WOMEN AND HEALTH IN AMERICA

WOMEN AND HEALTH IN AMERICA

Historical Readings

Second Edition

Edited by

Judith Walzer Leavitt

The University of Wisconsin Press

The University of Wisconsin Press
2537 Daniels Street
Madison, Wisconsin 53718

3 Henrietta Street
London WC2E 8LU, England

5   4   3   2   1

Printed in the United States of America

Library of Congress Cataloging-in-Publication Data
Women and health in America : historical readings / edited by Judith Walzer Leavitt — 2nd ed.
   702 pp.   cm.
      Includes bibliographical references.
      ISBN 0-299-15960-4 (alk. paper).
      ISBN 0-299-15964-7 (pbk. : alk paper).
      1. Women—Health and hygiene—United States—History.   2. Women—United States—Sexual
   behavior—History.   3. Women—Diseases—United States—History.   4. Women in medicine—
   United States—History.   I. Leavitt, Judith Walzer.
   RA778.W744   1999
   362.1'082'0973—DC21   98-14446

Contents

Contents

Contents

Childbirth and Motherhood   327

Women and Mental Illness   389

Health Care Providers: Midwives   423

Contents

Contents

WOMEN AND HEALTH IN AMERICA

Introduction to the Second Edition

Judith Walzer Leavitt

In the introduction to the first edition of this book, published in 1984, I noted that a "quiet revolution" had occurred within the field of medical history. Responding in large part to the contributions of social historians, medical history had broadened beyond the study of intellectual and technical accomplishments to encompass the social, economic, and political contexts in which medical theories emerged and medical practices occurred. The articles included in the first edition of *Women and Health in America* represented what I judged at that time to be the best new scholarship in the history of women's health.

Fourteen years later *quiet* is no longer a word that applies to what has indeed revolutionized the history of medicine. Scholarship has been transformed and is now firmly embedded in the wider view that medicine and health can be fully understood only within the social complexity in which they existed. In the history of women's health the broad arenas in which women lived their lives now routinely inform the research questions and writing. New understandings of women's roles in society and theoretical analyses of the meanings of historical change have shaped recent scholarship. Along with these multifaceted understandings, the quantity of scholarship in the area continues to expand. All this intellectual ferment, as exciting as it is, makes the selection of any group of representa-

tive articles for this second edition exceedingly difficult. As readers will no doubt notice when they look in vain for one of their favorite recent articles, I was forced to reduce the field in significant ways. I kept only six articles from the initial collection. In adding to them my selection of articles published since 1985, I strived to present a comprehensive story of the history of women's health in the United States and offer a taste of the diversity of the new work, hoping that by whetting intellectual appetites, this volume would send readers out to discover for themselves all that is missing here.

One way to understand the effect of the new historical writing is to compare it with what preceded it. American historians have shown interest in women and health at least since the 1940s. At that time scholars concentrated on famous people and medical advances. Articles with titles like "Louisiana's Contributions to Obstetrics and Gynecology" (1948), "Grace Revere Osler: Her Influence on Men of Medicine" (1949), and "The First Woman Dentist, Lucy Hobbs Tayler, D.D.S. (1833–1910)" (1951) indicate the range of topics.[1] Such early articles can be characterized as encompassing the "great lady" or the "great men conquering women's diseases" approaches, both parallel to medicohistorical scholarship in general during the middle years of the twentieth century. Historians struggled to identify significant people and to record major contributions of the past, and they saw as their job the documentation of the march of medical progress. Where women fit into this model, if they were pioneers like Elizabeth Blackwell or wives of famous physicians like Grace Osler, they became relevant subjects for historical writing. Similarly, the male doctors who conquered women's health problems were likely targets of study. Historians in the second half of the century, as both editions of this book testify, have moved beyond this celebratory stage to more comprehensive analyses of the meaning of women's experiences.

As the field blossomed in the early 1970s, some literature continued to identify and celebrate women achievers. Because women had been virtually absent from medical histories, names such as Elizabeth Blackwell, Rebecca Lee Crumpler, Alice Hamilton, Harriet Hunt, Mary Putnam Jacobi, Lilly Rosa Minoka-Hill, Mary Gove Nichols, Susan Smith McKinney Steward, and Bertha Van Hoosen had to be looked for, found, and put on a pedestal. Institutions such as the Woman's Medical College of Pennsylvania, New York Infirmary for Women and Children, New England Female Medical College, and Mary Thompson Hospital had to be recovered. We wanted to know they had existed and to celebrate what they represented. This was a necessary first step and served future scholarship well.

The early 1970s were also characterized by a more analytic strain of writing, one that was quite polemical. For example, Barbara Ehrenreich and Deirdre English, in two widely circulated and important pamphlets, "Complaints and Disorders: The Sexual Politics of Sickness" (1973), and "Witches, Midwives, and Nurses: A History of Women Healers" (1974), staked out the territory. Both concentrated on identifying a single villain, the medical profession, as a—if not *the*—major oppressor of women. "Doctors pass judgment on who is sick and who is well," they wrote. In this telling of the history of

women and health, women were the victims of an oppression that was both socially determined by male power and, in medical thinking, biologically determined by women's bodies. Women were victims of social forces and of their reproductive systems, neither of which they could completely escape.

In accepting the biology-as-destiny theme as driving the male medical profession, the historians writing in this period also rejected the idea that women's biology might actually have limited women's activities and abilities. Thus the writing of the 1970s, while celebrating early heroines and defining the sources of women's oppression, avoided topics that might be identified as the very sources of women's oppression. Subjects like childbirth and motherhood, which too easily reminded women of their historic subjugation, were essentially taboo, except as life models to overcome and move beyond. The very things that defined biology's hold over women—pregnancy, labor and delivery, lactation—could only be seen as instruments of limitation and oppression, instruments to be abandoned and rejected in the politics of the period. So scholarship reflected a major dilemma: Biology should have been a mainstay of a field concentrating on women and health, yet historians, like other scholars trying to reject the meaning of biology as women's destiny, felt constrained to tiptoe around what biology really meant for women.

By the late 1970s an integration of traditional with newer ideas occurred in the political world of feminism and was reflected in the scholarly world of women's studies and the history of women and health. The notion of women as hapless victims, either of their bodies or of the medical profession's powers, subsided. Women's agency in determining their lives, even historically, became the new emphasis. If biology had been used to confine women to the domestic sphere, scholars saw the need to study it and understand it as a factor that also enriched, liberated, and broadened women's experiences. Recognition and celebration continued in this period too because so many women and institutions remained to be recovered. While continuing to recognize the social forces working against women in this society, past and present, scholars added the layer of women's activities and agendas. Scholars in this period and since have recognized how and why women participated in their subordination, studied the effects of culture on behavior, and learned to value women's world. Carroll Smith-Rosenberg's now-classic article, "The Female World of Love and Ritual," which was reprinted in the first edition of this reader, was most representative of this kind of work.[2] It made women's culture an important point for historians by recognizing the value of women's institutions and private lives.

Smith-Rosenberg and two important books of the late 1970s opened an important era of scholarship that emphasized women's agency, women's control over their bodies, and women's power. Most important was Linda Gordon, *Woman's Body, Woman's Right,* a rich contextual history of birth control. Adrienne Rich, in *Of Woman Born,* saw women's experience as powerful in itself and helped to make motherhood an acceptable subject of study. Thus was launched an incredibly rich period of active research in the area of women's health history, uncovering and celebrating collective and private experience, much of it mirrored in the first edition of *Women and Health in America.* Scholarship from that period examined questions about the relationship between biology and women's life experiences and explored the meanings of the biological parts of women's social status. It no longer avoided the hard issues.

Studying women's medical schools and hospitals made historians face separatism as a complex historical issue. Were women's institutions fundamentally different, and if so, how were they different from their male counterparts? How did the feminist movement that led to their creation influence their ideology and daily activity? Did women practice medicine any differently from men? Scholars wrote about how women were bound in their culture in many of the same ways men were; they saw how domestic values gave women both an element of liberation and an element of subjugation. They got beyond glorification of women pioneers and into analyses of motivations, needs, and dreams of women of the past. Historical writing recognized that women were not merely victims throughout history. Women had some powers, exercised those powers, and were actors in history.

Using a sample of four journals, the graph (see fig. I.1) illustrates the periods of interest in the history of women and health, which correspond to times of increased participation and visibility of women in American society.[3] Although men wrote most articles published during the 1940s, they were probably influenced by the cultural and patriotic value of women's participation in war-related work in areas previously dominated by men. Similarly, the second wave of American feminism (since the late 1960s) spawned and revitalized interest in women's history. This time women wrote most of the historical articles and continue to lead scholarship in women's health history in new directions. After a brief lull in the mid-1980s (for which I have no ready explanation) the scholarly production in women's health history has been more active than ever.

The historical writing in the most recent period has tried to make the gender analyses more inclusive and representative of the diversity of women's experiences. Instead of addressing "difference" as a lens through which to differentiate men's and women's lives, recent scholarship has sought to understand and study the differences among women. Leisured white

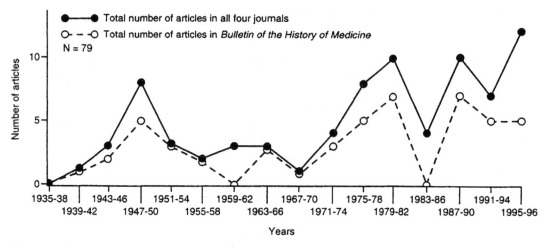

Fig. I.1 Number of articles on women and health in America appearing in *Bulletin of the History of Medicine, Journal of the History of Medicine and Allied Sciences, Journal of American History,* and *Journal of Social History* from 1935–1996.

women of the middle and upper classes had been the focus of much earlier historical work; now historians have begun to explore more comprehensively the experiences of African-American women, ethnic working-class and poor women, lesbians, Native American, and Asian-American women and Latinas. Race, ethnicity, and class are now more equal partners with gender in the analysis of historical literature on women and health. Historians now more readily understand and address the importance of differences in religion, region, age, wartime experiences, and other variables.

At the end of the twentieth century, biology and social context still define the core of the field. Historians have gone well beyond a simple rejection of the biology-is-destiny analysis. We have gained some understanding of how biology influenced women's lives, how biology has been used against women, and how women have used biology in self-definition and in strength. The recognition of women's agency, encompassing biology and society, has been an important corrective to past interpretations. We have learned how biology has been mediated in many different social contexts over time. We no longer talk about one social context but of many contexts, understanding that women are a varied group, living in many worlds over time and within time, and that all these contexts need analysis in order to understand the diversity that is Woman.

The articles in the second edition include six that were in the first edition and originally were published between 1973 and 1984. The additions represent newer work, all published since 1986. The articles are first in rough chronological order and then, for the late nineteenth and twentieth centuries—for which there is a superabundance of material—by topic. Three articles provide studies of women and health in the colonial and revolutionary periods. Four follow that concentrate on the first half of the nineteenth century through the Civil War period. The remainder of the articles concern the late nineteenth and twentieth centuries and address body image and physical fitness, sexuality, fertility, abortion and birth control, childbirth and motherhood, mental illness, women health care providers (midwives, nurses, physicians), health reform and public health, and the medicalization and the construction of woman.

The articles selected address women's physical and mental capacities and analyze how perceptions of these have changed over time. Theories about what could be expected from women often determined how society and the medical profession defined and reacted to women's health problems. In the first half of the nineteenth century, doctors believed that women's bodies were regulated by the uterus—the organ that Philadelphia physician Charles Meigs called the "disturbing radiator" of women's constitution—and their diagnoses frequently reverted to finding fault with the reproductive system.[4] As a result, when women tried to move into jobs outside the home, they found the way blocked by the biological argument that their reproductive systems governed their lives and limited their capacity for other activity. The argument of women's limited physical and mental capacity and the centrality of reproduction for understanding women's bodies thus defined medical treatments and views of women's health and sickness and sustained traditional roles for women in the culture at large. As compelling as the "anatomy-is-destiny" conception might have been, however, biology's influence over women's lives never was total. Although most women did bear children, they were not perpetually confined by their bodies. American women participated widely in domestic activities but also in nondomestic and nonreproductive activity, providing factory and farm labor for America's burgeoning economy, creating active service organizations, and entering the professions in substantial numbers. Women's lives might have been affected by cultural notions of their reproductive capacities, but physiology alone did not define womanhood.

The articles in this book also address the question of women's control over their bodies and seek to analyze how ideas about the importance of control influenced women's health and women's participation in the health occupations. As today, women in the past sought to govern their bodies, and they went to extreme lengths at various times to get this sometimes elusive authority. They actively sought birth control measures against the wishes of their husbands and religious advisers; they aborted their fetuses, putting their lives in jeopardy with knitting needles and toxic substances; they used potentially dangerous drugs to alleviate their painful and out-of-control childbirth; and they did these things both as individuals and as members of organizations formed to increase women's power over their lives. How much control women wanted and what form it took varied from group to group and over time, but the issue was as relevant to women's lives in the past as it is today.

Scholars today recognize gender as a crucial factor in understanding health status and health care. It is my hope that the historical articles in this book can further that understanding by helping to inform current health policy debates. Women's health agenda for the twenty-first century can best be built on a solid examination of the past. By looking back while we plan ahead, the issues at the turn of the millenium, such as mandatory inclusion of women in clinical studies, alarming differentials in health care access and outcomes among various groups of women, and the continuing limitations on women's social and economic opportunities, can be put in a perspective and context that can maximize future development.

I want to thank the students in my course "Women and Health in American History" who over the years have helped shape my conceptualization of this field. I especially want to thank

Karen Walloch, who helped me locate and then offered her candid opinion about most articles included in the second edition. Tomomi Kinukawa stepped in at the end with splendid insights. Rima Apple, Susan Friedman, Vanessa Northington Gamble, Linda Gordon, Maneesha Lal, Gerda Lerner, Mariamne Whatley, and Nancy Worcester, all colleagues at the University of Wisconsin, offered special contributions to this endeavor.

Notes

1. Thomas Benton Sellers, "Louisiana's Contributions to Obstetrics and Gynecology," *Bulletin of the History of Medicine* 22 (1948): 196–207; John F. Fulton, "Grace Revere Osler: Her Influence on Men of Medicine," *Bulletin of the History of Medicine* 23 (1949): 341–51; and Ralph W. Edwards, "The First

Woman Dentist, Lucy Hobbs Taylor, D.D.S. (1833–1910)," *Bulletin of the History of Medicine* 25 (1951): 277–83.

2. Her article originally appeared in *Signs* 1 (1975): 1–29.

3. I want to thank Sarah Kate Benton, Sarah Webber, and Susan Mobley for their help in collecting articles from the four journals, the *Bulletin of the History of Medicine* (1932–1996), *Journal of the History of Medicine and Allied Sciences* (1946–1996), *Journal of American History* (1913–1996), and the newer *Journal of Social History* (1967–1996). We chose these journals because they offer the longest runs of articles in the field, although in the 1990s we could have used many other journals that show the vibrancy of scholarship in the area.

4. Charles Meigs, *Woman: Her Diseases and Remedies: A Series of Letters to His Class*, 3d ed. (Philadelphia: Blanchard and Lea, 1854).

PART I
Women and Health in the Seventeenth and Eighteenth Centuries

IN THE SEVENTEENTH AND EIGHTEENTH CENTURIES, AS IN OTHER HISTORICAL PERI-ods, context and culture defined women's health status and participation as much as did biology or medicine. While the colonial relationship with England shaped much of the economic life of the European settlers, and the practice of medicine among the colonists in many ways mimicked their English counterparts', the American environment also shaped their daily experiences. As the three articles in this section demonstrate, social practices and values constrained many options available to women who needed medical attention and set parameters on how much and what women practitioners could accomplish. Yet women—part of their culture yet sometimes in defiance of it—often found ways to manipulate events and circumvent some restraints to become proactive agents in their own lives.

Through a case study of abortion and death in a New England village Cornelia Hughes Dayton vividly portrays the social situation surrounding a young woman's attempt to end her pregnancy, an action taken under coercion from her lover, and helps readers understand the community and family responses to the event. In this excellent example of microhistory we become privy to the medical alternatives and punishments available to small-town inhabitants in 1742; we come to understand the taboos and silences under which young people, and young women in particular, lived out their lives. Dayton powerfully demonstrates how gender and generational relationships shaped events that led to a postabortion infection and the death of a woman. Dayton's contribution forces a recognition of the limitations of female agency in such instances.

Next, turning to Native American childbirth practices through two centuries, Ann Marie Plane describes how birthing experiences of different groups of women of New England and southern Canada collectively differed from their colonial French or English neighbors. Readdressing questions about the "ease" of labor and delivery among native women and whether they actually went off alone into the forest to have their babies, and rereading some often-used European descriptions of native practices, Plane demonstrates how cultural diversity shaped health experiences and practices of women in colonial America.

Finally, Laurel Thatcher Ulrich examines a period in American history when a midwife such as Martha Moore Ballard was, contrary to earlier historical conclusions, the single most important practitioner in town. Not only did Ballard skillfully deliver 814 babies in her small Maine town, she also interacted and competed with the town's physicians and successfully treated numerous general medical ailments. Ulrich brilliantly situates one woman's practice in the full life of the community in which it occurred and in the context of women's history and experiences. She provides us with a model for how the careful application of textual analysis, in this case of one woman's diary, can lead to a complete transformation of historical knowledge.

1 Taking the Trade: Abortion and Gender Relations in an Eighteenth-Century New England Village

Cornelia Hughes Dayton

In 1742 in the village of Pomfret, perched in the hills of northeastern Connecticut, nineteen-year-old Sarah Grosvenor and twenty-seven-year-old Amasa Sessions became involved in a liaison that led to pregnancy, abortion, and death. Both were from prominent yeoman families, and neither a marriage between them nor an arrangement for the support of their illegitimate child would have been an unusual event for mideighteenth-century New England. Amasa Sessions chose a different course; in consultation with John Hallowell, a self-proclaimed "practitioner of physick," he coerced his lover into taking an abortifacient. Within two months, Sarah fell ill. Unbeknown to all but Amasa, Sarah, Sarah's sister Zerviah, and her cousin Hannah, Hallowell made an attempt to "Remove her Conseption" by a "manual opperation." Two days later Sarah miscarried, and her two young relatives secretly buried the fetus in the woods. Over the next month, Sarah struggled against a "Malignant fever" and was at-

CORNELIA HUGHES DAYTON is Associate Professor of History at the University of Connecticut, Storrs.
Reprinted from *The William and Mary Quarterly: A Magazine of Early American History and Culture*, Third Series, 48 (1991): 19–49, by permission of the author and the Omohundro Institute of Early American History and Culture.

tended by several physicians, but on September 14, 1742, she died.[1]

Most accounts of induced abortions among seventeenth- and eighteenth-century whites in the Old and New Worlds consist of only a few lines in a private letter or court record book; these typically refer to the taking of savin or pennyroyal—two common herbal abortifacients. While men and women in diverse cultures have known how to perform abortions by inserting an instrument into the uterus, actual descriptions of such operations are extremely rare for any time period. Few accounts of abortions by instrument have yet been uncovered for early modern England, and I know of no other for colonial North America.[2] Thus the historical fragments recording events in a small New England town in 1742 take on an unusual power to illustrate how an abortion was conducted, how it was talked about, and how it was punished.

We know about the Grosvenor-Sessions case because in 1745 two prominent Windham County magistrates opened an investigation into Sarah's death. Why there was a three-year gap between that event and legal proceedings, and why justices from outside Pomfret initiated the legal process, remain a mystery. In November 1745 the investigating magistrates offered their preliminary opinion that Hallowell, Amasa Sessions, Zerviah Grosvenor, and Hannah Grosvenor were guilty of Sarah's murder, the last three as accessories. From the outset, Connecticut legal officials concentrated not on the act of abortion per se but on the fact that an abortion attempt had led to a young woman's death.[3]

The case went next to Joseph Fowler, king's attorney for Windham County. He dropped charges against the two Grosvenor women, probably because he needed them as key witnesses and because they had played cover-up roles rather than originating the scheme. A year and a half passed as Fowler's first attempts to get convictions against Hallowell and Sessions failed, either before grand juries or before the Superior Court on technical grounds. Finally, in March 1747, Fowler presented Hallowell and Sessions separately for the "highhanded Misdemeanour" of attempting to destroy both Sarah Grosvenor's health and "the fruit of her womb."[4] A grand jury endorsed the bill against Hallowell but rejected a similarly worded presentment against Sessions. At Hallowell's trial before the Superior Court in Windham, the jury brought in a guilty verdict, and the chief judge sentenced the physician to twenty-nine lashes and two hours of public humiliation standing at the town gallows. Before the sentence could be executed, Hallowell managed to break jail. He fled to Rhode Island; as far as records indicate, he never returned to Connecticut. Thus, in the end, both Amasa Sessions and John Hallowell escaped legal punishment for their actions, whereas Sarah Grosvenor paid for her sexual transgression with her life.

Nearly two years of hearings and trials before the Superior Court produced a file of ten depositions and twenty-four other legal documents. This cache of papers is extraordinarily rich, not alone for its unusual chronicle of an abortion attempt but for its illumination of the fault lines in Pomfret dividing parents from grown children, men from women, and mideighteenth-century colonial culture from its seventeenth-century counterpart.

The depositions reveal that in 1742 the elders of Pomfret, men and women alike, failed to act as vigilant monitors of Sarah Grosvenor's courtship and illness. Instead, young married householders—kin of Sarah and Amasa—pledged themselves in a conspiracy of silence to allow the abortion plot to unfold undetected. The one person who had the opportunity to play middleman between the generations was Hallowell. A man in his forties, dogged by a shady past and yet adept at acquiring respectable connections, Hallowell provides an intriguing and rare portrait of a socially ambitious rural medical practitioner. By siding with the young people of Pomfret and keeping their secret, Hallowell betrayed his peers and elders and thereby opened himself to severe censure and expulsion from the community.

Beyond depicting generational conflict, the Grosvenor-Sessions case dramatically highlights key changes in gender relations that reverberated through New England society in the eighteenth century. One of these changes involved the emergence of a marked sexual double standard. In the midseventeenth century, a young man like Amasa Sessions would have been pressured by parents, friends, or the courts to marry his lover. Had he resisted, he would most likely have been whipped or fined for the crime of fornication. By the late seventeenth century, New England judges gave up on enjoining sexually active couples to marry. In the 1740s, amid shifting standards of sexual behavior and growing concern over the evidentiary impossibility of establishing paternity, prosecutions of young men for premarital sex ceased. Thus fornication was decriminalized for men but not for women. Many of Sarah Grosvenor's female peers continued to be prosecuted and fined for bearing illegitimate children. Through private arrangements, and occasionally through civil lawsuits, their male partners were sometimes cajoled or coerced into contributing to the child's upkeep.[5]

What is most striking about the Grosvenor-Sessions case is that an entire community apparently forgave Sessions for the extreme measures he took to avoid accountability for his bastard child. Although he initiated the actions that led to his lover's death, all charges against him were dropped. Moreover, the tragedy did not spur Sessions to leave town; instead, he spent the rest of his life in Pomfret as a respected citizen. Even more dramatically than excusing young men from the crime of fornication, the treatment of Amasa Sessions confirmed that the sexually irresponsible activities of men in their youth would not be held against them as they reached for repute and prosperity in their prime.[6]

The documents allow us to listen in on the quite different responses of young men and women to the drama unfolding in Pomfret. Sarah Grosvenor's female kin and friends, as we shall see, became preoccupied with their guilt and with the inevitability of God's vengeance. Her male kin, on the other hand, reacted cautiously and legalistically, ferreting out information in order to assess how best to protect the Grosvenor family name. The contrast reminds us yet again of the complex and gendered ways in which we must rethink conventional interpretations of secularization in colonial New England.

Finally, the Grosvenor case raises more questions than it answers about New Englanders' access to and attitudes toward abortion. If Sarah had not died after miscarriage, it is doubtful that any word of Sessions's providing her with an abortifacient or Hallowell's operation would have survived into the twentieth century. Because it nearly went unrecorded and because it reveals that many Pomfret residents were familiar with the idea of abortion, the case supports historians' assumptions that abortion attempts were far from rare in colonial America.[7] We can also infer from the case that the most dangerous abortions before 1800 may have been those instigated by men and performed by surgeons with instruments.[8] But both abortion's frequency and the lineaments of its social context remain obscure. Did cases in which older women helped younger women to abort unwanted pregnancies far outnumber cases such as this one in which men initiated the process? Under what circumstances did family members and neighbors help married and unmarried women to hide abortion attempts?

Perhaps the most intriguing question centers on why women and men in early America acted *covertly* to effect abortions when abortion before quickening was legal. The Grosvenor case highlights the answer that applies to most known incidents from the period: abortion was understood as blameworthy because it was an extreme action designed to hide a prior sin, sex outside of marriage.[9] Reading the depositions, it is nearly impossible to disentangle the players' attitudes toward abortion itself from their expressions of censure or anxiety over failed courtship, illegitimacy, and the dangers posed for a young woman by a secret abortion. Strikingly absent from these eighteenth-century documents, however, is either outrage over the destruction of a fetus or denunciations of those who would arrest "nature's proper course." Those absences are a telling measure of how the discourse about abortion would change dramatically in later centuries.

The Narrative

Before delving into the response of the Pomfret community to Sarah Grosvenor's abortion and death, we need to know just who participated in the conspiracy to cover up her pregnancy and how they managed it. The following paragraphs, based on the depositions, offer a reconstruction of the events of 1742. A few caveats are in order. First, precise dating of crucial incidents is impossible, since deponents did not remember events in terms of days of the week (except for the Sabbath) but rather used phrases like "sometime in August." Second, the testimony concentrated almost exclusively on events in the two months preceding Sarah's death on September 14. Thus we know very little about Sarah and Amasa's courtship before July 1742.[10] Third, while the depositions often indicate the motivations and feelings of the principals, these will be discussed in subsequent sections of this article, where the characters' attitudes can be set in the context of their social backgrounds, families, and community. This section essentially lays out a medical file for Sarah Grosvenor, a file that unfolds in four parts: the taking of the abortifacient, Hallowell's operation, the miscarriage, and Sarah's final illness.

The case reveals more about the use of an abortifacient than most colonial court records in which abortion attempts are mentioned. Here we learn not only the form in which Sarah received the dose but also the special word that Pomfret residents applied to it. What the documents do not disclose are either its ingredients or the number of times Sarah ingested it.[11]

The chronicle opens in late July 1742 when Zerviah Grosvenor, aged twenty-one, finally prevailed upon her younger sister to admit that she was pregnant. In tears, Sarah explained that she had not told Zerviah sooner because "she had been taking [the] trade to remove it."[12] *Trade* was used in this period to signify stuff or goods, often in the deprecatory sense of rubbish and trash. The *Oxford English Dictionary* confirms that in some parts of England and New England the word was used to refer to medicine. In Pomfret trade meant a particular type of medicine, an abortifacient, thus a substance that might be regarded as "bad" medicine, as rubbish, unsafe and associated with destruction.

What is notable is that Sarah and Zerviah, and neighboring young people who also used the word, had no need to explain to one another the meaning of "taking the trade." Perhaps only a few New Englanders knew how to prepare an abortifacient or knew of books that would give them recipes, but many more, especially young women who lived with the fear of becoming pregnant before marriage, were familiar with at least the *idea* of taking an abortifacient.

Sarah probably began taking the trade in mid-May when she was already three-and-a-half-months pregnant.[13] It was brought to her in the form of a powder by Amasa.[14] Sarah understood clearly that her lover had obtained the concoction "from docter hollowel," who conveyed "directions" for her doses through Amasa. Zerviah deposed later that Sarah had been "loath to Take" the drug and "Thot it an Evil," probably because at three and a half months she anticipated quickening, the time from which she knew the law counted abortion an "unlawful measure."[15] At the outset, Sarah argued in vain with Amasa against his proposed "Method." Later, during June and July, she sometimes "neglected" to take the doses he left for her, but, with mounting urgency, Amasa and the doctor pressed her to comply. "It was necessary," Amasa explained in late July, that she take "more, or [else] they were afraid She would be greatly hurt by what was already done." To calm her worries, he assured her that "there was no life [left] in the Child" and that the potion "would not hurt her."[16] Apparently, the men hoped that a few more doses would provoke a miscarriage, thereby expelling the dead fetus and restoring Sarah's body to its natural balance of humors.

Presumably, Hallowell decided to operate in early August because Sarah's pregnancy was increasingly visible, and he guessed that she was not going to miscarry. An operation in which the fetus would be removed or punctured was now the only certain way to terminate the pregnancy secretly.[17] To avoid the scrutiny of Sarah's parents, Hallowell resorted to a plan he had used once before in arranging a private examination of Sarah. Early one afternoon he arrived at the house of John Grosvenor and begged for a room as "he was weary and wanted Rest."[18] John, Sarah's thirty-one-year-old first cousin, lived with his wife, Hannah, and their young children in a homestead only a short walk down the hill but out of sight of Sarah's father's house. While John and Hannah were busy, the physician sent one of the little children to fetch Sarah.[19]

The narrative of Sarah's fateful meeting with Hallowell that August afternoon is best told in the words of one of the deponents. Abigail Nightingale had married and moved to Pomfret two years earlier, and by 1742 she had become Sarah's close friend.[20] Several weeks after the operation, Sarah attempted to relieve her own "Distress of mind" by confiding the details of her shocking experience to Abigail. Unconnected to the Grosvenor or Sessions families by kinship, and without any other apparent stake in the legal uses of her testimony, Abigail can probably be trusted as a fairly accurate paraphraser of Sarah's words.[21] If so, we have here an unparalleled eyewitness account of an eighteenth-century abortion attempt.

This is how Abigail recollected Sarah's deathbed story:

On [Sarah's] going down [to her cousin John's], [Hallowell] said he wanted to Speake with her alone; and then they two went into a Room together; and then sd. Hallowell told her it was necessary that something more should be done or else she would Certainly die; to which she replyed that she was afraid they had done too much already, and then he told her that there was one thing more that could easily be done, and she asking him what it

was; he said he could easily deliver her. but she said she was afraid there was life in the Child, then he asked her how long she had felt it; and she replyed about a fortnight; then he said that was impossible or could not be or ever would; for that the trade she had taken had or would prevent it: and that the alteration she felt Was owing to what she had taken. And he farther told her that he verily thought that the Child grew to her body to the Bigness of his hand, or else it would have Come away before that time. and that it would never Come away, but Certainly Kill her, unless other Means were used.[22] On which she yielded to his making an Attempt to take it away; charging him that if he could percieve that there was life in it he would not proceed on any Account. And then the Doctor openning his portmantua took an Instrument[23] out of it and Laid it on the Bed, and she asking him what it was for, he replyed that it was to make way; and that then he tryed to remove the Child for Some time in vain putting her to the Utmost Distress, and that at Last she observed he trembled and immediately percieved a Strange alteration in her body and thought a bone of the Child was broken; on which she desired him (as she said) to Call in some body, for that she feared she was a dying, and instantly swooned away.[24]

With Sarah's faint, Abigail's account broke off, but within minutes others, who would testify later, stepped into the room. Hallowell reacted to Sarah's swoon by unfastening the door and calling in Hannah, the young mistress of the house, and Zerviah, who had followed her sister there. Cold water and "a bottle of drops" were brought to keep Sarah from fainting again, while Hallowell explained to the "much Surprized" women that "he had been making an Attempt" to deliver Sarah. Despite their protests, he then "used a further force upon her" but did not succeed in "Tak[ing] the Child . . . away."[25] Some days later Hallowell told a Pomfret man that in this effort "to distroy hir conception" he had "either knipt or Squeisd the head of the Conception."[26] At the time of the attempt, Hallowell explained to the women that he "had done so much to her, as would Cause the Birth of the Child in a Little time." Just before sunset, he packed up his portmanteau and went to a nearby tavern, where Amasa was waiting "to hear [the outcome of] the event."[27] Meanwhile, Sarah, weak-kneed and in pain, leaned on the arm of her sister as the young women managed to make their way home in the twilight.

After his attempted "force," Hallowell fades from the scene, while Zerviah and Hannah Grosvenor become the key figures. About two days after enduring the operation, Sarah began to experience contractions. Zerviah ran to get Hannah, telling her "she Tho't . . . Sarah would be quickly delivered." They returned to find Sarah, who was alone "in her Father's Chamber," just delivered and rising from the chamber pot. In the pot was "an Untimely birth"—a "Child [that] did not Appear to have any Life In it." To Hannah, it "Seemed by The Scent . . . That it had been hurt and was decaying," while Zerviah later remembered it as "a perfect Child," even "a pritty child."[28] Determined to keep the event "as private as they Could," the two women helped Sarah back to bed, and then "wr[ap]ed . . . up" the fetus, carried it to the woods on the edge of the farmstead, and there "Buried it in the Bushes."[29]

On learning that Sarah had finally miscarried and that the event had evidently been kept hidden from Sarah's parents, Amasa and Hallowell may have congratulated themselves on the success of their operation. However, about ten days after the miscarriage, Sarah grew feverish and weak. Her parents consulted two college-educated physicians who hailed from outside the Pomfret area. Their visits did little good, nor were Sarah's symptoms—fever, delirium, convulsions—relieved by a visit from Hallowell, whom Amasa "fetcht" to Sarah's bedside.[30] In the end, Hallowell, who had decided to move from nearby Killingly to more distant Providence, washed his hands of the case. A few days before Sarah died, her cousin John "went after" Hallowell, whether to bring him back or to express his rage, we do not know. Hallowell predicted "that She woul[d] not live."[31]

Silence seems to have settled on the Grosvenor house and its neighborhood after Sarah's death on September 14. It was two and a half years later that rumors about a murderous abortion spread through and beyond Pomfret village, prompting legal investigation. The silence, the gap between event and prosecution, the passivity of Sarah's parents—all lend mystery to the narrative. But despite its ellipses, the Grosvenor case provides us with an unusual set of details about one young couple's extreme response to the common problem of failed courtship and illegitimacy. To gain insight into both the mysteries and the extremities of the Grosvenor-Sessions case, we need to look more closely at Pomfret, at the two families centrally involved, and at clues to the motivations of the principal participants. Our abortion tale, it turns out, holds beneath its surface a complex trail of evidence about generational conflict and troubled relations between men and women.

The Pomfret Players

In 1742 the town of Pomfret had been settled for just over forty years. Within its central neighborhood and in homesteads scattered over rugged, wooded hillsides lived probably no more than 270 men, women, and children.[32] During the founding decades, the fathers of Sarah and Amasa ranked among the ten leading house-holders; Leicester Grosvenor and Nathaniel Sessions were chosen often to fill important local offices.

Grosvenor, the older of the two by seven years, had inherited standing and a choice farmstead from his father, one of the original six purchasers of the Pomfret territory.[33] When the town was incorporated in 1714, he was elected a militia officer and one of the first selectmen. He was returned to the latter post nineteen times and eventually rose to the highest elective position—that of captain—in the local trainband. Concurrently, he was appointed many times throughout the 1710s and 1720s to ad hoc town committees, often alongside Nathaniel Sessions. But unlike Sessions, Grosvenor went on to serve at the colony level. Pomfret freemen chose him to represent them at ten General Assembly sessions between 1726 and 1744. Finally, in the 1730s, when he was in his late fifties, the legislature appointed him a justice of the peace for Windham County. Thus, until his retirement in 1748 at age seventy-four, his house would have served as the venue for petty trials, hearings, and recordings of documents. After retiring from public office, Grosvenor lived another eleven years, leaving behind in 1759 an estate worth over £600.[34]

Nathaniel Sessions managed a sizable farm and ran one of Pomfret's taverns at the family homestead. Town meetings were sometimes held there. Sessions was chosen constable in 1714 and rose from ensign to lieutenant in the militia—always a step behind Leicester Grosvenor. He could take pride in one exceptional distinction redounding to the family honor: in 1737 his son Darius became only the second Pomfret resident to graduate from Yale College, and before Sessions died at ninety-one he saw Darius elected assistant and then deputy gover-

nor of Rhode Island.[35]

The records are silent as to whether Sessions and his family resented the Grosvenors, who must have been perceived in town as more prominent, or whether the two families—who sat in adjoining private pews in the meetinghouse—enjoyed a close relationship that went sour for some reason *before* the affair between Sarah and Amasa. Instead, the signs (such as the cooperative public work of the two fathers, the visits back and forth between the Grosvenor and Sessions girls) point to a long-standing friendship and dense web of interchanges between the families. Indeed, courtship and marriage between a Sessions son and a Grosvenor daughter would hardly have been surprising.

What went wrong in the affair between Sarah and Amasa is not clear. Sarah's sisters and cousins knew that "Amasy" "made Sute to" Sarah, and they gave no indication of disapproving. The few who guessed at Sarah's condition in the summer of 1742 were not so much surprised that she was pregnant as that the couple "did not marry."[36] It was evidently routine in this New England village, as in others, for courting couples to post banns for their nuptials soon after the woman discovered that she was pregnant.

Amasa offered different answers among his Pomfret peers to explain his failure to marry his lover. When Zerviah Grosvenor told Amasa that he and Sarah "had better Marry," he responded, "That would not do," for "he was afraid of his Parents . . . [who would] always make their lives [at home] uncomfortable."[37] Later, Abigail Nightingale heard rumors that Amasa was resorting to the standard excuse of men wishing to avoid a shotgun marriage—denying that the child was his.[38] Hallowell, with whom Amasa may have been honest, claimed "the Reason that they did not marry" was "that Sessions Did not Love her well a nough for [he] saith he did not believe it was his son and if he Could Cause her to gitt Red of it he would not Go near her again."[39] Showing yet another face to a Grosvenor kinsman after Sarah's death, Amasa repented his actions and extravagantly claimed he would "give All he had" to "bring Sarah . . . To life again . . . and have her as his wife."[40]

The unusual feature of Amasa's behavior was not his unwillingness to marry Sarah but his determination to terminate her pregnancy before it showed. Increasing numbers of young men in eighteenth-century New England weathered the temporary obloquy of abandoning a pregnant lover in order to prolong their bachelorhood or marry someone else.[41] What drove Amasa, and an ostensibly reluctant Sarah, to resort to abortion? Was it fear of their fathers? Nathaniel Sessions had chosen Amasa as the son who would remain on the family farm and care for his parents in their old age. An ill-timed marriage could have disrupted these plans and threatened Amasa's inheritance.[42] For his part, Leicester Grosvenor may have made it clear to his daughter that he would be greatly displeased at her marrying before she reached a certain age or until her older sister wed. Rigid piety, an authoritarian nature, an intense concern with being seen as a good household governor—any of these traits in Leicester Grosvenor or Nathaniel Sessions could have colored Amasa's decisions.

Perhaps it was not family relations that proved the catalyst but Amasa's acquaintance with a medical man who boasted about a powder more effective than the herbal remedies that were part of women's lore. Hallowell himself had fathered an illegitimate child fifteen years earlier, and he may have encouraged a rakish attitude in Amasa, beguiling the younger man with the promise of dissociating sex from its possible consequences. Or the explanation may have been that classic one: another woman. Two years after Sarah's death, Amasa married Han-nah Miller of Rehoboth, Massachusetts. Perhaps in early 1742 he was already making trips to the town just east of Providence to see his future wife.[43]

What should we make of Sarah's role in the scheme? It is possible that she no longer loved Amasa and was as eager as he to forestall external pressures toward a quick marriage. However, Zerviah swore that on one occasion before the operation Amasa reluctantly agreed to post banns for their nuptials and that Sarah did not object.[44] *If* Sarah was a willing and active participant in the abortion plot all along, then by 1745 her female kin and friends had fabricated and rehearsed a careful and seamless story to preserve the memory of the dead girl untarnished.

In the portrait drawn by her friends, Sarah reacted to her pregnancy and to Amasa's plan first by arguing and finally by doing her utmost to protect her lover. She may have wished to marry Amasa, yet she did not insist on it or bring in older family members to negotiate with him and his parents. Abigail Nightingale insisted that Sarah accepted Amasa's recalcitrance and only pleaded with him that they not "go on to add sin to sin." Privately, she urged Amasa that there was an alternative to taking the trade—a way that would enable him to keep his role hidden and prevent the couple from committing a "Last transgression [that] would be worse then the first." Sarah told him that "she was willing to take the sin and shame to her self, and to be obliged never to tell whose Child it was, and that she did not doubt but that if she humbled her self on her Knees to her Father he would take her and her Child home." Her lover, afraid that his identity would become known, vetoed her proposal.[45]

According to the Pomfret women's reconstruction, abortion was not a freely chosen and defiant act for Sarah. Against her own desires, she reluctantly consented in taking the trade only because Amasa "So very earnestly perswaided her." In fact, she had claimed to her friends that she was coerced; he "would take no denyal."[46] Sarah's confidantes presented her as being aware of her options, shrinking from abortion as an unnatural and immoral deed, and yet finally choosing the strategy consistent with her lover's vision of what would best protect their futures. Thus, if Amasa's hubris was extreme, so too was Sarah's internalization of those strains of thought in her culture that taught women to make themselves pleasing and obedient to men.

While we cannot be sure that the deponents' picture of Sarah's initial recoil and reluctant submission to the abortion plot was entirely accurate, it is clear that once she was caught up in the plan she extracted a pledge of silence from all her confidantes. Near her death, before telling Abigail about the operation, she "insist[ed] on . . . [her friend's] never discovering the Matter" to anyone.[47] Clearly, she had earlier bound Zerviah and Hannah on their honor not to tell their elders. Reluctant when faced with the abortionist's powder, Sarah became a leading co-conspirator when alone with her female friends.

One of the most remarkable aspects of the Grosvenor-Sessions case is Sarah and Amasa's success in keeping their parents in the dark, at least until her final illness. If by July Sarah's sisters grew suspicious that Sarah was "with child," what explains the failure of her parents to observe her pregnancy and to intervene and uncover the abortion scheme? Were they negligent, preoccupied with other matters, or willfully blind?[48] Most mysterious is the role of forty-eight-year-old Rebecca Grosvenor, Grosvenor's second wife and Sarah's stepmother since 1729. Rebecca is mentioned only once in the depositions,[49] and she was not summoned as a witness in the 1745–1747 investigations into Sarah's death. Even if some extraordinary circumstance—an invalid condition or an implacable hatred between Sarah and her stepmother—explains Rebecca's abdication of her role as guardian, Sarah had two widowed aunts living in or near her household. These matrons, experienced in childbirth matters and concerned for the family reputation, were just the sort of older women who traditionally watched and advised young women entering courtship.[50]

In terms of who knew what, the events of summer 1742 in Pomfret apparently unfolded in two stages. The first stretched from Sarah's discovery of her pregnancy by early May to some point in late August after her miscarriage. In this period a determined collective effort by Sarah and Amasa and their friends kept their elders in the dark.[51] When Sarah fell seriously ill from the aftereffects of the abortion attempt and miscarriage, rumors of the young people's secret activities reached Leicester Grosvenor's neighbors and even one of the doctors he had called in.[52] It is difficult to escape the conclusion that by Sarah's death in mid-September her father and stepmother had learned of the steps that had precipitated her mortal condition and kept silent for reasons of their own.

Except for Hallowell, the circle of intimates entrusted by Amasa and Sarah with their scheme consisted of young adults ranging in age from nineteen to thirty-three.[53] Born between about 1710 and 1725, these young people had grown up just as the town attracted enough settlers to support a church, militia, and local market. They were second-generation Pomfret residents who shared the generational identity that came with sitting side by side through long worship services, attending school, playing, and working together at children's tasks. By 1740 these sisters, brothers, cousins, courting couples, and neighbors, in their visits from house to house—sometimes in their own households, sometimes at their parents'—had managed to create a world of talk and socializing that was largely exempt from parental supervision.[54] In Pomfret in 1742 it was this group of young people in their twenties and early thirties, *not* the cluster of Grosvenor matrons over forty-five, who monitored Sarah's courtship, attempted to get Amasa to marry his lover, privately investigated the activities and motives of Amasa and Hallowell, and, belatedly, spoke out publicly to help Connecticut juries decide who should be blamed for Sarah's death.

That Leicester Grosvenor made no public move to punish those around him and that he avoided giving testimony when legal proceedings commenced are intriguing clues to social changes underway in New England villages in the mideighteenth century. Local leaders like Grosvenor, along with the respectable yeomen whom he represented in public office, were increasingly withdrawing delicate family problems from the purview of their communities. Slander, illegitimacy, and feuds among neighbors came infrequently to local courts by midcentury, indicating male householders' growing preference for handling such matters privately.[55] Wealthy and ambitious families adopted this ethic of privacy at the same time that they became caught up in elaborating their material worlds by adding rooms and acquiring luxury goods. The "good feather bed" with all of its furniture that Grosvenor bequeathed to his one unmarried daughter was but one of many marks of status by which the Grosvenors differentiated themselves from their Pomfret neighbors.[56] But all the fine accoutrements in the world would not excuse Justice Grosvenor from his obligation to govern his household effectively. Mortified no doubt at his inability to monitor the young people in his extended family, he responded, ironically, by extending their conspiracy of silence. The best way for him to shield the family name from scandal and protect his political reputation in the county and colony was to keep the story of Sarah's abortion out of the courts.

The Doctor

John Hallowell's status as an outsider in Pomfret and his dangerous secret alliance with the town's young adults may have shaped his destiny as the one conspirator sentenced to suffer at the whipping post. Although the physician had been involved in shady dealings before 1742, he had managed to win the trust of many patients and a respectable social standing. Tracking down his history in northeastern Connecticut tells us something of the uncertainty surrounding personal and professional identity before the advent of police records and medical licensing boards. It also gives us an all-too-rare glimpse into the fashion in which an eighteenth-century country doctor tried to make his way in the world.

Hallowell's earliest brushes with the law came in the 1720s. In 1725 he purchased land in Killingly, a Connecticut town just north of Pomfret and bordering both Massachusetts and Rhode Island. Newly married, he was probably in his twenties at the time. Seven months before his wife gave birth to their first child, a sixteen-year-old Killingly woman charged Hallowell with fathering her illegitimate child. Using the alias Nicholas Hallaway, he fled to southeastern Connecticut, where he lived as a "transient" for three months. He was arrested and settled the case by admitting to paternity and agreeing to contribute to the child's maintenance for four years.[57]

Hallowell resumed his life in Killingly. Two years later, now referred to as "Dr.," he was arrested again; this time the charge was counterfeiting. Hallowell and several confederates were hauled before the governor and council for questioning and then put on trial before the Superior Court. Although many Killingly witnesses testified to the team's suspect activities in a woodland shelter, the charges against Hallowell were dropped when a key informer failed to appear in court.[58]

Hallowell thus escaped conviction on a serious felony charge, but he had been tainted by stories linking him to the criminal subculture of transient, disorderly, greedy, and manually skilled men who typically made up gangs of counterfeiters in eighteenth-century New England.[59] After 1727 Hallowell may have given up dabbling in money-making schemes and turned to earning his livelihood chiefly from his medical practice. Like two-thirds of the male medical practitioners in colonial New England, he probably did not have college or apprentice training, but his skill, or charm, was not therefore necessarily less than that of any one of his peers who might have inherited a library of books and a fund of knowledge from a physician father. All colonial practitioners, as Richard D. Brown reminds us, mixed learned practices with home or folk remedies, and no doctor had access to safe, reliable pharmacological preparations or antiseptic surgical procedures.[60]

In the years immediately following the counterfeiting charge, Hallowell appears to have made several deliberate moves to portray himself as a sober neighbor and reliable physician. At about the time of his second marriage, in 1729, he became a more frequent attendant at the Killingly meetinghouse, where he renewed his covenant and presented his first two children for baptism.[61] He also threw himself into the land and credit markets of northeastern Connecticut, establishing himself as a physician who was also an enterprising yeoman and a frequent litigant.[62]

These activities had dual implications. On the one hand, they suggest that Hallowell epitomized the eighteenth-century Yankee citizen—a man as comfortable in the courtroom and countinghouse as at a patient's bedside; a man of restless energy, not content to limit his scope to his fields and village; a practical, ambitious man with a shrewd eye for a good deal.[63] On the other hand, Hallowell's losses to Boston creditors, his constant efforts to collect debts, and his farflung practice raise questions about the nature of his activities and medical practice. He evidently had clients not just in towns across northeastern Connecticut but also in neighboring Massachusetts and Rhode Island. Perhaps rural practitioners normally traveled extensively, spending many nights away from their wives and children.[64] It is also possible, however, either that Hallowell was forced to travel because established doctors from leading families had monopolized the local practice or that he chose to recruit patients in Providence and other towns as a cover for illicit activities.[65] Despite his land speculations and his frequent resort to litigation, Hallowell was losing money. In the sixteen years before 1742, his creditors secured judgments against him for a total of £1,060, while he was able to collect only £700 in debts.[66] The disjunction between his ambition and actual material gains may have led Hallowell in middle age to renew his illicit money-making schemes. By supplying young men with potent abortifacients and dabbling in schemes to counterfeit New England's paper money, he betrayed the very gentlemen whose respect, credit, and society he sought.

What is most intriguing about Hallowell was his ability to ingratiate himself throughout his life with elite men whose reputations were unblemished by scandal. Despite the rumors that must have circulated about his early sexual dalliance, counterfeiting activities, suspect medical remedies, heavy debts, and shady business transactions,[67] leading ministers, merchants, and magistrates welcomed him into their houses. In Pomfret such acceptance took its most dramatic form in September 1739 when Hallowell was admitted along with thirty-five other original covenanters to the first private library association in eastern Connecticut. Gathering in the house of Pomfret's respected conservative minister, Ebenezer Williams, the members pledged sums for the purchase of "useful and profitable English books." In the company of the region's scholars, clergy, and "gentlemen," along with a few yeomen—all "warm friends of learning and literature"—Hallowell marked himself off from the more modest subscribers by joining with thirteen prominent and wealthy signers to pledge a sum exceeding £15.[68]

Lacking college degree and family pedigree, Hallowell traded on his profession and his charm to gain acceptability with the elite. In August 1742 he shrewdly removed himself from the Pomfret scene, just before Sarah Grosvenor's death. In that month he moved, possibly without his wife and children, to Providence, where he had many connections. Within five years Hallowell had so insinuated himself with town leaders such as Stephen Hopkins that fourteen of them petitioned for mitigation of what they saw as the misguided sentence imposed on him in the Grosvenor case.[69]

Hallowell's capacity for landing on his feet, despite persistent brushes with scandal, debt, and the law, suggests that we should look at the fluidity of New England's eighteenth-century elite in new ways.[70] What bound sons of old New England families, learned men, and upwardly mobile merchants and professionals in an expanded elite may partly have been a reshaped, largely unspoken set of values shared by men. We know that the archetype for white New England women as sexual beings was changing from carnal Eve to resisting Pamela and that the calculus of accountability for seduction was shifting blame solely to women.[71] But the simultaneous metamorphosis in cultural images and values defining manhood in the early and mid-eighteenth century has not been studied. The scattered evidence we do have suggests that, increasingly, for men in the more secular and anglicized culture of New England, the lines between legitimate and illegitimate sexuality,

between sanctioned and shady business dealings, and between speaking the truth and protecting family honor blurred. Hallowell's acceptability to men like minister Ebenezer Williams and merchant Stephen Hopkins hints at how changing sexual and moral standards shaped the economic and social alliances made by New England's male leadership in the 1700s.[72]

Women's Talk and Men's Talk

If age played a major role in determining who knew the truth about Sarah Grosvenor's illness, gender affected how the conspiring young adults responded to Sarah's impending death and how they weighed the issue of blame. Our last glimpse into the social world of eighteenth-century Pomfret looks at the different ways in which women and men reconstructed their roles in the events of 1742.

An inward gaze, a strong consciousness of sin and guilt, a desire to avoid conflict and achieve reconciliation, a need to confess—these are the impulses expressed in women's intimate talk in the weeks before Sarah died. The central female characters in the plot, Sarah and Zerviah Grosvenor, lived for six weeks with the daily fear that their parents or aunts might detect Sarah's condition or their covert comings and goings. Deposing three years later, Zerviah represented the sisters as suffering under an intensifying sense of complicity as they had passed through two stages of involvement in the concealment plan. At first, they were passive players, submitting to the hands of men. But once Hallowell declared that he had done all he could, they were left to salvage the conspiracy by enduring the terrors of a first delivery alone, knowing that their failure to call in the older women of the family resembled the decision made by women who committed infanticide.[73] While the pain and shock of miscarrying a five-and-one-half-month fetus through a possibly lacerated vagina may have been the experience that later most grieved Sarah, Zerviah would be haunted particularly by her stealthy venture into the woods with Hannah to bury the shrouded evidence of miscarriage.[74]

The Grosvenor sisters later recalled that they had regarded the first stage of the scheme—taking the trade—as "a Sin" and "an Evil" not so much because it was intended to end the life of a fetus as because it entailed a protracted set of actions, worse than a single lie, to cover up an initial transgression: fornication.[75] According to their religion and the traditions of their New England culture, Sarah and Zerviah knew that the proper response to the sin of "uncleanness" (especially when it led to its visible manifestation, pregnancy) was to confess, seeking to allay God's wrath and cleanse oneself and one's community. Dire were the consequences of hiding a grave sin, so the logic and folklore of religion warned.[76] Having piled one covert act upon another, all in defiance of her parents, each sister wondered if she had not ventured beyond the pale, forsaking God and in turn being forsaken.

Within hours after the burial, Zerviah ran in a frenzy to Alexander Sessions's house and blurted out an account of her sister's "Untimely birth" and the burying of the fetus. While Alexander and Silence Sessions wondered if Zerviah was "in her right mind" and supposed she was having "a very bad fit," we might judge that she was in shock—horrified and confused by what she had done, fearful of retribution, and torn between the pragmatic strategy of silence and an intense spiritual longing to confess. Silence took her aside and demanded, "how could you do it?—I could not!" Zerviah, in despair, replied, "I don't Know; the Devil was in us." Hers was the characteristic refuge of the defiant sinner: Satan made her do it.[77]

Sarah's descent into despondency, according to the portrait drawn in the women's depositions, was not so immediate. In the week following the miscarriage she recovered enough to be up and about the house. Then the fever came on. Bedridden for weeks yet still lucid, she exhibited such "great Concern of mind" that Abigail, alone with her, felt compelled to ask her "what was the Matter." "Full of Sorrow" and "in a very affectionate Manner," Sarah replied by asking her friend "whether [she] thought her Sins would ever be pardoned?" Abigail's answer blended a reassuringly familiar exhortation to repent with an awareness that Sarah might have stepped beyond the possibility of salvation. "I answered that I hoped she had not Sinned the unpardonable Sin [that of renouncing Christ], but with true and hearty repentance hoped she would find forgiveness." On this occasion, and at least once more, Sarah responded to the call for repentance by pouring out her troubled heart to Abigail—as we have seen—confessing her version of the story in a torrent of words.[78]

Thus visions of judgment and of their personal accountability to God haunted Sarah and Zerviah during the waning days of summer—or so their female friends later contended. Caught between the traditional religious ethic of confession, recently renewed in revivals across New England, and the newer status-driven cultural pressure to keep moral missteps private, the Grosvenor women declined to take up roles as accusers. By focusing on their own actions, they rejected a portrait of themselves as helpless victims, yet they also ceded to their male kin responsibility for assessing blame and mediating between the public interest in seeing justice done and the private interests of the Grosvenor family. Finally, by trying to keep the conspiracy of silence intact and by allowing Amasa frequent visits to her bedside to lament his role and his delusion by Hallowell, Sarah at once endorsed a policy of private repentance and forgiveness *and* indicated that she wished her lover to be spared eventual public retribution for her death.

Talk among the men of Pomfret in the weeks

preceding and following Sarah's death centered on more secular concerns than the preoccupation with sin and God's anger that ran through the women's conversations. Neither Hallowell nor Sessions expressed any guilt or sense of sin, as far as the record shows, *until* Sarah was diagnosed as mortally ill.[79] Indeed, their initial accounts of the plot took the form of braggadocio, with Amasa (according to Hallowell) casting himself as the rake who could "gitt Red" of his child and look elsewhere for female companionship, and Hallowell boasting of his abortionist's surgical technique to Sarah's cousin Ebenezer. Later, anticipating popular censure and possible prosecution, each man "Tried to Cast it" on the other. The physician insisted that "He did not do any thing but What Sessions Importuned him to Do," while Amasa exclaimed "That he could freely be Strip[p]ed naked provided he could bring Sarah . . . To life again . . . , but Doct Hollowell had Deluded him, and Destroyed her."[80] While this sort of denial and buck passing seems very human, it was the antithesis of the New England way—a religious way of life that made confession its central motif. The Grosvenor-Sessions case is one illustration among many of how New England women continued to measure themselves by "the moral allegory of repentance and confession" while men, at least when presenting themselves before legal authorities, adopted secular voices and learned self-interested strategies.[81]

For the Grosvenor men—at least the cluster of Sarah's cousins living near her—the key issue was not exposing sin but protecting the family's reputation. In the weeks before Sarah died, her cousins John and Ebenezer each attempted to investigate and sort out the roles and motives of Amasa Sessions and John Hallowell in the scheme to conceal Sarah's pregnancy. Grilled in August by Ebenezer about Sarah's condition, Hallowell revealed that "Sessions had bin Interseeding with him to Remove her Consep-

tion." On another occasion, when John Grosvenor demanded that he justify his actions, Hallowell was more specific. He "[did] with her [Sarah] as he did . . . because Sessions Came to him and was So very earnest . . . and offered him five pounds if he would do it." "But," Hallowell boasted, "he would have twenty of[f] of him before he had done." John persisted: did Amasa know that Hallowell was attempting a manual abortion at John's house on that day in early August? Hallowell replied that Amasa "knew before he did anything and was at Mr. Waldo's [a Pomfret tavernkeeper] to hear the event."[82]

John and Ebenezer, deposing three or four years after these events, did not mention having thrown questions at Amasa Sessions at the time, nor did they explain why they did not act immediately to have charges brought against the two conspirators. Perhaps these young householders were loath to move against a male peer and childhood friend. More likely, they kept their information to themselves to protect John's wife, Hannah, and their cousin Zerviah from prosecution as accessories. They may also have acted, in league with their uncle Leicester, out of a larger concern for keeping the family name out of the courts. Finally, it is probable that the male cousins, partly because of their own complicity and partly because they may have believed that Sarah had consented to the abortion, simply did not think that Amasa's and Hallowell's actions added up to the murder of their relative.

Three years later yet another Grosvenor cousin intervened, expressing himself much more vehemently than John or Ebenezer ever had. In 1742 John Shaw at age thirty-eight may have been perceived by the younger Grosvenors as too old—too close to the age when men took public office and served as grand jurors—to be trusted with their secret. Shaw seems to have known nothing of Sarah's taking the trade or having a miscarriage until 1745 when "the Storys" suddenly surfaced. Then Hannah and Zerviah gave him a truncated account. Shaw reacted with rage, realizing that Sarah had died not of natural causes but from "what Hollowell had done," and he set out to wring the truth from the doctor. Several times he sought out Hallowell in Rhode Island to tell him that "I could not look upon him otherwise Than [as] a Bad man Since he had Destroyed my Kinswoman." When Hallowell countered that "Amasa Sessions . . . was the Occasion of it," Shaw's fury grew. "I Told him he was like old Mother Eve When She said The Serpent beguild her; . . . [and] I Told him in my Mind he Deserved to dye for it." [83]

Questioning Amasa, Shaw was quick to accept his protestations of sincere regret and his insistence that Hallowell had "Deluded" him. [84] Shaw concluded that Amasa had never "Importuned [Hallowell] . . . to lay hands on her" (that is, to perform the manual abortion). Forged in the men's talk about the Grosvenor-Sessions case in 1745 and 1746 appears to have been a consensus that, while Amasa Sessions was somewhat blameworthy "as concerned in it," it was only Hollowell—the outsider, the man easily labeled a quack—who deserved to be branded "a Man of Death." Nevertheless, it was the stories of *both* men and women that ensured the fulfillment of a doctor's warning to Hallowell in the Leicester Grosvenor house just before Sarah died: "The Hand of Justice [will] Take hold of [you] sooner or Later." [85]

The Law

The hand of justice reached out to catch John Hallowell in November 1745. The warrants issued for the apprehension and examination of suspects that autumn gave no indication of a single informer or highly placed magistrate who had triggered the prosecution so long after the events. Witnesses referred to "those Stories Concerning Amasa Sessions and Sarah Grosvenor" that had begun to circulate beyond the inner circle of Pomfret initiates in the summer

of 1745. *Something* had caused Zerviah and Hannah Grosvenor to break their silence. [86] Zerviah provided the key to the puzzle, as she alone had been present at the crucial series of incidents leading to Sarah's death. The only surviving account of Zerviah's belated conversion from silence to public confession comes from the stories told by Pomfret residents into the nineteenth century. In Ellen Larned's melodramatic prose, the "whispered" tale recounted Zerviah's increasing discomfort thus: "Night after night, in her solitary chamber, the surviving sister was awakened by the rattling of the rings on which her bed-curtains were suspended, a ghostly knell continuing and intensifying till she was convinced of its preternatural origin; and at length, in response to her agonized entreaties, the spirit of her dead sister made known to her, 'That she could not rest in her grave till her crime was made public.' " [87]

Embellished as this tale undoubtedly is, we should not dismiss it out of hand as a Victorian ghost story. In early modern English culture, belief persisted in both apparitions and the supernatural power of the guiltless victim to return and expose her murderer. [88] Zerviah in 1742 already fretted over her sin as an accomplice, yet she kept her pledge of silence to her sister. It is certainly conceivable that, after a lapse of three years, she could no longer bear the pressure of hiding the acts that she increasingly believed amounted to the murder of her sister and an unborn child. Whether Zerviah's sudden outburst of talk in 1745 came about at the urging of some Pomfret confidante, or perhaps under the influence of the revivals then sweeping Windham County churches, or indeed because of her belief in nightly visitations by her dead sister's spirit, we simply cannot know. [89]

The Pomfret meetinghouse was the site of the first public legal hearing into the facts behind Sarah Grosvenor's death. We can imagine that townsfolk crowded the pews over the course of two November days to watch two prominent county magistrates examine a string of witnesses before pronouncing their preliminary judgment.[90] The evidence, they concluded, was sufficient to bind four people over for trial at the Superior Court: Hallowell, who in their opinion was "Guilty of murdering Sarah," along with Amasa Sessions, Zerviah Grosvenor, and Hannah Grosvenor as accessories to that murder.[91] The inclusion of Zerviah and Hannah may have been a ploy to pressure these crucial, possibly still reluctant, witnesses to testify for the crown. When Joseph Fowler, the king's attorney, prepared a formal indictment in the case eleven months later, he dropped all charges against Zerviah and Hannah. Rather than stand trial, the two women traveled frequently during 1746 and 1747 to the county seat to give evidence against Sessions and Hallowell.

The criminal process recommenced in September 1746. A grand jury empaneled by the Superior Court at its Windham session first rejected a presentment against Hallowell for murdering Sarah "by his Wicked and Diabolical practice." Fowler, recognizing that the capital charges of murder and accessory to murder against Hallowell and Sessions were going to fail before jurors, changed his tack. He presented the grand jury with a joint indictment against the two men, not for outright murder but for endangering Sarah's health by trying to "procure an Abortion" with medicines and "a violent manual opperation"; this time the jurors endorsed the bill. When the Superior Court trial opened in November, two attorneys for the defendants managed to persuade the judges that the indictment was faulty on technical grounds. However, upon the advice of the king's attorney that there "appear reasons vehemently to suspect" the two men "Guilty of Sundry Heinous Offenses" at Pomfret four years earlier, the justices agreed to bind them over to answer charges in March 1747.[92]

Fowler next moved to bring separate indictments against Hallowell and Sessions for the "highhanded misdemeanour" of endeavoring to destroy Sarah's health "and the fruit of her womb." This wording echoed the English common law designation of abortion as a misdemeanor, not a felony or capital crime. A newly empaneled grand jury of eighteen county yeomen made what turned out to be the pivotal decision in getting a conviction: they returned a true bill against Hallowell and rejected a similarly worded bill against Sessions.[93] Only Hallowell, "the notorious physician," would go to trial.[94]

On March 20, 1747, John Hallowell stepped before the bar for the final time to answer for the death of Sarah Grosvenor. He maintained his innocence, the case went to a trial jury of twelve men, and they returned with a guilty verdict. The Superior Court judges, who had discretion to choose any penalty less than death, pronounced a severe sentence of public shaming and corporal punishment. Hallowell was to be paraded to the town gallows, made to stand there before the public for two hours "with a rope visibly hanging about his neck," and then endure a public whipping of twenty-nine lashes "on the naked back."[95]

Before the authorities could carry out this sentence, Hallowell escaped and fled to Rhode Island. From Providence seven months after his trial, he audaciously petitioned the Connecticut General Assembly for a mitigated sentence, presenting himself as a destitute "Exile." As previously noted, fourteen respected male citizens of Providence took up his cause, arguing that this valued doctor had been convicted by prejudiced witnesses and hearsay evidence and asserting that corporal punishment was unwarranted in a misdemeanor case. While the Connecticut legislators rejected these petitions, the language used by Hallowell and his Rhode Island patrons is yet another marker of the distance separating many educated New England men at midcentury from their more God-fearing predecessors. Never mentioning the words "sin" or "repentance," the Providence men wrote that Hallowell was justified in escaping the lash since "every Person is prompted [by the natural Law of Self-Preservation] to avoid Pain and Misery."[96]

In the series of indictments against Hallowell and Sessions, the central legal question became who had directly caused Sarah's death. To the farmers in their forties and fifties who sat as jurors, Hallowell clearly deserved punishment. By recklessly endangering Sarah's life he had abused the trust that heads of household placed in him as a physician.[97] Moreover, he had conspired with the younger generation to keep their dangerous activities secret from their parents and elders.

Several rationales could have been behind the Windham jurors' conclusion that Amasa Sessions ought to be spared the lash. Legally, they could distinguish him from Hallowell as not being *directly* responsible for Sarah's death. Along with Sarah's male kin, they dismissed the evidence that Amasa had instigated the scheme, employed Hallowell, and monitored all of his activities. Perhaps they saw him as a native son who deserved the chance to prove himself mature and responsible. They may have excused his actions as nothing more than a misguided effort to cast off an unwanted lover. Rather than acknowledge that a culture that excused male sexual irresponsibility was responsible for Sarah's death, the Grosvenor family, the Pomfret community, and the jury men of the county persuaded themselves that Sessions had been ignorant of the potentially deadly consequences of his actions.

Memory and History

No family feud, no endless round of recriminations followed the many months of deposing and attending trials that engaged the Grosvenor and Sessions clans in 1746 and 1747. Indeed, as Sarah and Amasa's generation matured, the ties

between the two families thickened. In 1748 Zerviah married a man whose family homestead adjoined the farm of Amasa's father. Twenty years later, when the aging Sessions patriarch wrote his will, Zerviah and her husband were at his elbow to witness the solemn document. Amasa, who would inherit "the Whole of the Farm," was doubtless present also.[98] Within another decade the third generation came of age, and despite the painful memories of Sarah's death that must have lingered in the minds of her now middle-aged siblings, a marriage directly joining the two families finally took place. In 1775 Amasa's third son, and namesake, married sixteen-year-old Esther Grosvenor, daughter of Sarah's brother, Leicester, Jr.[99]

It is clear that the Grosvenor clan was not willing to break ranks with their respectable yeoman neighbors and heap blame on the Sessions family for Sarah's death. It would, however, be fascinating to know what women in Pomfret and other Windham County towns had to say about the outcome of the legal proceedings in 1747. Did they concur with the jurors that Hallowell was the prime culprit, or did they, unlike Sarah Grosvenor, direct their ire more concertedly at Amasa, insisting that he too was "a Bad man?" Several decades later, middle-class New England women would organize against the sexual double standard. However, Amasa's future career tells us that female piety in the 1740s did not instruct Windham County women to expel the newly married, thirty-two-year-old man from their homes.[100]

Amasa, as he grew into middle age in Pomfret, easily replicated his father's status. He served as militia captain in the Seven Years' War, prospered in farming, fathered ten children, and lived fifty-seven years beyond Sarah Grosvenor. His handsome gravestone, inscribed with a long verse, stands but twenty-five feet from the simpler stone erected in 1742 for Sarah.

After his death male kin remembered Amasa fondly; nephews and grandsons recalled him as a "favorite" relative, "remarkably capable" in his prime and "very corpulent" in old age. Moreover, local storytelling tradition and the published history of the region, which made such a spectacular ghost story out of Sarah's abortion and death, preserved Amasa Sessions's reputation unsullied: the *name* of Sarah's lover was left out of the tale.[101]

If Sarah Grosvenor's life is a cautionary tale in any sense for us in the late twentieth century, it is as a reminder of the historically distinctive ways in which socialized gender roles, community and class solidarity, and legal culture combine in each set of generations to excuse or make invisible certain abuses and crimes against women. The form in which Sarah Grosvenor's death became local history reminds us of how the excuses and erasures of one generation not unwittingly become embedded in the narratives and memories of the next cultural era.

Notes

The author is grateful to the members of the Early American Seminar at the Huntington Library, to her colleagues in the Women's History Reading Group at UCI for helpful criticisms of earlier versions of this article, and to Carol Karlsen, Stanley Katz, Robert Moeller, and Laurel Ulrich for close readings and advice.

1.   The documentation is found in the record books and file papers of the Superior Court of Connecticut: *Rex v. John Hallowell et al.,* Superior Court Records, book 9, 113, 173, 175, and Windham County Superior Court Files, box 172, Connecticut State Library, Hartford. Hereafter all loose court papers cited are from *Rex v. Hallowell,* Windham County Superior Court Files, box 172, unless otherwise indicated. For the quotations see Security bond for John Hallowell, undated; Deposition of Ebenezer Grosvenor, probably April 1746; Indictment against John Hallowell and Amasa Sessions, Sept. 20, 1746; Deposition of Parker Morse.

2.   One such abortion was reported in *Gentleman's Magazine* (London) 2, 20 (August 1732): 933–34; see Audrey Eccles, *Obstetrics and Gynaecology*

in Tudor and Stuart England (London, 1982), 70. On the history of abortion practices see George Devereux, "A Typological Study of Abortion in 350 Primitive, Ancient, and Preindustrial Societies," in Harold Rosen, ed., *Abortion in America: Medical, Psychiatric, Legal, Anthropological, and Religious Considerations* (Boston, 1967), 97–152; Angus McLaren, *Reproductive Rituals: The Perception of Fertility in England from the Sixteenth Century to the Nineteenth Century* (London, 1984), ch. 4; Linda Gordon, *Woman's Body, Woman's Right: A Social History of Birth Control in America* (New York, 1976), 26–41, 49–60; and Edward Shorter, *A History of Women's Bodies* (New York, 1982), ch. 8.

For specific cases indicating use of herbal abortifacients in the North American colonies see Julia Cherry Spruill, *Women's Life and Work in the Southern Colonies* (New York, 1972; orig. pub. Chapel Hill, N.C., 1938), 325–26; Roger Thompson, *Sex in Middlesex: Popular Mores in a Massachusetts County, 1649–1699* (Amherst, Mass., 1986), 11, 24–26, 107–8, 182–83; and Lyle Koehler, *A Search for Power: The "Weaker Sex" in Seventeenth-Century New England* (Urbana, Ill., 1980), 204–5. I have found two references to the use of an abortifacient in colonial Connecticut court files. Doubtless, other accounts of abortion attempts for the colonial period will be discovered.

3.   Abortion before quickening (defined in the early modern period as the moment when the mother first felt the fetus move) was not viewed by the English or colonial courts as criminal. No statute law on abortion existed in either Britain or the colonies. To my knowledge, no New England court before 1745 had attempted to prosecute a physician for carrying out an abortion.

On the history of the legal treatment of abortion in Europe and the United States see McLaren, *Reproductive Rituals,* ch. 5; Gordon, *Woman's Body, Woman's Right,* ch. 3; James C. Mohr, *Abortion in America: The Origins and Evolution of National Policy, 1800–1900* (New York, 1978); Michael Grossberg, *Governing the Hearth: Law and the Family in Nineteenth-Century America* (Chapel Hill, N.C., 1985), ch. 5; and Carroll Smith-Rosenberg, "The Abortion Movement and the AMA, 1850–1880," in *Disorderly Conduct: Visions of Gender in Victorian America* (New York, 1985), 217–44.

4. Indictment against John Hallowell, March 1746/47.

5. The story of the decriminalization of fornication for men in colonial New England is told most succinctly by Carol F. Karlsen, *The Devil in the Shape of a Woman: Witchcraft in Colonial New England* (New York, 1987), 194–96, 198–202, 255. Laurel Thatcher Ulrich describes a late eighteenth-century Massachusetts jurisdiction in *A Midwife's Tale: The Life of Martha Ballard, Based on Her Diary, 1785–1812* (New York, 1990), 147–60. For New Haven County see Cornelia Hughes Dayton, "Women Before the Bar: Gender, Law, and Society in Connecticut, 1710–1790" (Ph.D. diss., Princeton University, 1986), 151–86. See also Zephaniah Swift, *A System of Laws of the State of Connecticut*, 2 vols. (Windham, Conn., 1795–1796), 1:209. A partial survey of fornication prosecutions in the Windham County Court indicates that here too the local JPs and annually appointed grand jurymen stopped prosecuting men after the 1730s. The records for 1726–1731 show that fifteen men were prosecuted to enjoin child support and twenty-one single women were charged with fornication and bastardy, while only two women brought civil suits for child maintenance. Nearly a decade ahead, in the three-year period 1740–1742, *no* men were prosecuted, while twenty-three single women were charged with fornication, and ten women initiated civil paternity suits.

6. Such also was the message of many rape trials in the mid and late eighteenth century. See Dayton, "Women Before the Bar," 112–43; trial of Frederick Calvert, Baron Baltimore, as reported in the *Connecticut Journal*, New Haven, June 10, 1768, and in other colonial newspapers and separate pamphlets; and the Bedlow-Sawyer trial discussed by Christine Stansell in *City of Women: Sex and Class in New York, 1789–1860* (New York, 1986), 23–30.

7. For a recent summary of the literature see Brief for American Historians as *Amicus Curiae* Supporting the Appellees 5–7, *William L. Webster et al. v. Reproductive Health Services et al.*, 109 S. Ct. 3040 (1989).

8. In none of the cases cited in n. 2 did the woman ingesting an abortifacient die from it. If abortions directed by male physicians in the colonial period were more hazardous than those managed by midwives and laywomen, then, in an inversion of the mid-twentieth century situation, women from wealthy families with access to, and preferences for, male doctors were those most in jeopardy. For a general comparison of male and female medical practitioners see Ulrich, *A Midwife's Tale*, 48–66, esp. 54.

9. Married women may have hidden their abortion attempts because the activity was associated with lewd or dissident women.

10. Conception must have occurred sometime in the months of January through March, most probably in late January. Sarah had been pregnant nearly seven months at her delivery in early August, according to one version offered later by her sister.

11. Hallowell's trade may have been an imported medicine or a powder he mixed himself, consisting chiefly of oil of savin, which could be extracted from juniper bushes found throughout New England. For a thorough discussion of savin and other commonly used abortifacients see Shorter, *History of Women's Bodies*, 184–88.

12. Deposition of Zerviah Grosvenor. In a second deposition Zerviah used the word "Medicines" instead of "trade"; Testimony of Zerviah Grosvenor in Multiple Deposition of Hannah Grosvenor et al.: hereafter cited as Testimony of Zerviah Grosvenor. Five times out of eight, deponents referred to "the trade" instead of simply "trade" or "some trade."

13. So her sister Zerviah later estimated. Testimony of Rebecca Sharp in Multiple Deposition of Hannah Grosvenor et al.

14. After she was let into the plot, Zerviah more than once watched Amasa take "a paper or powder out of his pockett" and insist that Sarah "take Some of it." Deposition of Zerviah Grosvenor.

15. Deposition of John Grosvenor; Deposition of Zerviah Grosvenor; Testimony of Zerviah Grosvenor in Multiple Deposition of Hannah Grosvenor et al. "Unlawful measure" was Zerviah's phrase for Amasa's "Method." Concerned for Sarah's well-being, she pleaded with Hallowell not to give her sister "any thing that should harm her"; Deposition of Zerviah Grosvenor. At the same time, Sarah was thinking about the quickening issue. She confided to a friend that when Amasa first insisted she take the trade, "she [had] feared it was too late"; Deposition of Abigail Nightingale.

16. Deposition of Zerviah Grosvenor; Testimony of Zerviah Grosvenor.

17. Hallowell claimed that he proceeded with the abortion in order to save Sarah's life. If the powder had had little effect and he knew it, then this claim was a deliberate deception. On the other hand, he may have sincerely believed that the potion had poisoned the fetus and that infection of the uterine cavity had followed fetal death. Since healthy babies were thought at that time to help with their own deliveries, Hallowell may also have anticipated a complicated delivery if Sarah were allowed to go to full term— a delivery that might kill her. On the operation and variable potency of herbal abortifacients see Gordon, *Woman's Body, Woman's Right,* 37, 40; Shorter, *History of Women's Bodies,* 177–88; and Mohr, *Abortion in America,* 8–9.

18. Testimony of Hannah Grosvenor in Multiple Deposition of Hannah Grosvenor et al. Hannah may have fabricated the account of Hallowell's deception to cover her own knowledge of and collusion in Hallowell and Sessions's scheme to conceal Sarah's pregnancy.

19. Deposition of Zerviah Grosvenor. Hallowell attended Sarah overnight at John Grosvenor's house once in July; Multiple Deposition of Sarah and Silence Sessions.

20. On Abigail's husband, Samuel, and his family see Clifford K. Shipton, *Biographical Sketches of Those Who Attended Harvard College in the Classes 1731– 1735* (Boston, 1956), 9:425–28; Pomfret Vital Records, Barbour Collection, Connecticut State Library. All vital and land records cited hereafter are found in the Barbour Collection.

21. Hearsay evidence was still accepted in many eighteenth-century Anglo-American courts; see J. M. Beattie, *Crime and the Courts in England, 1660–1800* (Princeton, N.J., 1986), 362–76. Sarah's reported words may have carried special weight because in early New England persons on their deathbeds were thought to speak the truth.

22. Twentieth-century obstetrical studies show an average of six weeks between fetal death and spontaneous abortion; J. Robert Willson and Elsie Reid Carrington, eds., *Obstetrics and Gynecology,* 8th ed. (St. Louis, Mo., 1987), 212. Hallowell evidently grasped the link between the two events but felt he

could not wait six weeks, either out of concern for Sarah's health or for fear their plot would be discovered.

23. A 1746 indictment offered the only other point at which the "instrument" was mentioned in the documents. It claimed that Hallowell "with his own hands as [well as] with a certain Instrument of Iron [did] violently Lacerate and . . . wound the body of Sarah"; Indictment against John Hallowell, endorsed "Ignoramus," Sept. 4, 1746.

24. Deposition of Abigail Nightingale.

25. Joint Testimony of Hannah and Zerviah Grosvenor in Multiple Deposition of Hannah Grosvenor et al.; Deposition of Hannah Grosvenor; Deposition of Zerviah Grosvenor.

26. Deposition of Ebenezer Grosvenor.

27. Deposition of John Grosvenor; Deposition of Hannah Grosvenor; Deposition of Ebenezer Grosvenor.

28. Testimony of Hannah Grosvenor, Alexander Sessions, and Rebecca Sharp in Multiple Deposition of Hannah Grosvenor et al. In a second statement Hannah said that "the head Seemed to be brused"; Deposition of Hannah Grosvenor.

29. Testimony of Rebecca Sharp, Hannah Grosvenor, and Alexander Sessions in Multiple Deposition of Hannah Grosvenor et al.; Testimony of Silence Sessions in Multiple Deposition of Sarah and Silence Sessions.

30. Joint Testimony of Hannah and Zerviah Grosvenor in Multiple Deposition of Hannah Grosvenor et al.; Deposition of Parker Morse of Woodstock, April 1746. Although Pomfret had had its own resident physician (Dr. Thomas Mather) since 1738, Sarah's family called in young Dr. Morse of Woodstock, who visited twice (he later admitted he was not much help), and a Dr. Coker of Providence (who I assume was Theodore Coker). On Mather see Ellen D. Larned, *History of Windham County, Connecticut* (Worcester, Mass., 1874), 1:354. On Morse: Shipton, *Biographical Sketches,* 9:424. On Coker: *Biographical Sketches,* 8:19, and Eric H. Christianson, "The Medical Practitioners of Massachusetts, 1630–1800: Patterns of Change and Continuity," in *Medicine in Colonial Massachusetts, 1620–1820,* Publications of the Colonial Society of Massachusetts, 57 (Boston, 1980), 123.

31. Deposition of John Grosvenor.

32. I am using a list of forty heads of household in the Mashamoquet neighborhood of Pomfret in 1731, presuming five persons to a household, and assuming a 2.5 percent annual population growth. See Larned, *History of Windham County*, 1:342, and Bruce C. Daniels, *The Connecticut Town: Growth and Development, 1635–1790* (Middletown, Conn., 1979), 44–51. Pomfret village had no central green or cluster of shops and small house lots around its meeting-house. No maps survive for early Pomfret apart from a 1719 survey of proprietors' tracts. See Larned, *History of Windham County* (1976 ed.), 1, foldout at 185.

33. Leicester's father, John Grosvenor, a tanner, had emigrated from England about 1670 and settled in Roxbury, Mass., whence the first proprietors of Pomfret hailed. John died in 1691 before he could re-settle on his Connecticut tract, but his widow, Esther, moved her family to their initial allotment of 502 acres in Pomfret in 1701. There she lived until her death at eighty-seven in 1738, known in the community as a woman of energy and "vigorous habits," "skillful in tending the sick," and habitual in "walking every Sunday to the distant meeting-house." See Daniel Kent, *The English Home and Ancestry of John Grosvenor of Roxbury, Mass.* (Boston, 1918), 10–13, and Larned, *History of Windham County*, 1:353–55.

34. Kent, *The English Home of John Grosvenor*, 10–13; Larned, *History of Windham County*, 1:200–202, 204, 208–9, 269, 354, 343–44; Charles J. Hoadly and J. Hammond Trumbull, eds., *The Public Records of the Colony of Connecticut, 1636–1776*, 15 vols. (Hartford, Conn., 1850–1890), vols. 5–9; Inventory of Leicester Grosvenor, Oct. 29, 1759, Pomfret District Probate Court Records, 2:260.

35. Larned, *History of Windham County*, 1:201, 204, 206, 208–9, 344; Ellen D. Larned, *Historic Gleanings in Windham County, Connecticut* (Providence, R.I., 1899), 141, 148–49; Francis G. Sessions, comp., *Materials for a History of the Sessions Family in America; The Descendants of Alexander Sessions of Andover, Mass., 1669* (Albany, N.Y., 1890), 34–35, hereafter cited as Sessions, *Sessions Family*. Nathaniel's inheritance from his father Alexander of Andover (d. 1687) was a mere £2.14.5.

36. Deposition of Hannah Grosvenor; Deposi-

tion of Ebenezer Grosvenor; Deposition of Anna Wheeler, Nov. 5, 1745; Deposition of Zerviah Grosvenor; Testimony of Zerviah Grosvenor.

37. Deposition of Zerviah Grosvenor; Testimony of Zerviah Grosvenor.

38. Deposition of Abigail Nightingale. Contradicting Amasa's attempt to disavow paternity were both his investment in Hallowell's efforts to get rid of the fetus and his own ready admission of paternity privately to Zerviah and Sarah.

39. Deposition of Ebenezer Grosvenor. Hallowell revealed this opinion in an Aug. 1742 conversation with Sarah's twenty-eight-year-old cousin Ebenezer at Ebenezer's house in Pomfret. In a study of seventeenth-century Massachusetts court records Roger Thompson finds evidence that when pregnancy failed to pressure a couple into marriage, it was often because love "had cooled"; Thompson, *Sex in Middlesex*, 69.

40. Testimony of John Shaw in Multiple Deposition of Hannah Grosvenor et al.

41. For one such case involving two propertied families see Kathryn Kish Sklar, "Culture Versus Economics: A Case of Fornication in Northampton in the 1740's," *The University of Michigan Papers in Women's Studies* (1978), 35–56. For the incidence of illegitimacy and premarital sex in families of respectable yeomen and town leaders see Dayton, "Women Before the Bar," 151–86, and Ulrich, *A Midwife's Tale*, 156.

42. Two years later, in Feb. 1744 (9 months before Amasa married), the senior Sessions deeded to his son the north part of his own farm for a payment of £310. Amasa, in exchange for caring for his parents in their old age, came into the whole farm when his father died in 1771. Pomfret Land Records, 3:120; Estate Papers of Nathaniel Sessions, 1771, Pomfret Probate District. On the delay between marriage and "going to housekeeping" see Ulrich, *A Midwife's Tale*, 138–44.

43. Sessions, *Sessions Family*, 60; Pomfret Vit. Rec., 1:29.

44. The banns never appeared on the meeting-house door. Sarah may have believed in this overdue betrothal. She assured her anxious sister Anna that "thay designed to mary as soone as thay Could and that Sessions was as much Concarned as she." Deposi-

tion of Zerviah Grosvenor; Testimony of Zerviah Grosvenor; Deposition of Anna Wheeler.

45. Deposition of Abigail Nightingale. I have argued elsewhere that this is what most young New England women in the eighteenth century did when faced with illegitimacy. Their parents did not throw them out of the house but instead paid the cost of the mother and child's upkeep until she managed to marry. Dayton, "Women Before the Bar," 163–80.

46. Deposition of John Grosvenor; Deposition of Abigail Nightingale. Amasa Sessions, "in his prime," was described as "a very strong man," so it is possible that his physical presence played a role in intimidating Sarah. See Sessions, *Sessions Family*, 31.

47. Deposition of Abigail Nightingale.

48. Like his wife, Leicester was not summoned to testify in any of the proceedings against Hallowell and Sessions.

49. Zerviah testified that, a day or two after Sarah fell sick for the first time in July, the family heard "that Doctor Hallowell was at one of our Neighbors [and] my Mother desired me to go and Call him." Deposition of Zerviah Grosvenor.

Sarah's mother had died in May 1724, when Sarah was eleven months old. Perhaps Sarah and Zerviah had a closer relationship with their grandmother Esther (see n. 33) than with their stepmother. Esther lived in their household until her death in 1738, when Zerviah was seventeen and Sarah fifteen.

50. Laurel Thatcher Ulrich, *Good Wives: Image and Reality in the Lives of Women in Northern New England, 1650–1750* (New York, 1982), ch. 5, esp. 98.

51. In Larned's account, the oral legend insisted that Hallowell's "transaction" (meaning the abortion attempt) and the miscarriage were "utterly unsuspected by any . . . member of the household" other than Zerviah. *History of Windham County*, 1:363.

52. Deposition of Parker Morse.

53. Within days of Sarah's miscarriage, the initial conspirators disclosed their actions to others: Hallowell talked to two of Sarah's older male cousins, John (age thirty-one) and Ebenezer (age twenty-eight), while Zerviah confessed to Amasa's brother Alexander (age twenty-eight) and his wife Silence. Anna Wheeler (age thirty-three), Sarah's older sister, knew of Sarah's pregnancy before the abortion operation and thus must have guessed or secured information

about the miscarriage. As we have seen, Sarah would soon confess privately to Abigail Nightingale, recently married and in her twenties. Others in the peer group may also have known. Court papers list seven witnesses summoned to the trials for whom no written testimony survives. At least four of those witnesses were in their twenties or early thirties.

54. The famous "bad books" incident that disrupted Jonathan Edwards's career in 1744 involved a similar group of unsupervised young adults ages twenty-one to twenty-nine. See Patricia J. Tracy, *Jonathan Edwards, Pastor: Religion and Society in Eighteenth-Century Northampton* (New York, 1980), 160–64. The best general investigation of youth culture in early New England is Thompson's *Sex in Middlesex*, 71–96. Thompson discusses the general ineffectiveness of parental supervision of courtship (52–53, 58–59, 69–70). Ellen Rothman concludes that in New England in the mid to late eighteenth century "parents made little or no effort to oversee their children's courting behavior"; Rothman, *Hands and Hearts: A History of Courtship in America* (New York, 1984), 25.

55. Helena M. Wall, *Fierce Communion: Family and Community in Early America* (Cambridge, Mass., 1990); Bruce H. Mann, *Neighbors and Strangers: Law and Community in Early Connecticut* (Chapel Hill, N.C., 1987).

56. Leicester Grosvenor's Will, Jan. 23, 1754, Pomfret Dist. Prob. Ct. Rec., 1:146. For recent studies linking consumption patterns and class stratification see Richard L. Bushman, "American High-Style and Vernacular Cultures," in Jack P. Greene and J. R. Pole, eds., *Colonial British America: Essays in the New History of the Early Modern Era* (Baltimore, 1984), 345–83; T. H. Breen, "'Baubles of Britain': The American and Consumer Revolutions of the Eighteenth Century," *Past and Present*, 119 (May 1988), 73–104; and Kevin M. Sweeney, "Furniture and the Domestic Environment in Wethersfield, Connecticut, 1639–1800," in Robert Blair St. George, ed., *Material Life in America, 1600–1860* (Boston, 1988), 261–90.

57. Killingly Land Records, 2:139; *Rex v. John Hallowell and Mehitable Morris*, Dec. 1726, Windham County Court Records, bk. 1:43, and Windham County Court Files, box 363. Hallowell paid the £28 he owed Mehitable, but there is no evidence that he

took any other role in bringing up his illegitimate namesake. Just before his death, Samuel Morris, the maternal grandfather of John Hallowell, Jr., out of "parentiall Love and Effections," deeded the young man a three-hundred-acre farm "for his advancement and Settlement in the World"; Killingly Land Rec., 4:261.

58.   Hallowell was clearly the mastermind of the scheme, and there is little doubt that he lied to the authorities when questioned. According to one witness, Hallowell had exclaimed that "If he knew who" had informed anonymously against him, "he would be the death of him tho he ware hanged for it the next minit"; Letter of Joseph Leavens, Sept. 1727, Windham Sup. Ct. Files, box 170. The case is found in Windham Sup. Ct. Files, box 170; Sup. Ct. Rec., bk. 5:297–98; and *Public Records Conn. Colony,* 7:118. One associate Hallowell recruited was Ephraim Shevie, who had been banished from Connecticut for counterfeiting four years earlier. See Kenneth Scott, *Counterfeiting in Colonial America* (New York, 1957), 41–45.

59.   The authority on counterfeiting in the colonies is Kenneth Scott. His 1957 general book on the subject emphasizes several themes: the gangs at the heart of all counterfeiting schemes, the ease with which counterfeiters moved from colony to colony (especially between Connecticut and Rhode Island), "the widespread co-operation between" gangs, "the readiness of [men of all ranks] ... to enter such schemes," the frequent use of aliases, the irresistible nature of the activity once entered into, and "the extreme difficulty of securing the conviction of a counterfeiter"; *Counterfeiting in Colonial America,* esp. 123, 35, 10, 36. See also Scott's more focused studies, *Counterfeiting in Colonial Connecticut* (New York, 1957), and *Counterfeiting in Colonial Rhode Island* (Providence, R.I., 1960).

For an illuminating social profile of thieves and burglars who often operated in small gangs see Daniel A. Cohen, "A Fellowship of Thieves: Property Criminals in Eighteenth-Century Massachusetts," *Journal of Social History* 22 (1988): 65–92.

60.   Richard D. Brown, "The Healing Arts in Colonial and Revolutionary Massachusetts: The Context for Scientific Medicine," in Col. Soc. Mass., *Medicine in Colonial Massachusetts,* esp. 40–42. For detailed

analysis of the backgrounds and training of one large sample of New England practitioners see Christianson, "Medical Practitioners of Massachusetts," in *Medicine in Colonial Massachusetts,* 49–67, and Eric H. Christianson, "Medicine in New England," in Ronald L. Numbers, ed., *Medicine in the New World: New Spain, New France, and New England* (Knoxville, Tenn., 1987), 101–53. That the majority of colonial physicians made "free use of the title 'doctor'" (ibid., 118) and simply "taught themselves medicine and set up as doctors" is reiterated in Whitfield J. Bell, Jr., "A Portrait of the Colonial Physician," *Bulletin of the History of Medicine* 44 (1970): 503–4.

61.   Hallowell's sons, baptized between 1730 and 1740, were named Theophilus, Bazaleel, Calvin, and Luther. Killingly Vital Records, 1:3, 24; Putnam First Congregational Church Records, 1:5–7, 14–15. Hallowell may have been one of the "'horse-shed' Christians" whom David D. Hall describes as concerned to have their children baptized but more interested in the men's talk outside the meetinghouse than in the minister's exposition of the Word. Hall, *Worlds of Wonder, Days of Judgment: Popular Religious Belief in Early New England* (New York, 1989), 15–16.

62.   Between 1725 and 1742, Hallowell was a party to twenty land sales and purchases in Killingly; he also assumed two mortgages. During the same period he was involved in county court litigation an average of three times a year, more often as plaintiff than defendant, for a total of forty-six suits.

63.   For example, in early 1735 Hallowell made a £170 profit from the sale of a sixty-acre tract with mill and mansion house that he had purchased two months earlier. Killingly Land Rec., 4:26, 36.

64.   Evidence of Hallowell's widespread clientele comes from his 1727–1746 suits for debt, from his traveling patterns as revealed in the depositions of the abortion case, and from a petition written in 1747 on his behalf by fourteen male citizens of Providence. They claimed that "Numbers" in Rhode Island "as well as in the Neighbouring Colonies" had "happily experienc'd" Hallowell's medical care. Petition of Resolved Waterman et al., Oct. 1747, Connecticut Archives, Crimes and Misdemeanors, series 1, 4:109.

65.   For a related hypothesis about the mobility of self-taught doctors in contrast to physicians from established medical families see Christianson, "Medi-

cal Practitioners of Massachusetts," in Col. Soc. Mass., *Medicine in Colonial Massachusetts*, 61.

In autumn 1745 Hallowell was in jail in Providence, for what reason (debt or crime) I have yet to discover. Counterfeiters, including a Killingly woman, were apprehended in Rhode Island in that year. See Scott, *Counterfeiting in Colonial Rhode Island*, 26–27, and Scott, *Counterfeiting in Colonial America*, ch. 6.

66. These figures apply to suits in the Windham County Court record books, 1727–1742. Hallowell may, of course, have prosecuted debtors in other jurisdictions.

67. In Dec. 1749 Samuel Hunt, "Gentleman" of Worcester County, revoked the power of attorney he had extended to Hallowell for a Killingly land sale. Hunt claimed that the physician had "behaved greatly to my hindrance [and] Contrary to the trust and Confidence I Reposed in him." Killingly Land Rec., 5:151.

68. Larned, *History of Windham County*, 1:356–59.

69. The petition's signers included Hopkins, merchant, assembly speaker, and Superior Court justice, soon to become governor; Daniel Jencks, judge, assembly delegate, and prominent Baptist; Obadiah Brown, merchant and shopkeeper; and George Taylor, justice of the peace, town schoolmaster, and Anglican warden. Some of the signers stated that they had made a special trip to Windham to be "Earwitnesses" at Hallowell's trial. The petition is cited in n. 64.

70. For discussions of the elite see Jackson Turner Main, *Society and Economy in Colonial Connecticut* (Princeton, N.J., 1985), esp. 317–66, and Joy B. and Robert R. Gilsdorf, "Elites and Electorates: Some Plain Truths for Historians of Colonial America," in David D. Hall, John M. Murrin, and Thad W. Tate, eds., *Saints and Revolutionaries: Essays on Early American History* (New York, 1984), 207–44.

71. Ulrich, *Good Wives*, 103–5, 113–17.

72. Compare the seventeenth-century case of Stephen Batchelor (Charles E. Clark, *The Eastern Frontier: The Settlement of Northern New England, 1610–1763* [New York, 1970], 43–44) with eighteenth-century Cape Cod, where ministers retained their posts despite charges of sexual misconduct (J. M. Bumsted, "A Caution to Erring Christians: Ec-

clesiastical Disorder on Cape Cod, 1717 to 1738," *William and Mary Quarterly*, 3d Series, 28 [1971]: 412–38). I am grateful to John Murrin for bringing these references to my attention. For a prominent Northampton, Massachusetts, man (Joseph Hawley) who admitted to lying in civil and church hearings in the 1740s and yet who suffered no visible damage to his career see Sklar, "A Case of Fornication," *Michigan Papers in Women's Studies* (1978), 46–48, 51.

73. See Ulrich, *Good Wives*, 195–201, and Cornelia Hughes Dayton, "Infanticide in Early New England," unpub. paper presented to the Organization of American Historians, Reno, Nev., March 1988.

74. Burying the child was one of the key dramatic acts in infanticide episodes and tales, and popular beliefs in the inevitability that "murder will out" centered on the buried corpse. For two eighteenth-century Connecticut cases illustrating these themes see Dayton, "Infanticide in Early New England," n. 31. For more on "murder will out" in New England culture see Hall, *Worlds of Wonder*, 176–78, and George Lyman Kittredge, *The Old Farmer and His Almanack . . .* (New York, 1920), 71–77.

75. Testimony of Zerviah Grosvenor.

76. Hall, *Worlds of Wonder*, 172–78.

77. Testimony of Silence Sessions in Multiple Deposition of Sarah and Silence Sessions; Testimony of Alexander Sessions in Multiple Deposition of Hannah Grosvenor et al.; Testimony of Silence Sessions; Hall, *Worlds of Wonder*, 174. Alexander and Silence may have had in mind their brother Amasa's interests as a criminal defendant when they cast doubt on Zerviah's reliability as the star prosecution witness.

78. Deposition of Abigail Nightingale.

79. Testimony of Zerviah Grosvenor; Deposition of John Grosvenor. Abigail Nightingale recalled a scene when Sarah "was just going out of the world." She and Amasa were sitting on Sarah's bed, and Amasa "endeavour[ed] to raise her up &c. He asked my thought of her state &c. and then leaning over her used these words: poor Creature, I have undone you[!]"; Deposition of Abigail Nightingale.

80. Deposition of Ebenezer Grosvenor; Testimony of John Shaw in Multiple Deposition of Hannah Grosvenor et al. See also Deposition of John Grosvenor. For discussions of male and female speech patterns and the distinctive narcissistic bravado of

men's talk in early New England see Robert St. George, "'Heated' Speech and Literacy in Seventeenth-Century New England," in David Grayson Allen and David D. Hall, eds., *Seventeenth-Century New England,* Publications of the Colonial Society of Massachusetts, 43 (Boston, 1984), 305–15; Dayton, "Women Before the Bar," 248–51, 263–83, 338–41; and John Demos, "Shame and Guilt in Early New England," in Carol Z. Stearns and Peter N. Stearns, eds., *Emotion and Social Change: Toward a New Psychohistory* (New York, 1988), 74–75.

81.   On the centrality of confession see Hall, *Worlds of Wonder,* 173, 241. The near-universality of accused men and women confessing in court in the seventeenth century is documented by Gail Sussman Marcus in "'Due Execution of the Generall Rules of Righteousnesse': Criminal Procedure in New Haven Town and Colony, 1638–1658," in Hall, Murrin, and Tate, eds., *Saints and Revolutionaries,* esp. 132–33. For discussions of the increasing refusal of men to plead guilty to fornication (the most frequently prosecuted crime) from the 1670s on see Thompson, *Sex in Middlesex,* 29–33; Karlsen, *Devil in the Shape of a Woman,* 194–96, 198–202; and Dayton, "Women Before the Bar," 168–69. On the growing gap between male and female piety in the eighteenth century see Mary Maples Dunn, "Saints and Sisters: Congregational and Quaker Women in the Early Colonial Period," *American Quarterly* 30 (1978): 582–601. For the story of how the New England court system became more legalistic after 1690 and how lawyerly procedures subsequently began to affect religious practices and broader cultural styles see Mann, *Neighbors and Strangers,* and John M. Murrin, "Anglicizing an American Colony: The Transformation of Provincial Massachusetts" (Ph.D. diss., Yale University, 1966).

82.   Deposition of Ebenezer Grosvenor; Deposition of John Grosvenor. Although a host of witnesses testified to the contrary, Hallowell on one occasion told Amasa's brother "That Sessions never applied to him for anything, to cause an abortion and that if She was with Child he did not Think Amasa knew it"; Testimony of Alexander Sessions in Multiple Deposition of Hannah Grosvenor et al.

83.   Testimony of John Shaw in Multiple Deposition of Hannah Grosvenor et al. One of these confrontations took place in the Providence jail, probably in late 1745 or early 1746.

84.   It is interesting to note that Sessions claimed to have other sources for strong medicines: he told Shaw that, had he known Sarah was in danger of dying, "he tho't he could have got Things that would have preserved her Life"; ibid.

85.   Ibid. Shaw here was reporting Dr. [Theodore?] Coker's account of his confrontation with Hallowell during Sarah's final illness. For biographical data on Coker, see n. 30.

86.   Testimony of Rebecca Sharp, Zebulon Dodge, and John Shaw in Multiple Deposition of Hannah Grosvenor et al.; Deposition of Ebenezer Grosvenor.

87.   Larned reported that, according to "the legend," the ghostly visitations ceased when "Hallowell fled his country." *History of Windham County,* 1:363.

88.   For mideighteenth-century Bristol residents who reported seeing apparitions and holding conversations with them see Jonathan Barry, "Piety and the Patient: Medicine and Religion in Eighteenth-Century Bristol," in Roy Porter, ed., *Patients and Practitioners: Lay Perceptions of Medicine in Preindustrial Society* (Cambridge, 1985), 157.

89.   None of the depositions produced by Hallowell's trial offers any explanation of the three-year gap between Sarah's death and legal proceedings. Between 1741 and 1747, revivals and schisms touched every Windham County parish except Pomfret's First Church, to which the Grosvenors belonged; see Larned, *History of Windham County,* 1:393–485, esp. 464.

90.   One of the magistrates, Ebenezer West, had been a justice of the county court since 1726. The other, Jonathan Trumbull, the future governor, was serving both as a county court justice and as an assistant. The fact that the two men made the twenty-four mile trip from their hometown of Lebanon to preside over this Inferior Court, rather than allow local magistrates to handle the hearing, may indicate that one or both of them had insisted the alleged crime be prosecuted.

91.   Record of the Inferior Court held at Pomfret, Nov. 5–6, 1745. Hallowell was the only one of the four persons charged who was not examined at this time. He was in jail in Providence for debt. Appre-

hended in Connecticut the following March, he was jailed until the Pomfret witnesses could travel to Windham for a hearing before Trumbull and West. At the second hearing, the magistrates charged Hallowell with "murdering Sarah ... *and* A Bastard Female Child with which she was pregnant" (emphasis added). See Record of an Inferior Court held at Windham, April 17, 1746.

92. Indictment against Hallowell, Sept. 4, 1746; Indictment against Hallowell and Sessions, Sept. 20, 1746; Pleas of Hallowell and Sessions before the adjourned Windham Superior Court, Nov. [18], 1746; Sup. Ct. Rec., bk. 12, 112–17, 131–33.

93. Sup. Ct. Rec., bk. 12, 173, 175; Indictment against John Hallowell, March 1746/47; *Rex v. Amasa Sessions,* Indictment, March 1746/47, Windham Sup. Ct. Files, box 172. See William Blackstone, *Commentaries on the Laws of England* (Facsimile of 1st ed. of 1765–69) (Chicago, 1979), 1:125–26, 4, 198.

94. Larned, *History of Windham County,* 1:363.

95. Even in the context of the inflation of the 1740s, Hallowell's bill of costs was unusually high: £110.2*s*.6*d.* Sessions was hit hard in the pocketbook too; he was assessed £83.14*s*.2*d.* in costs.

96. Petition of John Hallowell, Oct. 1747, Conn. Archives, Crimes and Misdemeanors, ser. 1, 4:108; Petition of Resolved Waterman et al., ibid., 109. Specifically, Hallowell and his supporters asked that his sentence be reduced to a fine in an amount "adequate to his reduced Circumstances." Such requests for reduced sentences were increasingly submitted by convicted felons in eighteenth-century Connecticut, and some were granted. See ibid., Ser. 1 and 2. I have not been able to peruse local Rhode Island records to see what became of Hallowell there after 1747. His name disappears from Connecticut records after 1749.

97. Note Blackstone's discussion of the liability of "a physician or surgeon who gives his patient a potion ... to cure him, which contrary to expectation kills him." *Commentaries,* 4:197.

98. Killingly Land Rec., 3:99; Estate papers of Nathaniel Sessions, 1771, Pomfret Prob. Dist. Although Zerviah bore five daughters, she chose not to name any of them after the sister she had been so close to. In 1747, the final year of the trials, Sarah's much older sister, Anna, gave birth to a daughter whom she named Sarah.

99. Pomfret Vit. Rec., 2:67.

100. Carroll Smith-Rosenberg, "Beauty, the Beast and the Militant Woman: A Case Study in Sex Roles and Social Stress in Jacksonian America," *American Quarterly* 23 (1971): 562–84. There were branches of the Female Moral Reform Society in several Connecticut towns.

101. Sessions, *Sessions Family,* 31, 35; Larned, *History of Windham County,* 1:363–64.

2   Childbirth Practices Among Native American Women of New England and Canada, 1600–1800

Ann Marie Plane

When seventeenth-century European men wrote descriptions of the New World, they often included detailed passages on Indian life.[1] Almost without exception, these authors marveled at the ease of childbirth among the "savages." A native woman went off alone into the forest and returned in a short while with a new baby, resuming her activities as if nothing had happened. Yet, to modern readers, these accounts seem rather incredible.[2] Did Indian women really lie down under any "Bush" or "Tree" that they fancied, as John Josselyn believed?[3] Was native childbirth actually so easy and painless that a woman might be "merry in the House, and delivered and merry againe"[4] in only a quarter of an hour, as Roger Williams wrote? Indeed, would a Pokanoket or a Micmac woman even recognize her experience in the descriptions made by French priests and English gentrymen?

This paper examines these accounts afresh. Although written by men and circumscribed by literary conventions of the day,[5] these sources do reveal part of the native woman's experience.

The scope of this study is limited to the natives of New England and eastern Canada, an area which includes the Narragansett, Massachusett, Nipmuc, Mohegan, Pequot, and Nauset Indians of southern New England, and their Abenaki, Passamaquoddy, Penobscot, Algonquin, Malecite, and Micmac neighbors to the north. Birthing practices may have varied from group to group, especially between the hunting bands of the north and the agricultural peoples to the south.[6] Unfortunately, European authors did not leave sufficient evidence to explore such distinctions. Therefore, while this essay groups these many people together, in no way does it suppose that all native peoples of this vast region found the same meanings in childbirth, or even that they all shared the same practices. Yet the accounts show that Native American women constructed a different sort of childbirth than their French or English neighbors. While European women would have called four or five female friends over to help during the birth, native women apparently preferred to be alone or attended by only a few people. The required sociability of Euro-Americans, what historians have called "social childbirth,"[7] was absent from the native world. To look at the experiences of Amerindian peoples, then, allows the recovery

ANN MARIE PLANE is Assistant Professor of History at the University of California at Santa Barbara.

Reprinted from *Medicine and Healing: The Dublin Seminar for New England Folk Life*, edited by Peter Benes (1992): 13–24, by permission of Boston University.

of a piece of the cultural diversity which shaped colonial America.

To understand what Indian women of colonial New England and Canada experienced, we must look at the cultural and environmental factors which made their childbirths different. While not necessarily a medical event, childbirth is certainly a biological one. Yet it is also a profoundly cultural act.[8] Since the pathbreaking work of Arnold van Gennep in the first decade of this century, anthropologists have described childbirth as a "liminal" experience—a fluid, transitional state which must be processed within each culture by various rites of passage.[9] How then did the Amerindian women of New England and eastern Canada experience this liminal period? And how did their beliefs and practices fit within the logic of their culture? Reading the accounts together, we can take a close look at a composite native birth and attempt to answer these questions.

Following traditional directives for proper diet and care during pregnancy, the Indian mother-to-be continued her daily routines until overcome by active labor. Then she left her village, either alone or in the care of one or two older women. Indian women did not give birth in isolation from their community, even though they may have often been physically separated from the village. In Rhode Island Roger Williams recorded the Narragansett phrases for "She is falling into Travell," "She is in Travell," "She is already delivered," and "She was just now delivered." All of these suggest that the woman's kin kept close tabs on the progress of her labor, even if they themselves were not present.[10] Women usually gave birth in small huts away from the village. A French missionary, Father Le Jeune, remarked that the wife of a male Malecite convert "was delivered of a child alone, and without the assistance of any one. She was confined in the morning, and at noon I saw her working." This woman "had withdrawn into a

miserable bark hut, which did not shelter her at all from the wind."[11] While it may have seemed miserable to the French priest, it probably was familiar to the native woman. Like all the New England and eastern Canadian Indians, Malecite women lived in separate small houses during their menstrual periods.[12] For this reason and others, separate huts were the preferred place of birth.

Because of the nature of the sources, few accounts describe the actual moment of birth. Both infrequent references in the accounts and modern ethnographic data suggest that the birthing woman probably remained in a vertical posture—either hanging, standing, kneeling, or squatting.[13] A 1691 description relates that a Micmac woman who had a difficult labor was helped into a hanging position in hopes of hastening the birth.[14] One early twentieth-century narrative, told by a Fox woman, details the use of similar practices by this closely related midwestern Indian group. During her labor her attendant put a strap above her. With each contraction she held on to the strap and sat up on her knees.[15] Many of the European commentators remarked that Indian women did not make noise during labor. The French missionary, Father Jouvency, noted that the women concealed "the pangs of childbirth," and "if a tear or a groan should escape any one of them, she would be stigmatized by everlasting disgrace."[16] Roger Williams noted similarly that most of the Indians "count it a shame for a woman in Travell to make complaint, and many of them are scarcely heard to groane."[17] Indeed, some Europeans thought that Indian women did not even feel pain during birth.[18]

Several explanations come to mind. One is that Indian women did not feel so much pain as European women—whether due to physical robustness, small fetal head size, or other physiological factors. Physical anthropologists have demonstrated that overall pelvic size does vary for different human populations, although variations in some traits are not obstetrically relevant. Yet they have also shown the tremendous variety of contingent factors, including diet, disease, and climate that, along with cultural practices, influence birth outcomes.[19] This author cannot begin to address these complex issues in any satisfactory way. Therefore this discussion will set these biosocial explanations aside and focus primarily on cultural practices and their influence. Second, perhaps Indian women had means of relaxation or were not so afraid of childbirth as their European contemporaries. Modern natural birth advocates have argued that reducing fear through education and relaxation can be effective in minimizing pain.[20] No evidence suggests that Indian women used such methods, although the European authors seemed to have found little fear of birth. A third possibility may be most likely. Perhaps, as Jouvency and Williams suggest, native women faced cultural constraints on their expression of pain. While pain seems to be an expected part of birth in all societies, medical anthropologists usually look at the ways in which it is processed—whether it is "highlighted or discounted." Indian women kept quiet during labor; they felt pain, but they did not let it overwhelm them. Modern ethnographic evidence supports this interpretation. Among the Cree, an Algonkian people of northern Ontario, ideal social behavior consists of reticence, emotional control, and self-reliance. A study of Cree women who gave birth in the 1940s and 1950s found that, in keeping with the Cree social model, both they and their native attendants were very careful to minimize the moments in which fear or pain interfered with self-control.[21] Perhaps natives of the seventeenth century valued a similar self-reliance.

Native women may also have taken steps to alleviate labor pains. These remedies probably varied from group to group, as the range of climates and the different subsistence patterns of the northeastern groups would have influenced their availability. Chrestien Le Clercq noted that the Micmac Indians of the Gaspé Peninsula would take a small bone found in the heart of a moose, make a powder of it, and then drink it in a broth.[22] He also recorded that the beaver's "kidneys" (probably the castor glands) "are sought by apothecaries, and are used with effect in easing women in childbirth."[23] (Le Clercq probably means native apothecaries, not European ones, although the statement is ambiguous.) While evidence is sketchy for other groups, various herbal remedies seem to have been commonly administered during childbirth throughout northern New England.[24] With the use of remedies, cultural attitudes toward birth, and the constraints on expression of pain, native women of New England and eastern Canada appear to have deemphasized the pain of childbirth.

After the birth the baby was usually washed. European observers reported that natives did this to make the child hardy,[25] and this may well be an accurate report of native beliefs. Nevertheless, we can decode other meanings in this practice as well. In his classic work on transition rituals van Gennep has argued that all such acts have social meaning. Although washing the baby is certainly a "hygienic practice," in van Gennep's view it also often serves as a rite to announce the baby's new separation from its mother.[26] Further, in native culture water was seen as a powerful element.[27] In 1674 a missionary, Daniel Gookin, wrote down a creation myth told by the Indians of Massachusetts. In this story all the people of the earth came from just two young women. These women were wading in the water when some of its foam "touched their bodies, from whence they became with child"; one had a boy, the other a girl, and these two begat the rest of the Indian people.[28] Such a story suggests the link between water and fecundity in the fertility beliefs of some New England native peoples. Washing the newborn may have been a symbolic expression of fecundity, at the same time as it both cleaned the baby and served as a rite to make the baby fully alive.

After the washing the mother or an attendant wrapped the baby in skins, often of beaver,[29] and placed it on a cradleboard. Among some peoples the baby was fed some oil—of seal, bear, or other respected animals.[30] Perhaps they felt this would secure the protection of the animal over the new baby. Among other peoples the mother kept a piece of the navel cord in a little sack which she put around the baby's neck. Le Jeune reported that Algonkian people believed "if they were to lose it, their children would all be dolts and lacking in sense."[31] Van Gennep argues that such practices might protect the baby from any "diminution of his personality" and serve as a means of affirming his ties to his family.[32] Yet with such limited evidence it is difficult to assess the exact meaning of these practices in the native cultures.

Most accounts report that the mother returned to normal life right away. But a few suggest that the mother remained at the birth house for a prescribed period after the birth, sometimes as much as month or more.[33] Perhaps, when a woman gave birth during travel, she had to press on, as described by Le Clercq and others.[34] But even when she did, she may have observed special transitional rituals. For example, Susannah Johnson, an English woman taken captive in 1754, gave birth during her forced march to Canada. For at least six nights after the birth her captors built her a new hut or "booth" to sleep in each night.[35] Probably, then, most native women stayed in a special house for a minimum of a few days after childbirth. However, even after they appeared to resume their normal activities, they may have continued to observe special proscriptions. Father Le Clercq noted that Micmac women would not eat from the usual family dishes for a month or two because

they believed that this might cause animals no longer to allow themselves to be hunted.[36] Others would stay separate from their husbands during the transitional period.[37]

In most cultures childbirth, as a liminal period, requires some sort of event to signal the safe return of the mother to her normal life. These "rites of incorporation" may have been more elaborate for first-time mothers and observed less dramatically for those bearing subsequent children.[38] At this distance the exact events of incorporation among the natives of New England and eastern Canada may never be identified. Indeed, the rites of incorporation probably varied tremendously from group to group. These proscriptions and other aspects of native childbirth practice are closely related to taboos which governed the behavior of menstruating women. As in many tribal societies, menstruation probably occurred relatively infrequently, due to later puberty, heavy physical labor, diets low in fat, and lengthy infant nursing.[39] Like childbirth, the infrequent menstrual periods had a liminal status, and native women similarly withdrew to special houses. They remained there for the duration of their courses.[40] Father Biard reported that the Acadian natives "always put up a separate cabin for the women when they have their menses, for then they believe them to be infectious."[41] Roger Williams wrote that the Narragansetts of southern New England called this house *Wetuomemese,* or "Little House." He noted, "Their women and maids live apart in [this dwelling], four, five, or six dayes, in the time of their monethly sicknesse, which custome in all parts of the Countrey they strictly observe, and no *Male* may come into that house."[42] Both menstruating and parturient women removed themselves from the larger community for the duration of these powerful liminal states.

These proscriptions have often been interpreted as taboos designed to prevent the "impure" woman from polluting the rest of the community. Yet they should also be read as a way of protecting the women from contact with the everyday world of village life.[43] Menstruating and pregnant women shared special powers which could effect good or ill.[44] They could cause sickness or bad fortune in hunting,[45] but the pregnant woman could also bring good results under certain circumstances. According to François du Peron, her presence was required to make a certain root effective in removing an arrow from a wounded man.[46] Menstruating, pregnant, and postpartum women shared a similar status in these native cultures as conduits of change, whether for good or ill.

Why, then, did native women choose to leave their villages during such powerful states as menstruation and childbirth? In many human communities the boundaries between forest and village, water and land seem to serve as "thresholds," or transition points between various worlds.[47] The Indians of New England and eastern Canada believed that passage across these zones allowed humans to come into contact with natural manifestations of spiritual power, what one anthropologist has termed the "grandfathers" or the "dragon man-beings." These powerful beings, especially those inhabiting the underwater world, were "the sources of precious gifts, medicines, and charms."[48] Perhaps this was one reason that native women preferred to give birth near the water: several accounts mention birthsites near brooks or streams.[49] Possibly they believed sites close to these protectors might ensure a successful childbirth. Certainly, though, both the village and the woman had best be protected from the ill effects of contact between them.[50] For this reason the woman left the quotidian world of the village for that of the forest.

Difficult births offer further insight into the symbolic world of these peoples. Most native births probably proceeded normally.[51] By all accounts the Indians were healthy and strong, ex-

cept for their curious vulnerability to contagious diseases—European diseases to which they had no immunities. Native women also may have had fewer children than their European neighbors. They commonly nursed for three years, and they practiced infanticide and abortion.[52] Such practices would have contributed to a greater spacing between the children of Amerindian women than of English women. This might have meant that Indian women had fewer of the complications associated with multiple pregnancies,[53] experiencing better birth outcomes than European women. In the European tradition midwives used manual version—manipulating the abdomen to turn the baby into a favorable birth position. In rare cases obstetrical specialists might reach inside the vagina, using knives and hooks to dismember the baby in hopes of saving the mother's life.[54]

By contrast, native peoples seem to have been less intrusive in their interventions, although they may not have achieved any better outcomes. As Pierre de Charlevoix recounted, among some Canadian peoples, younger members of the community aided a mother in a long labor by startling her with loud noises, "when the surprize occasions a sudden fright, which procures her an immediate delivery."[55] Le Clercq reported that Micmacs attached the woman's arms "above to some pole, her nose, ears, and mouth being stopped up. After this she is pressed strongly on the sides, in order to force the child to issue from the belly of its mother."[56] Similar practices were noted in 1931 among the Cherokee of North Carolina, suggesting their wide distribution and persistence among native peoples of eastern North America.[57] As a last resort, native women would call on the shamans—the traditional healers—to effect a cure.[58] Sickness could occur if an individual offended either a shaman or any one of the spirits of the land, air, or water.[59] As one Englishman perceived it, native healers cured their patients by means of "extraordinary strange motions of their bodies, insomuch that they will sweat until they foam; and thus continue for some hours together, stroking and hovering over the sick."[60] In addition to these treatments, the shaman used "external applications of herbs" and roots, as well as sweatbaths.[61] As with some illnesses, tobacco may have been used as an offering to the other-than-human spirits during labor in order bring on the birth.[62]

Of course, these practices changed after contact with the Europeans. While Native American childbirth was certainly structured by social interaction, it was not the same "social childbirth" as that of the European colonists. The agricultural peoples of New England probably faced greater changes than their Canadian neighbors. Tremendous European immigration and other factors forced southern New England's Indians into dependency by the late seventeenth century.[63] Codes of conduct for Christian Indians mandated the abandonment of many native practices, including menstrual separation and the use of shamans.[64] English missionaries praised the Christian Indian couple who refused to call upon a shaman even after the woman had suffered "great pains and sorrowful throes for sundry days, and could not be delivered." The couple prayed to the Christian God instead, and the baby was soon born.[65] At the Natick community the 1646 codes prohibited menstrual seclusion, stating that "the old Ceremony of the Maide walking alone and living apart so many dayes" was now subject to a 20-shilling fine.[66] Such prohibitions of native practice are a common facet of colonialism; indigenous peoples often abandoned practices found objectionable in the face of European pressure, demographic tragedies, and changes in subsistence patterns.[67] But even as they abandoned many censured practices, Amerindians may have come up with a creative synthesis of old and new ways of childbirth, just as they did for other aspects of

life.[68] Certainly, the complex changes in native childbirth need fuller investigation.

This examination of native birth practices recovers much of the lost world of Indian women. It has shown that the Indians of New England and eastern Canada had a fully elaborated system of childbirth belief and practice during the early years of the colonial period. While European authors recorded important details of native fertility systems, they were often blind to the meaning of practices so different from their own system of "social childbirth."[69] Yet such practices served native women well. Behavior that looked bestial to European eyes fitted logically into native culture. Rites and proscriptions governing behavior during this liminal state helped both the new mother and her community to process the dramatic transition to motherhood. But more important, this study has shown that the European system of social childbirth did not apply to Native American women. In order to understand childbirth in colonial America, we must go beyond the concept of social childbirth to include the reproductive practices of Amerindian peoples as well.

Notes

1. The author would like to thank Laurel Thatcher Ulrich, Robert B. St. George, Alice Nash, Katsi Cook, David W. Murray, David Hackett Fischer, Gregory Button, Ruth Holmes Whitehead, Bill Wicken, and Thomas Buckley for their guidance and criticisms.

2. See Paul J. Lindholdt, ed., John Josselyn, *Colonial Traveler: A Critical Edition of "Two Voyages to New England"* (1674; reprint ed., Hanover: University Press of New England, 1988), 91–92; Pierre Biard, "Relation of New France," in *Jesuit Relations and Allied Documents* [Acadia: 1611–1616], 72 vols., ed. Reuben Gold Thwaites (Cleveland: Burrows Bros. Co., 1896–1901), 3:109; William Wood, *New England's Prospect* (London: Thomas Cotes, 1634; reprint ed., Boston: n. p., 1897), 101; and Roger Wil-

liams, *A Key Into the Language of America,* ed. R. C. Alston (London: Gregory Dexter, 1643; reprint ed., English Linguistics, 1500–1800 series, 299, Menston, England: Scolar Press, Ltd., 1971), 141.

3. Lindholdt, ed., *Colonial Traveler,* 91–92.

4. Williams, *A Key,* 141.

5. See Margaret T. Hodgen, *Early Anthropology in the Sixteenth and Seventeenth Centuries* (Philadelphia: University of Pennsylvania Press, 1964), esp. 111–54.

6. William Cronon, *Changes in the Land: Indians, Colonists, and the Ecology of New England* (New York: Hill and Wang, 1983), 42, 53.

7. Richard W. Wertz and Dorothy C. Wertz, *Lying-In: A History of Childbirth in America* (New Haven: Yale University Press, 1989), 1–28; Ellen Fitzpatrick, "Childbirth and an Unwed Mother in Seventeenth-Century New England," *Signs: Journal of Women in Culture and Society* 8, 4 (Summer 1983): 744; Judith Walzer Leavitt, *Brought to Bed: Childbearing in America, 1750–1950* (New York: Oxford University Press, 1986), 37–38; Laurel Thatcher Ulrich, *A Midwife's Tale: The Life of Martha Ballard, Based on Her Diary, 1785–1812* (New York: Alfred A. Knopf, 1990), 12.

8. See Margaret Mead and Niles Newton, "Cultural Patterning of Perinatal Behavior," in *Childbearing—Its Social and Psychological Aspects,* ed. Stephen A. Richardson and Alan F. Guttmacher (n.p.: Williams and Wilkins Co., 1967), 147; Brigitte Jordan, *Birth in Four Cultures: A Crosscultural Investigation of Childbirth in Yucatan, Holland, Sweden, and the United States* (Montreal: Eden Press Women's Publications, 1980), 34, and throughout; Carole P. MacCormack, "Adaptation in Human Fertility and Birth," in *Ethnography of Fertility and Birth,* ed. Carole P. MacCormack (New York: Academic Press, 1982), 1–2.

9. Arnold van Gennep, *The Rites of Passage,* ed. Solon T. Kimball, trans. Monika B. Vizedom and Gabrielle L. Caffee (Chicago: University of Chicago Press, 1960), 50; Victor Turner, *The Forest of Symbols: Aspects of Ndembu Ritual* (Ithaca: Cornell University Press, 1967), 93–95; Sheila Kitzinger, "The Social Context of Birth," in *The Ethnography of Fertility and Birth,* ed. Carole P. MacCormack (New York: Academic Press, 1982), 182; Marla N. Powers, "Menstruation and Reproduction: An Oglala Case," *Signs: Jour-nal of Women in Culture and Society* 6, 1 (Autumn 1980): 55–56.

10. Williams, *A Key,* 140–41.

11. Paul le Jeune, "Relation de ce qui c'est passé en la Nouvelle France, en l'année 1639," in *Jesuit Relations and Allied Documents* [*Quebec and Hurons 1639*], 16:107.

12. Chrestien Le Clercq, *New Relation of Gaspesia: with the Customs and Religion of the Gaspesian Indians, 1691,* Champlain Society Publication, 3, trans. and ed. William F. Ganong (reprint ed., New York: Greenwood Press, 1968), 227.

13. Frada Naroll, Raoul Naroll, and Forrest H. Howard, "Position of Women in Childbirth: A Study in Data Quality Control," *American Journal of Obstetrics and Gynecology* 82 (1961): 953.

14. Le Clercq, *New Relation,* 90.

15. Truman Michelson, ed. and trans., "The Autobiography of a Fox Indian Woman," U.S. Bureau of American Ethnology, *Annual Report,* 40 (Washington, D.C., 1925), 315–21, quoted in James Axtell, ed. *The Indian Peoples of North America: A Documentary History of the Sexes* (New York: Oxford University Press, 1981), 29. The experiences of the Fox, who inhabited the northern Midwest, are relevant to this study, because they belonged to the Algonkian language group and shared many similarities with the Algonkian peoples of New England and eastern Canada. James Axtell is one of the few scholars who has addressed the issue of sexuality, pregnancy, and birth as part of this collection of primary documents pertaining to the life cycle in Algonkian-speaking societies.

16. Joseph Jouvency, "Concerning the Country and Manners of the Canadians, or the Savages of New France," in *Jesuit Relations and Allied Documents* [*Acadia 1610–1613*], 1:277.

17. Williams, *A Key,* 141. Scholars have similarly attributed this quiet to stoicism, likening it to the conduct of men under torture. See Bruce Trigger, *Children of the Aataentsic I: A History of the Huron People to 1660* (Montreal: McGill-Queen's University Press, 1976), 47; Axtell, *Indian Peoples,* 17.

18. Adriaen Van der Donck, *A Description of the New Netherlands,* 2d ed. (Amsterdam, 1656), trans. Jeremiah Johnson, *Collections of the New-York Historical Society,* 2d ser. 1 (1841): 200–201, quoted in Axtell,

Indian Peoples, 22. Van der Donck reported that "some persons" believed that Indians "do not suffer the pains of sin in bringing forth their children," although he himself did not espouse this view.

19. Personal communication, Dr. Robert G. Tague, Dept. of Geography and Anthropology, Louisiana State University at Baton Rouge, November 1, 1990. Indeed, there are also significant variations in pelvic size between differing Amerindian populations, dispelling any lingering notions of a "racial" (European/Indian) difference; for a related discussion, see also Dr. Robert G. Tague, "Variation in Pelvic Size Between Males and Females," *American Journal of Physical Anthropology* 80 (1989): 59–71.

20. Flora Hommel, "Twelve Years Experience in Psychoprophylactic Preparation for Childbirth," in *Psychosomatic Medicine in Obstetrics and Gynaecology,* ed. Norman Morris (New York: S. Karger, 1972), 49–51. There does seem to be evidence supporting the notion that seventeenth-century European women feared birth. See Wertz and Wertz, *Lying-In,* 23–24.

21. Jordan, *Birth in Four Cultures,* 36; Sarah Preston, "Competent Social Behavior Within the Context of Childbirth: A Cree Example," 211–17 in *Papers of the Thirteenth Algonquian Conference,* ed. William Cowan (Ottawa: Carleton University, 1982), 211–12. Also, as we see in narratives of white captives, Indian children were taught at very early ages to remain quiet. Possibly the women's restraint in childbirth stemmed from the same desire—which, after all, was an adaptation necessary for a hunter-gatherer society.

22. Le Clercq, *New Relation,* 275. This may be an error on the part of Le Clercq or in the process of translation or transcription, as the heart itself is a muscle and contains no bone.

23. Le Clercq, *New Relation,* 279.

24. Nicholas N. Smith, "The Adoption of Medicinal Plants by the Wabanaki," in *Papers of the Tenth Algonquian Conference,* ed. William Cowan (Ottawa: Carleton University, 1979): 168–69.

25. Le Clercq, *New Relation,* 88–89; he notes that the Micmac wash newborns "in the river as soon as they are born"; Van der Donck, *A Description,* 22, notes that they immerse the males in water in order to make them "strong brave men and hardy hunters."

26. Van Gennep, *The Rites of Passage,* 52.

27. According to one present-day Mohawk midwife, water served as a metaphor for both sperm and vaginal fluids among all northeastern Indian peoples. Katsi Cook (Mohawk), midwife, personal communication, November 20, 1989. It may have been important to the new mother to complete the circle of life— water, which served as a metaphor for the conception which began pregnancy, had to be used to make the baby fully alive after birth. Other areas of Indian life were also enacted as cyclical relationships. Jouvency explained that Canadian Indians buried their dead of all ages in a fetal position with the limbs pressed against the body "in order that, as they say, he [the dead person] may be committed to the earth in the same position he once lay in his mother's womb." Jouvency, "Concerning the Country," 265.

28. Daniel Gookin, "Historical Collections of the Indians in New England, of their several nations, numbers, customs, manners, religion and government, before the English planted there," 1674, *Collections of the Massachusetts Historical Society for 1792,* vol. 1 (1806), 146.

29. Josselyn and Le Clercq both report beaver skin, and the Huron sachem, Tonkhratacouan, also mentions being wrapped in a beaver skin. See Lindholdt, *John Josselyn,* 92; Le Clercq, *New Relation,* 88–89; and Francois Joseph le Mercier, "Relation de ce qui s'est passé en la Nouvelle France en l'année 1637," in *Jesuit Relations and Allied Documents* [*Hurons: 1637*], 13:107.

30. Le Clercq, *New Relation,* 88–89. This perhaps also had the practical benefit of helping to expel the meconium of the baby's first bowel movements.

31. Le Jeune, "Relation 1639," 16:197.

32. Van Gennep, *The Rites of Passage,* 51.

33. Marc LesCarbot, *The History of New France,* vol. 3, ed. W. L. Grant (Toronto: Champlain Society, 1914), 200. The Fox woman remained away from her home for thirty-three days. Michaelson, "Autobiography," 30.

34. Le Clercq, *New Relation,* 89.

35. Susannah Willard Johnson Hastings, *Narrative of the Captivity of Mrs. Johnson . . . Containing an account of her Sufferings with the Indians and French* (Walpole, N.H.: David Carlisle, 1796; reprint, Microfilm ed., Evans, *Early American Imprints*), 38.

36. Le Clercq, *New Relation,* 228.

37. LesCarbot, *History of New France,* 3:200.

38.   Van Gennep, *The Rites of Passage*, 45–46; Kitzinger, "The Social Context of Birth," 182–83; and Jordan, *Birth in Four Cultures*, 30–31.

39.   Thomas Buckley and Alma Gottlieb, "A Critical Appraisal of Theories of Menstrual Symbolism," in *Blood Magic: The Anthropology of Menstruation*, ed. Thomas Buckley and Alma Gottlieb (Berkeley: University of California Press, 1988), 44–45.

40.   Axtell, *Indian Peoples*, 58–59. Axtell notes an "eastern pattern of menstrual seclusion." Also, ethnographic accounts of the nineteenth century report widespread menstrual seclusion. See Editor's Note, *Jesuit Relations and Allied Documents*, 9:308–9.

41.   Biard, "Relation of New France," 105.

42.   Williams, *A Key*, 32.

43.   Buckley and Gottlieb, "A Critical Appraisal," 32–34.

44.   Axtell, *Indian Peoples*, 3.

45.   Paul Le Jeune, "Relation de ce qui s'est passé en la Nouvelle France, en l'année 1636," in *Jesuit Relations and Allied Documents* [*Quebec: 1636*], 9:123; François du Peron, Lettre au P. Joseph-Imbert du Peron, in *Jesuit Relations and Allied Documents* [*Hurons and Quebec: 1638–1639*], 15:181.

46.   Du Peron, Lettre, 181.

47.   The threshold idea comes from van Gennep, *The Rites of Passage*, 15–21, but a similar concept has recently been used by Alma Gottlieb as an organizing concept in her explanation of the prohibition against menstruating Beng women (Ivory Coast, Africa) entering the forest. Alma Gottlieb, "Menstrual Cosmology Among the Beng of the Ivory Coast," in *Blood Magic*, 65–66; George Hamell also points to the separation between village and forest, or village and ocean, as experienced by northeastern woodland American natives. George R. Hamell, "Mythical Realities and European Contact During the Sixteenth and Seventeenth Centuries," *Man in the Northeast* 33 (1987): 69.

48.   Hamell, "Mythical Realities," 69–70.

49.   Hastings, *Narrative*, 26; [Ezra Stiles], *Extracts from the Itineraries and other Miscellanies of Ezra Stiles . . . 1755–1794*, ed. Franklin B. Dexter (New Haven: Yale University Press, 1916), 146.

50.   For the classic analysis of the concept of taboo and defilement see Mary Douglas, *Purity and Danger: An Analysis of Concepts of Pollution and Taboo*

(1966; reprint ed., London: ARK Paperbacks, 1988), esp. 94–113, 140–58.

51.   Given adequate nutrition, it has been estimated that 96 percent of all births will proceed normally; 1 percent of these involve tearing, hemorrhages, and so forth; only 4 percent of all births should require intervention to relieve obstructions. See Ulrich, *A Midwife's Tale*, 170.

52.   Jouvency, "Concerning the Country," 257–58; Stiles, *Extracts*, 144–45.

53.   James Trussell and Anne R. Pebley, *The Potential Impact of Changes in Fertility on Infant, Child, and Maternal Mortality*, World Bank Staff Working Papers, 698; Population and Development Series, 23 (Washington, D.C.: World Bank, 1984), 3–4.

54.   Wertz and Wertz, *Lying-In*, 18.

55.   P[ierre] de Clarlevoix, *Journal of a Voyage to North America* (London, 1761), 2:54–57, 109–10, quoted in Axtell, *Indian Peoples*, 15.

56.   Le Clercq, *New Relation*, 90.

57.   Frans M. Olbrechts, "Cherokee Belief and Practice with Regard to Childbirth," *Anthropos* 26 (1931): 27.

58.   Gookin, "Historical Collections," 155. These shamans were usually male, but occasionally women filled the role. See Robert Grumet, "Sunksquaws, Shamans, and Tradeswomen: Middle Atlantic Coastal Algonkian Women during the Seventeenth and Eighteenth Centuries," in *Women and Colonization: Anthropological Perspectives*, Praeger Special Studies, ed. Mona Etienne and Eleanor Leacock (New York: J. F. Bergin Publishers, 1980), 53–54. These shamans would rid the sufferer of *tcipai* or *cheepi*—spirits which inhabited the sick person's body. See William S. Simmons, *Spirit of the New England Tribes: Indian History and Folklore, 1620–1984* (Hanover: University of New England Press, 1986), 101–2. Le Clercq's account uses the French word *ver* (literally: worm) to describe that which the shamans removed from their patients. Le Clercq, *New Relation*, 90, n. 2.

59.   Smith, "The Adoption of Medicinal Plants," 167.

60.   Gookin, "Historical Collections," 154.

61.   Gookin, "Historical Collections," 154; Simmons, *Spirit of the New England Tribes*, 91.

62.   Le Clercq, *New Relation*, 90, wrote: "If she feels it a little too severely, she calls on the jugglers

[Indian shamans], who come with joy, in order to extort some smoking tobacco, or other things of which they have need." This suggests the importance of tobacco in effecting cures; Hammell suggests that offerings were made at significant places to ask for favors from the otherworldly beings. Hamell, "Mythical Realities," 69–70.

63. Neal Salisbury, "Social Relationships on a Moving Frontier: Natives and Settlers in Southern New England, 1638–1675," *Man in the Northeast* 33 (1987): 95.

64. Gookin, "Historical Collections," 154.

65. Gookin, "Historical Collections," 155. Also recounted in Experience Mayhew, *Indian Converts: or, some Account of the Lives and dying speeches of the Christianized Indians of Martha's Vineyard, in New England* (London: for Samuel Gerrish, 1727), 13.

66. "Conclusions and Orders made and agreed upon by divers Sachims and other principall men amongst the Indians at Concord [1646]" in Thomas Shepard, *The Cleare Sun-shine of the Gospel, Breaking forth upon the Indians in New-England* (reprint ed., *Collections of the Massachusetts Historical Society*, 4, 3, Cambridge, Mass.: Charles Folsom, 1834), 40.

67. Buckley and Gottlieb, "A Critical Appraisal," 12. Among the Mackenzie Dene Indians of western Canada, nineteenth- and twentieth-century census data revealed no marked decline in population despite repeated epidemics of disease. This finding led one researcher to conclude that natives abandoned selective female infanticide to offset the deaths from disease. See June Helm, "Female Infanticide, European Diseases, and Population Levels Among the MacKenzie Dene," *American Ethnologist* 7, 2 (May 1980): 271, 274. Studies of the James Bay Cree Indians reveal a slight increase in fertility after modernization, associated with greater sedentarism, leading to more frequent coitus. See A. Romaniuk, "Modernization and Fertility: The Case of the James Bay Indians," *Canadian Review of Sociology and Anthropology* 11, 4 (November 1974): 344.

68. Indians in New England sometimes preserved material and cultural practices long after conversion to Christianity. For example, even relatively prosperous members of the eighteenth-century Natick Indian community rarely had more than one room in their English-style framed houses, organizing their households as their ancestors would have designed wigwams. See Kathleen J. Bragdon, "Probate Records as a Source for Algonquian Ethnohistory," in *Papers of the Tenth Algonquian Conference*, ed. William Cowan (Ottawa: Carleton University, 1979), 137–38.

69. This paper has, of necessity, not gone into the possible changes in Euro-American social childbirth in the seventeenth and eighteenth centuries. Undoubtedly there were many.

3 "The Living Mother of a Living Child": Midwifery and Mortality in Postrevolutionary New England

Laurel Thatcher Ulrich

Forty-one years ago, Richard Harrison Shryock could summarize the history of early American midwifery in a few sentences. "The history of obstetrics and of pediatrics," he wrote, "affords other illustrations of the way in which inadequate medical science affected the public health. Maternity cases were left, in English-speaking lands, almost entirely to midwives.... And since midwives lacked any scientific training, obstetrics proceeded on the level of folk practice, and with consequences which may be easily imagined."[1] The consequences could be imagined because few persons in 1948 doubted the superiority of medical science over folk practice.

The advent of "natural" childbirth, culminating in recent years in the revival of lay midwifery, has changed historical judgments as well as obstetrics. In revisionist histories of childbirth the pleasant story of scientific progress has been replaced by a darker tale of medical competitiveness and misplaced confidence in an imperfect science. Medical science did not on the whole increase women's chances of surviving childbirth until well into the twentieth century, the new histories argue, and may actually have increased the dangers. As Richard W. and Dorothy C. Wertz explain, puerperal fever, the dreaded infection that killed so many women in the nineteenth century, "is probably the classic example of iatrogenic disease—that is, disease caused by medical treatment itself."[2]

Although historians trace to the eighteenth century the gradual supplanting of midwives by physicians, most detailed studies have concentrated on the nineteenth century or later. The few discussions of childbirth in early America have dealt with urban centers and with the work of prominent physicians such as William Shippen of Philadelphia.[3] Almost nothing is known about rural obstetrics or about the activities and attitudes of midwives. This essay begins to fill that gap. Its central document is the manuscript diary of a Maine midwife, Martha Moore Ballard, who lived at Augusta (then part of Hallowell) from 1778 to 1812. It also uses English obstetrical literature, scattered physicians' and midwives' records from Maine and New Hampshire, and the papers of Dr. Jeremiah Barker of Gorham, Maine.

LAUREL THATCHER ULRICH is James Duncan Phillips Professor of Early American History at Harvard University, Cambridge, Massachusetts.

Reprinted from *The William and Mary Quarterly: A Magazine of Early American History and Culture*, Third Series, 46 (1989): 27–48, by permission of the author and the Omohundro Institute of Early American History and Culture.

Martha Ballard performed her first delivery in 1778, though her diary does not begin until six years later. Between 1785 and 1812 she recorded 814 deliveries. The expansiveness of her record is unusual not only among midwives (few of whom left any written evidence of their practice) but among country physicians as well. Yet the diary has received little scholarly attention. Historians who have used it have relied on an abridged version published in Charles Elventon Nash's *History of Augusta.* For most, the details of Ballard's practice have seemed less important than her symbolic image as a "traditional" midwife. One work portrays her as an untrained, intensely religious, and poorly paid practitioner, who nevertheless shared some of the attitudes of contemporary physicians. Another associates her with nineteenth-century controversies between midwives and physicians, emphasizing her helplessness when accused by a local physician of "meddling by giving her opinion of a disease."[4]

Serious study of the entire diary shatters such stereotypes. Although physicians were delivering babies in Hallowell as early as 1785, Martha Ballard was clearly the most important practitioner in her town. Because her record documents traditional midwifery at a moment of strength, it allows us to shift the focus of inquiry from the eventual triumphs of medical science to the immediate relations of doctors and midwives in an era of transition. What is most apparent on close examination is the *success* of Ballard's practice, measured on its own terms or against contemporary medical literature. Although elements of the new obstetrics had begun to filter into the region, the old rituals of childbirth remained powerful. In her record, it is the physicians—particularly the young physicians—who appear insecure and uncomfortable.

The diary also extends and deepens recent discussions concerning eighteenth-century modes of delivery. A number of historians have argued that English innovations, such as William Smellie's improved forceps, encouraged an interventionist obstetrics that eventually displaced the gentler practices of midwives. Edward Shorter has countered that eighteenth-century English midwives were themselves "wildly interventionist" and that physicians, not midwives, introduced the notion of "natural" childbirth. This new obstetrics, he argues, was in part a response to general cultural trends—a medical reflection of Enlightenment respect for nature—and a consequence of the work of pioneering physicians like Charles White, whose textbook published in London in 1773 was the first example of a fully noninterventionist obstetrics.[5]

By shifting the balance of attention from obstetrical prescriptions to obstetrical results, Ballard's diary provides a new vantage point for assessing this controversy. Although it reveals little about the particulars of Ballard's methods (we do not know, for example, whether she applied hog's grease to the perineum or manually dilated the cervix), it offers compelling evidence of her skill. Maternal and fetal mortality rates extracted from the diary compare favorably with those for physicians in both England and America, countering the horror stories of eighteenth-century literature as well as the casual assumptions of twentieth-century historians. The consequences of Ballard's practice need not be imagined.

In most respects Martha Ballard's is a typical eighteenth-century rural diary—a laconic record of weather, sermon texts, family activities, and visits to and from neighbors. Obstetrical and general medical entries are interwoven with this larger accounting of ordinary life, although she gave birth records a special significance by summarizing them in the margins, numbering each year's births from January to December.

Each delivery entry gives the father's name, the child's sex, the time of birth, the condition of mother and infant, and the fee collected. Many also include the time of the midwife's arrival and departure, the names of the attendants who assisted her, and the arrival of the "afternurse," who cared for the woman during lying-in (the week or two following delivery). Succeeding entries record follow-up visits or hearsay reports about the mother and child.

The account of Tabitha Sewall's delivery on November 12–13, 1790, is typical:

> I was Calld by Colonel Sewall to see his Lady who was in Labour. Shee was not so ill as to Call in other assistance this day. I slept with her till about 1 hour morn when shee calld her Neighbours to her assistance. Mrs Sewal was ill till 3 hour pm when shee was thro divine asistance made the Living Mother of a Living Son her 3d Child. Mrs Brooks, Belcher, Colman, Pollard & Voce assisted us . . . Colonel Sewall gave me 6/8 as a reward. Conducted me over the river.

The only unusual thing about this account is the reference to "divine asistance," suggesting that Mrs. Sewall or her midwife encountered some difficulty along the way. Everything else about the description is routine. The father or a near neighbor summoned the midwife. The woman remained "ill" for several hours. Just before the birth she called her female neighbors. The child was delivered—safely. The father paid the midwife and escorted her home. In the eight deliveries Martha Ballard performed for Tabitha Sewall, the description of one differs very little from another. Mrs. Sewall "was safe delivered at 7 hour morn of a fine Daughter and is Cleverly," Ballard wrote, or "Mrs Suall Delivard at 1 this morn of a son & is Cleverly."[6]

Ballard performed her first delivery at the age of forty-three shortly after moving to the District of Maine from Oxford, Massachusetts. Al-

though she had no doubt assisted in many births in Oxford (she was herself the mother of nine children), she seems not to have practiced alone until she came to Hallowell. Demographics may explain her entry into the profession. In Oxford she had been surrounded by older women; her maternal grandmother was still alive in 1777. In Hallowell she was one of the older women in a young and rapidly growing town. The diary opens in January 1785, the year she turned fifty. It ends with her death in May 1812, just ten days after she performed her last delivery at the age of seventy-seven.

The diary tells us nothing of how she acquired her skills, though genealogical data suggest that her family had something of a medical bent. Two of her sisters married doctors; a maternal uncle was a physician. Certainly, her family demonstrated an unusual commitment to education. Her uncle Abijah Moore was Oxford's first college graduate. Her younger brother, Jonathan, was the second.[7] She probably learned midwifery in the same way her husband, Ephraim, learned milling or surveying—by practice, by observation, and by working alongside someone who knew more than she.

Ballard's assurance as a midwife is the best evidence we have of her training. In almost one thousand births she did not lose a single mother at delivery, and only five women died in the lying-in period. Infant deaths were also rare. The diary lists fourteen stillbirths in 814 deliveries and five infant deaths within an hour or two of birth. When Mrs. Claton and her infant both died in the autumn of 1787, a week after delivery, Ballard noted the singularity of the event: "I asisted to Lay her out, her infant Laid in her arms, the first such instance I ever saw & the first woman that died in Child bed which I delivered."[8] The sight was as unusual as it was affecting. Under Martha Ballard's care, a woman could expect to become "the living mother of a living child."[9]

By twentieth-century standards, of course,

both maternal and infant mortality were high. The diary records one maternal death for every 200 births. Today the rate for the United States is one per 10,000. But as Judith Walzer Leavitt has demonstrated, such dramatic gains in obstetrical safety have come in the past fifty years; as late as 1930 there was one maternal death for every 150 births in the United States. A 1986 study of early twentieth-century births in a Portsmouth, New Hampshire, hospital gives stillbirth rates five times as high as Ballard's. The turning point for fetal as well as maternal deaths was the 1940s.[10]

The appropriate question is how Martha Ballard's record compared with those of her contemporaries, particularly with New England physicians who began the regular practice of obstetrics in the eighteenth and early nineteenth centuries. Direct comparisons are difficult, in part because physicians' records tend to be organized much differently from hers. Most are simply a record of fees collected. Some doctors kept notes on unusual cases; a few compiled mortality tables for their towns. Account books, obstetrical case notes, and mortality tables seldom overlap, however, so that we know the numbers of deliveries performed by one physician but not the results, the management of extraordinary cases by another but not the overall caseload, and the incidence of stillbirths for a given town but not the numbers of maternal deaths or the names of practitioners. Comparison with midwives' registers is easier, since midwives typically listed all births, live as well as stillbirths, chronologically from the beginning to the end of their careers. Few such lists survive, however, and none that I have found offers the kind of narrative detail available in the Ballard diary.

Despite the difficulties, it is nevertheless possible to construct some comparisons. Table 3.1 gives stillbirth ratios derived from Ballard's diary, two physicians' records, two midwives' registers, and several published mortality tables. At

Table 3.1 Comparative stillbirth rates

	Total births	Total stillbirths	Stillbirths per 100 live births
Martha Ballard Augusta, Maine 1785–1812	814	14	1.8
Hall Jackson Portsmouth, N.H. 1775–1794	511	12	2.4
Lydia Baldwin Bradford, Vt. 1768–1819	926	26	2.9
James Farrington Rochester, N.H. 1824–1859	1,233	36	3.0
Jennet Boardman Hartford, Conn. 1815–1849	1,113	36	3.3
Portsmouth, N.H. 1809–1810	541	14	2.7
Marblehead, Mass. 1808	222	7	3.3
Exeter, N.H. 1809	53	1	1.9
United States* 1942			2.0

*Fetal death ratio, defined as fetal deaths of 28 weeks' or more gestation per 1,000 live births.

Sources: Martha Moore Ballard Diary, 2 vols., Maine State Library, Augusta, Maine; J. Worth Estes, *Hall Jackson and the Purple Foxglove: Medical Practice and Research in Revolutionary America, 1760–1820* (Hanover, N.H., 1979), 120; A Copy of Records from an Original Memorandum Kept by Mrs. Lydia (Peters) Baldwin, typescript, Baker Library, Dartmouth College, Hanover, N.H.; James Farrington Medical Record Books, 1824–1859, Special Collections, Dimond Library, University of New Hampshire, Durham; "Midwife Records, 1815–1849, Kept by Mrs. Jennet Boardman of Hartford," Connecticut Historical Society, *Bulletin,* XXXIII (1968), 64–69; Lyman Spalding, *Bill of Mortality for Portsmouth,* Broadside (Portsmouth, N.H., 1809, 1810); John Drury, *Bill of Mortality for Marblehead, 1808,* Broadside (Marblehead, Mass., 1809); Joseph Tilton, M.D., *Bill of Mortality for Exeter, New Hampshire,* Broadside ([Exeter, N.H., 1809]); Helen M. Wallace, "Factors Associated with Perinatal Mortality and Morbidity," in Helen M. Wallace, Edwin M. Gold, Edward F. Lis, eds., *Maternal and Child Health Practices: Problems, Resources, and Methods of Delivery* (Springfield, Ill., 1973), 507.

first glance it is the success of Ballard's practice that stands out. Whether her record is compared to that of Hall Jackson, a prominent eighteenth-century physician, or to Lydia Baldwin's, a contemporary Vermont midwife, it is eminent.[11] Yet none of the mortality ratios is as high as impressionistic accounts would lead us to believe, nor are there clear differences between midwives and physicians.

Most obstetrical treatises published in the first three-quarters of the eighteenth century emphasized the terrors of obstructed birth. Even authors who mistrusted "man-midwifery" and the use of forceps acknowledged the problems. Sarah Stone, an English midwife writing in 1737, described a breech delivery in which it took her an hour and a half to turn and extract the fetus. When she reached for the child, it "suck'd my fingers in the Womb, which concern'd me, fearing it impossible for the poor Infant to be born alive." Writing two decades later Dr. Edmund Chapman, an English physician, included more gruesome tales. Among cautionary examples he cited one ignorant doctor who, not knowing "the Method of *Turning* a Child, made frequent use of the *Hook* and the *Knife*, and several other shocking and barbarous instruments, even while the Child was *Living*." Dr. William Smellie, the London physician whose improved forceps supposedly solved such problems as these, included vivid case studies in his published works, evenhandedly distributing the blame for mismanaged deliveries among superstitious midwives and poorly trained physicians. In comparison, Ballard's delivery descriptions are remarkably bland: "the foet[u]s was in an unnatural posetion but I Brot it into a proper direction and shee was safe delivered." Usually she said even less: "removed obstructions" or "used means."[12]

Just as striking, given the tenor of the prescriptive literature, is her independence of Hallowell's physicians. Although the English authors agreed that midwives were capable of handling routine deliveries, authorities differed on the question of their ability to negotiate emergencies. Most publishing physicians argued that the sign of a good midwife was her willingness to call for help when needed. As Brudenell Exton put it, "the more knowledge they have, the readier they are to send for timely Assistance in Cases of Danger." Sarah Stone, the English midwife, disagreed, as did Nicholas Culpeper, a seventeenth-century herbalist and astrologer whose books were still being reprinted in New England in the early nineteenth century. Culpeper told the "Grave Matrons" who followed his advice that "the Lord will build you Houses as he did the Midwives of the *Hebrews,* when *Pharaoh* kept their Bodies in as great bondage as *Physitians* of our times do your Understandings." Both authors believed that experienced midwives were better equipped to handle difficult deliveries than officious but poorly prepared physicians.[13]

Ballard's philosophy was closer to Culpeper's than to Exton's. Although she had cordial relations with Hallowell's physicians, several of whom occasionally officiated at routine births, she seldom needed their help. A handful of her patients called *both* a doctor and a midwife at the onset of labor, but even in those cases she usually handled the delivery. Only twice in her entire career did she summon a doctor in an emergency, once in 1785 and again in 1792. She was not herself responsible for the first emergency. Arriving late, she found the patient "greatly ingered by some mishap," though the midwife or neighbor who had delivered the child did "not allow that shee was sencible of it." Calling the doctor may have been Ballard's way of resolving a disagreement over the severity of the injury.[14]

In the other case she described her feelings in vivid language, though characteristically she offered little obstetrical detail:

My patients illness Came on at 8 hour morning. Her women were Calld, her Case was Lingering till 7 p.m. I removd dificulties & waited for Natures opperations till then, when shee was more severly atackt with obstructions which alarmed me much. I desird Doct Hubard might be sent for which request was Complid with, but by Divine assistance I performed the oppration, which was blisst with the preservation of the lives off mother and infant. The life of the latter I dispard off for some time.

In the margin of the day's entry she added, "The most perelous sien I Ever past thro in the Course of my practice. Blessed be God for his goodness."[15] Whether Dr. Hubbard's emergency skills included the forceps delivery of a living child or only the dismemberment of a dead one, we do not know. Fortunately, in this case as in all the others, Ballard and her patient got along without him.

In difficult deliveries she typically gave God the credit for her success. The phrases are formulaic: "Her illness was very sever a short space but Blessed be God it terminated in Safety and the infant is numbered among the living," or "She had a Laborious illness but Blessed be God it terminated in safety. May shee and I ascribe the prais to the Great Parent of the universe."[16] One should not assume from such language, however, that Ballard lacked confidence in her own ability or that she relied on faith to the exclusion of skill. She knew that God worked through her hands.

Her confidence may actually have increased with the arrival of Dr. Benjamin Page in Hallowell in 1791. Page is remembered in local history as an extraordinarily successful physician. When he died in 1844, after more than fifty years of practice, the *Boston Medical and Surgical Journal* published an eleven-page biography proclaiming his skills as a general practitioner, surgeon, and gentleman. According to the anonymous author, Page was also "unequalled in the success of his obstetric practice. . . . [H]e at-tended upwards of *three thousand females in their confinement, without the loss of a single life from the first year of his practice!* This is almost miraculous, and may challenge the professional records of Europe or America for anything to compare with it."[17]

This is not the picture of Page preserved in Ballard's diary. Her first encounter with the young doctor was at the delivery of his near neighbor, Mrs. Benjamin Poor, the wife of a printer newly arrived in the town. Perhaps the woman intended medical delivery; perhaps she was simply worried that her midwife would not arrive in time. "I Extracted the child," Ballard wrote. "He Chose to close the Loin." The language is opaque here, suggesting either a friendly division of duties or an officious takeover by the doctor. The second encounter was more troublesome. Ballard had been sitting up all night with twenty-year-old Hannah Sewall, who had recently arrived in Hallowell from the town of York. "They were intimidated," she wrote, "& Calld Dr. Page who gave my patient 20 drops of Laudanum which put her into such a stupor her pains (which were regular & promising) in a manner stopt till near night when she pukt & they returned & shee delivered at 7 hour Evening of a son her first Born."[18] Hannah Sewall's intimidation, so called, may have had something to do with the fact that she had grown up in an elite family in a coastal town and was already familiar with medical delivery. As for Ballard, she was openly annoyed. Thereafter she was unmerciful in reporting Page's mistakes.

"Sally Cocks went to see Mrs. Kimball," she wrote. "Shee was delivered of a dead daughter on the morning of the 9th instant, the operation performed by Ben Page. The infants limbs were much dislocated as I am informed." She even questioned the doctor's judgment on nonobstetrical matters. Called to treat an infant's rupture, she recommended the application of brandy. "They inform me that Dr. Page says it must be opined [opened], which I should think improper from present appearance," she added. In June 1798, while she was engaged in another delivery, the doctor again delivered a stillborn child. Her report of the event was blunt: "Dr Page was operator. Poor unfortunate man in the practice."[19]

Page was unfortunate, but in eighteenth-century terms he was also ill prepared, as his administration of laudanum at Hannah Sewall's delivery suggests. The prescriptive literature recommended the use of opiates for *false* pains but not for genuine labor; Page was apparently having difficulty telling one from the other. Experience was the issue here as in so many other aspects of midwifery. Ballard had sat through enough lingering labors to know promising pains from false ones. Her reference to the doctor's dislocation of an infant's limbs also suggests lack of familiarity with the difficult manual operation required in breech births. The English midwife Sarah Stone had warned against doctors like him, "boyish Pretenders," who having attended a few dissections and read a few books professed to understand the manipulative arts so important to midwifery. Even Henry Bracken, an author who insisted that midwives should call in a doctor in difficult births, cautioned, "I would never advise any one to employ a *young* physician."[20]

After 1800, Page's misadventures disappear from the diary. Presumably, he eventually learned the obstetrical art in the way Ballard did—by experience.

Extracting the child was only part of the problem. Toward the end of the eighteenth century, English writers began to give as much attention to the dangers of the lying-in period, particularly the problem of childbed fever, as to delivery itself. Puerperal fever may in fact have been rare in England in the early years of the eighteenth century; obstetrical treatises published before 1760 rarely comment on its treatment.[21] Thomas Denman's *Essays on the Puerperal Fever* appeared in London in 1768. Four years later Charles White appended a detailed account of puerperal mortality in British hospitals to his *Treatise on the Management of Pregnant and Lying-In Women.*[22]

Puerperal fever is a wound infection caused by bacterial invasion of the uterine cavity. The infectiousness of the disorder was first suggested in the 1840s by Dr. Oliver Wendell Holmes in the United States and Dr. Ignaz Semmelwis in Austria, though the bacteriology of the disease was not settled until the 1880s, when Louis Pasteur demonstrated the presence of what is now known as streptococcus in patients suffering from the affliction. The symptoms—elevated temperature, headache, malaise, and pelvic pain—usually do not appear until several days after delivery. With certain strains of bacteria there is a profuse and foul-smelling discharge.[23]

At least one of Martha Ballard's patients probably died of puerperal infection. Mrs. Craig was "safe Deliverd of a very fine Daughter" on March 31, 1790, but after five days finding her "not so well as I could wish," Ballard administered a "Clister [enema] of milk, water, & salt" and applied an "ointment & a Bath of Tansy, mugwort, Cammomile & Hysop which gave Mrs. Cragg great relief." A week later the woman was still "Exceeding ill." Someone (perhaps a physician) prescribed rhubarb and Peruvian bark but without effect. A day or two later Dr. Cony "plainly told the famely Mrs. Cragg must die." She expired that night. Ballard helped dress her body for burial. "The Corps were Coffined & sett in the west room," she wrote. "Purge & smell very ofensive." Meanwhile, neighbors came by turns to "give the infant suck."[24]

Although Ballard attempted no diagnosis in this case, the symptoms fit the clinical description of puerperal fever. Perhaps one or two others among the five maternal deaths in her practice can also be attributed to infection. One woman was "safe delivered," fell ill a few days later, and died two weeks after delivery. Another died four days after giving birth at a time when scarlet fever, a form of streptococcus infection, was present in the town. In the two remaining maternal deaths, however, other symptoms were apparent. One woman was suffering from measles.[25] The other was in convulsions when delivered of a stillborn daughter and was still experiencing "fitts" four days later when she died. She was no doubt a victim of eclampsia, the most severe stage of an acute toxemia of pregnancy, a condition that is still considered one of the gravest complications of childbirth today.[26]

The Ballard diary suggests that puerperal infection was present in late eighteenth-century Maine, but the random appearance of the disease shows why it was seldom identified and discussed. In contrast, contemporary English physicians were encountering a truly alarming phenomenon. Charles White reported mortality rates for several London and Dublin hospitals that at midcentury were losing one out of every thirty or forty patients to puerperal fever. In 1770, in one London hospital, one of every four women died, most from infection (see Table 3.2). White was astonished that two hospitals that had been established at the same time, were an equal distance from the center of London, were directed by eminent physicians, and treated the same number of patients should have markedly different death rates. In true Enlightenment fashion he concluded that one hospital smothered patients with an artificial regimen, while the more successful one was not only less crowded and closer to fields and fresh air but obliged patients to do more for themselves.

White believed that bad habits led to childbed fevers. "Violence used either by instruments or by the hand, in the extraction of the child or the placenta," might bring on an inflammation of

Table 3.2
Comparative maternal mortality rates

Place	Total births	Maternal deaths	Deaths per 1,000 births
London A			
1767–1772	653	18	27.5
1770	63	14	222.2
London B			
1749–1770	9,108	196	21.5
1770	890	35	39.3
London C			
1747–"present"	4,758	93	19.5
1771	282	10	35.4
London D	790	6	7.5
Dublin A			
1745–1754	3,206	29	9.0
Dublin B			
1757–1775	10,726	152	14.1
1768	633	17	26.8
1770	616	5	8.1
Martha Ballard			
1777–1812	998	5	5.0
1785–1812	814	5	6.1
United States			
1930			6.7
1935			5.8
1940			3.8
1945			2.1

Sources: Charles White, *A Treatise on the Management of Pregnant and Lying-In Women* (Worcester, Mass., 1793 [orig. publ. London, 1772]); Ballard Diary; Wallace, Gold, and Lis, eds., *Maternal and Child Health Practices*, 285.

the womb, a condition made worse by the custom of pampering women in childbed. A woman should not be delivered in a hot room, or have her child or placenta dragged from her, or lie in a horizontal position in a warm bed drinking warm liquids for a week after delivery. Physicians and midwives were both to blame for practices that all too frequently led to maternal death. He suspected that lower-class women, who could not afford pampering, did better in childbirth than their more affluent neighbors, and he cited christening and death ratios from London and Manchester parish records to prove his point.[27]

Had White known about Martha Ballard, he would have had a ready explanation for her success: she practiced among frontier women who lived close to nature. In fact, Ballard was probably guilty of one of the practices White deplored—using hot drinks laced with alcohol. Still, there is plenty of evidence in the diary of the kind of vigor he admired. Ballard's patients were not all as sturdy as Mrs. Walker, who was "sprigh about house till 11 [and] was safe delivrd at 12," or as courageous (or foolhardy) as Mrs. Herriman, who "wrode in a sleigh 13 miles after her illness was on her"; but few Hallowell women could afford to lie in bed.[28] Ballard's own daughter, Dolly Lambert, was "so well as to be helpt up and sett at table for breakfast" twenty-four hours after giving birth to her fourth child. Ballard generally left her patients in the care of an afternurse a few hours after delivery, but when she stayed overnight she helped to get the woman out of bed in the morning. "Got my patient up, Changd her Lining and came home," she wrote (in this case, twelve hours after delivery), and "help[ed] Mrs Williams up & maid her Bed and returned home" (twenty-four hours after birth).[29]

Modern epidemiology confirms Charles White's belief that environment affected mortality, though, of course, the theoretical explanations differ from his.[30] Because Ballard was a part-time practitioner who delivered women at home and shared their postpartum care with nurses and family members, she had little opportunity to spread puerperal infection from one patient to another. The opposite conditions existed in the London hospitals, where as White himself suspected, the use of instruments in delivery probably increased the lacerations and tears that encouraged septicemia. Higher incidence of venereal disease in London may also have been a factor.[31]

That childbearing was safer in rural Maine than in London hospitals hardly seems surprising. The more interesting question for our purposes is how the literature emanating from those hospitals affected obstetrical practice in country places. Here the writings of Dr. Jeremiah Barker of Gorham, Maine, are particularly revealing. In February 1785 Barker initiated a discussion in the Falmouth [Maine] Gazette over the causes of an unusual "mortality among child-bed women, which has prevailed of late." Dr. Nathaniel Coffin, whose practice was in Falmouth (now Portland), submitted an angry response that was published in the next issue of the paper. Yes, several women had died in childbed in and about the town, but since the cause was unknown there was nothing that could have been done to save them. He denied that there was an epidemic, and he accused Barker of awakening "all those fears and apprehensions, which are but too often cherished by the sex." The debate continued through four issues of the newspaper, Barker insisting that an excess of bile characterized all the cases of puerperal fever he had studied, Coffin retorting that Barker had misread the symptoms.[32]

Barker included additional detail on the puerperal fever controversy in "History of Diseases in the District of Maine," a manuscript that he wrote after his retirement from active practice in 1818. Taken together, the newspaper stories and the "History" tell us a great deal about how medical reforms, initiated in London, were received in America. In his letters to the Falmouth Gazette Barker appears as a bold empiric asserting the power of direct experimentation against the dated theories of academic physicians. In his manuscript he reveals that the source of his ideas was a work by Thomas Denman, presumably his 1768 essays on puerperal infection.[33]

According to Barker's history, the puerperal fever outbreak began at the same time as an equally troubling rash of wound infections. In the spring of 1784, he recalled, "some unusual appearances took place in wounds & bruises, even trivial ones, which baffled the skill of the Surgeon, and issued in the death of the patient. . . . Local inflammations chiefly from injuries were more frequent and untractable during the year than I ever knew them to be before or since. The subjects of these complaints were chiefly males and apparently of good constitutions."[34] At the same time, several women in Gorham, Falmouth, and adjoining towns contracted puerperal fever. Although Barker made no connection between the two phenomena, it is difficult for a twentieth-century reader to avoid doing so.

Since Barker gave no statistics on the number of men who died from infected wounds or of women who suffered from childbed fever, and since birth and death records for the region are incomplete, it is impossible to know how serious the problem really was. Barker simply tells us that few women who suffered childbed fever survived and that he attended autopsies in three different towns. Yet his description confirms the rarity of the disorder in the region. "The ill success which attended my practice," he wrote, "induced me to write to several aged & experienced physicians in different portions of Massachusetts, for advice, as puerperal fever had never appeared among us excepting in a few sporadic cases, which yielded to common means." His correspondents had never seen such an epidemic themselves, but they referred him to the works of Denman and other unnamed British authors. It was from Denman's book, apparently, that Barker got his notions about bile and the use of "the bark" (quinine) as a remedy. He also wrote to Dr. Ammi Ruhamah Cutter of Portsmouth, New Hampshire, who had reportedly experienced high mortality from childbed fever. Cutter suggested applying "fermenting cataplasms to the abdomen composed of flower & yeast."[35]

Barker credited none of these sources in his 1785 newspaper letters, however, nor did he elaborate on the problems in his own practice. Alluding to an unusual childbed mortality, "especially in the town of Falmouth" (where Nathaniel Coffin practiced), he offered his remedies as a disinterested effort to "secure the happiness of mankind." Although he claimed to have "taken the opinion of the Massachusetts Medical Society," he gave no names.[36] Whom was he addressing? Surely not his fellow doctors. If that had been his intent, he would have limited himself to the private correspondence he had already begun. Instead he reached beyond the medical fraternity to the literate public of his region. The very form of his argument suggests that some part of his intended audience was female.

When Barker asserted that his patients could testify to the effectiveness of his methods, Coffin countered, "I am sorry the Dr. is obliged to have recourse to the female sex for a vindication of them." He suggested that the young doctor read "Astruc, Brooks, and others" to correct his faulty diagnosis. Barker retorted that the proposed authors were not only "Obsolete" but "esteemed of less consequence, in many respects, than the opinion of some of the female sex, founded on experience, in this more enlightened age."[37] The reliance on experience was, of course, a staple of Enlightenment medicine. Whereas earlier physicians had relied on theoretical learning, English reformers like William Smellie had emphasized the necessity for practical training in the manual arts of midwifery. Ironically, the obstetrical Enlightenment encouraged physicians to assume women's work in the very act of celebrating its importance. As Thomas Denman expressed it, "A natural labour was the last thing well understood in the practice of midwifery, because scientific men, not being formerly employed in the management of common labours, had no opportunity of making observations upon them."[38]

Barker's regard for female experience was conditional. He praised enlightened women who sought his care but mistrusted traditional midwives and nurses. His case notes from 1774 describe his efforts to deliver a woman with an imperforate vagina after the ministrations of her "friends" had failed. "I found that nothing could be done but to dilate the Perineum for the egress of the Child," he wrote, "and 'tho the operation is simple, yet fearing the sensure of the Vulgar (if any misfortune should befall the patient, afterwards) advised to send for Dr. Savage as an assistant." As it happened, the dead fetus was delivered before the second physician arrived.[39] Barker's concern about the censure of "the Vulgar" suggests the difficulties many physicians had in establishing credibility in the region, not only in obstetrical but in general medical cases. One young man entering practice in Waterville, Maine, in the 1790s even signed contracts with prospective patients, promising not to charge them if his remedies failed.[40]

Like Benjamin Page of Hallowell, Barker had begun his medical career after a brief apprenticeship with a Massachusetts physician. In 1774 he was twenty-two and in his second year of practice. The newspaper debate suggests that, ten years later, he had grown tired of his practice in Gorham and adjoining towns and perhaps hoped to attract the attention of prosperous families in the port of Falmouth.[41] Jeremiah Barker knew that women, whether vulgar or enlightened, were guardians of a doctor's reputation.

For his part, Coffin was furious at Barker for questioning the skills of other physicians. He was also dismayed that the younger doctor should invoke the authority of the Massachusetts Medical Society, even though he was not a member. When Coffin wrote to the society in 1803 recommending a number of new members from Maine, he explicitly excluded Barker, partly on the basis of the 1785 affair, which still rankled. (The society ignored his advice and elected Barker anyway.)[42] Thus a young physician moving into obstetrical practice in the 1780s and 1790s had two obstacles to overcome—folk reliance on traditional midwifery and the mistrust of older, more conservative physicians.

For our purposes, however, the more important issue is the way in which the puerperal fever incident of 1784–1785 began to shape Barker's practice. All of the cases of childbed fever described in his history came from that outbreak, yet he used them to support a long, detailed discussion of the cure and prevention of the disorder. Even by his own account, puerperal fever cannot have been a serious problem in the region. Most of the physicians to whom he wrote had seen only scattered cases; all of them referred him to British authors for an understanding of the subject. Coffin even doubted that the deaths could be attributed to a single disease, and he questioned whether he or any other physician could have done anything to prevent them. In Barker's own practice the trouble also faded away. There were additional cases during the winter of 1784–1785, he wrote, yet the disease showed "decreasing malignancy and mortality. Since which it has not appeared among us, excepting in a few sporadic cases, which seldom proved fatal."[43] Yet by 1818 his interpretation of the 1784–1785 cases had expanded to encompass citations from medical literature published as late as 1817.[44] Barker measured his entire career against that single early disaster. Since it was never repeated, he assumed that his preventative practices were successful.

Barker combined the noninterventionist prescriptions of the late eighteenth century—better ventilation, lighter food, avoidance of alcohol—with more heroic measures. "The means of prevention may be reduced to two," he argued. First, the physician should treat the patient during labor as though she were already a victim of the disease, drawing blood, administering emetics and cathartics, debarring her "entirely from spirits," and keeping her "on a low diet, without any animal food, in a well ventilated apartment without any curtains, on a mattress or straw bed." The second method involved "facilitating or rather hastening, by artificial means, the termination of labour."[45] Presumably, this meant using forceps and possibly ergot, a powerful and dangerous drug that, when given orally, stimulates uterine contractions.[46] In this Barker departed from the advice of his 1784 mentor, Thomas Denman, who, like Charles White, believed that forceps should be used rarely and that hastening labor led to postpartum complications.[47]

What we have, then, is a clear example of the way in which medical literature in combination with local experience came to define a practice. Barker's need to differentiate himself from other practitioners, as well as his desire to apply the latest in scientific knowledge to the management of his practice, made it impossible for him to see the 1784–1785 outbreak as an anomaly. Thereafter, he was convinced that it was his own intervention that had prevented a similar disaster from occurring. In contrast, Martha Ballard's nonscientific, even providential interpretation of events enabled her to treat each case on its own terms. For every patient she did what she knew how to do and let God determine the outcome. This is not to say that she was incapable of experimentation or that she never wondered why one infant died and another lived. It is simply to argue that her craft was oriented toward practical results rather than theoretical explanation. The death of Mrs. Claton or Mrs. Craig did not destroy her confidence in the soundness of her methods. Hers was not an approach that encouraged innovation, but neither did it promote ill-considered intervention.

Adrian Wilson has estimated that, in nature, 96 percent of births occur spontaneously. Approximately 4 percent involve serious obstruction of some kind and cannot be delivered without intervention. An additional 1 percent, though spontaneous, result in complications—minor ones such as fainting, vomiting, and tearing of the perineum, or major events like hemorrhaging or convulsions.[48] Martha Ballard's records fit Wilson's typology well. Approximately 95 percent of her entries simply say "delivered" or "safe delivered." In the remaining 5 percent some sort of complication is indicated, by explicit reference to obstructions, an oblique comment on the severity of the labor, or simply an acknowledgment that the delivery was accomplished through the mercy of God. Her records thus attest to the relative safety of childbearing as well as to her skill in managing difficult labors. Her ministrations no doubt improved the conditions of birth, but, perhaps even more important, she did little to augment the dangers.

In this regard it is interesting to compare her records with those of James Farrington of Rochester, New Hampshire, a nineteenth-century physician whose caseload was similar to hers and whose records, unlike those of his eighteenth-century predecessors, are extraordinarily complete. Dr. Farrington began the study of medicine in 1814, two years after Ballard's death. His manuscript records include a systematic register of 1,233 deliveries performed between 1824 and 1859. At first glance, his stillbirth and mortality ratios confirm the conclusions of revisionist histories—that childbirth became more dangerous in the nineteenth century. Farrington's stillbirth ratios are higher than any of the eighteenth-century practitioners and closer to those of the nineteenth-century midwife Jennet Boardman (see Table 3.1). Even more striking is the number of maternal deaths at delivery. That he was occasionally called to complete someone else's mismanaged delivery is certain, though those few cases that include extended descriptions suggest that, regardless of practitioner, nineteenth-century obstetrical practice added new dangers to the old problems of obstructed birth. Curiously, there is no indication of puerperal fever in Farrington's records. One might have expected at least a few cases of infection over such a long career. Since his tightly organized accounts, with one exception, list deliveries *only*, it is possible that such cases, usually arising a week or so after delivery, appeared in another set of more general medical records.[49]

Farrington recorded five maternal deaths. One woman, he wrote, was "enfeebled by intemperance." Another had a severe cold and "spoke but few words after delivery, but sunk away without a groan." The most dramatic case had been abandoned by another physician. Farrington described it as "preternatural labor requiring in the end the dissection of the infant," adding details that might have come from English obstetrical literature a hundred years before: "the external parts of generation much lacerated and mangled by *hooks, pincers, and knives*." The woman survived Farrington's extraction of the dismembered fetus but died five days later. A fourth woman died of bleeding after an unidentified attendant failed to extract the placenta. The fifth woman suffered a ruptured uterus: "in a few minutes the whole child could be felt expelled from the Uterus within the abdominal cavity." The woman lived about an hour.[50]

The numbers are small, however, and, without more detail on postpartum infection, inconclusive. The most striking contrast between Farrington's and Ballard's records is not in mortality rates themselves but in their characterizations of delivery. The process of labor was biologically the same, yet their descriptions differ markedly. Whereas Ballard thought in terms of the general outcome ("left mother and child cleverly"), Farrington focused on theoretical categories. Labors were "natural," "tedious," "premature," "preternatural," "complicated," or,

after 1838, "instrumental," regardless of whether the mother and child survived.[51] Twenty percent of the deliveries in his records are listed as something other than "natural."[52]

Here the telltale category may be his 102 cases of "tedious" labor, defined in the medical literature as lasting longer than twenty-four hours. In one case, which terminated safely at twenty-six hours, Farrington reported taking blood from the woman's arm, then giving an opiate. Four hours before the birth he gave her "Ergot in Infusion" and was pleased when he was able to deliver the child "without Instruments though for several hours no alteration was made by the force of the Pains."[53] Reading such an account, one finds it difficult not to think of Ben Page's administration of laudanum at the delivery of Hannah Sewall. Ironically, the remedy that so dismayed Martha Ballard was by now a standard part of the physician's arsenal. The three remedies—laudanum, ergot, and forceps—went together, accomplishing, as the physicians and perhaps many of their patients thought, an artificial hastening of labor.[54]

Judith Walzer Leavitt has argued that women chose medical intervention. Sally Drinker Downing, for example, sought out the services of the Philadelphia physician William Shippen, who administered opium during her 1795, 1797, and 1799 deliveries. Leavitt concludes that "the prospect of a difficult birth, which all women fearfully anticipated, and the knowledge that physicians' remedies could provide relief and successful outcomes led women to seek out practitioners whose obstetric armamentarium included drugs and instruments."[55] Leavitt may be right about Downing, yet Martha Ballard's diary adds a new dimension to the question of choice. At ten o'clock on the evening of October 21, 1794, she was summoned to the house of Chandler Robbins, a Harvard graduate and new resident of Hallowell. "Doctor Parker was calld," she wrote, "but shee did not wish to see him when he Came & he returnd home. Shee was safe delivered of a son her first Born at 10 hour 30 minutes Evening"—that is, twenty-four and one-half hours after summoning the midwife. Ballard's reward for officiating at this "tedious labor" was eighteen shillings and the satisfaction of knowing that God and the parents were pleased.

This brief survey of Martha Ballard's diary and related documents supports the reformist point that birth is a natural process rather than a life-threatening event. It suggests that rural midwives were capable of managing difficult as well as routine births, that the need for medical intervention was by no means obvious, and that puerperal infection, though present, was still only a random problem in the last years of the eighteenth century. For midwives like Martha Ballard or Lydia Baldwin, experience defined competence, yet in the years following the Revolution a number of brash young men with more confidence than experience took up the practice of delivering babies. Not content with the more restrained role of older doctors, they consulted British literature and sought advice from other physicians to solve their problems and validate their skills. That they gravitated toward works that emphasized the necessity of intervention is hardly surprising. In a competitive environment no bright young physician could embrace Charles White's advice that the less done in childbirth the better. Employing forceps, letting blood, administering opiates and ergot, they set themselves apart from the manual skills and the providential faith of the midwives.

During the earlier years of Martha Ballard's midwifery in Hallowell, however, the success of such physicians was by no means assured. In 1800, when age, ill health, and a move to a more distant part of the town forced her to cut back her practice, she was the single most important practitioner in her town, and she knew it.

Notes

Versions of this article were presented at meetings of the Benjamin Waterhouse Medical Society (Boston University), the Maine Society for the History of Medicine, the American Antiquarian Society Seminar in Political and Social History, and the comparative history seminar at the University of New Hampshire. Acknowledgments: I am grateful to those groups and to Worth Estes, Judith Walzer Leavitt, Janet Polasky, and Cornelia Dayton for helpful comments. Some parts of this essay appear in my book A Midwife's Tale: The Life of Martha Ballard, Based on Her Diary, 1785–1812 (Knopf, 1990).

1. Shryock, The Development of Modern Medicine: An Interpretation of the Social and Scientific Factors Involved, rev. ed. (London, 1948), 77–78.

2. Judith Walzer Leavitt, Brought to Bed: Childbearing in America, 1750–1950 (New York, 1986), 56–57, and "'Science' Enters the Birthing Room: Obstetrics in America Since the Eighteenth Century," Journal of American History 70 (1983): 281–304; Wertz and Wertz, Lying-In: A History of Childbirth in America (New York, 1977), xi, x, 128.

3. Catherine M. Scholten, "'On the Importance of the Obstetrick Art': Changing Customs of Childbirth in America, 1760 to 1825," William and Mary Quarterly 34, 3 (1977): 429–31, and Childbearing in American Society, 1650–1850 (New York, 1985), ch. 2; Wertz and Wertz, Lying-In, ch. 2; Leavitt, Brought to Bed, 36–44, 263–65.

4. Nash, The History of Augusta: First Settlements and Early Days as a Town, Including the Diary of Mrs. Martha Moore Ballard (1785–1812) (Augusta, Me., 1904); Wertz and Wertz, Lying-In, 9–10; quotation from Scholten, Childbearing, 45; Leavitt, Brought to Bed, 37.

5. Wertz and Wertz, Lying-In, 34–43; Scholten, Childbearing, 34–36; Leavitt, Brought to Bed, 38–40; Shorter, "The Management of Normal Deliveries and the Generation of William Hunter," in W. F. Bynum and Roy Porter, eds., William Hunter and the Eighteenth-Century Medical World (Cambridge, 1985), 371–83.

6. Martha Moore Ballard Diary, 2 vols., Maine State Library, Augusta, Me., April 2, 1788, Dec. 31,

1786. According to the *Oxford English Dictionary*, *cleverly* means 'well' or 'in health' in some dialects. This is obviously the meaning Ballard intended. Henry Sewall, Tabitha's husband, also kept a diary. He mentioned Martha Ballard's presence on only four of the eight occasions, never recorded paying a fee, and only twice mentioned the presence of other birth attendants. Henry Sewall Diary, Massachusetts Historical Society, Boston, Mass.

7. The medical tradition continued into the nineteenth century. Ballard's diary was inherited and preserved by her great-granddaughter, Dr. Mary Hobart, who practiced obstetrics at New England Hospital in Boston. Clara Barton, the Civil War nurse and founder of the American Red Cross, was Ballard's grandniece.

8. Ballard Diary, Aug. 16, 20, 1787. Since the first fatality occurred during the diary period, I have included the 177 prediary births in arriving at the total of 991 births.

9. The phrase was conventional, and it persisted into the nineteenth century. Leavitt, for example, quotes a woman who gave thanks for having become "the living mother of a living and perfect child" (*Brought to Bed*, 34). Ballard's version of the statement was usually gender specific, as in "the living mother of a living son" or "the living mother of a fine Daughter" (Ballard Diary, Dec. 30, 1789).

10. Leavitt, *Brought to Bed*, 23–26; Helen M. Wallace, Edwin M. Gold, and Edward F. Lis, eds., *Maternal and Child Health Practices: Problems, Resources, and Methods of Delivery* (Springfield, Ill., 1973), 185; J. Worth Estes and David M. Goodman, *The Changing Humors of Portsmouth: The Medical Biography of an American Town, 1623–1983* (Boston, 1986), 298. In 2.3 percent of Ballard's deliveries the child was stillborn or died in the first twenty-four hours of life. For Portsmouth Hospital the figures were 11.4 percent (1915–1917), 4.8 percent (1925–1941), 1.2 percent (1954–1957), and 0.8 percent (1971–1983). Because methods of compiling statistics vary markedly over time, these numbers must be considered approximations. Stillbirth ratios, for example, might be affected by abortions, spontaneous or induced. On the development of obstetrical record keeping in general see James H. Cassedy, *American Medicine and Statistical Thinking, 1800–1860* (Cambridge, Mass., 1984), 80–83.

11. The lack of detail in the other sources makes it difficult to know whether the data are precisely comparable. Ballard's diary distinguishes between stillbirths and deaths within a few minutes or hours of birth. If other records melded those two categories, her record would look better by comparison. Still, adding the five very early deaths in her practice to stillbirths results in a ratio of only 2.3, almost identical with Jackson's and slightly lower than Baldwin's. Jennet Boardman's register includes three categories: "born dead," "died," and "died at age ——— or on ———." I list all those infants described as "dead" or "born dead" as stillborn but exclude the "died" entries, some of which deaths may have occurred immediately after birth.

12. Stone, *A Complete Practice of Midwifery . . .* (London, 1737), 76–77; Chapman, *A Treatise on the Improvement of Midwifery, Chiefly with Regard to the Operation . . . ,* 3d ed. (London, 1759), xiv; Smellie, *A Collection of Cases and Observations in Midwifery,* 3d ed. 3 (London, 1764), for example, 1–69, 416–27; Ballard Diary, Aug. 29, 1797, July 19, 1794, Feb. 18, 1799.

13. Nich[olas] Culpeper, *A Directory for Midwives; or, A Guide for Women, in Their Conception, Bearing, and Suckling Their Children . . .* (London, 1651), "Epistle Dedicatory"; Stone, *Complete Practice,* ix; Henry Bracken, *The Midwife's Companion; or, A Treatise of Midwifery, Wherein the Whole Art Is Explained . . .* (London, 1737), 146; Chapman, *Improvement of Midwifery,* vii–xiii; Brudenell Exton, *A New and General System of Midwifery . . .* (London, 1751), 11. The library of the College of Physicians and Surgeons, Philadelphia, has an autographed and annotated copy of Exton owned by Dr. John McKechnie, who emigrated from Scotland to Maine in 1755 and apparently practiced medicine until his death in 1782. Martha Ballard may have known him; three of his married daughters were among her patients (James W. North, *The History of Augusta* [Augusta, Maine, 1870], 913–914).

14. Ballard Diary, Nov. 11, 1785.

15. Ibid., May 19, 1792.

16. Ibid., June 30, 1807, March 31, 1800.

17. "Memoir of Benjamin Page, M.D.," *Boston Medical and Surgical Journal* 33 (1845): 9, 173.

18. Ballard Diary, Nov. 17, 1793, Oct. 9–10, 1794. For additional detail on relations between midwives

and physicians in Hallowell see Laurel Thatcher Ulrich, "Martha Moore Ballard and the Medical Challenge to Midwifery," in James Leamon and Charles Clark, eds., *From Revolution to Statehood: Maine in the Early Republic, 1783–1820* (Hanover, N.H., 1988), 165–83.

19. Ballard Diary, July 8, Aug. 14, 1796, June 14, 1798.

20. Thomas Denman, *An Introduction to the Practice of Midwifery* (New York, 1802 [orig. London, 1794, 1795]), 179; Stone, *Complete Practice*, 76–77, xiv; Bracken, *Midwife's Companion*, 194.

21. Exton, for example, gives no more attention to childbed fever than to afterpains (*System of Midwifery*, 150). In addition to the English works cited, I have read the Worcester, 1794, edition of Alexander Hamilton, *Outlines of the Theory and Practice of Midwifery*, first published in Edinburgh in 1784. It also ignores the problem.

22. White, *A Treatise on the Management of Pregnant and Lying-In Women* . . . (London, 1772).

23. Erna Ziegel and Carolyn Conant Van Blarcom, *Obstetric Nursing*, 6th ed. (New York, 1972), 522–26; Wertz and Wertz, *Lying-In*, 119–28; Leavitt, *Brought to Bed*, 154–55.

24. Ballard Diary, March 31, April 4, 5, 10, 11, 12, 13, 15, 16, 1790.

25. Ibid., Oct. 18, 21, 24–29, 1802.

26. Ibid., Feb. 26, 27, March 1, 2, 4, 1789; Ziegel and Van Blarcom, *Obstetric Nursing*, 208–213.

27. White, *A Treatise on the Management of Pregnant and Lying-In Women* (Worcester, Mass., 1793 [orig. publ. London, 1772]), 17–31, 219, 236–240. White's estimates for London and Manchester work out to maternal mortality rates of 13/1,000 and 6/1,000 respectively. For a modern effort to compute maternal mortality ratios from parish christening and death records, see B. M. Willmott Dobbie, "An Attempt to Estimate the True Rate of Maternal Mortality, Sixteenth to Eighteenth Centuries," *Medical History* 26 (1982): 79–90. Dobbie believes that maternal mortality in England may have been as high as 29/1,000, as compared with earlier estimates of 10–15/1,000.

28. Ballard Diary, March 11, 1790, Jan. 19, 1800.

29. Ibid., April 17, 1801, May 31, 1799, Nov. 28, 1787; see also June 30, 1794, June 3, 1795, Aug. 10–11, 1799.

30. Some nineteenth-century Americans debating the causes of childbed fever used the same environmental argument, anticipating the conclusions but not the logic of twentieth-century historians (Charles E. Rosenberg, *The Care of Strangers: The Rise of America's Hospital System* [New York, 1987], 124–26, 376, n. 10, n. 11).

31. Dorothy I. Lansing, W. Robert Penman, and Dorland J. Davis, "Puerperal Fever and the Group B Beta Hemolytic Streptococcus," *Bulletin of the History of Medicine* 57 (1983): 70–80. On the complexities of the puerperal fever debate in the nineteenth and early twentieth centuries see Leavitt, *Brought to Bed*, ch. 6.

32. *Falmouth Gazette and Weekly Advertiser*, Feb. 12, 26, March 5, 12, 1785.

33. Jeremiah Barker, "History of Diseases in the History of Maine," ch. 3, Barker Papers, Maine Historical Society, Portland.

34. Ibid.

35. Ibid.

36. *Falmouth Gazette*, Feb. 12, 1785.

37. Ibid., Feb. 26, March 5, 12, 1785. Coffin was perhaps referring to Richard Brookes, *The General Dispensatory* . . . (London, 1753), or *The General Practice of Physic* . . . (London, 1754) and to Jean Astruc, *A Treatise of the Diseases of Women* . . . , 2 vols. (London, 1762), or *Elements of Midwifery, containing the Most Modern and Successful Method of Practice in Every Kind of Labor* . . . (London, 1766). Astruc's works were translated from the French.

38. Smellie, *Collection* 3:533–43; Denman, *Introduction* 1:171.

39. Jeremiah Barker, Medical Cases, 1771–1796: Barker Papers.

40. Loose paper dated April 29, 1802, Moses Appleton Papers, Waterville Historical Society, Waterville, Me. On the larger question of lay resistance see William G. Rothstein, *American Physicians in the Nineteenth Century: From Sects to Science* (Baltimore, 1972), 128–38, and Joseph F. Kett, *The Formation of the American Medical Profession: The Role of Institutions, 1780–1860* (New Haven, Conn., 1968), 101–7.

41. Barker was born in Scituate, Mass., began his practice in Gorham, Me., in 1772, removed to Barnstable on Cape Cod after a year, returned to Gorham in 1779, and finally went to the Stroudwater section of Falmouth in 1796 (James Alfred Spalding, "Jere-

miah Barker, M.D., Gorham and Falmouth, Maine, 1752–1835," reprinted from *Bulletin of the American Academy of Medicine* 10 ([1909]: 1–2). Barker had an indirect link to British medicine. In midcareer his mentor, Dr. Bela Lincoln of Hingham, Mass., had spent a year studying in London hospitals and acquiring an M.D. from King's College, Aberdeen. Spalding, *Barker*, 1–2; Clifford K. Shipton, *Sibley's Harvard Graduates: Biographical Sketches of Those Who Attended Harvard College* 13 (Boston, 1965): 456.

42. Nathaniel Coffin to Massachusetts Medical Society, May 8, 1803, and Jeremiah Barker to Joseph Whipple, July 12, 1803, Countway Medical Library, Boston. In the long run Barker may have been more forgiving than Coffin. His manuscript history describes Coffin as a physician "who commanded an extensive practice in physic, surgery and obstetrics, with good success" ("History," ch. 2).

43. Barker, "History," ch. 2.

44. Ibid. The citations on puerperal fever are, as he gave them, "Dr. Terriere, 1789; Dr. Biskell, *Medical Papers*, v. 2, 1798; *London Medical Repository*, May 1815; *New England Journal*, v. 4, 5; Dr. Channing, *New England Journal*, vol. 6, 1817; *Medical Repository*, vol. II."

45. Barker, "History," ch. 2.

46. Leavitt, *Brought to Bed*, 144–45.

47. On some things Denman had changed his own mind by 1794. Although he continued to oppose intervention in labor, he did accept bloodletting as a cure for puerperal fever, something he had dismissed in his earlier treatise, as had Barker in his *Falmouth Gazette* letters. Denman, *Introduction* 1:184–90, 2:253–54; *Falmouth Gazette*, Feb. 26, 1785.

48. Adrian Wilson, "William Hunter and the Varieties of Man-midwifery," in Bynum and Porter, eds., *William Hunter*, 344–45.

49. Franklin McDuffie, *History of the Town of Rochester, New Hampshire, from 1722 to 1890*, ed. Silvanus Hayward (Manchester, N.H., 1892), 1:345–46; James Farrington Medical Record Books, 1824–1859, Special Collections, Dimond Library, University of New Hampshire, Durham. Farrington added an entry about the woman dying five days after delivery in different colored ink at the end of his delivery record. On the general pattern of listing childbed deaths under other causes see Wertz and Wertz, *Lying-In*, 125–26.

50. Farrington, Medical Record, Case #451, Sept. 9, 1835, #118, Feb. 24, 1825, #442, May 28, 1835, #292, Jan. 30, 1831.

51. Farrington used forceps before 1838; he just did not have a separate category to cover instrumental labors.

52. Joan M. Jensen's analysis of 109 deliveries by an early nineteenth-century Chester, Pa., physician shows no maternal deaths at delivery, 7 percent stillbirths, and 30 percent difficult labors (*Loosening the Bonds: Mid-Atlantic Farm Women, 1750–1850* [New Haven, Conn., 1986], 30–33). The low caseload of this physician, roughly fourteen deliveries a year, suggests the presence of other practitioners, probably including midwives.

53. Denman, *Introduction*, 171; Farrington, Medical Record, Case #539, Aug. 8, 1839.

54. Leavitt, *Brought to Bed*, 43–44.

55. Ibid., 40.

PART II
Women and Health in the Nineteenth Century

WOMEN'S LIVES AND HEALTH IN THE NINETEENTH CENTURY CONTINUED TO REFLECT the gendered culture in which they lived as well as the evolving medical practices of the period. We see a particularly dramatic portrayal of both medicine and culture in Susan Garfinkel's article about one woman's experiences with breast cancer in 1814. Garfinkel's description of unanaesthetized breast surgery serves as a vivid reminder of some benefits of modern medicine. Susanna Emlen's response to her cancer—that "this trial was sent in love and mercy for my refinement"—reflects her class and her Quaker and female upbringing. Personal experience in the context of full descriptions of what medicine could offer provides a compelling portrait of illness in early nineteenth-century America.

The next article, by Nancy Schrom Dye and Daniel Blake Smith, provides insight into an even more common part of eighteenth- and nineteenth-century women's lives, namely, motherhood and the seemingly ubiquitous reality of infant death. The authors focus on the historical debate about maternal feelings and perceptions toward their children, plumbing women's personal writings to illuminate the maternal experience and how it changed over time. They conclude that until the early twentieth century, when public health and nutritional improvements finally lowered the risks of infant death, high child mortality and what they call the "incessant anxiety that babies might die" remained central to women's experiences.

Carroll Smith-Rosenberg and Charles E. Rosenberg analyze nineteenth-century medical views about women's bodies and roles. They describe a medical discourse that created a biological argument for women's traditional domestic and childbearing roles. Many male physicians and scientists challenged women's attempts to expand their educational opportunities and limit their fertility, which the men perceived as potentially dangerous social changes, by insisting that "the female animal" was limited by her reproductive biology.

The actual state of nineteenth-century women's health remains of major interest to historians. Many contemporary observers noted that America seemed to be breeding physically inferior women. One of Catharine Beecher's friends wrote her at mid-century that among all her female acquaintances in Milwaukee she did "not know one perfectly healthy woman in the place." Until recently, historians assumed that women's invalidism, real and constructed, affected middle- or upper-class white women. But as Diane Price Herndl investigates here, African-American women, including enslaved, working, and middle-class women, suffered similar experiences of ill health. Using fictional alongside medical narratives, Herndl re-creates a world more whole and complex than we have recognized.

4 "This Trial Was Sent in Love and Mercy for My Refinement": A Quaker Woman's Experience of Breast Cancer Surgery in 1814

Susan Garfinkel

But why should I longer dwell on the remembrance of this season of inexpressible conflict? Suffice it to say, and I hope I may never forget that in sometimes the midst of it, my feeble mind was permitted to repose itself in the sweet persuasion that this trial was sent in love and mercy for my refinement.
— Susanna Dillwyn Emlen to her father 5th of 11th Month, 1814[1]

Breast cancer rates have increased dramatically in recent decades, yet the disease remains under-studied and incompletely understood. Women's advocates have argued that lack of attention to breast cancer is due to the institutional prejudices of a male-oriented medical profession. This issue has historic roots. From the classical rise of Hippocratic medicine to the development of the Halsted radical mastectomy in 1891, Western culture has equated the occurrence of breast cancer with the strange physiology of the female body. While surgery for breast cancer was known since ancient times, it was not much attempted before the sixteenth century, and the

procedure was still surprisingly rare by the Victorian era. More often, palliative measures were used to ease the patient through her final disease. As one historian of breast cancer writes, "Surgical literature of the eighteenth century abounds with case reports; a mastectomy was apparently still of sufficient interest to warrant a publication." Not only the pain suffered in a preanaesthesia environment but theories of female disease and the health of the female body guided this circumstance.[2]

In a detailed letter to her father written in the 11th Month, 1814 from her home in Burlington, New Jersey, Susanna Dillwyn Emlen described the set of events that led to her own decision to undergo surgery for breast cancer earlier that year.[3] This surgery was performed in her home, without anaesthesia, under the care of five doc-

SUSAN GARFINKEL is Visiting Assistant Professor in American Studies at George Washington University, Washington, D.C.
Reprinted from *New Jersey Folklife* 15 (1990): 18–31.

tors and a nurse.

It was about the middle of the last twelfth Month, that I first perceived a tumour in my left breast, irregularly shaped, about the size of a partridge egg.... My terror when I had fully ascertained the fact, is not to be express'd.

After several weeks Emlen told her husband Samuel of the situation, and though the couple's brother-in-law was a leading surgeon of the day, she settled first upon a course of self-treatment recommended by her aunt, using applications of "Logan's Salve."[4] When the salve proved ineffective, the Emlens reconsidered a consultation with their relative Dr. Philip Syng Physick but decided against the long trip to Philadelphia due to the severity of the winter. Susan Emlen next settled on "a light mild diet, with little exercize."

In 2d Month came a series of family illnesses, including that of the sickly Dr. Physick, which postponed the matter of her tumor further. Looking back, Emlen muses that

It now appears strange to me that I should have been unwilling to consult any Physician, but I knew Dr Physicks preference of a surgical operation in such cases, and I had not yet sufferd enough to endure the thought of so terrible a measure.

When Physick was at last consulted in Philadelphia late in 4th Month, "he said there was no time to be lost." She writes:

For three weeks I was confined to my bed, in nearly one posture on at first a very low diet, but as I bore it but poorly, being often affected with faintness, I was afterward indulged with milk, weak chocolate, and the liquor from Oysters and clams—I was twice bled with a lancet, three times with

leeches, and had a blister, and then a mercurial plaister on the tumour.

This treatment was found unyielding and after a consultation among four doctors it was decided that only a surgical operation might help. "This awful sentence I had anticipated," she says,

but it filled me with great distress.... I was candidly informed by Dr Physick, that there was some danger in it, and that till the part was laid open no certainty could be attaind, whether the disease had not proceeded beyond the reach of the knife.

Emlen returned to her own home to make the decision, "and in a short time it appear'd best to me to endeavour to submit to an operation." That event happened three weeks later: after twenty-five minutes of surgery Susan Emlen lay in undiminished pain for fourteen hours. She survived the operation and lived for five years longer; the cause of her death is as yet undiscovered.[5]

The detailed description of Susan Emlen's illness and its treatment is part of an ongoing correspondence between Susan and her husband Samuel Emlen of Burlington, and Susan Emlen's father William Dillwyn, resident in England for many years. The bulk of these letters survives in the Library Company collection at the Historical Society of Pennsylvania, spanning the years 1770–1824. One of perhaps a dozen letters referring to the breast cancer surgery, Susan Emlen's account of 11th Month, cited earlier, is extraordinary and compelling. Within a single narrative she is able to integrate into a descriptive chronology of external events her own personal experience of those events, with the emotions, attitudes and beliefs that inform them. The early nineteenth century has often been called a transitional period—in medical history, women's history, the history of religion, and beyond. Yet *transition* is a concept that gestures toward the future, while people experience the immediacy of their lives in the present. Susan Emlen's experience of breast cancer surgery is set in a time when the shifting paradigms we can now identify meant that multiple choices and multiple interpretations existed side by side for her to draw upon. Situated as it is within a group of family letters, Susan Emlen's breast cancer story opens to us a rare and expanded view into a historically specific ethnographic community, where ongoing issues can be viewed as they affect individual lives in a single, bounded time and place.

What guided Susan Emlen's decisions, those of her family, friends, and doctors? What assumptions did this upper-class Quaker woman, childless and middle aged, carry about herself, her body, its treatability, and her capacity for suffering that led her to go through with the difficult surgical procedure? How did others respond? Susan Emlen's breast cancer experience, when viewed in the contexts both of her letter-writing community and of a larger historical-cultural perspective, can be used to draw out and unlock the layers of cultural networks, communities, and modes of thought and action upon which it draws. Her bout with cancer is revealing at several levels: individual experience and the narratives used to recount it, family and community responses to a difficult illness, the state of health care and the medical profession in early nineteenth-century America, and societal attitudes toward femaleness at that time. The letters also reveal a strong overlay of specifically Quaker culture—from Quaker involvement in progressive medicine to spiritual attitudes toward bodily affliction to the ordering of relationships in everyday life.

Susanna Dillwyn was born the 31st of 3d Month, 1769 near Burlington, New Jersey, the only child of William Dillwyn and his first wife, Sarah Logan Smith.[6] In England her father's second family provided four sisters and a brother, but Susan Emlen—as she was called by her family—was raised in America by her aunt Susanna Dillwyn Cox at Oxmead, near Burlington. Her uncle George Dillwyn and his wife Sarah Hill Dillwyn, noted Quaker ministers, lived at Green Bank near Burlington as well.[7] Though Susan Emlen lived separately from her father for most of her life, the correspondence between them demonstrates a close and caring relationship.

Susanna Dillwyn was married to Samuel Emlen, Jr. (1766–1837), on the 16th of 4th Month, 1795 at Burlington. The couple lived at first in Philadelphia but hastened a planned return to Burlington in 1797 to escape the city's yellow fever epidemic. Their house at West Hill, adjoining the Oxmead property where Susan Emlen had been raised, was built at this time.[8] As the son of a noted Quaker minister, Samuel Emlen, Jr.—an influential Friend himself—concerned himself particularly with the issues of slavery and Indian affairs within the Society of Friends. By 1800 Samuel Emlen had been appointed assistant clerk to Burlington Monthly Meeting; he served as assistant until 1806 and as clerk of the meeting from 1807 to 1814.[9] The Emlen couple was childless, and Susan Emlen died in 1819, almost twenty years before her husband. In his will, Samuel Emlen established a trust of $20,000 for an agricultural school, The Emlen Institute for the Benefit of Children of African and Indian Descent, located first in Carthagena, Ohio, and later transferred to Bucks County, Pennsylvania. The Institute, according to Emlen's will "as near as may be in the manner of the Manual Labor School of Emanuel von Fellenberg of Hofwyl in Switzerland," was eventually merged into the Colored Normal School at Cheyney, Pennsylvania.[10]

It was Samuel Emlen's half-sister and only sibling, Elizabeth, who married the surgeon Dr. Physick of Philadelphia in 1800. Philip Syng Physick (1768–1837) has been called the father of American surgery. Having chosen the profession of doctor early on, he began his studies at Philadelphia's Friends' Academy. Physick next received an A.B. from the College (1785) and an A.M. (1788) from the University of Pennsylvania. Three years of study under Dr. Adam Kuhn in Philadelphia were followed by three years under the renowned London surgeon John Hunter. In a single year at the medical school of the University of Edinburgh he earned an M.D., graduating in 1792. Returning to Philadelphia, Physick became well known through his efforts in the yellow fever epidemics of 1793 and 1797 and as a strong ally of Dr. Benjamin Rush's in favor of the controversial use of bleeding and purging as a medical treatment. By 1794 Physick was on the staff of Pennsylvania Hospital, where he began to offer lectures to students in 1800, the same year that he married. In 1805 he was named to the newly created chair of surgery at the University of Pennsylvania. As one biographer notes, "his technical contributions to medicine were many and varied," including the development of new procedures and new surgical instruments for treating a wide range of medical conditions.[11] Despite his own poor health, Physick performed numerous surgical operations over many years, most notably a lithotomy on Chief Justice John Marshall near the end of both of their lives.[12] The father of two daughters and two sons, Philip Syng Physick was legally separated from his wife Elizabeth Emlen in 1815, a highly unusual circumstance at that time.[13]

Inhabiting the Dillwyn-Emlen letters are a close-knit circle of family and friends, radiating outward from Susan Emlen's home at West Hill to Burlington, Philadelphia, and England. Most of those found in the Emlens' circle were members of the Society of Friends and were related through blood or marriage as well.[14] Beyond these obvious connections is a more intricate network of ties that includes both mutual business or professional concerns and common

agendas, attitudes, and activities within the Society of Friends. Drs. Joseph Parrish, John Syng Dorsey, and Philip Syng Physick are frequently mentioned by Susan Emlen in her letters, less for their shared expertise than for their respective places in her daily life.[15] Yet Dorsey's position as Physick's nephew only enhanced a strong professional relationship shared by the two. In the same way, George and Sarah Dillwyn's role as traveling ministers in the Society of Friends meant that a one-time companion of theirs on a religious trip to England had been Samuel Emlen, Sr., Susan Emlen's father-in-law. Writing in 1854, John Jay Smith recalled the extended family circle at Burlington:

> George and Sarah Dillwyn returned to America in 1802, and cast anchor *near* in every respect their attached relatives John and Ann Cox, Margaret Morris, Milcah Martha Moore, Samuel and Susan Emlen, my father and mother John and Gulielma M. Smith, and Deborah Smith. A most united band of brothers and sisters, nephews and nieces, continued for some time in the enjoyment of as much happiness as often falls to the lot of mortals . . . there was a regular correspondence kept up with England and English friends and relations, which I well remember. William Dillwyn was intimately connected with Clarkson and Wilberforce, and all the philanthropic movements of the day were known and discussed in this amiable and virtuous circle. New and good books found their way to our quiet neighborhood, and a personal intercourse with the world was kept up by visits to Philadelphia, and from visitors abroad.[16]

Many of these same names are found in the minutes of Burlington Monthly Meeting; Samuel Emlen was at one time a business partner with Susan Emlen's cousin Benjamin Smith.[17]

The progressive, often literary interests of George and William Dillwyn were indeed an influence on the wider circle. A set of three volumes inscribed "William Dillwyn/to his son/Samuel Emlen" is titled, *History of the British and Foreign Bible Society*. This nondenominational Christian group promoted the distribution of the Bible as translated into indigenous languages, including in America the native Delaware and Mohawk.[18] Exposure to these volumes and others, along with George Dillwyn's published writings such as *Occasional Reflections, Offered Principally for the Use of Schools of 1815*, were no doubt an important influence on Samuel Emlen's bequest for the Emlen Institute which, despite his Quakerism, was defined as a nondenominational Christian school.[19] A slim volume featuring William Dillwyn's name, *Letter from Doctor Edward Jenner, to William Dillwyn, on the Effects of Vaccination, in Preserving from the Small Pox*, printed in Philadelphia in 1818, suggests a familywide interest in medical advancement and the use of medical cures.[20]

Susan Emlen's operation lasted only twenty-five minutes, but weeks of preparation led up to the event. While timeliness was a priority in treatment, it was neither as centrally stressed, nor as possible, as would be the case today. Following three weeks of preliminary treatment in Philadelphia during 4th Month, and a journey back to Burlington, Susan Emlen decided to undergo the operation. Her husband Samuel next wrote to Dr. Physick, who in turn made arrangements with the other attending doctors. It was the 18th of the 5th Month before Physick actually arrived at the Emlen home "with his sister Abby, his daughter Sally and the nurse" in tow. Physick was to perform the operation the following day with Drs. Dorsey and Parrish, scheduled to arrive that morning, as assistants.

On the day of the operation unexpected complications arose. Susan Emlen explains:

I went into the chamber over the kitchen, that some preparations might be made in my own, and thro the attention of the kind friends surrounding me, I was ignorant for several hours, that Dr Physick had soon after dinner, been attackd with a fit of cholick more violent than any he had ever had. I was however before night in part acquainted with it, as my dear Aunt Cox staid with me, while S Emlen, Abby Physick & Dr Tucker, found employment in endeavouring to alleviate a state of torture—The next day he was in some measure relieved.

Physick's health, rather than Emlen's operation, now became the focus of familial attention and concern.

Our solicitude continued to be exercised for some days on account of Dr Physick, who after great weakness began now to amend, and was frequently visited by his nephew [Dr. Dorsey], and on the 25th by his wife and daughter Susan, escorted by Dr Dorsey—they staid till the 31st, when Sister left Susan to wait on her Father, and took Sally with her.

Physick's health, in fact, led to Susan Emlen's being kept in a protracted state of waiting, what she later refers to as "a time of complicated affliction." She explains that

the other Physicians came up, and I was willing the business should proceed, but the young men hesitating to perform it, it was judged best to delay the matter for the present till Dr Physick should be a little better.[21]

Over two weeks passed in focused attention to Physick's rather than his patient's health, until at last "on the evening of the 3d of 6 month, Dr Dorsey once more came up to perform the dreaded operation." A mastectomy—at that time without anaesthesia—was an event to be dreaded, and anticipation approached overwhelming. Susan Emlen writes, "the distress of this evening I shall never forget, yet my resolution remained firm."[22]

Five doctors—Physick, Dorsey, Wistar, Parrish and Tucker—attended the operation in the Emlen home, with Physick to supervise and—because of his poor health—Dorsey acting as principal surgeon.[23] Relatives came to stay with Susan Emlen's husband, and, she reports, "Aunt G M Smith, a person in whom fortitude and tenderness are remarkably united, was willing to stand by me, during my trial." With the five doctors, one nurse, and an aunt in attendance, over six months after the tumor was first detected, Susan Emlen's operation at last began. Her narrative of 11th Month continues:

About one oclock, Nurse Hooke came into the chamber I was in to tell me the Doctors waited my coming in my own room. She cover'd my head with a handkerchief, as she led me in, hoping it might save me the sight of the preparations—I however saw Dr Dorsey with his sleeves tuckd up and his cloaths coverd with a large apron, and had a slight view of the other four: of whom Dr Tucker held my arm, while my Aunt supported my head, and at times gave me something to smell.

And here Susan Emlen's detailed description breaks down, painful experience overwhelming her ability to narrate the scene. Her next words are referential more than descriptive.

My suffering was severe beyond expression, my whole being seemd absorbd in pain—The tumour was taken out in 25 minutes, but it was an hour before I was in bed where I lay 14 hours, before I became easy enough to sleep.

The details of experience are erased in the face of the pain that Emlen endured. In her narrative the surgery ends abruptly, with detailed description reserved for the recovery to follow. Though the focused subject of her suffering and the motive for her narrative, the operation itself is the most underdescribed section of her letter.[24]

Susan Emlen's narrative spares a description of the surgical process and most of the surgical experience. In retrospect she writes:

The Physicians all agreed that the disease was completely eradicated—as it was thought to weigh no less than a pound, it left a great hollow (since completely fill'd up) where a number of arteries taken up with needles and tied with wax'd thread, were for some days in danger of bleeding.[25]

Her husband's account only adds that "the diseased parts both of the breast & the Arm-pit, they believe entirely extracted."[26] It is their uncle George Dillwyn, writing also to his brother William Dillwyn, who provides the most detailed description of the scene.

What I could not but suppose a very dangerous operation ... was perform'd the 4th instant, with a dexterity that does credit to the surgeon, in about 25 minutes; during which she was favor'd with such unshaken constancy, that hardly a sigh escap'd her. . . . [T]he operation was intended to be perform'd by Docr Physick, & delayed a considerable time on that acct, but the enfeebled state of his health continuing, his Nephew Doctor Dorsey took his place, between Susan & him there is an uncommon attachment; & while the knife was in motion, *she* expressed a fear that he "would stand too long," and *he*, "admiration at the fortitude, attainable thro' resignation."[27]

This description of Susan Emlen's courage is moving, but details of the surgical procedure itself are still unclear.

Yet surgical technique is an important aspect of what happened that day, and operations for breast cancer were not then as standardized as they later would become. It was not until 1891 and the development of the Halsted radical mastectomy that a single and maximally effective protocol was developed.[28] Without medical records or detailed descriptions by Emlen or her family, we cannot be sure how much of her breast was removed. Drs. Physick and Dorsey, however, were closely involved in the development of improved and standardized surgical procedures within their own time, working in particular to bring American practices up to the standards of European knowledge. In 1813, just a year before Susan Emlen's operation, John Syng Dorsey, Physick's nephew and principal assistant, authored and published *Elements of Surgery;* it was the first technical book on surgery to be published in America. While *Elements of Surgery* does not deal specifically with cancers (readers are refered to John Bell's *Principles of Surgery*), Dorsey would publish an expanded American version of Samuel Cooper's *Dictionary of Practical Surgery* three years later, where the detailed procedure for a mastectomy is well described. As principal surgeon for Susan Emlen's operation, Dorsey may well have followed these procedures himself.[29]

To begin the operation, Cooper says, "the patient is usually placed in a sitting position, well supported by pillows and assistants," although it is equally or more convenient to place the patient lying down, especially if a long operation or considerable loss of blood is expected. Next,

> The arms should be confined back, by placing a stick between them and the body, by which means, the fibres of the great pectoral muscle will be kept on the stretch.

This practical measure makes dissection easier and also prevents the patient from disruptively moving her arms. Both the logistics and the protocol of beginning the incision are next discussed. Cooper explains that the initial cut through the skin of the breast—the most painful stage of the surgery—should always be longer than the tumor or later difficulties will result.

> The fear, however, of giving pain has probably led many operators to err, by not making their first incision through the integuments large enough . . . [so] that there was not room enough to get at the tumour so as to dissect it out with facility.

The unfortunate result of this mistake, he notes, is that "the patient has been kept nearly an hour in the operating room, instead of five minutes, and the surgeon censured by the spectators, as awkward and tedious."

Continuing this theme of expeditious competence, Cooper advocates a three-step method for the actual tumor's removal. Once the skin is cut, the tumor is next detached from surrounding tissue on all sides. The location of cuts is again important:

> The operator must not dissect close to the swelling, but make his incisions on each side, at a prudent distance from it, so as to be sure to remove with the diseased mass, every atom of morbid mischief in its vicinity.

Finally, with the same degree of prudence, the tumor is separated at its base. Operating according to these "three methodological stages . . . in each of which there is a distinct object to be fulfilled," Cooper writes, provides confidence to the surgeon, and leads to both "expedition and adroitness."[30]

In her letter of 11th Month Susan Emlen records for her father the tedious and trying nature of recovery, noting with precision the markers of each small improvement in her health. The first days following surgery were spent in fear of impending crisis; Emlen reports that although in danger of severe bleeding for several days, "I was however watch'd day and night with the tenderest anxiety." At the end of four days Emlen's wound was dressed, a sign that the doctors felt that healing had begun. Monitoring her diet was also seen as an important factor in aiding recovery.

> At first, to prevent fever I was allow'd only strawberries and toast and water, but that being found too low, I was allowd milk and rich chocolate, and my pains and restlessness were mitigated by laudanum and rubbing my limbs.

Laudanum, a tincture of opium (i.e., opium suspended in an alcohol-based solution), was the only form of pain relief provided. The doctors, and assorted family guests, remained at the Emlen home until eight days had passed; at day thirteen the last of the ligatures was removed from Emlen's wound. Sixteen days after surgery Susan Emlen first "rode [by carriage] about half a mile. After this," she explains, "my daily rides were gradually lengthn'd, and my strength very slowly returnd."

A month after surgery Susan Emlen was visibly recovered as "our dear SN Dickinson came and spent a week of congratulation with us." The end of this visit marked the hoped-for end of medical attention as well. "The day after her departure Dr Tucker who had daily dressed my wound for several weeks took his leave, saying he thought it perfectly heal'd." Although a small discharge from the wound continued for two more weeks, she reports that it was attended to by the nurse alone. Yet the trial was not completely behind her even then. Writing five months after the surgery, Susan Emlen explains that "a degree of weakness long remain'd, and I think will never be wholly removed as Dr Dorsey told me he was obliged to cut a muscle which moves the arm." [31]

More of the anxiety and pain of Susan Emlen's recovery is expressed by those around her. Her husband Samuel Emlen writes to her father six days after the event that

> Since the preceeding date my dear Susan has had occasion for the exercise of that patience with which she has been so eminently furnished, for altho ever since the performance of the operation our Physicians have encouraged us to believe the symptoms were as favourable as could be expected, yet there has daily occurred a nervous effection producing a restlessness which has been very trying; this has generally taken place in the afternoon, & been allayed by taking opium of which she now takes daily two grains one at 6 & the other about 10 oClock in the evening; last night she had rather less sleep than usual, but appeared considerably refreshed by the portion which she did obtain.

The next day, however, there is more cause for concern. Despite improvements in the wound itself, "There is a debility about the dear Patient that is not so pleasant." Because of the weather and his own poor health, Dr. Physick had not left Burlington according to plan, a circumstance "of which we are now glad, as we see his mind is in some anxiety in consequence of this state of weakness that Susan has been in for a considerable part of the day." [32] In letters written throughout the summer and fall Samuel Emlen continues to register his distress at the extreme effects of pain, weakness, and debility in his wife. To his prominent cousin Roberts Vaux he confides that, "Most of the time since [the operation] a train of consequent apprehensions has kept my fears awake & my mind in a state of painful tension." Samuel Emlen keeps from his wife the news of the death of Vaux's mother, "until she acquires a greater degree of strength than she has yet recovered," forgoing himself a funeral visit to Philadelphia from fear evoked by minor changes in her daily state.[33]

Historians of medicine turn to the nineteenth century as a time of great change in conceptions of disease and the human body, a medical revolution of sorts. Marking this theoretical reconfiguration was a changing relationship between medical and therapeutic practices within the local community and the rise of an international community of professional practitioners. This was also a time when—under the influence of conflicting paradigms—ideas about the body itself shifted in the professional, and by extension, the popular mind. The medical historian Charles Rosenberg points out, however, that the theoretical debates of the early part of the century

> all served the same explanatory function relative to therapeutics; all related local to systemic ills; they described all aspects of the body as interrelated; they tended to present health or disease as general states of the total organism.

The early nineteenth century is noted as a time of transition from highly evolved humoral theories and an anecdotal practice to the start of a "scientific" medical approach.[34]

Philip Syng Physick—trained in London and Edinburgh, first chair of surgery, innovator of new operative techniques—was clearly part of the professional medical elite.[35] Yet, like the salve used by Susan Emlen to self-treat her newly found tumor, the weeks of therapeutic measures employed by Physick when he first took on her case conform to the widely shared view of a systemic body that Rosenberg describes. Of the bleedings Susan Emlen explains:

> Dr Physick had recently succeeded in reducing a large tumour not unlike mine, in a case when other remedies had been assisted by a strict confinement to a recumbent posture, keeping the diseased part higher than the rest of the body, that the humours might fall from it, for six weeks.[36]

Likewise, Susan Emlen's frequent reference to diet and Samuel Emlen's ongoing attention to the surface of her wound and the quality of her nightly sleep both suggest the desire to once again observe a bodily equilibrium. Treatment in the home—even a surgery such as Susan Emlen experienced—was in fact not just a luxury afforded to the rich. In the early nineteenth century, surroundings were seen as an integral part of overall health; hospitals existed only for those without the benefit of a stable, healthy home environment. As one historian explains:

> For most patients, the best therapeutic environment remained the home. Frequent house calls by the doctor were not simply for the comfort and convenience of the patient; they allowed the physician to observe and modify the home environment as an integral part of therapy.[37]

Treatment in her home best suited Susan Emlen's spiritual needs but suited the perceptions and needs of her doctors as well.

By placing herself in the care of her professionally prominent brother-in-law, however, Susan Emlen automatically took herself beyond the realm of ordinary medical attention. She recognizes as much when, in a short attachment to her husband's letter of 3d Month, 1814, she writes to reassure her father that "Dr Physick has been so successful in these cases that we believe no other person could offer me equal advantages."[38] Still, the choice of partial or full mastectomy, like any amputation, did not fit well the holistic view of her body that the average patient held.[39] We see Susan Emlen taking her time to make this extraordinary decision but should realize the severity of such a step for the doctor as well. Surgery, with its focus on a localized, tissue-based cure, was somewhat at theoretical odds with the physician's own focus on systemic treatment. Thus Physick routinely turned to therapeutics first.[40] Safe operative anaesthesia would become available only in the 1840s; during Physick's lifetime the pain involved in surgery had also to be a serious consideration with each patient he treated.[41]

Susan Emlen was forty-five years old and childless when her cancer first appeared; these factors alone would place her at high risk according to modern demographic studies.[42] For centuries breast cancer has been associated with the end of menstruation, which since classical times was thought to bring on a dangerous imbalance due to the cessation of the necessary flow of blood from a woman's body. The concept of the four humors—blood, phlegm, yellow bile, and black bile—as the essential components of the bodily system makes up the basic philosophy of classic Hippocratic medicine. Breast cancer, like other forms of the disease, was thought to arise from an imbalance of the humors. At first, the end of menstruation was thought to lead to engorgement of blood in the breasts. Later theories suggested that the entrapment of different humors led to different sorts of cancer, with the accumulation of black bile in the breast as the particular source of breast cancer. Both versions led to attempts to purge and bleed or to restart menstruation with medicines, herbs, exercise, or hot baths. A still later idea attributed the rise of cancer to mother's milk that curdled in the breast. Childlessness was also given as a cause.[43]

Theories of breast cancer have historically indicated a general conception of the female body as strange and different. If the ability to bear children was a positive benefit, diseases such as breast cancer were all-too-common negative consequences of being born a woman. According to Galen, the Greek compiler of medical

knowledge who remained an unchallenged authority on cancer into the sixteenth century, cancers arise in many parts of the body but most often in the female breast.[44] Breasts, in fact, have come to be in some ways equated with cancer. One scholar, tracing an outline of changing tumor theory over time, adds as an aside that "the reader will find that cancer of the female breast, which is the prototypical cancer, plays a central role in the story from the beginning."[45] Eighteenth-century conventional wisdom also suggested that female breasts were naturally inclined toward cancer. In one of his early accounts of Susan Emlen's illness Samuel Emlen attributes the cause of his wife's breast cancer: "owing to a bruise rec[eive]d in reaching for something."[46] Yet the operative condition is ultimately one of femaleness. In 1666 the Reverend John Ward of Strafford-on-Avon observed a mastectomy performed on one of his parishioners and reported that "Dr Needham hath affirmed that a cancer is as much within as without the breast, and hee hath seen a string, as I was told, going from the breast to the uterus."[47] Being a woman was itself a pathological condition, or at least a systemically unbalanced one in therapeutic terms. This is seen, for example, in the emphasis of hydropathy, an alternate nineteenth-century medical movement. In hydropathy, "Adolescence, puberty, menstruation, childbearing and menopause *were not* seen as illnesses but as natural physiological processes" [emphasis added]. Noted for its "nonpathological" view of women's health, the feminist hydropathy movement highlights the attitude of mainstream medicine by contrast.[48]

Because she does not tell us, it is hard to know what Susan Emlen's emotional or psychological reactions to losing a breast might have been. While her letters sometimes evoke a sense of the trials of being female, Emlen writes more consistently of surgery, illness, and spiritual understanding. In one instance, when she writes of "the disease being completely eradicated—as it was thought to weigh no less than a pound," she separates the idea of the cancer from the reality of her breast. Her husband, by contrast, refers to "the diseased parts both of the breast & the Arm-pit."[49] Because the underlying issues were so unspeakable, historians have found few references to the emotional reactions of women forced to deal with breast cancer's potential effects on their femininity. Anne of Austria, stricken with breast cancer in 1664, reported feelings of guilt about her illness, interpreting it as a consequence of former vanities.[50] The psychological effects of mastectomy would seemingly have been even more intense. As one historian suggests, the physical intrusion of the mastectomy operation becomes a violation of female privacy, the "private body [is] violated and made public through the experience of surgery."[51] Some Renaissance depictions of St. Agatha—the patron saint of breasts, whose own breasts were publically ripped from her body and later miraculously restored—show her carrying her breasts before her on a salver, removable body parts giving symbolic notice of their important but potentially impermanent role as markers of feminine identity.[52]

Breast cancer surgery was more common historically than we realize; it is firsthand narratives such as Susan Emlen's that are so rare.[53] The existence of a mastectomy narrative startlingly similar to Susan Emlen's own provides an opportunity for comparative analysis. In 1812 the British novelist Fanny Burney sent a letter to her sister containing her own version of the breast cancer surgery experience. Living then with her husband in France, Burney wrote her letter some six months after the event, apologizing for the need to cause her distant family any potential grief. "Nevertheless," she says, "if they should hear that I have been dangerously ill from any hand but my own, they might have

doubts of my perfect recovery which my own alone can obviate."[54] Burney's story traces many of the same preparatory steps as Susan Emlen's: initial hesitance to consult a doctor, eventual consultations and attempts at lesser cures, the difficult pronouncement of the need to operate, heightened anticipation of an event scheduled for an uncertain date and time. Burney's operation, attended by seven surgeons and a nurse, lasted twenty minutes; unlike Susan Emlen she chose to have no female friend at her side and later lamented this as a miscalculation. Like Susan Emlen, Burney's face was covered by a veil that allowed her still to see the proceedings, and like Susan Emlen, she engaged in conversation with the surgeons. Yet, in a spirit quite different from that which Susan Emlen expressed, when she saw the cut lines being indicated by the surgeon, Burney sat up and started to argue with the doctors that only a more modest removal was required.

Fanny Burney, unlike Susan Emlen, provides intense descriptive detail of the surgical procedure in her narrative—experienced detail that only the patient could recount—this, despite reference to "a terror that surpasses all description, & the most torturing pain." While Susan Emlen was reportedly nearly silent during her ordeal, Burney writes that she screamed and screamed. Her description of the surgical experience is graphic:

> When the wound was made, & the instrument was withdrawn, the pain seemed undiminished, for the air that suddenly rushed into these delicate parts felt like a mass of minute but sharp & forked poniards, that were tearing the edges of the wound—but when again I felt the instrument—describing a curve—cutting against the grain, if I may say so, while the flesh resisted in a manner so forcible as to

oppose & tire the hand of the operator who was forced to change from the right to the left—then, indeed, I thought I must have expired.

Yet Burney remained conscious and goes on to describe vividly the feeling of knife scraping against breast bone and "the finger of Mr. Dubois—which I literally *felt* elevated over the wound, though I saw nothing, & though he touched nothing, so indescribably sensitive was the spot." Having relived on paper the painful detail of experience, she concludes: "I fear this is all written—confusedly, but I cannot read it—& now I can write it no more." [55]

In a 1986 article the scholar Julia Epstein analyzes Burney's mastectomy account as a multilayered narrative, designed by an accomplished author to provide herself with a sense of detachment from the overwhelming event. Epstein points out that a synthesis of medical case-history language and the "'feminine' narrative of euphemism and disguise . . . remove Burney from direct contact with her body's representations and its fears and turn her to the narrative options of fiction." [56] By drawing on the genre of the "scientific" medical account, Burney removes the narrative of illness from the flow of customary everyday life, making it less of a threat to the continuation of the normal. "By focusing on her own narration as narration and self-representation, Burney simultaneously gives and takes her own medical history." [57]

Unlike Fanny Burney, Susan Emlen is not a professional writer. She does not share Burney's literary aims, and other factors seem to motivate both her letter writing in general and her account of breast cancer surgery in particular. The Emlen breast cancer narratives, in fact, seem primarily shaped by the forms and circumstances of family letter writing, the problems of maintaining close ties over long distances. While

Susan Emlen's letter of 11th Month, 1814 is the fullest version of her story, the first letter she wrote to her father following the operation was dated 24th of 7th Month. "Once more," Susan Emlen begins, "thro the goodness of Providence, I am enabled to resume my part of a correspondence, that has been the source of some of my sweetest pleasures." Here she reflects not only on the trial she has suffered but on the necessity of writing about it to loved ones far away.

> My dear husband has been careful to let our dear relations know by every conveyance he could find how we have lately fared; some things he had to say, were said with reluctance, and only because it seemed necessary to save you, the greater pain of hearing of our affliction from the transient report of strangers. [58]

Although Susan Emlen's letter of 11th Month is the most detailed that she wrote, the story of her illness is significantly amplified beyond any single narrative by the variety of other letter accounts that survive. One reason for the existence of all these letters is a practical matter—the uncertainty of prompt, or even eventual, delivery of any one letter in particular. In this case, the vagaries of delivery worked out for the best. When William Dillwyn responds to his children upon hearing of the successful surgery, he writes of the

> occasion for Thankfulness that the Miscarriage of my Son's Letter written during your Y[early] Meeting, had spared me the Poignancy of Distress with which I must have anticipated a less favourable Event. [59]

Providing an accounting of letters received was a periodic practice, but concern was no doubt heightened during Susan Emlen's illness because of the quick turn of events and the slowness of the mails. In a letter written at the end of 6th

Month, 1814 Samuel Emlen lists for his father-in-law five letters received from him over the last nine months.[60] William Dillwyn, in fact, reports back on nine letters relating to the illness sent to him or his friends in England and written by seven different American correspondents.[61] International events added a further worry to the transmission of personal news. In a letter of 9th Month, 1814 Susan Emlen musters her strength to write directly to her father, saying that due to the impending war, "this letter may be the last one we shall have an opportunity of sending for some time."[62]

Although the multiplicity of letters, and therefore narratives, is based on the uncertainty of delivery, the multiple accounts highlight the narrative qualities of Susan Emlen's breast cancer story. By 11th Month William Dillwyn would have had a full picture of the illness and recovery, based on the many letters received by that point. It is at this point, however, that Susan Emlen sits down to write out and reflect upon her own most complete account. In closing this narrative, she frames the letter as written for her father's benefit:

> After all that I know of thy inclination for minutia, I will not apologize, as I really should feel it necessary to do, to almost any one else, for all this prolixity. On a review of the whole, my dear affectionate Father will I believe join me in gratitude to Heaven, for conducting me thro so great a trial, and restoring me to bear again my little part in the offices and duties of life—Oh! that it may be with increased diligence, with added circumspection!

Yet her apology is for sending the letter, not writing it. Like Burney, Susan Emlen seems to need the existence of this final self-contained version of her story for her own peace of mind.

Susan Emlen's familial focus contrasts with Julia Epstein's foregrounding of Fanny Burney's literary aims. The work on personal narrative by scholars of folklore has looked at narratives like Susan Emlen's as a form of personal storytelling, used by members of a community to build and maintain group cohesiveness around the sharing of experience. While the focus of scholarship has been on spoken narrative in face-to-face contexts, the circumstances of the Dillwyn-Emlen letters (like many comparable bodies of letters that survive) suggest a community-based need for narrative experience in written forms as well. Whether spoken or written, the personal narrative account helps to define unruly and perhaps overwhelming experience as a definable event.[63]

Quakerism's focus on the individual and inward nature of experience as the source of understanding would only contribute further to Susan Emlen's interest, and that of her fellow Friends, in the wisdom of inner reflection on the event. Thus Susan Emlen turns away from the details of physical suffering that Burney, by contrast, so eloquently describes. Fiction, like that which Burney wrote, was not much read by Friends. Other types of narratives were more influential and likewise more familiar to Susan Emlen and her circle. Among these narratives are the often anecdotal Quaker sermons, heard in meeting and subsequently published, and the spiritual journals kept by many Friends in the manner of founder George Fox's original. With a focus on explaining and interpreting inward experience, these chronological narratives were often shared privately with family and friends but were also published more widely and formed a mainstay of Quaker literary life.[64] A comment by Emlen in an 1811 letter to her father suggests such an orientation. While narratives are often constructed around unusual or special circumstances or events, Susan Emlen suggests that inner experience, more than the enticement of worldly events, is the best source for communication among those who care for each other.

> I believe the case is this, that to those who love each other no news is more agreeable than an account of what is passing in "the little world within" and when nothing from without, or of more promising aspect presents itself, then we are writing to look into it and see what it has to produce, forgetting before or overlooking what we have so often experienced.[65]

"The little world within" is Quakerism's Inner Light, here reexpressed at the customary level of daily life. With their stronger component of spiritual orientation, Emlen's letters show less of the gender consciousness that motivated Burney's writing.

In fact, references to Quaker practice punctuate the various accounts by members of Susan Emlen's daily community. The use of numerical dating in the letters is the most obvious but taken-for-granted example. Regular attendance at various meetings for worship and business is central to the rhythm of daily life, as is clear when that rhythm gets disrupted. Susan Emlen first uses the opportunity of the Yearly Meeting of Friends in Philadelphia to consult with her surgeon brother-in-law. In 3d Month her husband explains that "Susan hoped to attend the settings of the Yearly Meeting before it might be necessary to use any remedies, but on examining the tumor the Docr thought no time ought to be lost." When she is immediately confined to her room for treatment, Susan Emlen writes that "Our dear Sally Sharpless who came down with us, expecting to go and spend the Yearly Meeting Week mostly with her sister at Darby, now changed her plan, and commenced my nurse." Three weeks later, when the need for an operation was first pronounced, it is a religious visit of the type favored by the Society of Friends that first influences Susan Emlen's personal spiritual state. She says: "Immediately after the Drs were gone, Nathan Hunt comforted us by a sympethizing visit, during which in a religious opportunity, my poor mind was once more brought into a state of quiet resignation to the divine Will."[66] Letters written following the operation anticipate when Susan Emlen will be able to attend Friends' weekly meeting for worship once again. In 7th Month Susan Emlen writes, "I hoped today to have gone to meeting for the first time but a restless night, or one partly so[,] disabled me." A few weeks later her husband reports, "This day for the first time the dear Sufferer has been at our Meeting." Susan Emlen's last reference to the operation in the surviving correspondence, in a letter to her father dating from the following 5th Month, is again placed in relation to the event of Yearly Meeting:

> Now it is over, I cannot but wish that I had written thee during the week of our Yearly Meeting. It was a very interesting one, but . . . the fatigue of long sittings failed not to affect my left side, long a weak, but now a very weak part, owing perhaps to so long a wound. The Doctor, my Bro: examined the scar, for I hope the last time.

These references tie the Emlens into a community-based cycle of shared daily life—serving too, in the recounting, to strengthen the cultural bonds shared by Susan Emlen and her father abroad.[67]

Finally, then, the Dillwyn-Emlen breast cancer letters are permeated by a marked spirituality, showing a strong influence of beliefs and attitudes associated with the Society of Friends. Based in a theology that characterized all aspects of life as equally sacred, Quakerism lacked the sense of negative fatalism found in some of its contemporary Protestant denominations. In the early stages of her illness Susan Emlen hesitated to seek treatment in part because of Dr. Physick's own poor health. In the 1st Month of 1814, when the presence of the tumor was known only to those closest around her, she wrote to her father in reflection on Physick's health. Her own impending condition was no doubt in her thoughts.

> How inestimable is the comfort of believing ourselves under the protection of an infinitely wise and good Father and friend, who afflicts his poor creatures not willingly, and knows how to extract a benefit out of every sorrow. I wish we may all profit by this dispensation—and abating our confidence in the skill and wisdom of man, seeing how little he can do for himself, repose it on the physicians of value, who can either heal the maladies of the body, or make them subservient to the more important welfare of the nobler parts, as it best pleases them.[68]

The parallel here between the metaphoric "physicians of value," capable of healing the body beyond the powers of man, and Emlen's own impending encounter with physicians bound by human weakness, is striking. It suggests her willingness to submit to manmade cures with an understanding that these fit a larger spiritual scheme. "He who afflicts not willingly," a quotation from Lamentations, is a common refrain in letters written by both Susan and Samuel Emlen during Susan Emlen's illness and recovery. In her letter of 11th Month, when Susan Emlen looks back to describe the start of her difficult decisions, she first recounts a "willingness to leave the matter at present and endeavour to trust in that ever gracious power, who afflicts not willingly." In 9th Month, having braved the worst of his wife's illness and recovery, Samuel Emlen writes that "I have never before as during the last few months felt so firm a confidence that we are in the hands of him who afflicts not willingly."[69] Disease is not punishment, but rather a challenge for the soul.

Ideas about breast cancer, whether we call them folk inheritance, customary wisdom, or professional knowledge, whether attributed to famous doctors or ordinary people, are also ideas about woman, femaleness, reproductive function, the overall strangeness and relative treatability of the female body. Susan Emlen's experience suggests that transitions in medical theory and in women's role served to heighten the multiple meanings of her illness. In this period before "The Cult of Domesticity" women were not as trapped by their bodies as they later would become. Later in the century a woman's body became a site for discussing and asserting her female disposition, for arguing the restrictions of her possible female roles.[70] The early nineteenth century, however, was a period of change, and societal change through industrialization, religious revivalism, abolitionism and the early feminist movement all affected women and the place of women in American society. For a woman like Susan Emlen, the constraints of traditional female propriety were balanced against the newly available and multiple definitions of the female self. Starting from the expectation that little could be said, there was actually more freedom to choose what to say.[71] Susan Emlen's letters display a frankness about her illness, and an interpretation of its effects and meanings, that move beyond stereotypes to evoke a strong sense of her personal experience.

Quakerism is noted for the value it places on family and for the relative equality with which it treated women members in this period. Nineteenth-century feminism—through such leaders as Susan B. Anthony, Alice Paul, and Lucretia Mott—shows significant Quaker influence.[72] Quakerism's progressive version of domestic and familial relationships shines clearly through the accounts of Susan Emlen's illness, in the close involvement of her husband and father, in her own autonomy as a decision maker throughout the ordeal.[73] Note that Susan Emlen's letters are written for her father, while Fanny Burney hoped that everyone but her father might know of her similar ordeal. Yet Burney dwells on a frankly physical experience while Emlen does not. In Susan Emlen's writing there is a certain silence of the flesh, a silence in keeping with the ideals of silence in Quaker worship.[74] Constraints of the body are meant to be overcome, and she does not dwell on her suffering or pain.

Susan Dillwyn Emlen died the 24th of 11th Month, 1819.[75] The reoccurrence or spread of her cancer as the cause seems likely but is not as yet confirmed. A lifelong correspondent with her distant father, she did not die without a chance to meet him again. As a biographer of her uncle describes it:

> In the summer of 1816 George Dillwyn's affectionate heart was put to a severe test by the departure for England, on a visit to her beloved father, of his dear niece, Susanna Emlen, who with her husband, Samuel Emlen, and Sarah Sharpless, remained there for two years, returning in the summer of 1818. He feared that his tenderly attached niece, upon whom he had looked to fulfil the claims of declining age, would never again enliven the scene she had so often cheered; but alas! though thus permitted to re-mingle with many loved ones, it was only for a season, which mostly embraced deep bodily suffering; and at length, in the 11th month, 1819, her said uncle formed part of the circle who surrounded her dying bed. At the grave he emphatically exclaimed, "Pre-

cious is the sight of the Lord in the death of his saints."[76]

Perhaps it was her ability to live within yet beyond the constraints of medical knowledge and female role, to embody the spiritual expectations of those around her while experiencing so vividly the details of more mundane daily life, that marks Susan Emlen as extraordinary. Twenty years after her death her cousin Hannah Logan Smith would write:

> A bright and beautiful example she has indeed left us of patience and resignation under the most severe sufferings and long protracted anguish. She had a very solid judgment united to great delicacy of taste. Her disposition was remarkably kind and tender; her mind serious, but her temper cheerful and social, and her countenance beamed united intelligence and softness.[77]

Susan Emlen's serious mind but cheerful temper, her solid judgment and delicacy of taste, and her patience and resignation in the face of bodily pain were what her family chose to remember her for. Her ability to expressively negotiate the spiritual and physical aspects of an excruciating illness are also memorable to us, for the insight they provide into the experience of women as patients within a medical system where such voices are mostly lost.

Notes

My attention was first drawn to the Dillwyn-Emlen letters during research on the Philadelphia home of Dr. Philip Syng Physick, now a historical house museum. I wish to thank Andrea Henderson, Andrew T. Miller, and members of the Gender and Theory dissertation workshop sponsored by the Program for Assessing and Revitalizing the Social Sciences at the University of Pennsylvania, for their helpful suggestions and comments.

1. Dating here follows the Quaker style, using

numbers to refer to both days and months (i.e., January becomes 1st Month, February becomes 2nd Month, etc.). Quotations from letters are given as written; minor changes have been made only when necessary to provide sense to the reader.

2. Daniel de Moulin, *A Short History of Breast Cancer* (The Hague: Martinus Nijhoff, 1983), 49. See also Carroll Smith-Rosenberg and Charles Rosenberg, "The Female Animal: Medical and Biological Views of Women in Nineteenth-Century America," *Journal of American History* 60 (1973): 332–56; Carroll Smith-Rosenberg, "Puberty to Menopause: The Cycle of Femininity in Nineteenth-Century America," *Feminist Studies* 1 (1973): 58–72. Inhalation anaesthesia was first introduced in the 1840s.

3. Susan Emlen to William Dillwyn, 5th of 11th Month, 1814. Unless otherwise noted, all letters cited are from the Dillwyn-Emlen Letters, Library Company Collection, Historical Society of Pennsylvania. Letters referring to Susan Emlen's operation are:

Susan and Samuel Emlen to William Dillwyn, 11th of 3d Month, 1814.
Samuel Emlen to William Dillwyn, 6th of 6th Month, 1814.
Samuel Emlen to William Dillwyn, 15th of 6th Month, 1814.
Susan and Samuel Emlen to William Dillwyn, 27th of 6th Month, 1814.
Susan Emlen to William Dillwyn, 24th of 7th Month, 1814.
Samuel Emlen to William Dillwyn, 4th of 8th Month, 1814.
Susan Emlen to William Dillwyn, 14th of 9th Month, 1814.
Susan Emlen to William Dillwyn, 5th of 11th Month, 1814.
Susan Emlen to William Dillwyn, 16th of 5th Month, 1815.
George Dillwyn to William Dillwyn (copied extract), 7th of 6th Month, 1814.
William Dillwyn to Susan and Samuel Emlen, 24th of 7th Month, 1814.
William Dillwyn to Susan and Samuel Emlen, 16th of 8th Month, 1814.

Letters in the Vaux Papers, Historical Society of Pennsylvania, also refer to Susan Emlen's mastectomy:

Samuel Emlen to Roberts Vaux, 13th of 6th Month, 1814.
Samuel Emlen to Roberts Vaux, 22d of 6th Month, 1814.
John Cox to Roberts Vaux, 22d of 6th Month, 1814.
Samuel Emlen to Roberts Vaux, 29th of 6th Month, 1814.
Susan Emlen to Roberts Vaux, 7th of 8th Month, 1814.

While her full name is Susanna, both Emlen and her family use the shorter Susan interchangeably with Susanna in the letters.

4. Surviving letters reveal three versions of the initial decision-making process that Susan Emlen followed as she considered family consultation, self-treatment, and eventual medical attention. They are Samuel Emlen to William Dillwyn, 11th of 3d Month, 1814:

It is now rather more than three months since she first made known to me her apprehensions of the formation of a tumor (*owing to a bruise recd in reaching for something) in her left breast, my alarm excited the wish to place her immediately under the care of Docr Physick & we proceeded as far as Burlington in our way to this City for that purpose, but finding herself much affected by the cold, added to the reluctance expressed by her Uncle Dillwyn & also felt by herself to submitting to surgical dissection before having made a trial of the Logan Plaster which had often proved affectual in apparently similar cases, induced her to relinquish the design for the present, & before the effect of this application could be ascertained Docr Physick was taken ill, so as to preclude any application to home.

Samuel Emlen to William Dillwyn, 15th of 6th Month, 1814:

the affecting situation in which my beloved Susan then was, in consequence of the existence of a cancerous tumor in her left breast, of which I was first informed in the latter part of last Autumn, tho she had suspected it about two months before; On being made acquainted with this alarming circumstance my first thought was to take her immediately to Philada to be under the care of Docr Physick, but some of her friends here whose affectionate interest in her welfare gave weight to their opinions, were desirous that she should make a trial of the Logan Plaster which had often been found efficacious in cases apparently similar; & with this, her own sentiments coincided (she according wore it constantly for 10 or 11 weeks, without any reduction of the size of the tumor, but by this time the illness of our Brother deprived us of the aid we should have hoped for from his advice.

Susan Emlen to William Dillwyn, 5th of 11th Month, 1814:

the lump was at first insensible but in a few weeks a slight pressure produced a dull pain, and sometimes shooting pains occurred without any external cause. After some weeks of anxiety I discovered the circumstance to my beloved S Emlen, and my dear Aunt Cox: the latter advised me to try for the present the Logans salve, but the application did not suit it, producing heat and irritation—We then thought of going to Philada to consult our Brother Physick, but other considerations discouraged the measure—the season was cold, passing back and forward out of the question, and the thought of leaving our comfortable home and going into lodgings as we first contemplated; was unpleasant to me, tho my dear Husband was desirous of making every possible sacrifice of his convenience to my comfort.

5. Susan Emlen to William Dillwyn, 5th of 11th Month, 1814.

6. William Dillwyn's estate, called Houghton, was located in Springfield Township. E. M. Woodward and John F. Hageman, *History of Burlington and Mercer Counties, New Jersey, with Biographical Sketches of Many of Their Pioneers and Prominent Men* (Philadelphia: Everts & Peck, 1883), 436.

7. Dillwyn Parrish, *The Parrish Family, Including Related Families of Cox—Dillwyn—Roberts—Chandler—Mitchell—Painter—Pussey* (Philadelphia: George H. Buchanan, 1925), 319. A miniature portrait of Susan Emlen is reproduced on 318.

8. William Wade Hinshaw, *Encyclopedia of American Quaker Genealogy,* vol. 2 (Ann Arbor, Mich.: Edwards Brothers, 1938), 217. Burlington Monthly Meeting of the Religious Society of Friends, men's minutes and women's minutes. In a letter to her sister Margaret Hill Morris, dated 22d of 10th Month, 1797, Sarah Hill Dillwyn writes from London that

We are informed by brother and sister Cox, and our dear Susan Emlen, that the report we had before heard of the yellow fever being in Philadelphia is true ... We understand that S. & S. Emlen's present situation [in Philadelphia] was thought rather damp, on which account perhaps they may think it right to hasten their intended building at West Hill.

(quoted in John Jay Smith, *Letters of Doctor Richard Hill and his Children* [Philadelphia: for the author, 1854], 293.) The West Hill property was purchased by the Emlens from John Cox in 1797. See George DeCou, *Burlington: A Provincial Capital; Historical Sketches of Burlington, New Jersey, and Neighborhood* (Philadelphia: Harris and Partridge, 1945), 162.

9. Samuel Emlen was the only son of Samuel Powel Emlen. See Frank Willing Leach, "Emlen Family," Written for the Philadelphia North American, reprint (Philadelphia: The Historical Publication Society, 1932), 13; Burlington Monthly Meeting, men's minutes, 1800–1815.

10. On the Emlen Institute see Ulrich F. Mueller, *Red, Black, and White* (Carthagena, Ohio: n.p. [1935]), particularly 64–65.

11. L. R. C. Agnew and G. F. Sheldon, "Philip Syng Physick (1768–1837): 'The Father of American Surgery,'" *Journal of Medical Education* 35 (1960): 541–49, 546.

12. Philip Syng Physick's unusual last name is coincidental. It is surprising that there is still no comprehensive biography of Physick or analysis of his medical career. See Agnew and Sheldon, "Philip Syng Physick"; William S. Middleton, "Philip Syng Physick: Father of American Surgery," *Annals of Medical History* 1 (1929): 562–82; Alexander Randall, "Philip Syng Physick's Last Major Operation," reprint, *Annals*

of Medical History 9 (1937): 133–42; J. Randolph, *A Memoir on the Life and Character of Philip Syng Physick, M.D.* (Philadelphia: 1839); George F. Sheldon and Ruth Guy Sheldon, "Philip Syng Physick, MD, 1768–1837, The Father of American Surgery," *Bulletin, American College of Surgeons* 64, 5 (May 1979): 16–27.

Lithotomy is surgery for the removal of calculi: hard, calcium-based deposits in internal organs, such as kidney, gall bladder, and urinary bladder stones.

13.   The reasons behind the legal separation of Philip Syng and Elizabeth Emlen Physick remain a mystery. Family legend suggests that she held undue affections for her nephew Dr. John Syng Dorsey, while he was involved in a relationship with the family maid (personal communication, Hill-Physick Keith House staff, summer 1987), Elizabeth Physick's mental condition or irreconcilable differences within the couple seem more likely causes for the separation. On this subject Susan Emlen writes:

> The fact is, that a long continued uneasiness between our sister and her Husband has at length resulted in a formal separation, by mutual agreement.—Referees were chosen, who assign'd a certain income to Betsey, and ... Trustees, who act for her, in matters of business. The Doctor has gone to Henry Hills house in fourth street, where the girls are with him, and the boys to be soon, when they leave us. [T]heir Mother is still in her fathers late dwelling, but expects in a few days to remove into a new handsome house in Pine Street, near the hospital, where her Aunt Huldah Mott is to be her companion.—Thou will easily believe we have not been unconcernd spectators in this most calamitous state of things, but we sometimes hope good of some kind, may arise out of this terrible evil. [P]ossibly the solitude to which poor Betsey has thus reduced herself may give time and occasion for useful reflection, and thus be a means of leading her back to the better way. [I]t is not to be told how much conversation, what variety of sentiments, what a profusion of censure, on all the connections as different views are taken of the subject, have been occasiond by a measure, which in this country has been almost without

example. The kindest and best of brothers has not wholly escaped blame because of the silence he has thought it right as much as possible to preserve on the sad occasion, but this is a very small part of the bad consequences to be expected.

(Susan Emlen to William Dillwyn, 29th of 9th Month, 1815.)

14.   At the time of his marriage, Susan Emlen writes to her father of Philip Syng Physick's ambiguous relationship to Quakerism:

> I cannot remember whether we have yet inform'd thee that our sister has given us a new brother of the pun-provoking name of Philip Syng Physick.... As it did not seem likely Betsy would marry in the society [of Friends], and as her situation, and gaiety of disposition exposed her to the danger of a more improper match, none of her friends seem'd to have much to say against this, particularly as the young man is attach'd to friends, and frequents their meetings, and likewise all his family.

(Susan Emlen to William Dillwyn, 16th of 10th Month, 1800.)

15.   Susan Emlen took great interest in the personal character of the physicians in her circle. She recounts to her father, for example, John Syng Dorsey's marriage of 1807 and Joseph Parrish's marriage to her cousin Susan Cox of 1808. In the letter of 1808 she compares the two doctors:

> Sister told me she had heard Dr Parrish much commended for setting a good example of moderation, for it has become a custom too general in this country for young persons who have their fortune to make, to make no kind of difference in their manner of living from those, whose fortunes are made. Dr Dorsey for instance, in every other respect than keeping carriages and horses, lives just as his Uncle [Physick] ... he is a young man I have much respect for, but tis hard to stem the torrent of fashion.

In a letter of 1809 Susan Emlen compares Drs. Parrish and Physick, commenting on how their shared profession affects their respective dispositions:

[Dr. Parrish] is one of the happiest beings I have ever known. [H]ow he can retain that even cheerfulness, amounting to vivacity thro the dismal scenes, rather I should say, notwithstanding the dismal scenes with which a Physician must continually be conversant is often my admiration. [T]here is a wonderful variety among persons who share perhaps pretty equally the indispensable qualities for ensuring esteem and affection—Dr Physick, a person I sincerely love, is very opposite in his character, naturally grave, constantly engaged in occupations that call for his whole attention, and daily a spectator of evils he can only at best, partially relieve, he has acquired what seems to me so gloomy a view of life, that I tell him, I could not live if I did not entertain more cheerful sentiments.

(Susan Emlen to William Dillwyn, 30th of 5th Month, 1807; Susan Emlen to William Dillwyn, 12th of 11th Month, 1808; Susan Emlen to William Dillwyn, 5th of 7th Month, 1809.)

16. John Jay Smith, *Letters of Doctor Richard Hill and his Children* (Philadelphia: for the author, 1854), 322.

17. Ibid., 390.

18. John Owen, *History of the British and Foreign Bible Society*, 3 vols. (London: vol. 1 printed by Tilling and Highes, 1816; vols. 2 and 3 printed by Richard Watts, 1820), inscription on flyleaf page of third volume, list of languages provided in appendix, volume 3. Rare Book Collection, University of Pennsylvania. These volumes eventually became part of the Philip Syng Physick library at the University.

19. George Dillwyn, *Occasional Reflections, Offered Principally for the Use of Schools* (Burlington, NJ: David Allinson, 1815). Other books bearing William Dillwyn's inscription now at the Rare Book Collection, University of Pennsylvania are Joseph Lancaster, *A Letter to John Foster, Esq. Chancellor of the Exchequer for Ireland, on the Best Means of Educating and Employing the Poor, in that Country* (London: Darton and Harvey, 1805), inscribed "William Dillwyn/to Geo: Dillwyn"; and "Rules and Orders for the Government of Friends School and Workhouse at Clerkenwell" (London: James Phillips, 1780), inscribed "W-Dillwyn to GD."

20. Edward Jenner, *Letter from Doctor Edward Jenner, to William Dillwyn, on the Effects of Vaccination, in Preserving from the Small Pox* (Philadelphia: Philadelphia Vaccine Society, 1818) (microopaque Early American Imprints, 2d ser., no. 44464). The introduction to the slim volume states:

The Importance of the following letter, lately received from London, has induced the Philadelphia Vaccine Society to publish it in its present form. The communication was made in answer to an enquiry of one of our fellow citizens, who was desirous of knowing Dr. Jenner's opinion of his truly interesting discovery, after it had stood the test of twenty years' experience. The friends of humanity will, no doubt, be highly gratified by this additional confirmation of the security to be obtained against the ravages of the Small-Pox, one of the most loathsome and desolating of all diseases. (n.p.)

21. Susan Emlen to William Dillwyn, 5th of 11th Month, 1814. Samuel Emlen explains that Physick nearly died the first night of his attack and that over four hundred drops of laudanum were required to sustain him. Samuel Emlen to William Dillwyn, 6th of 6th Month, 1814.

22. Susan Emlen to William Dillwyn, 5th of 11th Month, 1814.

23. Dr. Casper Wistar was a prominent physician, holding the chair of anatomy at the University of Pennsylvania Medical School. Dr. Tucker was a young assistant of Physick's. See James Thacher, *American Medical Biography: or Memories of Eminent Physicians Who Have Flourished in America*, 2 vols. (Boston: 1828), reprint (New York: DaCapo Press, 1967); S. D. Gross, *Lives of Eminent American Physicians and Surgeons of the Nineteenth Century* (Philadelphia: Lindsay & Blakiston, 1861).

24. Susan Emlen to William Dillwyn, 5th of 11th Month, 1814.

25. Susan Emlen to William Dillwyn, 5th of 11th Month, 1814.

26. Samuel Emlen to William Dillwyn, 6th of 6th Month, 1814.

27. George Dillwyn to William Dillwyn (copied extract), 7th of 6th Month, 1814.

28. The Halsted radical mastectomy, a standard amputation of the entire breast and surrounding tis-

sue, has survived as customary practice into the present and is the operation that women have objected to as unnecessarily disfiguring in cases of early detection of malignancy. William Stewart Halsted's groundbreaking article of 1894, "The Results of Operations for the Cure of Cancer of the Breast Performed at the Johns Hopkins Hospital from June, 1889 to January, 1894" is reprinted in A. Scott Earle, ed., *Surgery in America: From the Colonial Era to the Twentieth Century*, 2d ed. (New York: Praeger, 1983), 356–68.

29.   John Syng Dorsey (b. 1783), nephew of Physick, received his medical degree from the University of Pennsylvania in 1802 and was later on the faculty. He died prematurely in 1818. John Syng Dorsey, *Elements of Surgery* 2 (Philadelphia: Kimber & Conrad, 1813), 268 (microopaque, *Early American Imprints*, 2d ser., no. 28351). A. Scott Earle calls *Elements of Surgery* "the first text by an American author that attempted to encompass the entire field of surgery as it was then practiced." The book, which went through three editions in a decade, was based on Physick's own procedures combined with teachings of the European masters. See Earle, *Surgery in America*, 66. Samuel Cooper, *Dictionary of Practical Surgery* ["with notes and additions by John Syng Dorsey, M.D."] 2 vols. (Philadelphia: B. & T. Kite, et. al., 1810) (microopaque, *Early American Imprints*, 2d ser., no. 19857). Physick published only short journal articles during his career, two of which are reprinted in Earle, *Surgery in America*.

30.   Quotations are taken from a second American edition of 1816, based on an expanded London edition. Cooper, *Dictionary of Practical Surgery*, 196 (microopaque, *Early American Imprints*, 2d ser., no. 37338).

31.   Susan Emlen to William Dillwyn, 5th of 11th Month, 1814.

32.   Samuel Emlen to William Dillwyn, 6th of 6th Month, 1814. Samuel Emlen continued this first postoperative letter over the course of several days.

33.   Samuel Emlen to Roberts Vaux, 13th of 6th Month, 1814; Samuel Emlen to Roberts Vaux, 22d of 6th Month, 1814; Samuel Emlen to Roberts Vaux, 29th of 6th Month, 1814; See also Susan Emlen to Roberts Vaux, 7th of 8th Month, 1814.

34.   Charles E. Rosenberg, "The Therapeutic Revolution: Medicine, Meaning and Social Change in Nineteenth-Century America," in *Sickness and Health in America: Readings in the History of Medicine and Public Health*, 2d ed., ed. Judith Walzer Leavitt and Ronald L. Numbers (Madison: University of Wisconsin Press, 1985), 40. See also Lester S. King, *Transformations in American Medicine: Benjamin Rush to William Osler* (Baltimore: Johns Hopkins University Press, 1991); Henry Burnell Shafer, *The American Medical Profession, 1783 to 1850* (New York: AMS Press, 1968); Martin S. Pernick, "A House Divided: An Interpretive Overview of Nineteenth-Century American Medicine," ch. 2 in *A Calculus of Suffering: Pain, Professionalism, and Anaesthesia in Nineteenth-Century America* (New York: Columbia University Press, 1985), 9–31. On reactions to medical professionalism see the essays in Norman Gevitz, ed., *Other Healers: Unorthodox Medicine in America* (Baltimore: Johns Hopkins University Press, 1988).

35.   The spectrum of physicians ranged from the elite practicing in large cities—perhaps with European training—serving wealthy clients and associated with medical schools to the more common class of physicians trained through apprenticeship who tended to be more individualistic than the leading professionals and more likely to share therapeutic outlook of patients. At the other end, far away from this rising professionalism, were lay healers and alternate practitioners, such as the nineteenth-century homeopaths. William G. Rothstein, "The Botanical Movements and Orthodox Medicine," in *Other Healers*, 35. Homeopathy sought an alternative to the heroic treatments—such as bleeding and purging—taught and practiced by the medical establishment. See Martin Kaufman, "Homeopathy in America: The Rise and Fall and Persistence of a Medical Heresy," in *Other Healers*, 99–123.

36.   Susan Emlen to William Dillwyn, 5th of 11th Month, 1814.

37.   Pernick, "House Divided," 15–16.

38.   Samuel Emlen to William Dillwyn, 11th of 3d Month, 1814.

39.   Martin S. Pernick, "The Calculus of Suffering in Nineteenth-Century Surgery," in Leavitt and Numbers, eds., *Sickness and Health in America*, 98–112.

40.   Middleton quotes an anecdote about another breast cancer case that Physick supervised. "An opera-

tion was agreed upon, but was deferred for six weeks in order, by a process of rigid dieting, as was then the custom, to put the system in a proper condition." Physick's convictions about the necessary sparseness of diet are portrayed as comically extreme. Middleton, "Philip Syng Physick," 577–78.

41. Pernick, "Calculus of Suffering," 99.

42. See, for example, Judy A. Breen, "Epidemiological Review of Cancer in Women," in *Women with Cancer: Psychological Perspectives,* ed. Barbara L. Anderson (New York: Springer-Verlag, 1986), 73–74.

43. See de Moulin, *Short History,* especially 1–50. For a widely read contemporary example of humoral based theory, see Alexander Monro's "Histories of Collections of Bloody Lymph in Cancerous Breasts," 484–91 in *The Works of Alexander Monro, MD* (Edinburgh: Charles Elliot; London: George Pobinson, 1781). Julia L. Epstein's summary is also quite useful, "Writing the Unspeakable: Fanny Burney's Mastectomy and the Fictive Body," *Representations* 16 (1986): 131–66.

44. L. J. Rather, *Genesis of Cancer: A Study in the History of Ideas* (Baltimore: Johns Hopkins University Press, 1978), 16; de Moulin, *Short History,* 6–9.

45. Rather, *Genesis of Cancer,* 6.

46. Samuel Emlen to William Dillwyn, 11th of 3d Month, 1814.

47. De Moulin, *Short History,* 26.

48. Susan E. Cayleff, "Gender, Ideology, and the Water-Cure Movement," in *Other Healers,* 82–98, 85. For an insightful study of these issues, focused on childbirth rather than cancer, see Emily Martin's *The Woman in the Body: A Cultural Analysis of Reproduction* (Boston: Beacon Press, 1987).

49. Susan Emlen to William Dillwyn, 5th of 11th Month; Samuel Emlen to William Dillwyn, 6th of 6th Month, 1814.

50. Ruth Kleinman, "Facing Cancer in the Seventeenth Century: The Last Illness of Anne of Austria, 1664–1666," *Advances in Thanatology* 4, 1 (1977): 37–55. See also Epstein, "Writing the Unspeakable"; de Moulin, *Short History,* 34.

51. Epstein, "Writing the Unspeakable," 131.

52. Edward F. Lewison, "Saint Agatha: The Patron Saint of Diseases of the Breast in Legend and Art," *Bulletin of the History of Medicine* 24 (1950): 414–19.

53. See Epstein, "Writing the Unspeakable," 157.

54. Fanny Burney, "A Mastectomy," in *The Journals and Letters of Fanny Burney (Madame D'Arblay),* 6, "France 1803–1812," ed. Joyce Hemlow (Oxford: Clarendon Press, 1975), 596–616, esp. 598.

55. Ibid., 612–14.

56. Epstein, "Writing the Unspeakable," 143.

57. Ibid., 150–53.

58. Susan Emlen to William Dillwyn, 24th of 7th Month, 1814.

59. William Dillwyn to Susan and Samuel Emlen, 16th of 8th Month, 1814.

60. Susan and Samuel Emlen to William Dillwyn, 27th of 6th Month, 1814.

61. William Dillwyn to Susan and Samuel Emlen, 16th of 8th Month, 1814.

62. Susan Emlen to William Dillwyn, 14th of 9th Month, 1814.

63. Barbara Kirshenblatt-Gimblett, "Authoring Lives," *Journal of Folklore Research* 26 (1989): 123–49; Richard Bauman and Charles Briggs, "Poetics and Performance as Critical Perspectives on Language and Social Life," *Annual Review of Anthropology* 19 (1990): 59–88. Keith H. Basso, "The Ethnography of Writing," in *Explorations in the Ethnography of Speaking,* ed. Richard Bauman and Joel Sherzer (Cambridge: Cambridge University Press, 1974), 425–32; Carroll Smith-Rosenberg, *Disorderly Conduct: Visions of Gender in Victorian America* (New York: A. A. Knopf, 1985). On letter-writing see Bruce Redford, *The Converse of the Pen: Acts of Intimacy in the Eighteenth-Century Familiar Letters* (Chicago: University of Chicago, 1986).

64. Some of George Dillwyns's sermons are published, including one in William Savery, *Three Sermons Preached at the Meetinghouse of the People Commonly Called Quakers in Houndsditch,* 3d. ed. (Dublin: 1796). See also William Savery, *Some Remarks on the Practice of Taking Down and Publishing the Testimonies of Ministering Friends,* "with some additional observations, by George Dillwyn" (Philadelphia: Kimber & Conrad, 1809). John Woolman's *Journal,* along with Fox's, was already widely read by early nineteenth-century Friends. See Howard T. Brinton, *Quaker Journals: Varieties of Religious Experience Among Friends* (Wallingford, Pa.: Pendle Hill, 1972).

65. Susan Emlen to William Dillwyn, 20th of 2d Month, 1811.

66. Samuel Emlen to William Dillwyn, 11th of 3d Month, 1814; Samuel Emlen to William Dillwyn, 5th of 11th Month, 1814. In a letter of 1812 Samuel Emlen reports to his father-in-law on a religious visit he participated in, to the home of his sister and brother-in-law Physick. (Samuel Emlen to William Dillwyn, 11th of 2d Month, 1812.)

67. Susan Emlen to William Dillwyn, 24th of 7th Month, 1814; Samuel Emlen to William Dillwyn, 4th of 8th Month, 1814; Susan Emlen to William Dillwyn, 16th of 5th Month, 1815.

68. Susan Emlen to William Dillwyn, 30th of 1st Month, 1814.

69. Susan Emlen to William Dillwyn, 5th of 11th Month, 1814; Samuel Emlen to William Dillwyn, 14th of 9th Month, 1814. Lamentations 3:31–33, in the King James Version, reads:

> For the Lord will not cast off for ever:
> But though he cause grief, yet will he have
>   compassion
> according to the multitude of his mercies.
> For he doth not afflict willingly nor grieve
>   the children of men.

70. See Rosenberg and Smith-Rosenberg, "The Female Animal"; Smith-Rosenberg, "Puberty to Menopause."

71. See, for example, Nancy Cott, *The Bonds of Womanhood: "Woman's Sphere" in New England, 1780–1835* (New Haven: Yale University Press, 1977).

72. See Barry Levy, *Quakers and the American Family: British Settlement in the Delaware Valley* (New York: Oxford University Press, 1988); J. William Frost, *The Quaker Family in Colonial America: A Portrait of the Society of Friends* (New York: St. Martin's Press, 1973); Margaret Hope Bacon, *Mothers of Feminism: The Story of Quaker Women in America* (San Francisco: Harper & Row, 1986).

73. Samuel Emlen explains his wife's decision-making process during treatment.

On being informed of the opinion of the Physicians on dear Susans case, her Uncle Dillwyn, Aunt & Uncle Cox came to Philada but like myself were unable to give the dear sufferer any advice on the momentous question, whether relief was to be sought by undergoing an extensive surgical operation, or await the issue of the disease & endeavour to counteract its attendant pains by palliatives were therefore obliged, altho very desirous of affording her all the aid in our power, to leave the decision with all it attendant weight upon herself[.] this she thought she should be better qualified to come to at her own home than anywhere else.

(Samuel Emlen to William Dillwyn, 6th of 6 Month, 1814.)

74. See Richard Bauman, *Let Your Words Be Few: Symbolism of Speaking and Silence Among Seventeenth-Century Quakers* (Cambridge: Cambridge University Press, 1983), especially 20–31.

75. An extract from Susan Emlen's death notice reads:

Possessed of all the mild and endearing virtues, gentle, benevolent, good, she was the delight of her friends, and a treasure of inestimable worth to her husband and relatives. The spotless purity of her mind and the sweetness of her whole character, appeared so entirely without alloy, that she seemed like an inhabitant of a more blessed sphere.

(*National Recorder*, December 18, 1819, quoted in Parrish, *Parrish Family*, 319).

76. *Gathered Fragments: Briefly Illustrative of the Life of George Dillwyn of Burlington, West New Jersey, North America* (London: Alfred W. Bennett, 1858), 42–43.

77. Hannah Logan Smith, *Memorial and Reminiscences in Private Life* (Philadelphia: 1839), 264.

5   Mother Love and Infant Death, 1750–1920

Nancy Schrom Dye and Daniel Blake Smith

American motherhood has only recently begun to acquire a past. Mothering is far more than a biological constant; it is an activity whose meaning has altered considerably over time. Changes in cultural values, maternal self-perceptions, and attitudes toward children—all these factors underscore the historical dimensions of motherhood. Indeed, the shifting status of women and the changing nature of the family in history cannot be fully understood without close study of the experience of mothering.

Yet despite all the valuable work on women and the family in recent years, this central dimension in women's past has remained largely unexamined. Almost all previous research has either examined parental roles through the prescriptive lens of sermons and childrearing manuals, or it has addressed the question of how changing childrearing styles affected children's personality development.[1] As a result, a good deal is known about normative ideals of motherhood, particularly in the antebellum period, but comparatively little is known about women's own emotions and experiences of mothering.

This essay suggests elements of both change and continuity in the history of American motherhood by addressing two related questions. First, what were American women's experiences and perceptions as mothers, and how did maternal feelings and perceptions change over time? Second, how did the persistent reality of high infant mortality affect mothers and the relationships they established with their children? How did mothers come to terms with the ever-present reality that their children might die? These questions are explored primarily by examining women's personal writings—diaries, journals, and letters—to illuminate maternal experience.

Women's writings suggest that the experience of infant death formed a constant backdrop against which mothers' experiences and emotions must be set. Infant mortality remained high throughout the entire period under consideration. During the nineteenth century as much as 40 percent of the total death rate was comprised of the deaths of children under the age of five. Not until near the turn of the twentieth century, with improvements in public health, nutrition, and general standard of living, was there a dramatic downturn in the infant death rate.[2] Despite the continuous reality of infant death, mothers' responses to sickness and death in their families changed significantly over time as cultural explanations for infant death

NANCY SCHROM DYE is President of the College and Professor of History at Oberlin College, Oberlin, Ohio.
DANIEL BLAKE SMITH is Professor of History at the University of Kentucky, Lexington.

Reprinted from *The Journal of American History* 73 (Sept. 1986): 329–53, by permission of the publisher.

and definitions of maternal roles changed. Women's personal reflections on and descriptions of day-to-day mothering—fragmentary as they sometimes are—point to three central modes in mothering during the American past.

Throughout the colonial period, as Laurel Thatcher Ulrich and others have suggested, mothering was "extensive" in nature.[3] The experience of motherhood was shaped largely by the permeable household structure of colonial America, in which neighbors, friends, and kin played significant caretaking roles, and by a general dependence on divine providence for interpreting maternal experience, especially in times of infant death or illness.

During the second half of the eighteenth century, Americans began to move away from the well-ordered, father-dominated family of early America, with its emphasis on paternal control, obedience, and emotional restraint, toward a strikingly affectionate, self-consciously private family environment in which children gradually became the focus of indulgent attention and mothers emerged as guardians of their moral and physical well-being.[4] By 1800, and even earlier in some instances, one can trace an increasing focus on the individual mother as the most influential force in shaping and preserving a child's life. Reliance on God gradually gave way to a more secular belief that a child's welfare lay primarily in the hands of loving watchful mothers. The belief in the centrality of mothers to infant well-being has, of course, persisted to the present day. It carried particular force, however, throughout the nineteenth century.

By the turn of the twentieth century, women, working through women's clubs, municipal reform groups, and social welfare organizations, began to give public voice to their concerns about the deaths of so many children and to turn infant death—once seen solely as a private tragedy—into a major social and political issue. The first decades of the twentieth century,

then, witnessed a movement away from the confined, privatized mother-child relationship of nineteenth-century America toward a more broadly based "social motherhood" in which mothers shared responsibility for child welfare with public health officials, the medical profession, and ultimately the state.

Such a general overview drawn from women's personal documents requires certain caveats. By virtue of their literacy the eighteenth- and nineteenth-century mothers studied here belonged to middle- and upper-class families; their relatively privileged status may well have colored their maternal experiences and social values. No claim is made, therefore, that working-class women shared their particular perspective on motherhood.[5] In addition, the changes in women's maternal values discussed here, like much else in family history, are gradual and complex, rarely displaying dichotomous shifts. They reflect not so much abrupt transformations at certain specific times as blends of changing experiences and attitudes. Finally, the almost complete absence of female diaries and letters before 1750 severely hinders historians' understanding of motherhood in the colonial era. Despite such limitations, focusing on the writings of mothers themselves highlights central threads in the changing fabric of American family life and illustrates the nature and direction of change in American mothering.

Colonial Americans inhabited an uncertain world. Periodic epidemics of diphtheria, smallpox, scarlet fever, measles, whooping cough, and a host of other infectious diseases swept through early American communities, striking babies and young children with special ferocity. Given the dearth of writings by women, it is necessary to turn to men's writings to gain some sense of how seventeenth- and early eighteenth-century women experienced motherhood in the midst of almost continuous disease and death.

The diary of Massachusetts clergyman Ebenezer Parkman provides a vivid picture of the high child mortality that was central to mothers' experience. In early 1739 Parkman recorded the death of his own infant dauther. "A Morning of great Trouble!" he wrote. ". . . About nine o'Clock I was called down from my study with the Alarm that the Child was dying! About 10 She ceas'd to breathe. The will of the Lord be done! . . . O that we might have a due sense of the Divine Mind concerning us!" On February 5 he noted the death of another baby: "Mrs. Samuel Fay junior Infant Child bury'd which bled to death at Naval." And on May 18 there was yet another infant death to record: "Molly Hicks (infant) dyed—My wife there all night." In the first months of 1740 the "throat distemper"—a virulent strain of diphtheria that killed as many as half the children in some New England communities—appeared in Parkman's community of Marlborough. Parkman noted the mounting death toll among infants and children. In a typical entry he wrote, "A.M. I went over to the Funeral of widow Tomlin's Child and P.M. to the Funeral of Deacon Newton's Child both of which Dy'd of the Throat Distemper. A Third Grave was open'd this Day . . . for a Stillborn Infant of Daniel Stone."[6]

An incident Cotton Mather recorded in his diary suggests the parental response to such continuous illness and death among children. "Alas, for *my Sin*," Mather wrote after his daughter was badly burned in a household accident, "the just God throwes *my Child* into the *Fire!*"[7] Mather's comment reveals the inescapable power attributed to divine providence, especially in determining the fate of infants and children. It is one of the paradoxes of colonial family life that parents cared deeply for their children and yet expected neither conscientious care nor the best medical attention to cure their children's illnesses, prevent dangerous accidents, or forestall death. Children were God's tempo-

rary gift to parents; what He had freely given, He could just as freely—and suddenly—take away.[8]

The belief that the ultimate responsibility for a child's welfare lay in the hands of God is one of the cultural norms that must have shaped women's experiences as mothers. Ebenezer Parkman, for example, gives this description of a mother's response to her newborn's dangerously ill condition: "As to her dear Infant She had given it to God before it was born; She gave it up to him when it was born and I give it up to him now, Said She, and Shd be glad to do it in . . . Baptism." Reflecting on the loss of eight out of eleven children born to her between 1717 and 1736, a New England mother observed, "[S]o it pleased God to take away one after another of my dear children, I hope, to himself."[9]

By the mideighteenth century the beginning of a major transformation in American family values was underway through which mothers assumed increasingly important responsibilities for the nurturance and moral development of children. Consequently, the women who left maternal diaries during the late eighteenth century were transitional figures in the history of American mothering, sharing some of the values and attitudes typically associated with the colonial and the antebellum eras. Their accounts reflect a growing affection and concern for children alongside a continued reliance on divine providence.

Despite the growth of increasingly private, child-centered families beginning in the Revolutionary era, child care remained essentially a cooperative venture in many late eighteenth-century households, involving not only parents but servants, relatives, and friends as well. Children were born into large and busy households with continually changing casts of characters. Abigail Adams, for example, referred to her "constant family" of eighteen, "ten of which make my own family." Elizabeth Porter Phelps and Mary Vial Holyoke, far from remaining closeted at home with their children, thrust themselves into rounds of visiting, attending births, baptisms, and funerals, and running errands in their Massachusetts communities. A typical entry from Phelps's diary in the 1780s: "Monday I a visit at Mr. Hop[kins] took the two Little girls. Left them at Mr. Gaylords. Tuesday I down again—Mrs. Gaylord and I a visit at Mrs. Colts, the widow Warner went with us—at night we brought home Betty—left Thankful to go do school. Thursday Mrs. Crouch a visit here, the Majors wife and her new Daughter Eleazers wife a visit here. Fryday Mother up to see Rosel Smiths wife and her mother Prat is there too." [10]

Elizabeth Cranch Norton, of Weymouth, Massachusetts, likewise raised her children in an open inclusive household. Norton and her husband were usually in residence, although they frequently traveled to visit relatives elsewhere in Massachusetts and Rhode Island. In turn, relatives and friends—and their children—came for lengthy visits. Servants, whose ranks included children, also came and went. For example, in January 1795, "Solomon Porter, a boy about 12 years old came to live with us." The following year, "Mrs. Bates came and brought her daughter Nancy to spend some time with us—to be instructed in needle work, reading, writing, etc." In April of the same year, "Becca Clevely came to live with me . . . aged 13 years . . . am to cloathe her if she continues with me till she is 18." [11]

Such continuous coming and going did not slow down while women were pregnant, nursing, or caring for infants. On December 1, 1799, shortly before the birth of her fifth child, Elizabeth Norton recorded that "Mrs. Bicknell came and spent the evening and night." On December 11 Norton took her oldest son to her parents' home in Quincy, Massachusetts, while "Mrs. Bicknell staid and took care of my baby for me." Three days later Elizabeth visited "young Mr. Tufts. Betty Burrell took care of my baby." On

Christmas day the Nortons left for Quincy. Once again their servant Betty Burrell took care of baby Thomas, then ten months old. With all this bustling movement in and out of the Norton home, Elizabeth Norton was rarely alone. Indeed, a day such as December 25, 1794, about which she wrote "Staid at home with children all day—had to do all my housework—no help," stands out amid constant travel and visiting. [12]

Children in such households, even when very young, also seem to have spent considerable time elsewhere. Elizabeth Phelps's diary indicates that she frequently left her daughters, Betty and Thankful, with friends and relatives, and Mary Vial Holyoke often transported her children from Salem, Massachusetts, to Cambridge, where they stayed with her parents. All of Elizabeth Norton's children, beginning in infancy, stayed for extended periods in their grandparents' home in Quincy, a two-hour journey from Weymouth. This practice may have served several purposes. The timing of children's visits sometimes coincided with a particularly difficult pregnancy or illness. Children sometimes left their own households to go to school elsewhere. And their lengthy stays in the households of relatives may have served to foster close ties with members of an extended-kin network. [13] The frequent movement of both parents and children in and out of households contributed to the permeable quality of late eighteenth-century family life and suggests that mothers' relationships with their children formed part of a larger, continuously changing tapestry of multiple personal ties.

Given that extensive mothering style, it is perhaps not surprising that children—especially infants and toddlers—rarely figured in mothers' daily accounts of their own activities. When children were mentioned, the notations were spare and brief, often related to matters of birth and death. Between 1760 and 1782, for instance, Mary Vial Holyoke bore twelve children, eight of whom died in infancy. In the following entries she describes the birth and death of a daughter:

> Sept. 5, 1767. I was brought to bed about 2 o'clock AM of a daughter.
> Sept. 6. The Child Baptized Mary.
> Sept. 7. The Baby very well till ten o'clock in the evening & then taken with fits.
> Sept. 8. The Baby remained ill all day.
> Sept. 9. It died about 8 o'clock in the morning.
> Sept. 10. Was buried.

The diary of Sarah Snell Bryant also reflects this cryptic style. The wife of a Massachusetts physician, Bryant bore her third child on July 12, 1798. "A son born a little before sun set. Mrs. White and [Mrs.] Otis here." There were only two brief, indirect references to the baby until September 17, 1798: "babe sick—very rainy night—kept on [illegible] with the babe." Bryant did not refer to the infant by his name, Cyrus, until the following March, when he was eight months old.[14]

The late eighteenth-century maternal diaries tell little about how children looked or behaved. Although a child's teething and walking experiences were sometimes mentioned, most diarists did not reflect on developmental milestones. More important, the diaries do not reveal children's individual personalities. Some journals provide considerable detail about mothers' household work—when and how much they spun, wove, sewed, and ironed—and their daily social life—with whom they dined and visited, who stayed with them and how long—but their children, particularly infants, remain obscure. One wonders whether this silence testifies to something more significant—not to maternal indifference but to an unspoken acknowledgment that until infants had grown into children (and perhaps hearty ones at that) they remained something less than complete persons and

their well-being remained beyond a mother's control.[15]

This absence of maternal control is suggested in the accounts of the numerous home accidents that appear in diaries and letters. While few late eighteenth-century mothers expressed the divine determinism so evident in Puritan parents such as Cotton Mather, they nonetheless displayed a similar feeling of powerlessness. Consider Elizabeth Phelps's description of a toddler's accident in 1773: "Thursday morning our babe was left alone in the Room, crept to a teakettle of scalding water—turned it over scalded one hand very bad, the other a little. Lord what a great mercy twas no worse. thou are our constant benefactor, O may this providence serve to put me upon consideration that the Child is thine. Let me never forget it." When one of Elizabeth Norton's children was bitten by a dog, she responded with characteristic passivity: "Richard was a little bitten by a Dog which gave me some anxiety, as many dogs lately run mad and occasioned the deaths of several persons—in God's hands is his life—let him do as it seemeth to him good!"[16]

In many respects Elizabeth Norton's experiences were emblematic of the complex, sometimes contradictory attitudes of mothers in the late eighteenth century. Although her prose is spare, even cryptic, Norton's diary and letters reveal a mother who was warm and affectionate, who worried over her children's physical and spiritual welfare, and who missed them when they were away. "I long to get my little Thomas home," she wrote her mother in 1801. "I must have him (upon trial at least) a week or a fortnight. I think of him a great deal and have lived hoping every week I should find it convenient to go for him." As her children grew older, she wrote of them more frequently and with more description, perhaps indicative of her growing expectations for them. And yet she stood ready to surrender her children to God. On her son

William's seventh birthday, she noted: "May he be made a good and useful member of society if he should be continued to years of manhood—but may I be prepared to part with him, if so infinite wisdom should determine."[17]

Few maternal writings equal Louisa Park's sad account of her son Warren's death. Unlike many eighteenth-century women's personal writings, Louisa Park's diary effusively detailed the development and death of her child. Park's husband, a Newburyport, Massachusetts, physician, was away on shipboard throughout the period from Warren's birth in 1800 until his death one year later. Perhaps the father's absence drew mother and child unusually close together; indeed, Park seemed to luxuriate in caring for Warren—nursing him, playing with him, watching his growth, giving him medicine during his several bouts with influenza and other illnesses. "Stayed at home all day with my babe," she wrote in December 1800, "reading, writing, playing with my little innocent." One senses both the quiet pleasure and the tedium of Park's daily life as a mother: "My days have become so regular that when I have told the business of one, I have for the week. I rise at eight—eat breakfast at nine—then sew, what time I have besides the care of Warren, 'till after dinner. If no company presents, writing and reading in the afternoon. In the evening, knitting, if circumstances permit."[18]

In the midst of a virulent influenza epidemic Park began to worry openly about Warren, even though he was not yet ill. On January 29, 1801, when Warren was nine months old, she declared in her diary, "I hope Warren is not going to be sick. I begin to love him much." By March her son was dangerously ill and despite numerous consultations with physicians and heroic efforts to save him, he died on April 25. A week after Warren's death (there are no entries during the final stages of the boy's illness), Park began to reflect on her loss. "Until today, I found it im-

possible to compose myself sufficiently to make the attempt [to write]. At bedtime, instead of my charming boy, my lovely babe . . . instead of my laughing cherub to receive the caresses of a tender mother—I found a lifeless corps—laid out in the white robes of innocence and death. Though I wept and pressed him, he could not look at me. How could I endure it—much less compose myself—but by believing him gone to perfect rest and happiness—there to wait for his father and mother." Her son's death, without the consoling presence of her husband, left Louisa Park desperately alone. "I know not what to do with myself, now I have no Warren to care for, attend to, caress and love." By the next day, though, she was able to let go of her child. In a single sentence she gave voice to a range of powerful emotions and values, touching on both her love for Warren and her religious resignation: "Yet I do not wish thee back again, my lovely innocent. No—I will bless my God who has taken thee to Himself before thou couldst offend him, and has saved thee from a life of sickness, sorrow, and woe, although it has been at the expense of my health and happiness."[19]

Louisa Park's account of her emotions and experiences suggests that by the turn of the nineteenth century, American concepts of mothering were changing. In her detailed description of her son's life Park seems to have developed a richer, more intense relationship with her child than was characteristic of earlier generations of American mothers. As the nineteenth century progressed, such mother-child bonds became the norm. Over the same years ministers, educators, physicians, and social commentators produced a flood of childrearing literature advising women that motherhood was a full-time responsibility for which they were fitted by natural instinct and inclination. Central to that emerging concept of maternity was the belief that women stood as guardians of traditional moral values in an increasingly secular, commercialized society.[20]

Other social and cultural developments over the course of the century influenced women's perceptions of themselves as mothers and of the relationships they established with their children. The increasingly private and isolated nature of the family in nineteenth-century America seems to have encouraged the establishment of intense mother-infant bonds. And in all but the most evangelical of families Americans gradually supplemented reliance on divine providence with faith in the power of human agency. To be sure, mothers continued to turn to God in times of sickness and death and tried to resign themselves to God's will when children died. But increasingly they took active and sustained measures to ensure their children's health and well-being. Despite a continuously high infant death rate—indeed, the mortality rate may have risen in the first half of the nineteenth century as a result of urbanization—reactions to infant death gradually changed.[21] Individual mothers slowly came to replace God as the most important guarantors of their children's welfare—a role that created great anxiety for many nineteenth-century American women.

Women's accounts of mothering reflect a sense of high purpose and responsibility so characteristic of antebellum advice literature on maternity. "I will be obliged to consider his wants first, dear husband," Bessie Huntting Rudd wrote in 1860, shortly after giving birth to her first child. "How strange it seems to think of a *pet* coming into our thoughts requiring the *first* care and attention." For some women childrearing seems to have been more than a full-time commitment; it meant assuming a new identity altogether. As Elizabeth Sedgwick wrote after the birth of her first child, "At 6 o'clock on the 7th of January 1824 I was a *mother* and experienced that delightful transition from suffering, danger, and anxiety to happiness and that intense delight, that unspeakable sentiment which pervades the heart at its first maternal throb." For Fanny Appleton Longfellow the transformation from young wife to mother was signaled by her decision to give up the diary she had kept since she was a young girl. "With this day my Journal ends," she noted a few months after her first child's birth, "for I now have a living one to keep faithfully, more faithfully than this."[22]

Middle-class mothers in the nineteenth century cared for their children in relative isolation, compared with the large sociable households of early America. Left alone for the greater part of each day with their babies, mothers assumed primary responsibility for their care. Servants and nurses, although often present, were relegated to secondary roles. "I have been employed, and very happily so, with the child all day," Mary Lee noted in her journal in 1813. In the first months of their babies' lives many mothers rarely left home. Her firstborn was nearly three months old before Caroline White ventured out for an evening band concert. "It was a much longer time to leave baby than I have ever done before," White noted in her journal. Elizabeth Sedgwick found fulfillment in her hours of solitude with her children. Writing in 1827, when she was the mother of three children under three years old, Sedgwick declared, "My situation this winter confining me wholly to the House, excepting a daily walk for exercise, enabled me to devote myself very much to my children. Indeed my home was so happy a one that I found my Interest in general society constantly lessening."[23]

The secluded nature of the middle-class household and the amount of time that mothers spent in the company of their babies encouraged intense attachments between mothers and infants.[24] There is much evidence of such early attachment. Writing in 1810, Mary Lee declared that she had "passed a most delightful day in tending my child; she has been unusually pleasant and I have enjoyed the true comfort of a *little baby.*" Writing several generations later, in 1863, Elizabeth Child described her first sight of her baby daughter: "I remember that while I looked you opened your eyes . . . and as they met mine I thought they mutely recognized the new tie." "The dear little fellow," Caroline White wrote of her newborn in 1856, "I did not think I should love him so well so soon."[25]

Some mothers devoted their journals exclusively to detailed accounts of their children's development. Elizabeth Sedgwick, for instance, kept a daily account of her daughter's life. "Long has been my intention to keep a journal of my child's life extending even to the minutest action and the slightest unfolding of her character," she noted in the first entry. "The smallest events of her life have had their peculiar interest for us, who have been watching her as parents always watch their heart's treasure." When the child Lizzie Sedgwick herself became a mother, she continued the tradition. Fanny Longfellow was the mother of three children when she resumed a journal of 1847. In place of the introspective musings and accounts of social activities that had filled her earlier journals, her new diary, aptly titled "Chronicle of the Children of Craigie Castle," was given over to the unfolding of her children's personalities and to notes on their activities and behavior.[26]

Infants and toddlers appear in nineteenth-century journals as full-fledged, albeit physically fragile, individuals. Children are almost always referred to by name or, commonly, by affectionate nicknames: They are "birdies," "chicks," "kitties," "pets." Mothers also adopted as nicknames the names children called themselves. Thus Mary Lee's daughter Betty is "Beppa" in her writings; Caroline White's son Charlie is "Larlie"; Fanny Longfellow's daughter is "Sipsie."

Children's distinct personalities and physical characteristics emerge clearly. Fanny Longfel-

low's four-year-old son Charley was a "Man of action" who "promises to have a rich and noble nature." Erny, at two, was "an angelic little child" who "loves to nestle in one's lap with his thumb in his mouth . . . as timid and tender as a young bird. He promises to be the poet." And "Little Fan, not a year old, is only a round rosy merry plaything with dark blue eyes, a cunning little mouth and a very intelligent eager air." Calista Hall presents a similarly lively description of her baby daughter, "Little Frances grows more interesting every day. She will try to say everything you tell her too. Ask her whose baby she is, she will say Pa and Ma. . . . She is a little rogue. She will pull of her shoes and stokings as fast as I can put them on."[27]

Nineteenth-century mothers carefully recorded developmental milestones. On January 9, 1848, for instance, Fanny Longfellow noted that her daughter had "learned to creep." By February 3, "Little Fan has a tooth and climbs by chairs upon her feet." On March 8, "Baby . . . looked at me and smiling said, 'Mama' then put her finger in my mouth." Elizabeth Sedgwick gave a precise description of her daughter's walking: "[At] 14 months and 6 days old, you rose from your chair with great deliberation and walked across the nursery. . . . You seemed neither pleased nor surprised."[28]

Such detailed description of children's behavior and the many expressions of affection that fill these writings attest to the richly textured emotional bonds between middle-class women and their children. Mothers clearly delighted in their children. "Took Charley to village," Fanny Longfellow noted in a typical entry. "Love to feel his hand in mine for a walk and he is so glad to go with me." Or as Bessie Rudd wrote about her infant son, "He grows into our hearts every day, . . . he is becoming an idol with me, and his sweet face, as he looks up so innocently, seems to steal your heart immediately." But beneath this delight lay the ever-present fear that their children might die. In the first days and weeks after giving birth many mothers appear to have been frightened by the intensity of their feelings for infants whose lives seemed so fragile. "He is so tender," Caroline White wrote of her son in 1856, "I feel as though a very slight injury or sickness could carry him off—and I do not feel ready to part with him—short time as our relation to each other has existed."[29]

As their infants grew, mothers became more anxious and appear to have regarded serious illness as inevitable. Accounts of children's illnesses and their treatment consume a significant part of nineteenth-century maternal writings. "Edwin was quite sick Sunday night," Calista Hall wrote in 1849 in a typical description. "He could not keep any thing down and had a high fever. I set holding him. . . . I was afraid he would die, he seemed so sick." Elizabeth Cleveland's detailed record of her baby daughter's fatal illness is also characteristic of such accounts. In the spring of 1848 Cleveland wrote, "my dear little Lucretia had her first sickness. I was very anxious—as the Dr. considered it a bad state of the digestive organs—cankered bowels." The little girl improved temporarily but became seriously ill in July. For more than a month Cleveland's journal was given over to daily entries describing her daughter's symptoms and the medical efforts made to cure her. Late in August, despite Cleveland's increasingly frantic efforts to save her daughter, the child died. Only then did Cleveland invoke God: "I must show my love in my resignation that she has gone to God," she wrote some days after Lucretia died.[30]

Infancy and early childhood emerge as periods of successive physical crises. Feeding, teething, weaning, and walking all possessed their peculiar dangers. Feeding crises were particularly common. A breast infection and phlebitis forced Caroline White to wean her first child prematurely, and "the poor little fellow has pained and

been ailing ever since." Mothers who weaned their infants early or who bottlefed them had good reason to be apprehensive, for at a time when water and milk supplies could not be counted on to be clean, babies brought up "by hand" were at far greater risk of disease. For months after she weaned her son in 1856, White devoted her journal to the daily fluctuations in her son's weight and health. A gain in weight, however slight, was cause for rejoicing. But there were frequent setbacks. "Called on friend with six-week old baby bigger than mine of sixteen weeks. Poor little thing—he now weighs ten pounds—six weeks ago he weighed thirteen." Charlie White survived and developed into a healthy, thriving child, but his mother dreaded the possibility of raising another baby "by hand." She anxiously kept track of each infant's weight and despaired of her ability to nurse them successfully. "I am not strong—I feel quite discouraged about having nourishment enough for baby," she wrote shortly after the birth of her daughter in 1863. "I have to feed her considerably now—am hoping for a better state of things. She does not grow as well as she ought."[31]

Children's teething, long regarded as a critical period in infants' lives, was also a time of apprehension for mothers. Both mothers and physicians believed that dentition was a major cause of infant death. "The period of teething is filled with terror to a mother's imagination and I had looked forward to it with unceasing anxiety," Elizabeth Sedgwick wrote. "Poor little Erny was found . . . in his crib in convulsions from teeth," Fanny Longfellow noted in early 1848. "Little fragile flower. I tremble as I look at him and am devoured by anxiety." Longfellow believed that teething caused her daughter's fatal illness later that year. When little Fanny was eighteen months old, her mother noted that the child was "having a hard time teething." Less than a week after the first symptoms appeared, what had seemingly begun as difficult teething had developed into a life-threatening illness. "A very anxious day," Fanny Longfellow wrote on September 6. "Poor little baby seemed to have much trouble in her head and the Doctor feared congestion of the brain." Heroic medical measures, including large doses of mercury, were employed in a futile attempt to save her life. On September 11 Fanny died. "Sinking, sinking away from us," Longfellow wrote on the day of her daughter's death. "Felt a terrible desire to seize her in my arms and warm her to life again at my breast. Oh for one look of love, one mood or smile. . . . Heard her breathing shorten, then cease, without a flutter." [32]

Fanny Longfellow's profound grief over the death of her daughter plunged her into a lengthy depression. "I seem to have lost all interest in the future and can enjoy my children only from hour to hour. I feel as if my lost darling were drawing me to her—as I controlled her life before birth so does she me now." Almost all of Longfellow's diary notations for the next year dwelt on little Fanny's death and the fear that her remaining children would die. She could not look at them, she wrote, without imagining them in their own small graves. And for months after her loss Longfellow found herself reliving the final days of her daughter's life, "haunted by thoughts of what might have been avoided, the most pitiless of all." [33]

The child's death was a tragedy from which Fanny Longfellow never fully recovered. In 1850 she bore a second daughter, Alice, who seemed "an especial grace of God." But Alice's birth coincided with her son Erny's serious bout with pleurisy. "We were terribly anxious," Fanny wrote in the same entry in which she recorded Alice's birth. "I felt as if a great stone were hurled back upon my heart from which I had just been relieved. I have never been without anxiety for him but now it will be increased tenfold." Her daughter's death remained the central event in Fanny Longfellow's life. [34]

Why did nineteenth-century mothers appear to fear child illness and death so much more intensely than parents in early America? What factors account for growing intimacy between mother and child? To be sure, early American parents feared death too and grieved over the loss of loved ones. [35] But one detects in those parents a more passive attitude of Christian resignation. By the early decades of the nineteenth century, except among evangelical families, divine providence was giving way to exalted motherhood in the care and protection of children. [36] Parents, as always, tried to submit to God's will in the face of infant illness and death, but it was an effort increasingly undermined by the growing belief that a child's health depended mainly on maternal dedication and appropriate medical treatment. Mothers may well have worried so incessantly about the health of their children because they were coming to believe—despite persistently high infant death rates—that good mothering could somehow ensure a baby's survival.

That gradual shift in the locus of control reflected new assumptions about the mother's centrality in the family and her principal responsibility for child nurture. At a deeper level, however, it seems closely linked to the growing belief that human agency could shape and control the natural order. Just as Americans were beginning to reject pain and death as necessary concomitants of childbearing—as the growing popularity of ether and chloroform in childbirth attest—they embraced the idea of maternal efficacy in overseeing the health of their infants. [37] As the author of one popular home medical guide stated, "No one can for a moment believe that the excessive and increasing infant mortality among us, is part of the established order of nature, or of the systematic arrangements of Divine Providence." Or as another leading maternal advisor observed in 1844, "The first and most important truth . . . to be impressed on mothers, is, that the constitution of their offspring depends on natural consequences, many of which are under their own control.[38]

Physicians were concerned about good mothering but paid little attention to infant mortality. Many doctors were well aware of the fact that infant deaths constituted a very large percentage of total mortality, but medical literature throughout the century devoted strikingly little attention to this problem. Specialized medical interest in pediatrics was just emerging in the 1830s and 1840s. The few physicians who wrote on children's diseases believed that infancy was a period of extreme physical debility and weakness; inevitably, given infants' natural susceptibility to disease, many would perish. As D. Francis Condie, one of the first writers on pediatrics, explained, "During infancy and childhood, there exists a very strong predisposition to disease. . . . During the first few weeks of existence, the imperfect organization of the body, and the deficiency in vigour of most of its functions, render it particularly liable to the actions of various agents, the impression of which . . . produces in the delicate organs of the infant, the most serious disturbance, resulting in the greater number of cases, in a rapid extinction of life."[39] Infancy, like old age, was a time to die.

Significantly, doctors and other writers of medical guides did not recognize the social and economic dimensions of infant diseases and their prevention. Concerns about health lay entirely within individual mothers' domain. As one physician declared, the chief cause of infant death was the "ignorance and false pride of the mothers. Children are killed by the manner in which they are dressed, and by the food that is given them, as much as any other cause." Doctors agreed that there was much individual mothers could do to ensure their children's well-being. Mothers could exercise constant vigilance over their children's day-to-day care—bathing, dressing, playing, and ministering to them during illness. Most important, mothers had a duty

to nurse their own infants. A mother "must not delegate to any being the sacred and delightful task of suckling her child." Mothers could school themselves in the symptoms and etiology of childhood diseases and the essentials of child hygiene. They could learn to distinguish between ordinary innocuous "snuffles" and the "morbid snuffles" that portended serious illness or death. And they could treat their vulnerable infants gently by refraining from cold baths and harsh purgatives and medications and by sheltering their delicate nervous systems from bright lights, extreme temperatures, and loud noises.[40] Just as Americans assigned mothers the responsibility for their children's moral welfare, so too did they vouchsafe to them their babies' very survival.

American culture, then, as reflected in both professional and popular medical literature, sent mothers a mixed message. On the one hand, death in infancy was recognized as commonplace, even expected. On the other, "good" mothers were encouraged to believe that careful nurturing could ensure the well-being of their offspring. It is not known how many mothers read such medical guides or how they interpreted them. Still, the perception of infancy as a time of extreme physical vulnerability and the belief that only mothers could provide proper care help explain maternal anxiety. Certainly, women looked on infancy and early childhood as precarious times. Given the limited knowledge of infectious disease, it was impossible for a mother to gauge the severity of a child's illness. Thus any illness appeared potentially fatal. Alone at home with children whose fragile health they believed lay largely in their own hands, mothers fell prey to anxiety and fear.

The private agonies of watchful and worried mothers bear witness to the special intimacies of their increasingly separate sphere at home. While no doubt mothers could share some of their fears with caring husbands, for the most part their strong maternal feelings found expression in sentimental literature, maternal associations, and women's networks of friends and relatives.[41] In this context the idea of the family as a calm and soothing refuge from a busy world takes on a peculiarly patriarchal cast. For men away at work, hearth and home may have seemed like a place of repose and comfort; for many women, the family was the seat of apprehension and turmoil.[42]

Children in early nineteenth-century America were gradually slipping out of the hands of God and into their mothers' warm, if nervous, embrace. And infant death, like so much else in the increasingly intimate world of the family, became a private tragedy. Whereas the world beyond the household paid little notice to infant mortality, for mothers the lives and deaths of their infants were matters that took a powerful emotional toll. Living beyond the time of unquestioned confidence in the providential power of an omnipotent God and before that of public health advances and reliable medical help, women experienced motherhood in solitude—sole possessors of all the delights and fears that come with raising children.

When it came time to have her baby in Madison, Wisconsin, in 1906, Dorothy Reed Mendenhall expected a safe and normal delivery. A physician herself, she had been well trained in obstetrics at Johns Hopkins University Medical School in the 1890s and, thanks to several years at New York's Babies' Hospital, she had unusually thorough grounding in pediatrics as well. As soon as she learned she was pregnant, Mendenhall placed herself in the care of the surgeon reputed to be the best in the city. What followed was a disaster. Her physician proved careless and incompetent. After two days of lingering labor, the doctor manually turned the baby and delivered it feet first, a hazardous obstetrical procedure. "All I remember was turning my head into the pillow and thinking this will be the end; I shall

die, he doesn't know what to do." A few hours later Mendenhall became aware of her baby's labored and abnormal breathing. She tried to resuscitate her daughter with artificial respiration, but a few hours later "Margaret, my firstborn and only girl, died of cerebral hemorrhage and bad obstetrics."[43]

Mendenhall had three other children (the second of whom also died in early childhood) and for several years devoted herself to childrearing rather than to her career as a physician. But in 1914, on the urging of a home economist at the University of Wisconsin, she began to travel to small Wisconsin towns to lecture to women on maternal and infant welfare. What she found—women desperately looking for medical assistance and information on baby care—launched Mendenhall on a long public career as a maternal and infant health reformer.[44]

Mendenhall's medical training enabled her, unlike nineteenth-century mothers, to translate her private grief into public action. "My own part in [maternal and infant welfare] work is a great satisfaction to me," she reflected some years later. "It, in a way, was a compensation reaction to help me bear the bitter frustration that the deaths of my babies gave me. Helping another mother have her child safely and advising countless mothers how to care for their babies was a real outlet for my grief."[45]

Mendenhall's experience reflects a new development in the history of American mothering: By the first decades of the twentieth century, mothers' attitudes toward infant death had undergone a sea change. No longer was the loss of a child to be viewed passively as the will of God or to be endured silently in the privacy of the home. Instead, women in communities throughout the United States made the nation's high infant mortality a matter of serious public, political concern and called on doctors, public health officials, and the state to take action to reduce the number of infant deaths. Social reformer Florence Kelley articulated the new consciousness: "So long as mothers did not know that children need not die. . . . [W]e strove for resignation, not intelligence. A generation ago we could only vainly mourn. Today we now know that every dying child accuses the community. For knowledge is available for keeping alive and well so nearly all, that we may justly be said to sin in the light of the new day when we let any die."[46]

Such a dramatic change in attitude was due, in part, to the new views of infancy reflected in the child-study movement so popular with American mothers at the turn of the century. More important, however, during the Progressive Era women worked in a variety of ways to make visible and public concerns that had in earlier generations been private and to document connections between the quality of life in the home and the social and political institutions of the larger community. At the same time, Americans continued to identify mothers as the primary guarantors of children's well-being and to emphasize the crucial importance of mother-infant bonds. Thus both change and continuity—the new emphasis on the social and public aspects of mothering and the traditional belief in the sanctity of the mother-infant bond and maternal responsibility—characterize the history of American mothering in the modern period.

The child-study movement illustrates how nineteenth-century concepts of intensive mothering continued to influence twentieth-century mothers' perceptions of their roles and responsibilities. The movement, a coalition of psychologists interested in defining and charting infancy as a critical, distinct stage in human development and mothers anxious to raise their children according to scientific principles and to enhance the status of motherhood, produced a unique literature aptly called "baby biography."[47] Purportedly scientific in method and in-

tent, baby biographies were meticulously detailed accounts, usually kept by mothers, of the day-to-day development of individual infants. For example, Louise E. Hogan, who published her journal of her son's development, *A Study of a Child*, in 1898, made careful daily observations of her baby's behavior. A typical entry reads: "He began to build with blocks to-day, placing five or six on top of each other with great care and precision. The words he has learned since November, when he was nine months old, are as follows, given in the order of acquirement: 'Oh, mammam,' 'hab'em,' 'gib'em,' 'ups-a-dada,' 'wow wow,' 'bow wow,' 'ba' and 'baba' for papa (he generally says 'ba'), 'by-bye.'" Such works attested to the primacy of individual mothers in guiding infant development: Because infancy was a crucial time on which future emotional and intellectual growth depended, mothers' close relationships with their infants were critical. As Elizabeth Harrison explained in *A Study of Child-Nature from the Kindergarten Standpoint* (1907), "One of the greatest lines of the world's work lies here before us: the understanding of little children, in order that they may be properly trained. Correctly understood, it demands of woman her highest endeavor, the broadest culture, the most complete command of herself, and the understanding of her resources and environments. It demands of her that she become a physician, an artist, a teacher, a poet, a philosopher, a priest." Authors of baby biographies exhorted other mothers to keep systematic records of their own children's development, either by following the general outlines of published studies or by making notations in one of the commercially marketed, highly detailed "baby books" that appeared in the United States by the 1880s.[48]

By 1910 child study had lost its momentum as a social movement, but its popularity, however brief, provides insight into the ways that mothers, educators, and behavioral scientists placed new emphasis on the importance of infancy as a developmental stage and simultaneously reaffirmed traditional concepts of motherhood: Much as antebellum Americans entrusted mothers with the responsibility of imbuing children with sound moral values and republican ideals, Americans at the turn of the twentieth century assigned to mothers the responsibility of overseeing their children's cognitive and emotional development.

During the same years a generation of women progressive reformers also played a central role in reshaping American views of childhood and in making the issue of infant welfare the focus of social concern. Women such as Julia Lathrop and Grace Abbott, both of whom headed the federal Children's Bureau; Florence Kelley; Dorothy Mendenhall; S. Josephine Baker, who headed New York's Bureau of Child Hygiene; and Elizabeth Putnam, who pioneered in the development of prenatal care, made common cause with public health officials and physicians to create influential maternal- and child-welfare movements. Equally important, women in communities throughout the United States acted through women's clubs, mothers' associations, and civic reform organizations to link child health to municipal politics.

Women active in the infant-welfare movement differed among themselves as to the most efficacious solutions to the problem of an infant death rate that in 1900 remained as high as 159 per 1,000 population under one year old and soared to as high as 235 per 1,000 infant population in some industrial cities.[49] Some reformers stressed overarching social and economic problems such as poverty; others emphasized the importance of individual solutions such as maternal education and breastfeeding. Many women believed that both municipal reform and maternal education were necessary.[50] Collectively, however, women's groups and organizations such as the American Association for the Study and Prevention of Infant Mortality, established

in 1909, made infant mortality a highly visible public issue. In cities and towns all over the country, women's clubs organized "baby weeks," "baby-saving campaigns," and child-welfare exhibits. Those who visited the exhibits viewed graphic displays of the ways poverty, overcrowding, inadequate municipal services, substandard housing, and impure milk supplies affected the infant death rate.[51]

At one level, the child-welfare activities were designed to reach individual mothers, particularly working-class, immigrant, and rural women, to teach them about infant feeding and child care. But at a deeper level, the intentions of reformers cut across class lines, for their activities were meant to serve a broad educational function: The dedication of individual mothers, the reformers stressed, was futile in communities that failed to provide adequate sanitation, clean water, or pure milk. Nor could maternal efforts alone succeed in a society that paid lip service to the importance of mothering and the family but did not actually value children sufficiently to commit resources to try to keep them alive. Physician Emma DeVries put the matter bluntly in a letter to Julia Lathrop: "Here's to the future day when our dear old Uncle Sam will enable the Children's Bureau to do as much for the babies as has been done for the pigs."[52]

Women's growing conviction that mothers alone could not reduce infant mortality and that increased medical help and government involvement were necessary to preserve the well-being of their babies is illustrated most vividly in the hundreds of thousands of letters mothers wrote to the federal Children's Bureau after its establishment in 1912. Women from every region of the country and every social class wrote to the bureau, pouring out details of their mothering experiences.[53] Their letters provide graphic, often eloquent testimony to mothers' search for knowledge and assistance.

Many letters reflect both mothers' frustration

at knowing so little about childbirth and infant care and their anger at the lack of available medical expertise and social support. After losing her four-month-old son, for example, an Illinois woman wrote the bureau that "it was then that I knew how helpless I was when it came to knowing what a mother should know. My baby was sacrificed thru mere ignorance. . . . I felt my own baby had every chance. Our home was clean and sanitary and far more luxurious than lots of children. But when I had to stand by and see my baby slowly starve I made up my mind I'd fight the world but what I'd find out some way to teach people more about babies." The angry mother, writing two years after her child's death, was now living in Boston and had just attended a lecture series on baby welfare. She was trying to "study every word I [can] find" on baby care, for she was convinced that it was neither God nor bad mothering, but the "lack of proper care" that had killed her son. "I soon found that not only mothers of large families knew nothing about the scientific care of babies but the best doctors of the city knew less."[54]

An Alabama woman who had been isolated from reliable medical help when her twin sons, born prematurely, died at the age of thirteen days wrote to the bureau for information when she became pregnant again. "I want to find out the reason and how to prevent anything like that this time. . . . What could have been done during those thirteen days and what was possibly done that should not have been done? Is there any reason why the next child will be born before the full term on account of the twins coming too soon? is there anything that can be done to prevent it? . . . We are reading and trying to prepare but as this is such a poor place to get help of the right sort I am afraid that we still will know too little to handle a similar situation should we be placed that way again. My doctor seems to know nothing about premature births and I want some information there."[55] Such urgent requests for information reflect mothers'

continued belief that they were responsible for their children's welfare and a strong conviction that, through knowledge and proper care, infants' deaths could be prevented.

Many mothers who wrote to the bureau criticized what they perceived as widespread public indifference to children's well-being. "With the Baby Week Campaign in progress and knowing the importance of taking good care of babies," a Maryland woman wrote, "I cannot help making a few remarks." Neither pasteurized milk nor ice was available in her community, she continued, although "we are right close to the Naval Academy Dairy where they pasteurize milk and make ice every day, still the Government refuses to sell either article. If babies are to be saved these things are necessary." "I write to see if the US Department [Children's Bureau] will see that a pregnant women gets medical attention when needed," a Virginia woman stated. "Will not our government soon do something for the poor mothers of America to help her raise her babies? Now that I am soon to become a mother again will I have to see my child pass out of my arms to the Great beyond or will the Government help us poor mothers by just seeing we get physicians at the time of birth?" Or as an Ohio woman declared in 1920, "it seems to me that the progress in the treatment of obstetrical care is almost inexcusably slow or treated with remarkable indifference. Could not this matter be more seriously stressed by the government?" The federal government ought to hire physicians "to doctor these women and babies," a South Dakota woman wrote. "When that is done, the efforts of the government to save the mothers and babies will be partly accomplished."[56]

Perhaps no mother expressed so well the modern frustration and impatience with needless infant deaths as a Mississippi woman. "Here in Jackson we have been recently having such sad deaths," she wrote to the bureau in 1915. "I have just come from a funeral this afternoon, the baby was buried two or three days ago, the Mother to-day." She demanded to know what was responsible for such deaths. "Is it the corsets we wear? Is it the food we eat? Is it the strain we live under? What is it? It makes me mad to hear the preachers say 'It's God's will' 'She has fulfilled her mission,' etc. Something's got to be done—and done quickly."[57]

For mothers who wrote to the Children's Bureau, both maternal education, which would enable individual mothers to carry out their responsibilities more effectively, and social reform were essential. Underlying those maternal concerns were two assumptions that distinguished mothering in the twentieth century from that of previous generations. First, infant deaths were preventable. And second, society at large—rather than divine providence or individual mothers alone—must assume responsibility for children's survival and well-being.

For the first three centuries of American history, infant death was the central reality of maternal experience. Although mothers' interpretations of and responses to the ever-present possibility that their children might die changed considerably—from the seventeenth- and eighteenth-century conviction that God alone determined the fate of infants to the Progressive Era belief that civic reform and public health measures could ensure babies' survival—women's writings attest to the importance of infant death in shaping maternal experience and consciousness. Only as infant mortality began to decline significantly after 1900 did the deaths of infants and young children cease to be common events in middle-class families. By the 1920s maternal consciousness was no longer shaped primarily by incessant anxiety that babies might die. In this crucial respect the first decades of the twentieth century are a watershed in the history of mothering and of the family.

In another important respect, however, maternal experience has shown greater continuity than change. Since the turn of the nineteenth century, women's perceptions of themselves as mothers have been influenced by a cultural definition of motherhood that stresses the intense, essentially private nature of the mother-child bond and the primary responsibility of mothers for the well-being of their children. Americans redefined specific maternal responsibilities over time, but the new definitions supplemented traditional beliefs rather than supplanted them. In those beliefs the essential continuity in definitions of motherhood can be traced from the early nineteenth century to the present day. Early twentieth-century women articulated a new public dimension to motherhood that became institutionalized in the acceptance of some degree of government responsibility for child welfare. American society, however, has continued to define mothering almost entirely as an individual, private experience and to assign to individual mothers the primary responsibility for their children's care and welfare.

Notes

1. For an interpretive discussion of prescriptive maternal values see Ruth H. Bloch, "American Feminine Ideals in Transition: The Rise of the Moral Mother, 1785–1815," *Feminist Studies* 4 (June 1978): 101–26; Nancy Cott, "Notes Toward an Interpretation of Antebellum Childrearing," *Psychohistory Review* 6 (Spring 1978): 4–20; Mary P. Ryan, *Womanhood in America: From Colonial Times to the Present* (New York, 1975), esp. 45–49, 106–14, 140–48, 150–72, 180–91; Mary P. Ryan, *The Empire of the Mother: American Writing About Domesticity, 1830–1860* (New York, 1982), esp. 19–70, 97–114. The best historical analysis of the effects of childrearing patterns on personality development is Philip J. Greven, *The Protestant Temperament: Patterns of Childrearing, Religious Experience, and the Self in Early America* (New York, 1977).

2. R. S. Meindl and A. C. Swedlund, "Secular Trends in Mortality in the Connecticut Valley, 1700–1850," *Human Biology* 49 (Sept. 1977): 389–414; Maris A. Vinovskis, "Mortality Rates and Trends in Massachusetts Before 1860," *Journal of Economic History* 32 (March 1972): 195–201; Henry H. Hibbs, Jr., *Infant Mortality: Its Relation to Social and Industrial Conditions* (New York, 1916), 3–16; C.-E. A. Winslow and Dorothy Holland, "The Influence of Certain Public Health Procedures upon Infant Mortality," *Human Biology* 9 (May 1937): 133–74. For contemporary documentation compiled from bills of mortality see Gouverneur Emerson, "Vital Statistics of Philadelphia, for the Decennial Period from 1830 to 1840," *American Journal of Medical Sciences* 16 (July 1848): 13–32; Lemuel Shattuck, "On the Vital Statistics of Boston," *American Journal of Medical Sciences* 17 (April 1841): 369–99; Gouverneur Emerson, "Observations upon the Mortality of Philadelphia Under the Age of Puberty," *American Journal of Medical Sciences* 17 (Nov. 1835): 56–59; and Charles A. Lee, "Medical Statistics: Comprising a Series of Calculations and Tables Showing the Mortality in New York and Its Immediate Causes, During a Period of Sixteen Years," *American Journal of Medical Sciences* 18 (Nov. 1836): 25–51.

3. Laurel Thatcher Ulrich, *Good Wives: Image and Reality in the Lives of Women in Northern New England, 1650–1750* (New York, 1982), 146–63; Michael Zuckerman, "William Byrd's Family," *Perspectives in American History* 12 (1979): 255–311.

4. Carl N. Degler, *At Odds: Women and the Family in America from the Revolution to the Present* (New York, 1980), 8–25; John Demos, "The American Family in Past Time," *American Scholar* 43 (Summer 1974): 422–46; Joan Hoff Wilson, "The Illusion of Change: Women and the American Revolution," in *The American Revolution: Explorations in the History of American Radicalism,* ed. Alfred F. Young (De Kalb, Ill., 1976), 383–445; Mary Beth Norton, *Liberty's Daughters: The Revolutionary Experience of American Women, 1750–1800* (Boston, 1980), 92–105, 243–50; Daniel Scott Smith, "Parental Power and Marriage Patterns: An Analysis of Historical Trends in Hingham, Massachusetts," *Journal of Marriage and the Family* 35 (Aug. 1973): 419–28; Daniel Blake Smith, *Inside the Great House: Planter Family Life in Eighteenth-Century Chesapeake Society* (Ithaca, 1980), 285–99.

5. On the experiences and perceptions of working-class women see esp. Mary Christine Stansell, "Women of the Laboring Poor in New York City, 1820–1860" (Ph.D. diss., Yale University, 1979); Sharon Ann Burnston, "Babies in the Well: An Underground Insight into Deviant Behavior in Eighteenth-Century Philadelphia," *Pennsylvania Magazine of History and Biography* 106 (April 1982): 151–86; and Lois Green Carr and Lorena S. Walsh, "The Planter's Wife: The Experience of White Women in Seventeenth-Century Maryland," *William and Mary Quarterly* 34 (Oct. 1977): 542–71.

6. Francis G. Walett, ed., "The Diary of Ebenezer Parkman," *Proceedings of the American Antiquarian Society* 72 (part 1, 1962): 31, 33, 36, 50. See also Anne Bradstreet, "In Memory of My Dear Grandchild Anne Bradstreet Who Deceased June 20, 1669, Being Three Years and Seven Months Old," in *The Works of Anne Bradstreet,* ed. Jeannine Hensely (Cambridge, Mass., 1967), 236.

7. "Diary of Cotton Mather, 1681–1708," *Massachusetts Historical Society Collections,* 7th series, 7 (Boston, 1911), 283. See also Peter G. Slater, "'From the *Cradle* to the *Coffin*': Parental Bereavement and the Shadow of Infant Damnation in Puritan Society," *Psychohistory Review* 6 (Fall–Winter 1977–78): 4–24.

8. Peter Gregg Slater, *Children in the New England Mind: In Life and in Death* (Hamden, Conn., 1977); David E. Stannard, *The Puritan Way of Death: A Study in Religion, Culture, and Social Change* (New York, 1977), 57–71.

9. Walett, ed., "Diary of Ebenezer Parkman," 101; Ulrich, *Good Wives,* 161. See also Joy Day Buel and Richard Buel, Jr., *The Way of Duty: A Woman and Her Family in Revolutionary America* (New York, 1984), 30–32, 39, 61–64.

10. Stewart Mitchell, ed., *New Letters of Abigail Adams, 1788–1801* (Boston, 1947), 33; Thomas Eliot Andrews, ed., "The Diary of Elizabeth (Porter) Phelps," *New England Historical and Genealogical Register* 119 (April 1965): 136; "Diary of Mary (Vial) Holyoke, 1760–1800," in *The Holyoke Diaries, 1709–1865,* ed. George Francis Dow (Salem, Mass., 1911), 58, 59, 60, 69, 70, 75. See also Buel and Buel, *Way of Duty,* 42.

11.   Elizabeth Cranch Norton Diary, Jan. 2, 1795, Aug. 22, April 19, 1796, Norton Family Papers (Massachusetts Historical Society, Boston). The practice of taking in children is amply documented in eighteenth-century diaries and journals. See, for example, Thomas Eliot Andrews, ed., "The Diary of Elizabeth (Porter) Phelps," *New England Historical and Genealogical Register* 118 (April 1964): 119; *New England Historical and Genealogical Registers* 118 (Oct. 1964): 300, 305; *New England Historical and Genealogical Registers* 119 (Jan. 1965): 43; *New England Historical and Genealogical Registers* 119 (April 1965): 135, 137, 140; *New England Historical and Genealogical Registers* 119 (July 1965): 208, 212, 215, 221; *New England Historical and Genealogical Registers* 119 (Oct. 1965): 289, 304; and Elizabeth Drinker's diary excerpted in Cecil K. Drinker, *Not So Long Ago: A Chronicle of Medicine and Doctors in Colonial Philadelphia* (New York, 1937), 37–39, 63–64.

12.   Norton Diary, Dec. 2, 3, 11, 14, 1799, Dec. 25, 1794.

13.   Andrews, ed., "Diary of Elizabeth (Porter) Phelps," *New England Historical and Genealogical Register* 119 (Jan. 1965): 51; *New England Historical and Genealogical Register* 119 (April 1965): 128; "Diary of Mary (Vial) Holyoke, 1760–1800," 67–68; Norton Diary, Aug. 5, Nov. 11, Dec. 30, 1794, March 2, 1795, Sept. 5, Dec. 16, 1796, Dec. 15, 1797, Feb. 9, March 29, 1799, Aug. 20, Sept. 13, 1800, Jan. 26, May 28, 1802.

14.   "Diary of Mary (Vial) Holyoke," 67; Sarah Snell Bryant Diary, July 12–Sept. 17, 1798, Bryant Family Papers (Houghton Library, Harvard University, Cambridge, Mass.).

15.   John F. Walzer, "A Period of Ambivalence: Eighteenth-Century American Childhood," in *The History of Childhood*, ed. Lloyd deMause (New York, 1974), 380n100; Norton, *Liberty's Daughters*, 85–86; Daniel Scott Smith, "Child-Naming Practices, Kinship Ties, and Changes in Family Attitudes in Hingham, Massachusetts, 1641 to 1880," *Journal of Social History* 18 (Summer 1985): 541–66.

16.   Andrews, ed., "Diary of Elizabeth (Porter) Phelps," *New England Historical and Genealogical Register* 118 (April 1964): 125; Norton Diary, April 10, 1798. For a discussion of the relationship between frequent accidents and extensive mothering, see Ulrich, *Good Wives,* 157–58.

17.   Elizabeth Cranch Norton to Mary Cranch, Jan. 7, 1800, Norton Family Papers; Norton Diary, Dec. 29, 1798.

18.   Louisa Park Diary, Dec. 14, Dec. 24, 1800 (American Antiquarian Society, Worcester, Mass.).

19.   Ibid., Jan. 29, May 2–May 4, 1801.

20.   Barbara Welter, "The Cult of True Womanhood: 1820–1860," *American Quarterly* 18 (Summer 1966): 151–74; Bernard Wishy, *The Child and the Republic: The Dawn of Modern American Child Nurture* (Philadelphia, 1968), esp. 22–66.

21.   Meindl and Swedlund, "Secular Trends in Mortality," 389–96; Vinovskis, "Mortality Rates and Trends," 195–201.

22.   Bessie Huntting Rudd to Edward Rudd, n.d. [summer 1860], Bessie Huntting Rudd Correspondence, Huntting-Rudd Papers (Schlesinger Library, Radcliffe College, Cambridge, Mass.); Elizabeth Dana Ellery Sedgwick Journal, 1824 (Houghton Library); Fanny Longfellow Diary, July 13, 1844, Fanny Appleton Longfellow Papers (Longfellow Historical Site, Cambridge, Mass.).

23.   Frank Rollins Morse, ed., *Henry and Mary Lee: Letters and Journals, with Other Family Papers, 1802–1860* (Boston, 1926), 178–79; Caroline White Diary, Aug. 1, 1856, White Family Papers (American Antiquarian Society, Worcester, Mass.); Sedgwick Journal, n.d. [winter 1827].

24.   For a discussion of the privatization of the household see Mary P. Ryan, *Cradle of the Middle Class: The Family in Oneida County, New York, 1790–1865* (Cambridge, Eng., 1981); Kirk Jeffrey, "The Family as a Utopian Retreat from the City: The Nineteenth Century," *Soundings* 55 (Spring 1972): 24–41; Edward Shorter, *The Making of the Modern Family* (New York, 1975), 168–204, 242, 250; and Robert V. Wells, "Family History and Demographic Transition," *Journal of Social History* 9 (Fall 1975): 1–19.

25.   Morse, ed., *Henry and Mary Lee,* 91; Elizabeth Sedgwick Child Journal, n.d. [1865] (Houghton Library); White Diary, June 12, 1856.

26.   Sedgwick Journal, n.d. [1824]; Child Journal, 1865–1874; Longfellow Diary, 1848. Most other nineteenth-century maternal diaries examined contained daily comments on children.

27.   Longfellow Diary, Jan. 1848; Carol Kammen, ed., "The Letters of Calista Hall," *New York History* 63 (April 1982): 218.

28. Longfellow Diary, Jan. 9, Feb. 3, March 8, 1848; Sedgwick Journal, March 13, 1826.

29. Longfellow Diary, June 14, 1848; Bessie Huntting Rudd to Edward Rudd, n.d. [summer 1860], Bessie Huntting Rudd Correspondence, Huntting-Rudd Papers; White Diary, June 13, 1856.

30. Kammen, ed., "Letters of Calista Hall," 226; Elizabeth Cleveland Journal, May–August 1848 (Essex Institute, Salem, Mass.).

31. White Diary, Sept. 14, Sept. 15, 1856, June 30, 1863.

32. Sedgwick Journal, n.d. [1825]; Longfellow Diary, Feb. 18, Aug. 18, Sept. 6, Sept. 11, 1848.

33. Longfellow Diary, Oct. 14, Sept. 13, 1848.

34. Ibid., n.d. [1850], Feb. 22, 1851. For a discussion of death in nineteenth-century families see Lewis Saum, "Death in the Popular Mind of Pre-Civil War America," *American Quarterly* 26 (Dec. 1974): 477–95; and Jan Lewis, *The Pursuit of Happiness: Family and Values in Jefferson's Virginia* (New York, 1983), 98–102.

35. See, for example, David Stannard, "Death and the Puritan Child," *American Quarterly* 26 (Dec. 1974): 456–76; Smith, *Inside the Great House*, 249–80; and Lewis, *Pursuit of Happiness*, 72–76.

36. On evangelical childrearing see William G. McLoughlin, "Evangelical Childrearing in the Age of Jackson: Francis Wayland's Views on When and How to Subdue the Willfulness of Children," *Journal of Social History* 9 (Fall 1975): 21–43; and Greven, *Protestant Temperament*, 21–61.

37. Judith Walzer Leavitt, "'Science' Enters the Birthing Room: Obstetrics in America since the Eighteenth Century," *Journal of American History* 70 (Sept. 1983): 281–304. Historians have just begun to examine fathers' responses to these developments. For a good example of a father whose attitudes toward parenting reflected changes in thinking about human agency, see Carol E. Hoffecker, ed., "The Diaries of Edmund Canby, A Quaker Miller, 1822–1848," *Delaware History* 16 (Oct. 1974): 79–121; *Delaware History* 16 (Spring–Summer 1975): 184–243.

38. An American Matron [pseud.], *The Maternal Physician: A Treatise on the Nurture and Management of Infants, From the Birth Until Two Years Old, Being the Result of Sixteen Years' Experience in the Nursery* (Philadelphia, 1810), 7; Mrs. [Louisa Mary Bacon] Barwell, *Infant Treatment: With Directions to Mothers for Self-Management Before, During, and After Pregnancy* (New York, 1844), 15.

39. D. Francis Condie, *A Practical Treatise on the Diseases of Children* (Philadelphia, 1844), 85–86.

40. "Infant Mortality and Fashionable Dress," *New York Medical Journal* 10 (Jan. 1870): 424–25; John W. Thrailkill, *An Essay on the Causes of Infant Mortality; Being a Brief Account of the Origins of the Feebleness and Diseases Which Afflict and Destroy So Many Children Under Five Years of Age* (St. Louis, 1869), 6–7; William P. Dewees, *A Treatise on the Physical and Medical Treatment of Children* (Philadelphia, 1836), 48, 72, 108; W. M. Ireland, *Advice to Mothers on the Management of Infants and Young Children, with Directions on How to Distinguish and Prevent Their Complaints* (New York, 1820), esp. 5–38; American Matron [pseud.], *Maternal Physician*, 60–61.

41. See, for example, Welter, "Cult of True Womanhood"; and Carroll Smith-Rosenberg, "The Female World of Love and Ritual: Relations Between Women in Nineteenth-Century America," *Signs* 1 (Autumn 1975): 1–30. For a different interpretation of early nineteenth-century maternal values that is based on literary sources, see Ann Douglas, *The Feminization of American Culture* (New York, 1977), esp. 1–13, 50–93, 240–72.

42. For the view of the early nineteenth-century family as refuge see Jeffrey, "Family as a Utopian Retreat"; and Ryan, *Cradle of the Middle Class*, 146–55, 191–98.

43. Dorothy Reed Mendenhall, unpublished autobiography, section H, p. 13, Dorothy Reed Mendenhall Papers, Sophia Smith Collection (Smith College, Northampton, Mass.).

44. Ibid., section 1: 2.

45. Ibid.

46. Florence Kelley, "Children in the Cities," *National Municipal Review* 4 (April 1915): 199.

47. On the child-study movement see Steven L. Schlossman, "Before Home Start: Notes Toward a History of Parent Education in America, 1897–1929," *Harvard Educational Review* 46 (Aug. 1976): 436–67; Sheila M. Rothman, *Woman's Proper Place: A History of Changing Ideals and Practices, 1870 to the Present* (New York, 1978), 97–106; and Joseph F. Kett, *Rites of Passage: Adolescence in America, 1790 to the Present* (New York, 1977), 228–30.

48. Louise E. Hogan, *A Study of a Child* (New

York, 1898), 1–14, 30–31; Elizabeth Harrison, *A Study of Child-Nature from the Kindergarten Standpoint* (Chicago, 1907), 11; Milicent Washburn Shinn, *The Biography of a Baby* (Boston, 1900), esp. 1–6, 10–19; Jessie Chase Fenton, *Practical Psychology of Babyhood: The Mental Development and Mental Hygiene of the First Two Years of Life* (Boston, 1925), esp. 315–41. See also Winifred S. Hall, "The First 500 Days of a Child's Life," *Child-Study Monthly* 2 (Nov. 1896): 330–42; *Child-Study Monthly* 2 (Dec. 1896): 394–407; *Child-Study Monthly* 2 (Jan. 1897): 458–73; *Child-Study Monthly* 2 (Feb. 1897): 522–37.

49. Hibbs, *Infant Mortality*, 12. As late as the first decades of the twentieth century, neither the federal government nor most state governments provided for complete and accurate registration of births. The lack of such vital statistics renders infant mortality figures imprecise. See James H. Cassedy, "The Registration Area and American Vital Statistics: Development of a Health Research Resource, 1885–1915," *Bulletin of the History of Medicine* 39 (May–June 1965): 221–32.

50. Harvey Levenstein, "'Best for Babies' or 'Preventable Infanticide'? The Controversy over Artificial Feeding of Infants in America, 1880–1920," *Journal of American History* 70 (June 1983): 75–94; J. Stanley Lemons, "The Sheppard-Towner Act: Progressivism in the 1920s," *Journal of American History* 55 (March 1969): 776–86; Judith Walzer Leavitt, *The Healthiest City: Milwaukee and the Politics of Health Reform* (Princeton, 1982), 156–89; Constance D. Leupp, "Campaigning for Babies' Lives," *McClure's Magazine*, 38 (Aug. 1912): 361–73; "For the Babies of Philadelphia," *Survey*, July 10, 1909, 533; "Survey of Sickness," *Survey*, Oct. 16, 1915, 65–69; Helen Worthington Rogers, "A Modest Experiment in Foster-Motherhood; The Work of the Pure Milk Commission of the Children's Aid Association of Indianapolis," *Survey*, May 1, 1909, 176–83; Wilbur C. Phillips, "The Mother and the Baby," *Survey*, Aug. 7, 1909,

623–31; "Coordinated Child-Saving," *Charities and the Commons*, Feb. 20, 1909, 1010; "The Right View of the Child," *Charities and the Commons*, April 25, 1908, 123–25; "To Reduce Infant Mortality," *Charities and the Commons*, May 30, 1908, 285; S. Josephine Baker, "The Value of Municipal Control of Child Hygiene," *American Journal of Obstetrics and the Diseases of Women and Children* 65 (June 1912): 1061–68; Mrs. William Lowell Putnam, "The Importance of Prenatal Care," *American Journal of Obstetrics and the Diseases of Women and Children* 78 (July 1918): 103–107.

51. Mary Ritter Beard, *Woman's Work in Municipalities* (New York, 1915), 56–68; Anna Louise Strong, "Child Welfare Exhibits," *National Municipal Review* 1 (April 1912): 248–52; S. Josephine Baker, *Fighting for Life* (New York, 1939).

52. Emma DeVries to Julia Lathrop, March 6, 1916, Records of the U.S. Department of Labor Children's Bureau, RG 102 (National Archives).

53. Nancy Weiss, "Mother, the Invention of Necessity: Dr. Benjamin Spock's *Baby and Child Care*," *American Quarterly* 29 (Winter 1977): 519–46.

54. Mrs. WRD, Cambridge, Mass., to Children's Bureau, June 22, 1918, Records of the U.S. Department of Labor Children's Bureau.

55. MH, Bladon Springs, Alabama, to Children's Bureau, July 8, 1919, Records of the U.S. Department of Labor Children's Bureau.

56. Mrs. IL, Annapolis, to Children's Bureau, May 12, 1915; Mrs. DEH, Staunton, Va., to Children's Bureau, July 3, 1920; Mrs. HBC, Steubenville, Ohio, to Children's Bureau, Feb. 10, 1920; Mrs. RN, South Dakota, to Children's Bureau, Aug. 12, 1919, Records of the U.S. Department of Labor Children's Bureau.

57. Mrs. CFH, Jackson, Miss., to Children's Bureau, April 28, 1915, Records of the U.S. Department of Labor Children's Bureau.

6 The Female Animal: Medical and Biological Views of Woman and Her Role in Nineteenth-Century America

Carroll Smith-Rosenberg and Charles E. Rosenberg

Since at least the time of Hippocrates and Aristotle, the roles assigned women have attracted an elaborate body of medical and biological justification. This was especially true in the nineteenth century as the intellectual and emotional centrality of science increased steadily. Would-be scientific arguments were used in the rationalization and legitimization of almost every aspect of Victorian life and with particular vehemence in those areas in which social change implied stress in existing social arrangements.

This essay is an attempt to outline some of the shapes assumed by the nineteenth-century debate over the ultimate bases for women's domestic and childbearing role.[1] In form it resembles an exercise in the history of ideas; in in-

tent it represents a hybrid with social and psychological history. Biological and medical views serve as a sampling device, suggesting and illuminating patterns of social continuity, change, and tension.

The relationships between social change and social stress are dismayingly complex and recalcitrant to both psychological theorists and to the historian's normal modes of analysis. In an attempt to gain insight into these relationships the authors have chosen an analytic approach based on the study of normative descriptions of the female role at a time of widespread social change; not surprisingly, emotion-laden attempts to reassert and redefine this role constitute one response to the stress induced by such social change.

This approach was selected for a variety of reasons. Role definitions exist on a level of prescription beyond their embodiment in the individuality and behavior of particular historical persons. They exist rather as a formally agreed-upon set of characteristics understood by and acceptable to a significant proportion of the

CARROLL SMITH-ROSENBERG is Professor of History, Women's Studies, and American Culture at the University of Michigan, Ann Arbor.
CHARLES E. ROSENBERG is Janice and Julian Bers Professor of History and Sociology of Science at the University of Pennsylvania, Philadelphia.

Reprinted from *The Journal of American History* 60 (Sept. 1973): 332–56, by permission of the publisher.

population. As formally agreed-upon social values they are, moreover, retrievable from historical materials and thus subject to analysis. Such social role definitions, however, have a more than platonic reality, for they exist as parameters with which and against which individuals must either conform or define their deviance. When inappropriate to social, psychological, or biological reality, such definitions can themselves engender anxiety, conflict, and demands for change.

During the nineteenth century, economic and social forces at work within Western Europe and the United States began to compromise traditional social roles. Some women at least began to question—and a few to challenge overtly— their constricted place in society. Naturally enough, men hopeful of preserving existing social relationships, and in some cases threatened themselves both as individuals and as members of particular social groups, employed medical and biological arguments to rationalize traditional sex roles as rooted inevitably and irreversibly in the prescriptions of anatomy and physiology. This essay examines the ideological attack mounted by prestigious and traditionally minded men against two of the ways in which women expressed their dissatisfaction and desire for change: women's demands for improved educational opportunities and their decision to resort to birth control and abortion. That much of this often emotionally charged debate was oblique and couched in would-be scientific and medical language and metaphor makes it even more significant, for few spokesmen could explicitly and consciously confront those changes which impinged upon the bases of their particular emotional adjustment.

The Victorian woman's ideal social characteristics—nurturance, intuitive morality, domesticity, passivity, and affection—were all assumed to have a deeply rooted biological basis. These medical and scientific arguments formed an ideological system rigid in its support of tradition yet infinitely flexible in the particular mechanisms which could be made to explain and legitimate woman's role.

Woman, nineteenth-century medical orthodoxy insisted, was starkly different from the male of the species. Physically, she was frailer, her skull smaller, her muscles more delicate. Even more striking was the difference between the nervous system of the two sexes. The female nervous system was finer, "more irritable," prone to overstimulation and resulting exhaustion. "The female sex," as one physician explained in 1827,

> is far more sensitive and susceptible than the male, and extremely liable to those distressing affections which for want of some better term, have been denominated nervous, and which consist chiefly in painful affections of the head, heart, side, and indeed, of almost every part of the system.[2]

"The nerves themselves," another physician concurred a generation later, "are smaller, and of a more delicate structure. They are endowed with greater sensibility, and, of course, are liable to more frequent and stronger impressions from external agents on mental influences."[3] Few if any questioned the assumption that in males the intellectual propensities of the brain dominated, while the female's nervous system and emotions prevailed over her conscious and rational faculties. Thus it was only natural, indeed inevitable, that women should be expected and permitted to display more affect than men; it was inherent in their very being.

Physicians saw woman as the product and prisoner of her reproductive system. It was the ineluctable basis of her social role and behavioral characteristics, the cause of her most common ailments; woman's uterus and ovaries controlled her body and behavior from puberty through menopause. The male reproductive system, male physicians assured, exerted no parallel degree of control over man's body. Charles D. Meigs, a prominent Philadelphia gynecologist, stated with assurance in 1847 that a woman is "a moral, a sexual, a germiferous, gestative and parturient creature."[4] It was, another physician explained in 1870, "as if the Almighty, in creating the female sex, had taken the uterus and built up a woman around it."[5] A wise deity had designed woman as keeper of the hearth, as breeder and rearer of children.

Medical wisdom easily supplied hypothetical mechanisms to explain the interconnection between the female's organs of generation and the functioning of her other organs. The uterus, it was assumed, was connected to the central nervous system; shocks to the nervous system might alter the reproductive cycle—might even mark the gestating fetus—while changes in the reproductive cycle shaped emotional states. This intimate and hypothetical link between ovaries, uterus, and nervous system was the logical basis for the "reflex irritation" model of disease causation so popular in middle and late nineteenth-century medical texts and monographs on psychiatry and gynecology. Any imbalance, exhaustion, infection, or other disorders of the reproductive organs could cause pathological reactions in parts of the body seemingly remote.[6] Doctors connected not only the paralyses and headaches of the hysteric to uterine disease but also ailments in virtually every part of the body. "These diseases," one physician explained, "will be found, on due investigation, to be in reality, no disease at all, but merely the sympathetic reaction or the symptoms of one disease, namely, a disease of the womb."[7]

Yet despite the commonsensical view that such ailments resulted from childbearing, physicians often contended that far greater difficulties could be expected in childless women. Motherhood was woman's normal destiny, and those females who thwarted the promise immanent in their body's design must expect to suffer. The maiden lady, many physicians argued, was fated to a greater incidence of both physical and emotional disease than her married sisters and to a shorter life span.[8] Her nervous system was placed under constant pressure, and her unfulfilled reproductive organs—especially at menopause—were prone to cancer and other degenerative ills.

Woman was thus peculiarly the creature of her internal organs, of tidal forces she could not consciously control. Ovulation, the physical and emotional changes of pregnancy, even sexual desire itself were determined by internal physiological processes beyond the control or even the awareness of her conscious volition.[9] All women were prisoners of the cyclical aspects of their bodies, of the great reproductive cycle bounded by puberty and menopause, and by the shorter but recurrent cycles of childbearing and menstruation. All shaped her personality, her social role, her intellectual abilities and limitations; all presented as well possibly "critical" moments in her development, possible turning points in the establishment—or deterioration—of future physical and mental health. As the president of the American Gynecological Society stated in 1900: "Many a young life is battered and forever crippled in the breakers of puberty; if it crosses these unharmed and is not dashed to pieces on the rock of childbirth, it may still ground on the ever-recurring shallows of menstruation, and lastly, upon the final bar of the menopause ere protection is found in the unruffled waters of the harbor beyond the reach of sexual storms."[10]

Woman's physiology and anatomy, physicians habitually argued, oriented her toward an "inner" view of herself and her worldly sphere. (Logically enough, nineteenth-century views of heredity often assumed that the father was responsible for a child's external musculature and skeletal development, the mother for the internal viscera, the father for analytical abilities, the

113

mother for emotions and piety.)[11] Their secret internal organs, women were told, determined their behavior; their concerns lay inevitably within the home.[12] In a passage strikingly reminiscent of some midtwentieth-century writings, a physician in 1869 depicted an idealized female world, rooted in the female reproductive system, sharply limited socially and intellectually, yet offering women covert and manipulative modes of exercising power:

> Mentally, socially, spiritually, she is more interior than man. She herself is an interior part of man, and her love and life are always something interior and incomprehensible to him. . . . Woman is to deal with domestic affections and uses, not with philosophies and sciences. . . . She is priest, not king. The house, the chamber, the closet, are the centres of her social life and power, as surely as the sun is the centre of the solar system. . . . Another proof of the interiority of woman, is the wonderful secretiveness and power of dissimulation which she possesses. . . . Woman's secrecy is not cunning; her dissimulation is not fraud. They are intuitions or spiritual perceptions, full of tact and wisdom, leading her to conceal or reveal, to speak or be silent, to do or not to do, exactly at the right time and in the right place.[13]

The image granted women in these hypothetical designs was remarkably consistent with the social role traditionally allotted them. The instincts connected with ovulation made her by nature gentle, affectionate, and nurturant. Weaker in body, confined by menstruation and pregnancy, she was both physically and economically dependent upon the stronger, more forceful male, whom she necessarily looked up to with admiration and devotion.

Such stylized formulae embodied, however, a characteristic yet entirely functional ambiguity. The Victorian woman was more spiritual than man, yet less intellectual, closer to the divine, yet prisoner of her most animal characteristics, more moral than man, yet less in control of her very morality. While the sentimental poets placed woman among the angels and doctors praised the transcendent calling of her reproductive system, social taboos made woman ashamed of menstruation, embarrassed and withdrawn during pregnancy, self-conscious and purposeless during and after menopause. Her body, which so inexorably defined her personality and limited her role, appeared to woman often degrading and confining.[14] The very romantic rhetoric which tended to suffocate nineteenth-century discussions of femininity only underlined with irony the distance between behavioral reality and the forms of conventional ideology.

The nature of the formalistic scheme implied as well a relationship between the fulfilling of its true calling and ultimate social health. A woman who lived "unphysiologically"—and she could do so by reading or studying in excess, by wearing improper clothing, by long hours of factory work, or by a sedentary, luxurious life—could produce only weak and degenerate offspring. Until the twentieth century it was almost universally assumed that acquired characteristics in the form of damage from disease and improper lifestyles in parents would be transmitted through heredity; a nervous and debilitated mother could have only nervous, dyspeptic, and undersized children.[15] Thus appropriate female behavior was sanctioned not only by traditional injunctions against individual sin in the form of inappropriate and thus unnatural modes of life but also by the higher duty of protecting the transcendent good of social health, which could be maintained only through the continued production of healthy children. Such arguments were to be invoked with increasing frequency as the nineteenth century progressed.

In midnineteenth-century America it was apparent that women—or at least some of them—were growing dissatisfied with traditional roles. American society in midnineteenth century was committed—at least formally—to egalitarian democracy and evangelical piety. It was thus a society which presumably valued individualism, social and economic mobility, and free will. At the same time it was a society experiencing rapid economic growth, one in which an increasing number of families could think of themselves as middle class and could seek a lifestyle appropriate to that station. At least some middle-class women, freed economically from the day-to-day struggle for subsistence, found in these values a motivation and rationale for expanding their roles into areas outside the home. In the Jacksonian crusades for piety, for temperance, for abolition, and in pioneering efforts to aid the urban poor, women played a prominent role, a role clearly outside the confines of the home. Women began as well to demand improved educational opportunities—even admission to colleges and medical schools. A far greater number began, though more covertly, to see family limitation as a necessity if they would preserve health, status, economic security, and individual autonomy.

Only a handful of nineteenth-century American women made a commitment to overt feminism and to the insecurity and hostility such a commitment implied. But humanitarian reform, education, and birth control were all issues which presented themselves as real alternatives to every respectable church-going American woman.[16] Contemporary medical and biological arguments identified, reflected, and helped to eliminate two of these threats to traditional role definitions: demands by women for higher education and family limitation.

Since the beginnings of the nineteenth century, American physicians and social commentators generally had feared that American women were physically inferior to their English and Continental sisters. The young women of the urban middle and upper classes seemed in particular less vigorous, more nervous than either their own grandmothers or European contemporaries. Concern among physicians, educators, and publicists over the physical deterioration of American womanhood grew steadily during the nineteenth century and reached a high point in its last third.

Many physicians were convinced that education was a major factor in bringing about this deterioration, especially education during puberty and adolescence. It was during these years that the female reproductive system matured, and it was this process of maturation that determined the quality of the children which American women would ultimately bear. During puberty, orthodox medical doctrine insisted, a girl's vital energies must be devoted to development of the reproductive organs. Physicians saw the body as a closed system possessing only a limited amount of vital force; energy expended in one area was necessarily removed from another. The girl who curtailed brain work during puberty could devote her body's full energy to the optimum development of its reproductive capacities. A young woman, however, who consumed her vital force in intellectual activities was necessarily diverting these energies from the achievement of true womanhood. She would become weak and nervous, perhaps sterile, or more commonly, and in a sense more dangerously for society, capable of bearing only sickly and neurotic children—children able to produce only feebler and more degenerate versions of themselves.[17] The brain and ovary could not develop at the same time. Society, mid-century physicians warned, must protect the higher good of racial health by avoiding situations in which adolescent girls taxed their intellectual faculties in academic competition. "Why," as one physician pointedly asked, "spoil a good mother by making an ordinary grammarian?"[18]

Yet where did America's daughters spend these years of puberty and adolescence, doctors asked, especially the daughters of the nation's most virtuous and successful middle-class families? They spent these years in schools; they sat for long hours each day bending over desks, reading thick books, competing with boys for honors. Their health and that of their future children would be inevitably marked by the consequences of such unnatural modes of life.[19] If such evils resulted from secondary education, even more dramatically unwholesome was the influence of higher education upon the health of those few women intrepid enough to undertake it. Yet their numbers increased steadily, especially after a few women's colleges were established in the East and state universities in the Midwest and Pacific Coast began cautiously to accept coeducation. Women could now, critics agonized, spend the entire period between the beginning of menstruation and the maturation of their ovarian systems in nerve-draining study. Their adolescence, as one doctor pointed out, contrasted sadly with those experienced by healthier, more fruitful forebears: "Our great-grandmothers got their schooling during the winter months and let their brains lie fallow for the rest of the year. They knew less about Euclid and the classics than they did about housekeeping and housework. But they made good wives and mothers, and bore and nursed sturdy sons and buxom daughters and plenty of them at that."[20]

Constant competition among themselves and with the physically stronger males disarranged the coed's nervous system, leaving her anxious, prey to hysteria and neurasthenia. One gynecologist complained as late as 1901:

the nervous force, so necessary at puberty for the establishment of the menstrual function, is wasted on what may be compared as trifles to perfect health, for what use are they without health? The poor sufferer only adds another to the great army of neurasthenia and sexual incompetents, which furnish neurologists and gynecologists with so much of their material . . . bright eyes have been dulled by the brain-fag and sweet temper transformed into irritability, crossness and hysteria, while the womanhood of the land is deteriorating physically.

She may be highly cultured and accomplished and shine in society, but her future husband will discover too late that he has married a large outfit of headaches, back-aches and spine aches, instead of a woman fitted to take up the duties of life.[21]

Such speculations exerted a strong influence upon educators, even those connected with institutions which admitted women. The state universities, for example, often prescribed a lighter course load for females or refused to permit women admission to regular degree programs. "Every physiologist is well aware," the regents of the University of Wisconsin explained in 1877, "that at stated times, nature makes a great demand upon the energies of early womanhood and that at these times great caution must be exercised lest injury be done. . . . Education is greatly to be desired," the regents concluded:

but it is better that the future matrons of the state should be without a University training than that it should be produced at the fearful expense of ruined health; better that the future mothers of the state should be robust, hearty, healthy women, than that, by over study, they entail upon their descendants the germs of disease.[22]

This fear for succeeding generations born of educated women was widespread. "We want to have body as well as mind," one commentator noted, "otherwise the degeneration of the race is inevitable."[23] Such transcendent responsibilities made the individual woman's personal ambitions seem trivial indeed.

One of the remedies suggested by both educators and physicians lay in tempering the intensely intellectualistic quality of American education with a restorative emphasis on physical education. Significantly, health reformers' demands for women's physical education were ordinarily justified not in terms of freeing the middle-class woman from traditional restrictions on bodily movement but rather as upgrading her ultimate maternal capacities. Several would-be physiological reformers called indeed for active participation in housecleaning as an ideal mode of physical culture for the servant-coddled American girl. Bed making, clothes scrubbing, sweeping, and scouring provided a varied and highly appropriate regimen.[24]

Late nineteenth-century women physicians, as might have been expected, failed ordinarily to share the alarm of their male colleagues when contemplating the dangers of coeducation. No one, a female physician commented sardonically, worked harder or in unhealthier conditions than the washerwoman; yet would-be saviors of American womanhood did not inveigh against this abuse—washing, after all, was appropriate work for women. Women doctors often did agree with the general observation that their sisters were too frequently weak and unhealthy; however, they blamed not education or social activism but artificialities of dress and slavery to fashion, aspects of the middle-class woman's lifestyle which they found particularly demeaning. "The fact is that girls and women can bear study," Alice Stockham explained, "but they cannot bear compressed viscera, tortured stomachs and displaced uterus," the results

of fashionable clothing and an equally fashionable sedentary life. Another woman physician, Sarah Stevenson, wrote in a similar vein: "'How do I look?' is the everlasting story from the beginning to the end of woman's life. Looks, not books, are the murderers of American women."[25]

Even more significant than this controversy over woman's education was a parallel debate focusing on the questions of birth control and abortion. These issues affected not a small percentage of middle- and upper-middle-class women, but all men and women. It is one of the great and still largely unstudied realities of nineteenth-century social history. Every married woman was immediately affected by the realities of childbearing and child rearing. Though birth control and abortion had been practiced, discussed—and reprobated—for centuries, the midnineteenth century saw a dramatic increase in concern among spokesmen for the ministry and medical profession.[26]

Particularly alarming were the casualness, doctors charged, with which seemingly respectable wives and mothers contemplated and undertook abortions and how routinely they practiced birth control. One prominent New York gynecologist complained in 1874 that well-dressed women walked into his consultation room and asked for abortions as casually as they would for a cut of beefsteak at their butcher.[27] In 1857 the American Medical Association nominated a special committee to report on the problem, then appointed another in the 1870s; between these dates, and especially in the late 1860s, medical societies throughout the country passed resolutions attacking the prevalence of abortion and birth control and condemning physicians who performed and condoned such illicit practices. Nevertheless, abortions could in the 1870s be obtained in Boston and New York for as little as ten dollars, while abortifacients could be purchased more cheaply or through the mail. Even the smallest villages and rural areas provided a market for the abortionist's services; women often aborted any pregnancy which occurred in the first few years of marriage. The Michigan Board of Health estimated in 1898 that one-third of all the state's pregnancies ended in abortion. From 70 to 80 percent of these were secured, the board contended, by prosperous and otherwise respectable married women who could not offer even the unmarried mother's "excuse of shame." [28] By the 1880s English medical moralists could refer to birth control as the "American sin" and warn against England's women following in the path of America's faithless wives. [29]

So general a phenomenon demands explanation. The only serious attempts to explain the prevalence of birth control in this period have emphasized the economic motivations of those practicing it—the need in an increasingly urban, industrial, and bureaucratized society to limit numbers of children so as to provide security, education, and inheritance for those already brought into the world. As the nineteenth century progressed, it has been argued, definitions of appropriate middle-class lifestyles dictated a more and more expansive pattern of consumption, a pattern—especially in an era of recurring economic instability—particularly threatening to those large numbers of Americans only precariously members of the secure economic classes. The need to limit offspring was a necessity if family status was to be maintained. [30]

Other aspects of nineteenth-century birth control have received much less historical attention. One of these needs only to be mentioned, for it poses no interpretative complexities; this was the frequency with which childbirth meant for women pain and often lingering incapacity. Death from childbirth, torn cervixes, fistulae, prolapsed uteri were widespread "female complaints" in a period when gynecological practice was still relatively primitive and pregnancy every

few years common indeed. John Humphrey Noyes, perhaps the best-known advocate of family planning in nineteenth-century America, explained poignantly why he and his wife had decided to practice birth control in the 1840s:

> The [decision] was occasioned and even forced upon me by very sorrowful experiences. In the course of six years my wife went through the agonies of five births. Four of them were premature. Only one child lived. . . . After our last disappointment, I pledged my word to my wife that I would never again expose her to such fruitless suffering. [31]

The Noyeses' experience was duplicated in many homes. Young women were simply terrified of having children. [32]

Such fears, of course, were not peculiar to nineteenth-century America. The dangers of disability and death consequent upon childbirth extended back to the beginning of time, as did the anxiety and depression so frequently associated with pregnancy. What might be suggested, however, was that economic and technological changes in society added new parameters to the age-old experience. Family limitation for economic and social reasons now appeared more desirable to a growing number of husbands; it was, perhaps, also more technically feasible. Consequently married women could begin to consider, probably for the first time, alternative lifestyles to that of multiple pregnancies extending over a third of their lives. Women could begin to view the pain and bodily injury which resulted from such pregnancies not simply as a condition to be borne with fatalism and passivity but as a situation that could be avoided. It is quite probable, therefore, that, in this new social context, increased anxiety and depression would result once a woman, in part at least voluntarily, became pregnant. Certainly, it could be argued, such fears must have altered women's attitudes toward sexual relations generally. Indeed the decision to practice birth control must necessarily have held more than economic and status implications for the family; it must have become an element in the fabric of every marriage's particular psychosexual reality.[33]

A third and even more ambiguous aspect of the birth control controversy in nineteenth-century America relates to the way in which attitudes toward contraception and abortion reflected role conflict within the family. Again and again, from the 1840s on, defenders of family planning—including individuals as varied and idealistic as Noyes and Stockham, on the one hand, and assorted quack doctors and peddlers of abortifacients, on the other—justified their activities not in economic terms but under the rubric of providing women with liberty and autonomy. Woman, they argued with remarkable unanimity, must control her own body; without this she was a slave not only to the sexual impulses of her husband but also to endless childbearing and rearing. "Woman's equality in all the relations of life," a New York physician wrote in 1866, "implies her absolute supremacy in the sexual relation. . . . It is her absolute and indefeasible right to determine when she will and when she will not be exposed to pregnancy." "God and Nature," another physician urged, "have given to the female the complete control of her own person, so far as sexual congress and reproduction are concerned."[34] The assumption of all these writers was clear and unqualified: women, if free to do so, would choose to have sexual relations less frequently and to have far fewer pregnancies.

Implied in these arguments as well were differences as to the nature and function of sexual intercourse. Was its principal and exclusively justifiable function, as conservative physicians and clergymen argued, the procreation of children, or could it be justified as an act of love, of tenderness between individuals? Noyes argued that the sexual organs had a social, amative function, separable from their reproductive

function. Sex was justifiable as an essential and irreplaceable form of human affection; no man could demand this act unless it was freely given.[35] Nor could it be freely given in many cases unless effective modes of birth control were available to assuage the woman's anxieties. A man's wife was not his chattel, to be violated at will, and forced—ultimately—to bear unwanted and thus almost certainly unhealthy children.

Significantly, defenders of women's right to limit childbearing employed many of the same arguments used by conservatives to attack women's activities outside the home; all those baleful hereditary consequences threatened by over-education were seen by birth control advocates as resulting from the bearing of children by women unwilling and unfit for the task, their vital energies depleted by excessive childbearing. A child, they argued, carried to term by a woman who desired only its death could not develop normally; such children proved inevitably a source of physical and emotional degeneracy. Were women relieved from such accustomed pressures, they could produce fewer but better offspring.[36]

Many concerned midnineteenth-century physicians, clergymen, and journalists failed to accept such arguments. They emphasized instead the unnatural and thus necessarily deleterious character of any and all methods of birth control and abortion. Even coitus interruptus, obviously the most common mode of birth control in this period, was attacked routinely as a source of mental illness, nervous tension, and even cancer. This was easily demonstrated. Sex, like all aspects of human bodily activity, involved an exchange of nervous energy; without the discharge of such accumulated energies in the male orgasm and the soothing presence of the male semen "bathing the female reproductive organs," the female partner could never, the reassuring logic ran, find true fulfillment. The nervous force accumulated and concentrated in sexual excitement would build up dangerous levels of undischarged energy, leading ultimately to a progressive decay in the unfortunate woman's physical and mental health. Physicians warned repeatedly that condoms and diaphragms—when the latter became available after midcentury—could cause an even more startlingly varied assortment of ills. In addition to the mechanical irritation they promoted, artificial methods of birth control increased the lustful impulse in both partners, leading inevitably to sexual excess. The resultant nervous exhaustion induced gynecological lesions, and then through "reflex irritation" caused such ills as loss of memory, insanity, heart disease, and even "the most repulsive nymphomania."[37]

Conservative physicians similarly denounced the widespread practice of inserting sponges impregnated with supposedly spermicidal chemicals into the vagina immediately before or after intercourse. Such practices, they warned, guaranteed pelvic injury, perhaps sterility. Even if a woman seemed in good health despite a history of practicing birth control, a Delaware physician explained in 1873 that "as soon as this vigor commences to decline ... about the fortieth year, the disease [cancer] grows as the energies fail—the cancerous fangs penetrating deeper and deeper until, after excruciating suffering, the writhing victim is yielded up to its terrible embrace."[38] Most important, this argument followed, habitual attempts at contraception meant—even if successful—a mother permanently injured and unable to bear healthy children. If unsuccessful, the children resulting from such unnatural matings would be inevitably weakened. And if such grave ills resulted from the practice of birth control, the physical consequences of abortion were even more dramatic and immediate.[39]

Physicians often felt little hesitation in expressing what seems to the historian a suspiciously disproportionate resentment toward such unnatural females. *Unnatural* was of course the operational word, for woman's presumed maternal instinct made her primarily responsible for decisions in regard to childbearing.[40] So frequent was this habitual accusation that some medical authors had to caution against placing the entire weight of blame for birth control and abortion upon the woman; men, they reminded, played an important role in most such decisions.[41] In 1871, for example, the American Medical Association Committee on Criminal Abortion described women who patronized abortionists in terms which conjured up fantasies of violence and punishment:

> She becomes unmindful of the course marked out for her by Providence, she overlooks the duties imposed on her by the marriage contract. She yields to the pleasures—but shrinks from the pains and responsibilities of maternity; and, destitute of all delicacy and refinement, resigns herself, body and soul, into the hands of unscrupulous and wicked men. Let not the husband of such a wife flatter himself that he possesses her affection. Nor can she in turn ever merit even the respect of a virtuous husband. She sinks into old age like a withered tree, stripped of its foliage; with the stain of blood upon her soul, she dies without the hand of affection to smooth her pillow.[42]

The frequency with which attacks on family limitation in midnineteenth-century America were accompanied by polemics against expanded roles for the middle-class woman indicates with unmistakable clarity something of one of the motives structuring such jeremiads. Family limitation necessarily added a significant variable within conjugal relationships generally; its successful practice implied potential access for women to new roles and a new autonomy.

Nowhere is this hostility toward women and

the desire to inculcate guilt over women's desire to avoid pregnancy more strikingly illustrated than in the warnings of "race suicide" so increasingly fashionable in the late nineteenth century. A woman's willingness and capacity to bear children was a duty she owed not only to God and husband but to her "race" as well.[43] In the second half of the nineteenth century, articulate Americans forced to evaluate and come to emotional terms with social change became, like many of their European contemporaries, attracted to a world view which saw racial identity and racial conflict as fundamental. And within these categories birthrates became all-important indices to national vigor and thus social health.

In 1860 and again in 1870 Massachusetts census returns began to indicate that the foreign born had a considerably higher birthrate than that of native Americans. Indeed, the more affluent and educated a family, the fewer children it seemed to produce. Such statistics indicated that native Americans in the Bay State were not even reproducing themselves. The social consequences seemed ominous indeed.

The Irish, though barely one-quarter of the Massachusetts population, produced more than half of the state's children. "It is perfectly clear," a Boston clergyman contended in 1884, "that without a radical change in the religious ideas, education, habits, and customs of the natives, the present population and their descendants will not rule that state a single generation."[44] A few years earlier a well-known New England physician, pointing to America's still largely unsettled western territories, had asked: "Shall they be filled by our own children or by those of aliens? This is a question that our own women must answer; upon their loins depends the future destiny of the nation." Native-born American women had failed themselves as individuals and society as mothers of the Anglo-Saxon race. If matters continued for another half-century in the same manner, "the wives who are to be

mothers in our republic must be drawn from trans-Atlantic homes. The Sons of the New World will have to re-act, on a magnificent scale, the old story of unwived Rome and the Sabines."[45]

Such arguments have received a goodly amount of historical attention, especially as they figured in the late nineteenth and early twentieth centuries as part of the contemporary rationale for immigration restriction.[46] Historians have interpreted the race suicide argument in several fashions. As an incident in a general Western acceptance of racism, it has been seen as a product of a growing alienation of the older middle and upper classes in the face of industrialization, urbanization, and bureaucratization of society. More specifically, some American historians have seen these race suicide arguments as rooted in the fears and insecurities of a traditionally dominant middle class as it perceived new and threatening social realities.

Whether or not historians care to accept some version of this interpretation—and certainly such motivational elements seem to be suggested in the rhetorical formulae employed by many of those bemoaning the failure of American Protestants to reproduce in adequate numbers—it ignores another element crucial to the logical and emotional fabric of those arguments. This is the explicit charge of female sexual failure. To a significant extent contemporaries saw the problem as in large measure woman's responsibility; it was America's potential mothers, not its fathers, who were primarily responsible for the impending social cataclysm. Race suicide seemed a problem in social gynecology.

Though fathers played a necessary role in procreation, medical opinion emphasized that it was the mother's constitution and reproductive capacity which most directly shaped her offspring's physical, mental, and emotional attributes. And any unhealthy mode of life—anything, in short, which seemed undesirable to contemporary medical moralists, including both education and birth control—might result in a woman's becoming sterile or capable of bearing only stunted offspring. Men, it was conceded, were subject to vices even more debilitating, but the effects of male sin and imprudence were, physicians felt, "to a greater extent confined to adult life; and consequently do not, to the same extent, impair the vitality of our race or threaten its physical destruction." Women's violation of physiological laws implied disaster to "the unborn of both sexes." [47]

Though such social critics tended to agree that woman was at fault, they expressed some difference of opinion as to the nature of her guilt. A few felt that lower birthrates could be attributed simply to the conscious and culpable decision of American women to curtail family size. Other physicians and social commentators, while admitting that many women felt little desire for children, saw the roots of the problem in somewhat different—and perhaps even more apocalyptic—terms. It was not, they feared, simply the conscious practice of family limitation which resulted in small families; rather the increasingly unnatural lifestyle of the "modern American woman" had undermined her reproductive capacities so that even when she would, she could not bear adequate numbers of healthy children. Only if American women returned to the simpler lifestyles of the eighteenth and early nineteenth centuries could the race hope to regain its former vitality; women must from childhood see their role as that of robust and self-sacrificing mothers. If not, their own degeneration and that of the race were inevitable.

Why the persistence and intensity of this masculine hostility, of its recurring echoes of conflict, rancor, and moral outrage? There are at least several possible, though by no means exclusive, explanations. One centers on the hostility implied and engendered by the sexual deprivation—especially for the male—implicit in many of the modes of birth control employed at this time. One might, for example, speculate—as Oscar Handlin did some years ago—that such repressed middle-class sexual energies were channeled into a xenophobic hostility toward the immigrant and the black and projected into fantasies incorporating the enviable and fully expressed sexuality of these alien groups. [48] A similar model could be applied to men's attitudes toward women as well; social, economic, and sexual tensions which beset late nineteenth-century American men might well have caused them to express their anxieties and frustrations in terms of hostility toward the middle-class female. [49]

Such interpretations are, however, as treacherous as they are inviting. Obviously, the would-be scientific formulations outlined here mirror something of postbellum social and psychic reality. Certainly some middle-class men in the late nineteenth century had personality needs—sexual inadequacies or problems of status identification—which made traditional definitions of gender roles functional to them. The hostility, even the violent imagery expressed toward women who chose to limit the number of children they bore indicates a significant personal and emotional involvement on the part of the male author. Some women, moreover, obviously used the mechanisms of birth control and, not infrequently, sexual rejection as role-sanctioned building blocks in the fashioning of their particular adjustment. Their real and psychic gains were numerous: surcease from fear and pain, greater leisure, a socially acceptable way of expressing hostility, and a means of maintaining some autonomy and privacy in a life which society demanded be devoted wholeheartedly to the care and nurturance of husband and children. Beyond such statements, however, matters become quite conjectural. At this moment in the development of both historical methodology and psychological theory great caution must be exercised in the development of such hypotheses—especially since the historians of gender and sexual behavior have at their disposal data which from a psychodynamic point of view are at best fragmentary and suggestive.[50]

What the nineteenth-century social historian can hope to study with a greater degree of certainty, however, is the way in which social change both caused and reflected tensions surrounding formal definitions of gender roles. Obviously, individuals as individuals at all times and in all cultures have experienced varying degrees of difficulty in assimilating the prescription of expected role behavior. When such discontinuities begin to affect comparatively large numbers and become sufficiently overt as to evoke a marked ideological response, one can then speak with assurance of having located fundamental cultural tension.[51]

Students of nineteenth-century American and Western European society have long been aware of the desire of a growing number of women for a choice among roles different from the traditional one of mother and housekeeper. It was a theme of Henry James, Henrik Ibsen, and a host of other, perhaps more representative if less talented, writers. Women's demands ranged from that of equal pay for equal work and equal education for equal intelligence to more covert demands for abortion, birth control information, and sexual autonomy within the marriage relationship. Their demands paralleled and were in large part dependent upon fundamental social and economic developments. Technological innovation and economic growth, changed patterns of income distribution, population concentrations, demographic changes in terms of life expectancy and fertility all affected woman's behavior and needs. Fewer women married; many were numbered among the urban poor. Such women had to become self-supporting and at the same time deal with the changed self-image that self-support neces-

sitated. Those women who married generally did so later, had fewer children, and lived far beyond the birth of their youngest child. At the same time ideological developments began to encourage both men and women to aspire to increased independence and self-fulfillment. All these factors interacted to create new ambitions and new options for American women. In a universe of varying personalities and changing economic realities, it was inevitable that some women at least would—overtly or covertly—be attracted by such options and that a goodly number of men would find such choices unacceptable. Certainly for the women who did so the normative role of homebound nurturant and passive woman was no longer appropriate or functional but became a source of conflict and anxiety.

It was inevitable as well that many men, similarly faced with a rapidly changing society, would seek in domestic peace and constancy a sense of the continuity and security so difficult to find elsewhere in their society. They would—at the very least—expect their wives, their daughters, and their family relationships generally to remain unaltered. When their female dependents seemed ill disposed to do so, such men responded with a harshness sanctioned increasingly by the new gods of science.

Notes

1. For historical studies of women's role and ideological responses to it in nineteenth-century America, see William L. O'Neill, *Everyone Was Brave: The Rise and Fall of Feminism in America* (Chicago, 1969); William Wasserstrom, *Heiress of All the Ages: Sex and Sentiment in the Victorian Tradition* (Minneapolis, 1959); Eleanor Flexner, *Century of Struggle: The Woman's Rights Movement in the United States* (New York, 1968); Aileen S. Kraditor, *The Ideas of the Woman Suffrage Movement, 1890–1920* (New York, 1965). For studies emphasizing the interaction between social change and sex role conflict, see Carroll

Smith-Rosenberg, "Beauty, the Beast, and the Militant Woman: a Case Study in Sex Roles and Social Stress in Jacksonian America," *American Quarterly* (Oct. 1971): 562–84; Carroll Smith-Rosenberg, "The Hysterical Woman: Sex Roles and Role Conflict in 19th-century America," *Social Research* 39 (Winter 1972): 652–78. The problem of sexuality in the English-speaking world has been a particular subject of historical concern. Among the more important, if diverse, attempts to deal with this problem are Peter T. Cominos, "Late-Victorian Sexual Respectability and the Social System," *International Review of Social History* 8 (1963): 18–48, 216–50; Stephen Nissenbaum, "Careful Love: Sylvester Graham and the Emergence of Victorian Sexual Theory in America, 1830–1840," Ph.D. diss., University of Wisconsin, 1968; Graham J. Barker-Benfield, "The Horrors of the Half Known Life: Aspects of the Exploitation of Women by Men," Ph.D. diss., University of California, Los Angeles, 1968; Nathan G. Hale, Jr., *Freud and the Americans: The Beginnings of Psychoanalysis in the United States, 1876–1917* (New York, 1971), 24–46; David M. Kennedy, *Birth Control in America: The Career of Margaret Sanger* (New Haven, 1970), 36–71; Steven Marcus, *The Other Victorians: A Study of Sexuality and Pornography in Midnineteenth-Century England* (New York, 1966). See also Charles E. Rosenberg, "Sexuality, Class, and Role in 19th-century America," *American Quarterly* 25 (May 1973): 131–54.

2. Marshall Hall, *Commentaries on Some of the More Important of the Diseases of Females,* in three parts (London, 1827), 2. Although this discussion centers on the nineteenth century, it must be understood that these formulations had a far longer pedigree.

3. Stephen Tracy, *The Mother and Her Offspring* (New York, 1860), xv; William Goodell, *Lessons in Gynecology* (Philadelphia, 1879), 332; William B. Carpenter, *Principles of Human Physiology: With Their Chief Applications to Pathology, Hygiene, and Forensic Medicine,* 4th ed. (Philadelphia, 1850), 727. In mid-nineteenth century many of these traditional views of woman's peculiar physiological characteristics were restated in terms of the currently fashionable phrenology. For example, see Thomas L. Nichols, *Woman, in All Ages and Nations: A Complete and Authentic History of the Manners and Customs, Character and*

Condition of the Female Sex in Civilized and Savage Countries, from the Earliest Ages to the Present Time (New York, ca. 1849), xi.

4. Charles D. Meigs, *Lecture on Some of the Distinctive Characteristics of the Female, Delivered before the Class of the Jefferson Medical College, January 5, 1847* (Philadelphia, 1847), 5.

5. M. L. Holbrook, *Parturition without Pain: A Code of Directions for Escaping from the Primal Curse* (New York, 1882), 14–15. See also Edward H. Dixon, *Woman, and her Diseases, from the Cradle to the Grave: Adapted Exclusively to Her Instruction in the Physiology of Her System, and All the Diseases of Her Critical Periods* (New York, 1846), 17; M. K. Hard, *Woman's Medical Guide: Being a Complete Review of the Peculiarities of the Female Constitution and the Derangement to Which It Is Subject, with a Description of Simple Yet Certain Means for Their Cure* (Mt. Vernon, Ohio, 1848), 11.

6. In the hypothetical pathologies of these generations the blood was often made to serve the same function as that of the nerves; it could cause general ills to have local manifestations and effect systemic changes based on local lesions. By midcentury, moreover, physicians had come to understand that only the blood supply connected the gestating mother to her child.

7. M. E. Dirix, *Woman's Complete Guide to Health* (New York, 1869), 24. So fashionable were such models in the late nineteenth century that America's leading gynecologist in the opening years of the twentieth century despaired of trying to dispel such exaggerated notions from his patients' minds. "It is difficult," he explained, "even for a healthy girl to rid her mind of constant impending evil from the uterus and ovaries, so prevalent is the idea that woman's ills are mainly 'reflexes' from the pelvic organs." Gynecological therapy was the treatment of choice for a myriad of symptoms. Howard A. Kelly, *Medical Gynecology* (New York, 1908), 73.

8. [Dr. Porter], *Book of Men, Women, and Babies: The Laws of God Applied to Obtaining, Rearing, and Developing the Natural, Healthful, and Beautiful in Humanity* (New York, 1855), 56; Tracy, *Mother and Offspring,* xxiii; H. S. Pomeroy, *The Ethics of Marriage* (New York, 1888), 78.

9. On the involuntary quality of female sexuality,

see Alexander J. C. Skene, *Education and Culture as Related to the Health and Diseases of Women* (Detroit, 1889), 22.

10. George Engelmann, *The American Girl of To-day: Modern Education and Functional Health* (Washington, 1900), 9–10.

11. Alexander Harvey, "On the Relative Influence of the Male and Female Parents, in the Reproduction of the Animal Species," *Monthly Journal of Medical Science* 19 (Aug. 1854): 108–18; M. A. Pallen, "Heritage, or Hereditary Transmission," *St. Louis Medical and Surgical Journal* 14 (Nov. 1856): 495. William Warren Potter, *How Should Girls be Educated? A Public Health Problem for Mothers, Educators, and Physicians* (Philadelphia, 1891), 9.

12. As one clerical analyst explained, "All the spare force of nature is concerned in this interior nutritive system, unfitting and disinclining the woman for strenuous muscular and mental enterprise, while providing for the shelter and nourishment of offspring throughout protracted periods of embryo and infancy." William C. Conant, "Sex in Nature and Society," *Baptist Quarterly* 4 (April 1870): 183.

13. William H. Holcombe, *The Sexes Here and Hereafter* (Philadelphia, 1869), 201–2. William Holcombe was a Swedenborgian, and these contrasting views of the masculine and feminine also reflect New Church doctrines.

14. In regard to pregnancy many middle-class women "sought to hide their imagined shame as long as possible," by tightening corsets and then remaining indoors, shunning even the best of friends—certainly never discussing the impending event. Henry B. Hemenway, *Healthful Womanhood and Childhood: Plain Talks to Non-Professional Readers* (Evanston, Ill., 1894); Elizabeth Evans, *The Abuse of Maternity* (Philadelphia, 1875), 28–29.

15. For a brief summary of late nineteenth-century assumptions in regard to human genetics, see Charles E. Rosenberg, "Factors in the Development of Genetics in the United States: Some Suggestions," *Journal of the History of Medicine* 22 (Jan. 1967): 31–33.

16. Since both male and female were ordinarily involved in decisions to practice birth control, the cases are not strictly analogous. Both, however, illustrate areas of social conflict organized about stress on traditional role characteristics. This discussion emphasizes only those aspects of the birth control debate which placed responsibility on the woman. Commentators did indeed differ in such emphases; in regard to abortion, however, writers of every religious and ideological persuasion agreed in seeing the matter as woman's responsibility.

17. "The results," as Edward H. Clarke put it in his widely discussed polemic on the subject, "are monstrous brains and puny bodies; abnormally active cerebration, and abnormally weak digestion; flowing thought and constipated bowels; lofty aspirations and neuralgic sensations." Edward H. Clarke, *Sex in Education; or, A Fair Chance for Girls* (Boston, 1873), 41. Thomas A. Emmett, in his widely used textbook of gynecology, warned in 1879 that girls of the better classes should spend the year before and two years after puberty at rest. "Each menstrual period should be passed in the recumbent position until her system becomes accustomed to the new order of life." Thomas Addis Emmett, *The Principles and Practice of Gynecology* (Philadelphia, 1879), 21.

18. T. S. Clouston, *Female Education from a Medical Point of View* (Edinburgh, 1882), 20; Potter, *How Should Girls Be Educated?*, 9.

19. The baleful hereditary effects of woman's secondary education served as a frequent sanction against this unnatural activity. Lawrence Irwell, "The Competition of the Sexes and its Results," *American Medico-Surgical Bulletin* 10 (Sept. 19, 1896): 319–20. All the doyens of American gynecology in the late nineteenth century—Emmett, J. Marion Sims, T. Gaillard Thomas, Charles D. Meigs, William Goodell, and Mitchell—shared the conviction that higher education and excessive development of the nervous system might interfere with woman's proper performance of her maternal functions.

20. William Goodell, *Lessons in Gynecology* (Philadelphia, 1879), 353.

21. William Edgar Darnall, "The Pubescent Schoolgirl," *American Gynecological and Obstetrical Journal* 18 (June 1901): 490.

22. Board of Regents, University of Wisconsin, *Annual Report, for the Year Ending, September 30, 1877* (Madison, 1877), 45.

23. Clouston, *Female Education*, 19.

24. James E. Reeves, *The Physical and Moral Causes of Bad Health in American Women* (Wheeling,

W.Va., 1875), 28; John Ellis, *Deterioration of the Puritan Stock and Its Causes* (New York, 1884), 7; George Everett, *Health Fragments; or, Steps toward a True Life: Embracing Health, Digestion, Disease, and the Science of the Reproductive Organs* (New York, 1874), 37; Nathan Allen, "The Law of Human Increase; or, Population Based on Physiology and Psychology," *Quarterly Journal of Psychological Medicine* 2 (April 1868): 231; Nathan Allen, "The New England Family," *New Englander* (March 1882): 9–10; Pye Henry Chavasse, *Advice to a Wife on the Management of Her Own Health, And on the Treatment of Some of the Complaints Incidental to Pregnancy, Labour and Suckling with an Introductory Chapter Especially Addressed to a Young Wife* (New York, 1886), 73–75.

25. Sarah H. Stevenson, *The Physiology of Woman, Embracing Girlhood, Maternity and Mature Age*, 2d ed. (Chicago, 1881), 68, 77; Alice Stockham, *Tokology: A Book for Every Woman*, rev. ed. (Chicago, 1887), 257. Sarah H. Stevenson noted acidly that "the unerring instincts of woman have been an eloquent theme for those who do not know what they are talking about." Stevenson, *Physiology of Woman*, 79. The dress reform movement held, of course, far more significant implications than one would gather from the usually whimsical attitude with which it is normally approached; clothes were very much a part of woman's role. Health reformers, often critical as well of the medical establishment whose arguments we have—essentially—been describing, were often sympathetic to women's claims that not too much, but too little, mental stimulation was the cause of their ills, especially psychological ones. M. L. Holbrook, *Hygiene of the Brain and Nerves and the Cure of Nervousness* (New York, 1878), 63–64, 122–23; James C. Jackson, *American Womanhood: Its Peculiarities and Necessities* (Dansville, N.Y., 1870), 127–31.

26. For documentation of the progressive drop in the white American birthrate during the nineteenth century, and some possible reasons for this phenomenon, see Yashukichi Yasuba, *Birth Rates of the White Population in the United States, 1800–1860: An Economic Study* (Baltimore, 1862); J. Potter, "American Population in the Early National Period," in *Proceedings of Section V of the Fourth Congress of the International Economic History Association*, Paul De-

prez, ed. (Winnipeg, Canada, 1970), 55–69. For a more general background to this trend, see A. M. Carr-Saunders, *World Population: Past Growth and Present Trends* (London, 1936).

27. A. K. Gardner, *Conjugal Sins against the Laws of Life and Health* (New York, 1874), 131. H. R. Storer of Boston was probably the most prominent and widely read critic of such "conjugal sins." Abortion had in particular been discussed and attacked since early in the century, though it was not until the postbellum years that it became a widespread concern of moral reformers. Alexander Draper, *Observations on Abortion, With an Account of the Means both Medicinal and Mechanical, Employed to Produce that Effect . . .* (Philadelphia, 1839); Hugh L. Hodge, *On Criminal Abortion: A Lecture* (Philadelphia, 1854). Advocates of birth control routinely used the dangers and prevalence of abortion as one argument justifying their cause.

28. *Report of the Suffolk District Medical Society on Criminal Abortion and Ordered Printed . . . May 9, [1857]* (Boston, 1857), 2. The report was almost certainly written by Storer. The Michigan report is summarized in William D. Haggard, *Abortion: Accidental, Essential, Criminal,* Address before the Nashville Academy of Medicine, Aug. 4, 1898 (Nashville, Tenn., 1898), 10. For samples of contemporary descriptions of prevalence, cheapness, and other aspects of abortion and birth control in the period, see Ely Van de Warker, *The Detection of Criminal Abortion, and a Study of Foeticidal Drugs* (Boston, 1872); Evans, *Abuse of Maternity;* Horatio R. Storer, *Why Not? A Book for Every Woman*, 2d ed. (Boston, 1868); N. F. Cook, *Satan in Society: By a Physician* (Cincinnati, 1876); Discussion, *Transactions of the Homeopathic Medical Society of New York,* 1866, 4:9–10; H. R. Storer and F. F. Heard, *Criminal Abortion* (Boston, 1868); H. C. Ghent, "Criminal Abortion, or Foeticide," *Transactions of the Texas State Medical Association at the Annual Session, 1888–89* (1888–89), 119–46; Hugh Hodge, *Foeticide, or Criminal Abortion: A Lecture Introductory to the Course on Obstetrics, and Diseases of Women and Children, University of Pennsylvania* (Philadelphia, 1869), 3–10. Much of the medical discussion centered on the need to convince women that the traditional view that abortion was no crime if per-

formed before quickening was false and immoral and to pass and enforce laws and medical society proscriptions against abortionists.

29. Compare the warning of Pomeroy, *Ethics of Marriage*, v, 56, with the editorial "A Conviction for Criminal Abortion," *Boston Medical and Surgical Journal* 106 (Jan. 5, 1882): 18–19. It is significant that discussions of birth control in the United States always emphasized the role and motivations of middle-class women and men; in England, following the canon of the traditional Malthusian debate, the working class and its needs played a far more prominent role. Not until late in the century did American birth control advocates tend to concern themselves with the needs and welfare of the working population. It is significant as well that English birth control advocates often used the prevalence of infanticide as an argument for birth control; in America this was rarely discussed. And one doubts if the actual incidence of infanticide was substantially greater in London than New York.

30. For a guide to literature on birth control in nineteenth-century America, see Norman Himes, *Medical History of Contraception* (Baltimore, 1936). See also J. A. Banks, *Prosperity and Parenthood: A Study of Family Planning Among the Victorian Middle Classes* (London, 1954), and J. A. Banks and Olive Banks, *Feminism and Family Planning in Victorian England* (Liverpool, 1964); Margaret Hewitt, *Wives and Mothers in Victorian Industry* (London, ca. 1958). For the twentieth century, see David M. Kennedy, *Birth Control in America.*

31. John Humphrey Noyes, *Male Continence* (Oneida, N.Y., 1872), 10–11.

32. It is not surprising that the design for a proto-diaphragm patented as early as 1846 should have been called "The Wife's Protector." J. B. Beers, "Instrument to prevent conception, patented Aug. 28th, 1846," design and drawings (Historical Collections, Library of the College of Physicians of Philadelphia).

33. In some marriages, for example, even if the male had consciously chosen, indeed urged, the practice of birth control, he was effectively deprived of a dimension of sexual pleasure and of the numerous children which served as tangible and traditional symbols of masculinity as well as the control over his wife which the existence of such children implied. In

some marriages, however, birth control might well have brought greater sexual fulfillment because it reduced the anxiety of the female partner. Throughout the nineteenth century, withdrawal was almost certainly the most common form of birth control. One author described it as "a practice so universal that it may well be termed a national vice, so common that it is unblushingly acknowledged by its perpetrators, for the commission of which the husband is even eulogized by his wife." [Cook], *Satan in Society*, 152. One English advocate of birth control was candid enough to argue that "the real objection underlying the opposition, though it is not openly expressed, is the idea of the deprivation of pleasure supposed to be involved." Austin Holyyoake, *Large or Small Families* (London, 1892), 11.

34. R. T. Trall, *Sexual Physiology: A Scientific and Popular Exposition of the Fundamental Problems in Sociology* (New York, 1866), xi, 202. As women awoke to a realization of their own "individuality," as a birth control advocate explained it in the 1880s, they would rebel against such "enforced maternity." E. B. Foote, Jr., *The Radical Remedy in Social Science; or, Borning Better Babies* (New York, 1886), 132. See also Stevenson, *Physiology of Women*, 91; T. L. Nichols, *Esoteric Anthropology* (New York, 1824). E. H. Heywood, *Cupid's Yokes; or, the Binding Force of Conjugal Life* (Princeton, Mass., 1877); Stockham, *Tokology*, 250; Alice Stockham, *Karezza: Ethics of Marriage* (Chicago, 1896); E. B. Foote, *Medical Commonsense Applied to the Causes, Prevention and Cure of Chronic Diseases and Unhappiness in Marriage* (New York, 1864), 365; J. Soule, *Science of Reproduction and Reproductive Control: The Necessity of Some Abstaining from Having Children; the Duty of All to Limit Their Families According to Their Circumstances Demonstrated; Effects of Continence Effects of Self-Pollution—Abusive Practices; Seminal Secretion—Its Connection with Life; with All the Different Modes of Preventing Conception, and the Philosophy of Each* (n.p., 1856), 37; L. B. Chandler, *The Divineness of Marriage* (New York, 1872). To radical feminist Tennie C. Claflin, man's right to impose his sexual desires upon woman was the issue underlying all opposition to woman suffrage and the expansion of woman's role. Tennie C. Claflin, *Constitutional Equality: A Right of Woman; or, A Consideration of the*

Various Relations Which She Sustains as a Necessary Part of the Body of Society and Humanity; with Her Duties to Herself—together with a Review of the Constitution of the United States, Showing That the Right to Vote Is Guaranteed to All Citizens; Also a Review of the Rights of Children (New York, 1871), 63. Particularly striking are the letters from women desiring birth control information. Margaret Sanger, *Motherhood in Bondage* (New York, 1928); E. B. Foote, Jr., *Radical Remedy*, 114–20; Henry C. Wright, *The Unwelcome Child; or The Crime of an Undesigned and Undesired Maternity* (Boston, 1858). This distinction between economic, "physical," and role consideration is, quite obviously, justifiable only for the sake of analysis; these considerations must have coexisted within each family in particular configuration.

35. Noyes, *Male Continence*, 16; Frederick Hollick, *The Marriage Guide; or, Natural History of Generation; A Private Instructor for Married Persons and Those about to Marry, Both Male and Female* (New York, ca. 1860), 348; Trall, *Sexual Physiology*, 205–6.

36. Indeed, in these post-Darwinian years it was possible for at least one health reformer to argue that smaller families were a sign of that higher nervous evolution which accompanied civilization. [M. L. Holbrook], *Marriage and Parentage* (New York, 1882). For the eugenic virtues of fewer but better children, see E. R. Shepherd, *For Girls: A Special Physiology: Being a Supplement to the Study of General Physiology,* 20th ed. (Chicago, 1887), 213; M. L. Griffith, *Ante-Natal Infanticide* (n.p. [1889]), 8.

37. See Louis François Etienne Bergeret, *The Preventive Obstacle; or, Conjugal Onanism*, trans. P. de Marmon, (New York, 1870); C. H. F. Routh, *Moral and Physical Evils Likely to Follow If Practices Intended to Act as Checks to Population Be Not Strongly Discouraged and Condemned,* 2d ed. (London, 1879), 13; Goodell, *Lessons in Gynecology*, 371, 374; Thomas Hersey, *The Midwife's Practical Directory; or, Woman's Confidential Friend: Comprising, Extensive Remarks on the Various Casualties and Forms of Diseases Preceeding, Attending and Following the Period of Gestation, with Appendix,* 2d ed. (Baltimore, 1836), 80; William H. Walling, *Sexology* (Philadelphia, 1902), 79.

38. J. R. Black, *The Ten Laws of Health; or, How Disease is Produced and Can Be Prevented* (Philadelphia, 1873), 251. See also C. A. Greene, *Build Well:*

The Basis of Individual Home, and National Elevation, Plain Truths Relating to the Obligations of Marriage and Parentage (Boston, ca. 1885), 99; E. P. LeProhon, *Voluntary Abortion; or Fashionable Prostitution, with Some Remarks upon the Operation of Craniotomy* (Portland, Me., 1867), 15; M. Solis-Cohen, *Girl, Wife, and Mother* (Philadelphia, 1911), 213.

39. There is an instructive analogy between these ponderously mechanistic sanctions against birth control and abortion and the psychodynamic arguments against abortion used so frequently in the twentieth century; both served precisely the same social function. In both cases the assumption of woman's childbearing destiny provided the logical basis against which a denial of this calling produced sickness, in the nineteenth century through physiological, and ultimately pathological processes—in the twentieth century through guilt and psychological but, again, ultimately pathological processes.

40. A. K. Gardner, for example, confessed sympathy for the seduced and abandoned patron of the abortionist, "but for the married shirk, who disregards her divinely-ordained duty, we have nothing but contempt." Gardner, *Conjugal Sins*, 112. See also E. Frank Howe, *Sermon on Ante-Natal Infanticide Delivered at the Congregational Church in Terre Haute, on Sunday Morning, March 28, 1869* (Terre Haute, Ind., 1869); J. H. Tilden, *Cursed before Birth* (Denver, ca. 1895); J. M. Toner, *Maternal Instinct, or Love* (Baltimore, 1864), 91.

41. It must be emphasized that this is but one theme in a complex debate surrounding the issue of birth control and sexuality. A group of more evangelically oriented health reformers tended to emphasize instead the responsibility of the "overgrown, abnormally developed and wrongly directed amativeness of the man" and to see the woman as victim. John Cowan, Henry C. Wright, and Dio Lewis were widely read exemplars of this point of view. This group shared a number of assumptions and presumably psychological needs, and represents a somewhat distinct interpretive task. John Cowan, *The Science of a New Life* (New York, 1874), 275.

42. W. L. Atlee and D. A. O'Donnell, "Report of the Committee on Criminal Abortion," *Transactions of the American Medical Association* 22 (1871): 241.

43. The most tireless advocate of these views was

Nathan Allen, a Lowell, Massachusetts, physician and health reformer. Nathan Allen, "The Law of Human Increase; or, Population based on Physiology and Psychology," *Quarterly Journal Psychological Medicine* 2 (April 1868): 209–66; Nathan Allen, *Changes in New England Population, Read at the Meeting of the American Social Science Association, Saratoga, September 6, 1877* (Lowell, Mass., 1877); Nathan Allen, "The Physiological Laws of Human Increase," *Transactions of the American Medical Association* 21 (1870): 381–407; athan Allen, "Physical Degeneracy," *Journal of Psychological Medicine* 4 (Oct. 1870): 725–64; Nathan Allen, "The Normal Standard of Woman for Propagation," *American Journal of Obstetrics* 9 (April 1876): 1–39.

44. Ellis, *Deterioration of Puritan Stock,* 3; Storer, *Why Not?,* 85.

45. Clarke, *Sex in Education,* 63. For similar warnings, see Henry Gibbons, *On Feticide* (San Francisco, 1878), 4; Charles Buckingham, *The Proper Treatment of Children, Medical or Medicinal* (Boston, 1873), 15; Edward Jenks, "The Education of Girls From a Medical Stand-Point," *Transaction of the Michigan State Medical Society* 13 (1889): 52–62; Paul Paquin, *The Supreme Passions of Man* (Battle Creek, Mich., 1891), 76.

46. These arguments, first formulated in the 1860s, had become clichés in medical and reformist circles by the 1880s. See Barbara Miller Solomon, *Ancestors and Immigrants: A Changing New England Tradition* (Cambridge, Mass., 1956); John Higham, *Strangers in the Land: Patterns of American Nativism, 1860–1925* (New Brunswick, N.J., 1955). Such arguments exhibited a growing consciousness of class as well as of ethnic sensitivity; it was the better-educated and more sensitive members of society, anti-Malthusians began to argue, who would curtail their progeny, while the uneducated and coarse would hardly change their habits. H. S. Pomeroy, *Is Man Too Prolific? The So-Called Malthusian Idea* (London, 1891), 57–58.

47. Ellis, *Deterioration of Puritan Stock,* 10.

48. Oscar Handlin, *Race and Nationality in American Life,* 5th ed. (Boston, 1957), 139–66.

49. One might postulate a more traditionally psychodynamic explanatory model, one which would see the arguments described as a male defense against their own consciousness of sexual inadequacy or ambivalence or of their own unconscious fears of female sexual powers. These emphases are quite distinct. The first—though it also assumes the reality of individual psychic mechanisms such as repression and projection—is tied very much to the circumstances of a particular generation, to social location, and to social perception. The second kind of explanation is more general, time-free, and based on a presumably ever-recurring male fear of female sexuality and its challenge to the capacity of particular individuals to act and live an appropriately male role. For the literature on this problem, see Wolfgang Lederer, *The Fear of Women* (New York, 1968).

50. At this time, moreover, most psychiatric clinicians and theoreticians would agree that no model exists to extend the insights gained from individual psychodynamics to the behavior of larger social groups such as national populations or social classes.

51. Most societies provide alternative roles to accommodate the needs of personality variants—as, for example, the shaman role in certain Siberian tribes or the accepted man-woman homosexual of certain American Indian tribes. In the nineteenth-century English-speaking world, such roles as that of the religious enthusiast and the chronic female invalid or hysteric may well have provided such modalities. But a period of peculiarly rapid or widespread social change can make such available role alternatives inadequate mechanisms of adjustment for many individuals. Others in the same society may respond to the same pressures of change by demanding an undeviating acceptance of traditional role prescriptions and refusing to accept the legitimacy of such cultural variants. The role of the hysterical woman in late nineteenth-century America suggests many of the problems inherent in creating such alternative social roles. While offering both an escape from the everyday duties of wife and mother, and an opportunity for the display of covert hostility and aggression, this role inflicted great bodily (though nonorganic) pain, provided no really new role or interest, and perpetuated—even increased—the patient's dependence on traditional role characteristics, especially that of passivity. The reaction of society, as suggested by the writings of most male physicians, can be described as at best an unstable compromise between patronizing

tolerance and violent anger. See Carroll Smith-Rosenberg, "The Hysterical Woman: Sex Roles and Role Conflict in 19th-Century America," 652–78. For useful discussions of hysteria and neurasthenia, see Ilza Veith, *Hysteria: The History of a Disease* (Chicago, 1965); Henri F. Ellenberger, *The Discovery of the Unconscious: The History and Evolution of Dynamic Psychiatry* (New York, 1970); Charles E. Rosenberg, "The Place of George M. Beard in Nineteenth-Century Psy-

chiatry," *Bulletin of the History of Medicine* 36 (May–June 1962): 245–59; John S. Haller, Jr., "Neurasthenia: the Medical Profession and the 'New Woman' of Late Nineteenth-Century," *New York State Journal of Medicine* 71 (Feb. 15, 1971): 473–82. Esther Fischer Homberger has recently argued that these diagnostic categories masked an endemic male-female conflict: "Hysterie und Mysogynie—ein Aspekt der Hysteriegeschichte," *Gesnerus* 26 (1969): 117–27.

7 The Invisible (Invalid) Woman: African-American Women, Illness, and Nineteenth-Century Narrative

Diane Price Herndl

In the last several years feminist scholars have begun writing the history of women, illness, and narrative, but from Barbara Ehrenreich's and Deirdre English's *Complaints and Disorders,* Haller and Haller's *The Physician and Sexuality in Victorian America,* and Carroll Smith-Rosenberg's *Disorderly Conduct* to Elaine Showalter's *The Female Malady,* Mary Poovey's *Uneven Developments,* and my own *Invalid Women,* that history of female illness is white and middle class. Since I finished my study, however, I have begun to notice that the illness of African-American women is not nearly so invisible in writings by nineteenth-century African-American women. It remains invisible, though, not only in the medical writing of the time but in the historical and critical accounts of both slave and free black women.[1] I've found an amazing lack of historical information about nineteenth-century African-American women's health and no critical commentary on the representations of illness in literary texts by African-American women. Almost everything one can find approaches the question from a distinctly ideological perspective, from overtly racist claims that slaves were treated well to the progressive perspective of writing a positive history of black women; in emphasizing the black woman's strength, however, the historian sometimes overlooks that with which [the black woman] had to struggle. This paper will begin the work of recovering that invisible history and looking at the ways that nineteenth-century African-American women writers dealt with a cultural discourse that not only refused to recognize their illnesses but in many ways denied their physicality, even as it paradoxically refused to grant them an existence that went beyond that physicality. While the absence of ill black women in medical and historical texts springs from racist assumptions about black women's bodies, the seeming invisibility of African-American women's illnesses in their own texts, even when they mention real illnesses, may well have been a rhetorical strategy aimed at defusing this racist medical-cultural discourse.

DIANE PRICE HERNDL is Associate Professor of English at New Mexico State University, Las Cruces.

Reprinted from *Women's Studies* 24 (1995): 553–72, by permission of Gordon and Breach Publishers.

African-American Women and Health in the Nineteenth Century

I'm not going to rehearse at length the history of nineteenth-century medicine and women, since I and many others have done that elsewhere.[2] Suffice it to say that in Victorian medical thought women were believed to be naturally weak and liable to all sorts of illness but also guilty of exacerbating the dangers to their bodies by inappropriate activities. A passage from Gunning S. Bedford's 1856 *Clinical Lectures on the Diseases of Women and Children* summarizes the dominant medical view succinctly:

> There are numerous causes which conspire to the frequent production of functional and organic derangements of the uterus; but numerous as these causes are, experience proves very conclusively how unequally they operate under different circumstances. Child-bearing, unrestrained sexual intercourse, abortions, precocious nervous excitement from the perusal of prurient books, the lascivious polka, and the various exciting scenes of city life, ... Add to these the uninterrupted rounds of excitement consequent upon balls, parties, the opera ... and more than all, the fact that these disastrous influences ... are exercised on a *physique* too often without a single attribute of solidity—and you will at once have explained why it is that the females in the higher classes of our large cities decay long before they have attained the meridian of life. (author's emphasis, 7)

Bedford goes on to assert that, in fact, illness does "not fall with equal force on all" and that "the lower classes who reside in the city enjoy, comparatively at least, an immunity from these special diseases" (7). For Bedford, and for the midnineteenth-century medical profession as a whole, women's illnesses were the province chiefly of the upper class.

When we factor race into this equation of illness and class, it becomes clear that much nineteenth-century medicine was explicitly racist. For example, Dr. Columbat wrote in his widely translated and quoted 1850 text, *A Treatise on the Diseases and Special Hygiene of Females,* that girls from southern climates matured sexually at much earlier ages than did girls from northern climates. Indeed, throughout the first two-thirds of the nineteenth century, the medical theory of "specificity—an individualized match between medical therapy and the specific characteristics of a particular patient and of the social and physical environments—was an essential component of proper therapeutics" (Warner 58). One of the chief of these "specifics" was race (and this held as much for people of different ethnic origins as for those of different racial origins), but none of the ideas about physicality and race was more damaging than that directed at the African-American.

The ideas about African-Americans and illness were deeply influenced by the racist anthropometric studies of the nineteenth century and the religious beliefs that often fueled them. Even years after emancipation physicians and ministers—sometimes working together, often just drawing on each other's work—continued to claim that blacks were fundamentally physically and morally different, and different, of course, meant inferior. This often took the form of claiming that physical labor and difficult working conditions that would be harmful to whites were in fact beneficial for blacks and that slavery was the ideal condition for maintaining the African-American's health.[3] In 1867 a contributor to a book called *The Land We Love* claimed that slavery left the slaves "happy, thriving, contented" because "they spent their days in healthful easy employment, their powers never overtasked, their nights under good shelters in healthful sleep, with plentiful supply [sic] of food, with no thought of the past, no care for the morrow" (quoted in Smith 42). These claims—at their most vitriolic when they joined science and religion—often were extended to the point of claiming that blacks were not even human. "Ariel" (Nashville clergyman Buckner H. Payne), in his pamphlet *The Negro: What Is His Ethnological Status*, "maintained that God disapproved of any attempt to equalize the races. 'Ariel' based his theory on the supposition 'That the negro being created before Adam, consequently . . . is a *beast* in God's nomenclature'" (Smith 43). Two prominent antiabolitionist physicians, Samuel A. Cartwright and Josiah C. Nott, believed that the only alternatives for the future of the African-American race were perpetual enslavement or genocide (Smith 44). According to John David Smith, author of *An Old Creed for the New South: Proslavery Ideology and Historiography, 1865–1918*, this blend of biblical and scientific arguments for black inferiority went for the most part unquestioned by whites at least until the end of World War I.

In these classist and racist contexts the conclusion seems to have been that black women never suffered any illness at all. No nineteenth-century medical text that I have been able to find actually mentions treating black women, even though the theory of specificity would have convinced physicians that [black women's] bodies and illnesses were fundamentally different from white women's and even though we know they received some medical care.[4] (In fact no textbook on treating African-American patients was published in the U.S. until 1975 [Savitt 16].) Few texts that deal with the health of slaves discuss the care of women, except to discuss lying-in and birthing. As Deborah White notes in her

history of slave women, *Ar'n't I a Woman?: Female Slaves in the Plantation South*, histories of slavery tend to be histories of the enslavement of black men.[5] As she explains,

> In short, it is very difficult, if not impossible, to be precise about the effect of any single variable on female slaves. Source material on the general nature of slavery exists in abundance, but it is very difficult to find source material about slave women in particular. Slave women were everywhere, yet nowhere. . . . A consequence of the double jeopardy and powerlessness is the black woman's invisibility. . . . To both [whites and African-American males] the female slave's world was peripheral. The bondwoman was important to them only when her activities somehow involved them. (23)

Unfortunately, even White's text does not really illuminate the medical condition, treatment, or beliefs of slave women.[6]

When we turn to histories of medicine, we find slaves in general and slave women in particular strikingly absent. The only book-length study of the medical condition of slaves, Todd Savitt's *Medicine and Slavery*, unfortunately bears out White's contention that the slave represented in these texts is male. Savitt's book very admirably addresses the lack of information about the health and health care of slaves—material which is woefully lacking in other histories of slavery (as he notes, 309–10)—but women merit only five pages of Savitt's book. The black invalid woman remains invisible even in this revisionist account.

The histories we have are largely contradictory and unhelpful. Many of the histories that rely on the evidence of plantation owners claim that slave women were fairly well treated. When slave-holders realized that female slaves could be profitable producers of more slaves, so this argument goes, they protected their investment and even indulged their female slaves' desires not to work. As White explains, "In an age when women's diseases were still shrouded in mystery, getting the maximum amount of work from women of childbearing age while remaining confident that no damage was done to their reproductive organs was a guessing game that few white slave owners wanted to play or could afford to lose" (80).[7] White claims that many female slaves were able to use this situation to their advantage and could use illness as a strategy for resistance. She quotes a Virginia planter who complained that slave women were "nearly valueless" for work because of the excuses for not working they could come up with:

> 'they don't come to the field and you go to the quarters and ask the old nurse what's the matter and she says, "Oh, she's not ... fit to work sir"; and ... you have to take her word for it that something or other is the matter with her, and you dare not set her to work; and so she will lay up till she feels like taking the air again, and plays the lady at your expense.' (White, quoting from Olmstead, *Seaboard Slave States*, 80)

Other histories, based on the reports of slave women themselves, contradict such a view. One slave woman told an interviewer in 1862, "You neber 'lowed to drop you hoe till labor 'pon you, neber! no matter how bad you feel, you neber 'lowed to stop till you go in bed, neber!" (quoted in Sterling 40; from Austa French, *Slavery in South Carolina*, New York, 1862). Other women reported beatings and overwork well into the final trimester of pregnancy: "When women was with child they'd dig a hole in the groun' and put their stomach in the hole, and then beat 'em. They'd allus whoop us" (quoted in Sterling 39; from George P. Rawick, ed., *The American Slave:*

A Composite Autobiography, Vol. 4, Westport, Conn., 1972).

What is clear from the data available is that slaves in general were quite unhealthy; pneumonia, cholera, diarrhea, and smallpox were rampant in the slave population, in part because the slave diet—generally high in calories but lacking protein and other infection-fighting nutrients—was poor, as were their living and working conditions. Slave women suffered a good deal from gynecological problems attendant on poor and nonexistent prenatal and natal care.[8] "Still," as Deborah White points out, "it is difficult to ascertain whether bondwomen who claimed to be ill were actually sick or whether they were practicing a kind of passive resistance. They certainly had more leverage in the realm of feigning illness than men, but they also perhaps had more reason than men to be ill" (84).

Understanding the reality of African-American women's physical history is further complicated by the racist ideologies and histories of the nineteenth century. Black women were represented in both scientific and historical tracts as being either excessively sexualized or as enduringly maternal and strong. Patricia Morton, in her historiographical study *Disfigured Images: The Historical Assault on Afro-American Women,* characterizes these two stereotypes of the black woman as the oversexed and lascivious "Jezebel" and the strong, maternal workhorse "Mammy" and points out that they dominate even contemporary representations of black women.[9] In the effort to combat these destructive images, however, contemporary historians seem reluctant to approach the unpleasant pieces of history to look at either the representations or the realities of African-American women's illness.

The result of this confused and incomplete history is that we cannot really determine the "truth" about slave and free-black women's illnesses in the nineteenth century from currently available sources. What this incomplete history does show us is the peculiar double bind in which black women writers found themselves. Their texts are shaped by this medical-political ideology and can be read as responses to this specific form of oppression. We can read their representations of ill black women as strategic rhetorical responses to a pervasive proslavery argument that claimed that blacks were better off as slaves than as free persons. In this context black women who used the sentimental trope of illness to assert their likeness to white women and their humanity ran the risk of supporting the proslavery arguments against which they were fighting. If they followed the conventions of nineteenth-century white fiction and represented their heroines as invalids, the dominant culture could interpret this as evidence of their unfitness for freedom. If they did not represent their heroines as ill, however, they not only risked the lack of sympathy such representations could evoke but also fueled the ideology that black women's bodies were fundamentally different and indestructible.

Invalid African-American Women

White women engaged in writing antislavery narratives did, from time to time, mention black women's illnesses. In *Uncle Tom's Cabin* Harriet Beecher Stowe describes Eva's mammy as having frequent headaches (because she is awakened twenty times a night and never allowed to make up her sleep), details the near-insanity of Cassy, and depicts the alcoholism and general weakness of Prue. Still, in a novel which focuses so exhaustively on white women's illnesses—Eva's consumption and death, her mother Marie's hypochondriac whining—these illnesses are distinctly understated, if not all but invisible. There is a more significant invalid black character in

E. D. E. N. Southworth's 1848 novel, *Retribution;* in a very involved subplot an heiress is robbed of her inheritance because she is in fact her father's illegitimate quadroon daughter by a former slave. Minny falls into hysterical fits and paralysis when she is sold to traders who plan to make her a prostitute. She is treated by a woman doctor who makes [Minny's] body well by dulling her mind. She is eventually restored to her freedom, family, inheritance, and health. Involved as this subplot is, it is still a minor piece of the novel, which focuses instead on a white woman's illness, her betrayal by her husband and best friend, and the sorrow of her young, soon-orphaned daughter.[10] Again, as in Stowe, the black woman's illness is trivialized by comparison to white women's. In Stowe's and Southworth's texts both white and black women suffer illness that results from the evils of the patriarchal institutions of slavery and marriage, but white women's sufferings take on a more spiritual cast; black women's sufferings are either slight, easily put to rights, or purely physical. The racism of nineteenth-century stereotypes of women even here, in abolitionist fictions, marks black women as more physical than white. These representations are fraught with what Carolyn Karcher has noted as the central problem facing the whole genre of sentimental antislavery fiction: "Without dangerous concessions to the prejudices of its intended audience, the genre could not arouse compassion for the slaves; with those concessions, it inevitably perpetuated an ideology relegating people of color and white women to subordinate status" (71–72). The result is that black women's illnesses are all but invisible in white women's texts.

If we look at narratives written by African-American women, we get a different version of the role of illness in the lives of black women. I will turn here to three narratives: two novels—

Harriet Wilson's autobiographical *Our Nig* and Frances Harper's *Iola Leroy*—and one slave narrative—Harriet Jacob's *Incidents in the Life of a Slave Girl*. Although all three represent the illness of a black woman, they are nonetheless somewhat reticent. Even these narrators pass over their illnesses quickly, not dwelling on sickroom or deathbed scenes at length, and, more pointedly, not really drawing fully on the sentimental tradition of empowering illness, what Jane Tompkins calls the "sentimental power" of these deaths. Still, their reticence differs from Stowe's or Southworth's in strategic ways and demonstrates sentimental rhetoric's inadequacy to deal with bodily representation. It is not the relative unimportance of the African-American woman's illness that these writers' reticence demonstrates but the difficulty of claiming any physicality at all without being subsumed by an ideological view that sees black women as *only* physical, in their roles as lascivious Jezebels or workhorse Mammies.

Frances Harper probably comes closest to the standard model of white sentimental fiction in *Iola Leroy, or Shadows Uplifted,* which was published fairly late in the century, in 1892. Harper echoes and revises Stowe's *Uncle Tom's Cabin* in a number of interesting ways, one of which is to rewrite Eva St. Clare's beatific death scene by featuring a young girl of African descent.[11] Stricken by the awful news not only of the father's death but also of the family's return to slavery, Marie and Gracie Leroy, the title character's mother and younger sister, succumb to brain fever. Like Eva St. Clare, Gracie is beloved in the household and is completely self-forgetting; she tends to her mother's illness, ignoring her own wasting away. Finally, though, she cannot continue to nurse her mother and is herself taken to bed to die:

Swiftly the tidings went through the house that Gracie was dying. The servants gath-ered around her with tearful eyes, as she bade them all good-bye. When she had finished, and Mammy had lowered the pillow, an unwonted radiance lit up her eye, and an expression of ineffable gladness overspread her face, as she murmured: "It is beautiful; so beautiful!" Fainter and fainter grew her voice, until, without a struggle or sigh, she passed away beyond the power of oppression and prejudice. (108)

However sentimental such a death may seem to a twentieth-century reader, for its time it is distinctly understated. Gracie's death is led up to by only a few pages, and, as we've just seen, the death itself and her leave-taking comprise only a paragraph.

Harper uses the same tradition Stowe had used for Eva's death, but she tempers it a good deal. Harper employs many of the tropes of what Welter calls "the Cult of True Womanhood" in the service of emphasizing the Leroy women's superiority, but Gracie's death is more than overbalanced by Iola's strength; instead of succumbing to brain fever, Iola becomes a nurse in a Civil War hospital and later a teacher. Harper manages to use the sentimental representation of death to emphasize (for a nineteenth-century audience) Gracie's nobility, her delicacy, her femininity, and, frankly, her *humanity* and, by extension, to alert us to those qualities in Iola and her mother, Marie, but she chooses not to offer us a narrative in which such an invalid is the central figure. The rhetorical gesture here is double; it assures readers of the Leroy sisters' inherent value and virtue (two things of which illness was a sign for the nineteenth-century reader), and it rebuts arguments that African-Americans are better off in slavery by representing the Leroy women as strong, capable, and *well* when they are free.

The dangers of representing the postbellum

African-American woman as an invalid were tremendous. As I discussed earlier, one of the strongest lines taken by proslavery writers was the argument that without the "healthful benefits" of slavery the African-American race would face extinction. Therefore any representation of a free black woman who suffered an illness was liable to a kind of terrible misconstrual. To show how extreme this difficulty could prove, we can look at the life and death of Olivia Davidson Washington, the second wife of Booker T. Washington. A few years ago, while teaching Washington's *Up From Slavery,* I was struck by the fact that, although it was clear that Olivia had been absolutely instrumental in helping to found, fund, and run Tuskegee, Washington notes her death almost as an afterthought and in only a few lines: "In 1889 she died, after four years of happy married life and eight years of hard and happy work for the school. She literally wore herself out in her never ceasing efforts in behalf of the work that she so dearly loved" (199). At the time I took the brevity of this narrative of illness and death as a sign of Washington's self-absorption, egotism, and sexism. However, I recently ran across a short biography of Olivia Davidson Washington that sheds a different light on her husband's reticence.

Olivia Davidson had suffered debilitating illness for many years, even before coming to Tuskegee and marrying its founder, but she remained "uncertain" about her own condition and worked amazingly long hours whenever it was possible to stay on her feet. In April of 1886, a few months before she married Washington, she spoke to the Alabama State Teachers' Association on the subject of "How shall we make the women of our race stronger?" (This seems a significant topic, if only as evidence that for some the question of black women's weakness *was* a serious issue.) In line with the general principles of Tuskegee she emphasized physical development above other development and argued that poor health could be disastrous for the race "for though a strong, earnest spirit may rise above, and inspire a weak body, generally the weakness of the body will crop the wings and keep the soul from soaring" (quoted in Dorsey, 355). Small wonder, then, given Booker Washington's own emphasis on health and hygiene, and the concurrent threat that any illness might be taken as evidence of the unfitness of the African-American to stand up to the rigors of freedom, that her own ill health would be ignored by her and understated by her husband.

To return, then, to the representation of Iola and Gracie Leroy, we see that Harper is caught between discourses: if the Leroy sisters suffer no physical ill effects of their enslavement, then they will not be delicate, feminine, and sympathetic enough to enlist Harper's white readers on their side. On the other hand, Iola must live to be healthy and strong as a free woman, or Harper's novel might act against the very cause she was writing to promote—freedom, education, and equality under the law for African-Americans.

We can see a similar rhetorical dilemma at work in Harriet Jacob's *Incidents in the Life of a Slave Girl.* No one can read this narrative without wonder at Jacobs's physical strength and stamina and awe at her ability to withstand the cold, heat, and physical strain of living for seven years in an unheated, uncooled attic. Jacobs *does,* however, describe much of the illness that she suffered in her years of running and imprisonment. When she first escapes, she suffers fevers from fright, sleeping on damp ground, and hundreds of mosquito bites (168, 171). She describes one serious illness, which came on during her second winter of imprisonment, during which she is in a coma for sixteen hours and seriously ill for over six weeks. But even during her description of this long illness Jacobs dwells on the suffering of her soul more than the suffering of her body: "Dark thoughts passed through my mind as I lay there day after day. I tried to be thankful . . . Sometimes I thought God was a compassionate Father. . . . At other times, it seemed to me there was no justice or mercy in the divine government. I asked why the curse of slavery was permitted to exist" (186). Jacobs notes that she suffered tremendously from the effects of this imprisonment, but even so she does it so briefly—almost in passing—that, unless readers are looking for it, they will hardly notice that she has been ill at all. She mentions once that "my friends feared I should become a cripple for life" (192) and notes, toward the end of the description of her confinement, that her "body still suffers from the effects of that long imprisonment" (224). But that is almost all there is of the descriptions of her physical suffering. Her emphasis throughout is on the spiritual suffering she endures because of slavery and on the heartrending pain she and her family feel at being separated. In a narrative which otherwise employs sentimental tropes to gain the reader's sympathy, Jacobs rejects the sentimental emphasis on illness and weakness used so effectively in sentimental novels by and about white women.

Frances Smith Foster argues that in rejecting the representation of herself as ill and weak, Jacobs rejects the ideal of "True Womanhood" with its privileging of pale, invalid women and promotes instead what Frances Cogan calls the ideal of "Real Womanhood," which privileges health, strength, and women's domination of their sphere. This contrasts markedly with many abolitionist writers who used the image of the delicate slave woman to convince their audiences of her femininity.[12] The difference observable in Jacobs's text, however, seems less an opposing of True and Real Womanhood than a rewriting of the nineteenth-century definitions of what it is to be a woman. Jacobs does reject the model of "True Womanhood," but she rewrites the ideal of "Real Womanhood" because

it involves (beyond claims to health and strength) adherence to domestic ideals of the two-parent nuclear family and women's working exclusively in the home—options that were closed to Harriet Jacobs—and because it insists on a kind of robust leisure-time athleticism at odds with the lives of working-class women who suffered illnesses caused by overwork and exposure. While Harper has been both condemned and praised for her unrealistic adherence to standards of white, bourgeois behavior for her black heroine (see Christian and Tate, respectively), Jacobs seems to be constructing a different model of womanhood altogether—one that both acknowledges the body and its weaknesses but that ties that body directly to the mind and soul.

In this insistence on the unity of mind and body, of spiritual and physical suffering, Jacobs constructs a model of womanhood which both claims her humanity and does not disclaim her body. If Karen Sanchez-Eppler, in "Bodily Bonds," is correct in identifying as "the essential dilemma of both feminist and abolitionist projects," the contradiction that "the recognition of ownership of one's own body as essential to claiming personhood is matched by the fear of being imprisoned, silenced, deprived of personhood by that same body" (104), then we can see Jacobs as finding a resolution to the conflict. Sanchez-Eppler notes that the dilemma is usually dealt with through a denial of the body in favor of a focus on the spiritual crises of the woman or the slave, a decorporealization that denies the texts' own political message (112–13). In other words, if the body does not matter, and the spirit can be free under any circumstances, then why bother to free the body? But in her insistence on keeping the African-American woman's body in the foreground of her representation, Jacobs avoids the decorporealizing of much sentimental fiction that troubles Sanchez-Eppler.[13] Jacobs insists on a discourse in which both her soul *and her body* could be free, a discourse in which the two are not separate: bodily suffering *is* spiritual suffering.

In doing so, Jacobs's text upholds Carla Peterson's contention, in "'Doers of the Word,'" that African-American women in the antebellum North were involved in a project to redefine the ideals of womanhood as contained in the private sphere, a redefinition that would allow them a clear say in public civic debates (184). Peterson argues that the way black women entered the arena of public discourse was through "'achieving' an additional 'oppression,' ... a self-marginalization that ... paradoxically allowed empowerment" (191–2). These women became so "other," she asserts, that they were allowed to engage in political debate in a way that women who aspired to white bourgeois respectability were not. The "self-marginalization" that Peterson examines took the form of religious evangelicalism that manifested itself in the ability to become public speakers and to travel: "In the process, these women entered into various hybrid spaces that freed them from the fixed social and economic hierarchies determined by capitalism" (192). Because illness in Jacobs's text functions as the expression of a political complaint against the system of slavery, it becomes another of these liminal spaces, a space in between "True" and "Real" womanhood, in between racial and class lines, in between body and spirit.[14]

(In representing the connections between physical and spiritual sufferings Jacobs finds a way to recognize her corporeality and yet avoid either of the stereotypes of African-American women's bodies as lascivious or as impervious to pain.)Cynthia Davis argues that Harriet Wilson's *Our Nig* similarly challenges the "simplified, monological representation of black bodies as truly and innately lascivious." She argues that by testifying to the black woman's ability to feel pain, Wilson "intervenes in the racist attempt to

classify blacks as bestial" (392). Jacobs's insistence on Linda Brent's susceptibility to illness functions in the same way. But while torture and pain, as Davis explains, only metonymically displace sexuality into violence and pleasure into pain, illness works to deconstruct the separation of mind and body and the identification of African-American women as *only* body. Jacobs insists on both her physical existence as suffering body *and* her mental existence as thinker and writer.

In *Our Nig* the emphasis on illness is much more pronounced and yet still somehow oddly muted. From what we know of the novel, Wilson wrote it because she was ill and needed a source of income that did not depend on physical labor. She hoped that writing the book would provide her with enough income to be able to rejoin her young son and to supplement her small income from hat making.[15] Even though the autobiographical novel focuses on the events that broke her health, and on the difficulties she has faced as a result of debilitating illness, Wilson, like Jacobs, Harper, and Olivia Davidson Washington, remains reticent and vague about her illness. She also describes in detail how Frado (her heroine) continues to work despite this illness and frequently quotes friends' comments on Frado's strength.

In contrast to nineteenth-century physicians who argued that black people were ideally suited to difficult physical labor, Wilson insists that Frado's sickness is a direct result of overwork and exposure:

From early dawn until after all were retired, was she toiling, overworked, disheartened, longing for relief.

Exposure from heat to cold or the reverse, often destroyed her health for short intervals. She wore no shoes until after frost, and snow even, appeared; and bared her feet again before the last vestige of winter disappeared. These sudden changes she was so illy guarded against, nearly conquered her physical system. (65–66)

Wilson argues that Frado's constitution is like other women's, that she is subject to the same forces that make other women ill. To a twentieth-century reader this does not seem all that amazing, given that Frado *is* another woman like any other; for an antebellum African-American woman, however, asserting her *likeness* to white women is a declaration of equality and humanity that rings like Frederick Douglass's avowal of his intelligence and self-control.

Still, Wilson seems well aware that she faces the danger of the proslavery argument that, had Frado been enslaved, she would have been well. First, she emphasizes the similarities between Frado's years at the Bellmont house and slavery (the subtitle of the novel states that slavery also exists in the North); second, she is adamant about the fact that it was the toil at the Bellmont house and not her subsequent freedom that led to Frado's illness. Frado's illnesses usually follow a scene of torture (she is beaten, tied, left with blocks of wood in her mouth, and kicked several times in the narrative), and, when her health breaks after she leaves the Bellmont household, Wilson points out that the *community* blames the Bellmonts: "All felt that the place where her declining health began, should be the place of relief" (121). Wilson also guards against any accusations that Frado is trying to "play the lady" or feign illness to avoid work by recounting the ways that Frado continues to try to work despite her illness. When her illness first comes on she continues her housework even when she has to sit down to do it (64, 82), and later, when she is bedridden, Frado is reluctant to take charity and attempts to be self-supporting despite her invalidism (122).

Despite the fact that illness seems to be the motivating factor in Wilson's writing the narrative, and one of the central dramatic events in Frado's life, Wilson remains guarded about depicting a black woman's illness. Where a white woman's narrative might dwell on her character's illness and sickbed, these writers are very circumspect about mentioning the illness, even when, as in Wilson's case, it is the chief factor in her *need* to write. At the end of *Our Nig* Wilson does attempt to enlist the reader's sympathy on the grounds of her and Frado's illness; writing of herself in the third person, she entreats: "Still an invalid; she asks your sympathy, gentle reader. Refuse not" (130). Even so, Wilson doesn't dwell on details, except to tell us that for three years she has been on the public's support because of illness and would like to be self-sufficient, passing over years of suffering as will Jacobs later. In doing this Wilson, like Jacobs, tries to reconfigure the tropes of sentimental illness to suit the peculiar rhetorical problems of combining gender and race in nineteenth-century America.

Rhetoric and the African-American Invalid Woman

Obviously, illness appears in these African-American women's texts because, like everyone else, these women suffered from illnesses and ailments, and sickness is a natural part of their attempt to provide a truthful account of their lives. But the question of how they present it, and why they seem too reticent to represent it in the same style as nineteenth-century white authors, is a question about the rhetoric of the representation. One could argue that including it at all is a sign of the influence of white female authors. Nineteenth-century African-American male authors of autobiographical and fictional narratives usually do not mention illness, describing instead injuries and physical abuse.[16] The appearance, then, of illness in African-American women's narratives may be a matter of their participation in a female tradition as well as the tradition of slave narratives.

An unsympathetic reader might well see illness in these narratives as a sign of the writer "playing the lady," a rhetorical gesture like the making of the heroine's skin light, to elevate the heroine's sympathy-getting ability with a white reading public and to make her appearance more in line with white middle-class values. There may well be something to this explanation, but in all three texts the heroine more or less explicitly denies being a "lady." Iola Leroy is the clearest about her sense that she must work and avoid either the appearance or the reality of being pampered, in part because her background makes *being* a lady possible in a way that is closed off to the other two heroines. Both Wilson and Jacobs emphasize their heroines' strength and resilience. These are not women who faint away at every opportunity but women who finally break after truly unendurable hardships. When we recognize the dangers for African-American women of presenting themselves as too weak, it becomes clear that the rhetoric of "playing the lady" really does not fit the representation or the circumstances.

Further, in these three narratives the writers resist describing invalidism as an *ontological* state for their characters, describing it instead as a temporary one. This temporality of illness signals a real departure from the white sentimental tradition and from the dominant medical-cultural view of women in general. These narratives represent, in fact, a different theory of gender and a different theory of illness which resist the theory of dominant white medicine: illness in these narratives is not a matter of a state of being (female) but a result of circumstances, shock, exposure, and overwork. In a culture where hard work is believed to be beneficial to African-Americans' health and where it is "the dissipations of society" and the "constant excitement of fashionable life" which are believed to cause illness (Bedford 9), these black women rely on their life experience to refute such a claim and argue instead that overwork, fatigue, and the constant stress of enslavement are the causes of *their* illness.

There is yet another dynamic that these writers must confront when writing a narrative of an African-American woman's life. In writing such a narrative, the black woman has to prove not only that she is a woman but that she is human. Further, she has to show that blacks and women are unfairly characterized in the popular conceptions of them. Wilson's novel offers a good example of how this works. During the time that Frado is suffering from increasingly severe illness, she is also undergoing a religious awakening. The evil Mrs. Bellmont simultaneously questions both Frado's possession of a soul and the reality of her illness: Mrs. Bellmont "did not feel responsible for [Frado's] spiritual culture, and hardly believed she had a soul" (86).[17] When her son suggests that Frado should be allowed to go to church meetings, and argues that making her work so hard will ruin her health, Mrs. Bellmont replies: "you know these niggers are just like black snakes; you *can't* kill them. If she wasn't tough she would have been killed long ago. There was never one of my [white serving] girls could do half the work" (88). Mrs. Bellmont's reference to African-Americans' being "black snakes" would have a particular resonance for nineteenth-century readers; one of the arguments used by proslavery religious writers was that the serpent that tempted Eve had been black and was, in fact, an African (see Smith 43). Here Wilson not only directly confronts that racist contention but rejects the claim that her very strength and stamina make her inferior. This representation challenges the dominant views of women's weakness, refutes the sexism and classism of medical constrictions on women, and rebuts racist religiomedical tracts that likened blacks to animals. Like Jacobs, she replaces these ideas with a new model of womanhood in which physicality and spirituality are linked. The rhetoric of this rebuttal turns the self-effacement of the narrator—"I am just like everyone else"—into a direct self-promotion.[18] That is, it becomes an argument that the African-American woman *has* a "self."[19]

All three of the texts I've examined reflect an attitude toward illness that rejects the dominant nineteenth-century ideology in which illness is sentimental and beautiful; they represent it instead as *just* suffering. Given the cultural status of illness as both a class and status marker, it may be that this alternative symbolic system is not a matter of a healthier model for black women; it was just that their very invalid*ity* in the culture made any emphasis on their invalid*ism* dangerous. The way that these women assert their heroines' liability and resistance to illness suggests a rhetoric of *self*-assertion; in asserting their "invalidism/invalidity," African-American women assert just the opposite. In a culture in which white women claimed power through illness and black women were not considered "cultured" enough to suffer, they declare their very *humanity* through invalidism. Whereas the black man (as James Olney demonstrates) asserted his existence as human and not animal through the claim "I was born," the black woman may have done so through the claim "I was ill." These three texts, then, take part in what Deborah McDowell calls "the consistent preoccupation of black female novelists throughout their literary history," the "imaging of the black woman as a 'whole' character or 'self'" (283). Read against medical texts which denied her humanity, the African-American writer's assertion of her own suffering becomes a qualitatively different kind of assertion than either the black man's or the white woman's. Seen in this light, illness becomes, ironically, a way to imagine wholeness.

Notes

1. Even in histories of women of color, the history of nineteenth-century African-American illness is absent, including Cunningham and the recent excellent collection by Bair and Cayleff, *Wings of Gauze: Women of Color and the Experience of Health and Illness.*

2. See my *Invalid Women.* See also Ehrenreich and English, Carroll Smith-Rosenberg, Haller and Haller, Martha Verbrugge, Mary Poovey, and Elaine Showalter.

3. Drs. Josiah Nott and Samuel Cartwright not only argued that blacks were inferior but that they were naturally immune to certain diseases which devastated whites: "Slaveowners, they said, did not sacrifice blacks every time they sent them into the rice fields or canebreaks" (Savitt, 8).

4. My search included several university medical libraries and the Library of Congress. We know that the slave women who were the subjects of J. Marion Sims's gynecological experiments were originally his patients and that, while his experiments on them were chiefly a means to enhance his treatment of white women, he did perform surgery on other slave women after he "perfected" his treatment for vesicovaginal fistula. (We also know that he believed that slave women did not need anaesthesia, even though he used ether for his white patients.) For more information on Sims's experiments on slave women, see McGregor and Axelson.

5. I am focusing here on the health of slave women for two reasons. First, 90 percent of the antebellum African-American population in the United States was enslaved, and second, while there is little information about slave women, there is even less about free black women.

6. As a kind of confirmation of White's point, William L. Van Deburg's *Slavery and Race in American Popular Culture,* which is a study of "how American novelists, historians, dramatists, poets, filmmakers,

and songwriters have *perceived* and *interpreted* the Afro-American slave experience" not of the "reality" of it (xi), not only does not list illness, health, condition of slaves, medicine, sickness or disease in the index but also does not list "women," "slave women," Frances Harper, Harriet Wilson, Harriet Jacobs, or the titles of any of their books.

Using the male slave as the type for all slaves has a long history. Hoganson argues that "Since categorizing a race as feminine could be used to justify reduced rights, if not outright bondage, then Garrisonian [abolitionists] who demanded liberation highlighted the manliness of slaves, even when they discussed slaves in the abstract" (580).

7.   Savitt claims that "Whites found it impossible to separate the sick from the falsely ill; as a result they often indulged their breeding-age women rather than risk unknown complications" and argues that "white Virginians regarded pregnancy as almost holy. In addition to the fact that slave women received time off from work, avoided whippings, and were even able to feign pregnancy to gain their ends, expectant women were protected from execution in capital offenses until after parturition" (116).

There were, though, only three cases of postponed executions for pregnancy between the Revolution and the Civil War, and all three women were put to death after delivery, certainly some cause to suspect Virginians' reverence for the motherhood of slaves.

8.   Savitt asserts that "Female slaves probably lost more time from work for menstrual pain, discomfort, and disorders than for any other cause. Planters rarely named illness in their diaries or daybooks, but the frequency and regularity with which women of childbearing age appeared on sick lists indicates that menstrual conditions were a leading complaint" (115).

9.   For a discussion of the scientific representation of black women as purely sexual, see Gilman.

10.   I discuss both Stowe and Southworth's representations of the white invalid at length in Chapter 2 of *Invalid Women*.

11.   For example, one of Harper's heroes is named Tom and sacrifices himself to save the lives of others. Like Stowe, Harper draws on strong coincidence to reunite separated slave families. And also like Stowe, she names the mother of her sentimental female victim "Marie." What is interesting is how, in every case, Harper signifies on Stowe's text, altering the character and circumstances of the African-American people.

12.   Several critics have noted the use of these tropes in abolitionist writings; see Sanchez-Eppler (esp. 96) and Hoganson (esp. 574). For a discussion of these tropes and their relation to whiteness, see Christian (22, 27–30).

13.   Samuels also reads this decorporealizing of the slave in abolitionist writings.

14.   Berlant suggests that Jacobs', Harper's, and Anita Hill's testimony against sexual abuse functions similarly to break down the barriers between public and private; in testifying about abuses allowed by law, these women demonstrate that what looks like a private conflict can be an issue of public and national concern.

15.   For a discussion of *Our Nig* as a product which can be produced by an invalid and traded for humanity, see Holloway, 129.

16.   Olney discusses the usual form of the African-American slave narrative, characterizing the conventions of the slave narrative by listing the twelve most common narrative components; these accounts almost always contain accounts of abuse and poor treatment but rarely mention illness. In many ways Olney's description doesn't accord well with slave narratives and other autobiographies by women, which usually include accounts of specifically sexual violence and illness that are left out of male narratives.

17.   The relationship between Mrs. Bellmont and Frado is strongly reminiscent of that between Mrs. Reed and Jane Eyre; here, however, we see that the added factor of racial difference results in the torture of the little girl being even more marked and more cruel.

18.   In a forthcoming essay, "*Apologia Pro Vitae Nostra*: Sickness and Community in Jarena Lee's *Religious Experience and Journal*," Joycelyn Moody argues that Lee uses this rhetoric of self-effacement as an assertion of her election by God to be a minister. Moody suggests that this rhetoric characterizes many African-American women's spiritual narratives of the nineteenth century. See also Moody's fine discussion

of nineteenth-century black women's autobiography, "Twice Other, Once Shy," and Tate's assertion that the authorial modesty of *Our Nig* serves as a kind of minstrelsy that allows Wilson to fulfill nineteenth-century ideals of subservience and maternal obligation (116).

Elizabeth Young also reads *Iola Leroy* as a text about the difficulties of self-assertion and argues that, for Harper, the Civil War serves as a metaphor for self-conflict.

19.   Cynthia Davis argues that the representation of torture allows Wilson to "assume the position of authoritative speaking subject. . . . One could say, then, that the project of *Our Nig* is essentially a humanist one, designed to clear a space in which 'our nig' can assert her essential humanity" (400).

References

Axelson, Diana E. Women as Victims of Medical Experimentation: J. Marion Sims' Surgery on Slave Women, 1845–1850. *Sage* 2.2 (1985): 10–13. Rpt. *Black Women in American History,* ed. Darlene Clark Hine. Brooklyn: Carlson, 1990. 1–4: 51–9.

Bair, Barbara, and Susan E. Cayleff. *Wings of Gauze: Women of Color and the Experience of Health and Illness.* Detroit: Wayne State University Press, 1993.

Bedford, Gunning S. *Clinical Lectures on the Diseases of Women and Children.* New York: Samuel S. and William Wood, 1856.

Berlant, Lauren. The Queen of America Goes to Washington City: Harriet Jacobs, Frances Harper, Anita Hill. *American Literature* 65.3 (1993): 549–74.

Christian, Barbara. *Black Feminist Criticism: Perspectives on Black Women Writers.* New York: Pergamon Press, 1985.

Cogan, Frances B. *All-American Girl: The Ideal of Real Womanhood in Midnineteenth-Century America.* Athens: University of Georgia Press, 1989.

Columbat, Marc. *A Treatise on the Diseases and Special Hygiene of Females.* Philadelphia: Lea and Blanchard, 1850.

Cunningham, Constance A. The Sin of Omission: Black Women in Nineteenth Century American History. *Journal of Social and Behavioral Sciences* 33.1 (1987): 35–46. Rpt. *Black Women in American History,* ed. Darlene Clark Hine. Brooklyn: Carlson, 1990. 1–4: 275–86.

Davis, Cynthia J. Speaking the Body's Pain: Harriet Wilson's *Our Nig. African-American Review* 27.3 (1993): 391–403.

Dorsey, Carolyn A. Despite Poor Health: Olivia Davidson Washington's Story. Orig. pub. *Sage* 2 (Fall 1985): 69–72. Reprinted: *Black Women's History: Theory and Practice,* ed. Darlene Clark Hine. Brooklyn, N.Y.: Carlson, 1990, 351–57.

Ehrenreich, Barbara, and Deirdre English. *Complaints and Disorders: The Sexual Politics of Sickness.* New York: Feminist Press, 1973.

Foster, Frances Smith. Harriet Jacobs's *Incidents* and the 'Careless Daughters' (and Sons) Who Read It. In Warren, 92–107.

Gilman, Sander L. *Difference and Pathology: Stereotypes of Sexuality, Race, and Madness.* Ithaca: Cornell University Press, 1985.

Haller, John, and Robin Haller. *The Physician and Sexuality in Victorian America.* New York: Norton, 1974.

Harper, Frances E. W. *Iola Leroy, or, Shadows Uplifted* (1892). New York: Oxford University Press, 1988.

Hoganson, Kristin. Garrisonian Abolitionists and the Rhetoric of Gender, 1850–1860. *American Quarterly* 45.4 (1993): 558–95.

Holloway, Karla F. C. Economies of Space: Markets and Marketability in *Our Nig* and *Iola Leroy.* In Warren, 126–40.

Jacobs, Harriet. *Incidents in the Life of a Slave Girl* (1861). New York: Oxford University Press, 1988.

Karcher, Carolyn. Rape, Murder, and Revenge in 'Slavery's Pleasant Homes': Lydia Maria Child's Antislavery Fiction and the Limits of Genre. In Samuels, 58–72.

McDowell, Deborah. 'The Changing Same': Generational Connections and Black Women Novelists. *New Literary History* 18 (1987): 281–302.

McGregor, Deborah Kuhn. *Sexual Surgery and the Origins of Gynecology: J. Marion Sims, His Hospital, and His Patients.* New York: Garland, 1989.

Moody, Joycelyn. *Apologia Pro Vitae Nostra:* Sickness and Community in Jarena Lee's *Religious Experience and Journal.* Forthcoming.

Moody, Joycelyn. Twice Other, Once Shy: Nineteenth-Century Black Women Autobiographers and the American Literary Tradition of Self-effacement. *Auto/biography Studies: a/b.* 7.1 (1992): 46–61.

Morton, Patricia. *Disfigured Images: The Historical Assault on Afro-American Women.* New York: Greenwood, 1991.

Olney, James. 'I Was Born': Slave Narratives, Their Status as Autobiography and as Literature. In *The Slave's Narrative.* Charles T. Davis and Henry Louis Gates, Jr., eds. New York: Oxford University Press, 1985.

Peterson, Carla L. 'Doers of the Word': Theorizing African-American Women Writers in the Antebellum North. In Warren, 183–202.

Poovey, Mary. *Uneven Developments: The Ideological Work of Gender in Mid-Victorian England.* Chicago: University of Chicago Press, 1988.

Price Herndl, Diane. *Invalid Women: Figuring Feminine Illness in American Fiction and Culture, 1840–1940.* Chapel Hill: University of North Carolina Press, 1993.

Samuels, Shirley, ed. *The Culture of Sentiment: Race, Gender, and Sentimentality in Nineteenth-Century America.* New York: Oxford University Press, 1992.

Samuels, Shirley. The Identity of Slavery. In Samuels, 157–71.

Sanchez-Eppler, Karen. Bodily Bonds: The Intersecting Rhetorics of Feminism and Abolition. In Samuels, 92–114.

Savitt, Todd. *Medicine and Slavery: The Diseases and Health Care of Blacks in Antebellum Virginia.* Urbana: University of Illinois Press, 1978.

Showalter, Elaine. *The Female Malady: Women, Madness, and English Culture, 1830–1980.* New York: Pantheon, 1985.

Smith, John David. *An Old Creed for the New South: Proslavery Ideology and Historiography, 1865–1918.* Westport, Conn.: Greenwood, 1985.

Smith-Rosenberg, Carroll. *Disorderly Conduct: Visions of Gender in Victorian America.* New York: Oxford University Press, 1985.

Southworth, E. D. E. N. *Retribution.* Chicago: M. A. Donahue, 1849.

Sterling, Dorothy, ed. *We Are Your Sisters: Black Women in the Nineteenth Century.* New York: Norton, 1984.

Stowe, Harriet Beecher. *Uncle Tom's Cabin.* 1852. Rpt. New York: Penguin American Library, 1981.

Tate, Claudia. Allegories of Black Female Desire; or, Rereading Nineteenth-Century Sentimental Narratives of Black Female Authority. In *Changing Our Own Words: Essays on Criticism, Theory, and Writing by Black Women.* Cheryl A. Wall, ed. New Brunswick, N.J.: Rutgers University Press, 1989. 98–126.

Tompkins, Jane. *Sensational Designs: The Cultural Work of American Fiction.* New York: Oxford University Press, 1985.

Van Deburg, William L. *Slavery and Race in American Popular Culture.* Madison: University of Wisconsin Press, 1984.

Verbrugge, Martha. *Able-Bodied Womanhood: Personal Health and Social Change in Nineteenth-Century Boston.* New York: Oxford University Press, 1988.

Warner, John Harley. *The Therapeutic Perspective: Medical Practice, Knowledge, and Identity in America, 1820–1885.* Cambridge: Harvard University Press, 1988.

Warren, Joyce W., ed. *The (Other) American Traditions: Nineteenth-Century Women Writers.* New Brunswick, N.J.: Rutgers University Press, 1993.

Washington, Booker T. *Up from Slavery* (1901). New York: Penguin, 1986.

Welter, Barbara. The Cult of True Womanhood, 1820–1860. *American Quarterly* 18 (1966): 151–74.

White, Deborah Gray. *Ar'n't I a Woman?: Female Slaves in the Plantation South.* New York: Norton, 1985.

Wilson, Harriet E. *Our Nig; or, Sketches from the Life of a Free Black, In a Two-Story White House, North. Showing That Slavery's Shadows Fall Even There* (1859). New York: Random House, 1983.

Young, Elizabeth. Warring Fictions: *Iola Leroy* and the Color of Gender. *American Literature* 64.2 (1992): 273–97.

PART III
Women and Health in the Late Nineteenth and Twentieth Centuries: Issues

Body Image and Physical Fitness

In the nineteenth century, as in other historical periods, women's lives and health were affected by many factors that were cultural, biological, and medical. But to many members of the American medical profession in the nineteenth century, the reproductive life cycle, beginning with puberty and the onset of the menstrual cycle and ending with menopause and the years "beyond the reach of sexual storms," defined the productive boundaries of women's lives. The centrality of reproductive capability in medical thinking was based on more than simple biological theory. The articles in this section examine the broad social context in which the theory of reproductive control thrived, some of the ways in which women responded to it, and how they tried to shape the medical and cultural forces concerning their bodies and physical abilities.

Joan Jacobs Brumberg examines how women reacted to their monthly menstrual periods and how these periodic episodes shaped women's activities and self-perceptions. Historical investigation has revealed women's strong opinions about Dr. Edward H. Clarke's 1873 theory that women could not simultaneously experience puberty and gain an education. But historians know far less about how women coped with their periods each month and how the lack of disposable sanitary napkins may have limited women's activities. How, for example, did Elizabeth Blackwell manage her menstrual periods when she attended an all-male medical school in 1848 and 1849, where her presence was required for long hours each day and where it must have been difficult to arrange for privacy and impossible to consult with another woman? We know, of course, that she found a solution, but the personal costs involved in her pioneer triumph remain speculation. Similarly, how did women workers on farms or in factories contend with their long hours away from home? Not until the post–World War I years, when nurses returning from the battlefields showed industry and women that the surplus bandages had a domestic use, did women have an easy way to collect and dispose of the monthly discharges. Brumberg provides insight into menstrual management, the cultural forces that influenced it, and how the experiences affected women's lives.

Margaret Lowe's article furthers our understanding of how cultural imperatives affected women's lives and health through a case study of dieting among Smith College students in the 1920s. Social attitudes about body image shaped young women's ideas and actions, but, as Lowe demonstrates, young women also used dieting as a tool to shape their appearance and define their identity. Together the articles in this section provide a telling portrait of some pressures young women felt as they faced their adult lives and help us understand how their responses affected their health.

8 "Something Happens to Girls": Menarche and the Emergence of the Modern American Hygienic Imperative

Joan Jacobs Brumberg

On September 3, 1959, Ruth Teischman, a twelve-year-old girl living in suburban Long Island, New York, bought herself a small imitation leather diary, which she embraced immediately as her special friend. In extremely neat handwriting, the kind that girls use self-consciously at the start of important literary projects, she wrote expectantly: "I am going to tell you loads of personal things. . . . I am going to write in you all my feelings and emotions." Two weeks later she noted: "Today I got my period for the third time. I thought I had it at lunchtime but I wasn't sure. I went to the bathroom and there it was. So, I put on a napkin and told my mother."[1]

Although the diary of Ruth Teischman was primarily about her friends and family, it also provided her with a way of acknowledging and keeping track of a newly acquired bodily function, menstruation. The first menstruation or menarche is an inevitable developmental milestone that has social as well as personal meaning. Because it is the bellwether of female fertility, menarche confirms the reproductive potential of every young woman and marks her as essentially different from males. Although menarche is only one of a number of physical changes experienced by adolescent girls, it is generally given greater cultural significance than budding breasts or the appearance of body hair. "Something happens to girls," wrote a twentieth-century physician, and it usually initiates a new phase of life.[2]

Ruth Teischman began her diary only a few months after her periods began. The nature of her entries suggests that she was probably prepared for menstruation in a combination of ways that were commonplace by the 1950s—conversations with her mother and with her peers, as well as reading materials and corporate-sponsored films provided for her at home and in school. Because she had been "coached" on the logistics of what is called "sanitary protection," she knew exactly what to do when she saw her menstrual blood. For Ruth, menarche clearly involved a certain drama: it occurred suddenly, it involved a display of

Joan Jacobs Brumberg is Stephen H. Weiss Presidential Fellow and Professor of Human Development, Women's Studies and History at Cornell University, Ithaca, New York.

Reprinted from *Journal of the History of Sexuality* 4 (1993): 99–127, by permission of The University of Chicago. Copyright © 1993 by The University of Chicago.

blood, and it required some kind of immediate hygienic response in order to reduce her fear of disclosure, or soiling her clothes. In this respect Ruth was absolutely typical; clinical studies demonstrate that in the United States both pre- and postmenarcheal girls regard menarche as a hygienic crisis rather than as a maturational event.[3]

In the days afterward Ruth continued a normal round of school and extracurricular activities, phone conversations with friends, disagreements with her mother, and earnest ruminations about death, boys, and blackheads—all the while carefully noting the density of her flow and the number of sanitary napkins that she used. Although her diary is short-lived (it covers only four months), it documents Ruth's active social life along with her careful attention to her changing body and her desire for "regularity," a condition that she regarded as a sure sign of maturity. On November 9 she wrote: "Soon I'll be getting my period again. I'm glad. I hope I either get it on the 15, 16 or 17 of this month because then I'll be sure I'm regular." Menstruation was obviously new to Ruth, and it was a source of both anxiety and power: "Today I got my period for the fourth time. It is very light and black. But I think it is light because it's just begun and will get heavier later. I hope so. Not that I like it. I don't like it at all. I just want to have it a long time and heavy. So I can have something good to talk about."

For Ruth Teischman menarche implied new hygienic concerns and some degree of attention from her peers, but it did not imply adult sexuality or adult status. Ruth's attention was clearly riveted on what her body was doing rather than on the larger issue of what it meant to become a woman. Although something very important was happening to her, which she eagerly shared with her girlfriends, the larger society provided no overt recognition of Ruth's entrance into a new stage of life. In the 1930s anthropologist

Margaret Mead took this lack of public ac knowledgment to mean that menarche in America involved "no rituals," although she noted that it generated a great deal of "apprehensiveness" on the part of parents and "tension" on the part of sexually maturing girls.[4]

What Mead failed to recognize was the manner in which menarche could still be an important rite of passage even though it involved no flamboyant rituals of initiation or exclusion. Although our contemporary response to menarche seems rational and altogether commonplace, the American girl experiences her first display of menstrual blood in a way that is shaped by her own body but also by family, friends, economic forces, and values at work in the larger society. So even though menarche appears to be one of those inevitable, natural, and developmental progressions that is "hardwired" into the female organism, its meaning, and even its pattern, is derived from the particular culture that surrounds the body of the sexually maturing girl.[5] Because menarche is a biological event that men do not experience, it has always loomed large in rhetorical strategies and ideologies about female difference.[6]

Ruth Teischman's diary, with its careful notation of the quantity of blood and the number of napkins, confirms the centrality of hygienic concerns in the mind of the modern adolescent. It also raises a number of other questions about the historic evolution of our contemporary "menstrual script." At what age do American girls generally first menstruate? Have they always been prepared for menstruation beforehand and what have they been told? Were mothers or other adult women consistently involved? If not, how do sexually maturing girls in different classes learn about becoming a woman? And why does "personal hygiene" loom so large in contemporary thinking about adult womanhood?

Anthropological studies provide a model of

ness in terms of describing
of menstrual education in
in most ethnographic studies
turing girls learn about cultural con-
of their bodies from older women; in these
small traditional societies young women acquire
the skills necessary to manage "women's mysteries"—that is, menstruation and childbirth—
within intergenerational single-sex groups in
which mothers and other adult women play a
primary role.[7] Sitting at the feet of older women,
girls absorb practical information, cultural values, and a larger sense of their female identity.
But my research suggests that in the United
States this model for female education about the
body was eclipsed during the midnineteenth
century precisely because menarche and puberty, like so many other aspects of the life
course, were transformed by industrialization,
urbanization, and secularization, all critical dimensions of modern mass society.

In moving what has been an essentially anthropological concern into the realm of American social and cultural history, I hope to dispel
the notion that our treatment of menarche is
culturally neutral because it is simply a matter
of "hygiene."[8] Although we do not isolate, segre-
gate, or formally restrict our girls at menarche,
we do have a set of socially shared beliefs and
practices about menstruation that are sufficiently consistent to resemble a "menstrual taboo," and these have consequences for girls at
menarche.[9] Part of this taboo is the idea that
women should not talk about their monthly
bleeding (although women and girls clearly do)
and that, when menarche occurs, it should remain a private concern without public acknowledgment. But the trouble with a great deal of
thinking on this subject is that it portrays our
menstrual taboo—and menarche itself—as
fixed and universal when, in fact, both are
changeable, subject to reformulation, and highly
specific to time and place. There are societies

where no such taboo exists and some where
there is no requirement at all for what we call
"sanitary protection." Among the Kayapo of the
Amazon, for example, napkins are not used, and
the word for menstruation means literally
"stripe down the leg."[10]

Cultural variability exists within the United
States as well, as this essay reveals. Although the
nature of historical sources tends to focus our
attention on middle-class girls and their mothers, my narrative also suggests how cultural
knowledge about menarche and menstruation is
structured by class and ethnicity as well as cohort or generation. When newly arrived ethnic
immigrant women were presented with "modern scientific" information about menstruation
and personal hygiene, it was not always accepted
or utilized, despite the claims of Progressive Era
reformers that sex education programs inevitably improved the quality of life and mitigated
class differences. Their American-born daughters, however, were generally more responsive to
the middle-class sanitary ideal and accepted the
notion that there was an "American way" to
menstruate.

In addition to cultural variation, biological
variability complicates the history of menarche
and raises questions about our wholesale adoption of the social construction model—that is,
how a biological event or biomedical phenomenon becomes subject to cultural values and constraints. In the case of menarche there is a
changing material body to consider along with
its figurative and ideological representations.
For example, in 1780 the average age at menarche in the United States and western Europe
was probably about 17; by 1877 the average age
had declined to almost 15; by 1901 it was 13.9;
by 1948 it was 12.9. Today menarche generally
occurs at about twelve and a half years old, and
this developmental material reality makes it a
different experience from the time when it occurred typically in the late teens, closer to the

M

management of menarche ↓ changing

time of marriage.[11] As I began to reconstruct the history of this experience, I tried to consider how this changing female body interacted with my "social construction" story. Although my explication of this relationship is in no way complete, I suggest in this essay something that most deconstructionist scholarship has not considered sufficiently: that is, how a changing biological pattern could provoke a new kind of discourse about menarche and the bodies of adolescent girls.[12]

The Inadequate Mother

In the traditional communities of eighteenth-century America sexually maturing girls learned about menstruation within a single-sex environment, or "women's culture." Mothers and other adult women were the primary and most regular source of information about the biological realities of womanhood. Adult women also served as accessible role models for young girls in a society where there was little discontinuity between the experience of mothers and of their daughters. As far as we can tell, learning about being a woman was a kind of "integrated, core curriculum" that happened organically as part of day-to-day life. In the absence of books and pamphlets on the subject of female biology, young women in the eighteenth century learned by word of mouth from older women about their menstrual "flowers."[13]

There is a parallel between what happened to women in childbirth and what has happened to girls at menarche. According to Judith Walzer Leavitt, as the nineteenth century progressed, the experience of childbirth moved from the world of women—what has been called "social childbirth"—to the world of the hospital and the terrain of scientific medicine.[14] Women stopped learning about their parturient bodies from other women and instead began to rely on medical professionals who were, by and large,

males. The management of menarche seems to have followed the same general scenario: as the century progressed, and as age-segregated schools and activities increased, girls were less and less likely to garner knowledge about their menstruating bodies from naturally occurring events and discussions within an age-heterogeneous community of women. The decline of traditional communities and the rise of a highly individualistic style of domesticity in the nineteenth century relegated girls to the instruction of their mothers alone.

By the midnineteenth century, evidence began to appear that American mothers were not adequately preparing their daughters for menarche. In 1852 a physician who pioneered in the field of obstetrics and gynecology, Edward John Tilt, found that of one thousand girls 25 percent were totally unprepared for their first menstruation. Many, he said, were frightened and thought that they were wounded.[15] Tilt's report is important because it signals a new concern about preparedness that was echoed by many individuals throughout the nineteenth and early twentieth centuries. This new concern probably was based on changing expectations as well as actual experience; for at least one hundred years after Tilt, all kinds of observers—physicians, educators, and women themselves—lamented the fact that so many girls did not know what was happening to them when they started to menstruate because mothers failed to provide adequate information.[16]

By the 1870s, at the moment when young women began to enter residential collegiate institutions in significant numbers, mothers were blamed for what appeared to be a pervasive lack of preparation for womanhood. In his well-known 1873 diatribe against coeducation Harvard University's Dr. Edward Clarke suggested that American girls knew very little about their essential "rhythmic periodicity." But even supporters of educational equity, such as Cornell

University biologist Burt Wilder, marveled at how parents could send their daughters away to a university without telling them about "the generative function," including menstruation. Wilder cried out for more parental instruction in his popular 1875 advice book, *What Young People Should Know,* claiming that many sixteen- and seventeen-year-old college girls "thought their first menses was a hemorrhage."[17]

In *For Girls: A Special Physiology* (1884) Mrs. E. R. Shepherd confirmed that female advisers and analysts identified the same problem. "I have met numbers of women and some of them young who knew nothing of their coming 'courses' until they were upon them. One in particular I remember who said: 'It has taken me nearly a lifetime to forgive my mother for sending me away to boarding school without telling me about it.'" Yet the same was true for girls living at home; evidence drawn from Helen Kennedy's 1895 study of Boston area high school girls reveals that 60 percent were ignorant at the time of menarche. And among the college women surveyed by Dr. Clelia Mosher at Stanford University in the years between 1892 and 1920, over 25 percent gave responses such as "Not one thing, no knowledge" or "Not the least in the world" when asked their degree of preparation for their first period. (Another 50 percent had some "slight" knowledge of menarche when it happened to them, apparently meaning that they knew the names of some of the reproductive organs and had been alerted to the prospect of bleeding.) Even a daughter whose mother was a doctor told Dr. Mosher: "[I received] vague ideas from fellow pupils at school. My mother was a physician but refused to instruct me when I asked questions. I remember well the first time I asked a question which showed that I already had the idea there was something shameful about child bearing. Yet she told me I would read books about it when I was older and I never asked again." Another woman reported that,

while her mother did tell her about "the facts" of menstruation, she simultaneously "taught her that such things were not talked about [and] also not thought of."[18] By all accounts the mother-daughter dialogue was a terribly painful process characterized by great awkwardness and maternal reserve.

While this may seem like an abdication of maternal responsibility and a clear indication of Victorian sexual repression, the situation was really more complicated because it involved the interaction of cultural ideology with biological development and changing social life. The reluctance to talk appears to have been a pervasive maternal strategy related to the middle-class mother's desire to preserve her daughter's innocence for as long as possible. Most middle-class Victorian mothers believed that menarche initiated their daughter's sexuality; thus they attempted to keep it at bay for as long as they could. As Constance Nathanson rightly observed, "the supression of sexuality" was defined in the nineteenth century "as necessary to the healthy development of a young woman's reproductive capabilities."[19]

But the preservation of innocence became axiomatic because of a new "biosocial" development in the female life course: middle-class girls were menstruating earlier but marrying later. By the end of the century the average age at menarche had dropped to thirteen or fourteen but the age at marriage was approximately twenty-two or twenty-three.[20] For most mothers the most salient effect of earlier menstruation was that a daughter's adolescence was prolonged and that, in turn, increased the prospect of social dangers. Because knowledge about menstruation was considered the first step on the slippery slope to loss of innocence, many Victorian mothers simply avoided the subject altogether, believing that it was in the best interest of their young daughters. In this way, an earlier age at menarche may have contributed to the ongoing disaggregation of information about sex and the body that was characteristic of the decline of traditional women's culture. Where "innocence" was revered, as it was in middle-class Victorian America, menstruation would be taught only when necessary and the "sexual connection" left until "later."

Talking about menstruation actually may have become harder rather than easier in the late nineteenth century as a consequence of the parallel processes of secularization and medicalization. Among liberally educated middle-class mothers, menarche and menstruation were less and less likely to be regarded simply as events ordained by God. Although the menstrual process was still characterized by most people as one of the great wonders of nature, there was an increasing realization that menstruation was something more than the punishment of Eve and that there was some actual physiology involved. Most mothers, however, did not understand physiology, and they often did not know what to call the participating organs. But menstruation was not clearly understood even by the experts. In 1907 G. Stanley Hall, the architect of modern adolescence, stated frankly: "Precisely what menstruation is, is not very well known," and his work reveals that he was still uncertain as to whether or not the monthly period was analogous to estrus in animals.[21] A modern understanding of ovarian function would have to wait until the 1930s.[22] Despite a lack of definitive information from the experts, women in the middle class felt the need to turn to physicians for explanations of normal female life experiences such as menarche.[23]

The idea of the "inadequate mother" was a powerful one, because it served as a critical justification for medical and other expert interventions in what had once been a strictly female domain. Fueled by the idea that mothers did not tell their daughters about menstruation and did not know how, nineteenth-century physicians began to take an increased role in defining and treating the experience of menarche (and menstruation), and women of the middle class were more than willing to accept their "expert" help. The medicalization of menarche was the result of a joint effort between physicians and middle-class mothers, albeit with different motives. Both groups, however, were confronted with a new biological reality that may help to explain their alliance: the changing shape of the female life course. Because of a series of demographic and social changes that were neither linear nor uniformly distributed (lower age at menarche, later marriage, fewer children, improved nutrition, and the decline of infectious diseases), nineteenth-century women were experiencing more ovulatory cycles in a lifetime than ever before in human history. This robust improvement in women's health is an important context for understanding medical intervention in traditionally female domains, as well as the nineteenth-century physician's preoccupation with the reproductive capabilities of women.[24]

The Hygiene of Puberty

Among middle- and upper-class girls who came of age between 1890 and World War I, reading about menstruation became the common pathway from girlhood to womanhood. The American mother's sense of inadequacy about how to explain menstruation, combined with a general fear of "vulgar" information, contributed to a reliance on health and hygiene guides for domestic use. Good books, monitored by sensitive middle-class mothers, were the best way to teach adolescent girls about changes in their bodies. In the Mosher survey more than half of the respondents achieved their sexual education through books, including those read surreptitiously in parental libraries, health and hygiene manuals handed to girls by their mothers, and, later in life, college textbooks. "I learned everything I knew from good sources and in a pure and sacred way," said one of Dr. Mosher's respondents. (John Cowan's *Science of a New Life* and George Naphey's *Physical Life of Woman* repeatedly were cited as influential, along with Wilder and Shepherd and a category of books known as "tokologies," which were intended as guides to childbirth and maternity for married women.)[25]

Even though mothers and their advisers understood the utility of published explanations of menstruation, they consistently worried about how to keep the scope of the discussion from moving on to graphic discussions of sexual intercourse, particularly among younger adolescents: "[We] would like to put books treating of these topics in the hands of [our] girls; there are plenty of medical works which present them well enough, but there are so many other themes introduced that [we] hesitate to give them the book at all."[26] Still, books were almost always regarded as a better option than learning from peers. Informal social learning about female biology was soundly condemned by those who were advising the American middle-class mother. This approach implied a coarser, rougher way of life associated with the working class and the poor. Learning about menstruation in the middle class increasingly was cast as a systematic, scientific, and essentially private undertaking that required the acquisition of an anatomical vocabulary as well as appropriate moral understandings.

For these reasons middle-class mothers and daughters turned to popular health books that codified the "hygiene of puberty" and provided a narrative explanation of the cyclical process of "ripened ovules." This was done in a mix of scientific and romantic language; although the uterus was referred to as the "mother-room" by some writers, most tried to teach the correct anatomical vocabulary. Popular hygiene guides also reported that there was an average or "normative" age at menarche, and they elabo-

rated a full range of what could go wrong: amenorrhea and "suppression"; dysmenorrhea; leucorrhea; and imperforate hymen. Implicit in this catalogue of pubertal pathologies was the suggestion that menarche required medical as well as maternal management.[27]

In this new division of labor the doctor was the biomedical strategist, but the mother was the chief operative. Her role was to monitor those habits and behaviors that would impact on menstrual functioning and subsequent reproductive capacity (that is, bathing, dress, and exercise). Thus mothers became the source of a persistent litany of things to be avoided during menstruation: excessive exercise, hot and cold baths, wet feet. These directives encouraged mothers to fixate on things they could control— matters of habit and behavior, as well as school attendance—that were outside the day-to-day supervision of physicians.

Hygiene writers seemed to assume, and they probably were correct, that if mothers spoke about nothing else, they at least would tell their daughters of ways to "fix themselves" in order to prevent displays of menstrual blood and soiled clothes. Most women made something absorbent out of pieces of cotton or chambray found in the ever-present "rag bag," or they purchased gauze and cheesecloth for assembling their own pads. (Mass-produced napkins were available by the 1890s, but they do not appear to have been used widely, probably because of their cost.) Although advice writers generally did not give explicit descriptions of middle-class menstrual practices, one text confidently claimed that "napkins" were "almost universally worn" and set out specifications for the "best" napkins: they were made of linen (as opposed to cotton), at least one-half yard square, folded, secured to the clothing at the front and back, and "worn between the limbs." Soiled napkins were to be left soaking for a few hours before washing.[28]

The model of a linen "napkin" obviously was

meant to suggest genteel middle-class dining rather than defecation, but the recommendation that menstrual rags be left to soak would soon be regarded as a dangerous, if not offensive, practice. Because of a set of increasingly powerful ideas about the sources of disease, long-standing feminine hygiene practices were reexamined in the late nineteenth century. Beginning in the 1860s the middle class became conversant with Joseph Lister's concept of "antisepsis" and the idea that there was something alive and dangerous in human waste, air, and water; in the 1880s public health officials, motivated by the new "germ theory," began to advocate "antiseptic cleanliness" of the home and the person. These new ideas about the sources of disease and contamination stimulated a revolution in the hygiene of women's bodies, as well as their homes.[29]

As a result, in the middle and upper classes feminine washing and menstrual discharge were subjected to more intense hygienic standards disseminated by physicians, popular books, and by word of mouth among women.[30] "Every part of the body [should be] as clean as the face," wrote Dr. Joseph H. Greer in his well-known guide, *The Wholesome Woman* (1902). At menstruation time this meant that "the napkins should be changed at least every morning upon dressing and at night upon retiring."[31] Greer's recommendation that napkins be changed at least twice a day was typical, and it was based on the notion that "absorption" of the menstrual blood simply was not "wholesome" and that soiled napkins generated unpleasant odors. (Odors, of course, were a sure sign of "noxious effluvia" and breeding bacteria.) Greer believed that menstrual odor resulted from "uncleanliness" of the external genitals and not from women's normal "glandular secretions," which, he said, were in fact "as pure as the synovial fluids which oil the joints." But he firmly recommended a vaginal douche of tepid water in con-

junction with the daily bath, a clear indication of his belief that the vagina, like the middle-class home, required antisepsis.

Menstrual blood, which had long been taboo, was now suspect on additional scientific grounds as a potential contaminant. Because menstrual rags facilitated a dangerous mixing of germs and gases in a warm place, the soaking of cloth rags was rejected. Consequently, by the end of the nineteenth century, the long-standing reliance on cloth rags was in question, particularly among women of privilege—those who were in the closest contact with physicians who espoused the new "sanitary science." Middle- and upper-class American women and their daughters began to cope with menstrual blood in a new way, putting aside their ancient cloths for disposable, purchasable materials such as gauze and surgical cotton.[32]

For middle-class women with this new hygienic sensibility, commercially produced napkins became a personal necessity. The 1895 catalogue from Montgomery Ward advertised the "Faultless Serviette or Absorbent Health Napkin," which was "antiseptic . . . required no washing . . . [and was] burned after using." The 1897 Sears Roebuck catalogue featured two different menstrual aids: a "Ladies Elastic Doily Belt" (made of silk trimmings and elastic) and "Antiseptic and absorbent pads" (made from cotton, gauze, and bandage). Advertisers claimed that doctors endorsed these products because of their antiseptic and absorbent properties and suggested that commercially made napkins could revolutionize the menstrual experience by eliminating heat and chafe.

As the hygiene practices of privileged women were transformed, so was the instruction of their daughters. In 1913 the American Medical Association presented its formula for the correct mix of science, maternal nurturance, and sanitary hygiene that should occur at menarche. Their publication, *Daughter, Mother, and Father:*

A Story for Girls, stands as a model of middle-class sensibilities and the ideal of preparedness immediately before World War I.[33] (I stress the word *ideal* precisely because we have so many indications that middle-class mothers did not talk with their daughters in the ways that were prescribed.)

In an episode entitled "Life Problems," the Dawson family responded to the ongoing physical maturation of their thirteen-year-old daughter, Margaret. In the idealized middle-class household enlightened parents anticipated their daughter's need for menstrual preparation and spoke to one another about it. In this case the father, quite conveniently, was a physician who had educated Margaret as a young girl about "female anatomy and the function of the ovaries"; he was also familiar with the problem of preparedness since he was "frequently asked" to explain menstruation to "other people's daughters." While the physician-father provided the basic scientific background (and probably orchestrated the discourse), it was Mrs. Dawson who appropriately initiated the subject "which every girl should know, preferably from her own mother's lips."

In the cosy serenity of her sewing room on a beautiful April day Mrs. Dawson used the example of the lilacs outside her window to begin talking with Margaret about cycles of blooming and fading foliage, cycles that suggested the natural "periodicity" of women. Because she was concerned about her daughter's reaction to this heady information, the pace of the conversation was slow and deliberate; there was no maternal unease. Mrs. Dawson warned Margaret that, as a rule, the sign that a person had "crossed the threshold from girlhood to womanhood" was "the sight of a stain on her garment"—particularly, "the garment that is next to her body, perhaps her nightdress." But she pointed out, in a reassuring manner, that staining was perfectly natural and that it happens to "all women in all

lands." When Margaret asked with some foreboding about the amount of blood she could expect to lose, her mother assured her that it was only "a few spoonfuls" and, though periods may seem "difficult" at first, "they are your Creator's preparation of you for future motherhood." Margaret responded enthusiastically to this linkage between menstruation and motherhood. To her mother's pleasure she proclaimed happily: "Oh, that's different. It isn't really a sickness at all then." In the remainder of the model dialogue Mrs. Dawson underscored for Margaret the importance of cultivating a "normal attitude" toward such "sacred things" and suggested ways to manage possible discomfort.

Three weeks later, when Margaret noticed a "mark" on her bed linen, she immediately sought out her understanding and prescient mother, who put an arm around her daughter's waist and then showed her how the lower drawer of the dresser in her bedroom had been prepared with a canvas package containing two or three dozen handmade napkins. Margaret, who claimed that she had not opened her dresser drawer for a number of weeks, was surprised and grateful to have such a responsible and well-organized mother who anticipated her basic needs at this stressful juncture. In the privacy of Margaret's bedroom Mrs. Dawson predictably took the opportunity to instruct Margaret in the latest methods of sanitary protection. "Here is the whole outfit," she explained. "Some girls wear little folded napkins made from old linen or cotton, but such napkins have to be washed. As a rule, the girl washes them herself. But by using thoroughly laundered cheese-cloth and absorbent cotton, materials which a girl may get in any department store at very reasonable price, she is able to make for herself these little 'sanitary napkins' as we call them. They need not be washed; after they are soiled they are rolled up in a paper and thrown into the furnace." And in an appeal to her daughter's generational identification she noted that "sanitary napkins" were the wave of the future: "They are very much less trouble and more satisfactory than the others and are now generally used, especially by school and college girls and by the young women in the business world."

Although commercially made napkins would not be mass-marketed and adopted until after World War I, the turn to "sanitary napkins" in the years before the war reflects a heightened sensibility to feminine hygiene among middle-class mothers and daughters. It was certainly apparent to these women that the new "antiseptic" napkins promised a number of real benefits: less work, improved comfort, greater mobility, and a germ-free environment. The benefits of the new hygiene also provided middle-class mothers with a subject that furnished an organizing focus for their private conversations about what it meant to grow up female. In those intimate moments, as mothers taught daughters how to "fix themselves," middle-class mothers surely indicated that cleanliness was as much a social concern as a matter of health. "Sanitary protection" had become an article of middle-class faith because, in addition to everything else, it was an important means of class differentiation.

A Different Sensibility

Among the immigrant working class, sexually maturing girls were not necessarily any better prepared for menarche. Evidence drawn from oral history interviews with Slavic, Italian, and Jewish American women in western Pennsylvania suggest that, in the years before 1920, only 10 percent of immigrant mothers provided any preparation for menarche. When asked what they knew about menstruation before it occurred, almost 40 percent of the Pittsburgh sample replied, "Nothing."[34]

In 1918, when she menstruated for the first time at age fourteen, a Yugoslavian immigrant asked her mother: "Why I bleed?" "She give me a big rag," the daughter observed nearly sixty years later, "and then she said, 'Put that you know . . . in the bottom.'" A Jewish girl who got her first period at age eleven also did not know what was happening and remembered her mother scavenging for cloth: she "went and tore up an old underwear suit or something from . . . father." In an Italian immigrant household a mother of three adolescent daughters regularly used fifty-pound flour sacks, which she bleached and cut into napkins for her girls, well into the late 1930s.[35]

Before World War I working-class women simply did not have the money or the inclination to adopt commercially made, easily disposable "sanitary napkins." As a result, mothers and their adolescent daughters were bound together by a common need, napkin washing. The rhythm of menstruation meant that households went through this domestic ritual on a fairly regular basis. In the working class, where mothers and daughters lived in extremely close proximity to one another, the biological realities of women's lives were more obvious because there were common wash basins, no servants, and a central location for hanging wash to dry. Despite the sanitarians' injunctions against it, napkins were left to soak so that stains would not set and the fabric could be reused. "Every once in a while when the washing would come up, you would see them soaking. So after awhile you would ask what that was all about."[36]

Small stained rags left to soak were a regular and provocative visual source of information for young daughters, but bloodstained clothing or bedding also served as an important nonverbal form of communication between reticent daughters and their mothers. Watchful mothers monitored their family wash to see if a daughter's "changes" had come, or if her cycles were regular. Many simply waited until they saw

159

proof of menarche before they ventured any explanation or offered advice on what to do. One fourteen-year-old in Pittsburgh, who knew nothing of menstruation and thought that she was "hurt," treated her genitals with mercurochrome and cotton, until her stained laundry divulged her secret. Unlike the daughters of the middle class, however, her menstrual blood stimulated no special coddling or lengthy scientific explanation. Instead, she was provided with "a piece of material" to wrap around her and sent directly to school.[37] Other girls who had picked up information from their friends understood the meaning of the spots on their clothes but said nothing to their mothers; they waited for their news to come out in the wash.

Unlike mothers in the middle class, working-class mothers expected their girls to learn from their friends, older sisters, fellow workers, or even resident grandmothers or aunts. They relied on their daughters to pick up the information they needed at school or on the street in precisely the fashion that repelled the middle class and motivated their investment in health and hygiene books written by experts. A Finnish mother cuddled her newly menstruating daughter on her lap and said: "Poor dear. I thought the girls at school would tell you. I just didn't think it would happen so young."[38]

Learning about womanhood probably was less privatized and more social in the working class: poor girls experienced less age segregation (in adult-sponsored groups) and had more of an opportunity to mix with women of different ages at work, within the kitchen, or in a shared bedroom. Girls who developed early initiated others into "women's mysteries." Irene, a Jewish immigrant in Pittsburgh, menstruated for the first time in 1912 at age thirteen. Although her Rumanian-born mother had told her nothing about "women's things," she learned from a girlfriend at the Irene Kaufmann Settlement House: "[My friend] was mature; she was like a woman

... and she would tell me things that would just go over my head." When her periods came, Irene was able to take it "as a matter of course" because of her conversations with the friend who developed ahead of her.[39] "We didn't learn about [menstruation] at home," explained the daughter of an Austro-Hungarian coal miner. "We just had to learn it from each other ... because at that time, way back, they didn't discuss things like that. At school they didn't either."[40] For the naive daughter of Finnish immigrants in Minnesota the lavatory in her public school was the critical learning environment. There she heard the "whispers of older girls" that were "just loud enough so we younger girls could hear them complaining—and bragging—about 'the monthlies' and 'Grandma coming to visit.' Some of them didn't smell very good at those times. Once in a while one had an embarrassing stain on the back of her skirt. But they acted as if they were somehow superior to the rest of us, and ... I envied them their mysterious passage to adulthood."[41]

In ethnic families the immigrant mother was a reluctant participant in a dialogue that seemed to occur naturally among peers and only occasionally between mother and daughter. "We ashamed people. We don't tell mother nothing, mother no tell us," explained Mika, a Yugoslavian immigrant daughter, as she recalled her first menstruation at age fourteen. When her mother finally began to talk, she offered a common religious explanation: "God ... made a woman that way. That's why we, every month, going to be sick, till you go into trouble."[42]

There was silence between the generations even among Jews whose orthodox religious rules, set forth in the book of Niddah, required special menstrual practices—such as mikvah baths—to be observed by all menstruating women.[43] However, some Jewish mothers still acted in at least one ritualistic way that was not totally understood by their assimilating daughters: on the occasion of their first period some Jewish girls still received a maternal slap on the face, which signified the difficulties of life as a woman: "Mother did come in and slap me as the old Jewish custom. . . . I don't even know if it's a Jewish custom, but it's a custom. I was angry. I didn't want to have it to begin with and I was terribly insulted that mother hit me. But she said, 'I really didn't mean to do that, but that's the custom and you're supposed to.'"[44] Despite the emphasis in Jewish tradition on ritual cleanliness at menstruation, the oral history interviews reveal a striking lack of communication between generations of Jewish mothers and daughters.

Among other groups—notably the Italians—maternal explanations of menstruation and reproduction were particularly rare, since the culture put an extremely high value on female chastity and innocence. Mothers who were close to their old-world tradition clearly thought that they should not tell their daughters the facts of life. Italian immigrant mothers therefore were largely unresponsive to the call for an extensive program of "sex education" by middle-class educators and physicians after World War I. Born in Petilia, Italy, in 1892, Rose came to Pittsburgh at age fourteen and then married at age sixteen in 1908. Her first menstruation, however, occurred one week after her marriage. "I just got blood, that was it," she remembered many years later. Rose had been completely unprepared for both sexual intercourse and menstruation, but her mother felt totally justified. Later in life, when Rose complained, her mother explained her silence: "They say [we] weren't supposed to tell you."[45]

Italian immigrants also displayed resistance to middle-class efforts to sanitize the menstrual experience. Although they usually were silent beforehand, when their daughters reached menarche Italian mothers verbalized a set of ideas about menstrual blood that had consequence

for their daughters' personal hygiene. Because of their belief in the importance of the free flow of menstrual blood, Italian mothers worried about anything that diverted or interfered with that flow. (For example, they believed that an emotional shock at the time of menstruation could send blood to the lungs, causing tuberculosis, or to the brain, causing insanity.) A heavy steady flow was a symptom of health and fertility; the best indication of that flow was a soaked napkin rather than an unstained one. To the chagrin of physicians and health educators Italian mothers did not then encourage their daughters to change their menstrual rags often. They believed that a stained napkin was a good sign and that it stimulated a healthy flow (as opposed to a clean antiseptic one that might act as a suppressant).[46]

Ideas such as these, as well as smaller family budgets, help to explain the class differential in the long revolution from menstrual cloths to commercially produced napkins. After World War I young working women and college girls became the first mass market for the disposable "sanitary napkins" sold as Kotex and made from cellucotton by Kimberly-Clark.[47] But among Italian-Americans, as late as the 1930s and 1940s, daughters had to find a justification for their change to more modern forms of sanitary protection. In 1938, in a family where the mother still made "diapers" from flour sacks, the burden of producing enough for four daughters (plus herself) finally drove the family to enter the market for commercially made napkins. In 1945, when she menstruated for the first time, Lillian Petrillo was totally unprepared: "[My mother] didn't tell me anything before and she didn't tell me anything after." Her mother did show her, however, how to fold a cloth and stick it into her underwear between her legs. After three cycles it was clear that this method of protection gave Lillian a nasty rash. Only then did her mother reluctantly allow the purchase of napkins: "She figured that was one expense she was going to have."[48] For these women learning to menstruate the "American way" required participation in the larger consumer society, a critical step in the assimilation process.

A Pittsburgh physician's abrupt advice to an Italian girl—"Be clean. Don't do anything. Wash yourself."—was an abbreviated form of education about menstruation (and sexuality) that was in distinct contrast to the labored and carefully constructed mother-daughter discourse that was the ideal of the middle class. Although genital cleanliness and sexual purity were ideals for all girls regardless of their position in life, the sexually maturing daughter of the middle class was treated as a "hothouse" flower while her working-class sister was not. In fact, among the immigrant working class the daughter rather than the mother took increasing responsibility for her own—as well as her mother's—education on matters of feminine hygiene. A 1911 school-based physiology and sex-hygiene program at Technical High School in Cleveland, Ohio, provides a case in point. After the students learned "the physiology and hygiene of the pelvic organs" from a papier-mâché female figure, each girl was asked to take home "nine typewritten sheets containing the essential facts . . . and [they were] requested to have their mother read these pages with them."[49] Clearly, there were ways to get around the problem of inadequate maternal guidance, and these became increasingly apparent after World War I, as menstruation itself became less privatized and more sanitized.

Menarche After the War: A Summary

After 1918, as a result of wartime experience that heightened sensitivity to the problem of venereal disease, the nation embarked on a crusade to promote moral health in which all aspects of sexuality were sanitized, including menarche and menstruation. After the war young women of all classes were more and more likely to learn about menstruation in school and at an increasingly younger age.[50] Although there were critics who worried about "decency" and feared that teachers were trying to usurp the maternal role, the experience of the war, combined with the reports of urban social workers, confirmed that lack of preparation had social costs. The model of maternal initiation remained a nostalgic ideal, but physicians and educators recognized a new reality. "No one can quite take the place of the mother in instructing her daughter in the simple and beautiful truths of the reproductive life and its various manifestations," wrote Emil Novak in 1921. "However [when] such home instruction is out of the question . . . there is a legitimate field for the activity of various agencies now interested in 'sex education' of young people."[51]

Although there would be a debate among educators over where menstruation should be taught—in biology, physiology, physical education, or home economics classes—the teachers, physicians, and social workers who made up the social hygiene movement generally agreed that, as far as sex information was concerned, "better a year too early, than an hour too late."[52] Consequently, menstrual preparedness was linked to the imperatives of the social hygiene movement, and it became a regular part of Progressive Era curricula in the schools and in the programs of groups for girls, such as the Girl Scouts.[53] In these settings young women were formally instructed about female physiology and the hygiene of puberty. What they were taught reflected middle-class sensibilities: menstruating women and girls must be concerned about genital cleanliness; they should do only moderate exercise; and they could not afford to rely on makeshift solutions to the hygienic problem posed by menstrual blood. By the early 1920s more and more young women followed a predictable menstrual script that kept the focus on hygiene needs at that "special" time. These educational discussions of menarche and menstruation were, by and large, "desexualized," in that they did not initiate the natural sequelae: conversations about intercourse and female sexuality.

The emergence of mass-produced sanitary products also had a potent effect on the experience of menarche. After World War I the sanitary products industry saw the vast commercial implications of the "inadequate mother" and adapted it to their own purposes. In the early 1920s their advertising campaigns in women's periodicals relied on what Roland Marchand calls a "vacuum of advice" argument to sell their wares.[54] In the early ads, which constituted the first real public acknowledgment of menstruation, the industry targeted mothers and their difficulties with preparing their daughters. They hired female experts, such as Ellen J. Buckland, "a graduate nurse," to provide advice, answer letters, and send free samples in "a plain, unmarked wrapper." The ads usually featured an illustration depicting either the idealized intergenerational conversation or a brooding young woman; the texts provided mothers with the appropriate words and rationale for introducing the subject of menstruation. A generation of mothers who had themselves suffered the indignities of "old fashioned makeshifts" that were "unhygienic and dangerous to health" were urged to tell their daughters: "This new way is Kotex, widely urged by doctors and nurses. . . . Kotex is used by eight women in ten in the better walks of life."[55]

In the 1930s and 1940s the industry moved even further into the business of preparation for menarche, but by this time it was filling a longstanding void rather than supplanting an important cultural tradition or rite of passage.[56] Newly established educational divisions within the sanitary products industry began to supply mothers, teachers, parent-teacher associations, and the Girl Scouts with free, ready-made programs of instruction on "menstrual health." Oral history interviews with three generations of immigrants in Pittsburgh are definitive on the pervasiveness of this material; beginning in the 1930s, but especially in the 1940s, almost all the daughters of the Slovak, Italian, and Jewish families in Pittsburgh were given corporate-sponsored pamphlets—such as "Marjorie May's Twelfth Birthday" (1932)—either at school or by their mothers. And in the 1940s the industry developed, in conjunction with Walt Disney, the first corporate-sponsored educational film on the subject, *The Story of Menstruation* (1946), an animated cartoon that has been seen by approximately 93 million American women. Numerous pamphlets and films followed, often suggesting that mothers did not know how to talk with their daughters, and when they proceeded without instruction they could "scare them to death."[57]

Advertising, educational booklets, and films combined to generate a common vocabulary for thinking about menarche, talking about it to girlfriends, and writing about it in personal diaries. Entries such as Ruth Teischman's only began to appear in post–World War II America, as a new cohort of sexually maturing young women—in early rather than late adolescence—began to keep personal diaries and internalize the behavioral imperatives disseminated in pamphlets, books, and films. In 1950, after she went to see a "menstruation movie" with her mother at a Cleveland junior high school, Sandra Rubin wrote, "I'm afraid Mom doesn't understand me and laughs when I ask for a sanitary belt. I wish she wouldn't do it because it makes me embarrassed." When Lois Pollack got her first period in Dorchester, Massachusetts, in January 1956, she regarded it as "exciting" and immediately did what she had been instructed to do: "Today I got my period. I was over Pam's house when I got it. I called up my mother but at first she didn't believe me. When I came home my mother was surprised and I don't want to forget it either."[58]

Through its aggressive education (cum marketing) programs the sanitary products industry untied many tongues and stimulated a great deal of talk within the culture of girls about who had her period and who did not. This kind of talk was not "invented," however, in the postwar world, although it did become more frequent and more audible.[59] Because American girls were being taught about menstruation in latency, in anticipation of a more precocious arrival, some longed for this new sign of maturity for a number of years. At thirteen Sarah Vaughn was jubilant: "I got my period today! I'm so happy! The pads are very uncomfortable. It is weird to be bleeding. I'm still not used to it . . . I got it at approximately 5:45. . . . I told Kerry I had it. . . . She said oh my gosh!"[60]

In the 1950s and 1960s the industry initiated a deliberate campaign to further develop the youthful adolescent market, a strategy that successfully played on adolescent awkwardness and the embarrassing spectre of stained clothes. The identification of this niche market was a brilliant strategy because it capitalized on a continuous and increasingly affluent group of consumers: the expansive baby-boom generation. The industry's elaboration of even more exacting standards of feminine hygiene translated, for this group, into more and varied purchases. By the late 1940s and 1950s napkins had to be changed as often as six times a day; today women born in those decades, many of whom are still menstruating, routinely select from at least three generic types of protection (napkins, tampons, and panty liners), each with different absorbent capacities. And some use tampons and napkins simultaneously and panty liners almost continuously, throughout their cycle.[61]

Although the industry's educational efforts undoubtedly were part of the important demystification of menstruation, the long-term consequences for girls at puberty may not be so benign. In fact, surrendering menarche to Walt Disney probably contributed in some measure to the difficulties we face today in the realm of female adolescent sexuality. As the industry became an ever-present third party in mother-daughter, doctor-patient, and teacher-student discussions, personal experience and testimony from older women became even less authoritative. There was more information available, but it was increasingly abstracted from real-life experience; in the Disney case menstruation was reduced to a cartoon shown primarily in school or in girls' groups. The oral histories also suggest that the availability of free corporate-sponsored materials meant that many mothers and teachers simply gave out pamphlets and free samples rather than provide personal advice and counsel about growing up female. We know from personal reports that this was not a satisfying substitute and that young women wanted (and continue to want) meaningful exchanges about female sexuality and womanhood in addition to the best techniques for keeping their clothes and their genitals clean. Unfortunately, however, American girls in the twentieth century have grown up equating the experience of menarche and menstruation with a hygiene product. To wit: a woman who spent her childhood in Pittsburgh in the 1940s recalled that before any of her friends had gotten their periods, one of her fellow fifth-graders suddenly declared one day that she could not slide down a snowy hill. When asked why, this prepubertal girl said laughingly, "I can't, I'm practicing Kotex."[62]

This childish remark captures the extent to which the sanitary products industry was implicated in the long-term transformation of menarche from a maturational event into a hygienic crisis. But it was also a perceptive piece of cultural logic on the part of a young girl coming of age in a society where identity, particularly adult female identity, was inextricably linked to purchases in the marketplace.[63] In harnessing adolescent angst over menstruation to capitalist imperatives, the sanitary products industry paved the way for the commercialization of other areas of the body that were of great concern to developing girls—namely, their skin, their hair, and their "figures." By the late 1940s and 1950s menstruation was owned by neither mothers nor doctors; in fact, the rites of passage for American girls were clearly in the commercial realm, where they centered on consumer activities such as the purchase of sanitary products, high heels, lipsticks, and "training bras." Never before, with adult approval, had so many girls become followers and interpreters of "the fashion system."[64] As a result, mothers and their teenage daughters became absorbed in a discussion (or tug-of-war) about "good grooming," a dialogue that ultimately contributed little to the adolescent girl's understanding of her body and her sexuality. This state of affairs, combined with the fact that young women were menstruating earlier than ever before, left girls unprepared and unprotected for the new sexual liberalism that characterized American life in the 1960s and 1970s.

Notes

My work on this subject was supported by a National Endowment for the Humanities Fellowship for University Teachers (1990–91) as well as a residential fellowship at the American College of Obstetricians and Gynecologists. In writing this essay I have benefited particularly from the insightful commentary of Susan Bell and Nancy Tomes. In an earlier version, prepared for the 1990 Organization of American Historians meetings in Louisville, Ky. Allan Brandt, G. David Brumberg, Faye Dudden, Judith Walzer Leavitt, and Barbara Sicherman supplied helpful advice. Susan Matt assisted with research in Olin Library at Cornell University and Maria Cochran at the Countway Library, Harvard Medical School.

1. The quotations here are from Ruth Teischman (pseud.), manuscript diary, 83-M31, Schlesinger Library, Radcliffe College. Although menarche is the critical event that inevitably starts the "biological clock" of every woman, we have very few personal accounts that describe the experience as it was lived in the nineteenth or even twentieth centuries. "On-the-spot," authentic reporting such as Teischman's is extremely hard to find. Readers of *The Diary of a Young Girl* (New York, 1953) may recall that one of the pleasures of reading that diary as an adolescent was the author's description of her emotional reaction to menarche and menstruation. Anne Frank wrote at age fourteen: "Each time I have a period—and that has only been three times—I have the feeling that in spite of all the pain, unpleasantness, and nastiness, I have a sweet secret, and that is why, although it is nothing but a nuisance to me in a way, I always long for the time that I shall feel that secret within me again" (p. 143). More common are retrospective memories in autobiographies: see, for example, Emma Goldman, *Living My Life* (New York, 1931); Kate Simon, *Bronx Primitive* (New York, 1982), 139–41, 176; and Simone de Beauvoir, *Memoirs of a Dutiful Daughter* (New York, 1959), 101. Other autobiographical accounts of menarche can be found in Mavis Hiltunen Biesanz, ed., *Mavis Helmi: A Finnish American Girlhood* (St. Cloud, Minn., 1989); Dorothy Sterling, ed., *We Are Your Sisters: Black Women in the Nineteenth Century* (New York, 1984); and Gussie Kimball, *Gitele* (New York, 1960). Stories of first menstruation are also a popular part of contemporary American feminist writing; see Lisa Alther, *Kinflicks* (New York, 1976); Jamaica Kincaid, *Annie John* (New York, 1985); and Toni Morrison, *The Bluest Eye* (New York, 1970). There are also menarcheal memoirs in Janice Delaney, Mary Jane Lupton, and Emily Toth, eds., *The Curse: A Cultural History of Menstruation* (Urbana, Ill., 1976), especially ch. 17, "The Absent Literature: The Menarche."

2. Mary Chadwick, *The Psychological Effects of Menstruation* (New York, 1932), 31. Chadwick was among a number of psychoanalytic theorists, such as Helena Deutsch, Melanie Klein, and Karen Horney,

who regarded menarche as "the first pollution." These analysts are outside the scope of this essay.

3. Contemporary studies demonstrate that accidents and staining are a major concern of girls at menarche. See Lynn Whisant and Leonard S. Zegans, "A Study of Attitudes Toward Menarche in White Middle-Class American Adolescent Girls," *American Journal of Psychiatry* 132 (August 1975): 809–14; and Lynn Whisant, Leonard S. Zegans, and Elizabeth Brett, "Implicit Messages Concerning Menstruation in Commercial Educational Materials Prepared for Young Adolescent Girls," *Journal of American Psychiatry,* 815–20; and Jeanne Brooks-Gunn and Diane Ruble, "Menarche: The Interaction of Physiological, Cultural, and Social Factors," in *The Menstrual Cycle: A Synthesis of Interdisciplinary Research,* ed. Alice Dan (Urbana, Ill., 1980).

4. See Margaret Mead, "Adolescence in Primitive and Modern Society," in *The New Generation,* ed. Victor Francis Calverton and Samuel Schmalhausen (New York, 1930), 169–88. By "rituals" Mead meant something formal, explicit, and derived from religion, not the broad secular usage common today. Others regard the biological process as "a puberty rite cast upon women by nature itself." See, for example, Therese F. Benedek, "Sexual Functions in Women and Their Disturbance," in *American Handbook of Psychiatry* (New York, 1959), 1: 727–48.

5. In *Simians, Cyborgs, and Women: The Reinvention of Nature* (New York, 1991) Donna Haraway points out that nature is not really fixed or given. Menarche provides an excellent example because it depends on culture for its meaning, its pattern (timing), and even its existence. For example, when a girl becomes anorexic she does not menstruate; anorexia nervosa is both a cultural and a biomedical condition. See Joan Jacobs Brumberg, *Fasting Girls: The Emergence of Anorexia Nervosa as a Modern Disease* (Cambridge, Mass., 1988).

6. In the nineteenth century, menarche was the critical site for establishing female difference. In Victorian medicine menarche was the critical event in the life of a woman, and the way in which the girl traversed menarche was believed to determine her future life, health, and happiness—as well as the future of "the race." For some it provided a rationale for a separate educational program. For examples of this

kind of "ovarian determinism" see Edward Clarke, *Sex in Education; or, A Fair Chance for the Girls* (Boston, 1873); George Engelmann, "The American Girl of Today," *Transactions of the American Gynecological Society* 25 (1900): 8–44. Clarke and Engelmann were well-known physicians, but many other lesser-known practitioners followed their lead and made the management of schoolgirls at menarche a feature of their clinical practice. There was, of course, a female response to "ovarian determinism" by Clelia Mosher, Helen Kennedy, Mary Putnam Jacobi, and others. See, for example, Mary Putnam Jacobi, *The Question of Rest for Women During Menstruation* (New York, 1877); and Clelia Duel Mosher, "Respiration in Women" (M.A. thesis, Stanford University, 1894), "Normal Menstruation and Some Factors Modifying It," *Johns Hopkins Hospital Bulletin of Medicine* 12, 121 (1901): 178–89, and *Health and the Woman Movement* (New York, 1916).

7. For some examples of the anthropological model see Carol P. MacCormack, "Biological Events and Cultural Control," *Signs* 3 (1977): 93–100; Karen Paige and Jeffrey Paige, *The Politics of Reproductive Ritual* (Berkeley, Calif., 1981); Bruce Lincoln, *Emerging from the Chrysalis: Studies in Rituals of Women's Initiation* (Cambridge, Mass., 1981); N. N. Bhattacharyya, *Indian Puberty Rites* (Calcutta, 1968); Audrey Richards, *Chisungu: A Girls' Initiation Ceremony in Northern Rhodesia* (London, 1956); and Shirley Begay, *Kinaalda: A Navajo Puberty Ceremony* (Rough Rock, Ariz., 1983). In developmental psychology the emphasis is on the quality of the mother-daughter bond at menarche; see the discussion of this literature in Whisant and Zegans, "A Study of Attitudes," 809–10.

8. *Hygiene* is a nineteenth-century term. According to George Vigarello, in *Concepts of Cleanliness: Changing Attitudes in France Since the Middle Ages* (Cambridge, 1983), the term was not used in the eighteenth century. Hygiene means a collection of practices and knowledge that are used to preserve health.

9. See the excellent summary of anthropological literature in Thomas Buckley and Alma Gottlieb, eds., *Blood Magic: The Anthropology of Menstruation* (Berkeley, Calif., 1988).

10. Personal communication with Kathryn March and Jane Fajans about the fieldwork of Terry

Turner. According to a World Health Organization study done by Robert Snowden and Barbara Christian (*Patterns and Perceptions of Menstruation* [New York, 1983], 16), some women never use sanitary protection irrespective of the amount of blood loss, while others use protection even for the slightest manifestation of blood loss, or for ordinary vaginal discharges that involve no blood at all. An early, popular feminist analysis, which tended to flatten cultural and subcultural variation, is Delaney, Lupton, and Toth, eds., *The Curse* (see also the revised edition [Urbana, Ill., 1988] with new afterwords by Lupton and Delaney).

11. These data are from Henry P. Bowditch, "The Growth of Children," *Massachusetts State Board of Health, Eighth Annual Report* (January 1877), 284; Engelmann, 8–44; and Leona Zacharias, William Rand, and Richard Wurtman, "A Prospective Study of Sexual Development and Growth: The Statistics of Menarche," *Obstetrical and Gynecological Survey* 31 (1976): 336. On the "secular trend" see J. M. Tanner, *Growth at Adolescence* (Oxford, 1962); Grace Wyshak and Rose E. Frisch, "Evidence for a Secular Trend in Age at Menarche," *New England Journal of Medicine* 36 (1982): 1033–35; P. E. Brown, "The Age at Menarche," *British Journal of Preventive and Social Medicine* 29 (1966): 9–14; R. V. Short, "The Evolution of Human Reproduction," *Proceedings of the Royal Society,* series B, 195 (1976): 3–24; Leona Zacharias, Richard Wurtman, and Martin Schatzoff, "Sexual Maturation in Contemporary Girls," *American Journal of Obstetrics and Gynecology* 108 (1970): 833–46; and Marion E. Maresh, "A Forty-five Year Investigation for Secular Changes in Physical Maturation," *American Journal of Physical Anthropology* 36 (1972): 103–9. For a criticism of the idea of the declining age at menarche by a historian, see Vern Bullough, "Age at Menarche: A Misunderstanding," *Science* 213 (1981): 365–66. Despite Bullough's objections, there seems to be general agreement that improved nutrition in infancy and childhood and the decline of infectious diseases has produced larger, healthier young women who menstruate earlier. The prevailing theory is that in order to start menstruation a young woman must have a certain level of stored, easily metabolizable energy in the form of body fat. See Rose E. Frisch and R. Revelle, "Height and Weight at Menarche and a Hypothe-sis of Critical Body Weights and Adolescent Events," *Science* 169 (1970): 397–99; Rose E. Frisch, "A Method of Prediction of Age of Menarche from Height and Weight at Ages Nine through Thirteen Years," *Pediatrics* 53 (1974): 384–90. In psychology the current debate is over the influence of family stressors in early maturation and menarche. For a provocative argument see Jay Belsky, Laurence Steinberg, and Patricia Draper, "Childhood Experience, Interpersonal Development, and Reproductive Strategy: An Evolutionary Theory of Socialization," *Child Development* 62 (1991): 647–70. There are important critical commentaries by other psychologists in the same issue.

12. See, for example, Susan Suleiman, *The Female Body in Western Culture* (Cambridge, Mass., 1986); and Mary Jacobus, Evelyn Fox Keller, and Sally Shuttleworth, eds., *Body/Politics: Women and the Discourses of Science* (New York, 1990), which was an admirable and early collection of work on the female body and sexuality. However, it did not attend to the role of biology itself in the stimulation or disruption of discourses about the body.

13. "Flowers," from the French *fleurs,* was used to denote menstrual discharge until the nineteenth century. See *Oxford English Dictionary,* 2d ed., s.v. "flower."

14. Judith Walzer Leavitt, *Brought to Bed: Childbearing in America, 1750–1950* (New York, 1986). For a portrait of childbearing and women's culture in the colonial period see Laurel Ulrich, *A Midwife's Tale: The Life of Martha Ballard* (New York, 1990).

15. Edward Tilt, *On the Preservation of the Health of Women at Critical Periods of Life* (London, 1851), quoted in G. Stanley Hall, *Adolescence* (New York, 1907), 481. Tilt does not note anything about the race and ethnicity of these young women.

16. There is a developmental issue involved here that complicates the issue of preparedness: sometimes children are provided with information that they cannot absorb or that they forget because it has no relation to their experience at that point. Despite this caveat, there does appear to be a cultural pattern (or perception) of "not talking" about menstruation that emerges in the mid to late nineteenth century. With remarkably few exceptions the studies point in a sin-

gle direction—that is, the number of unprepared girls actually increased throughout the nineteenth century before it began to decrease in the twentieth. Although there are obvious problems with using a disparate group of studies, the overall direction is highly suggestive. I have extracted quantitative information about menarcheal preparation from all of the following: Helen P. Kennedy, "Effects of High School Work upon Girls During Adolescence," *Pedagogical Seminary* 3 (June 1896): 469–82; Clelia Duel Mosher, *The Mosher Survey: Sexual Attitudes of Forty-five Victorian Women,* ed. James Mahood and Christian Wenburg (New York, 1980); A. Louise Brush, "Attitudes, Emotional and Physical Symptoms Commonly Associated with Menstruation in One Hundred Women," *American Journal of Orthopsychiatry* 8 (1938): 286–301; Carney Landis, et al., *Sex in Development* (New York, 1940); Natalie Shainess, "A Reevaluation of Some Aspects of Femininity Through a Study of Menstruation: A Preliminary Report," *Comprehensive Psychiatry* 2 (1961): 20–26; W. G. Shipman, "Age at Menarche and Adult Personality," *Archives of General Psychiatry* 10 (1964): 155–59; and Frances Y. Dunham, "Timing and Sources of Information About, and Attitudes Toward, Menstruation Among College Females," *Journal of Genetic Psychology* 117 (1970): 205–17. Brooks-Gunn and Ruble, "Menarche," state that only 5 to 10 percent of young women in the 1970s had no advance preparation. Generational differences are also reported in Research and Forecasts, Inc., *Summary of Survey Results: Tampax, Inc.* (New York, 1981): 37 percent of women over age thirty-five remembered no advance preparation, as compared to 19 percent of women under age thirty-five. An interesting study of menstrual memories and adult experience is David P. Pillemer, Elissa Koff, Elizabeth Rhinehart, and Jill Rierdan, "Flashbulb Memories of Menarche and Adult Menstrual Distress," *Journal of Adolescence* 10 (1987): 187–99. There is general agreement in the literature on memory that women have accurate memories of their first menstruation.

17. Burt Wilder, *What Young People Should Know: The Reproductive Function in Man and the Lower Animals* (Boston, 1875), 168.

18. Mrs. E. R. Shepherd, *For Girls: A Special Physiology* (New York, 1884), 9–10; Kennedy, "Effects," 472–73; Mosher survey, nos. 41 and 35.

19. Constance Nathanson, *Dangerous Passage: The Social Control of Sexuality in Women's Adolescence* (Philadelphia, 1991), 82.

20. See Maris Vinovskis and Susan Juster, "Adolescence in Nineteenth-Century America," in Richard Lerner, Anne C. Petersen, and Jeanne Brooks-Gunn, eds., *Encyclopedia of Adolescence,* 2 vols. (New York, 1991), 2: 698–707.

21. Hall, *Adolescence,* 480.

22. See Patricia Crawford, "Attitudes to Menstruation in Seventeenth-Century England," *Past and Present* 91 (May 1981): 47–73; Vern Bullough and Martha Voght, "Women, Menstruation, and Nineteenth-Century Medicine," *Bulletin of the History of Medicine* 47 (1973): 66–82; Elaine Showalter and English Showalter, "Victorian Women and Menstruation," *Victorian Studies* 14 (1970): 83–91; Harvey Graham, *Eternal Eve* (London, 1950); R. O. Valdiserri, "Menstruation and Medical Theory: An Historical Overview," *Journal of the American Medical Women's Association* 38 (1983): 66–70; and Victor Cornelius Medvei, *A History of Endocrinology* (Hingham, Mass., 1982). So far as I know, there is no history of the dissemination of scientific information about menstrual function.

23. For the classic statement on medicalization see Irving K. Zola, "Medicine as an Institution of Social Control," *Sociological Review* 20 (1972): 487–504. The nineteenth-century case follows the model of medicalization provided by C. K. Reissman, "Women and Medicalization: A New Perspective," *Social Policy* 14 (1983): 3–18. Reissman argues that women's experiences are more easily medicalized for a variety of biological, social, and psychological reasons but that women have not been passive victims of this process. Medicalization has been the outcome of a joint effort between physicians and the dominant class of women, although undertaken for different motives. The expansion of medical control over menarche and menstruation is similar to medicine's expansion elsewhere and should be seen as part of the push for cultural authority described in Paul Starr, *The Social Transformation of American Medicine* (New York, 1982).

24. For an important overview of nineteenth-century medicine's view of the female life course, see Carroll Smith-Rosenberg, "From Puberty to Menopause: The Cycle of Femininity in Nineteenth-

Century America," *Feminist Studies* 1 (1973): 58–72. I think that Edward Clarke, George Engelmann, and others (see n. 6) all need to be considered against the backdrop of biological change that I have suggested; this is not, however, to excuse their misogyny but rather to understand the biomedical world in which they operated.

25. Mosher survey, no. 24; "tokology" was sometimes used instead of an exact title ("I read a Tokology") or, as one of Mosher's respondents noted, "I read Wilder's Tokology." Since there was no book of this name, she probably meant Wilder's *What Young People Should Know* (1875). *Tokology* was obviously used as a generic term for a book about the body and sex; the term is also used by Dr. Naphey and other physicians (see George Naphey, *The Physical Life of Women* [Philadelphia, 1890], 57).

26. Shepherd, *For Girls,* 8.

27. In this respect menarche follows the classification model described by Michel Foucault in *The History of Sexuality: An Introduction,* vol. 1 of *The History of Sexuality,* trans. Robert Hurley (New York, 1978).

28. Shepherd, *For Girls,* 129–30. Other rare descriptions are in Mary Wood-Allen, *What a Young Woman Ought to Know* (Philadelphia, 1905), 149; and Emma Frances Angell Drake, *What a Young Wife Ought to Know* (Philadelphia, 1908), 194–95. Wood-Allen describes how to make a reusable cloth envelope for holding napkins that was supported by shoulder straps; Drake describes a homemade disposable pad made of purchased cotton and cheesecloth. In the oral histories that follow, "flannelette" is the fabric most often mentioned. My Cornell University colleague Kathryn March explained to me that the protective materials used to absorb menstrual blood generally reflect the ecology of a region or group. The North American Indians, for example, used pads of shredded cedar bark or buckskin; the Toda of southern India used pads of moss wrapped in cloth.

29. The idea of "invisible cleanliness" and the notion that washing obliterated microbes actually began with Pasteur in the nineteenth century. See Viagrello, *Concepts of Cleanliness,* 202–9; and also Richard L. Bushman and Claudia Bushman, "The Early History of Cleanliness in America," *Journal of American History* 74 (1988): 1213–38. On the subject of domestic hygiene see Nancy Tomes, "The Private Side of

Public Health: Sanitary Science, Domestic Hygiene, and the Germ Theory, 1870–1900," *Bulletin of the History of Medicine* 64 (1990): 509–39, and "The Wages of Dirt Were Death: Women and Domestic Hygiene, 1870–1930" (Paper delivered at the annual meeting of the Organization of American Historians, Louisville, Ky., 1990). For the broad history of public health in the nineteenth and twentieth centuries, see Howard D. Kramer, "The Germ Theory and the Early Public Health Program in the United States," *Bulletin of the History of Medicine* 22 (1948): 233–47; Oswei Temkin, *The Double Face of Janus* (Baltimore, Md., 1977), 456–71; Charles E. Rosenberg, "Florence Nightingale on Contagion: The Hospital as a Moral Universe," in *Healing and History: Essays for George Rosen,* ed. Charles Rosenberg (New York, 1979), 116–36; Judith Walzer Leavitt, *The Healthiest City: Milwaukee and the Politics of Health Reform* (Princeton, N.J., 1982; rpt. Madison, Wis., 1996); John Duffy, *The Sanitarians: A History of American Public Health* (Urbana, Ill., 1990).

30. According to Vigarello, *Concepts of Cleanliness,* 107–11, the insistence on feminine cleanliness became explicit for the first time in the 1770s as part of a larger turn to intimate washing of body parts. The story of the elaboration of menstrual hygiene practices fits the general model suggested by Norbert Elias in *The Civilizing Process: The History of Manners,* trans. Edmund Jephcott (Oxford, 1978).

31. Joseph H. Greer, *The Wholesome Woman* (Chicago, 1902), 172.

32. On technological innovation in this area see Vern L. Bullough, "Technology and Female Sexuality and Physiology: Some Implications," *Journal of Sex Research* 16 (1980): 59–71; Fred E. H. Schroeder, "Feminine Hygiene, Fashion, and the Emancipation of American Women," *American Studies* 17 (Fall 1976): 101–11; and Delaney, Lupton, and Toth, eds., *The Curse.* Menstrual pads may have been one of the very first disposable products in our "throw-away" culture.

33. Winfield Scott Hall, *Daughter, Mother, and Father: A Story for Girls* (Chicago, 1913), 4–14. The author was a professor of physiology at the Northwestern University Medical School; the publication was the fourth in the Sex Education series issued by the Council on Health and Public Instruction of the American Medical Association.

34.   Corinne Azen Krause, *Grandmothers, Mothers, and Daughters: An Oral History of Ethnicity, Mental Health, and Continuity of Three Generations of Jewish, Italian, and Slavic American Women* (New York, 1978), 55–59.

35.   Krause Collection, Historical Society of Western Pennsylvania (hereafter cited as HSWP), S2A, J9A, I5A, I11A.

36.   HSWP, I9A.

37.   HSWP, I1B. The same scenario appears in S11B and S2B, but the mother provides some explanation after finding the telltale blood. Reduced exercise at menstruation obviously was an important issue: middle-class mothers did some pampering that was not evident in the working class. This issue was played out in the 1920s in the debate over whether menstruating girls should be excused from gym classes in high school and college.

38.   Biesanz, ed., *Mavis Helmi,* 185.

39.   HSWP, J3A.

40.   HSWP, S10A.

41.   Biesanz, *Mavis Helmi,* 179.

42.   HSWP, S2A.

43.   The biblical heroine Queen Esther said: "Thou knowest that I abhor the sign of my high estate [her crown] . . . as a menstrous rag." On niddah, the state of uncleanness, see *The Jewish Encyclopedia,* ed. Isidore Singer, 16 (New York, 1901), s.v. "Niddah."

44.   HSWP, J15B. See also the biographies of Emma Goldman, *Living My Life,* and Kate Simon, *Bronx Primitives,* for reports of slapping at menarche.

45.   Although marriage before menarche probably was unusual, a number of other Italian immigrant daughters reported that their sexual education had been so limited that they had to ask where the baby would come out at the time of their first delivery. The answer was frequently expressed in this way: "It come from where it entered" or "It come out where it goes in." See HSWP, I1A, I2B, and others.

46.   Theodora M. Abel and Natalie F. Joffe, "Cultural Backgrounds of Female Puberty," *American Journal of Psychotherapy* 4 (1950): 91–92. In America, where they undoubtedly led a more secular life, Jewish girls were not any better prepared than anyone else, but because of the ritual emphasis on cleanliness they may have been among the first immigrant groups to adopt commercial sanitary products.

47.   Cellucotton was the invention of Ernest Mahler, a chemist at Kimberly-Clark; it was derived from wood pulp and was twice as absorbent as surgical cotton. Shipped to the allied troops in World War I, cellucotton was "discovered" by Red Cross nurses as an effective form of sanitary protection. Kimberly-Clark went so far as to set up the International Cellucotton Production Company to market Kotex because they did not want to be associated publicly with their product. The Kimberly-Clark name was not in Kotex advertisements or promotions until the 1950s. See Margot Kennard, "The Corporation in the Classroom: The Struggles over Meanings of Menstrual Education in Sponsored Films, 1947–1983" (Ph.D. diss., University of Wisconsin–Madison, 1989).

48.   HSWP, I8B and I11A. See also I1B and I6B for cases where Italian-American girls were using rags well into the 1930s. This was the pattern, not the exception.

49.   Anna C. Arbuthnot, "Physiology and Sex Hygiene for Girls in the Technical High School, Cleveland, Ohio," *School Science and Mathematics* 11 (1911): 106.

50.   The fact that menstruation occurs earlier and is taught earlier confirms the general "death of childhood" argument advocated by Neil Postman, *The Disappearance of Childhood* (New York, 1983), and Marie Winn, *Children Without Childhood* (New York, 1983). However, it is incorrect to assume that age at first menstruation will continue to decline: the decline in the age at menarche is best understood as a function of change in the range of ages at which menstruation occurs. Over the past century it has become uncommon for girls to start menstruating at seventeen or eighteen, but the lower age range—nine or ten—has remained constant. According to both Brown, "Age at Menarche," and Zacharis, Rand, and Wurtman, "Sexual Maturation," there does seem to be a biological "bottom line."

51.   Emil Novak, *Menstruation and Its Disorders* (New York, 1921), 108. On sex education see Bryan Strong, "Ideas of the Early Sex Education Movement in America, 1890–1920," *History of Education Quarterly* 12 (1972): 129–61; and Wallace H. Maw, "Fifty Years of Sex Education in the Public Schools of the United States, 1900–1950: A History of Ideas" (Ph.D. diss., University of Cincinnati, 1953).

52. A German proverb quoted in C. F. Hodge, "Instruction in Social Hygiene in the Public Schools," *School Science and Mathematics* 11 (1911): 304. Hodge was a biology professor at Clark University. The discussion about where menstruation and sex should be taught can be found in Lo Ree Cave, "Domestic Science as an Opportunity for Sex Education," *Bulletin of the Kansas State Board of Health* 16 (April 1920): 67–72; and Benjamin Gruenberg, *High Schools and Sex Education: A Manual of Suggestions on Education Related to Sex* (Washington, D.C., 1922). On the social hygiene movement see Allan Brandt, *No Magic Bullet: A Social History of Venereal Disease in the United States* (New York, 1985); John Burnham, "The Progressive Era Revolution in American Attitudes Towards Sex," *Journal of American History* 59 (1973): 885–908; and David Pivar, *Purity Crusade: Sexual Morality and Social Control, 1868–1900* (Westport, Conn., 1973).

53. Menstrual education was incorporated into Girl Scouts of America programming beginning in the 1920s; to earn the "Health Winner Badge" (described in the *Girl Scout Handbook*) girls had to indicate an understanding of the process and hygiene of menstruation. Today menstrual hygiene is part of the curriculum in 88 percent of our elementary and junior high schools. See Whisant, Zegans, and Brett, "Implicit Messages," 815.

54. Roland Marchand, *Advertising the American Dream: Making Way for Modernity, 1920–1940* (Berkeley, Calif., 1985). Another stage in the industry's advertising strategy—the transformation of shame to liberation—is described in Ann Treneman, "Cashing in on the Curse: Advertising and the Menstrual Taboo," in *The Female Gaze: Women as Viewers of Popular Culture*, ed. Lorraine Gamman and Margaret Marshment (Seattle, 1989), 153–65.

55. Kotex advertisement, *Good Housekeeping* 80 (1925): 190. See also the popular ad for "Marjorie May's Twelfth Birthday," *Parents' Magazine* 14 (February 1939): 49.

56. I am referring here to the following corporations: Kimberly-Clark, Personal Products, Tampax, Inc., and Campana Corporation.

57. "Do You Scare Her to Death?" asked a Kimberly-Clark ad published in 1949 in *Parents' Magazine*, *Ladies' Home Journal*, and *Good Housekeeping*.

58. Sandra Rubin (pseud.), born in 1939, and Lois Pollock (pseud.), born in 1944, manuscript diaries, in the possession of the author.

59. Menarcheal competition among bourgeois girls was described in *A Young Girl's Diary: With a Letter by Sigmund Freud*, trans. Eden Paul and Cedar Paul (New York, 1921), with an introduction by Sigmund Freud.

60. Sarah Vaughn, born in 1968, manuscript diary, in the possession of the author.

61. This heightened differentiation or segmentation of the market is typical of American capitalism. For a discussion of these strategies in the domain of food, see Sidney Mintz, *Sweetness and Power: The Place of Sugar in Modern History* (New York, 1985), ch. 5.

62. HSWP, J15B.

63. There is a pervasive interdisciplinary feminist literature on past and present connections between adolescent feminine identity and the consumer culture. Some examples are Kathy Peiss, *Cheap Amusements: Working Women and Leisure in Turn-of-the-Century New York* (Philadelphia, 1986); Erica Carter, "Alice in the Consumer Wonderland," in Angela McRobbie and Mica Nava, eds., *Gender and Generation* (London, 1984), 185–214; Wendy Chapkis, *Beauty Secrets: Women and the Politics of Appearance* (Boston, 1986); Leslie G. Roman and Linda Christian-Smith, *Becoming Feminine: The Politics of Popular Culture* (New York, 1988).

64. Bernard Barber and Lyle S. Lobel, "Fashion in Women's Clothes and the American Social System," in *Class, Status, and Power: A Reader in Social Stratification*, ed. Reinhard Bendix and Seymour Martin Lipset (New York, 1953), 323–32; Roland Barthes, *The Fashion System*, trans. Matthew Ward and Richard Howard (New York, 1983); Joanne Finkelstein, *The Fashioned Self* (Philadelphia, 1991).

9   From Robust Appetites to Calorie Counting: The Emergence of Dieting Among Smith College Students in the 1920s

Margaret A. Lowe

On October 29, 1924, the *Smith College Weekly* published a letter to the editor entitled, "To Diet or Not to Die Yet?" Three students warned the campus community: "If preventive measures against strenuous dieting are not taken soon, Smith College will become notorious, not for the sylph-like forms but for the haggard faces and dull, listless eyes of her students."[1] In striking contrast to previous generations of Smith students, dieting to lose weight, or "reducing" as it was more commonly called, had infiltrated women's daily lives.[2] Although the prevalence of dieting among Smith students is difficult to determine in the post–World War I college environment, it clearly emerged as a tool utilized by Smith women to shape their appearance.

Few historians have directly analyzed questions of diet and body image, and none has specifically addressed dieting among college women, but historians have produced significant research on the history of women in higher education, the history of fashion, beauty, and health ideals, and the history of nutrition. Further, historians have recently turned their attention to the history of the body, and new scholarship does examine the history of dieting practices.

Historians of women's education have thoroughly documented women's entry into higher education and the challenges they posed to traditional notions of femininity. While some education historians have investigated the ways in which this challenge was filtered through debates about the female body, they have not analyzed the effect of those debates on college women's attitudes toward their bodies.[3] Scholars have chronicled idealized images of female beauty and costume historians have documented fashion trends, but their work has relied primarily on prescriptive literature.[4] They have documented idealized images of female beauty rather than the daily social and bodily practices that women employed to create their appearance. Historians of the body, on the other hand, have challenged scholars to understand the body as a social construct which changes over time, but they too have relied primarily on discourse anal-

MARGARET A. LOWE is Assistant Professor of History at State University of New York, College at Potsdam.

Reprinted from *Journal of Women's History* 7, 4 (Winter 1995): 37–61, by permission of Indiana University Press.

ysis.[5] My detailed study of Smith College students' dieting practices furthers the analysis: it delineates the relationship between cultural discourses about the female body and women's attitudes toward their bodies and traces the "historical moment" when dieting became integral to college women's conceptions of their bodies and a tool for changing them.

Most historians who have explored the history of dieting date its onset to the mid or late nineteenth century. Keith Walden and T. J. Jackson Lears have both suggested that its roots lay in nineteenth-century industrialization and modernization rather than gender or class differentiation.[6] As the United States became more urban, fast-paced, and homogeneous, they argued, men and women found that one way to stem anxiety amid disorder was to exert control over their bodies. Hillel Schwartz, in his wide-ranging cultural history of dieting, determined that while "each epoch has had different tolerances for weight and for fatness, since the 1880s, those tolerances have grown especially narrow."[7] Schwartz supported Walden's and Lears's theses that industrialization augmented the emergence of dieting. He concluded that "slimming . . . [was] the modern expression of an industrial society confused by its own desires and therefore never satisfied."[8]

Historians Roberta Seid and Joan Brumberg have most thoroughly documented dieting for aesthetic purposes as a twentieth-century phenomenon. Unlike scholars who have dated the proliferation of dieting to the mid or late nineteenth century, Roberta Seid located the "first significant thinness craze as between 1919 and 1935"—the same years that dieting emerged among Smith students.[9] Joan Brumberg dated the onset of modern dieting a bit earlier. She determined that "within the first two decades of the twentieth century, . . . the voice of American women revealed that the female struggle with weight was already underway."[10] While Brumberg demonstrated that dieting information permeated popular literature prior to World War I, she also suggested that "in the 1920s, the imperative to diet intensified."[11]

For Smith women it was in this period that dieting became a popular preoccupation. Seid's conceptualization of the twenties as "a period of transition" accurately describes Smith women's dieting practices.[12] Similar to Seid's findings, dieting became a "craze" at Smith College in the 1920s, but as she stated, "the standards of slenderness were not as extreme as ours . . . [and] the cult of slimness did not have the monolithic character it has today."[13] Though Smith students dieted, it was not their only relationship to food. In addition, dieting did not necessarily result in weight loss or a slender figure. The standards of slenderness and the expectations to diet were not as extreme in the 1920s as they are today. But the emergence of dieting in the twenties does signal a pivotal shift in the way white middle-class college women understood and shaped their bodies.

Prior to World War I, food played a central role in the lives of Smith College students but not as an element to manipulate in order to lose weight. While historians have documented widespread dieting in the late nineteenth and early twentieth centuries, this was not the case at Smith College. In contrast, students revered weight *gain*. Students, their families, and college officials perceived weight gain as a sign of health, and it was essential that early college women appear healthy. Critics of higher education for women had predicted that academic life would destroy female health and feminine appeal. Scientists and physicians argued that brainwork would usurp the body's finite resources, depleting the female reproductive system of blood, nutrients, and energy. They contended that women's generative organs would suffer, college women would not reproduce, and thus the "race" would be diminished. Not surprisingly, critics were most concerned about white, native, middle- and upper-class women. And it was these women who were most likely to attend Smith College. In this atmosphere weight gain demonstrated that the students' bodies were not "breaking down" under the strain of college life. Weight loss, perceived as a symptom of some of the most common female illnesses (neurasthenia, hysteria, and consumption), was troublesome. Weight gain, by contrast, signified a healthy adjustment to college life.[14]

Granted permission to gain weight, the students went about the business of eating. Smith student letters written between 1875 and 1910 teem with references to food. In contrast to scholars who have concluded that gendered social edicts encouraged privileged adolescent girls to hide their appetites in order to appear appropriately middle-class and feminine in the late nineteenth century, Smith students repeatedly described their daily pleasure in satisfying their appetites.[15] They wrote long letters to friends and family detailing the delights of college food, sent profuse thank-you notes for the cornucopia of candy, sandwiches, oranges, and fruitcake their mothers dispatched from home, and recounted luscious food parties or "spreads" they shared with newfound friends. They indulged with abandon.

In a letter to her mother, Josephine Wilkin recounted the excitement of a surprise birthday spread for her friend Daisy: "When Daisy came in she was *so* astonished. We had two kinds of crackers, chocolate, pine-apple cake, candy and nuts . . . such fun as we had."[16] Even after great indulgence the night before, Gertrude Barry wrote to her mother with mischievous glee that "Peetie and I did not arise with the larks but slept on and on. Then we got up and fixed the grapefruit which was *delicious, Butter,* then a fried-cake and the last of the fig-cake! I went to church but Azalia went back to bed again."[17]

If such pleasurable and abundant eating re-

sulted in weight gain, students such as Charlotte Wilkinson figured that was a good thing. Writing to her mother in February 1892, she stated somewhat cheekily, "It is my ambition to weigh 150 pounds."[18] In closing remarks of a letter written in April of that year, she wrote, "Now I must stop, dearest Mamma, with a heart full of love from your devoted and healthy daughter, Char." Then she added, "I put in healthy because I know you want me to be that, next to being good, as I am very well now, as I was all winter term. I weigh 135 1/2 pounds."[19] In June of that year she used weight gain once more to substantiate her health. For most Smith students, campus activities, or the "life" as they called it, offered constant temptations. Parents and administrators worried that students would become fatigued and ill. Wilkinson countered with her most common and effective form of reassurance. "I have never had so much going on in my life as this last month," she wrote. "But don't be afraid that I shall get tired out for I am bouncingly well. I weighed 137 pounds the other day."[20] She was slowly gaining on her goal of 150. Reports of hearty appetites and weight gain served to reassure parents and society that college life would not undermine female health. Neither weight gain, a healthy appetite, nor robust eating threatened femininity.

By the 1920s, however, attempts to "reduce" began to accompany comments about school food and thank-you notes for food sent from home. In the twenties, when students gained weight and wrote home about it, they expressed anxiety. Weight gain no longer symbolized health; it suggested weakened willpower and a potential loss of feminine appeal. Even Lucy Kendrew, who continued to write with delight about food and dining, expressed dismay after an encounter with the gymnasium scale. "I had the worst scare the other day, when I came down, I weighed 119 or 122, Wednesday I weighed myself on the gym scales, & weighed

136 1/2! Friday I got weighed on them in the same clothes & had lost 2 1/2 pounds."[21] Notwithstanding an unreliable scale, this speaks to her fear of weight gain and its message of apprehension rather than reassurance.[22]

Upper-class students introduced incoming students to dieting behavior on campus. Beginning in 1923, the "Hints to Freshmen" section of the Smith College student handbook advised: "Don't consider it necessary to diet before your first vacation. Your family will be just as glad to see you if you look familiar."[23] Though juniors and seniors chided freshmen not to succumb, enough Smith students dieted to warrant its mention in the handbook. From 1925 onward, the Mount Holyoke College student handbook warned students: "Beware of eating between meals. Freshmen traditionally gain ten pounds so patronize the 'gym' scales."[24] Although this allusion to the dreaded "freshman ten" might fit snugly into today's college parlance, it would have made little sense to first- or second-generation Smith students. The weight gain may have sounded familiar but not the students' response to it.

By the mid-1920s dieting had become part of many students' normative relationship to food and their bodies. Dorothy Dushkin's diary entry from May 5, 1922, revealed the conflicted feelings dieting could engender. She resented her classmates' constant preoccupation with dieting but also struggled with it herself. She reported:

> Resolved once more to cut down my diet. Betty & Fran's chief topic of conversation is dieting. It is extremely wearisome especially since they are both slender. I shall try once again to exert my will power. I'm not going to say a word about it. I'm not going to foolishly cut meals and starve on certain days & relax on others as they do—but attend all meals & refrain from eating between meals.[25]

Reducing, along with its attendant battles of the will, had become a familiar symbol of student adjustment to college life.

College administrators monitored the shifting dietary codes as well. William Neilson, president of Smith, was concerned about dieting, according to Anne Morrow Lindbergh, class of 1928. She wrote to her mother, "the President disconcerts me sometimes for he always asks me for the student opinion, which I am the last to know! . . . This time he asked me if they were still *dieting!*"[26] Health officials also documented the dieting trend. In 1923 Dr. Anna Richardson, while reassuring alumnae that the senior class exhibited good health, noted that 16 percent or one in six of the senior class was underweight.[27] She ambivalently rationalized: "This is not such a very high percent when we consider the present vogue for the slender figure."[28] Yet her ambivalence revealed itself in her following comment:

> Recent studies in weight in relation to longevity place much emphasis on the advantages of normal or overweight before the age of thirty. Concern for the underweight students thus becomes of very real importance so that the fad for dieting cannot be ignored. Fortunately this practice is interfered with by its inherent inconvenience as well as the fact that the students become frightened by certain physiological changes that happen to them during personal experimentation.[29]

Though we do not know what "physiological changes" (weight loss, amenorrhea, fatigue, headaches?) frightened the students, Dr. Richardson's comments suggest that dieting had become a pressing concern to college administrators.

Dieting on campus became one concern among many. By World War I the "healthy" success of college women had silenced early critics.

From the school's inception Smith officials constantly proclaimed that academic life posed no threat to female health or femininity. After World War I college administrators faced a different challenge: how to combat the harmful living habits of modern youth. In 1920 college physician Florence Gilman lamented the effects of such habits on student health: "Never have we had such a large proportion who [seniors] are tired, nervously tense, underweight, anemic, with a low blood pressure showing a condition of depressed vitality. These things are found in students who have not been ill for the most part."[30] Throughout the twenties college physicians bemoaned "the attitude of young people . . . the feverishness and unrest . . . the attitude toward pleasures . . . changes in the social conduct of undergraduates of both sexes [and the oft resulting] neglect of academic work."[31] Health officials responded with renewed health and hygiene education and continued physical examinations. College officials consistently reported that students were in good health, but they also admitted that "few students . . . do not need some direction in regard to their health. . . . Each student is again advised, as she undoubtedly has been before, regarding living habits she should follow."[32] Yet many students declined college administrators' sensible advice and instead joined their peers in, among other things, dieting.

In general, "reducing" meant exercise, cutting out sweets and starches, and no snacking between meals. Since health advisers constantly warned mothers to make sure their daughters did not join the dieting craze at college, students may have moderated or concealed more severe dieting practices.[33] For example, although the "Hollywood Eighteen Day Diet" enticed many a dieter in the twenties, students make no mention of following its 585-calorie regimen of grapefruits, oranges, melba toast, green vegetables, and hard-boiled eggs.[34] Yet some students did skip meals, "starve" themselves, and follow

exacting dietary regimens. Pauline Ames, who subsisted on a diet of just "fruit and milk until supper," suggested the restrictive nature of student dieting.[35] For her, hunger and anxiety were constant: "You can imagine how hungry I get," she pined to her mother. "The food is a great difficulty and I don't know how strict I ought to be."[36]

The "dieting craze" at Smith College emerged within a complex web of social and cultural factors that coalesced in the 1920s, prompting new eating behaviors and bodily practices. Dieting among Smith students reinforced their affluent, middle-class social status and the development of the new "youth culture" which emphasized heterosexual dating. It also revealed the influence of popular culture, especially flapper imagery, on women's bodily practices. Last and perhaps most important, dieting reflected the popularization of scientific nutrition. These postwar changes were transforming food activities generally; dieting among Smith students mirrored this large transformation.

As mostly white, Protestant, middle- to upper-middle-class women, Smith students represented and sometimes embodied the ideal body configuration of the modern woman.[37] In the 1920s both dieting and slenderness signified middle-class status as well as female youthfulness and modernity.[38] According to anthropologist Claire Cassidy, in twentieth-century Western society "slenderness symbolizes the freedom from want. . . . The wealthy . . . are able to switch the bodily metaphor of success from fat to thin because they do not need to worry about famine or infectious disease. They can go beyond the message of fat—'Look how much abundance I have' to a more etheric model—'I'm so safe, I can afford to ignore abundance.'"[39] By the 1920s fat was most closely associated with the old world: immigrants and the lower classes. As Roberta Seid has stated, "the poor and lower classes began to be seen as stocky and plump rather than as thin and undernourished."[40] At Smith, when depictions of postwar working-class, immigrant, or African-American women surfaced in student literary magazines or the college newspaper, they were usually described as having "sturdy" or "hearty" constitutions.[41] In addition, historian Harvey Levenstein has argued that dieting itself had become a "middle-class obsession."[42] Dieting was considered a middle-class activity and its desired result signaled middle-class status. However elusive, dieting offered Smith students a path to slenderness and all that it represented. Yet, while Smith women's dieting practices were laden with class meaning, Smith students did not articulate their motivations for dieting as class based. While viewed as models of middle-class femininity and perceived as trendsetters, they did not refer to dieting with such class consciousness. Instead, they reported that they dieted in order to appear attractive or because their friends did—to partake in one more fad of the 1920s.[43]

The switch from "spreads" to "bacon bats" introduces the nature of postwar campus life at Smith College and its effect on food practices.[44] Late nineteenth- and early twentieth-century students' lavish informal food parties contained the essential ingredients of campus life before 1910: a single-sex social life, clear links to parental love and approval, and on-campus entertainment. Only women attended spreads, they were hosted within the gates of Smith College, and the food consumed was abundant and homemade. By the 1910s, however, students emphasized the pleasures of "batting." On a typical bacon bat a group of students and guests (often including men) motored to a rural off-campus spot and picnicked on coffee, sweets, and bacon roasted over an open fire. While outdoor picnics had always been a part of student life at Smith, in the 1910s picnics were called "bats," and bats symbolized off-campus, heterosocial, unsupervised fun.[45]

Between 1910 and 1930 student-directed socializing still included food parties, but they had

shifted from predominantly single-sex campus activities to largely mixed-sex off-campus gatherings. While students expressed the same excitement and pleasure over "batting" that earlier students had over spreads, their pleasures derived from different sources. In the 1920s students relished their bonds with men, their separation from campus, and their freedom from adult supervision. One result of this shift was that students experienced all the pleasures and anxieties related to food and eating before a wider audience. On bacon bats, as well as at restaurants and inns, they presented their appetites and manners to not only their exacting female peers but to men whom they hoped to attract. Bacon bats reflected the relaxed sexual codes, affluence, and mixed-sex socializing of the 1920s college campus.

As women's participation in academic life became less threatening to society, Smith students began to exhibit all the complex signs of the modern youth.[46] Students preserved their beloved school traditions but added new ingredients: men, automobiles, and a complex peer culture. Smith students bobbed their hair with or without parental approval; donned shorter skirts; flocked to football games; loved jazz, dancing, and movies; and orchestrated active mixed-sex social lives. Until the early 1910s the ten o'clock rule (lights out at ten) was the only general campus rule. By the late 1910s Smith had established elaborate student regulations covering such matters as chaperonage, overnights away from the college, and dances. Students dined out so often that administrators created a list of approved "eating places outside Northampton" where students could dine without a chaperon.[47] And restaurateurs and innkeepers marketed their establishments as the answer to the "eternal question, 'Where shall we eat?'"[48] In the early 1920s President Nielsen warned parents that "the increase in the number of girls who bring automobiles to Northampton . . . is a mistake."[49] He cautioned that "supervision is rendered more difficult when a student is enabled in a short time to remove herself from the observation of the authorities."[50] Suggesting just how much fun college had become, the handbook forewarned freshmen: "Smith College is not a country club."[51] But historian Helen Horowitz has cautioned that "college women did not suddenly become hedonistic: they had been so for decades. What happened in the 1920s is that their hedonism turned its focus to men."[52]

In turn, college women catered their appearance to the possibilities of both planned and surprise encounters with men. References to heterosexual dating appear as early as 1908, but in the 1920s men became central figures on campus. Early Smith students avidly discussed their chapel and reception "dates," but such dates were women. In the 1920s one's "date" was invariably male. Historian Beth Bailey has demonstrated that dating "moved courtship into the public world, relocating it from family parlors and community events to restaurants, theaters and dance halls."[53] The goal of dating was not matrimony but popularity; women competed for success in peer culture by attracting men who 'rated.'"[54] While not the only element, an attractive appearance, at least in part defined by slenderness, facilitated successful dating.

In the twenties Smith College students assessed one another's dates, followed the latest engagements, organized mixed-sex campus events, and traveled to New York for weekend outings and football games. Such administration rules as "there is no dancing from house to house by individuals or couples" suggest just how casual heterosexual contact had become.[55] The *Smith College Weekly* carried a running debate about the "fussing [dating] problem."[56] Students and administrators wrangled over when and where students could socialize with men. A member of the class of 1926 responded with irritation, "Rainy Sundays are a problem for those who are 'fussing' if their callers do not possess cars. There is no place to go except to an already crowded living-room. . . . Wouldn't it be possible to allow bridge-playing or attendance at the movies under these circumstances?"[57] By 1930 the first item on the official statement regarding dances at Amherst College stated, "There will be seventeen dances a year at Amherst; one for each fraternity and Sophomore Hop, Senior Hop, cotillion club, and Sphinx club. This is exclusive of Amherst Junior Prom arrangements."[58] In a culture of "fussing," cotillion clubs, senior hops, and sojourns to the Northfield Inn, students continually envisioned and created their appearance in accordance with what they thought would be appealing or at least appropriate to interaction with young men.

To catch the eye of a college man an attractive appearance was essential. By the 1920s college men admired the slender flapper figure. At nearby Amherst College the student humor magazine, the *Lord Jeff*, was filled with references to Smith students, proms, dating, and petting. After each dance the names of all the students and their dates were published in the student newspaper. Many Amherst students cut these out and kept them in their scrapbooks. While the students did not specify the exact dimensions of their ideal date, they did express their preference for slenderness. For example, in a verse published in the *Lord Jeff*, an Amherst student wrote:

The secret of a girl's success; Is not her
    lips, her locks of gold;
or eyes so full of tenderness;—But its her
    figure, I've been told.
A girl may think she is a peach; She may
    be sweet & debonair;
But when she goes upon the beach; her
    secret is at last laid bare.[59]

Smith student Ruby Mae Jordon expressed some of the complexity men on campus engendered in regard to female appearance. To her mother she lamented, "I am quite the joke with them [housemates] because I am so quiet and insignificant looking. I guess they think I wouldn't dare look at a boy edgeways."[60] In another letter she recounted her relief that she "happened to look well" upon encountering a surprise male acquaintance on campus. She divulged, "I had my blue flannel dress on, with flesh stockings & suede shoes, and my hair had been freshly curled the night before. I never was so thankful for anything in my life as I was that Elsie saw me. She thought Ralph was awfully cute."[61] But alone on prom night, she lamented to her mother, "It sure makes me lonesome to sit here and listen to the gorgeous music. And the men! My goodness, chapel was crowded like it was *only* the first day.... And why I studied all morning." But then she went on, perhaps to regain her esteem, "I couldn't say much for the men in our house. There were only two attractive ones that I saw."[62] The students hoped to not only impress their dates with their appearance but also their classmates with the rewards of that appearance—their men.

Within this arena students dieted to cultivate a more favorable appearance and win popularity. In other words, dieting among Smith students signified not only their middle-class status but also efforts at "intraclass" differentiation. Dieting provided a means to distinguish the popular attractive students—those with lots of dates and the right "look"—from the less popular. Student publications let them know, for example, that the "chubby chester ... [a] tall, fat girl" filled the dance card only "to make [one] appear sylph-like, a fairy creature of remarkable charm in contrast."[63] Dieting boosted popularity in several ways. Most obviously, students dieted to cut a more fashionable figure and thus win the admiration of fellow classmates and men. In addition, dieting cemented friendships within a sophisticated peer culture. Dieting

offered the camaraderie of watching one's weight with friends. Last, the rewards of dieting were clearly visible, the transformation of the body attracted attention and admiration in and of itself.

In their attempts to forge an appealing feminine identity, students employed a variety of poses. In accordance with cultural historian Warren Susman's thesis, Smith students seem to have embraced the notion "that pleasure [and success] could be attained by making oneself pleasing to others."[64] In this "new culture of personality," according to Susman, "every American was to become a performing self."[65] Personality, "both the unique qualities of an individual and the performing self that attracts others,"[66] formed one's identity in the 1920s, Susman argued, not character—the solid, immutable self that nineteenth-century society held so dear. Dieting provided a simple means to manipulate self-presentation. Accordingly, the body itself, in contrast to fashion or elements placed upon the body, became one more variable to manipulate in the orchestration of a performing self.

Both students and administrators considered a woman's appearance critical in defining her "type" and ultimately her success—not only in the marriage market but in the job market.[67] In Personnel Department records vocational counselors regularly evaluated student appearance in the career assessment process. Such remarks as "very pronounced Southern drawl & hair drawn severely back from face"; "large, rather gawky in manner & appearance"; "well-bred, pleasant manner"; "makes a nice appearance, excellent public school material"; and "large overgrown girl, pleasant manners, attractive blond hair" accompanied each student file.[68] Counselors linked physical appearance to vocational direction. As the "largest women's college in the world," Smith sought to preserve its reputation. Faculty and administrators feared that certain "types" of girls, placed in prominent or public positions, might diminish Smith's status. While the link between appearance, "type," and vocation, and its impact on Smith students' dieting practices is difficult to trace, these records suggest that dieting became an important tool for young women, not only due to fads or fashion but also as a mechanism to facilitate an appropriate professional appearance and career success. After all, students depended on vocational counselors to match them with potential employers. Female appearance continued its long-standing influence on courtship and marital decisions in the 1920s, but it also came to signify occupational aptitude. While administrators discouraged student dieting, they rewarded the most attractive students with expanded career opportunities. Within this context dieting acted as one tool among many that students employed to create an image that they believed would best represent the identity they wished to embody.

The most prevalent popular image of youthful femininity that the students emulated was the flapper. It is perhaps ironic that in an era of unprecedented emphasis on individuality and style, one fashion ideal—the flapper—thoroughly dominated the fashion stage. The shift to one commanding ideal by the 1920s may have more strongly propelled dieting than its specific slender dimensions. The effect of the flapper's thin standard of femininity on college women's eating behavior was dramatically compounded because it was so dominant.

Before World War I students encountered multiple, overlapping, and competing images of feminine beauty. In the late nineteenth century, as Americans grappled with the New Woman, a flurry of new cultural images materialized.[69] The new images reflected women's changing social status. By 1900 women had not only entered the halls of higher education, but they had demanded the vote, organized national political movements, worked in settlement houses, and opted not to marry in unprecedented num-

bers.[70] As the Victorian era came to a close, so did its fainting, wispy image of womanhood. As historian Lois Banner has documented, once the frail, ethereal, pre–Civil War model of femininity faded, several new models emerged: the "natural woman," the "voluptuous woman," and the "Gibson Girl."[71] The new models exhibited more robust figures: more flesh, more color, and more movement. The voluptuous woman, as the name suggests, idealized the hourglass figure with accentuated ample breasts and hips.[72] The natural woman exhibited a curvaceous figure but shed its extreme hourglass shape and its artifice in favor of a "healthy" constitution. As the century closed, an athletic physique joined this group. The exercise boom of the 1890s increased the social acceptability of women in sport, which created new images of feminine beauty and new lines of fashionable attire. By the early 1900s the athletic Gibson Girl dominated advertisements, retail catalogues, and fashion magazines. The Gibson Girl dropped the fleshiness of the natural woman in favor of a linear but sensuous shapeliness. As New Women, Smith students created and interacted with the new definitions and models of middle-class womanhood. At the turn of the century no one knew quite what to expect as the slight, frail image of Victorian womanhood had become outmoded before new standards of female appearance were firmly established.

In the 1920s, however, the popular flapper image prevailed. The flapper represented above all youthful modernity. She was both slender and linear. In allegiance with an emphasis on youthfulness in the twenties, signs of female maturity and maternity were suppressed. According to historian Valerie Steele, "the ideal woman was no longer the voluptuous mother, but the young woman with the girlish figure."[73] The ideal flapper was flat chested and small hipped. Smith students decked out in full flapper attire did not always look slender, but they usually looked straight. Though the flapper image minimized breasts and hips, it radiated sensuality. The straight lines signaled adherence to the new sexual codes of twenties' youth: dating, petting, fast dancing, and freedom from parental supervision.[74]

The flapper style, marketed in ways unknown to earlier generations, permeated popular culture. As mass consumerism geared up and took off during their stay at Smith, postwar students viewed the latest fashions in more sophisticated and more rapidly disseminated advertisements, in darkened movie theaters on the big screen, and at traveling department-store fashion shows which featured live models. Earlier students and their mothers also purchased "ready made," sized clothes, but it was in the late 1910s that the sophisticated fashion network of mass production, consumption, and distribution coalesced and accelerated fashion conformity.[75]

Students relished the new consumerism and avidly followed flapper fashions. They urged one another to conform and buy. A Debating Society report entitled: "Are the Movies a Benefit to Smith College?" exemplifies the students' often creative responses to the new media. The affirmative team won by claiming that local movie showings saved students money by presenting the very latest fashions for students to imitate in ordering and tailoring their clothes.[76] Students viewed and purchased the latest fashions away from home and among their friends. Students often ordered items from Lord and Taylor, Bonwit Teller, and Chandler and Company and then wrote home for permission and money. By 1928 the handbook scolded new students: "Don't provide yourself with a college trousseau. Smith has a style of her own."[77] In other words, be prepared to "revamp" your wardrobe—but according to the predominant style.

While historians have pointed out that diet information permeated popular literature prior

to the slender ideal of the 1920s, the flapper image further encouraged student dieting. Losing weight would slim their bodies into a prized slender physique. College women who emulated popular taste made their food and fashion decisions within a strong peer culture that aligned itself with all the accoutrements of the flapper style. In the 1920s students defined themselves as "youth"—distinct from adult parents and administrators. They looked first to each other for approval. To gain approval they followed the latest fads. Dieting served as both a fad in and of itself and a means to achieve the flapper look. The dominant flapper image, combined with a persuasive peer culture, created a powerful impetus to diet.

On the internal side of the dieting ledger, an explosion of scientific research in food chemistry formulated in the late nineteenth century flooded the market by the 1920s. The fundamental principle, "eat to live, don't live to eat," was its common refrain, repeated in popular magazines and health books. According to historian Laura Shapiro, scientific cookery at the turn of the century "pursued the science of food, not the sensuality."[78] The new food scientists advocated a rational businesslike approach to eating rather than an unpredictable "messy" approach associated with gratification of the appetite. The new nutritionists advocated eating what was good for you rather than what you liked.[79]

Building on midnineteenth-century German research, American food scientists encouraged people to select food on the basis of its structural composition (nutritional value) rather than taste or appearance. Food scientists ranked food according to the amount of minerals, fats, proteins, and carbohydrates it contained. Similar to today's Food and Drug Administration guidelines, they recommended reduced fats and sugars, limited carbohydrates and proteins, and plenty of fruits and vegetables.[80] They also recommended calorie counting. In the late nineteenth century, Wilbur Atwater invented the "calorimeter." Its ability to measure the exact number of calories burned during various physical activities spawned the widely held belief that the human body required and burned food calories as fuel, much like a machine. To run efficiently the body required a specific amount of calories which were determined by food type and amount.[81] During World War I food scientists and home economists disseminated these new ideas under the auspices of the Food Administration. In the 1920s new nutrition tenets popularized by food retailers, advice columnists, and home economists became commonplace.[82]

By the early 1920s the calorie concept functioned as the structural basis for most reducing regimens. Dr. Lulu Peters, "the best known and loved physician in America," and her colleagues, along with food retailers and advertisers, popularized the notion that food contained a certain number of calories, invisible to the naked eye, which acted to either fuel or fatten the body.[83] One of many prolific diet promoters, Peters churned out numerous popular articles and books which recommended counting and restricting calories for weight reduction. These articles supplied daily and weekly diet menus with caloric equivalents attached. Many magazines, including *Hygeia*, the publication of the American Medical Association, began to run reader response advice columns devoted exclusively to dieting. In popular women's magazines doctors suggested that wives and mothers prepare specific daily calorie-coded menus to keep their family healthy and at the "right" weight.[84] Standardized height and weight charts, based on insurance table ideals rather than averages, often accompanied such articles.[85] The popularization of newer nutrition tenets propagated a dietary system which advocated body regulation via the calorie, specific nutritional guidelines, and ideal target weights.

Newer nutrition principles both reflected and augmented the growing social disdain for "fat." By World War I fat had lost its positive value for the middle class. It no longer symbolized health, abundance, or joviality as it had during the nineteenth century. Fat was displeasing on both aesthetic and medical grounds.[86] According to Harvey Levenstein, "by 1918, . . . the idea that being overweight was unhealthy had caught up with the traditional idea that being underweight denoted poor health."[87] By the 1920s female illnesses associated with weight loss had diminished, removing the stigma of ill health from thin women. On the other hand, fat had become associated not only with laziness and gluttony but with death. Insurance company literature increasingly linked corpulence to mortality. The accumulation of fat over a lifetime, once recommended, was now viewed as dangerous.[88]

Newer nutrition guidelines were not lost on Smith students. Health records and student comments suggest that students adjusted their diets in accordance with the new nutritional paradigm. The students' integration of newer nutrition principles had modified school menus. In 1930 campus warden Laura Scales indicated that dieting among students had modified the school's food purchases. Until the early 1920s individual "heads of house" ordered and prepared their own food. By the mid-1920s a central purchasing office coordinated the ordering and distribution of food, while individual house matrons planned menus and prepared meals. In 1930 Scales reported that "last year 6000 meals were served per day with food purchased through our offices." She added that the $253,000 spent was less than usual. "Perhaps," she surmised, "the girls eat less too. The reducing fad shows itself particularly in the amount of potatoes consumed . . . the increase in amounts of lettuce, tomatoes, and celery may also be an indirect result of the craze for the 'boyish figure' and 'that schoolgirl complexion.'"[89] Students dieted in large enough numbers to alter bulk food purchases.

In 1917 Smith College physician Dr. Goldthwait's dietary recommendations to the student body began, "Food to the human being is in part similar to fuel in a furnace." He continued, "The amount of food required each day is figured in heat units called calories, and from 2500 to 3000 are required each day."[90] A detailed calorie table was enclosed which provided not only the number of calories in such food items as corn muffins, lamb stew, and vanilla ice cream but specified the protein, fat, and carbohydrate composition of each food. The obligatory height and weight chart was distributed as well. Not surprisingly, when students dieted they adhered to newer nutrition tenets: they counted calories, limited carbohydrates, and exercised to burn excess fat. Unlike earlier generations of Smith women who consumed food with little regard to its structural composition, students in the 1920s incorporated detailed knowledge of caloric and nutritional values into their food decisions. In the 1920s, just as the food scientists had hoped, elaborate and overt nutritional guidelines stood between students and their appetites. The new guidelines did not incite dieting in and of themselves, but they dispensed the requisite tools for building and implementing dieting habits. Newer nutrition concepts helped students "watch their weight."

In the 1920s Smith students encountered a dominant slender model of beauty, a popular scientific discourse which extolled the benefits of dieting, a modern heterosexual dating environment, and a cultural ethos which linked and deemed mutable both identity and appearance. It is within this cultural context that Smith students began dieting for aesthetic purposes.

On the one hand, dieting was a blow to women in the triumphant wake of suffrage. Dieting required vigorous self-restraint, constant suppression of hunger, and intensified competition, if not among women then at least between women and their scales. The "internalization" that seems most important here is not the concept of a thinner ideal but the perception that a normal relationship to food and the body included dieting. Women learned to exert internal control over their appetites, counterbalancing an era of unprecedented external social freedoms. In addition, the female body itself, not just its adornment, began to be considered a telling representation of one's identity. The physical dimensions of the body accrued more and more power to signify the nature of the person.[91] At the same time, dieting did not dominate the cultural landscape as it does today. As Roberta Seid has argued, "this was a period of transition, and the craze did not create the hysteria or the terrible effects on women that we know today."[92] Yet, in comparison to prewar students' voluptuous enjoyment of food, the tension and anxiety that twenties' students expressed stands in sharp and disheartening contrast.

Notes

I would like to thank Smith College, Amherst College, and Oakes Ames Plimpton for permission to publish material quoted in this article. I would also like to thank Maida Goodwin, Amy Hague, and Margery Sly, archivists at Smith College, for their generous help in locating materials for this essay. My thanks to Professors Joyce Berkman, David Glassberg, Helen Lefkowitz Horowitz, Kathy Peiss, and Patricia Warner for their always constructive comments. Thank you also to the anonymous reviewers for the *Journal of Women's History* for their thorough and insightful reading.

1. "To Diet or Not to Die Yet?" *Smith College Weekly* 15 (October 29, 1924): 2.

2. Although dieting practices reflect various motivations (gaining, losing, disease-related, for pregnancy, lactation, allergies, and so on), in this essay dieting refers to efforts to lose weight for aesthetic purposes.

3. For histories of women's education, many of which include sections on college women's athletics, physical culture, and health, see Lynn D. Gordon, *Gender and Higher Education in the Progressive Era* (New Haven, Conn.: Yale University Press, 1990); Helen Lefkowitz Horowitz, *Alma Mater: Design and Experience in the Women's Colleges from their Nineteenth-Century Beginnings to the 1930s* (New York: Alfred K. Knopf, 1984); Mary Kelley, ed., *Woman's Being, Woman's Place: Female Identity and Vocation in American History* (Boston: K. Hall, 1979); Elaine Kendall, *Peculiar Institutions: An Informal History of the Seven Sister Colleges* (New York: Putnam, 1975); Mabel Newcomber, *A Century of Higher Education for American Women* (New York: Harper, 1959); Barbara Solomon, *In the Company of Educated Women: A History of Women in Higher Education in America* (New Haven, Conn.: Yale University Press, 1985); David and Sheila M. Rothman, eds., *The Dangers of Education: Sexism and the Origins of Women's Colleges* (New York: Garland, 1987); Thomas Woody, *A History of Women's Education in the United States* (New York: Science Press, 1929). Specific to Smith College, see Rosalind Cuomo, "Student Relationships at Smith College and Mount Holyoke College" (Masters thesis, University of Massachusetts, 1988); Sarah H. Gordon, "Smith College Students: The First Ten Classes, 1879–1888," *Journal of Higher Education* 15, 2 (1975): 147–65; Eleanor Terry Lincoln, *Through the Grecourt Gates: Distinguished Visitors to Smith College, 1875–1975* (Northampton, Mass.: Smith College, 1978); Thomas C. Mendenhall, *Chance and Change in Smith College's First Century* (Northampton, Mass.: Smith College, 1976); L. Clark Seelye, *The Early History of Smith College, 1875–1910* (Boston, Mass.: Houghton Mifflin, 1923); Jacqueline Van Voris, *College: A Smith Mosaic* (W. Springfield, Mass.: M. J. O'Malley, 1975). Historians who highlight issues of the body within the educational setting include Louise Michele Newman, *Men's Ideas/Women's Realities: Popular Science, 1870–1915* (New York: Pergamon, 1985); and Martha H. Verbrugge, *Able-Bodied Womanhood: Personal Health and Social Change in Nineteenth-Century Boston* (New York: Oxford University Press, 1988).

4. See Lois Banner, *American Beauty* (Chicago: University of Chicago Press, 1984); Martha Banta, *Imaging American Women: Idea and Ideals in Cultural History* (New York: Columbia University Press, 1987); Claudia Kidwell and Valerie Steele, *Men and Women: Dressing the Part* (Washington, D.C.: Smithsonian Institution Press, 1989); Claudia Kidwell and Margaret C. Christman, *Suiting Everyone: The Democratization of Clothing in America* (Washington, D.C.: Smithsonian Institution Press, 1974); Valerie Steel, *Fashion and Eroticism: Ideals of Feminine Beauty from the Victorian Era to the Jazz Age* (New York: Oxford University Press, 1985); Elisabeth Wilson, *Adorned in Dreams: Fashion and Modernity* (Berkeley: University of California Press, 1985).

5. The history of the body, a rapidly expanding field of research, offers new theoretical frameworks for understanding changes in bodily practices. See Susan Bordo, "Reading the Slender Body," in *Body/Politics: Women and the Discourses of Science*, ed. Mary Jacobus et al., (New York: Routledge, 1990), 83–112; Judith Butler, "Performative Acts and Gender Constitution: An Essay on Phenomenology and Feminist Theory" *Theater Journal* 40 (December 1988): 519–31; Joanne Finkelstein, *The Fashioned Self* (Philadelphia, Pa.: Temple University Press, 1991); Allison M. Jaggar and Susan R. Bordo, eds., *Gender/Body/Knowledge* (New Brunswick, N.J.: Rutgers University Press, 1989); Stephen Kearn, *Anatomy and Destiny: A Cultural History of the Human Body* (Indianapolis, Ind.: Bobbs-Merrill, 1974); Thomas Lacqueur, *Making Sex: Body and Gender from the Greeks to Freud* (Cambridge, Mass.: Harvard University Press, 1990); Anson Rabinbach, *The Human Motor: Energy, Fatigue, and the Origins of Modernity* (Berkeley: University of California Press, 1992); Allan Sekula, "The Body and the Archive," *October* 39 (Winter 1986): 3–64; Peter Stallybrass and Allan White, *The Politics and Poetics of Transgression* (Ithaca, N.Y.: Cornell University Press, 1986); Susan Rubin Suleiman, ed., *The Female Body in Western Culture: Contemporary Perspectives* (Cambridge, Mass.: Harvard University Press, 1985).

6. T. J. Jackson Lears, "American Advertising and the Reconstruction of the Body: Images of Health, Sport and the Body, 1880–1930," in *Fitness in American Culture*, ed. Katherine Grover (Amherst: University of Massachusetts Press, 1989), 47–66; Keith Walden, "The Road to Fat City: An Interpretation of the Development of Weight Consciousness in Western

Society," *Historical Reflexions* 12 (1985): 331–73; see also Bryan S. Turner, "The Discourse of Diet," *Theory, Culture, and Society* 1, 1 (1982): 23–32.

7. Hillel Schwartz, *Never Satisfied: A Cultural History of Fantasy and Fat* (New York: Free Press, 1986), 4; according to Schwartz, attitudes toward the body shifted between 1880 and 1920. Dieting emerged within this period, which included an emphasis on lightness and buoyancy in the late nineteenth century and body regulation and measurement in the early twentieth.

8. Ibid., 5.

9. Roberta Pollack Seid, *Never Too Thin: Why Women Are at War with Their Bodies* (New York: Prentice Hall, 1989), 102.

10. Joan Jacobs Brumberg, *Fasting Girls: A History of Anorexia Nervosa* (New York: New American Library, 1988), 238.

11. Ibid., 244.

12. Seid, *Never Too Thin*, 97.

13. Ibid., 98.

14. For the most famous attack on women's education by a physician see Edward Clarke, *Sex in Education: or, A Fair Chance for the Girls* (Boston, Mass.: Houghton, Mifflin, 1873). Charles Darwin's theories of evolution, sex selection, and species selection were manipulated by early critics of women's education to substantiate claims of "race suicide." Darwin did not make such claims. See Louise Newman, ed., *Men's Ideas/Women's Realities*, 1–53; Charles Darwin, *The Origin of the Species* (London: John Murray, 1859); *Descent of Man and Selection in Relation to Sex* (New York: D. Appleton, 1906, 1st ed., 1871). See also Barbara Rosenberg, *Beyond Separate Spheres: Intellectual Roots of Modern Feminism* (New Haven, Conn.: Yale University Press, 1982). For the official response from Smith College see Rev. L. Clark Seelye, *The Early History of Smith College, 1875–1910*, and "The Higher Education of Women: Its Perils and Benefits," c. 1888, public address, Smith College Archives, Northampton, Mass.

15. Several historians, based on their review of nineteenth-century prescriptive literature, have concluded that young women felt pressured to conceal their appetites because the open enjoyment of food was linked to sexual desire, carnality, and the "unrefined" working classes. According to Laura Shapiro, "there was a long-standing assumption that well-bred women were creatures with light, disinterested eating habits." *Perfection Salad: Women and Cooking at the Turn of the Century* (New York: Farrar, Straus and Giroux, 1986), 72. Joan Brumberg argued that "in the late nineteenth century . . . there was a wide spectrum of 'picky eating' and food refusal ranging from the normative to the pathological" among middle-class adolescent girls. *Fasting Girls*, 171. See also Haller and Haller, *The Physician and Sexuality in Victorian America*; Seid, *Never Too Thin*; Michelle Stacey, *Consumed: Why Americans Love, Hate, and Fear Food* (New York: Simon and Schuster, 1994). Several factors may account for the robust appetites and hearty eating of Smith students. First, I examined the food practices of Smith students as they recorded them rather than advice literature. The stark differences may reflect the distance between prescriptions and behavior. Second, I examined the letters of young women living away from home at a single-sex college, whereas other historians have focused on adolescent girls living within the confines and responding to the expectations of the bourgeois home (including courtship). Last, concerns about student health took precedence over other matters. Smith students could not afford dainty appetites if they expected to prove their health and remain in college.

16. Josephine Wilkin, October 15, 1891, Smith College Archives, Northampton, Mass.

17. Gertrude Martha Barry, January 26, 1908, Smith College Archives, Northampton, Mass.

18. Charlotte Wilkinson, February 16, 1892, Smith College Archives, Northampton, Mass.

19. Ibid., April 24, 1892.

20. Ibid., June 11, 1892.

21. Lucy Eliza Kendrew, December 15, 1924, Smith College Archives, Northampton, Mass.

22. According to Hillel Schwartz, bathroom and bedroom scales began to appear in private homes in the early twentieth century. By the 1920s kitchen scales for weighing and measuring food had arrived. Penny scales were plentiful by the 1910s. Schwartz connects the popularization of the scale to new interpretations of the meaning of one's weight. By the 1920s, Schwartz argued, "weight began to carry with it a moral imperative . . . braced by the truth-telling powers of the scale." *Never Satisfied*, 153.

23. "Hints to Freshmen," *Smith College Student Handbook* (1923–1930), Smith College Archives, Northampton, Mass.

24. "Words to the Wise," *Freshman Handbook of Mount Holyoke College* (1925–26), Mount Holyoke College Archives, South Hadley, Mass.

25. Dorothy Smith Dushkin, Papers, 1906–1988, Sophia Smith Collection, Smith College Archives, Northampton, Mass. My thanks to Cathy Verenti for this citation.

26. Anne Morrow Lindbergh, *Bring Me a Unicorn: Diaries and Letters of Anne Murrow Lindbergh* (New York: Harcourt Brace Javanovich, 1971), 110. My thanks to Maddie Cahill for this citation.

27. Anna M. Richardson M.D., "How Well Are the Seniors?" *Smith College Alumnae Quarterly* 19 (1923): 426, Smith College Archives, Northampton, Mass.

28. Ibid.

29. Ibid., 426–27.

30. Florence Gilman, "Report of the College Physician," *Bulletin of Smith College Annual Reports* (1919–1920): 23–24, Smith College Archives, Northampton, Mass.

31. For the general perceptions of college officials see "Reports of the College Physician," *Smith College Annual Bulletins* (1916–1931), Smith College Archives, Northampton, Mass. For specific references to these terms see Gilman, (1919–1920), 23; and (1921–22), 12; Florence Meredith, "Report of the College Physician," *Bulletin of Smith College Annual Report* (1925–1926), 21.

32. Florence Meredith, M.D., "Report of the College Physician," *Bulletin of Smith College Annual Report* (1923–1924), 20, Smith College Archives, Northampton, Mass. College physicians also catalogued "Disturbances of Digestion," which hovered between 12 and 16 percent of all complaints each year.

33. Women's magazines were filled with health and diet advice geared to the mothers of college students in the twenties. See, for example, William Emerson M.D., "The Health of the College Girl," *Woman's Home Companion* 56 (April 1929): 35; and Clarence Lieb, "That Schoolgirl Digestion," *Woman's Home Companion* (June 1929): 22–24.

34. Frances B. Floore, "An Analysis of The Hollywood Eighteen Day Diet," *Hygeia* 9 (March 1930): 245–46; Patricia Seid, *Never Too Thin*, 96.

35. Pauline Ames Plimpton, January 11, 1922, Smith College Archives, Northampton, Mass.

36. Ibid.

37. The number of women attending college increased dramatically after World War I, creating a more diverse student population, but the majority of those students did not attend the private women's colleges. By the mid-1920s about 10 percent of the Smith College student population was Catholic and Jewish, and the vast majority was Protestant. For religious and geographic breakdowns of the student population, see *Bulletin of Smith College Annual Reports* (1920–1930), Smith College Archives, Northampton, Mass. African-American and immigrant women did not attend Smith College in large numbers until the 1970s. While Smith College was interested to have "among its students those who are partially self-supporting," most students came from middle-class families who could afford to pay their daughter's college expenses. See Students and Society, Administration Box 34, Smith College Archives, Northampton, Mass. For general trends in the female college population see Dorothy Brown, *Setting a Course: American Women in the 1920s* (Boston, Mass.: Twayne, 1987); Mabel Newcomber, *A Century of Higher Education for Women;* and Barbara Solomon, *In the Company of Educated Women.*

38. Most historians agree that by the 1920s slenderness signified affluence. Building on Thorstein Veblen's thesis of conspicuous consumption, "that people do not use [a] surplus, . . . for useful purposes . . . but to impress other people with the fact that have a surplus," scholars have argued that dieting symbolized abundance, the luxury of refusing food for aesthetic purposes. Thorstein Veblen, *The Theory of the Leisure Class* (New York: Modern Library, 1899), xiv; Joan Brumberg, *Fasting Girls,* 244–48; Claire Cassidy, "The Good Body: When Big Is Better," *Medical Anthropology* 13 (1991): 181–213; Roberta Seid, *Never Too Thin.* Other scholars suggest that the emergence of a mass culture, including the mass-marketing of clothing, made class lines less distinct in the twenties. As a result, changing the body itself accrued more significance to indicate class status. See Stuart Ewen and Elizabeth Ewen, *Channels of Desire: Mass Images and the Shaping of American Consciousness* (New York: McGraw-Hill, 1982); Valerie Steele, *Fashion and Eroticism;* Claudia B. Kidwell and Margaret Christman,

Suiting Everyone. In *Adorned in Dreams* Elizabeth Wilson, on the other hand, contended that the thin ideal had more to do with modernist aesthetics, with a love of an angular form that suggested movement and speed.

39. Cassidy, "The Good Body," 203.

40. Seid, *Never Too Thin*, 91.

41. See student stories by Helen Josephy, "The Way of Man," *Smith College Monthly* (April 1921): 217–24; and Ethel Halsey, "The Buryin," *Smith College Monthly* (March 1921): 181–85.

42. Levenstein, *Revolution at the Table*, 163.

43. Many historians have documented the cultural changes that swept through middle America in the 1920s and changed the nature of economic, political, and social relations. See Loren Baritz, ed., *The Culture of the Twenties* (Indianapolis: Bobbs-Merrill, 1970); Joan Hoff, ed., *The Twenties: The Critical Issues* (Boston, Mass.: Little, Brown, 1972); William Leuchtenberg, *The Perils of Prosperity, 1914–1932,* rev. ed. (Chicago: University of Chicago Press, 1993); Elizabeth Stevenson, *Babbits and Bohemians: The American 1920s* (New York: Macmillan, 1967).

44. The meaning of *bat* most likely stems from an early definition: 'to go or move; to wander, to potter.' J. A. Simpson and E. S. C. Weiner, eds., *The Oxford English Dictionary*, rev. ed. (Oxford: Clarendon), 995. Among Smith College students, the word *batting*, slang for an outdoor picnic, came into vogue in the 1910s.

45. My thanks to Smith College archivists Margery Sly and Maida Goodwin for their help in sorting through the chronology and meaning of bats. Many collections of student letters and photographs include references to batting in the 1910s. See Agnes Betts, Class of 1916; Marjorie Stafford Root, Class of 1917; and Dorothy Atwill, Class of 1915. See also the *Smith College Weekly*, (May 16, 1923): 3.

46. America's "youth culture" was fully developed by the twenties. See Paula Fass, *The Damned and the Beautiful: American Youth in the 1920s* (Oxford: Oxford University Press, 1977); Beth Bailey, *From Front Porch to Back Seat: Courtship in Twentieth-Century America* (Baltimore, Md.: Johns Hopkins University Press, 1988); and Joseph Kett, *Rites of Passage: Adolescence in America, 1790 to the Present* (New York: Basic Books, 1977). For changes in women's education in the twenties see Dorothy M. Brown, *Setting*

a Course; Horowitz, *Alma Mater;* Kendall, *Peculiar Institutions;* Solomon, *In the Company of Educated Women.*

47. "Approved Eating Places," *Smith College Customs and Regulations,* 1915–1930, Smith College Archives, Northampton, Mass.

48. *Smith College Weekly*, April 11, 1923. Joanne Finkelstein has provocatively argued that modern dining out can be understood "as a means to expression of one's individuality." She suggested that "the event can summarize our knowledge of food and interests in pleasure, status, fashionability and entertainment." Considering the emphasis on individuality and style in the twenties, eating out may have connoted similar meaning to Smith students. Joanne Finkelstein, *Dining Out: A Sociology of Modern Manners* (New York: New York University Press, 1989), 4.

49. President Neilson, "Report of the President," *Bulletin of Smith College Annual Reports*, 10, Smith College Archives, Northampton, Mass.

50. Ibid.

51. "Hints to Freshman," *Smith College Handbook,* 92, Smith College Archives, Northampton, Mass.

52. Horowitz, *Alma Mater*, 285.

53. Bailey, *From Front Porch*, 13. See also Ellen Rothman, *Hand and Hearts: A History of Courtship in America* (New York: Basic Books, 1984).

54. Bailey, *From Front Porch*, 56.

55. Heads of House, "Regulations for Spring Dance 1928," box 541, Smith College Archives, Northampton, Mass.

56. *Fussing* is derived from its early meaning, 'a bustle or commotion' modified by the turn of the century into the slang 'a drinking spree, a binge, a spree of any kind.' J. E. Lighter, ed., *Random House Dictionary of American Slang*, vol. 1 (New York: Random House, 1994), 102.

57. "Fussing Problem Again," *Smith College Weekly*, March 11, 1925, Smith College Archives, Northampton, Mass.

58. Heads of House Meeting Minutes, 1930, Box 541, Smith College Archives, Northampton, Mass.

59. *Lord Jeff* 3, 1 (1921): 12, Amherst College Archives, Amherst, Mass.

60. Ruby Mae Jordon, October 3, 1922, Smith College Archives, Northampton, Mass.

61. Ibid., c. April 1923.

62. Ibid., April 18, 1923.

63. "Program for Prom," *Campus Cat* (January, 1924): 9, Smith College Archives, Northampton, Mass.

64. Warren Susman, *Culture as History: The Transformation of American Society in the Twentieth Century* (New York: Pantheon, 1984), 280–81.

65. Ibid.

66. Ibid.

67. Throughout the college literature, students categorized themselves and each other according to type. Fitting a type depended on family background, class, and ethnicity but also on physical style and adornment. For example, the *Campus Cat* characterized college types by highlighting physical appearance. The collegiate girl wore "blouses under her sweaters, brogues and no hat," while the off-campus type wore "short, flannel dresses, silk stockings and pumps." In another issue the editors contrasted "the all-round girl . . . trying to get thin," while the girl-of-moods lounged at vespers, "sleeping off her last exam." Smith students hoped to appear alluringly feminine or modern to eligible young men, the all-around college girl to classmates, and appropriately professional to faculty and vocational counselors.

68. Personnel Department, Class of 1926, box 1, Smith College Archives, Northampton, Mass.

69. For a discussion of the New Woman see Mary P. Ryan, *Womanhood in America*, 2nd ed. (New York: New Viewpoints, 1979), ch. 4; Carroll Smith-Rosenberg, "The New Woman as Androgyne: Social Disorder and the Gender Crisis, 1870–1936," in *Disorderly Conduct* (Oxford: Oxford University Press, 1985), 245–96. See also Martha Banta, *Imagining American Women*, 45–91.

70. See Mary E. Cookingham, "Blue Stockings, Spinsters, and Pedagogues: Women College Graduates, 1865–1910," *Population Studies* 38 (1984): 349–64; Gordon, "The First Ten Classes of Smith College"; Horowitz, *Alma Mater*; Newcomber, *A Century of Higher Education for American Women*; Newman, *Men's Ideas/Women's Realities*; Barbara Sicherman, "College and Careers: Historical Work Patterns of Women College Graduates," in *Women and Higher Education*, ed. Faraghar and Howe; Solomon, *In the Company of Educated Women*.

71. Lois Banner, *American Beauty*; see also Martha Banta, *Imaging American Women;* Valerie Steele, *Fashion and Eroticism;* Claudia Bush Kidwell and Valerie Steele, *Men and Women: Dressing the Part.*

72. According to Banner, the voluptuous woman originated in 1870s working-class culture which featured the "British Blondes" on stage and curvaceous prostitutes on the street. See Robert Allen for an analysis of how the burlesque stage challenged and subverted late nineteenth-century gender categories: Robert Allen, *Horrible Prettiness: Burlesque and American Culture* (Chapel Hill: University of North Carolina Press, 1991).

73. Steele, in *Men and Women Dressing the Part*, 20.

74. Scholars differ in their interpretation of the flapper. Some see the image as boyish or trivial while others consider it sophisticated and urbane. For Smith students the most resonant characteristics were its youthful rebellious style and modern sensibility that included sexual experimentation. In *Adorned in Dreams* Elizabeth Wilson argued that by the 1920s slenderness reflected several cultural changes: the effect of photography and film which accentuated width; the vogue for youth which rejected an "aging" Western society; and the modernist love of angular form that rejected "the natural." For fashion history see Banner, *American Beauty;* and Steele, *Fashion and Eroticism;* for more detail on the connections between 1920s' youth and the flapper, see Bailey, *From Front Porch;* Baritz, *The Culture of the Twenties;* Leuchtenberg, *The Perils of Prosperity.*

75. Kidwell and Christman, *Suiting Everyone;* Kidwell and Steele, *Men and Women;* Ewen and Ewen, *Channels of Desire.* Kidwell and Christman suggested that early sizing standards were based on college women's physiques, including Smith women. This would make sense since women's colleges kept detailed records of the students' physical measurements taken during physical exams and college women were often perceived as in the fashion vanguard.

76. *Smith College Weekly* (May 17, 1922): 5. Smith students were avid moviegoers. Regulars at the Calvin Theater and the Academy of Music, they rated movies and actors in letters home and debated the merits of moviegoing and movie etiquette in the *Weekly.* For two views of the relationship between the movies and women see Hansen Miriam, *Babel and*

Babylon: Spectatorship in American Silent Film (Cambridge, Mass.: Harvard University Press, 1991); and Majorie Rosen, *Popcorn Venus: Women, Movies, and the American Dream* (New York: Coward, McCann and Geoghegan, 1973).

77. "Hints to Freshmen," *Smith College Handbook* (1928): 96, Smith College Archives, Northampton, Mass.

78. Shapiro, *Perfection Salad,* 47.

79. Levenstein, *Revolution at the Table;* Shapiro, *Perfection Salad;* Michelle Stacey, *Consumed: Why Americans Love, Hate, and Fear Food* (New York: Simon and Schuster, 1994); James C. Whorton, "Eating to Win: Popular Concepts of Diet, Strength, and Energy in the Early Twentieth Century," in *Fitness in American Culture: Images of Health, Sport, and the Body, 1830–1940,* ed. Kathryn Grover (Amherst: University of Massachusetts Press, 1989), 86–122.

80. See *New York Times* (May 4, 1992), 1.

81. Levenstein, *Revolution at the Table,* 72–85; Seid, *Never Too Thin,* 87–88; Michelle Stacey, *Consumed,* 27–59; Green, *Fit for America.*

82. The field of home economics was well established by the 1920s. College women entered the profession in high numbers and also served as research subjects to their professors and fellow students. The *Journal of Home Economics* contained numerous nutrition studies that focused on college students during this period. For examples, see Katherine Blunt and Virginia Bauer, "The Basal Metabolism and Food Consumption of Underweight College Women," *Journal of Home Economics* 14, 1 (1922): 171–80; Martha Kramer and Edith Grundmeier, "Food Selection and Expenditure in a College Community," *Journal of Home Economics* 18, 1 (1926): 18–23. See also Hamilton Cravens, "Establishing the Science of Nutrition at the USDA: Ellen Swallow Richards and Her Allies," *Agricultural History* 64 (Spring 1990): 122–34.

83. Lulu Peters, *Diet and Health with the Key to the Calories* (Chicago: Reilly and Britton, 1918); see also Jean L. Bogert, *Diet and Personality: Fitting Food to Type and Environment* (New York: Macmillan, 1934); Kathryn Daum, M.D., "How to Change Your Weight," *Good Housekeeping* 80 (May 1925): 76–77.

84. See, for example, Guilielma Alsop, "Food for a Good Figure," *Woman Citizen* 9 (April 18, 1925): 26. Alsop was one of many doctors in the 1920s who regularly published her recommendations in women's magazines. Alsop promoted "the perfect diet for the perfect body."

85. For a discussion of the development of insurance height and weight tables see Brumberg, *Fasting Girls,* 232–33; Schwartz, *Never Satisfied,* 153–59; Seid, *Never Too Thin,* 90.

86. For examples of the popular discourse see Alonzo E. Taylor, "The National Overweight," *Scientific Monthly* 32 (1931): 393–97; and "The Cult of Slimness," *Living Age* 280 (March 1914): 572–75.

87. Levenstein, *Revolution at the Table,* 166.

88. See Banner, *American Beauty;* William Bennett and Joel Gurin, *The Dieter's Dilemma: Eating Less and Weighing More* (New York: Basic Books, 1982); Anne Scott Beller, *Fat and Thin: A Natural History of Obesity* (New York: Farrar, Straus and Giroux, 1977); Schwartz, *Never Satisfied.*

89. Buildings and Grounds, Administration of Houses, box 184, 1930 Smith College Archives, Northampton, Mass.

90. Dr. Goldthwait, "Health Information Distributed," 1916–1917, box 52.Sci, Smith College Archives, Northampton, Mass.

91. Such interpretational reading of the body, according to Joanne Finkelstein, perpetuates "physiognomic perspectives" which regard "physical appearance as the key to understanding human character." See Joanne Finkelstein, *The Fashioned Self,* 7.

92. Seid, *Never Too Thin,* 97.

Sexuality

According to Victorian ideology, men were passionate and sexual and women were passionless and uninterested in sex. As Carl Degler relates, one British woman advising her daughter before her marriage is said to have told her to submit to her husband's needs and to "lie still and think of the Empire." Most physicians, however, recognized that women could and did enjoy sexual activity. But the medical profession had its own ideas about what was healthy for women sexually and what produced problems. In this section three historians explore the ideology and the medical and women's response to it.

Degler examines the common medical notion that women were not sexual beings and counters it with data from a survey of forty-five women's descriptions of their sexual activity, information collected around the turn of the century. Turning to a different aspect of women's sexuality, Marylynne Diggs explores the dual "problem" of masturbation and homosexuality, which produced its own set of medical advice literature. Physicians and society in general certainly did not sanction what they believed was abnormal and dangerous behavior, and yet the society that did not condone homosexuality nonetheless harbored it. Diggs describes the historiographic debates about the extent to which intimate and erotic relationships between women might have been acceptable and critically scrutinizes the often repeated image of a "golden age" of romantic friendships between women. She concludes that, while romantic friendship was one current of nineteenth-century discourse, it was "neither the only nor the dominant signification for exclusive and erotic relations between women."

Elizabeth Lunbeck addresses the issue of women's sexuality through a study of early twentieth-century psychiatrists' discussions of the "hypersexual" woman. Lunbeck understands physicians' ideas about sexually assertive women within the social context of women who were moving into the public sphere and especially in terms of working-class women. She sees the medical idea of the psychopathic hypersexual as born partly from an imaginary worry brought about by the social changes and partly as a real but exaggerated notion of the threat that women could pose.

Sexuality, a sensitive and personal arena, was a subject, like menstruation, that most women did not write about explicitly or frequently. Thus the historian confronts the same dilemma of trying to understand a subject in the face of limited data. Degler's analysis of the Clelia Mosher survey provides some concrete evidence of a few women's perceptions of their sexuality, but the record remains too silent for most other women. Diggs and Lunbeck add greatly to our understanding of the range of choices available to women. Their clear delineation of what discourse or activity was acceptable and the sanctions that waited women who ventured too far illustrates how society and medicine together defined and limited women's sexual options.

10 What Ought to Be and What Was: Women's Sexuality in the Nineteenth Century

Carl N. Degler

As every schoolgirl knows, the nineteenth century was afraid of sex, particularly when it manifested itself in women. Captain Marryat, in his travels in the United States, told of some American women so refined that they objected to the word *leg*, preferring instead the more decorous *limb*. Marryat also reported seeing this delicacy carried to extremes in a girls' school where a schoolmistress, in the interest of protecting the modesty of her charges, had dressed all four "limbs" of the piano "in modest little trousers with frills at the bottom of them!"[1] Women's alleged lack of passion was epitomized too in the story of the English mother who was asked by her daughter before her marriage how she ought to behave on her wedding night. "Lie still and think of the Empire," the mother advised.

This view of Victorian attitudes toward sexuality is captured in more than stories. Steven Marcus, writing about the attitudes of English Victorians toward sexuality, and Nathan Hale, Jr., summarizing the attitudes of Americans on the same subject, both quote at length from Dr.

William Acton's *Functions and Disorders of the Reproductive Organs,* which went through several editions in England and the United States during the middle years of the nineteenth century.[2] Acton's book was undoubtedly one of the most widely quoted sexual advice books in the English-speaking world. The book summed up the medical literature on women's sexuality by saying that "the majority of women (happily for them) are not very much troubled with sexual feelings of any kind. What men are habitually, women are only exceptionally."[3] Theophilus Parvin, an American doctor, told his medical class in 1883, "I do not believe one bride in a hundred, of delicate, educated, sensitive women, accepts matrimony from any desire for sexual gratification; when she thinks of this at all, it is with shrinking, or even with horror, rather than with desire."[4]

Modern writers on the sexual life of women in the nineteenth century have echoed these contemporary descriptions. "For the sexual act was associated by many wives only with a duty," writes Walter Houghton, "and by most husbands with a necessary if pleasurable yielding to one's baser nature; by few, therefore, with any innocent and joyful experience."[5] Writing about late nineteenth-century America, David Ken-

CARL N. DEGLER is Margaret Byrne Professor Emeritus of American History at Stanford University, Stanford, California.

Reprinted from *American Historical Review* 79 (1974): 1467–90, by permission of the author.

nedy quotes approvingly from Viola Klein when she writes that "in the whole Western world during the nineteenth century and at the beginning of the twentieth century it would have been not only scandalous to admit the existence of a strong sex urge in women, but it would have been contrary to all observation."[6] Nathan Hale, Jr., sums up his review of the sexual advice literature at the turn of the century with a similar conclusion: "Many women came to regard marriage as little better than legalized prostitution. Sexual passion became associated almost exclusively with the male, with prostitutes, and women of the lower classes."[7] Most recently, Ben Barker-Benfield has argued that male doctors were so convinced that women had no sexual interest that when it manifested itself drastic measures were taken to subdue it, including excision of the sexual organs. "Defining the absence of sexual desire in women as normal, doctors came to see its presence as disease. . . . Sexual appetite was a male quality (to be properly channelled of course). If a woman showed it, she resembled a man."[8]

Despite the apparent agreement between the nineteenth-century medical writers and modern students of the period, it is far from clear that there was in the nineteenth century a consensus on the subject of women's sexuality or that women were in fact inhibited from acknowledging their sexual feelings. In examining these two issues I shall be concerned with an admittedly limited yet significant population, namely, women of the urban middle class in the United States. This was the class to which the popular medical advice books, of which William Acton's volume was a prime example, were directed. It is principally the women of this class upon whom historians' generalizations about women's lives in the nineteenth century are based. And though these women were not a numerical majority of the sex, they undoubtedly set the tone and provided the models for most women. The sources

drawn upon are principally the popular and professional medical literature concerned with women and a hitherto undiscovered survey of married women's sexual attitudes and practices that was begun in the 1890s by Dr. Clelia D. Mosher.

Let me begin with the first question or issue. Was William Acton representative of medical writers when he contended that women were essentially without sexual passion? Rather serious doubts arise as soon as one looks into the medical literature, popular as well as professional, where it was recognized that the sex drive was so strong in woman that to deny it might well compromise her health. Dr. Charles Taylor, writing in 1882, said, "It is not a matter of indifference whether a woman live a single or a married life. . . . I do not for one moment wish to be understood as believing that an unmarried woman cannot exist in perfect health for I know she can. But the point is, that *she must take pains for it.*" For if the generative organs are not used, then "some other demand for the unemployed functions, must be established. Accumulated force must find an outlet, or disturbance first and weakness ultimately results." His recommendation was muscular exercise and education for usefulness. He also described cases of women who had denied their sexuality and even experienced orgasms without knowing it. Some women, he added, ended up, as a result, with impairment of movement or other physical symptoms.[9]

Other writers on medical matters were even more direct in testifying to the presence of sexual feelings in women. "Passion absolutely necessary in woman," wrote Orson S. Fowler, the phrenologist, in 1870. "Amativeness is created in the female head as universally as in the male. . . . That female passion exists, is as obvious as that the sun shines," he wrote. Without woman's passion, he contended, a fulfilled love could not

193

occur.[10] Both sexes enjoy the sexual embrace, asserted Henry Chavasse, another popular medical writer, in 1866, but among human beings, as among the animals in general, he continued, "the male is more ardent and fierce, and . . . the desires of the female never reach that hight [sic] as to impel her to the commission of crime." Woman's pleasure, though it may be "less acute," is longer lasting than man's, Chavasse said. R. T. Trall, also a popular medical writer, counseled in a similar vein. "Whatever may be the object of sexual intercourse," he wrote, "whether intended as a love embrace merely, or a generative act, it is very clear that it should be as pleasurable as possible to *both parties*."[11]

If one can judge the popularity of a guide for women by the number of its editions, then Dr. George Napheys's *The Physical Life of Woman: Advice to the Maiden, Wife, and Mother* (1869) must have been one of the leaders. Within two weeks of publication it went into a second printing, and within two years sixty thousand copies were in print. Napheys was a well-known Philadelphia physician. Women, he wrote, quoting an unnamed "distinguished medical writer," are divided into three classes. The first consists of those who have no sexual feelings, and it is the smallest group. The second is larger and comprises those who have "strong passion." The third is made up of "the vast majority of women, in whom the sexual appetite is as moderate as all other appetites." He went on to make his point quite clear. "It is a false notion and contrary to nature that this passion in a woman is a derogation to her sex. The science of physiology indicates most clearly its propriety and dignity." He then proceeded to denounce those wives who "plume themselves on their repugnance or their distaste for their conjugal obligations." Napheys also contended that authorities agree that "conception is more assured when the two individuals who co-operate in it participate

at the same time in the transports of which it is the fruit." Napheys probably had no sound reason for this point, but the accuracy of his statement is immaterial. What is of moment is that as an adviser to women he was clearly convinced that women possessed sexual feelings, which ought to be cultivated rather than suppressed. Concerning sexual relations during pregnancy he wrote, "There is no reason why passions should not be gratified in moderation and with caution during the whole period of pregnancy." And since his book is directed to women, there is no question that the passion he is talking about here is that of women.[12]

In 1878 Dr. Ely Van de Warker of Syracuse, a fellow of the American Gynecological Society, described sexual passion in women as "the analogue of the subjective copulative sensations of man, ... the acme of the sexual orgasm in woman is the sensory equivalent of emission in man, observing the distinction necessarily implied between the sexes—that in woman it is psychic and subjective, and that in man it has also a physical element and is objective," that is, it is accompanied by seminal emission. The principal purpose of Van de Warker's article was to deplore the fact that some women lacked sexual feeling, a state which he called "female impotency."[13] What is striking about his article is that he obviously considered such lack of feeling in women abnormal and worthy of medical attention, just as impotency in a man would cause medical concern.

Van de Warker's remarks, as well as his use of the word, make it evident that physicians were well aware that normal women experienced orgasms. Lest there be any doubt that their meaning of the word was the same as ours today, let me quote from a physician in 1883 who described in some detail woman's sexual response. He began by describing the preparatory stage, which, he said, may be reached by any means, bodily or mental, which, in the opposite sex, cause erection. Following upon this, then, is a stage of pleasurable excitement, gradually increasing and culminating in an acme of excitement, which may be called the state of consummation, and the analogue of which in the male is emission. This is followed in both sexes by a degree of nervous prostration, less marked, however, in the female, and . . . by a relief to the general congestion of all the genital organs which has existed, and perhaps increased, from the beginning of the preparatory stage.[14]

All of this evidence, it seems to me, shows that there was a significant body of opinion and information quite different from that advanced on women's sexuality by William Acton and others of his outlook. Now it might be asked how widespread was this counter-Acton point of view? Was it not confined primarily to physicians writing for other physicians? Not at all. Napheys, Chavasse, and Fowler, to name three, were all writing their books for the large lay public that was interested in sexual matters. As we have seen, many of these marriage manuals, particularly Napheys's and Fowler's, were printed in several large editions.

Yet, in the end, there is a certain undeniable inconclusiveness in simply raising up one collection of writers against another, even if their existence does make the issue an open one, rather than the closed one that so many secondary writers have made it. It suggests, at the very least, that there was a sharp difference of medical opinion, rather than a consensus, on the nature of women's sexual feelings and needs. In fact, there is some reason to believe, as we shall see, that the so-called Victorian conception of women's sexuality was more that of an ideology seeking to be established than the prevalent view

of practice of even middle-class women, especially as there is a substantial amount of nineteenth-century writing about women that assumes the existence of strong sexual feelings in women. One of the historian's recognized difficulties in showing, through quotations from writers who assert a particular outlook, that a social attitude prevailed in the past is that one always wonders how representative and how self-serving the examples or quotations are. This is especially true in this case where medical opinion can be found on both sides of the question. When writers, however, assume the attitude in question to be prevalent while they are intent upon writing about something else, then one is not so dependent upon the tyranny of numbers in quoting from sources. For behind the assumption of prevalence lie many examples, so to speak. Such testimony, moreover, is unintended and therefore not self-serving. This kind of evidence, furthermore, helps us to answer the second question—to what extent were women in the nineteenth century inhibited from expressing their sexual feelings? For in assuming that women had sexual feelings, these writers are offering clear, if unintended, testimony to women's sexuality.

Medical writers like Acton may have asserted that women did not possess sexual feelings, but there were many doctors who clearly assumed not only that such feelings existed but that the repression of them caused illness. One medical man, for example, writing in 1877, traced a cause of insanity in women to the onset of sexuality. "Sexual development initiates new and extraordinary physical change," he pointed out. "The erotic and sexual impulse is awakened."[15] Another, writing ten years later, asserted that some of women's illnesses were due to a denial of sexual satisfaction. "Females feel often that they are not appreciated," wrote Dr. William McLaury in a medical journal, "that they have

no sexual feelings = disorder/illness

195

no one to confide in; then they become morose, angular, and disagreeable as a result of continual disappointment to their social and sexual longings. Even those married may become the victims of sexual starvation when the parties are mentally, magnetically, and physically antagonistic."[16] Henry Chavasse, writing for a popular audience, was also impressed by the need for sexual outlets for women. There may be some individuals "of phlegmatic temperament," he conceded, who are not injured by celibacy, but "absolute continence in the sanguine and ardent disposition predisposes to the gravest maladies." His listing of the resulting maladies, of which nymphomania was one, makes it clear that he was referring to women as well as men. These maladies, he went on, "are born as well of extreme restraint as of extreme excess. . . . Females seem to suffer even more than males . . . perhaps because their continence is more complete." (Presumably he was referring here to the absence of nocturnal emissions in women.) As a result, he continued, nunneries were notorious as places of fanaticism. "Hence the old proverb, 'The convent and the confessional are the cradles of hysteria and nymphomania.'"[17]

To Dr. Van de Warker women's sexuality was so obvious that he assumed men required it in order to achieve full sexual satisfaction for themselves. In marriage, he wrote, the husband

not only demands pleasure and satisfaction for himself, but he requires something much more difficult to give—the appearance, if not the real existence, of satisfaction and pleasure in the object of his attentions. Unhappiness and suspicion are often the result of the absence of this pleasure [in women], and are sure to work to the material disadvantage of the weaker party. To show that this is really the case, I need but to remind physicians how often they are approached by husbands upon

this subject; yet further, how often the coldness and indifference of wives are alleged as the excuse for conjugal infidelity.[18]

What is striking in this passage is that husbands complained to doctors about their wives' coldness, a fact that makes it quite evident that passion in wives was not only desired by men but expected—why, otherwise, would they complain of its lack? Van de Warker, it is worth pointing out, was writing for his fellow physicians, who were in a position to verify his assertions from their own experience with patients.

Van de Warker's explanation for "impotency" in women is revealing too. Ascribing it to "sexual incompatibility," he went on to say that "so far as my own observation extends, the husband is generally at fault. The more common cause is acute sexual irritability on the part of the husband."[19] Dr. William Goodell, writing in 1887, also asserted that mutual pleasure was essential to successful marital intercourse. In Goodell's mind, as in Van de Warker's, that meant men must recognize women's interests and sexual rhythm. "Destroy the reciprocity of the union," Goodell cautioned, "and marriage is no longer an equal partnership, but a sensual usurpation on the one side and a loathing submission on the other."[20] Another medical writer who also acknowledged women's pleasure in the sex act made the same point as Goodell and Van de Warker. Men must not force themselves upon women or "overpersuade, but await the wife's invitation at this time [during ovulation], when her husband is a hero in her eyes." In this way the husband "would enjoy more and suffer less," the physician predicted.[21] These writers, in short, were not only testifying to their knowledge that women possessed sexual feelings, they were also explaining how those feelings were sometimes denied legitimate satisfaction by inept husbands.

contraception

Masturbation
can happen by
1. touching
2. fantasize

DEGLER/What Ought to Be and What Was

The assumption that women had sexual feelings which required satisfaction also comes through in the course of discussions about contraception. Generally, physicians and other writers on this subject in the nineteenth century strongly opposed contraception, though all recognized that it was widely practiced. One of the methods in common use was coitus interruptus, or withdrawal by the male prior to ejaculation. This method was condemned for a variety of reasons, but for our purposes it is significant that among the objections was its harmful effects upon women. This method, wrote Henry Chavasse, is "attended with disastrous consequences, most particularly to the female, whose nervous system suffers from ungratified excitement."[22] Dr. John Harvey Kellogg, a popular writer on medical matters, also warned against the method because of its effects upon women. He quoted at length from a French authority. Whenever this method is practiced, the authority wrote, all of women's genital organs "enter into a state of orgasm, a storm which is not appeased by the natural crisis; a nervous super excitation persists" after the act. The authority then compared the unreleased tension to that evoked in presenting food to a "famished man" and then snatching it away. "The sensibilities of the womb and the entire reproductive system are teased to no purpose." It is evident that in the minds of both writers women were assumed to have sexual feelings that were normally aroused during sexual intercourse.[23] Dr. Augustus Gardner, writing in 1870 also for a popular audience, quoted from the same French authority and for the same purpose as Kellogg.[24]

Anyone who has looked into the sexual history of the nineteenth century is immediately struck by the deep and anxious concern physicians as well as other people felt about masturbation. Although it is often thought that boys were the principal objects of that concern, the fact is that girls were just as much fretted about.

That there were such concerns about girls' masturbating is in itself a sign and measure of the recognition of sexual feelings in women. In fact, in 1866 one popular medical writer on women defined masturbation as "the mechanical irritation of the sexual organs in order to excite the same voluptuous sensations attendant upon natural intercourse."[25] Mary Wood-Allen, a leader in the Women's Christian Temperance Union and a writer of advice books for young women, had no doubt that girls could be led into self-abuse. Even girls who would not use any mechanical means "to arouse sexual desire," she pointed out, nevertheless permitted themselves to fantasize or to have mental images that "arouse the spasmodic feelings of sexual pleasure."[26] Indeed, from Wood-Allen's book one receives the message that women's sexual feelings were not only present but dangerously easy to arouse.

Discussion about masturbation in women reveals in another way how widely accepted was the idea that women possessed sexual desires. One physician, in the course of an article on the subject, said that the worst thing about masturbation in women was that a climax and resolution of tension were generally not achieved; hence the vice was persisted in. In response another doctor agreed that masturbation indeed gave rise to all the physical harm alleged in the article. But he disagreed with the assertion that in a woman sexual excitation could stop short of orgasm. "A commencement of the act, either of masturbation or coition," the letter writer contended, "*naturally* leads to its consummation, viz., an orgasm." Furthermore, he persisted, if "in the *healthy* female, an orgasm is not produced in the act of coition, she is not satisfied, and either will continue the act herself or with her coadjutor till such consummation does take place."[27]

Women's sexuality is also assumed in another class of medical concerns. When Dr. J. Marion

concern about masturbation for girls and boys

to masturbate means to reach "O" if not, she is not satisfy.

197

women can be "impotent" in their sex lives because they don't like it

Sims, the "founding father" of American gyne-cology, published *Clinical Notes on Uterine Surgery* in 1866, conception was only dimly understood. In explaining how it took place Sims revealed, in passing, that most people took for granted that women experienced sexual feelings. "It is the vulgar opinion, and the opinion of many savants," Sims remarked, "that, to ensure conception, sexual intercourse should be performed with a certain degree of completeness, that would give an exhaustive satisfaction to both parties at the same moment." This sounds like twentieth-century ideas on optimum sexual performance, for Sims then went on to note, again in passing, that husbands and wives strove for such simultaneity and were unhappy when they failed to have simultaneous orgasms. "How often do we hear husbands complain of coldness on the part of the wives; and attribute to this the failure to procreate. And sometimes wives are disposed to think, though they never complain, that the fault lies with the hasty ejaculation of the husband."[28] Sims's point, of course, was that conception did not depend upon either sexual arousal or satisfaction in the women. The important point for us, however, is that Sims, the medical readers he was addressing, and the patients he treated all believed women were naturally capable of sexual feelings. Napheys in his popular book of advice for women also alluded to the prevalent idea that conception and pleasure were connected. He said that many people erroneously believed that conception could be known from the "more than ordinary degree of pleasure" on the part of the woman during the sexual act.[29]

In the course of discussing other kinds of women's illnesses, physicians often made it clear that they not only recognized the existence of sexual feelings in women but expected them in normal women. As we have observed already, Dr. Van de Warker considered the lack of sexual feelings in a woman as an abnormality to be cured. He called such women "impotent," just as one would denominate a man who failed to have adequate sexual responses. To Van de Warker, women had to learn how to dislike sex; enjoyment of it was natural.[30] Napheys too saw frigidity as abnormal; its removal, he thought, was "so desirable."[31] One physician in 1882, in discussing a case of excessive masturbation, wrote that during an examination his female patient experienced "the most intense orgasm that I have ever witnessed,"[32] implying that he had witnessed others. Another physician listed among the pathological symptoms of one patient "an absence of all sexual desire"[33]—as if its presence were the normal condition of a woman. One medical doctor, in trying to show how intense was the pain a married patient experienced during intercourse, said that both partners had given up sexual relations "although both had usually violent animal passions."[34] In arguing against birth control Dr. Augustus Gardner told of a wife who, fearing pregnancy since she had borne seven children in seven years, was "otherwise very ardent."[35]

During the 1880s and 1890s, as surgeons became more skillful and antisepsis made abdominal operations safer, a number of doctors sought to alleviate otherwise incurable or obscure pelvic pains and nervous conditions in women through the removal of ovaries. This medical development is a complex one, especially as to the attitudes it might reveal on the part of doctors and society in general. This is not the place to pursue that question, however. It serves to explain, though, why ovariodectomies were a subject of considerable interest among gynecologists. One consequence of that interest was a report in 1890 by a surgeon who had removed forty-six pairs of ovaries. Significantly, he related that "the sexual instinct was always preserved. Three patients, virginal before operation, married later and lived in happy wedlock. The passions persist particularly when the oper-

removal of ovaries because couples can be happy

Why not to have removal of ovaries?

ation is performed early on young persons," he concluded.[36] For us the significance of this report is not whether it is accurate; in fact, I suspect that it is not. For as Dr. Van de Warker remarked on a different occasion, many women who suffered the pain or nervousness that caused them to submit to the operation in the first place probably had never felt any sexual pleasure. Consequently, to ask them after the operation whether there was any diminution in sexual feeling generally brought a denial. Moreover, the removal of the ovaries may well have reduced or eliminated hormonal secretions that may contribute to normal sexual feelings in women. In short, the physician's report suffers from his clear wish to put his series of operations in a good light. But that very wish is revealing, for what it tells us is that women were expected to have sexual feelings and it was undesirable for a surgeon or, presumably, anyone else, to eliminate or even reduce those feelings.

evidence to say women have sexual feelings?

In the light of the foregoing it is difficult to accept the view that women were generally seen in the nineteenth century as without sexual feelings or drives. The question then arises as to how this widely accepted historical interpretation got established? Part of the reason, undoubtedly, is the result of the general reticence of the nineteenth century in regard to sex. The excessive gentility of the middle class has been read by historians as a sign of hostility toward sexuality, particularly in women. The whole cult of the home and women's allegedly exalted place in it was easily translated by some historians into an antisexual attitude.[37] But a good part of the explanation must also be attributable to the simple failure on the part of historians to survey fully the extant sources. The kind of statements quoted from medical writers in this article, for example, was either overlooked or ignored. Another important part of the explanation is that the sources that were surveyed and quoted were

taken to be descriptive of the sexual ideology of the time when in fact they were part of an effort by some other medical writers to establish an ideology, not to delineate an already accepted one. In other words, the medical literature that was emphasized by Steven Marcus, Oscar Handlin, or Nathan Hale, Jr., was really normative or prescriptive rather than descriptive.

This misinterpretation was easy enough to make since much nineteenth-century medical literature was often descriptive in form even though in fact it was seeking to set a new standard of sexual behavior. Sometimes, however, the normative concerns and purposes showed through the ostensible description. A close reading, for example, of William Acton's second edition of *The Functions and Disorders of the Reproductive Organs* reveals in several places his desire to establish a new and presumably "higher" standard of sexual attitude and behavior. After pointing out that publicists strongly condemn sexual relations outside marriage, he asks, "But should we stop there? I think not. The audience should be informed that, in the present state of society, the sexual appetites must not be fostered; and experience teaches those who have had the largest means of information on the matter, that self-control must be exercised." So far, he continues, no one has "dared publicly to advocate . . . this necessary regulation of the sexual feelings or training to continence." Or later, when he discusses women in particular, it is evident that he is arguing for a special attitude, not merely describing common practice. "The *best* mothers, wives, and managers of households know little or nothing of sexual indulgence. Love of home, children, and domestic duties are the only passions they feel," he writes.[38]

American writers of the time who followed the lead of Acton as well as quoting him display a similar mixture of prescription and description. Take Dr. John Kellogg's *Plain Facts for Old and Young,* which sold over three hundred thousand copies by 1910 and went through five editions. Kellogg, like Acton, made it clear that he thought sex was too dominant in the thoughts of people. As we look around us today, he wrote, "it would appear that the opportunity for sensual gratification has come to be, in the world at large, the chief attraction between the sexes. If to these observations," he continued, "we add the filthy disclosures constantly made in police court and scandal suits, we have a powerful confirmation of the opinion."[39] It was this excess that he warns against, drawing upon quotations from Acton to support his arguments. He is at pains to show too that continence, especially in men, is not deleterious to health, as some contended. He admits that the medical profession is not in agreement on the amount of sexual indulgence permitted in marriage. "A very few hold that the sexual act should never be indulged except for the purpose of reproduction, and then only at periods when reproduction will be possible. Others, while equally opposed to the excesses . . . limit indulgence to the number of months in the year." Human beings, he advised, should take their cue from animals, who have intercourse only for procreation and then at widely spaced intervals. Instead of heeding this counsel, he writes, loosely quoting from Acton, "the lengths to which married people carry excesses is perfectly astonishing."[40]

Kellogg's reference to the behavior of animals as a worthy guideline for human behavior was echoed by other writers who sought to control sexuality. William Acton and Orson S. Fowler, for example, also used that standard of sexual behavior. Kellogg even went so far as to make an overt defense of the analogy. He carefully explained to his readers that in the modern age of biology these analogies were extremely helpful in getting at nature's purpose. "It is by this method of investigation," he remarked, "that most of the important truths of physiology have been developed; and the plan is universally ac-

knowledged to be a proper and logical one." Then he launched into a condemnation of those men who use their wives as harlots, "having no other end but pleasure." For it was clear that among animals the end was reproduction only and then only at those one or two times a year when reproduction was possible. But by the time Kellogg reached the place in his book where he defended the analogy with animals, he had already revealed that his purpose in invoking the analogy was reformist and normative, not simply scientific and logical. For in the early pages of his book, in making a different normative point—the need to protect children from premature sexuality—he told of a parent whose adolescent children often played games in the nude. When admonished for permitting this practice, the parent replied that it was only natural. "Perfectly harmless; just like little pigs!" Kellogg quoted the parent as saying. Kellogg's comment, however, was quite different from what he would advise later in his book: "as though pigs were models for human beings!"[41]

In the end Kellogg himself virtually admitted that his "plain facts" were hardly facts at all but prescriptions and hopes. "There will be many," he wrote, "the vast majority, perhaps, who will not bring their minds to accept the truth which nature seems to teach, which would confine sexual acts to reproduction wholly." And so he was prepared to offer a compromise, that is, a method of contraception. It was not a very effective method, as he admitted—the so-called safe period—but again what is important is his frank recognition that only a minority among his readers confined their sexual activities to reproduction and that he hoped he would be able to induce more to do so.[42]

It would be a mistake, in short, to accept the prescriptive or normative literature, like that of Acton, Kellogg, and others,[43] as revealing very much about sexual behavior in the Victorian era. It may be possible to derive a sexual ideol-

200

ogy from such writers, but it is a mistake to assume that the ideology thus delineated is either characteristic of the society or reflective of behavior. On the contrary, it is the argument of this article that the attitudes and behavior of middle-class women were only peripherally affected by the ideology. Not only did many medical writers, as we have seen, encourage women to express their sexuality, but there is a further, even more persuasive reason for believing that the prescriptive literature is not a reliable guide to either the sexual behavior or the attitudes of middle-class women. It is the testimony of women themselves.

Any systematic knowledge of the sexual habits of women is a relatively recent historical acquisition, confined to the surveys of women made in the 1920s and 1930s and culminating in the well-known Kinsey report.[44] Until recently no even slightly comparable body of evidence for nineteenth-century women was known to exist. In the Stanford University Archives, however, are questionnaires completed by a group of women testifying to their sexual habits. The questionnaires are part of the papers of Dr. Clelia Duel Mosher (1863–1940), a physician at Stanford University and a pioneer in the study of women's sexuality. Mosher began her work on the sexual habits of married women when she was a student at the University of Wisconsin prior to 1892. That year she transferred for her senior year to Stanford, where she received an A.B. degree in 1893 and an M.A. in 1894. In 1900 she earned an M.D. degree from Johns Hopkins University. After a decade of private practice she joined the Stanford faculty as a member of the department of hygiene and medical adviser of women students. Her published work dealt with the physical capabilities of women; she was a well-known advocate of physical exercise for women. Mosher's questionnaires are carefully arranged and bound in vol-

ume 10 of her unpublished work, "Hygiene and Physiology of Women." Mosher, however, apparently never drew more than a few impressionistic conclusions from the highly revealing questionnaires. She did not even publish the fact of their existence, and so far as can be ascertained no use has heretofore been made of this manuscript source. Yet the amount and kind of information on sexual habits and attitudes of married women in the late nineteenth century contained in these questionnaires are unique.

The project, which spanned some twenty years, was begun at the University of Wisconsin when Mosher was a student of biology in the early 1890s. She designed the questionnaire when asked to address the Mother's Club at the university on the subject of marriage. In later years she added to her cases and used the information when giving advice to women about sexual and hygienic matters.[45] This initiative, as well as the kind of questions she asked, reveals that Mosher was far ahead of her time. She amassed information on women's sexuality that none of the many nineteenth-century writers on the subject studied in any systematic way at all.

The questionnaire itself is quite lengthy, comprising twenty-five questions, each one of which is divided into several parts. Much of the questionnaire, it is true, is taken up with ascertaining facts about the parents and even the grandparents of the respondents, but over half of the questions deal directly with women's sexual behavior and attitudes.[46] The information contained in the questionnaires not only supports the interpretation of women's sexuality that already has been drawn from the published literature, both lay and medical, but also provides us with a means of measuring the degree to which the prescriptive marriage literature affected women's sexual behavior.

Since the evidence in this questionnaire, which I call the Mosher survey, has never been used before, it is first worthwhile to examine the social background of the women who answered the questionnaires. All told there are forty-six usable questionnaires, but since two of the questionnaires seem to have been filled out by the same woman at an interval of twenty-three years, the number of women actually surveyed is forty-five.[47] In the aggregates that follow I have counted only forty-five questionnaires. The questionnaires, it ought to be said, were not administered at the same time but at three different periods at least; moreover, the date of administration of nine questionnaires cannot be ascertained. Of those that do provide that information, seventeen were completed before 1900, fourteen were filled out between 1913 and 1917, and five were answered in 1920.

More important than the date of administration of the questionnaires are the birth dates of the respondents. All but one of the forty-four women who provided their dates of birth were born before 1890. In fact, thirty-three, or 70 percent of the whole group, were born before 1870. And of these, seventeen, or slightly over half, were born before the Civil War. For comparative purposes it might be noted that in Alfred Kinsey's survey of women's sexuality the earliest cohort of respondents was born only in the 1890s. In short, the attitudes and practices to which the great majority of the women in the Mosher survey testify were those of women who grew up and married within the nineteenth century, regardless of when they may have completed the questionnaires.

An important consideration in evaluating the responses, of course, is the social origins of the women. From what class did they come, and from what sections of the country? The questionnaire, fortunately, provides some information here but not with as much precision as one might like. Since the great majority of the respondents attended college or a normal school (thirty-four out of forty-five, with the education of three unknown), it is evident that the group

is not representative of the population of the United States as a whole. The remainder of the group attended secondary school, either public or private, a pattern that is again not representative of a general population in which only a tiny minority of young people attended secondary school. But for purposes of evaluating the impact of the prescriptive or marital advice literature upon American women this group is quite appropriate. For inasmuch as their educational background identifies them as middle- or upper-class women, it can be said that they were precisely those persons to whom that advisory literature was directed and upon whom its effects ought to be most evident.

In geographical origin the respondents to the Mosher survey seem to be somewhat more representative, if the location of parents, birthplaces, and colleges attended can be taken as a measure, albeit impressionistic, of geographical distribution. Unfortunately, there is no other systematic or more reliable information on this subject. The colleges attended, for example, are located in the Northeast (Cornell [6], Smith, Wellesley, and Vassar [2]), in the Middle West (Ripon, Iowa State University, and Indiana), and in the Far West (Stanford [9], the University of California, and the University of the Pacific). The South is not represented at all among the colleges attended.

Although the emphasis upon prestigious colleges might make one think that these were women of the upper or even leisure class, rather than simply middle class, a further piece of information suggests that in fact they were not. One of the questions asked concerned working experience prior to marriage. Although seven of the respondents provided no data at all on this point, and eight reported that they had married immediately after completing their education, thirty of the women reported that they had worked prior to marriage. As a sidelight on the opportunities available to highly educated women in the late nineteenth century, it is worth adding that twenty-seven of the thirty worked as teachers. On the basis of their working experience it seems reasonable to conclude that the respondents were principally middle- or upper-middle-class women rather than members of a leisure class.

Despite the high level of education of these women, they confessed to having a pretty poor knowledge, by modern standards, of sexual physiology before marriage. Only eleven said that they had much knowledge on that subject, obtained from female relatives, books, or courses in college, while another thirteen said that they had some knowledge. The remainder—slightly over half—reported that they had very little or no knowledge. No guidelines were given in the questionnaire for estimating the amount of knowledge. The looseness of the definition is shown by the fact that three of the respondents who said that they had no knowledge at all named books on women's physiology that they had read. From other titles mentioned in passing it is clear that a number of these women had direct acquaintance with the prescriptive and advisory literature of the time. How did it affect their behavior? Did they repress their sexual impulses or deny them, as some of the prescriptive literature advised? Were they in fact without sexual desire? Or were they motivated toward personal sexual satisfaction as the medical literature quoted in this article advised?

The Mosher survey provides a considerable amount of evidence to answer these and other questions. To begin with, thirty-five of the forty-five women testified that they felt desire for sexual intercourse independent of their husband's interest, while nine said they never or rarely felt any such desire. What is more striking, however, is the number who testified to orgasmic experience. According to the standard view of women's sexuality in the nineteenth century, women

were not expected to feel desire and certainly not to experience an orgasm. Yet it is striking that in constructing the questionnaire Dr. Mosher asked not only whether the respondents experienced an orgasm during intercourse but whether "you *always* have a venereal orgasm?" (my italics). Although that form of the question makes quite clear Mosher's own assumption that female orgasms were to be expected, it unfortunately confuses the meaning of the responses. (Incidentally, only two of the forty-five respondents failed to answer this question.) Five of the women, for instance, responded no without further comment. Given the wording of the question, however, that negative could have meant, "not always but almost always" as well as "never" or any response in between these extremes. The ambiguity is further heightened when it is recognized that in answer to another question, three of the five negatives said that they had felt sexual desire, while a fourth said "sometimes but not often," and the fifth said sex was "usually a nuisance." Luckily, however, most of the women who responded to the question concerning orgasm made more precise answers. The great majority of them said that they had experienced orgasms. The complete pattern of responses is set forth in Table 10.1.

In sum, thirty-four of the women experienced orgasm, with the possibility that the figure might be as high as thirty-seven if those who

Table 10.1
Response to the query: "Do you always have a venereal orgasm?"

Response	Number	Percentage
No response	2	4.4
"No" with no further comment	5	11.1
"Always"	9	20.0
"Usually"	7	15.5
"Sometimes," "Not always," or "No" with instances	18	40.0
"Once" or "Never"	4	8.8

Table 10.2
Percentage of women experiencing orgasm
during intercourse (by decade of birth)

Women Born	Ages 21–25	Ages 26–30
Before 1900	72%	80%
1900–1909	80%	86%
1910–1919	87%	91%
1920–1929	89%	93%

Source: Alfred C. Kinsey et al., *Sexual Behavior in the Human Female* (Philadelphia, 1953), 397, table 97.

reported no but said they had felt sexual desire are categorized as "sometimes." (Interestingly enough, of nine women out of the forty-five who said they had never felt any sexual desire, seven said that they had experienced orgasms.) Moreover, sixteen or almost half of those who experienced orgasms did so either "always" or "usually." As we have seen, in the whole group of forty-five, all but two responded to the question asking if an orgasm was always experienced. Of those forty-three, thirty-four were born before 1875. Five answered no to that question without any further comment. One other woman responded "never," and two others said "once or twice." If the "noes" and the "never" are taken together, the proportion of women born before 1875 who experienced at least one orgasm is 82 percent. If the "noes" are taken to mean "sometimes" or "once or twice," as they might well be, given the wording of the question, then the proportion rises to 95 percent. For comparative purposes the figures for twentieth-century women provided in Kinsey's study are given in Table 10.2. Kinsey's proportions are arranged by age group and chronological period; hence they are not strictly comparable with those derived from Mosher's data. But the comparison is still suggestive, even when made with the women in the age group 26–30.[48]

Much more interesting and valuable than the bare statistics are the comments or rationales furnished by the women, which provide an insight into the sexual attitudes of middle-class women. As one might expect in a population by its own admission poorly informed on sexual physiology, the sexual adjustment of some of these women left something to be desired. Mosher, for example, in one of her few efforts at drawing conclusions from the survey, pointed out that sexual maladjustment within marriage sometimes began with the first intercourse. "The woman comes to this new experience of life often with no knowledge. The woman while she may give mental consent often shrinks physically." From her studies Mosher had also come to recognize that women's "slower time reaction" in reaching full sexual excitement was a source of maladjustment between husband and wife that could kill off or reduce sexual feelings in some women. Women, she recognized, because of their slower timing were left without "the normal physical response. This leaves organs of women over congested."[49] At least one of her respondents reported that for years intercourse was distasteful to her because of her "slow reaction," but "orgasm [occurs] if time is taken." On the other hand, the respondent continued, "when no orgasm, [she] took days to recover."[50] Another woman spoke of the absence of an orgasm during intercourse as "bad, even disastrous, nerve-wracking—unbalancing, if such conditions continue for any length of time." Still a third woman, presumably referring to the differences in the sexual rhythms of men and women, said, "Men have not been properly trained." One of the women in the Mosher survey testified in another way to her recognition of the differences in the sexuality of men and women. "Every wife submits when perhaps she is not in the mood," she wrote, "but I can see no bad effect. It is as if it had not been. But my husband was absolutely considerate. I do not think I could endure a man who forced it." And her response to a question about the effects of an orgasm upon her corroborate her remark: "a general sense of well being, contentment and regard for husband. This is true Doctor," she earnestly wrote.[51]

Mosher's probing of the attitudes of women toward their sexuality went beyond asking about orgasms. Several of her questions sought to elicit the reactions of women to sexual intercourse. What is the purpose of sex, she asked? Is it a necessity for a man or for a woman? Is it for pleasure, or is it for reproduction?[52] Only two of the women failed to respond in some fashion to these questions. Nine thought sex was a necessity for men, while thirteen thought it was a necessity for both men and women. Fifteen of the respondents thought it was not a necessity for either sex. Twenty-four of the forty-five thought that it was a pleasure for both sexes, while only one thought it was exclusively a pleasure for men. Given the view generally held about sexual attitudes in the nineteenth century, it comes as something of a surprise to find that only thirty marked "reproduction" as the primary purpose of sex. In fact, as we shall see in a moment, some of the women thought reproduction was not as important a justification for intercourse as love.

As one might expect, this particular series of questions was usually answered with a good deal of explanation. One woman who emphasized reproduction as the principal justification took the opportunity to condemn those couples she apparently had heard of who did not want children. "I cannot recognize as true marriage that relation unaccompanied by a strong desire for children." She thought it was close to "legalized prostitution." She admitted that because of her love for her husband she "cultivated the passion to effect the 'compromise' in this direction that must come in every other [area] when people marry." She went on to say that she did not experience orgasm until the fifth or sixth year of her marriage and that even at the time of her response to the questionnaire—the early

1890s—she still did not reach a sexual climax half of the time. A second woman was also apparently out of phase with her husband's sexual interests, for she thought a woman's needs for sex occurred "half as often as a man's." It is revealing of her own feelings that though she said "half as often," the figures she used to illustrate her point—twice a week for a man and twice a month for a woman—are actually in the ratio of one to four rather than of one to two as she said. Her true attitude was also summed up in the remark that since she was always in good health and intercourse "did not hurt me, ... I always meant to be obliging."[53]

But, as the earlier statistical breakdown makes evident, the women who only tolerated intercourse were in a decided minority. A frank and sometimes enthusiastic acceptance of sexual relations was the response from most of the women. Sexual intercourse "makes more normal people," said a woman born in 1857. She was not even sure that children were necessary to justify sexual relations within marriage. "Even if there are no children, men love their wives more if they continue this relation, and the highest devotion is based upon it, a very beautiful thing, and I am glad nature gave it to us." Since marriage should bring two people close together, said one woman born in 1855, sexual intercourse is the means that achieves that end. "Living relations have a right to exist between married people and these cannot exist in perfection without sexual intercourse to a moderate degree. This is the result of my experience," she added. A woman born in 1864 described sexual relations as "the gratification of a normal healthy appetite." The only respondent who was divorced and remarried testified in 1913 that at age fifty-three "my passionate feeling has declined somewhat and the orgasm does not always occur," but intercourse, she went on, was still "agreeable" to her.[54]

Several of the women even went so far as to

reject reproduction as sufficient justification for sex. Said one woman, "I consider this appetite as ranking with other natural appetites and like them to be indulged legitimately and temperately; I consider it illegitimate to risk bringing children into the world under any but most favorable circumstances." This woman was born before the Compromise of 1850 and made her comment after she had been married ten years. Another woman, also born a decade before the Civil War, denied that reproduction "alone warrants it at all; I think it is only warranted as an expression of true and passionate love. This is the prime condition for a happy conception, I fancy." To her too the pleasure derived from sexual intercourse was "not sensual pleasure, but the pleasure of love."[55]

A third woman born before 1861 doubted that sex was a necessity in the same sense as food or drink, but she had no doubt that "the desire of both husband and wife for this expression of their union seems to me the first and highest reason for intercourse. The desire for offspring is a secondary, incidental, although entirely worthy motive but could never to me make intercourse right unless the mutual desire were also present." She saw a clear conflict between the pleasure of intercourse and reproduction. "My husband and I," she said in 1893,

believe in intercourse for its own sake— we wish it for ourselves and spiritually miss it, rather than physically, when it does not occur, because it is the highest, most sacred expression of our oneness. On the other hand, even a slight risk of pregnancy, and then we deny ourselves the intercourse, feeling all the time that we are losing that which keeps us closest to each other.[56]

Another woman, in describing the ideal of sexual relations, said that she did not want intercourse to occur at any time when conception

was likely, for conception should not occur by accident. Instead, it ought to be the result of

deliberate design on both sides in time and circumstances most favorable physically and spiritually for the accomplishment of an immensely important act. It amounts to separating times and objects of intercourse into (a) that of expression of love between man and woman (that act is frequently simply the extreme caress of love's passion, which it would be a pity to limit . . . to once in two or three years) and (b) that of carrying on a share in the perpetuation of the race, which should be done carefully and prayerfully.[57]

It seems evident that among these women sexual relations were neither rejected nor engaged in with distaste or reluctance. In fact, for them sexual expression was a part of healthy living and frequently a joy. Certainly, the prescriptive literature that denigrated sexual feelings or expression among women cannot be read as descriptive of the behavior or attitude of these women. Nevertheless this is not quite the same as saying that the marriage handbooks had no effect at all. To be sure, there is no evidence that the great majority of women in the Mosher survey felt guilty about indulging in sex because of what they were told in the prescriptive literature. But in two cases the literature seems to have left feelings of guilt. One woman said that sexual relations were "apparently a necessity for the *average* person" and that it was "only [the] superior individuals" who could be "independent of sex relations with no evident ill-results." To her, as to St. Paul and some of the marriage advice books, it was better to indulge than to burn, but it was evidently even better to be free from burning from the beginning. A more blatant sign of guilt over sex came from the testimony of a woman who quite frankly thought the pleasure of sex was a justification for intercourse, but she added "not necessarily a legitimate one."[58]

Dr. Mosher herself obliquely testified to the effects of the prescriptive literature. She attributed the difficulties some women experienced in reaching orgasm to the fact that "training has instilled the idea that any physical response is coarse, common and immodest which inhibits [women's] proper part in this relation."[59] That was the same point that some of the medical writers in the nineteenth century had made in explaining the coldness of some women toward their husbands.

The advice literature, for men as well as for women, generally warned against excessive sexual activity.[60] This emphasis upon limits is reflected in the remarks of some of the women in the Mosher survey. One woman said, for example, that "the pleasure is sufficient warrant" for sexual relations but only if "people are extremely moderate and do not allow it to injure their health or degrade their best feelings toward each other." Another woman had concluded that "to the man and woman married from love," sexual intercourse "may be used temperately as one of the highest manifestations of love granted us by our Creator." A third woman, who had no doubt that sexual relations were "necessary to marital happiness," nonetheless said she believed in "temperance in it."[61] But temperance, another one of the women in the Mosher survey reminds us, should not be confused with repugnance or distaste. Although this respondent did not think the ideal sexual relation should occur more often than once a month, she did think it ought to take place "during the menstrual period . . . and in the daylight." The fact is this woman, in answer to other questions, indicated that she experienced sexual desire about once a week but with greatest intensity "before and during menses." She was, in short, restricting her own ideal to what she considered an acceptable frequency of indulgence. Her de-

scription of her feelings after orgasm suggests where she learned that limits on frequency might be desirable or expected: "Very sleepy and comfortable. No disgust, as I have heard it described."[62]

This examination of the literature, the popular advice books, and particularly the Mosher survey makes clear that the historians are ill advised to rely upon the marital advice books as descriptions either of the sexual behavior of women or of general attitudes toward women's sexuality. It is true that a literature as admittedly popular as much of the prescriptive or normative literature was could be expected to have some effect upon behavior as well as attitudes. But those effects were severely limited. Most people apparently did not follow the prescriptions laid down by the marriage and advice manuals. Indeed, some undoubtedly found that advice wrong or misleading when measured against experience. Through some error or accident the same woman was apparently interviewed twice in the Mosher survey, twenty-three years apart. As a result we can compare her attitudes at the beginning of her marriage in 1896 and her attitude in 1920. After one year of marriage she thought that sexual relations ought to be confined to reproduction only, but when asked the same question in 1920, she said that intercourse ought not to be confined to reproduction, though she thought it should be indulged in only when not pressed with work and when there was time for pleasure.[63] Another woman in the Mosher survey changed her mind about sexual relations even earlier in her sexual life. She said,

My ideas as to the reason for [intercourse] have changed materially from what they were before marriage. I then thought reproduction was the only object and that once brought about, intercourse should cease. But in my experience the habitual bodily expression of love has a deep psychological effect in making possible complete mental sympathy, and perfecting the spiritual union that must be the lasting "marriage" after the passion of love has passed away with years.

These remarks were made in 1897 by a woman of thirty after one year of marriage.[64]

Her comments make clear once again that historians need to recognize that the attitudes of ordinary people are quite capable of resisting efforts to reshape or alter them. That there was an effort to deny women's sexual feelings and to deny them legitimate expression cannot be doubted in the light of the books written then and later about the Victorian conception of sexuality. But the many writings by medical men who spoke in a contrary vein and the Mosher survey should make us doubt that the ideology was actually put into practice by most men or women of the nineteenth century, even among the middle class, though it was to this class in particular that the admonitions and ideology were directed. The women who responded to Dr. Mosher's questions were certainly middle- and upper-middle-class women, but they were, as a group, neither sexless nor hostile to sexual feelings. The great majority of them, after all, experienced orgasm as well as sexual desire. Their behavior in the face of the antisexual ideology pressed upon them at the time offers testimony to the truth of Alex Comfort's comment that "the astounding resilience of human commonsense against the anxiety makers is one of the really cheering aspects of history."[65]

Notes

1. Captain Frederick Marryat, *A Diary in America, with Remarks on Its Institutions* (London, 1839), 2: 244–47. The story of the trousers on piano legs is taken seriously in John Duffy, "Masturbation

and Clitoridectomy: a Nineteenth Century View," *Journal of the American Medical Association* 186 (1963): 246; G. Rattray Taylor, *Sex in History* (New York, 1954), 203; and Peter T. Cominos, "Innocent Femina Sensualis in Unconscious Conflict," in *Suffer and Be Still: Women in the Victorian Age,* ed. Martha Vicinus (Bloomington, 1972), 157.

2. William Acton, *The Functions and Disorders of the Reproductive Organs in Youth, in Adult Age, and in Advanced Life: Considered in Their Physiological, Social, and Psychological Relations* (1857; 2d ed., London, 1858; expanded American ed., Philadelphia, 1865). For references to Acton's writings see Steven Marcus, *The Other Victorians: A Study of Sexuality and Pornography in Midnineteenth-Century England* (New York, 1966), ch. 1; and Nathan G. Hale, Jr., *Freud and the Americans: The Beginnings of Psychoanalysis in the United States, 1876–1917* (New York, 1971), 36–37.

3. Acton, *Functions and Disorders* (1865), 133.

4. Theophilus Parvin, "Hygiene of the Sexual Functions," *New Orleans Medical and Surgical Journal,* n.s. 11 (1883–84): 607. Parvin also quotes at length from Acton's book.

5. Walter E. Houghton, *The Victorian Frame of Mind, 1830–1870* (New Haven, 1957), 353. Marcus presents a portrait of Victorian attitudes toward sex similar to that of Houghton, but he disclaims to be talking about behavior: "We need not pause to discuss the degree of truth or falsehood in these assertions. What is of more immediate concern is that these assertions indicate a system of beliefs." *Other Victorians,* 32. Yet it is not clear what point there is in detailing a system of beliefs unless it has some behavioral consequences. Peter T. Cominos also relies upon Acton, in "Late Victorian Sexual Respectability and the Social System," *International Review of Social History* 7/8 (1963): 18–48, 217–50. E. M. Sigsworth and T. J. Wyke doubt the pervasiveness in Victorian England of Acton's conception of women's sexuality. They write: "Victorian opinion on the innate sexuality of women was cloudy and divided"—a view about which more will be said in this article. "A Study of Victorian Prostitution and Venereal Disease," in Vicinus, *Suffer and Be Still,* 83.

6. Viola Klein, *The Feminine Character: History of an Ideology* (1946; reprint, Urbana, 1972), 85, as

quoted in David M. Kennedy, *Birth Control in America: The Career of Margaret Sanger* (New Haven, 1970), 56–57.

7. Hale, *Freud and the Americans*, 31. Elsewhere Hale sums up the medical view as he sees it: "By 1906 . . . some physicians regarded the asexual female as the norm: 'It may be offered that the sexual appetite in the majority of American females is evoked only by the purest love. In many the appetite never asserts itself and, indeed, the only impulse thereto is in the desire to gratify the object of affection'" (39–40; quotation from Ferdinand C. Valentine, "Education in Sexual Subjects," *New York Medical Journal* [Feb. 10, 1906], 276).

8. Ben (G. J.) Barker-Benfield, "The Spermatic Economy: a Nineteenth Century View of Sexuality," *Feminist Studies* 1 (1972): 54.

9. Charles Fayette Taylor, "Effect on Women of Imperfect Hygiene of the Sexual Function," *American Journal of Obstetrics* 15 (1882): 175–76, 168–71, italics in original.

10. Orson S. Fowler, *Sexual Science; Including Manhood, Womanhood, and Their Mutual Interrelations, etc. . . . as Taught by Phrenology* (Philadelphia, 1870), 680.

11. P. Henry Chavasse, *Physical Life of Man and Woman; or, Advice to Both Sexes* (1866; reprint, New York, 1897), 291–92; quotation from Trall in Michael Gordon, "From an Unfortunate Necessity to a Cult of Mutual Orgasm: Sex in American Marital Education Literature, 1830–1940," in *Studies in the Sociology of Sex*, ed. James M. Henslin (New York, 1971), 58, my italics.

12. George H. Napheys, *The Physical Life of Woman: Advice to the Maiden, Wife, and Mother* (1869; Philadelphia, 1871), 74–75, 180.

13. Ely Van de Warker, "Impotency in Women," *American Journal of Obstetrics* 11 (1878): 47.

14. J. Milne Chapman, "On Masturbation as an Etiological Factor in the Production of Gynic Diseases," *American Journal of Obstetrics* 16 (1883): 454.

15. Montrose S. Pallen, "Some Suggestions with Regard to the Insanities of Females," *American Journal of Obstetrics* 10 (1877): 209.

16. William M. McLaury, "Remarks on the Relation of the Menstruation to the Sexual Functions," *American Journal of Obstetrics* 20 (1877): 161.

17. Chavasse, *Physical Life of Man and Woman*, 372–73.

18. Van de Warker, "Impotency in Women," 38–39.

19. Ibid., 41. Today the complaint is called premature ejaculation.

20. William Goodell, *Lessons in Gynecology* (Philadelphia, 1887), 567, as quoted in Hale, *Freud and the Americans*, 40.

21. McLaury, "Remarks on the Relation of the Menstruation," 161.

22. Chavasse, *Physical Life of Man and Woman*, 424–25.

23. John Harvey Kellogg, *Plain Facts for Old and Young* (1879; Burlington, Iowa, 1881), 252.

24. Augustus K. Gardner, *Conjugal Sins against the Laws of Life and Health and Their Effects upon the Father, Mother, and Child* (New York, 1870), 98.

25. Chavasse, *Physical Life of Man and Woman*, 33.

26. Mary Wood-Allen, *What a Young Woman Ought to Know* (Philadelphia, 1905), 155.

27. Chapman, "On Masturbation"; letter from S. E. McCully, *American Journal of Obstetrics* 16 (1883): 844, my italics.

28. James Marion Sims, *Clinical Notes on Uterine Surgery* (London, 1866), 369.

29. Napheys, *Physical Life of Woman*, 104–5. This belief, which other writers also speak of, may well have affected some women's attitudes toward orgasm, for if a woman, under this view, could repress pleasure or climax, conception could be prevented.

30. Van de Warker, "Impotency in Women," 39.

31. Napheys, *Physical Life of Woman*, 86.

32. Horatio Bigelow, "Aggravated Instance of Masturbation in the Female," *American Journal of Obstetrics* 15 (1882): 437.

33. "A Case of Excision of Both Ovaries for Fibrous Tumors of the Uterus, and a Case of Excision of the Left Ovary for Chronic Oöphoritis and Displacement," reported by Dr. E. H. Trenholme in *Canada Lancet*, July 1876, *American Journal of Obstetrics* 9 (1876–77): 703.

34. "Case of Vaginismus," reported by Dr. George Pepper, *American Journal of Obstetrics* 3 (1871): 322–24.

35. Gardner, *Conjugal Sins*, 97.

36. Summary of paper by Dr. Keppler, "The Sexual Life of the Female after Castration," given at the 10th International Medical Congress, *American Journal of Obstetrics* 23 (1890): 1155–56.

37. Not all historians, it should be noted, have assumed that nineteenth-century concerns about sex meant hostility toward women's sexuality. In tracing the history of the social purity movement after 1870, David J. Pivar is careful to distinguish between a concern with the exploitation of women's sexuality and an opposition to women's sexual feelings. See his *Purity Crusade: Sexual Morality and Social Control, 1868–1900* (Westport, 1973).

38. Acton, *Functions and Disorders* (1858), 8–9; (1865), 134, my italics.

39. Kellogg, *Plain Facts,* 178. Hale gives the figures on Kellogg's sales in *Freud and Americans,* 37.

40. Kellogg, *Plain Facts,* 206, 209, 247, 225–26. Kellogg also quoted Acton.

41. Ibid., 217, 221–25, 118.

42. Ibid., 265–66.

43. It is true that some of the advice and medical literature that recognized women's sexual feelings and from which I have been quoting was also prescriptive rather than merely descriptive. But for convenience and economy of words in subsequent pages when I refer to "prescriptive or normative literature" I mean only that which minimized or denied women's sexuality.

44. Among the largest and most significant of such surveys were Katherine B. Davis, *Factors in the Sex Life of Twenty-two Hundred Women* (New York, 1929); Robert Latou Dickinson and Lura Beam, *A Thousand Marriages: A Medical Study of Sex Adjustment* (Baltimore, 1931); and Alfred C. Kinsey et al., *Sexual Behavior in the Human Female* (Philadelphia, 1953). The first chapter of Robert Latou Dickinson and Lura Beam, *The Single Woman* (Baltimore, 1934), concerns the sexual life of working girls in the 1890s, but it is based on forty-six cases, the typical patient being born "soon after 1870." I am indebted to David M. Kennedy of Stanford University for this reference.

45. Mosher, "Hygiene and Physiology of Woman," 10:xv, Mosher Papers, Stanford University Archives.

46. The principal questions dealing with women's sexual habits are number of conceptions; number of conceptions by choice and by accident; frequency of intercourse; whether intercourse is participated in during pregnancy; whether intercourse is "agreeable"; whether an orgasm occurs; what effects from orgasm, or from failure to have one; purpose of intercourse; the ideal habit of sexual relations; whether there is desire for intercourse other than during pregnancy; whether contraception is used and method employed; whether wife sleeps in same bed with husband; knowledge of sexual physiology prior to marriage; and the character of menses: age of onset, pain, and amount.

47. The small number of women queried in the Mosher survey may cause some readers to discount almost entirely the significance of any conclusions drawn from it. While such a response may be understandable as a first reaction, in the end I think it would be unwise. So far as I know, this is the only survey of sexual attitudes and practices in the nineteenth century; historians' standard conception of women's sexual practices and attitudes in the nineteenth century has been derived from no previous survey at all. Certainly the systematic questioning of forty-five women at considerable length and their rationales for their answers ought to be at least as significant in shaping historians' conceptions of women's sexuality as the scraps of information from interested writers at the time, novels, and recollections, which have been the bases of our traditional picture of women's sexual attitudes and behavior in the nineteenth century. It is true that we do not know at the present time who these women were or how random their selection was. But there seems little reason to believe that the women were specifically chosen by Mosher, if only because the purpose of the original questionnaire as well as the use of the information gained from it was to help her in advising women students. Moreover, as an unmarried woman herself, it is very likely that the information from the questionnaires was Mosher's most valuable source of knowledge on women's sexuality. It is probably true, given the general reluctance of nineteenth-century people to discuss sex, that some women whom Mosher approached refused to answer the questionnaire. But it is worth recalling that the value even of modern sex surveys, including Kinsey's, has been questioned on the grounds that the respondents were

largely self-selected. Obviously, the Mosher survey is not the final word on the sexual behavior and attitudes of women in the nineteenth century. But at the same time it ought not to be rejected because of its limited size; that would be applying a methodological standard quite inappropriate for a sensitive subject in which the evidence is always limited and fugitive.

48. A comparison of the sexual responses of the older and younger women in the Mosher survey did not reveal any greater interest in sex among the younger group, but the numbers involved were too small to be significant. The responses of fourteen women born before 1860 were compared with those of the eight women born after 1875. On the other hand, if the responses to the questions about desire for sex and about orgasmic experience are categorized by date at which the questionnaire was completed, regardless of the age of the respondent, there is a slight, if somewhat ambiguous, difference between the earlier and later respondents. Seventeen women completed the questionnaire before 1900; nineteen did so after 1912. Thirteen of the seventeen who completed before 1900 responded to the question of whether they had experienced orgasm; four of the thirteen said they had not. Eighteen of the nineteen who completed the questionnaire after 1912 answered that question; only one out of eighteen failed to experience an orgasm. In themselves these data suggest that women who answered the questionnaire in the twentieth century achieved somewhat more satisfaction in their sexual experience than those who completed the questionnaire in the nineteenth century. But when a similar division by century is made of the questionnaires in regard to another question, that conclusion is not so clear. One of the questions asked whether the respondent felt sexual desire. Fourteen women answered the question prior to 1900, of whom only two said they had failed to feel desire. But of the sixteen who responded to the same question after 1912, three said they lacked any feeling of desire. Here the proportion of sexuality was higher among the nineteenth- than the twentieth-century respondents.

49. Mosher, "Hygiene and Physiology of Women," 10:1. Twelve of the women were asked how soon after marriage they engaged in intercourse. Six said within the first three days, while six said from ten days to a year after the ceremony.

50. Ibid., case no. 51. The case numbers have been assigned by Mosher herself and appear on each page of each questionnaire. Hereafter the citation of cases will carry only "Hygiene and physiology of women" and case number.

51. Ibid., case nos. 47, 40, 41.

52. Since each respondent could legitimately answer yes to all three suggested justifications for sexual relations, the totals here can go beyond forty-five, though not all questions were always answered.

53. Ibid., case nos. 24, 19.

54. Ibid., case nos. 41, 18, 2. It is worth noting that here, as elsewhere in the survey, no mention was made of religious reasons for or against intercourse. These women had almost entirely secularized their sexual ideology.

55. Ibid., case nos. 14, 12.

56. Ibid., case no. 15.

57. Ibid., case no. 22.

58. Ibid., case nos. 47, 30. Marcus found a comparable example of guilt arising out of the prescriptive literature against masturbation. In discussing the Victorian sexual autobiography *My Secret Life*, Marcus observes that the anonymous author gave full credence to the dangers described in the literature, yet he masturbated nonetheless. After doing so, however, the anonymous author reported he suffered from depression, guilt, fatigue, and general feelings of debilitation though he felt none of these symptoms after sexual intercourse. Marcus ascribes these feelings to an internalizing of social attitudes, presumably derived from the prescriptive literature against masturbation. *Other Victorians*, 112. It is significant, however, that the prescriptions did not stop the practice. Why it did not stop is suggested by a more recent study of sexual behavior. Masters and Johnson report that most of their male subjects still believed the old tales of physical and psychical harm from masturbation, especially from "excessive" activity, but none of them desisted from the practice. The authors point out that no matter how active a subject was in this respect, he always defined "excessive" as more active than his own practice. William H. Masters and Virginia E. Johnson, *Human Sexual Response* (Boston, 1966), 201–2.

59. Mosher, "Hygiene and Physiology of Women," 10:1.

60. Hale cites sources ranging in origin from 1830 to 1910 on the concern for conserving sexual energy. *Freud and the Americans,* 35. Oscar Handlin sums up the advice in this fashion: "Abstinence, repression, and self-restraint thus were the law; and violations were punished by the most hideous natural consequences, described in considerable graphic detail." Handlin's conclusion, however, that the readers of that literature "were overwhelmed by the guilt and shame the necessities of self-control imposed," seems unwarranted on the basis of present evidence. *Race and Nationality in American Life* (Garden City, 1957), 122–23.

61. Mosher, "Hygiene and Physiology of Women," case nos. 33, 10, 13.

62. Ibid., case no. 11.

63. Ibid., case nos. 30, 33. Mosher gives no indication that she knew the two questionnaires were from the same person.

64. Ibid., case no. 22.

65. Alex Comfort, *The Anxiety Makers: Some Curious Preoccupations of the Medical Profession* (London, 1967), 113.

11 Romantic Friends or a "Different Race of Creatures"? The Representation of Lesbian Pathology in Nineteenth-Century America

Marylynne Diggs

The acceptability of intimate and erotic relationships between women prior to the twentieth century has long been a topic of critical debate. In 1975 Carroll Smith-Rosenberg's "Female World of Love and Ritual: Relations Between Women in Nineteenth-Century America" described an eighteenth- and nineteenth-century American culture in which passionate same-sex relationships between women were "socially acceptable and fully compatible with heterosexual marriage."[1] Smith-Rosenberg's groundbreaking article was followed by the influential work of Lillian Faderman, whose *Surpassing the Love of Men: Romantic Friendship and Love Between Women from the Renaissance to the Present* (1981) may be taken as a turning point after which the acceptability of "romantic friendship"

MARYLYNNE DIGGS teaches English at Clark College in Vancouver, Washington.

Reprinted from *Feminist Studies* 21, 2 (Summer 1995): 317–40, by permission of the publisher, *Feminist Studies*, Inc., c/o Department of Women's Studies, University of Maryland, College Park, MD 20742.

before the twentieth century became a scholarly "given" on which rested much subsequent historical and literary research. Faderman argued that the nineteenth century was a presexological culture in which women who maintained intimate relations with one another—relations that Faderman believed were not sexual—enjoyed complete freedom from disapprobation. The twentieth century, she argued, marked a turn toward internalization of sexological representations and the production of a literary culture by and about the now-stereotypical self-loathing invert.[2]

The insistence on a "golden age" of romantic friendship was perhaps generated out of several concerns common to lesbian and feminist scholarship in the United States. Much of the feminist research of the 1970s and early 1980s was devoted to reclaiming lost traditions and finding literary foremothers. As part of this project of reclamation, the evocation of an Edenic era, when the world might have been kinder to women who loved women, may have operated as a promise of a better world to come. The continued belief in an idyllic age of romantic friendships, especially in the cultural context of the United States, is also perhaps understandable as a historical transposition of the sex debates of the early 1980s.[3] Representing a presexual world where women's primary bonds were made with other women, the romantic friendship hypothesis provided a "purified," woman-centered notion of lesbian history that resonated with lesbian feminist theories of the 1970s and early 1980s.[4]

Although the romantic friendship model advanced by Smith-Rosenberg and Faderman went more or less unquestioned for over a decade, new scholarship on sexual relationships between women has led historians to complicate the relationship of romantic friendship to modern lesbian identity. Qualifying her earlier argument, Smith-Rosenberg has noted that the New Woman represented a threat to a social order dependent on the maintenance of rigid gender roles and gender inequalities. Some medical professionals, she noted, defined as pathological those who challenged the "natural" place of women, labeling them "atavistic throwbacks" who "'aped'" masculine gender roles.[5] Martha Vicinus has argued that the romantic friendship model may have survived well into the twentieth century, and more recently she has suggested that historians "may have exaggerated the acceptability of romantic friendships" in the nineteenth century.[6] She argues that medical and legal discourses began defining and pathologizing many forms of sexuality in the first half of the nineteenth century, precisely that period believed to have been a "golden age" of romantic friendship.

The debate over romantic friendship may seem overdone at this point. Vicinus has recently lamented that the debate has probably consumed "far too much energy" on what is a particularly "American concern."[7] Indeed, histo-

rians of European and Asian cultures, such as those whose work is included in *Hidden from History: Reclaiming the Gay and Lesbian Past,* have shown that, especially in Europe, lesbian and gay identities emerged much earlier than some historians had believed. Terry Castle, in *The Apparitional Lesbian: Female Homosexuality and Modern Culture,* argues that, although *lesbian* and *homosexual* are indeed recent terms, other terms for women who desired sexual intimacy with other women have abounded since the eighteenth century. Noting also the rumors about Marie Antoinette's female lovers, and the "homoerotic cult" which emerged in the wake of her execution, Castle indicates that "libidinal self-awareness" was more common than the romantic friendship hypothesis has suggested.[8] Lisa Moore, discussing early nineteenth-century England, also addresses the "contradictory status of the ideology of romantic friendship" and suggests that women readers of domestic fiction may have found in the interstices of these contradictions spaces in which to "construct themselves as female homosexual characters."[9] Among the texts Moore discusses are the diaries of Anne Lister, which describe her sexual relationships with other women and her doubts that the famed "Ladies of Llangollen" had a purely "platonic" relationship.[10] The diaries provide evidence that, at the turn of the nineteenth century, women were aware of the popular notion of nonsexual romantic friendships and engaged in sexual relationships under that cover.[11]

Nonetheless, these recent attempts to undercut the rigidity of the romantic friendship hypothesis coexist with scholarship about U.S. culture that continues to operate on the basis of its assumptions. Bonnie Zimmerman's *The Safe Sea of Women: Lesbian Fiction, 1969–1989* places the emergence of lesbian identity in the late nineteenth century and contends that most women who had romantic friendships before this time did not feel different from other women.[12] Lillian Faderman's *Odd Girls and Twilight Lovers: A History of Lesbian Life in Twentieth-Century America,* probably the most widely read study of lesbian history in recent years, reaffirms the positions she took in *Surpassing the Love of Men:* "women's intimate relationships," she argues, "were universally encouraged in centuries outside of our own."[13]

Although I agree with the historiographical principles implied in the work of Zimmerman and Faderman—that homosexuality is a historically specific category of sexual identity and that emotional and erotic intimacies with members of the same sex have held different meanings in different historical contexts—I disagree with the supersessionist claims of these histories and the resulting generalizations about nineteenth-century sexual culture in the United States. I agree with Vicinus that the romantic friendship hypothesis exaggerates the extent to which intimate and primary relationships between women were accepted. Mary E. Wood convincingly supports this argument in "'With Ready Eye': Margaret Fuller and Lesbianism in Nineteenth-Century American Literature." Wood locates constructions and deconstructions of lesbian identity in the first half of the nineteenth century, finding lesbian significations in the confusions of and resistances to the implicitly heterosexual ideology of separate female and male spheres.[14]

I would further argue that the continued reliance on the romantic friendship model overlooks both the pathologizing and the resistant discourses that emerged in the United States well before the turn of the century. Even before the middle of the nineteenth century a broad spectrum of discourses, including scientific discourses not directly related to sexuality and subjectivity, were constructing relationships between women in terms of sexual pathology. Moreover, even the intensified pathologizing of female homosexuality in the late nineteenth century did not necessarily lead, as scholars like Faderman have argued, to the internalization of sexological ideologies: discourses of lesbian resistance also emerged in the late nineteenth century, elucidating the beginnings of lesbian identity politics in U.S. culture.

In making these arguments, I am not suggesting that women's relationships were universally discouraged, but I am insisting that the representation of what we might now call homoerotic or lesbian relations was less monolithic than recent literary and social histories have implied—and that the tidy division between the nineteenth and twentieth centuries misrepresents the varieties of representation and, most important, resistance occurring in the United States throughout the nineteenth century. Moreover, I suggest that even those studies that have focused on scientific discourses have been too tentative about locating the production of sexual identities—and the influence of this production on American literature—before the late nineteenth century. My reading of nineteenth-century discourses about sexuality and identity, and the popular fiction informed by them, shows the emergence throughout the nineteenth century of a specific sexual identification built upon the pathologizing of erotic and exclusive relationships between women. Far from a presexological world in which women's intimacies were encouraged by a culture that perceived women to be innocent of sexuality, the nineteenth century was a period of contentious struggle over the definition and representation of a lesbian sexuality—a struggle in which writers of popular fiction were, by the 1850s, active participants.

Although a number of American literary texts do represent relatively uncomplicated and "Edenic" intimate attachments between women, an analysis that takes into account a broader examination of scientific discourse, including theories of pathology, Darwinian monstrosity, and heredity, yields new interpretations even of some texts formerly deemed paradigmatic of romantic friendship.[15] In Louisa May Alcott's *An Old-Fashioned Girl* (1870), Rebecca Jeffrey and Lizzie Small, appearing in a brief but important scene in the novel, are said to "live together, and take care of one another in true Damon and Pythias style." In keeping with Smith-Rosenberg's thesis, the two are determined never to part and, although Lizzie is about to be married, Rebecca plans to continue living with Lizzie and her husband. Despite these intimations of the acceptance of Rebecca and Lizzie's commitment and its compatibility with heterosexual marriage, the language of pathology and gender inversion creeps in. Fanny, the novel's young and naive spokesperson for fashion and conventionality, thinks Rebecca and Lizzie are a "different race of creatures" and is relieved to find that they are not "mannish and rough" as she feared they would be.[16] This suggests that although romantic friendship had a certain degree of visibility in nineteenth-century American literature, that visibility is not necessarily a sign of conventionality or social approbation.

More thoroughly influenced by the discourses of pathology are the novels of physician Oliver Wendell Holmes. Holmes's novels do not employ the language of sexology or Darwinism specifically, but they do present fictional case histories of his own somewhat eccentric theories of identity. Thus his work uses not only the rhetoric of sexual pathology but also the narrative structure of scientific inquiry. In *A Mortal Antipathy* (1885) exclusive same-sex friendships are sometimes represented as pathological conditions preventing heterosexual marriage. As one might expect, these relationships form the conflict which must be overcome in order for the narrative obligations of the marriage plot to be fulfilled. *A Mortal Antipathy* is replete with female and male characters who prefer the company of the same sex, but the novel focuses on a

young man's distaste for women, an "antipathy" caused by being dropped by a young woman when he was an infant. His "perverted instinct" is "cured" when he is saved from a burning house by Euthymia Tower, a strong Amazonian woman who then leaves her female companion to marry him. Euthymia's companion, Lurida, who has always been frail due to excessive study and an unfeminine interest in medicine, takes up translating the poems of Sappho, finding inspiration in the poet whom she calls "that impassioned lesbian."[17] Although "lesbian" seems to refer to a regional as much as a sexual identity in this passage, this specific usage in the context of a novel where same-sex attachments are an "antipathy" is an intriguingly early use of the term to refer generally to love between women. Ultimately, Lurida also regains her health when she marries and develops more domestic interests.

Holmes's "medicated novels," like *A Mortal Antipathy*, remind us that in order to understand the variety of discourses which participated in the construction and representation of lesbian pathology, one must attend to the profound cross-fertilization between advice literature, scientific discourse, and popular fiction in early nineteenth-century American culture. Until the last third of the nineteenth century, moreover, science and medicine were defined rather broadly and were highly influenced by theology, transcendentalism, mesmerism, and many of the other spiritual and eclectic movements of the period.[18] The distinctions between layperson, physician, scientist, and writer were much less rigid than they have since become. It was not uncommon for writers of literature to explore the mysteries of physiology or for physicians such as Oliver Wendell Holmes to write fictional representations of their scientific theories. The success of Holmes's "medicated novels," despite the protests of some who preferred their literature to provide amusement rather than lessons on heredity, attests to the fact that a variety of professional and popular discourses, including sexual advice manuals, popular fiction, and Darwinian theories of variation, participated in the specification of sexual identities.[19]

Thus, before the emergence of sexology as a specific professional discourse dedicated to studying sexual variation, there were other discourses of sexual advice, sexual health, and sexual pathology which began to construct and represent lesbian identity. A concept of sexuality in general as pathological, and pathology as constitutive of identity, is a part of the discourse of health reform as early as the 1830s.[20] Although much of the discourse of this period focused on male sexuality, by the 1850s women's sexuality was similarly subject to new taxonomies and definitions in the United States. Advice manuals on health, courtship, and marriage, and medical texts about sexuality and "generation" written between 1820 and 1860, often alluded to the appropriate limits of relations between women, of women's sexual desire, and of sexual and gender variations.[21] Granted, as late as 1868 William Alger could write of "The Friendships of Women" with few hints of any social disapprobation of the "enthusiastic and steadfast friendships" which women form with each other.[22] But other books warned against young girls maintaining intimacies with "polluted" friends.[23] Moreau de St. Méry, traveling in America shortly after the French Revolution and the circulation of scandalous rumors about Marie Antoinette's sexuality, was appalled by women's open displays of affection for each other and their willingness to "seek unnatural pleasures with persons of their own sex."[24]

As early as 1837 Eliza Farrar advises young women to avoid certain "intimacies" with each other which, she says, "do not deserve the name of friendship, but which consume much time, and expose [the girls] to ill-natured observation and misconstruction." Farrar recommends that all "kissing and caressing of your female friends should be kept for your hours of privacy, and never indulged in before gentlemen." As a measure of the ambivalence concerning the propriety of such relations, Farrar does not indicate that girls should give up "kissing and caressing" their female friends altogether, only that they should avoid such activity in front of men. The reasons she gives for avoiding such public displays are more provocative and intriguing than they are threatening: "There are some reasons for this, which will readily suggest themselves, and others, which can only be known to those well acquainted with the world."[25] Apparently well acquainted with the world herself, Farrar suggests that there may be more to these relationships than the women themselves know. Dinah Maria Craik, a British novelist whose advice book was published in 1858, considered women's intimate relationships with each other primarily as preparation for marriage. Craik acknowledges, and warns, that these relationships may be as intense as those with men. But she also claims that the two kinds of relationships cannot coexist; for the women's friendship to be compatible with the new "real" love, it must "change its character, temper its actions, resign its rights: in short, be buried and come to life again in a totally different form."[26]

These warnings about the appropriate limits of women's relationships are not unique to popular advice literature. Although the surge of sexological discourse which took place in the 1880s and 1890s does suggest that the late century was a pivotal time for the production of a fully professionalized and scientifically validated discourse on variant female sexualities, as early as 1850 medical science had begun discussing the implications of what Frederick Hollick called "doubtful or double sex": the presence of sexual "monstrosities" who, he determined, were females with masculine characteristics.[27]

As these increasingly scientific interpretations of women's relations emerged, women writers participated in the discourse but did not automatically internalize the premise of pathology. Two writers whose work can help us understand the complicated relationship between emerging representations of lesbian identity and resistance to the pathological model are Margaret J. M. Sweat, a little-known midcentury novelist whose work has rarely been discussed, and Mary E. Wilkins Freeman, a "regionalist" whose late nineteenth- and early twentieth-century writing has long been a site for debating the "romantic friendship" hypothesis.[28] Sweat's *Ethel's Love-Life: A Novel* and Freeman's "The Long Arm" are informed by midnineteenth-century popular and scientific discourses of pathology but reveal an ambivalence between identifying with and resisting the pathology model.

Ethel's Love-Life, published in 1859, represents erotic relationships which do not fit neatly into the model of either nineteenth-century romantic friendship or twentieth-century internalized pathology. Rather, the novel reveals a conflict over the possibilities for lesbian identification and representation, embodying the struggle between "romantic" and pathological representations and between resistance to scientific authority and submission to it. Although Sweat's novel is a fictional variation on the spiritual autobiography, a popular genre from the early colonial period through the nineteenth century, its reliance on midcentury medical terminology suggests that it is also a transitional text, combining the rhetoric of Calvinist self-examination with the medical rhetoric of biological healing, and pointing ahead as well to the case history.[29]

The novel takes the form of an epistolary narrative written by Ethel to her future husband, Ernest. Ethel begins the narrative by referring to herself as an organism constituted through a variety of influences: hereditary nature, early life circumstances, and exchanges of "magnetic force" with the atmosphere and other people. As a result of the luxurious circumstances of her childhood, the enervation of "desultory" learning, and her physical inactivity, Ethel says that she became "fragile in appearance," "requiring careful watching." The Calvinist roots of her self-surveillance are evident. She begins to keep vigilant watch over herself, explaining that "aimless reveries" and "morbid doubt and restless scepticism" had "gained complete mastery" over her. She tells Ernest that she longs for a time before such self-analysis became an obsession and thanks him for his influence, which has already begun "restoring" the "healthy and vigorous tone" to her heart and mind.[30]

Despite her professed desire to give up her obsession with self-analysis, the narrative is motivated by a healing impetus, the purpose of which is to exorcise Ethel's sexual history so that she may find her "true self" and bring that self into a healthy marriage with Ernest. Ethel writes about several episodes in her "love-life," involving both women and men, but the most evidently homoerotic episodes in the novel focus on her experience with Leonora, whom she describes as both an evil creature and an organism of imbalanced and contradictory nature. We get our first introduction to Leonora in what Ethel calls a dream creation, an erotic fantasy complete with "longing, burning looks" and "hot kisses." Ethel writes that the dream proves her theory that once you love a friend she is always a part of you and "clings to us as the monster to Frankenstein," an analogy that prepares us for the pathological figure she describes. In the dream Leonora enters the room with a "peculiar rushing step" in "contrast with her usual quiet movements" under which lies a "stormy, fiery, tempestuous nature." The layers of personality, each hiding another, more "essential" aspect of her "nature," represent Leonora alternately hiding and revealing her unhealthy nature. Only when under the influence of Ernest's healing gaze is Ethel able to discern Leonora's "traitorous and unworthy nature" surfacing from the "bewilderingly beautiful physical" nature to which she is so attracted. Leonora and Ethel's relationship, and the reasons for its dissolution, remain an enigma in subsequent letters to Ernest. We never do get a precise understanding of why Leonora's nature is so "unworthy." Her enigmatic qualities are only heightened by Ethel's claim to have received information about Leonora's other relationships from a man whose "revelations of every possible kind" were supplemented by his sister who said "what he dared not utter." [31]

The key to Leonora's "unworthiness" and to her relationship with Ethel lies in the representation of Leonora throughout the novel as a "contradiction." She has, as "an unfortunate gift from nature," an inherent "duplicity" by which "she was compelled to live a dual life." [32] This duplicity and doubleness, concepts abstract enough to incorporate any number of moral ineptitudes or biological "anomalies" of the nineteenth century, is associated with sexual variations in Frederick Hollick's 1850 *Marriage Guide*. Hollick's chapter on "Doubtful or Double Sex" addresses the problem of "sexual monstrosities," tricks of nature by which some persons, whom he claims are usually women, are seemingly of both sexes and can have sexual relations with other women. [33] Hollick's representation of the duality of the gender identification and the double life led by such women suggests that literary allusions to contradiction and duplicity were informed by an emerging concept of female homosexuality as pathological. [34] The lesbian and erotic implications of the enigmatic miscellany of psychological contradictions that Leonora represents become more explicit in the discussion of relations between women that follows. Saying that "[the] study and analysis of such an organism as [Leonora's] are full of interest to one who possesses the key to its contradictions," Ethel defines Leonora as a specific kind of organism or, to use the terms of Alcott's novel, a "different race of creature" identifiable by her love for the same sex. She goes on to discuss at length women's erotic relationships with each other, their difference from relations with men, and the secrecy that surrounds such relations. Ethel makes it clear that these relations are not free from passion or eroticism but, rather, represent a different kind of eroticism. Passion between women, she says,

> retains its energy, its abandonment, its flush, its eagerness, its palpitation and its rapture—but all so refined, so glorified, and made delicious and continuous by an ever-recurring giving and receiving from each to each. The electricity of one flashes and gleams through the other, to be returned not only in *degree* as between man and woman, but in *kind* as between precisely similar organizations.

Further highlighting the fact that these are not simply friendships, Ethel makes a distinction between these "passionate attachments" and the more common concept of intimate friendship: "these passions are of much more frequent occurrence than the world is aware of—generally they are unknown to all but the parties concerned, and are jealously guarded by them from intrusive comment." [35] That these relations are, in some way, unknown—explained away as something other than what they really are—indicates that they were not so "innocent" nor so conventional and "universally encouraged," as historians have suggested. [36] This becomes even more clear in Ethel's discussion of the potentially tenuous social position of such relationships and the value of silence as a form of resistance at a time when scientific representations threaten to influence their meaning. She says:

> silence and mystery help to guard the sacred spot where we go to meet our best-loved friends. The world sees only the ordinary appearances of an intimate acquaintanceship, and satisfies itself with a few common-place comments thereon—but the joy and beauty of the tie remain in sweet concealment—silent and inexpressive when careless eyes are upon it, but leaping in to the sunlight when free from cold and repelling influences.[37]

Although the very existence of Ethel's discussion of relations between women might suggest a kind of presexological acceptance—indeed, Sweat was able to publish the novel—her plea for keeping these relations secure from "intrusive comment" implies that "intrusive comment" did exist. In fact, this passage indicates that such relations were tolerated only to the extent that they were *mis*understood as "romantic friendship." The words of Richard von Krafft-Ebing further support this claim. Later in the century he explained the relatively scant medical attention to lesbianism by arguing that "inverted sexual intercourse among women is less noticeable, and by outsiders is considered mere friendship."[38]

Leonora is not the only "pathological" creature in Sweat's novel. As I suggested earlier, Sweat represents Ethel herself as a self-examining, convalescent organism. The novel's medical rhetoric suggests that she is informed by a concept of herself as a sexual organism whose unhealthy relations must be cured. That this cure comes through the influence of Ernest and through her marriage to him further supports this reading. But it is difficult to read the novel as simply representing lesbian pathology. Ethel contrasts the "serene loftiness" and "unselfish

devotion" of Claudia, another figure in her "love-life," with Leonora's "feminine beauty, bewitching grace," and ability "to fascinate and to subdue."[39] This difference in her descriptions of Claudia and Leonora indicates that, for Ethel, not all women who engage in same-sex relationships are monstrous contradictions of nature.

However, this intimation of the potential for a more "noble" relationship between women is overshadowed by the focus of the narrative on the restoration of Ethel's health through her marriage to Ernest. In a passage that calls into question the compatibility of women's intimate relations with heterosexual marriage, Ethel makes it clear that she had to choose between Leonora and Ernest. Indicating both the exclusiveness of the two relationships and their inherent differences, she tells Ernest that, having already promised herself to him, she could not also promise herself to Leonora: "I could crown but one king within my heart—to him and to him alone, could I reveal myself completely—yet though I wish not to worship but one, I could have loved her and enjoyed much with her on a different level."[40] That both Ernest and Leonora represent possible "kings" who could rule Ethel's heart suggests that her relations with the two are in some way analogous—both are gendered male—and that relationships between women were not compatible with marriage, since she could "crown but one." And yet Ethel also says that she could have loved Leonora on a "different level," suggesting, as Smith-Rosenberg and Faderman have argued, that there was a variety of possibilities for women's relations with each other and that they did not necessarily exclude the possibility of heterosexual marriage. This "different level" may seem to suggest compatibility with marriage, but it also echoes Craik's admonition that love between women must "resign its rights" and "come to life again in a totally different form."

As the audience for her epistolary self-analysis and as the figure providing her with the hope of marriage, Ernest represents health, happiness, and balance through a perfect match of complementarities. Earlier in the novel Ethel suggests that they are perfect for each other because he has "almost feminine intuitions" and "a woman's tenderness." But to reassure us of Ernest's masculinity, Ethel qualifies his feminine qualities: "I should love your tenderness less and rejoice in your gentleness with more of doubt, did I not know this other side of your strong manly nature." By the end of the novel it is his masculinity that, as a complement to her femininity, makes him so fitting a marital partner. In their marriage she finds the "consummation" of her "being" and the fulfillment of her "deepest yearnings" in his "noble and generous manliness." Refusing to paint a fairytale portrait of their love, however, Ethel acknowledges the difficulties which life will bring them and sees her past as preparation for these hardships: "we have lost no strength, but have rather been disciplined to a more skilful [sic] and successful warfare."[41] Thus, disciplined through a varied love life and a thorough self-analysis, Ethel happily accepts marriage as that relation which completes her own identity.

Sweat's novel indicates the tenuous position of women's relationships in midnineteenth-century American culture. Neither actively and institutionally encouraged nor absolutely disqualified by scientific discourses of pathology, same-sex erotic relations were possible to the extent that the eroticism was unsuspected or inconceivable. The ultimate function of the relationship must be preparation for marriage. Sweat implies that such relationships are able to flourish in a culture not yet aware of their difference from "intimate acquaintanceship," but her own medically informed rhetoric suggests just how tenuous this position is. By the midnineteenth century, a medical context for women's relationships was emerging in advice manuals, scientific discourse, and popular fiction, ultimately defining what constituted marital health and "normal" sexual desire and pathologizing anything representing a threat to heterosexual marriage. Sweat's exposure of these relationships and her participation in the medical discourse about them, combined with her critique of the sciences as a discourse of intrusive commentary, represents a sexual culture much more complex than historians and literary critics have suggested.

As many historians have indicated, by the 1890s a medical discourse about exclusive companionships between women was gaining momentum and achieving scientific authority.[42] Havelock Ellis and Continental physicians such as Richard von Krafft-Ebing were joined by American physicians J. C. Shaw and G. N. Ferris in examining and defining a range of sexual variations from a binary male-female gender identity entailing desire for the opposite sex. They called such "conditions" "sexual inversion," "antipathic instinct," or "homo-sexual feeling."[43] During the 1890s the medical establishment seemed to be reaching a consensus that such relations were indications of a congenital pathology in which gender identification was inverted; inverts chose as sexual partners members of the same sex—generally noninverts with a slight capacity for homosexual feeling.[44]

This period of consolidation in medical opinion is often cited as the historical moment when the figure of the modern homosexual was brought fully into both scientific and public discourse. Although at the turn of the century, and during the first two decades of the twentieth century, disagreement increased among physicians about the "etiology" and "prognosis" of homosexuality, the presence or absence of somatic abnormalities, and even the validity of homosexuality as a psychological abnormality, the authority of science and medicine to debate and decide these issues was not directly questioned. Thus the end of the nineteenth century was a period of consolidated redefinition of the terms with which one could represent intimate and erotic relations between members of the same sex. The medical vocabulary of somatic pathology became the paradigmatic idiom for discourse about homosexuality.

Born in 1852, and publishing short stories and a few novels by the 1880s and 1890s, Mary E. Wilkins Freeman came of age during this pivotal period. During her lifetime the concept of relationships between women as potentially unhealthy forms of psychological duplicity developed into a widely disseminated and culturally dominant medical determination that such relations were indicative of a pathological type. Even as this medical model was becoming dominant, Freeman both employed the rhetoric of pathology and attempted to resist its pejorative implications.

"The Long Arm," like other Freeman stories, represents the conflict between heterosexual marriage and women's exclusive relationships, but the story is unique in its use of sexological rhetoric and in its defense of those pathologized. One of Freeman's few remaining uncollected stories, this work has been overlooked by scholars, perhaps because it was cowritten by British detective fiction writer J. Edgar Chamberlin, or perhaps because it is a departure from Freeman's canon in a variety of ways. Nonetheless, it is an important text that requires us to reconsider both Freeman's treatment of same-sex relations and the history of lesbian representation and resistance in general. Unlike Sarah in "Two Friends," another Freeman story about a woman's unfounded fear that her female companion will marry and leave her, Phoebe Dole, in "The Long Arm," uses illicit means to preserve her relationship with Maria Woods, who wants to be free from Phoebe's domination. The story is composed of the recollections and daily journal

entries of Sarah Fairbanks after the murder of her father, Martin Fairbanks. It begins by revealing Martin's intent to leave Sarah out of his will if she marries her admirer, Henry. But the two characters whose relationship is the key to the detective mystery are Phoebe Dole and Maria Woods, "two old maiden ladies."[45]

The narrator describes Phoebe as overbearing and controlling: "Phoebe always does things her own way. All the women in the village are in a manner under Phoebe Dole's thumb." Although Phoebe is very calm in the wake of the murder, her companion of many years, Maria Woods, is terribly upset and weeps constantly. Phoebe chides her in front of Sarah Fairbanks: "'If you can't keep calmer, you'd better go upstairs, Maria. . . . You'll make Sarah sick. Look at her! She doesn't give way—and think of the reason she's got.'" Phoebe's comment about Sarah's composure so soon after the murder of her father seems to present evidence against her. But at this point in the story, Maria's enigmatic response, "'I've got reason too. . . . Oh, I've got reason,'" is as perplexing as Phoebe's sharp command that she leave the room. At first Maria's seemingly excessive emotion is understood by Sarah Fairbanks as just another example of her weak constitution. Sarah describes Maria's and Phoebe's roles in the relationship in terms similar to a traditional heterosexual marriage: "Maria Woods has always been considered a sweet, weakly, dependent woman, and Phoebe Dole is undoubtedly very fond of her. She has seemed to shield her, and take care of her nearly all her life."[46]

Although Phoebe and Maria are central to the early scenes of the story, they become less so as the search for the culprit ensues. The suspicion soon turns to Sarah, who, by murdering her father, would be able to marry Henry and still gain a substantial inheritance. The fact that her father was killed in the house where they lived together, and that the house was left locked from the inside, seems further to substantiate her guilt. But when Sarah is put on trial and acquitted, the town is ready to leave the murder unsolved. Their blindness to any but a heterosexual motive leads everyone to ignore the inconsistencies in the behavior of Phoebe Dole. But when Sarah divides the crime scene into easily inspected square yards and investigates every inch, and then hires a detective to assist in her investigation, the evidence against Phoebe Dole becomes conclusive.

The first clue, which Sarah is initially unable to understand, is a wedding ring inscribed with two dates exactly forty years apart. One of the dates is August of the year the story takes place. The second clue is an old letter from Maria Woods refusing Martin's offer of marriage. The clues add intrigue to the lives of Martin Fairbanks and Maria Woods and highlight the significance of heterosexual marriage to the murder, but their significance is not understood until much later. The clue which ultimately solves the mystery is the detective's discovery that the only way the murderer could have escaped the house and left it locked from the inside was to reach up through the cat door and hook the latch after leaving. The person who did this had to have an arm at least six inches longer than Sarah's or the detective's. As Sarah goes about town surreptitiously comparing the length of people's arms, she is shocked to discover that "Phoebe Dole's arm is fully seven inches longer than mine. I never noticed before, but she has an almost abnormally long arm."[47] Phoebe Dole, it seems, is the only person in the village physically capable of killing Martin Fairbanks and leaving the crime scene sealed from the inside.

Despite the detective's claim that "[c]rime has no sex," Sarah refuses to believe that Phoebe committed the murder; she remains blind to the lesbian motive and insists that a woman, particularly "a good woman—a church member" could not possibly commit murder. But Sarah must finally accept the evidence when Maria explains to Sarah "her long subordination to Phoebe Dole." As Sarah narrates Maria's story:

> This sweet child-like woman had always been completely under the sway of the other's stronger nature. The subordination went way back beyond my father's original proposal to her; she had, before he made love to her as a girl, promised Phoebe that she would not marry; and it was Phoebe who, by representing to her that she was bound by this solemn promise, had led her to write a letter to my father declining his offer, and sending back the ring.

Sarah need not prove that stopping Martin and Maria's second attempt to marry is the motive for the crime, for Phoebe, approaching Sarah's house with "rapid strides like a man," soon confesses that she killed Martin Fairbanks, in order to prevent the marriage: "'She's lived with me in that house for over forty years. There are other ties as strong as the marriage one, that are just as sacred. What right had he to take her away from me and break up my home.'"[48] Phoebe's assertion that her bond with Maria is as strong and sacred as any conventional marriage tie, and that Fairbanks had no right to "break up (her) home," is a plea for sympathy on the grounds that neither society nor Martin Fairbanks recognizes her home and partnership as valid. Of course, the fact that Phoebe forbids Maria to marry, and commits murder to prevent it, raises the issue of consent and reduces the power of her argument that the relationship should be recognized. This element of coercion in their relationship represents Phoebe more stereotypically as a dominant, controlling sexual invert who has seduced an unwitting, weak woman into a relationship.[49]

Freeman's representation of Phoebe's dominance and Maria's subordination is given a more clearly lesbian signification in the ridiculous image of the long arm, an obvious phallic representation signaling Phoebe's abnormal gender identity and sexual ambiguity. As Phoebe's identifying feature and the conclusive evidence in the detection of her crime, the long arm is also an inscription of sexual deviance informed by the medical construction of lesbians as somatically monstrous. Although theories assuming a connection between somatic irregularities and homosexuality had been largely discredited by the first decades of the twentieth century, and Freudian interpretations of homosexuality as a sign of disrupted psychosexual development or normal variation, rather than a degenerative biological monstrosity, had begun to gain credibility and authority, the much-noted distortions that medical science produced in describing the somatic peculiarities of "sexual inverts" reached the public imagination in spectacular forms during the late nineteenth century.[50] The idea of single women and men uninterested in the opposite sex as examples of a third sex, freaks of nature whose social irregularities were undoubtedly accompanied by physical aberrations, had received public currency, manifesting itself in a variety of fantastic representations. The "attitude," angle, and length of the arms were viewed medically as a sign of either femininity or masculinity and even, as works of scientific racism show, signs of either evolutionary progress or degeneration to an apelike state.[51] Particularly interesting in relation to Freeman's story is the significance Havelock Ellis attributes to arms. He quotes at length the measurements and observations of an obstetric physician whose examination of one lesbian, identified as "Miss M," shows that "with arms, palms up, extended in front of her with inner sides of hands touching, she cannot bring the inner sides of forearms together, as nearly every woman can, showing

that the feminine angle of arm has been lost."[52]

Thus Phoebe's arms represent more than her ability to have murdered Martin Fairbanks; they represent her status as a sexual transgressor, her anomalous sexual identity, and her degeneration as a member of the human species. According to Nancy Leys Stepan, the American physician James Weir believed that women who transgressed social gender boundaries were "in danger of degenerating into psychosexual hybrids."[53] Ellis, to his credit, attempts to ameliorate the stigma of the terms *abnormality* and *anomaly;* he quotes a physician who explains that somatic irregularities do not represent degeneracy but indicate an anomalous figure in the process of "transformation ... from one species into another."[54] But in an era where the transformation from one species to another is widely viewed as indicative of degeneration, this explanation hardly reduces the stigma. It does, however, present a more neutral paradigm which some writers would appropriate to claim that the new species was a sign of evolutionary advancement, rather than degeneration.[55]

Novels like Oliver Wendell Holmes's *A Mortal Antipathy* and *Elsie Venner: A Romance of Destiny* (1861) indicate that somatic notions of sexual difference were circulated in popular forms. In addition, the widely publicized murder case of Alice Mitchell, who apparently killed her companion Freda Ward, also established a context of pathology and criminality for understanding exclusive relationships between women.[56] Lillian Faderman explains Freeman's "The Long Arm" as an expression of the association between lesbianism and violence in the wake of the Lizzie Borden and Alice Mitchell trials.[57] Although Freeman would have been aware of these notorious murder cases, her representation of Phoebe seems also to be informed by a Darwinian model of variation within species. The connection to Darwinian variation is particularly evident in an untitled, unpublished, short story extensively excerpted by Edward Foster in his 1956 *Mary E. Wilkins Freeman.* Although a thorough reading of this story is beyond the scope of this article, it bears mention because of its use of the specific language of evolution and monstrosity. Jane Lennox, the first-person narrator of this story, claims to have been deprived of her birthright, what she calls "the character of the usual woman." She does not lament this self-image, however; in fact, she celebrates it: "I am a graft on the tree of human womanhood. I am a hybrid. Sometimes I think I am a monster, and the worst of it is, I certainly take pleasure in it."[58] Freeman clearly contextualizes Jane's self-perception within the nineteenth-century scientific discourse of human variation, employing the specific rhetoric of Charles Darwin's *Origin of Species,* in which he uses the image of the tree to talk about the mutability of species and the word *monstrosity* to talk about anomalies within species.[59] The fact that Jane Lennox claims to "take pleasure" in her anomaly presents a subversive twist that undermines the representation of sexual transgressors as tragic, monstrous degenerates.

The playfulness obvious in the Jane Lennox story makes it tempting to suggest that Freeman's use of the long arm may also have been tongue in cheek. Phoebe's defense of the validity of her "home" suggests that Freeman's sympathies may have been with Phoebe and Maria. Nonetheless, the pathological implications of Phoebe's arm, and her status as criminal, undercut the subversive elements of the story. What "The Long Arm" and Freeman's untitled story suggest is that by the end of the century, sexological discussion of somatic irregularities and the Darwinian language of hybridity and monstrosity inflected popular writing with a medically and scientifically based rhetoric. Freeman's inconsistent use of this rhetoric indicates how

complicated the relationship between identification, internalization, appropriation, and resistance to such constructions of identity can be. The pathological implications of Phoebe's long arm contrast sharply with Jane Lennox's proud affirmation of her hybridity and the pleasure she takes in it, suggesting that the rhetoric of evolution and sexology could be appropriated but only at the risk of reinscribing the premise of pathology.

The various discourses of sexuality and identity—advice manuals, treatises on medicine and heredity, and fictions such as *Ethel's Love-Life,* "The Long Arm," and the Jane Lennox story—suggest that the representation of same-sex relations between women as pathological occurred earlier in the nineteenth century than critics have believed and that some writers occasionally resisted them before the turn of the century. Thus romantic friendship, although certainly a part of nineteenth-century discourse, was neither the only nor the dominant signification for exclusive and erotic relations between women. For even in the midnineteenth century, figures of "freakish friendship," contradictory nature, and monstrous hybridity are a part of the cultural imagination through which same-sex relationships were represented.[60] The nineteenth and twentieth centuries can hardly be reduced to a presexological utopia and a postsexological nightmare of self-loathing and despair. But neither can we simply push back the postsexological analysis to the middle of the nineteenth century. Indeed, the supersessionist notions of presexological and postsexological cultures ignore the interactions between a variety of discourses of sexuality and identity throughout the nineteenth century. As I have suggested here, some writers produced woman characters who, constituted as a "different race of creatures," appropriated their newly created identities, fashioning themselves as a hybrid class with the subversive potential to challenge the binary construction of gender and the limitations of gender expectations. Thus we might better represent the sexual culture of the nineteenth century as a site of ongoing struggle over the politics of representation and resistance, a struggle which continues as we debate the efficacy of the politics of identity today.

Notes

1. Carroll Smith-Rosenberg, "The Female World of Love and Ritual: Relations Between Women in Nineteenth-Century America," *Signs* 1 (Autumn 1975): 8.

2. Lillian Faderman, *Surpassing the Love of Men: Romantic Friendship and Love Between Women from the Renaissance to the Present* (New York: William Morrow, 1981), 19–20. Faderman repeatedly asserts that it is unlikely that Victorian women expressed their love for each other sexually (18–19, 250–51, 414). She cites as evidence the dearth of explicit references to sex between women in correspondence or diaries, ignoring the similar lack of such discussions of heterosexual sex (19).

3. I am not the first to suggest this. Martha Vicinus, perhaps unintentionally, links the romantic friendship debate with the controversy over woman-identified women and butch-femme roles, in "'They Wonder to Which Sex I Belong': The Historical Roots of the Modern Lesbian Identity" (471–72). And Lisa Moore shows how Faderman's history is presented as an analog to the lesbian feminism of the late 1970s and early 1980s in "'Something More Tender Still than Friendship': Romantic Friendship in Early Nineteenth-Century England" (502). Both essays appear in The Lesbian Issue, *Feminist Studies* (Fall 1992). For a text key to the "sex wars," see Carole S. Vance, ed., *Pleasure and Danger: Exploring Female Sexuality* (Boston: Routledge & Kegan Paul, 1984).

4. Two important lesbian feminist essays which deemphasized the sexual component of lesbianism, or at least broadened the definition of lesbian beyond the sexual, are Adrienne Rich's "It Is the Lesbian in Us . . . ," in Adrienne Rich, *On Lies, Secrets, and Si-*lence: Selected Prose, 1966–1978* (New York: Norton, 1979), 199–202 and "Compulsory Heterosexuality and Lesbian Existence," *Signs* 5 (Summer 1980): 631–60.

5. Carroll Smith-Rosenberg, "Discourses of Sexuality and Subjectivity: The New Woman, 1870–1936," in *Hidden from History: Reclaiming the Gay and Lesbian Past,* ed. Martin Bauml Duberman, Martha Vicinus, and George Chauncey, Jr. (New York: New America Library, 1989), 270. Although Smith-Rosenberg located the emergence of a scientific and taxonomic discourse on sexual subjectivity in the midnineteenth century, she claims that only in the mid-1880s did these texts begin to pay attention to lesbianism (268–69).

6. Martha Vicinus, "Distance and Desire: English Boarding School Friendships, 1870–1920," in *Hidden from History,* 228, and Vicinus, "'They Wonder to Which Sex I Belong,'" 483.

7. Vicinus, "'They Wonder to Which Sex I Belong,'" 471.

8. Terry Castle, *The Apparitional Lesbian: Female Homosexuality and Modern Culture* (New York: Columbia University Press, 1993), 8–10.

9. Lisa Moore, "'Something More Tender Still than Friendship,'" 502–3. Moore uses the idea of "character" as a "historical antecedent to what we would now call identity," belying a reluctance to rehistoricize the construction of a specifically lesbian identity (520 n. 4).

10. Helena Whitbread, ed., *I Know My Own Heart: The Diaries of Anne Lister, 1791–1840* (New York: New York University Press, 1992), 210.

11. Lister's diaries also reveal that the case history was becoming a context for such discussions. On determining the nature of friendships between women, she writes: "But much, or all, depends upon the story of their former lives, the period passed before they lived together, that feverish dream called youth" (ibid.).

12. Bonnie Zimmerman, *The Safe Sea of Women: Lesbian Fiction, 1969–1989* (Boston: Beacon Press, 1990), 3–4.

13. Lillian Faderman, *Odd Girls and Twilight Lovers: A History of Lesbian Life in Twentieth-Century America* (New York: Columbia University Press, 1991), 1.

14. Mary E. Wood, "'With Ready Eye': Margaret Fuller and Lesbianism in Nineteenth-Century American Literature," *American Literature* 65 (March 1993): 3–4.

15. Henry Wadsworth Longfellow's *Kavanaugh* (1849) and Florence Converse's *Diana Victrix* (1897) show the extent to which the representation of romantic friendship was possible throughout the nineteenth century and could coexist with representations of lesbian pathology.

16. Louisa May Alcott, *An Old-Fashioned Girl* (1870; reprint, Boston: Little, Brown, 1911), 255. Faderman, in *Surpassing the Love of Men*, uses Alcott's and Holmes's novels as examples of the romantic friendship hypothesis, 262–63.

17. Oliver Wendell Holmes, *A Mortal Antipathy* (1885; reprint, Boston: Houghton Mifflin, 1892), 294.

18. Paul Starr, *The Social Transformation of American Medicine* (New York: Basic Books, 1982).

19. One reader protested that she did not want to read a "medicated novel." See the second Preface in Oliver Wendell Holmes, *Elsie Venner: A Romance of Destiny* (1861; reprint, New York: New American Library, 1961), xii.

20. Sylvester Graham, *A Lecture to Young Men* (1834; reprint, New York: Arno, 1974).

21. For example, see Catherine Esther Beecher, *Letters to the People on Health and Happiness* (New York, 1855); Eliza Farrar, *The Young Lady's Friend* (New York, 1837); Goss and Co., *Hygeiana, a Non-Medical Analysis of the Complaints Incidental to Females* (London, 1823); Frederick Hollick, *The Marriage Guide, or Natural History of Generation* (1850; reprint, New York: Arno, 1974); and Samuel B. Woodward, *Hints for the Young in Relation to the Health of Body and Mind* (Boston, 1856).

22. William Alger, *The Friendships of Women* (Boston, 1868), 27.

23. John S. Haller Jr. and Robin M. Haller, *The Physician and Sexuality in Victorian America* (Urbana: University of Illinois Press, 1974), 106.

24. Kenneth Roberts and Anna M. Roberts, trans. and eds., *Moreau de St. Méry's American Journey* (Garden City, N.Y.: Doubleday, 1947), 286.

25. Farrar, *Young Lady's Friend*, 254, 269.

26. Dinah Maria Craik, *A Woman's Thoughts about Women* (New York, 1858), quoted in Wood, "With Ready Eye."

27. Hollick, *Marriage Guide,* 202–3.

28. Margaret J. M. Sweat, *Ethel's Love-Life: A Novel* (New York: Rudd & Carleton, 1859). The only reference I have seen to Sweat's novel is in John D'-Emilio and Estelle B. Freedman's *Intimate Matters: A History of Sexuality in America* (New York: Harper & Row, 1988), 125–26. They use the novel as an example of the acceptability of passionate attachments between women. For analyses of Freeman see, especially, Josephine Donovan, *New England Local Color Literature: A Women's Tradition* (New York: Frederick Ungar, 1983); Susan Koppelman, "About 'Two Friends' and Mary Eleanor Wilkins Freeman," *American Literary Realism* 21 (Fall 1988): 43–57; and Mary R. Reichardt, "'Friend of My Heart': Women as Friends and Rivals in the Short Stories of Mary Wilkins Freeman," *American Literary Realism* 22 (Winter 1990): 54–68.

29. In *The Marriage Guide,* Hollick focuses on specific cases of sexual monstrosity but does not follow the form of case history—tracing environmental influences and family diseases—which would become infamous with Krafft-Ebing and Ellis. Carl Westphal published case histories in 1869 which were widely referred to by American physicians, including J. C. Shaw and G. N. Ferris in 1883. As mentioned earlier, Holmes's "medicated novels," which he began writing in the 1850s, were fictional case histories.

30. Sweat, *Ethel's Love-Life,* 16, 22.

31. Ibid., 68, 69, 72, 73.

32. Ibid., 76, 77.

33. Hollick, *Marriage Guide,* 292–94.

34. This duality and duplicity are not unique to the representation of homosexuality. Mixed-race African-Americans were also represented as having an inherent duplicity. Such duplicity was ascribed to those who did not perfectly identify with one pole of racial or gender binaries and were therefore capable of misrepresenting themselves not only to the common observer but to the medical professional as well. See, especially, J. C. Nott and Geo. R. Gliddon, eds., *Types of Mankind* (Philadelphia, 1854), 373; and Lothrop Stoddard, *The Rising Tide of Color Against White World-Supremacy* (New York: Scribner's, 1920), 166.

35. Sweat, *Ethel's Love-Life*, 82, 83.

36. Faderman, in *Surpassing the Love of Men*, refers to the nineteenth-century presexological culture as a period of "innocence" (297).

37. Sweat, *Ethel's Love-Life*, 83.

38. Richard von Krafft-Ebing, *Psychopathia Sexualis*, trans. F. J. Rebman, 12th ed. (1886; reprint, Chicago: Login Brothers, 1929), 396–97.

39. Sweat, *Ethel's Love-Life*, 85–86.

40. Ibid., 81.

41. Ibid., 123, 126, 230–32.

42. See Faderman, *Surpassing the Love of Men*; see also George Chauncey, Jr., "From Sexual Inversion to Homosexuality: Medicine and the Changing Conceptualization of Female Deviance," *Salmagundi* 58–59 (Fall–Winter 1983); and D'Emilio and Freedman, *Intimate Matters.*

43. See Krafft-Ebing; *Psychopathia Sexualis*, and J. C. Shaw and G. N. Ferris, "Perverted Sexual Instinct," *Journal of Nervous and Mental Disease* 10 (1883): 185–204.

44. The idea of gender inversion, which involves identification with the opposite sex, allowed homosexual relations to be explained within a heterosexual paradigm. In effect, the model did not describe two women who were sexually involved with each other but a man (albeit in a woman's body) and a woman (a noninvert with a slight capacity to desire inverts) who was attracted to the invert's manliness.

45. Mary E. Wilkins Freeman, "The Long Arm," in *American Detective Stories*, ed. Carolyn Wells (New York: Oxford University Press, 1927), 138.

46. Ibid., 147, 148.

47. Ibid., 169.

48. Ibid., 171–74.

49. For this stereotype see Havelock Ellis, *Sexual Inversion* (1897; reprint, New York: Arno, 1975), 87–88.

50. Chauncey, "From Sexual Inversion to Homosexuality," 132. For an example of a medical article arguing against somatic theories, see Trigant Burrow, "Genesis and Meaning of 'Homosexuality' and Its Relation to the Problem of Introverted Mental States," *Psychoanalytic Review* 4 (July 1917): 272–84.

51. Nancy Leys Stepan, "Race and Gender: The Role of Analogy in Science," in *Anatomy of Racism*, ed. David Theo Goldberg (Minneapolis: University of Minnesota Press, 1990), 42.

52. Ellis, *Sexual Inversion*, 90.

53. Stepan, "Race and Gender," 40.

54. Ellis, *Sexual Inversion*, 136 n. 1.

55. See, for example, Edward Carpenter, *The Intermediate Sex in Selected Writings*, vol. 1, *Sex*, ed. Noël Greig (London: GMP, 1984), esp. 186–88, 233–34, and 241.

56. For a discussion of the Alice Mitchell case and documentary evidence from the trial, see Jonathan Katz, *Gay American History: Lesbians and Gay Men in the U.S.A.* (New York: Avon, 1976), 82–90.

57. Faderman, *Surpassing the Love of Men*, 292. Reichardt also sees "The Long Arm" as influenced by the Mitchell case, but she misreads the story, saying that Phoebe kills Maria when it is actually Martin Fairbanks whom she murders ("'Friend of My Heart,'" 62).

58. Edward Foster, *Mary E. Wilkins Freeman* (New York: Hendricks House, 1956), 143.

59. Charles Darwin, *The Origin of Species* (1859; reprint, London: Penguin, 1985), 101.

60. Lisa Moore talks about "frekish friendship" in her discussion of Harriot Freke, a character in Maria Edgeworth's *Belinda*; see "'Something More Tender Still than Friendship,'" 503–13, and 517–18.

12 "A New Generation of Women": Progressive Psychiatrists and the Hypersexual Female

Elizabeth Lunbeck

In the early years of the twentieth century a number of prominent American psychiatrists identified the hypersexual female, the willfully passionate woman who could not control her desires for sexual pleasure, as an issue of pressing medical concern. Psychiatrists diagnosed women whose sexuality they deemed abnormally aggressive as "psychopaths"; these women, they explained, suffered from an inborn condition for which there was no remedy save institutionalization. Borrowing the term *psychopathic* from the nosologies of their German counterparts, who began using it in the late nineteenth century to classify male deviants (including vagabonds, criminals, anarchists, revolutionaries, and reformers), American psychiatrists employed it to account for what many agreed was the problem of women's sexual excess. Psychiatrists early identified prostitutes and female juvenile delinquents, most of whom had been charged with immoral behavior, as psychopaths; soon a variety of other young women whose sexuality violated medical and social conceptions of proper female deportment swelled their

ranks. "New women" of the working class, these hypersexual psychopaths sought a sexual freedom unknown to their mothers. "A new generation of women has arrived," proclaimed one of them, Ethel Hancock, a twenty-one-year-old waitress, "and the wrong of Grandma's day is the right of today."[1]

Confronted by a revolution in sexual mores among working-class youth, psychiatrists, like other social investigators—sociologists, social workers, criminologists—turned to the problem of female immorality. Psychiatrists, like many others, ascribed the lax sexual etiquette of the time to the deterioration of women's morals.[2] They explained that psychopathic hypersexuals, women who could not restrain their boundless desires, were to blame for slack sexual mores; men were at best the passive recipients or, at worst, the unwitting victims of their unwanted attentions.

In constructing the category of the female hypersexual, psychiatrists proclaimed themselves arbiters of rapidly changing sexual mores. They drew on their capital as medical experts in arguing for the indisputable fitness of an older Victorian morality that held women chaste and reticent. In so doing, they attempted to medicalize the discussion of the working girl's sexual-

ELIZABETH LUNBECK is Associate Professor of History at Princeton University, Princeton, New Jersey. Reprinted from *Feminist Studies* 13 (1987): 513–543.

ity, to replace the prostitute, who in the nineteenth century had embodied all that was base in women's nature, with the hypersexual. In diagnosing women as hypersexual, psychiatrists in part were responding to the many demands that others placed on them—the families, police, and social workers who brought young women to their attention—to address the issue of female immorality and to find explanations and remedies beyond the outworn moral categories that saw the promiscuous woman as bad and the sociological theory that cast her as a victim. Yet the particular solution psychiatrists chose—to diagnose these women as incurable psychopaths—was problematic. For these women manifested none of the usual symptoms of mental disorder, and psychiatrists readily conceded that psychopathic personality might be thought of as a conduct disorder rather than a disease. The significance of psychiatrists' intervention lies not in their success in controlling, or even defining, female sexuality, but in identifying and attempting to comprehend a social change of singular importance: the emergence of the independent, sexually assertive woman in American society at the turn of the century.

This essay examines a moment in the largely unwritten history of female sexuality when psychiatrists first attempted to control the discourse concerning women's erotic nature. Drawing on case records from the Boston Psychopathic Hospital, one of the most highly regarded of American mental hospitals in the early years of the century, I will examine how psychiatrists constructed the category of the hypersexual psychopath and explore the worlds of the young women they considered oversexed.[3] Psychiatrists saw these women as sick; middle-class social workers, bonded by gender but distanced by class, saw them as victims and sought both to protect and to discipline them. The women themselves fashioned their own interpretations of their predicaments from elements of both

perspectives; they sometimes admitted, when pressed, to fears that they were not normal, that they were sick or bad, that they needed help and supervision. But they could also reject psychiatrists' and social workers' concerns altogether and argue for an end to the double standard of sexual morality that marked them as diseased. From this welter of confused perspectives a female voice emerged to challenge the strictures governing sexual conduct. This working-class voice was often halting and unsure. At times, however, it spoke with a strength and clarity rare in the annals of the historically invisible.

A Rightful Adolescence

Psychiatrists argued that the sexual natures and needs of young women and men differed fundamentally. A woman was by nature wholly sexual, her life colored by barely controllable sexual impulses that first surfaced at adolescence. Adolescence was a time of rapid physical and mental growth, of strong, conflicting emotions, of storm and stress for both sexes; if difficult for boys, it was treacherous for girls. Subject to the caprice of her developing physique, a growing girl was ever in danger of straying from the narrow path of respectability into promiscuity; indeed, the slightest sign of interest in boys or sex might set her on the road to ruin. The growing boy's course was less hazardous. Not burdened like a girl by the efflorescence of strong sexual impulses, he could celebrate their acquisition with ardor and evoke only the bemused tolerance of psychiatrists; the curiosity, experimentation, and satiation of desire that was symptomatic of gross defect in a girl was but the commendable manifestation of the boy's natural drive for self-expression and mastery. Psychiatrists held that social convention, which tolerated and even encouraged the fulfillment of male desire, only mirrored the immutable dictates of human nature. If it was for a girl once soiled, forever spoiled, while a boy could "sow an unusually large crop of wild oats," straighten up, and become a good citizen thereafter, this was the natural order of things. Men, unlike women, could weather a phase of intense sexual activity without breaking down under it; among men, psychiatrists agreed, "a separate standard of moral and sexual life" could and did prevail.[4]

In its recognition of female desire the double standard that psychiatrists championed differed from that of the Victorian era. Middle-class Victorian sexual ideology had set the passionlessness of women against the lustfulness of men, elevating the former and excusing the latter, with tacit tolerance of prostitution as a necessary social evil. Although they paid a high price in their renunciation of passion, Victorian women managed to turn this ideology to their advantage; trapped in a society that offered them little outside marriage, they fell back on their supposed passionlessness to gain a measure of control over sexual relations with their husbands and thus over the timing and number of their pregnancies.[5] Progressive psychiatrists, as they ceded some ground to women by recognizing their capacity for passion, stripped away from them all the protections Victorian ideology had offered. They not only overturned sexual Victorianism, but they reversed its equation of desire as well, casting women as sexual predators, men as sexual victims. If women wanted passion, they would give it to them with a vengeance.

Psychiatrists' recasting of the Victorian sexual drama is most starkly evident in their elaboration of the category of the hypersexual. Hypersexuality, like its elusive counterpoint, normal sexuality, first became apparent at adolescence. It was then that overdeveloped girls, girls who prematurely developed the womanly contours that so enfeebled male resolve, first began to constitute a real social menace. Upstanding men were absolutely unable to resist these young women. As psychiatrists related it, particularly

attractive young women seduced many a hapless man over telephone lines, in automobiles, on public conveyances, and even in church; young temptresses led countless sailors astray. Invigorated by overwhelming desire, a young woman could haul a man through the windows of her residence or accost an innocent on the street and force him to submit, against his will, to intercourse on the spot! The city's abundant public places teemed with hypersexuals, ready to lure unsuspecting men into questionable establishments, to hire rooms for immoral purposes, to plague men with the demands of their insatiable immorality.[6]

Psychiatrists elaborated this theory of female desire as they confronted the sexual mores of a generation of working-class new women. These women, born between 1890 and 1905, were among the first to achieve for themselves some limited freedom from family obligations, some limited freedom to earn and spend, and some limited freedom to associate with whom they pleased—all freedoms we now see as the prerogatives of late adolescence. Juvenile experts, the psychologist G. Stanley Hall and the physician William Healy foremost among them, had only recently delineated adolescence as a socially and biologically constructed interlude between childhood and adulthood. Working-class boys, however, had been enjoying the privileges of adolescence for decades. For them, adolescence was a time to taste the pleasures of adulthood—independence, mastery, sexuality—without having to assume all its burdens. Parents who loosed boys from the bonds of the family economy as soon as they entered the workplace, together with juvenile authorities who tolerated the lapses of boys who went too far—who smoked, drank, or associated with fast companions—on the conviction that they could later go straight, smoothed the way for widespread acceptance of adolescent boyhood as an admittedly difficult but indisputably normal stage of life.[7] Adolescent girlhood was, by contrast, nearly unthinkable, the very phrase so oxymoronic in its coupling of the female sex with what was so essentially male that juvenile experts could barely imagine its contours. Female adolescence appears only fitfully in the works of Hall. On the one hand, he argues that women are by nature forever adolescent—dependent, childish, without ambition—and on the other hand, he advocates, in prose thick with sexual innuendo, that they forever teeter on the brink of knowledge and sexual maturity. In many other works, including Healy's, considerations of what might have been the pleasures of adolescent girlhood—growing independence, pride in a maturing body—are so conflated with the problem of juvenile delinquency as to make them one and the same; indeed, the girl who sought the independence that was a boy's by birthright was by her very desire delinquent. As parents and juvenile authorities conceived it, the years of a girl's life that corresponded to those of the boy's adolescence were filled with danger, not possibility. They were years in which she was to submerge, rather than to free, her yearnings for independence, years in which she was to reconcile herself to her dependence on men and the inevitability of marriage. They were years best avoided altogether; ideally, a girl would progress from childhood directly to the exalted state of motherhood.[8]

These young women of the early twentieth century were the first to live in large numbers on their own in cities. They challenged, if only for a few precious years, the familiar, patriarchal paradigm that saw them moving from their fathers' to their husbands' homes. The rapid entrance of young women into the urban workforce—into the new pink-collar clerical occupations, into factory work like candymaking, in which they had long predominated, and, to a lesser extent, into occupations like printing that had been the exclusive preserve of men—provided the context in which some of them staked their claim to a rightful adolescence. Living alone or in boardinghouses with others like themselves, working for meager wages, skimping on food to buy the fine clothing that conferred status and an air of sophistication, these young women, many country bred, chose to participate fully in the life of the city. With little cash, without family obligations, and with few concerns for the future, they worked by day and pursued pleasure by night through the exciting commercial amusements—movies, dance halls, and theaters—that were just appearing on the urban scene. In these public, anonymous establishments young women and their men, as they carried on the courtships formerly overseen by watchful parental and neighborly eyes, rewrote the code that governed their mutual relations. This code had long sanctioned sexual play—passionate kissing, petting, even intercourse—between young women and men who intended to marry.[9] After the turn of the century some young women sought, with varying degrees of self-consciousness, to engage in the same sorts of intimacies with men they did not intend to marry.

Sexual Ethics

Participating in the sexual sphere was no easy task. With no construction of respectability available to her, the working girl struggling to define a morality that would enable her to do more than sit alone in her room at night was frustrated at every turn by the seeming timeless equation of the working girl and the prostitute. Codified in countless late nineteenth- and early twentieth-century investigations that found her virtue wanting, this equation was assuming new and damning resonances as reformers campaigning to eradicate the necessary evil once and for all transformed the prostitute from a pitiable yet redeemable fallen woman into a hardened predator, a spreader of vicious disease.[10] It was as well an equation to whose strictures she might be subject in her day-to-day dealings with men. Too often the working girl was taken for an easy mark, complained one woman who, because she had worn silk stockings to work one day, had suffered the taunts of her male workmates. Surely only a kept woman could afford such luxuries, they teased: "No one has anything on me," she shot back.[11]

The working girl seeking a good time and respectability was frustrated too by the restrictions of poverty. Many girls worked long hours to earn between six and eight dollars a week, barely enough to cover room, board, and carfare. To gain entrance to the movies, shows, and dances that were the stuff of working-class leisure, some of them who claimed respectable status—who did not, that is, consider themselves prostitutes—chose to bargain with sexual favors, ranging from flirtation to intercourse, exchanged for men's "treats" to entertainments. Others thought such exchanges unworthy of a self-respecting young woman. Yet the urban working-class view of sex as a commodity shaped the heterosexual relations of even those working girls who thought "treating" beneath them. The woman who did not want to give herself to a man knew she must refuse any favors or money he might proffer; as one woman, pregnant, out of work, and desperate, explained, "I never approved of taking money from men. It places you in their obligations." The nature of those obligations was well understood by all; they were nakedly manifest in the many cases where men paid women one or two dollars in exchange for intercourse. The implicit bargain that structured such exchanges—that sex came at a price, for men as well as women—allowed women some leverage in their dealings with the opposite sex, enabling them to extract something of value, whether cash or charity, in return for their favors. But the bargain also put the

working girl at a disadvantage in her relations with men, for men knew women had little in the way of capital besides sex and taunted them with this knowledge. As one young woman, desperate for money, told social workers, "Time and time again I have had chances to go astray, large sums of money and flattering remarks." She offered as an example of such the fellow who wanted to give her fifteen dollars "to go to a room with him, but I told him money could not buy me, he had the wrong girl." She continued, "Then he asked me, 'would I go as a gift,' and I said no and not for charity either."[12]

If sex, or its promise, was the working girl's capital, to middle-class eyes it was capital she too readily squandered. It is hardly surprising that middle-class observers of the working-class sexual economy saw girls' behavior as promiscuous. Nor is it surprising that they focused on the dangers to which the young working woman was daily exposed: on her own in the anonymous metropolis, bereft of male protection, underpaid and overworked, the working girl was all too easy prey for male seducers, schemers, and white slavers. There was much that was true in this construction. Women in the case records I examined were often subject to exploitation at the hands of dishonest men; at least one woman was raped, by a man claiming to be an employment agent, as she looked for work in the city, and women regularly complained of having to rebuff unwanted advances.

Why, then, did psychiatrists and social workers absolve men of all blame? Why did women become the dangerous characters? And why did the sexual prowess of the young woman on her own assume such mythic proportions? When social workers reported that young women hauled men through windows, waylaid grocery boys, and tried to force unwilling, respectable young men to "sexual connection," they may have been reporting the facts as they were. It is more likely, however, that attraction, seduction, and conquest were reciprocal, if not, as some young women insisted, solely the work of men. The evolving conventions of working-class courtship celebrated easy familiarity between the sexes on the streets and in theaters and dance halls. The sexual behavior of working-class men was of little concern to social workers, who saw that women bore all the untoward consequences of their new sexual freedom. Ignorant of any means of birth control other than abortion (to which they often resorted), women faced the prospect of pregnancy with each encounter. Additionally, a man, free of the constraints of custom, family, and neighborhood, might renege on his promises, explicit or not, to support or marry a woman he impregnated. As one pregnant twenty-one-year-old domestic servant, abandoned by the man she had hoped would marry her, put it, "it is pretty mean of a fellow to get a girl into trouble and not to stand by her." Her only recourse was her vow "to make it hard for him," uttered "rather vindictively" in the social workers' estimation. This woman's predicament pointedly illustrates that an urban revolution in sexual mores might render women vulnerable to new forms of exploitation.[13]

Still, this concern for young women's vulnerability does not fully account for the fact that psychiatrists and social workers blamed girls, not boys; women, not men. Nor does it explain why they cast the problem in terms of the "uncontrolled sex impulses" of one sex and not the other. Psychiatrists could understand the girl from a bad home who got into trouble. Her upbringing poor and her material pleasures few, this girl—the classic delinquent—traded favors for money. Material need, not passion, motivated her; most likely she felt no passion at all, like the sexually active young woman whom psychiatrists judged merely delinquent, not psychopathic, because, as she told them, she was "entirely without sexual feeling." The woman

who did *not* receive money in exchange for her favors, who was "attracted to such acts by sexual passion alone," psychiatrists were at a loss to explain. Men were expected to seek sexual pleasure; indeed, a man's *failure* to do so might earn for him the designation psychopathic. It was, on the other hand, unseemly but increasingly all too common for a woman forthrightly to pursue sexual fulfillment. Women, according to conventional wisdom, properly relied on intrigue and feminine wile to attain their ends; men, on a "direct and open procedure," as one commentator put it. Women should be sly and devious, should seduce and tease. A woman who openly avowed passion, who could observe of her sexual exploits, like one putative psychopath, that "life is too short to worry. If you don't enjoy this life you might as well be dead," was, in psychiatrists' eyes, altogether without moral sensibility, altogether inexplicable, altogether pathological.[14]

The psychopath's forthright sexuality was the most visible and disturbing manifestation of her social autonomy; the right to actively seek sexual fulfillment was but the most salient of male prerogatives she assumed. She wanted the freedom to earn and spend, like a man, free of supervision; she wanted to enjoy the pleasures of the city without having her character impugned; she wanted to make her own choices and live independently. Her independence of what many held were the proper and fitting constraints of family and home was nearly as troubling as her hypersexuality. Many hypersexuals had nothing to do with their families. Employed in factories or as domestic servants they enjoyed a freedom from adult supervision that many social commentators agreed was the source of their troubles. Others, in their twenties or thirties, eschewed marriage and chose instead to work and to live singly or with other women. To be sure, some of the young women psychiatrists diagnosed psychopathic looked forward to becoming respectable wives and mothers. One unmarried mother, for example, told psychiatrists she was "not worried about the morality of her act as many other young women have gotten into the same trouble and they have turned out alright and later established good homes."[15] But others rejected the conventional female lifeprint that saw a woman passing from her father to her husband.

The concern over female autonomy that was implicit in the category of hypersexuality helps explain why psychiatrists considered failure to engage in heterosexual courtship—whether simple lack of interest or overtly lesbian behavior—just as psychopathic as a woman's too vigorous exercise of her seductive powers. The "spark of womanliness," for example, glowed but feebly in one psychopathic woman who lacked "ordinary feminine charm and appeal." Wrote one of the hospital's psychiatrists: "It would be difficult to imagine . . . even the most accomplished Lothario successfully drawing her into a flirtation or being able to keep a sustained interest in the pursuit." A more explicit rejection of heterosexuality, like that of a woman who lived with a female companion and questioned the institution of marriage, might mark a woman psychopathic as well. This twenty-year-old, who claimed that she had "never cared for boys" and had "never had anything to do with them," informed psychiatrists she did not intend to marry. Asked to account for this, she explained: "You don't know what you are getting into. It is just as well to live single and be happy. From what I have seen, they don't always get along in marriage." Twenty-eight-year-old Rose Butler too preferred "single blessedness." She did not lack offers; indeed, it seemed to her that men were "like flies around a sugar bowl, when they see a girl." Troubled by her indifference to men, one psychiatrist confronted her: "On the whole you have not much use for men?" Smiling slyly, she replied: "As much as I have for

women." Butler turned the tables on psychiatrists, asserting cheekily that "Adam made woman sin. They say it was Eve who gave him the apple, but he did not have to bite into it."[16]

The psychopath's disturbing assumption of male prerogative was carried to its literal extreme in the case of Julia Brown, alias Alfred Mansfield, a lesbian cross-dresser who for twelve years had lived as a man—smoking a pipe, drinking whiskey regularly, sporting men's suits, working as a printer, and, most puzzling to psychiatrists, escorting young women to dances, suppers, and shows. Brown originally adopted male attire expecting "to find it easier to get work and better pay as a man." But, she admitted to social workers, "I think I really did it because I was so crazy about the girls." She "got more pleasure in making love to women than she ever did in having men make love to her," she told psychiatrists. To their puzzled query as to how she avoided the problem of matrimony, she voiced their worst fears, averring that "all women cared for was a good time, they did not want to get married." The sexual activities of these women who refused to seduce men were of little interest to psychiatrists, who focused instead on their gender-inappropriate independence. Before the midtwenties, psychiatrists focused on the lesbian's supposed masculinity, not her sexual object choice; the lesbian's refusal to court men, not her preference for women, marked her as a sexual deviant.[17]

A Medical Diagnosis for Immorality

It was in this context of growing concern over female independence and sexual deviance that psychiatrists settled on psychopathic personality as an explanation for female immorality. Many correctional and psychiatric experts had earlier assumed that defective intelligence would prove responsible for immorality. Psychiatrists first considered a diagnosis of feebleminded, which could be established with certainty and precision, when faced with immoral behavior, and reformers cited feeblemindedness as the cause of prostitution until it became clear that the scientific tests that were to have proven prostitutes defective instead demonstrated they were normal. Many sexual delinquents, like prostitutes, scored too high on mental tests to be considered feebleminded, and they too rarely manifested any of the usual signs of mental disease. Psychiatrists observed that in contrast to the many feebleminded people they saw, the majority of psychopathic girls "rated high from an intellectual point of view." But if immoral women were too intelligent to be feebleminded, they were still too defective to be normal. Psychiatrists argued that they belonged to a group of subnormal individuals that intelligence tests were powerless to identify. The diagnosis of psychopathic personality satisfied their search for a medical diagnosis for immorality.[18]

The women whom psychiatrists diagnosed as psychopaths at the Boston Psychopathic Hospital were overwhelmingly young (75 percent were younger than twenty-one), single, native-born whites. One-half were Protestant, one-third were Catholic, and the rest were Jewish. Although a few worked at middle-class occupations, such as teaching or office work, most, if employed at all, worked in factories or as domestic servants. Families, police, or courts committed one-half of them to the Boston Psychopathic Hospital for a variety of reasons; state social workers, or visitors, committed the rest.

Twenty-two-year-old Lillian Thomas's background, sexual behavior, and path to the hospital are typical. Thomas lived on her own in a home for young working girls. Her ties to her family had been severed long before; at age ten she had come under the care of the city as a neglected child. Court workers placed her with a woman who "looked after her very carefully and would not allow young men to call on her regularly." In her eighteenth year Thomas began living on her own, working as a waitress and "discouraging advances from young men for fear they might lead her into temptations." Within the year, however, she fell in love with a man she judged honorable and agreed to "illicit relations" with him. Finding herself pregnant, she considered his proposal of marriage. But she discovered he drank and, she told psychiatrists, she "preferred the alternative of living single and fighting out her own battle rather than being the wife of a drunkard." Thomas entered a home for expectant single women; one month after her boy's birth the matron of the home had her committed to the hospital to determine whether she was capable of caring for him. Social workers noted that Thomas's reputation was less than exemplary. At her previous lodgings she had been accused of entertaining men in her room at night; she "went often to dances and came home very late"; one man told them that she "could make up to men quicker than any girl he knew"; and, the case record notes portentously, "it is said she was discharged from one restaurant for underchecking accounts to men."[19]

Thomas was a single working woman who adhered to a standard of sexual morality that many working-class women and men lived by, a standard that sanctioned sexual relations between those who intended to marry. Her only mistake, and the chief evidence of her hypersexuality, lay in withdrawing from the impending compact. Many of the other women committed to the hospital by their families, police, or courts, had, like Thomas, long been on their own. Some had run away from intolerable family situations, others had been orphaned, and still others, in their late twenties and early thirties, had chosen not to marry. Many lived in boardinghouses; in some cases the matrons of these homes, observing behavior they judged either bizarre or promiscuous, petitioned to have them committed to the hospital. The failed suicide attempts of lonely and despondent young women brought a few others under psychiatrists' purview; family members, deeming the behavior of a daughter or sister inappropriate, committed still other women.

The other group of women diagnosed as psychopathic were "state charges" brought to the Boston Psychopathic Hospital for observation as to their sanity. Courts had already found these girls delinquent, primarily on the basis of their sexual behavior, and had committed them to the care of the State Industrial School at Lancaster, Massachusetts. Many of their parents, underemployed and alcoholic, burdened with large families (many of these girls had six or seven siblings) they could barely support, had originally petitioned for their commitment because they could not control them. Their daughters, they complained, went with undesirable companions, were on the streets at all hours, had immoral relations with boys, and, in general, ran wild. Some of this behavior, like taking to the streets, was a form of protest against overly strict parental control. Many widowed fathers, for example, expected their young daughters to keep house for them, prepare their meals, and wait on them as their wives had; many mothers expected daughters to perform heavy household labor—laundry, scrubbing, cleaning—as well as to care for younger siblings. Some of it too was textbook delinquent behavior; more than a few girls had consented to intercourse in exchange for various sums of cash, ranging from twenty-five cents for sex among the barrels at a beach resort to fifteen dollars—a substantial sum—for six episodes in a cheap hotel. But the relatively innocuous driving in automobiles with immoral persons; the frequenting of pool halls, dance halls, and saloons; the carousing with evil-minded girls and boys; and the casual sexual play that figured so prominently in parental and professional accounts of girls gone wrong

point to the larger battle being waged between parents and daughters, between middle-class professionals and working-class girls, over the nature of working-class adolescent girlhood. Could a girl make her own life the same as a boy? To parents who expected their adolescent daughters to refrain from any sort of sexual activity, to hand over their wages without protest, and to assume the household duties of a wife, the girl who strayed undermined not only her own reputation but the fragile family economy, which depended on her services, as well. Her transgressions, then, provided sufficient cause for commitment to reform school.

At Lancaster delinquent girls were subjected to a heavy dose of domestic service—cooking, cleaning, sewing—designed to inculcate industrious habits in the slack and shiftless while exposing them to the pleasures of true domesticity. These pleasures, though, would forever remain beyond the reach of girls for whom domesticity was, and would always be, sheer drudgery. Further, the conditions under which they were paroled from Lancaster, as domestic servants in private homes, were sufficiently loathsome to quell any faintly felt domestic yearnings. State social workers justified the placement of young girls as domestic servants by arguing that only in private homes could troublesome girls be constantly supervised. Occupations other than housework were too fraught with peril for young women already convicted of immorality, they argued; such occupations as waitressing or factory work afforded too many enticing opportunities. As a visitor said authoritatively, discussing one girl's placement: "I do not think she should wait on table. It would lead to prostitution."[20]

As domestic servants, many girls began anew with their employers the sorts of battles that had landed them in reform school in the first place. The city's pleasure palaces beckoned, and lonely, underpaid, and overworked girls saw little reason to resist their allure. Many ran away from their employers for short periods during which they roamed the streets and indulged their cravings for music, dancing, and male companionship. Still worse, the private homes in which they were placed could themselves be problematic, for they offered ample opportunities for secret assignations with employers' husbands and sons. The matron of one home, for example, found her nineteen-year-old domestic servant, clad only in a kimono, perched on the lap of her son, a Harvard student. Another woman dismissed her state charge after she heard her husband suggest they have sex. And the girl who became pregnant after two years of sexual relations with her employer's thirty-five-year-old son was but one of many young women whose behavior mocked social workers' earnest efforts to ensconce them within the confines of a safe domesticity.[21]

The state social workers who committed these "Burleigh girls"—girls supervised by Edith Burleigh, superintendent of the Girls' Parole Department, and her staff of twelve visitors—deemed them bad tempered, irritable, untruthful, and saucy and complained of their waywardness and petty thievery.[22] Yet psychiatrists, charmed yet baffled by their good-natured, pleasant, and essentially normal dispositions, could find nothing remarkable in the mental examinations of most of them. As a psychiatrist observed in one case, "the contrast between the impression gained by the examination of the girl, and the report as given by the Parole Department is rather striking." Remarked another: "It does not seem possible that everyone who goes to Lancaster can be surly, sullen, have outbursts of temper, etc. etc. Certainly we never see a case whose description doesn't have nearly all those adjectives. On the whole when they come here they behave pretty well." Judging them "perfectly attractive sorts of girls," psychiatrists nevertheless diagnosed them as psycho-

pathic, noting that this was "the group to which most of Miss Burleigh's girls belong."[23]

The case history of eighteen-year-old Gertrude Blackstone is similar to those of the most troublesome and impulsive state girls psychiatrists considered psychopathic. Social workers noted that from earliest childhood, Blackstone had been a persistent liar who stole repeatedly, went with bad companions, and, in general, ran wild. Unable to control her, her widowed father had her committed to the care of the state. She proved unsatisfactory in each of the five homes to which state workers sent her. "Lazy, slack, and shiftless about her work," she also told vulgar stories to the children in these homes and continued to steal. Committed to Lancaster after confessing to immoral relations with two men, Blackstone, although "bad tempered and saucy at times," proved "not a difficult girl." She was paroled at home and allowed to study shorthand and typewriting. But she soon became "saucy, lazy and independent" and was placed out to do housework. At this point Blackstone's real trouble began. She paid visits to disreputable families, was immoral on several occasions, and stole from her employer. Next, she married a man whom social workers judged "poor stuff," quickly left him, began to work as a waitress and go out with other men. Eventually, she was returned to reform school. Noting that she was "good natured, pleasant company, but deceitful and unreliable," Burleigh had her committed to the Boston Psychopathic Hospital for observation. She "seems very much fascinated by the life of prostitutes," Burleigh noted. "Enjoys and seeks the attention of men whom she knows to be not good and voluntarily talks about the 'white slave traffic.'" Burleigh closed her report with the observation that Blackstone "says she cannot see the use of living unless you can live as you wish."[24]

For girls like Gertrude Blackstone, the widely shared desire to live life as they wished was well-nigh unattainable. Once identified as delinquent and enmeshed in the state's agencies of social and moral reform, they enjoyed few opportunities and had few choices. Confronted with the bleak reality of these girls' lives—the isolation, the meager wages, the endless round of domestic duties—psychiatrists worried that "a great many of these people don't have enough decent amusement." But the "sport, basketball and such things" they hoped to substitute for the moving pictures, automobiles, and dance halls that they saw as sources of the girls' immorality would hardly have appealed to psychopaths, like one Violet, whose recreational tastes ran to "noisy, frolicking times with jazz music, dancing and loud laughter," or to those like the fifteen-year-old who preferred "dancing and nice clothes and amusements of rather low character" to "interest in the healthy pursuits of life." The wholesome entertainments social workers advocated—the church concerts, choir rehearsals, and chaperoned dances—elicited only the disdain of girls who frequented dance halls, palm gardens, and moving picture shows. The much-vaunted young man of good reputation, who would carry on a courtship "by bi-weekly letters and semi-yearly visits," stood little chance of engaging their interest.[25]

Competing Immoralities

Psychiatrists and social workers did not distinguish among the many varieties of female sexual activity they subsumed under the rubric of immorality. In diagnosing women whose behavior appeared to violate accepted moral standards—girls who rode unchaperoned in automobiles with young men or who engaged them in lewd telephone conversations, as well as those who accepted money in exchange for intercourse—psychiatrists pointed to the prostitute, the archetypal sexually active woman, with whose sordid activities and fallen state they identified all

sexual activity outside marriage. So resonant with associations of voracious sexuality and irredeemable immorality was the prostitute that psychiatrists had only to invoke her for a young woman's sexual expression to be transformed into hypersexuality. The distinction between the young woman who received indecent letters from men and the prostitute who received money was of little consequence; psychiatrists assumed that reception of the former led inexorably to reception of the latter.

Girls and women who readily admitted to their uncontrollable sexual desires conformed to psychiatric notions of hypersexuality. Theresa Beauvais, for example, confessed "that she could not control herself with respect to sex relations." Bessie Dunston "seemed thoroughly ashamed" of the activities that had resulted in her pregnancy "but said she could not seem to help it, the desire was so strong." Other young women admitted the "temptations of sex were too much" for them, that they "did not have the strength to decline," and that they were "powerless to resist" themselves. One girl's declaration that she would choose to be shut up in an institution could she not "live the life" exemplified the defiantly immoral sexuality psychiatrists considered psychopathic.[26]

These women, however emblematic their declarations, comprised a minority. More commonly, women asserted they were not hypersexual and objected that the psychiatrists and social workers examining them assumed only the worst. Nineteen-year-old Ethel Townson, for example, admitted to a social worker that she had "acted very foolishly" but argued she had "not gone to the extremes as perhaps you and a good many others think." Never had she been, she protested, "so overwhelming with joy to devote my time to such indiscriminate lewdness." "I am no ————," she added, "for if I had ever been why should I ever be in want of money? I would have lots of clothes and an apartment." She was instead a girl having difficulty finding work. Psychiatrists argued to nineteen-year-old Alice Lawson that her difficulties resulted from her bad temper and sexual activities. She disagreed, telling them all her "troubles started with the social workers. They imagine a whole lot." "They say you are fond of boys," a psychiatrist teasingly challenged her. "I like boys but I am not crazy about them," she replied. "It is only nature" to be fond of men, contended another girl. "Are you so fond of them that you cannot get along without them?" a doctor queried. "No," she conceded, "I could get along without them." These girls argued that their activities were normal. A young girl naturally would, as one said, want "to be on the go, i.e. to see life, movies, young men and dances." But in the girls' activities, however limited, social workers saw only nascent immorality.[27]

Psychiatrists' and social workers' conceptions of respectable, moral behavior rigidly divided good from bad women. The specter of the prostitute informed their prescriptions about girls' proper dress and deportment. To social workers, working girls who spent their meager incomes on frivolous ribbons and silk stockings and who purchased clothing on installment plans were no better than the "dolled-up" psychopathic prostitutes who found satisfaction in lavish personal adornment, plucking and penciling their eyebrows, and flaunting their wares as they strutted about in striking, cheap-rich styles of dress. But flashy dress, so tellingly tawdry to middle-class eyes, was the norm among working girls. Fine clothing compensated for the daily drudgery of work; it could bolster self-esteem flagging under the double burden of overwork and underpay; it could also enhance a young woman's desirability in the marketplace of pleasures. Social workers held that the working girl's meager wages were best allotted to the necessities, not the frivolities, of life; the claim of room and board on a young woman's wages, ac-

cording to their calculus of economic rationality, was prior to that of clothing. Social workers might complain that working girls had "no idea of the value of money," but working girls knew better; money invested in clothes was money well spent, capable of yielding ample returns.[28]

Many observers agreed that girls' sartorial desires and recreational tastes exposed them to moral dangers. Young women may have made light of these concerns, but they were not necessarily inappropriate. As young women moved more visibly into the public sphere, they confronted new dangers as well as new opportunities. Many knew nothing of sex—its mechanics, its pleasures—until knowledge was forced on them. As one twenty-five-year-old woman who became pregnant after her first experience with men, automobiles, and drink ruefully noted, "I thought a few cigarettes, one or two cocktails were harmless, that a girl could always take care of herself, and there I was mistaken." Women did, however, rightfully object to the narrow construction of respectability that rigidly separated good from bad women, that blurred the distinction between prostitution and sexual activity outside marriage, and that saw in this woman's pregnancy—even though, as she emphasized, "I never had anything to do with men except that once"—the "start of a new career."[29]

It was the readiness of psychiatrists and social workers to forever consign them, on the slightest of evidence, to the ranks of the promiscuous that young women found so irksome. Psychiatrists measured them against a middle-class moral code, a code that ensured that the middle-class woman would not squander her virtue, her most marketable of assets. But the strict morality they advocated held little appeal for working girls for whom a willingness to play fast was of equal value to the middle-class girl's chastity. A number of putative psychopaths struggled to define standards of sexual morality appropriate to the circumstances of their lives.

Psychiatrists were puzzled that they could entertain any real self-respect—testified to in some cases by girls' refusal to accept money for their irregularities—while engaging in relations the psychiatrists judged promiscuous.[30] Yet this distinction was central to these women who were trying to forge another standard of sexual morality. They objected to the double standard that marked them defective. "Do you think there will ever be equality of sex, one moral standard for both?" Ethel Hancock, the "immoral" woman who heralded the arrival of a new generation of women, demanded of Mary C. Jarrett, the head of the Social Service at the hospital. "If so, I hope it comes in my day, I would be as good as a bad man, which is twice as good as a virtuous woman. Wouldn't it be great to do just as one pleased?" In response, psychiatrists warned that for women "a little knowledge is a dangerous thing" and urged them not to question the dictates of conventional morality. Jarrett had warned Hancock not to "read or discuss on any subject, especially sex, that [she] did not understand and had not the education and intelligence to grasp in just the manner the author intended." Yet young women did think about sex in spite of experts' warnings. Hancock's declaration that she would "simply love it, revel in it" could she indulge her desires and maintain respectability at the same time only highlighted how weakened were the foundations of conventional morality. Though she added that her revelry would be "not as a pig wallows in swill— because I'd surely have some good desires," her intentions were clear. And Rose Talbot's admission that she had "read Casqueline's (I think it is spelled) *Truth about Woman* and Havelock Ellis' *Psychology of Sex* which," she wrote, "quite knocked all my good resolutions to smash," only underlined the dangers of free thinking. "So you see it is not wicked to have sexual intercourse with other men than your husband. It is simply not the fashion."[31]

Even as they decried the injustices of the double standard, these women fell back on its very distinctions as they assessed their own behavior. Many lived by moral codes structured, like those of the largely female social workers, around considerations of good and bad. But they defined good behavior far more broadly than social workers, and their respectability was a complex thing, a measure not of chastity but of self-respect.

Social workers and their charges, for example, agreed that a woman could "go wrong" or "to the bad." While these phrases signified to social workers a debased state from which a girl could never recover, to the young women they denoted a process that could be reversed at any point. A girl could start to go to the bad and stop; one, for example, jilted by the lover whom she had hoped to marry, thought of herself as a now-respectable woman who had once "started to go wrong."[32] "Going wrong" had as much to do with a woman's attitude as with her activities; it was not the social fact of her surrender to temptation, whether occasioned by need or passion, that signified a woman's descent into debasement but, rather, the discouraged state of mind that ofttimes preceded it. Although she might stray, a respectable young woman only became truly bad, in the idiom of the working class, when she succumbed to despondency and thought of herself as bad.

From this female working-class perspective, material need was a familiar culprit. Young women implored social workers to confront this, but the latter were indifferent to economic arguments. Rose Butler, a highly paid compositor, emphasized that she was not fast nor did she wish to be. She pointed out, however, familiar as she was with the privations of single womanhood, that need could sorely tempt a girl: "Why is it girls get discouraged? Why do girls get tired of trying? Why should a girl be befriended and then scorned and snubbed and squashed when she does try? I have come to the point where I don't blame a girl to get discouraged." Similarly, a twenty-year-old tried to impress upon social workers that desperate circumstances like hers—for the past two years she had owned barely any clothes—"would drive anyone to the bad, no work, no money, and nothing to wear."[33] The road to ruin, then, could begin with discouragement and end with self-respect and reputation lost.

At the same time a woman could salvage her self-respect by dint of determined hard work. Self-respect regained, she could reclaim respectability, which pulled powerfully even on those women who most brazenly flouted its strictures. Ethel Townson, whose breezy nonchalance appeared the embodiment of the hypersexual's disdain for morality, responded to its call. After leaving home at age fifteen, Townson settled into a life of visiting cafes, going with men, tangling with police, and working but irregularly. Social workers judged her sullen, uncommunicative, and defiant. On one occasion she informed them ominously that she had "learned a great deal in the past few days"; on another she announced her intention to secure an abortion, claiming "quite calmly" that "lots of girls get in the same box and take the same way out." Yet Townson, for all her bravado, admitted something was wrong. "I can't control myself when I am once started," she wrote. "I do my self really know there is something the matter with me and I do want to grow and be a respectful young woman." Ethel Hancock too sought respectability. Sexual relations with her first husband disgusted her, she confided to social workers; it was only with her next husband, with whom she was not much in love, that she had experienced "strong sexual desires." Separated from him by a bitter quarrel, Hancock began living as a single woman, working as a waitress in a restaurant with a "fast crowd of girls." She quickly and deliberately went to the bad, she told social work-

ers. On one occasion she went with her workmates to a sporty place but left, intoxicated and disgusted, when their unspecified sexual debauches began, and on another she was found with a man in her room in the early hours of the morning by the matron of the home in which she lived. It seemed to psychiatrists that Hancock did not take her many transgressions very seriously. But over the course of her five-year epistolary relationship with social worker Mary Jarrett, Hancock struggled to make good. "It seems as if I'm always doing some fool thing and the harder I try the more of a muddle I get in," she observed. "I have not yet learned self control, not even a wee bit. Goodness knows I try hard enough." She wrote triumphantly to another social worker: "Oh, my dear, you can't imagine how proud I am to be able to write that I've actually been good for two whole weeks." Hancock desperately sought respectability. Her lament—"I've been so dreadfully bad and I will truly do something big if I can make something proud of out of my comparatively wretched life"—was echoed in those of the otherwise confident young women who found it difficult to manage the psychic conflicts their unconventional behavior engendered.[34]

Hancock did make good, although, as she put it, only "by force of circumstance." Working as a telegrapher at Western Union, she could find "no time at all for mere masculine sex." She derived some satisfaction from the knowledge, gained through experience, that she could "start out at anytime and make a good living" for herself. Yet she still wished—"when I am up to devilry," she added, "mischief is too mild"—she could "give rein to all [her] evil desires" without incurring the wrath of social workers.[35] Hancock's repentance redeemed her in social workers' eyes. As they told her tale—her fall, her struggle, her triumph—they cast her as both offender and victim. Both were familiar roles; the prostitute had played the two well. Need, bad fortune, poor company—all led women astray, all fit neatly within social workers' moral universe. True, this construction of Hancock's story overlooked her willful sexual experimentation, her professions of desire, and her own interpretation of her successful struggle for respectability as one in which she had paid an almost unacceptably high price. But the final, favorable outcome allowed social workers to mold it to fit the conventions of a familiar genre, that of the fallen woman redeemed through the timely, though lengthy, intervention of middle-class female rescuers. In casting hypersexuals as prostitutes, then, social workers were not just relegating them to the nether reaches of the moral spectrum but rendering them familiar as well.

The prostitute, though deplorable, was at least a known quantity. The young women who celebrated the pleasures of sex free of the moderating influence of moral concerns or social context—who, that is, perfectly exemplified the psychiatric notion of the hypersexual—were, by contrast, women wholly outside social workers' and psychiatrists' moral universe. Simple declarations of erotic desire unmediated by morality stunned them into silence. Prudence Walker, for example, a nineteen-year-old with a long history of sexual exploits, followed her desires even as they led her into trouble. "I must have someone all the time to love me," she told social workers. Sometimes, she confessed, "I get so excited that I don't know what to do—and it's then I consent to the things I do and have done." She had no excuse to offer "except the old one—that we don't know what we do—in that condition." In a similar vein eighteen-year-old Helen Perkins wrote to a friend of her latest adventure: "The old saying is go as far as you like the Sky is the limit But Jee I think we have gone our limit. don't you?" She had "fallen in love" with a man who had come to repair the furnace in the house she occupied. "Of course you know what that

means. I was so God Dam hot that I didn't know weather I was going home or to hell. I suppose you got yours from Fred," she added generously. "I hope so anyway." Psychiatrists and social workers focused on sexual practice, not pleasure. They were uncharacteristically reticent when confronted with evidence of the latter, tacit recognition, perhaps, that these declarations of female desire were indeed an altogether new thing.[36]

If white immorality was a symptom of disease, black immorality, on the other hand, was entirely normal. "On the whole I have been very chary in making the diagnosis of psychopathic personality in these colored girls," one psychiatrist explained. "The level of the negro regarding conduct, using that term in the broad sense, is decidedly different [from] the conduct of the white." Psychiatrists contended that the fooling with boys that was a definite symptom of psychopathy in white girls was in blacks only the expression of the natural immorality of the race. A "normal negress" was unintelligent but high spirited, and her immorality could be shrugged off with the assurance that there was "nothing abnormal about her delinquencies. Some of us get caught, some do not."[37] Adding a psychiatric twist to a very old set of beliefs concerning the sexual nature of black women, psychiatrists diagnosed as normal those with histories every bit as flamboyant as the supposedly psychopathic whites. Only black women whose immorality was of such proportions that it offended even the low standards of their race were psychopathic. Thus the psychiatrist had to determine first of all whether a woman was "normal for her race." The sexual ethics of blacks, for example, "for thousands of years [had] been other and peculiar"; it was questionable whether they could "get on under the marriage conventions of the Aryan race." One psychiatrist's cautionary admonition that immoderate sexual behavior could not be "merely racial" because "not every member of the race does this sort of thing" fell on deaf ears.[38]

It was a strange disease indeed that so respected social convention. Yet psychiatrists were firm in their conviction that psychopathic personality was an inborn and incurable condition. Unlike "simple" delinquents, who were "perfectly able to do otherwise if they wish" and who could learn from experience, the inborn defects of the psychopath rendered her unable to refrain from misdeeds. Some psychopaths thought differently, acknowledging their delinquent behavior but attributing it to poor upbringing, not inborn taint. Their mothers had been unable to care for them properly, several girls told psychiatrists; no one had taught them right living. Others, like Alice Lawson, insisted there was nothing at all wrong with them. As Lawson emphasized: "There is not a thing the matter with me. I eat and sleep well. I am perfectly normal. I have no hallucinations, illusions or delusions." She implored doctors to release her. "I am sure shutting a girl up—shutting her away from civilization does not help the girl any."[39]

Some young women rejected not only psychiatrists' etiologies but their pessimistic prognoses as well. They promised they would make good if psychiatrists would give them their freedom. But the policies of the parole department and social workers' distrust of the world of work ensured they would be trained for little. "There are a number of things I should like to do," one young woman told the psychiatrists, "but there is only one thing I will be able to do, and that is to go back to housework. I am very fond of study, and I should like to study more." Another said that she "wanted to do something different from housework, such as in an office of some kind," but added, "I don't think I am educated enough." Alice Lawson too did not want to do housework. "What else can you do?" one psychiatrist queried. "There are lots of things a girl can do besides housework," Lawson countered, but

to the psychiatrist's challenge, "What else can you do?" she could only reply, "Can't I wait on table?" "What else?" he demanded. "I have not thought of anything else," she feebly replied.[40]

Psychiatrists ridiculed these women's aspirations and made light of their resolve to go straight. The endeavors of Yvette Gagnon, a newspaper editor who had "many extreme ideas as to women's rights" and who was writing a book entitled "Daughters of Tomorrow," only testified, psychiatrists asserted, to "her delusional insanity re her mental genius in the art of book writing." Psychiatrists consigned these women, endowed with boundless sexual desire and limited occupational skills, to futures circumscribed by their sexuality. As a psychiatrist confidently predicted of one young woman, "I can see her becoming a manic-depressive or a prostitute."[41]

Although psychiatrists proposed that all psychopaths be segregated in rural colonies, they realized that this was easier proclaimed than accomplished. First, they would never be able to ensnare the many psychopaths running wild in their communities. Large numbers of defectives, "stunningly asocial in conduct," would "never see an institution." Second, the psychopaths that psychiatrists were able to supervise temporarily they could not commit as insane; a diagnosis of psychopathic personality was insufficient grounds for long-term commitment to a mental hospital. Psychiatrists duly released most of the women they diagnosed as psychopaths. They returned Miss Burleigh's girls to her; most of them continued to work as domestics until age twenty-one, when they gained their freedom. The rest gained their freedom immediately.[42]

In the twenties, as the sexual revolution reached the ranks of the middle class, and as behavior that psychiatrists labeled hypersexual became more prevalent and less easily ascribed to a deviant working-class minority, psychiatrists' interest in hypersexuality flagged. The sexual psychopath of the late twenties and beyond was male and most often homosexual; rapists, child molesters, and other sex offenders displaced working-class women as objects of psychiatric attention. As psychiatrists adjusted their theories to new sexual mores, they championed a sanitized heterosexuality that could be safely contained within marriage. Indeed, psychiatrists transformed the passion that had been such a mark of deviance into the very criterion of normality. Frigidity, not its obverse, marked the deviant woman; the lesbian, the all-too-independent woman who rejected men and patriarchy altogether, inherited the mantle of sexual deviance from the hypersexual who, however worthy of contempt, at least had played the game. As the hypersexual faded, the lesbian became the exemplar of female sexuality gone awry.[43]

It is possible to see in psychiatrists' rendering of the hypersexual an exemplary instance of Michel Foucault's suggestive observation that the science of sexuality has functioned as the *ars erotica* of the Western world. In the guise of a decent positivism, Foucault argues, the late nineteenth-century psychiatrists—Richard Krafft-Ebing, Auguste Tardieu, Havelock Ellis—who recorded and classified the manifold forms human sexual desire assumed, inadvertently created "a great archive of the pleasures of sex." There was pleasure to be derived from the very creating of this archive, Foucault goes on; the "obligatory and exhaustive confessions" psychiatrists elicited, and, we might add, the subjects from whom they were drawn, could themselves stimulate and titillate.[44]

The confrontations between psychiatrists and the young women they deemed hypersexual partook, at times, of the erotic. On at least one occasion a psychiatrist, dismayed at seeing his colleagues succumb to the blandishments of a particularly attractive young woman, burst out:

"You all say give her a chance. To say 'Give her a chance' is an emotional way of putting it. The question," he reminded them, "is whether this girl is hypersexual." An attractive girl who assumed the pose of injured innocence, she was, he argued, all the more dangerous in her ability to "get the sympathy of men."[45] Uncomfortable with their own vulnerability to young women's charms—which were considerable relative to those of the rest of the mental hospital population—psychiatrists, eager perhaps to establish the collective innocence of the male half of the species, fell into blaming women for seducing them. The overdrawn sexuality of the adolescent girl in psychiatric discourse is in part, by this interpretation, but the projection of these psychiatrists' own barely acknowledged desires.

But there was more to the hypersexual than this. Proclaiming women wholly sexual creatures, psychiatrists fixed on their sexuality in attempting to comprehend a larger process of social change that saw women moving out of the home and more visibly into the world of work. The psychopathic hypersexual was in part a product of psychiatric imagination; the confident hedonist who seduced men left and right was the working-class new woman, the woman with a little cash and a lot of savvy, seen through middle-class, mostly male, eyes. But the sexually assertive woman, the woman endowed with passion equal to that of a man, was real, and psychiatrists, however exaggerated their notions of her sexual prowess, were alone in recognizing her passion as dicey. For the moment psychiatrists stood back, a bit awed perhaps, and attempted to comprehend this new phenomenon, the passionate woman. Reflexively, they turned to the familiar categories of the pure woman and the prostitute. At the same time they recognized this old dichotomy would not do. The uneasiness of the construct of the female hypersexual embodied their confusion and ambivalence toward women and their sexuality. With it they enjoined women to be at once chaste and seductive, to tease but not to conquer. The psychopathic hypersexual was at times a pathological deviant in psychiatrists' discourse. More often, however, she was Everywoman. If, as one authority proposed, "a clean and protected moron was not far from corresponding to the ideal woman of the Victorian age," the psychopathic hypersexual was her redoubtable twentieth-century counterpart.[46]

Notes

I am grateful to Elizabeth Pleck, Molly Nolan, Barbara Rosenkrantz, Sherrill Cohen, Lorraine Daston, Peter Mandler, Estelle Freedman, Nancy Cott, Theodore Brown, Christopher Lasch, and Gary Gerstle for comments on earlier versions of this essay, which has since been published as a chapter of my book, *The Psychiatric Persuasion: Knowledge, Gender, and Power in Modern America* (Princeton University Press, 1994).

1. Case 6867, 1917, letter from Ethel Hancock to Mary C. Jarrett, head of the Social Service at the Boston Psychopathic Hospital. All cases cited are from this hospital; all case numbers and patient names have been altered. Original punctuation and spelling have been retained in all quotations. The essay is based on analysis of data drawn from a simple random sample of 1,290 cases drawn from a total of 17,000 admissions to the Boston Psychopathic Hospital from 1912 to 1921, as well as on examination of more than 100 cases (not included in the sample) of women diagnosed as psychopathic. Psychiatrists diagnosed 5 percent of all patients as psychopathic; three-quarters of these were women. I have also drawn on an extensive published psychiatric and correctional literature, some of which is cited here.

2. The literature on the post-Victorian sexual revolution is incomplete and largely inconclusive. Most historians agree that there was a middle-class revolution in mores. See, for example, James R. McGovern, "The American Woman's Pre–World War I Freedom in Manners and Morals," *Journal of American History* 55 (September 1968): 315–33. Several historians, however, have questioned our notions of sexual Victorianism; for example, see Carl Degler, "What Ought to Have Been and What Was: Women's Sexuality in the Nineteenth Century," *American Historical Review*, 79 (December 1974): 1467–90; and, more recently, Peter Gay, *Education of the Senses*, vol. 1, *The Bourgeois Experience, Victoria to Freud* (New York: Oxford University Press, 1984). For a thoughtful critique of these attempts to rehabilitate Victorian sexuality see Carol Zisowitz Stearns and Peter N. Stearns, "Victorian Sexuality: Can Historians Do It Better?" *Journal of Social History* 18 (Summer 1985): 625–34. Historians have focused too little on working-class sexuality; the problem of working-class sexual revolution has been even less adequately explored. For exceptions to this general neglect see Christine Stansell, *City of Women: Sex and Class in New York, 1789–1860* (New York: Alfred A. Knopf, 1986) on nineteenth-century working-class sexuality; and Kathy Peiss, "'Charity Girls' and City Pleasures: Historical Notes on Working-Class Sexuality," in *Powers of Desire: The Politics of Sexuality*, ed. Ann Snitow, Christine Stansell, and Sharon Thompson (New York: Monthly Review Press, 1983), 74–88.

3. The Boston Psychopathic Hospital was established in 1912 to diagnose and treat acute and curable mental disease; as used in the hospital's name *psychopathic* denoted mental disease without neurological basis.

4. Case 7025, 1916, case conference. On adolescent male sexuality see William Healy, *The Individual Delinquent: A Text-Book of Diagnosis and Prognosis for All Concerned in Understanding Offenders* (Boston: Little, Brown, and Company, 1915), 255–56, 588–89 esp.

5. Daniel Scott Smith, "Family Limitation, Sexual Control, and Domestic Feminism in Victorian America," *Feminist Studies* 1 (Winter–Spring 1973): 40–57; reprinted in *A Heritage of Her Own: Toward a New Social History of American Women*, ed. Nancy F. Cott and Elizabeth Pleck (New York: Simon and Schuster, 1979), 222–45.

6. On the relationship between overdevelopment and sexual precocity see Healy, *Individual Delinquent*, 244–54, 402–4. The extraordinary nature of his claim that "the overwhelmingly attraction which negro men occasionally have for white girls and women . . . is to

be explained by the hypersexualism of the female attracted" (p. 403) can best be appreciated when set
against the racism that pervades the literature of the
time.

7. See Leslie Woodcock Tentler, *Wage-Earning
Women: Industrial Work and Family Life in the United
States, 1900–1930* (New York: Oxford University
Press, 1979), ch. 4, for a discussion of the boy's relationship to the family economy. Both female and
male criminologists considered the delinquencies of
boys far less serious than those of girls. See, for one
example among many, Mary W. Dewson, *Conditions
That Make Wayward Girls: A Study Based on Last
Year's Commitments to the Massachusetts State Industrial School for Girls* (Lancaster, Mass., 1910), 1: "The
boy often gets into trouble because of an excess of animal spirits, because of energy which, rightly utilized,
should be his greatest asset."

8. In 1922, years after the appearance of his two-
volume *Adolescence: Its Psychology and Its Relation to
Physiology, Anthropology, Sociology, Sex, Crime, Religion, and Education* (New York: D. Appleton, 1904),
G. Stanley Hall noted that "of all the stages of human
life [the adolescent girl's] was *terra incognita*." The
pubescent boy was, by contrast, an open book. See
"Flapper Americana Novissima," *Atlantic Monthly*
129 (June 1922): 771. I found Carol Dyhouse's perceptive discussion of adolescent girlhood, although
primarily concerned with middle-class girls, very useful in thinking about working-class adolescent girls.
Dyhouse notes Hall's "leery" tone when writing of adolescent girls. See her *Girls Growing Up in Late Victorian and Edwardian England* (London: Routledge &
Kegan Paul, 1981), 115–22.

9. On the growth of commercial amusements see
Kathy Peiss, *Cheap Amusements: Working Women and
Leisure in Turn-of-the-Century New York* (Philadelphia: Temple University Press, 1986). Tentler, *Wage-
Earning Women*, 109–13, also discusses the appeal of
commercial amusements. On small-town courtship
see the essay by Ernest W. Burgess, "Sociological Aspects of the Sex Life of the Unmarried Adult," in *The
Sex Life of the Unmarried Adult: An Inquiry into and
an Interpretation of Current Sex Practices*, ed. Ira S.
Wile (New York: Garden City Publishing, 1934):
116–54.

10. Most noteworthy among the investigations

are Carroll D. Wright, *The Working Girls of Boston*
(Boston, 1889), 118–26 esp., where Wright attempts
to counter what he considers the prevailing view that
working girls are immoral; Robert A. Woods and Albert J. Kennedy, *Young Working Girls: A Summary of
Evidence from Two Thousand Social Workers* (Boston,
1913), 84–100; and William I. Thomas, *The Unadjusted Girl: With Cases and Standpoint for Behavior
Analysis* (1923; Montclair, N.J.: Patterson Smith,
1969), 98–150. On changing views of the prostitute
see Ruth Rosen, *The Lost Sisterhood: Prostitution in
America, 1900–1918* (Baltimore: Johns Hopkins University Press, 1982); and Allan M. Brandt, *No Magic
Bullet: A Social History of Venereal Disease in the
United States Since 1880* (New York: Oxford University Press, 1985).

11. Case 6062, 1917, letter to social worker.

12. See Peiss, "'Charity Girls' and City Pleasures," for a discussion of "treating." Quotations are
from case 10397, 1917, case report and case 6062,
1916, letter to social worker. My understanding of the
working-class sexual milieu owes much to Stansell's
City of Women, 76–101, 171–92. See Thomas, *Unadjusted Girl*, 98–150, for a contemporary's analysis of
this same milieu.

13. Case 8196, 1916, case record.

14. Elizabeth G. Evans and Mary W. Dewson,
Feeble-Mindedness and Juvenile Delinquency (Boston,
1909), 1; case 11046, 1918, case conference; case 8184,
1916, case conference; W. I. Thomas, *Sex and Society:
Studies in the Social Psychology of Sex* (Chicago, 1907),
313; case 5602, 1915, letter to social worker.

15. Case 3735, 1914, case record.

16. Martin W. Peck, "Psychopathic Personality:
Report of a Case," *Journal of Abnormal and Social Psychology* 17 (July–September, 1922): 192; case 5634,
1915, case conference; case 6062, 1915, case conference.

17. Case 5943, 1915, case conference. In their
published account of Brown's case E. E. Southard and
Mary C. Jarrett wrote that they had minimized the
issue of her sexual preference and had "purposely laid
little stress upon the sex delinquencies, heterosexual
and homosexual, which are suspected or implicit in
some parts of" her case. See *The Kingdom of Evils:
Psychiatric Social Work Presented in One Hundred
Case Histories Together With a Classification of Social*

Divisions of Evil (New York: Macmillan, 1922), 137. Southard was director of the Boston Psychopathic Hospital from 1912 until he died in 1919. For an analysis of the stress on gender deviance see George Chauncey, Jr., "From Sexual Inversion to Homosexuality: Medicine and the Changing Conceptualization of Female Deviance," *Salmagundi* 58–59 (Fall 1982–Winter 1983): 114–46.

18. Rosen, *Lost Sisterhood*, points out that prostitutes were among the first populations tested for feeblemindedness. She cites contemporary investigations of prostitution that found high proportions—around 30 percent—of prostitutes to be feebleminded (21–22). But the fact that the majority of prostitutes proved to be *not* feebleminded shocked and troubled reformers who quickly turned to psychopathy to account for prostitutes' immorality. See, for example, Mabel Ruth Fernald, Mary Holmes Stevens Hayes, and Almena Dawley, *A Study of Women Delinquents in New York State* (New York: Century, 1920). The authors concluded that their data failed "absolutely to justify the view expressed recently by certain propagandists that delinquency and defective intelligence are practically synonymous, and that, accordingly, solving the problem of mental deficiency will solve the problem of delinquency" (434). Steven Schlossman and Stephanie Wallach argue that Progressive courts began to define female delinquency almost entirely in sexual terms; see "The Crime of Precocious Sexuality: Female Juvenile Delinquency in the Progressive Era," *Harvard Educational Review* 48 (February 1978): 65–94. The quotation on psychopathic girls is from case 8487, 1917, case conference.

19. Case 3735, 1914, case record.

20. See Barbara M. Brenzel, *Daughters of the State: A Social Portrait of the First Reform School for Girls in North America, 1856–1905* (Cambridge: MIT Press, 1983), for a history of the Lancaster state school; case 6187, 1915, case conference. The visitor may have been right. Frances Donovan noted that the games waitresses played with men bordered "on prostitution, although not actual prostitution because the waitresses earn the necessaries of life for themselves." See her *The Woman Who Waits* (Boston: R. G. Badger, 1920), 211–12.

21. Case 6286, 1915, case report; case 10284, 1917, case record; case 10545, 1918, visitor's report.

22. One girl told psychiatrists, "They call it saucy if you turn round up there." Case 10869, 1918, case conference.

23. Ibid.; case 8184, 1916, case conference; case 7763, 1916, case conference.

24. Case 7853, 1916, Girl's Parole Department visitor's report.

25. Case 7643, 1917, case conference; Peck, 190; case 15784, n.d., case history; case 4760, 1915; case 10469, 1917, visitor's report.

26. Southard and Jarrett, *Kingdom of Evils*, 354; case 4760, 1915, visitor's report; case 7693, 1916, case report; case 8184, 1916, case conference; case 10869, 1918, case conference; case 5664, 1915, case conference.

27. Case 5844, 1915–19, letters to social workers; case 6187, 1915, case conference; case 10264, 1918, case conference; case 10701, 1919, case record.

28. See Augusta Scott, "Three Hundred Psychiatric Examinations Made at the Women's Day Court, New York City," *Mental Hygiene* 6 (April 1922): 362–64, on dolled-up prostitutes; quotation is from case 11046, 1918, visitor's report. Many contemporaries called attention to the working-class girl's near-obsession with clothes. See Thomas, *Unadjusted Girl*, 109ff. for one account. Stansell, *City of Women*, 92–101, 187–88; and Peiss, *Cheap Amusements*, 62–67, discuss the significance of clothing in working-class culture.

29. Case 10397, 1917, autobiography and case report.

30. Southard and Jarrett, *Kingdom of Evils*, 60, 70.

31. Case 6867, 1917, letter from Hancock to Jarrett; case 1272, 1913, case conference; letter from Hancock to Jarrett; Southard and Jarrett, *Kingdom of Evils*, 55. The book Talbot read was C. Gasquoine Hartley, *The Truth about Woman* (New York, 1914).

32. Case 6062, 1916, letter to social worker.

33. Ibid; Case 2202, 1916, letter to social worker.

34. Case 5844, 1915, letters and case record; case 6867, 1916–17, letters to social workers.

35. Ibid.

36. Case 5602, 1915, letter to social worker; case 2366, 1913, letter to friend.

37. Case 10264, 1918, case conference.

38. Case 10666, 1918, case conference.

39.   Case 8184, 1916, case conference; case 6187, 1915, case conference.

40.   Case 7763, 1916, case conference; case 8563, 1917, case conference; case 6187, 1915, case conference.

41.   Case 10397, 1917, case conference; case 1272, 1913, case conference.

42.   Case 10264, 1918, case conference.

43.   Estelle B. Freedman, "'Uncontrolled Desires': The Response to the Sexual Psychopath," *Journal of American History* 74 (July 1987): 83–106; Christina Simmons, "Companionate Marriage and the Lesbian Threat," *Frontiers* 4 (Fall 1979): 54–59; Michael Gordon, "From an Unfortunate Necessity to a Cult of Mutual Orgasm: Sex in American Marital Education Literature, 1830–1940," in *Studies in the Sociology of Sex,* ed. James M. Henslin (New York: Appleton-Century-Crofts, 1971), 53–77.

44.   Michel Foucault, *The History of Sexuality,* vol. 1, *An Introduction,* trans. Robert Hurley (New York: Random House, 1978), 53–73.

45.   Case 6187, 1915, case conference.

46.   Thomas, *Unadjusted Girl,* 166.

Fertility, Abortion, and Birth Control

Women have tried to limit the number of their offspring, using different methods and with varying intensity and success, throughout history. The historians in this section examine four parts of the American response. Linda Gordon analyzes a nineteenth-century feminist movement that aimed to increase women's control over their bodies through sexual restraint. Leslie Reagan analyzes some pressures and dangers faced by women who tried to abort their unwanted pregnancies. Jessie Rodrique places African-American women in what historians have previously described as a white birth-control movement. Andrea Tone focuses on the commercial aspects of birth control, opening an important new area of investigation.

As with their attitudes toward menstruation and sexuality, women did not find it easy to reveal their feelings about birth control and abortion. Yet the issue of family limitation and discussion of various methods by which it could be accomplished are recoverable in the historical record. The questions of which sexual partner made decisions about controlling the number of offspring and how this changed through history can be particularly revealing of women's power to control this important aspect of their lives. Some methods, such as withdrawal or condoms, rested in male control; others, like abortion, could be controlled by women; and still others, like abstinence, encouraged cooperation. In addition to individual decision making, however, as Tone reminds us, community interests and pressures significantly influenced the range of choices available to women. By examining the patterns of birth control and abortion practices, historians can learn the extent to which individual women and groups of women in any given historical period wanted to or could control their reproduction.

Abortion is the topic about which particular controversy rages. The historical record demonstrates that abortion was legal for much of our history and that in the period when abortion was illegal, women nonetheless continued to seek and procure abortions. In the eighteenth century, desperate women aborted pregnancies that had usually resulted from illicit relationships. James Mohr's book on the history of abortion demonstrates how this changed early in the nineteenth century when the number of abortions grew. At first mostly unmarried and poor women used abortion as the only way they knew to limit the number of their offspring, but increasingly more affluent, and frequently married, women continued the practice. In the nineteenth century, abortion played a significant role in the reduction of American birthrates.

As Leslie Reagan explains here, in the period when abortion was illegal, beginning in the last third of the nineteenth century and lasting until the *Roe v. Wade* decision in 1973, millions of women continued to get abortions. Despite laws against the practice, women found cooperative physicians and other health practitioners to help them abort their unwanted pregnancies. In this period, however, the state, sometimes in conjunction with physicians, played a crucial and punishing role in the policing of abortion. Women often found themselves the targets of prosecution, public humiliation, and loss of privacy.

In January 1973 the U.S. Supreme Court determined that the nation's century-old criminal abortion laws were unconstitutional. *Roe v. Wade* declared that a decision about whether to carry a pregnancy to term belonged to women in private consultation with their physicians. During the first trimester of pregnancy, the Court declared, "the attending physician, in consultation with his patient, is free to determine, without regulation by the State, that, in his medical judgment, the patient's pregnancy should be terminated." More recently, some erosion of the 1973 principles has occurred at the federal and state levels, and the threat remains that abortion could again become completely illegal in America. History demonstrates that, if this were to happen, the incidence of abortion would continue and women would again besiege their health care practitioners to accommodate their needs, dying or becoming maimed in the process. The history of abortion practices reveals the depth of the importance of fertility control to American women and men.

13 Voluntary Motherhood: The Beginnings of Feminist Birth Control Ideas in the United States

Linda Gordon

Voluntary motherhood was the first general name for a feminist birth control demand in the United States in the late nineteenth century.[1] It represented an initial response of feminists to their understanding that involuntary motherhood and child raising were important parts of woman's oppression. In this paper I would like to trace the content and general development of "voluntary motherhood" ideas and to situate them in the development of the American birth control movement.

The feminists who advocated voluntary motherhood were of three general types: suffragists; people active in such moral reform movements as temperance and social purity, in church auxiliaries, and in women's professional and service organizations (such as Sorosis); and members of small, usually anarchist, Free Love groups. The Free Lovers played a classically vanguard role in the development of birth control ideas. Free Love groups were always small and sectarian, and they were usually male dominated, despite their extreme ideological feminism. They never coalesced into a movement. On the contrary,

they were the remnants of a dying tradition of utopian socialist and radical Protestant religious dissent. The Free Lovers, whose very self-definition was built around commitments to iconoclasm and to isolation from the masses, were precisely the group that could offer intellectual leadership in formulating the shocking arguments that birth control in the nineteenth century required.[2]

The suffragists and moral reformers, concerned to win mass support, were increasingly committed to social respectability. As a result, they did not generally advance very far beyond prevalent standards of propriety in discussing sexual matters publicly. Indeed, as the century progressed the social gap between them and the Free Lovers grew, for the second and third generations of suffragists were more concerned with respectability than the first. In the 1860s and 1870s the great feminist theoreticians had been much closer to the Free Lovers, and at least one of these early giants, Victoria Woodhull, was for several years a member of both the suffrage and the Free Love camps. But even respectability did not completely stifle the mental processes of the feminists, and many of them said in private writings—in letters and diaries—what they were unwilling to utter in public.

LINDA GORDON is Florence Kelley Professor of History and Women's Studies at the University of Wisconsin–Madison.
Reprinted from *Feminist Studies* 1 (1973): 5–22.

The similar views of Free Lovers and suffragists on the question of voluntary motherhood did not bridge the considerable political distance between the groups but did show that their analyses of the social meaning of reproduction for the women were converging. The sources of that convergence, the common grounds of their feminism, were their similar experiences in the changing conditions of nineteenth-century America. Both groups were composed of educated, middle-class Yankees responding to severe threats to the stability, if not dominance, of their class position. Both groups were disturbed by the consequences of rapid industrialization—the emergence of great capitalists in a clearly defined financial oligarchy, and the increased immigration which threatened the dignity and economic security of the middle-class Yankee. Free Lovers and suffragists, as feminists, looked forward to a decline in patriarchal power within the family but worried too about the possible disintegration of the family and the loosening of sexual morality. They saw reproduction in the context of these larger social changes and in a movement for women's emancipation, and they saw that movement as an answer to some of these large social problems. They hoped that giving political power to women would help to reinforce the family, to make the government more just and the economy less monopolistic. In all these attitudes there was something traditional as well as something progressive; the concept of voluntary motherhood reflected this duality.

Since we all bring a twentieth-century understanding to our concept of birth control, it may be best to make it clear at once that neither Free Lovers nor suffragists approved of contraceptive devices. Ezra Heywood, patriarch and martyr, thought "artificial" methods "unnatural, injurious, or offensive." Tennessee Claflin wrote that the "washes, teas, tonics and various sorts of appliances known to the initiated" were a "stand-ing reproach upon, and a permanent indictment against, American women.... No woman should ever hold sexual relations with any man from the possible consequences of which she might desire to escape." *Woodhull and Claflin's Weekly* editorialized: "The means they [women] resort to for ... prevention is sufficient to disgust every natural man."[3]

On a rhetorical level the main objection to contraception[4] was that it was "unnatural," and the arguments reflected a romantic yearning for the "natural," rather pastorally conceived, that was typical of many nineteenth-century reform movements. More basic, however, particularly in women's arguments against contraception, was an underlying fear of the promiscuity that it could permit. The fear of promiscuity was associated less with fear for one's virtue than with fear of other women—the perhaps mythical "fallen" women—who might threaten a husband's fidelity.

To our twentieth-century minds a principle of voluntary motherhood that rejects the practice of contraception seems so theoretical as to have little real impact. What gave the concept substance was that it was accompanied by another, potentially explosive, conceptual change: the reacceptance of female sexuality. As with birth control, the most open advocates of female sexuality were the Free Lovers, not the suffragists; nevertheless, both groups based their ideas on the traditional grounds of the "natural." Free Lovers argued, for example, that celibacy was unnatural and dangerous—for men and women alike. "Pen cannot record, nor lips express, the enervating, debauching effect of celibate life upon young men and women."[5] Asserting the existence, legitimacy, and worthiness of female sexual drive was one of the Free Lovers' most important contributions to sexual reform; it was a logical correlate of their argument from the "natural" and of their appeal for the integration of body and soul.

Women's rights advocates too began to demand recognition of female sexuality. Isabella Beecher Hooker wrote to her daughter: "Multitudes of women in all the ages who have scarce known what sexual desire is—being wholly absorbed in the passion of maternity, have sacrificed themselves to the beloved husbands as unto God—and yet these men, full of their human passion and defending it as righteous & God-sent lose all confidence in womanhood when a woman here and there betrays her similar nature & gives herself soul & body to the man she adores."[6] Alice Stockham, a Spiritualist Free Lover and feminist physician, lauded sexual desire in men and women as "the prophecy of attainment." She urged that couples avoid reaching sexual "satiety" with each other, in order to keep their sexual desire constantly alive, for she considered desire pleasant and healthful.[7] Elizabeth Cady Stanton, commenting in her diary in 1883 on the Whitman poem "There Is a Woman Waiting for Me," wrote: "He speaks as if the female must be forced to the creative act, apparently ignorant of the fact that a healthy woman has as much passion as a man, that she needs nothing stronger than the law of attraction to draw her to the male."[8] Still, she loved Whitman, and, largely because of that, openness about sex that made him the Free Lovers' favorite poet.

According to the system of ideas then dominant, women, lacking sexual drives, submitted to sexual intercourse (and notice how Beecher Hooker continued the image of a woman "giving herself," never taking) in order to please their husbands and to conceive children. The ambivalence underlying this view was expressed in the equally prevalent notion that women must be protected from exposure to sexuality lest they "fall" and become depraved lustful monsters. This ambivalence perhaps came from a subconscious lack of certainty about the reality of the sexless woman, a construct laid only thinly on top of the conception of woman as highly sexed, even insatiably so, that prevailed

up to the eighteenth century. Victorian ambivalence on this question is nowhere more tellingly set forth than in the writings of physicians, who viewed woman's sexual organs as the source of her being, physical and psychological, and blamed most mental derangements on disorders of the reproductive organs.[9] Indeed, they saw it as part of the nature of things, as Rousseau had written, that men were male only part of the time, but women were female always.[10] In a system that deprived women of the opportunity to make extrafamilial contributions to culture, it was inevitable that they should be more strongly identified with sex than men were. Indeed, females were frequently called "the sex" in the nineteenth century.

The concept of maternal instinct helped to smooth the contradictory attitudes about woman's sexuality. In many nineteenth-century writings we find the idea that the maternal instinct was the female analogue of the male sex instinct; it was as if the two instincts were seated in analogous parts of the brain, or soul. Thus to suggest, as feminists did, that women might have the capacity for sexual impulses of their own automatically tended to weaken the theory of the maternal instinct. In the fearful imaginations of self-appointed protectors of the family and of womanly innocence, the possibility that women might desire sexual contact not for the sake of pregnancy—that they might even desire it at a time when they positively did not want pregnancy—was a wedge in the door to denying that women had any special maternal instinct at all.

Most of the feminists did not want to open that door either. Indeed, it was common for nineteenth-century women's rights advocates to use the presumed "special motherly nature" and "sexual purity" of women as arguments for increasing their freedom and status. It is no wonder that many of them chose to speak their subversive thoughts about the sexual nature of women privately, or at least softly. Even among the more outspoken Free Lovers, there was a certain amount of hedging. Lois Waisbrooker and Dora Forster, writing for a Free Love journal in the 1890s, argued that while men and women both had an "amative" instinct, it was much stronger in men and that women—only women—also had a reproductive, or "generative" instinct. "I suppose it must be universally conceded that men make the better lovers," Forster wrote. She thought that it might be possible that "the jealousy and tyranny of men have operated to suppress amativeness in women, by constantly sweeping strongly sexual women from the paths of life into infamy and sterility and death," but she thought also that the suppression, if it existed, had been permanently inculcated in woman's character.[11]

Modern birth control ideas rest on a full acceptance, at least quantitatively, of female sexuality. Modern contraception is designed to permit sexual intercourse as often as desired without the risk of pregnancy. Despite the protestations of sex counselors that there are no norms for the frequency of intercourse, in the popular view there are such norms. Most people in the midtwentieth century thought that "normal" couples have intercourse several times a week. By twentieth-century standards, then, the Free Lovers' rejection of artificial contraception and "unnatural" sex seems to preclude the possibility of birth control at all. Nineteenth-century sexual reformers, however, had different sexual norms. They did not seek to make an infinite number of sterile sexual encounters possible. They wanted to make it possible for women to avoid pregnancy if they badly needed to do so for physical or psychological reasons, but they did not believe that it was essential for such women to engage freely in sexual intercourse.

not for fun... for health

In short, for birth control they recommended periodic or permanent abstinence. The proponents of voluntary motherhood had in mind two distinct contexts for abstinence. One was the mutual decision of a couple. This could mean continued celibacy, or it could mean following a form of the rhythm method. Unfortunately, all the nineteenth-century writers miscalculated women's fertility cycle. (It was not until the 1920s that the ovulation cycle was correctly plotted, and until the 1930s it was not widely understood among American doctors.)[12] Ezra Heywood, for example, recommended avoiding intercourse from six to eight days before menstruation until ten to twelve days after it. Careful use of the calendar could also provide control over the sex of a child, Heywood believed: conception in the first half of the menstrual cycle would produce girls, in the second half, boys.[13] These misconceptions functioned, conveniently, to make practicable Heywood's and others' ideas that celibacy and contraceptive devices should *both* be avoided.

Some of the Free Lovers also endorsed male continence, a system practiced and advocated by the Oneida community, in which the male avoids climax entirely.[14] (There were other aspects of the Oneida system that antagonized the Free Lovers, notably the authoritarian quality of John Humphrey Noyes's leadership.[15]) Dr. Stockham developed her own theory of continence called "Karezza," in which the female as well as the male was to avoid climax. Karezza and male continence were whole sexual systems, not just methods of birth control. Their advocates expected the self-control involved to build character and spiritual qualities, while honoring, refining, and dignifying the sexual functions, and Karezza was reputed to be a cure for sterility as well, since its continued use was thought to build up the resources of fertility in the body.[16]

Idealizing sexual self-control was characteristic of the Free Love point of view. It was derived mainly from the thought of the utopian communitarians of the early nineteenth century,[17] but Ezra Heywood elaborated the theory. Beginning with the assumption that people's "natural" instincts, left untrammeled, would automatically create a harmonious, peaceful society—an assumption certainly derived from liberal philosophical faith in the innate goodness of man—Heywood applied it to sexuality, arguing that the natural sexual instinct was innately moderated, self-regulating. He did not imagine, as did Freud, a powerful simple libido that could be checked only by an equally powerful moral and rational will. Heywood's theory implicitly contradicted Freud's description of inner struggle and constant tension between the drives of the id and the goals of the superego; Heywood denied the social necessity of sublimation.

On one level Heywood's theory may seem inadequate as a psychology, since it cannot explain such phenomena as repression and the strengthening of self-control with maturity. It may, however, have a deeper accuracy. It argues that society and its attendant repressions have distorted the animal's natural self-regulating mechanism and have thereby created excessive and obsessive sexual drives. It offers a social explanation for the phenomena that Freud described in psychological terms and thus holds out the hope that they can be changed.

Essentially similar to Wilhelm Reich's theory of "sex-economy," the Heywood theory of self-regulation went beyond Reich's in providing a weapon against one of the ideological bastions of male supremacy. Self-regulation as a goal was directed against the prevalent attitude that male lust was an uncontrollable urge, an attitude that functioned as a justification for rape specifically and for male sexual irresponsibility generally. We have to get away from the tradition of "man's necessities and woman's obedience to them," Stockham wrote.[18] The idea that men's desires are irrepressible is merely the other face of the idea that women's desires are nonexistent. To-

gether the two created a circle that enclosed woman, making it her exclusive responsibility to say no and making pregnancy her God-imposed burden if she didn't, while denying her both artificial contraception and the personal and social strength to rebel against male sexual demands.

Heywood developed his theory of natural sexual self-regulation in answer to the common anti–Free Love argument that the removal of social regulation of sexuality would lead to unhealthy promiscuity: "In the distorted popular view, Free Love tends to unrestrained licentiousness, to open the flood gates of passion and remove all barriers in its desolating course; but it means just the opposite; it means the *utilization of animalism,* and the triumph of Reason, Knowledge, and Continence."[19] He applied the theory of self-regulation to the problem of birth control only as an afterthought, perhaps when women's concerns with that problem reached him. Ideally, he trusted, the amount of sexual intercourse that men and women desired would be exactly commensurate with the number of children that were wanted. Since sexual repression had had the boomerang effect of intensifying our sexual drives far beyond "natural" levels, effective birth control now would require the development of the inner self-control to contain and repress sexual urges. But in time he expected that sexual moderation would come about naturally.

Heywood's analysis, published in the mid-1870s, was concerned primarily with excessive sex drives in men. Charlotte Perkins Gilman, one of the leading theoreticians of the suffrage movement, reinterpreted that analysis two decades later to emphasize its effects on women. The economic dependence of woman on man, in Gilman's analysis, made her sexual attractiveness necessary not only for winning a mate but as a means of getting a livelihood too. This is the case with no other animal. In the human female it had produced "excessive modification to sex," emphasizing weak qualities characterized by humans as "feminine." She made an analogy to the milk cow, bred to produce far more milk than she would need for her calves. But Gilman agreed completely with Heywood about the effects of exaggerated sex distinction on the male; it produced excessive sex energy and excessive indulgence to an extent debilitating to the whole species. Like Heywood, she also believed that the path of progressive social evolution moved toward monogamy and toward reducing the promiscuous sex instinct.[20]

A second context for abstinence, in addition to mutual self-regulation by a couple, was the right of the wife unilaterally to refuse her husband. This idea is at the heart of voluntary motherhood. It was a key substantive demand in the midnineteenth century when both law and practice made sexual submission to her husband a woman's duty.[21] A woman's right to refuse is clearly the fundamental condition of birth control—and of her independence and personal integrity.

In their crusade for this right of refusal the voices of Free Lovers and suffragists were in unison. Ezra Heywood demanded "Woman's Natural Right to ownership and control over her own body-self—a right inseparable from Woman's intelligent existence."[22] Paulina Wright Davis, at the National Woman Suffrage Association in 1871, attacked the law "which makes obligatory the rendering of marital rights and compulsory maternity." When, as a result of her statement, she was accused of being a Free Lover, she responded by accepting the description.[23] Isabella Beecher Hooker wrote her daughter in 1869 advising her to avoid pregnancy until "you are prepared in body and soul to receive and cherish the little one."[24] In 1873 she gave similar advice to women generally, in her book *Womanhood*.[25] Elizabeth Cady Stanton had characteristically

used the same phrase as Heywood: woman owning her own body. Once asked by a magazine what she meant by it, she replied: "Womanhood is the primal fact, wifehood and motherhood its incidents. . . . Must the heyday of her existence be wholly devoted to the one animal function of bearing children? Shall there be no limit to this but woman's capacity to endure the fearful strain on her life?"[26]

The insistence on women's right to refuse often took the form of attacks on men for their lusts and their violence in attempting to satisfy them. In their complaints against the unequal marriage laws, chief or at least loudest among them was the charge that they legalized rape.[27] Victoria Woodhull raged, "I will tell the world, so long as I have a tongue and the strength to move it, of all the infernal misery hidden behind this horrible thing called marriage, though the Young Men's Christian Association sentence me to prison a year for every word. I have seen horrors beside which stone walls and iron bars are heaven."[28] Angela Heywood attacked men incessantly and bitterly; if one were to ignore the accuracy of her charges, she could well seem ill-tempered. "Man so lost to himself and woman as to invoke legal *violence* in these sacred nearings, *should have solemn meeting with, and look serious at his own penis until he is able to be lord and master of it, rather than it should longer rule, lord and master, of him and of the victims he deflowers*."[29] Suffragists spoke more delicately but not less bitterly. Feminists organized social purity organizations and campaigns, their attacks on prostitution based on a critique of the double standard, for which their proposed remedy was that men conform to the standards required of women.[30]

A variant of this concern was a campaign against "sexual abuses"—a Victorian euphemism for deviant sexual practices, or simply excessive sexual demands, not necessarily violence or prostitution. The Free Lovers, particularly, turned to this cause, because it gave them an opportunity to attack marriage. The "sexual abuses" question was one of the most frequent subjects of correspondence in Free Love periodicals. For example, a letter from Mrs. Theresa Hughes of Pittsburgh described

> a girl of sixteen, full of life and health when she became a wife.... She was a slave in every sense of the word, mentally and sexually, never was she free from his brutal outrages, morning, noon and night, up almost to the very hour her baby was born, and before she was again strong enough to move about.... Often did her experience last an hour or two, and one night she will never forget, the outrage lasted exactly four hours.[31]

Or from Lucinda Chandler, well-known moral reformer:

> This useless sense gratification has demoralized generation after generation, till monstrosities of disorder are common. Moral education, and healthful training will be requisite for some generations, even after we have equitable economics, and free access to Nature's gifts. The young man of whom I knew who threatened his bride of a week with a sharp knive in his hand, to compel her to perform the office of 'sucker,' would no doubt have had the same disposition though no soul on the planet had a want unsatisfied or lacked a natural right.[32]

From an anonymous woman in Los Angeles:

> I am nearly wrecked and ruined by ... nightly intercourse, which is often repeated in the morning. This and nothing else was the cause of my miscarriage ... he went to work like a man a-mowing, and instead of a pleasure as it might have been, it was most intense torture.[33]

Clearly, these remarks reflect a level of hostility toward sex. The observation that many feminists hated sex has been made by several historians,[34] but they have usually failed to perceive that feminists' hostility and fear of it came from the fact that they were women, not that they were feminists. Women in the nineteenth century were, of course, trained to repress their own sexual feelings, to view sex as a duty. But they also resented what they experienced, which was not an abstraction but a particular, historical kind of sexual encounter—intercourse dominated by and defined by the male in conformity with his desires and in disregard of what might bring pleasure to a woman. (That this might have resulted more from male ignorance than malevolence could not change women's experiences. Furthermore, sexual intercourse brought physical danger.) Pregnancy, childbirth, and abortions were risky, painful, and isolating experiences in the nineteenth century; venereal diseases were frequently communicated to women by their husbands. Elmina Slenker, a Free Lover and novelist, wrote, "I'm getting a host of stories (truths) about women so starved sexually as to use their dogs for relief, and finally I have come to the belief that a CLEAN dog is better than a drinking, tobacco-smelling, venereally diseased man!"[35]

"Sex-hating" women were not just misinformed, or priggish, or neurotic. They were often responding rationally to their material reality. Denied the possibility of recognizing and expressing their own sexual needs, denied even the knowledge of sexual possibilities other than those dictated by the rhythms of male orgasm, they had only two choices: passive and usually pleasureless submission, with high risk of undesirable consequences, or rebellious refusal. In that context abstinence to ensure voluntary motherhood was a most significant feminist demand.

What is remarkable is that some women recognized that it was not sex per se, but only their

women needed to find new husbands

husbands' style of making love, that repelled them. One of the women noted earlier who complained about her treatment went on to say: "I am undeveloped sexually, never having desires in that direction; still, with a husband who had any love or kind of feelings for me and one less selfish it *might* have been different, but he cared nothing for the torture to *me* as long as *he* was gratified."[36]

Elmina Slenker herself, the toughest and crustiest of all these "sex haters," dared to explore and take seriously her own longings, thereby revealing herself to be a sex lover in disguise. As the editor of the *Water-Cure Journal,* and a regular contributor to *Free Love Journal,*[37] she expounded a theory of "Dianaism, or Nonprocreative Love," sometimes called "Diana-love and Alpha-abstinence." It meant free sexual contact of all sorts except intercourse.

> We want the sexes to love more than they do; we want them to love openly, frankly, earnestly; to enjoy the caress, the embrace, the glance, the voice, the presence & the very step of the beloved. We oppose no form or act of love between any man & woman. Fill the world as full of genuine sex love as you can . . . but forbear to rush in where generations yet unborn may suffer for your unthinking, uncaring, unheeding actions.[38]

Comparing this to the more usual physical means of avoiding conception—coitus interruptus and male continence—reveals how radical it was. In modern history, public endorsement of nongenital sex, and of forms of genital sex beyond standard "missionary position" intercourse, has been a recent, post-Freudian, even post–Masters-and-Johnson phenomenon. The definition of sex as heterosexual intercourse has been one of the oldest and most universal cultural norms. Slenker's alienation from existing sexual possibilities led her to explore alter-

natives with a bravery and a freedom from religious and psychological taboos extraordinary for a nineteenth-century Quaker reformer.

In the nineteenth century, neither Free Lovers nor suffragists ever relinquished their hostility to contraception. But among the Free Lovers, free speech was always an overriding concern, and for that reason Ezra Heywood agreed to publish some advertisements for a vaginal syringe, an instrument the use of which for contraception he personally deplored, or so he continued to assure his readers. Those advertisements led to Heywood's prosecution for obscenity, and he defended himself with characteristic flair by making his position more radical than ever before. Contraception was moral, he argued, when it was used by women as the only means of defending their rights, including the right to voluntary motherhood. Although "artificial means of preventing conception are not generally patronized by Free Lovers," he wrote, reserving for his own followers the highest moral ground, still he recognized that not all women were lucky enough to have Free Lovers for their sex partners.[39]

> Since Comstockism makes male will, passion and power absolute to *impose* conception, I stand with women to resent it. The man who would legislate to choke a woman's vagina with semen, who would force a woman to retain his seed, bear children when her own reason and conscience oppose it, would waylay her, seize her by the throat and rape her person.[40]

Angela Heywood enthusiastically pushed this new political line.

> Is it "proper," "polite," for men, to go to Washington to say, by penal law, fines and imprisonment, whether woman may continue her natural right to wash, rinse, or wipe her own vaginal body opening—as

well legislate when she may blow her nose, dry her eyes, or nurse her babe. . . . Whatever she may have been pleased to receive, from man's own, is his gift and her property. Women do not like rape, and have a right to resist its results.[41]

Her outspokenness, vulgarity in the ears of most of her contemporaries, came from a substantive, not merely a stylistic, sexual radicalism. Not even the heavy taboos and revulsion against abortion stopped her: "To cut a child up in woman, procure abortion, is a most fearful, tragic deed; but *even that* does not call for man's arbitrary jurisdiction over woman's womb."[42]

It is unclear whether Heywood, in this passage, was actually arguing for legalized abortion; if she was, she was alone among all nineteenth-century sexual reformers in saying it. Other feminists and Free Lovers condemned abortion and argued that the necessity of stopping its widespread practice was a key reason for instituting voluntary motherhood by other means. The difference on the abortion question between sexual radicals and sexual conservatives was in their analysis of its causes and remedies. While doctors and preachers were sermonizing on the sinfulness of women who obtained abortions,[43] the radicals pronounced abortion itself an undeserved punishment and a woman who had one a helpless victim. Woodhull and Claflin wrote about Madame Restell's notorious abortion "factory" in New York City without moralism, arguing that only voluntary conception would put it out of business.[44] Elizabeth Cady Stanton also sympathized with women who had abortions, and [she] used the abortion problem as an example of women's victimization by laws made without their consent.[45]

Despite stylistic differences, which stemmed from differences in goals, nineteenth-century American Free Lovers and women's rights advocates shared the same basic attitudes toward birth control: they opposed contraception and abortion but endorsed voluntary motherhood achieved through periodic abstinence; they believed that women should always have the right to decide when to bear a child; and they believed that women and men both had natural sex drives and that it was not wrong to indulge those drives without the intention of conceiving children. The two groups also shared the same appraisal of social and political significance of birth control. Most of them were favorably inclined toward neo-Malthusian reasoning (at least until the 1890s, when the prevailing concern shifted to the problem of underpopulation rather than overpopulation).[46] They were also interested, increasingly, in controlling conception for eugenic purposes.[47] They were hostile to the hypocrisy of the sexual double standard and, beyond that, shared a general sense that men had become oversexed and that sex had been transformed into something disagreeably violent.

But above all their commitment to voluntary motherhood expressed their larger commitment to women's rights. Elizabeth Cady Stanton thought voluntary motherhood so central that on her lecture tours in 1871 she held separate afternoon meetings for *women only* (a completely unfamiliar practice at the time) and talked about "the gospel of fewer children & a healthy, happy maternity."[48] "What radical thoughts I then and there put into their heads & as they feel untrammelled, these thoughts are permanently lodged there! That is all I ask."[49] Only Ezra Heywood had gone so far as to defend a particular contraceptive device—the syringe. But the principle of woman's rights to choose the number of children she would bear and when was accepted in the most conservative sections of the women's rights movement. At the First Congress of the Association for the Advancement of Women in 1873, a whole session was devoted to the theme "Enlightened Motherhood," which had voluntary motherhood as part of its meaning.[50]

The general conviction of the feminist community that women had a right to choose when to conceive a child was so strong by the end of the nineteenth century that it seems odd that they were unable to overcome their scruples against artificial contraception. The basis for the reluctance lies in their awareness that a consequence of effective contraception would be the separation of sexuality from reproduction. A state of things that permitted sexual intercourse to take place normally, even frequently, without the risk of pregnancy, inevitably seemed to nineteenth-century middle-class women as an attack on the family, as they understood the family. In the mid-Victorian sexual system men normally conducted their sexual philandering with prostitutes; accordingly, prostitution, far from being a threat to the family system, was a part of it and an important support of it. This was the common view of the time, paralleled by the belief that prostitutes knew of effective birth control techniques. This seemed only fitting, for contraception in the 1870s was associated with sexual immorality. It did not seem, even to the most sexually liberal, that contraception could be legitimized to any extent, even for the purposes of family planning for married couples, without licensing extramarital sex. The fact that contraception was not morally acceptable to respectable women was, from a woman's point of view, a guarantee that those women would not be a threat to her own marriage.

The fact that sexual intercourse often leads to conception was also a guarantee that men would marry in the first place. In the nineteenth century, women needed marriage far more than men. Lacking economic independence, women needed husbands to support them, or at least to free them from a usually more humiliating economic dependence on fathers. Especially in the cities, where women were often isolated from communities, deprived of the economic and psychological support of networks of relatives, friends, and neighbors, the prospect of dissolving the cement of nuclear families was frightening. In many cases children, and the prospect of children, provided that cement. Man's responsibilities for children were an important pressure for marital stability. Women, especially middle-class women, were also dependent on their children to provide them with meaningful work. The belief that motherhood was a woman's fulfillment had a material basis: parenthood was often the only creative and challenging activity in a woman's life, a key part of her self-esteem.

Legal efficient birth control would have increased men's freedom to indulge in extramarital sex without greatly increasing women's freedom to do so. The pressures enforcing chastity and marital fidelity on middle-class women were not only fear of illegitimate conception but a powerful combination of economic, social, and psychological factors, including economic dependence, fear of rejection by husband and social support networks, internalized taboos and, hardly the least important, a socially conditioned lack of interest in sex that may have approached functional frigidity. The double standard of the Victorian sexual and family system, which had made men's sexual freedom irresponsible and oppressive to women, left most feminists convinced that increasing, rather than releasing, the taboos against extramarital sex was in their interest, and they threw their support behind social purity campaigns.

In short, we must forget the twentieth-century association of birth control with a trend toward sexual freedom. The voluntary motherhood propaganda of the 1870s was associated with a push toward a more restrictive, or at least a more rigidly enforced, sexual morality. Achieving voluntary motherhood by a method that would have encouraged sexual license was absolutely contrary to the felt interests of the very group that formed the main social basis for the cause—middle-class women. Separating these women from the early twentieth-century feminists, with their interests in sexual freedom, were nearly four decades of significant social and economic changes and a general weakening of the ideology of the Lady. The ideal of the Free Lovers—responsible, open sexual encounters between equal partners—was impossible in the 1870s because men and women were not equal. A man was a man whether faithful to his wife or not. But women's sexual activities divided them into two categories—wife or prostitute. These categories were not mere ideas but were enforced in reality by severe social and economic sanctions. The fact that so many, indeed most, Free Lovers in practice led faithful, monogamous, legally married lives is not insignificant in this regard. It suggests that they understood that Free Love was an ideal not to be realized in that time.

As voluntary motherhood was an ideology intended to encourage sexual purity, so it was also a promotherhood ideology. Far from debunking motherhood, the voluntary motherhood advocates consistently continued the traditional Victorian mystification and sentimentalization of the mother. It is true that at the end of the nineteenth century an increasing number of feminists and elite women—that is, still a relatively small group—were choosing not to marry or become mothers. That was primarily because of their increasing interest in professional work, and the difficulty of doing such work as a wife and mother, given the normal uncooperativeness of husbands and the lack of social provisions for child care. Voluntary motherhood advocates shared the general belief that mothers of young children ought not to work outside their homes but should make mothering their full-time occupation. Suffragists argued both to make professions open to women and to enno-

ble the task of mothering; they argued for increased rights and opportunities for women *because* they were mothers.

The Free Lovers were equally promotherhood; they wanted only to separate motherhood from legal marriage.[51] They devised promotherhood arguments to bolster their case against marriage. Mismated couples, held together by marriage laws, made bad parents and produced inferior offspring, Free Lovers said.[52] In 1870 *Woodhull and Claflin's Weekly* editorialized, "Our marital system is the greatest obstacle to the regeneration of the race."[53]

This concern with eugenics was characteristic of nearly all feminists of the late nineteenth century. At the time eugenics was mainly seen as an implication of evolutionary theory and was picked up by many social reformers to buttress their arguments that improvement of the human condition was possible. Eugenics had not yet become a movement in itself. Feminists used eugenics arguments as if they instinctively felt that arguments based solely on women's rights had not enough power to conquer conservative and religious scruples about reproduction. So they combined eugenics and feminism to produce evocative, romantic visions of perfect motherhood. "Where boundless love prevails. . . ." *Woodhull and Claflin's Weekly* wrote, "the mother who produces an inferior child will be dishonored and unhappy . . . and she who produces superior children will feel proportionately pleased. When woman attains this position, she will consider superior offspring a necessity and be apt to procreate only with superior men."[54] Free Lovers and suffragists alike used the cult of motherhood to argue for making motherhood voluntary. Involuntary motherhood, wrote Harriot Stanton Blatch, daughter of Elizabeth Cady Stanton and a prominent suffragist, is a prostitution of the maternal instinct.[55] Free Lover Rachel Campbell cried out that motherhood was being "ground to dust under the misrule of masculine ignorance and superstition."[56]

Not only was motherhood considered an exalted, sacred profession, and a profession exclusively woman's responsibility, but for a woman to avoid it was to choose a distinctly less noble path. In arguing for the enlargement of woman's sphere, feminists envisaged combining motherhood with other activities, not rejecting motherhood. Victoria Woodhull and Tennessee Claflin wrote:

Tis true that the special and distinctive feature of woman is that of bearing children, and that upon the exercise of her function in this regard the perpetuity of race depends. It is also true that those who pass through life failing in this special feature of their mission cannot be said to have lived to the best purposes of woman's life. But while maternity should always be considered the most holy of all the functions woman is capable of, it should not be lost sight of in devotion to this, that there are as various spheres of usefulness outside of this for woman as there are for man outside of the marriage relation.[57]

Birth control was not intended to open the possibility of childlessness but merely to give women leverage to win more recognition and dignity. Dora Forster, a Free Lover, saw in the fears of underpopulation a weapon of blackmail for women:

I hope the scarcity of children will go on until maternity is honored at least as much as the trials and hardships of soldiers campaigning in wartime. It will then be worthwhile to supply the nation with a sufficiency of children ... every civilized nation, having lost the power to enslave woman as mother, will be compelled to recognize her voluntary exercise of that

function as by far the most important service of any class of citizens.[58]

"Oh, women of the world, arise in your strength and demand that all which stands in the path of true motherhood shall be removed from your path," wrote Lois Waisbrooker, a Free Love novelist and moral reformer.[59] Helen Gardener based a plea for women's education entirely on the argument that society needed educated mothers to produce able sons (not children, sons).

Harvard and Yale, not to mention Columbia, may continue to put a protective tariff on the brains of young men: but so long as they must get those brains from the proscribed sex, just so long will male brains remain an "infant industry" and continue to need this protection. Stupid mothers never did and stupid mothers never will, furnish this world with brilliant sons.[60]

Clinging to the cult of motherhood was part of a broader conservatism shared by Free Lovers and suffragists—acceptance of traditional sex roles. Even the Free Lovers rejected only one factor—legal marriage—of the many that defined woman's place in the family. They did not challenge conventional conceptions of woman's passivity and limited sphere of concern.[61] In their struggles for equality the women's rights advocates never suggested that men should share responsibility for child raising, housekeeping, nursing, cooking. When Victoria Woodhull in the 1870s and Charlotte Perkins Gilman in the early 1900s suggested socialized child care, they assumed that only women would do the work.[62] Most feminists wanted economic independence for women, but most, too, were reluctant to recommend achieving this by turning women loose and helpless into the economic world to compete with men.[63] This attitude was conditioned by an attitude hostile to the egoistic spirit of capitalism, but the attitude was not transformed into a political position and usually appeared as a description of women's weakness, rather than an attack on the system. Failing to distinguish, or even to indicate awareness of a possible distinction between, women's conditioned passivity and their equally conditioned distaste for competition and open aggression, these feminists also followed the standard Victorian rationalization of sex roles, the idea that women were morally superior. Thus the timidity and self-effacement that were the marks of women's powerlessness were made into innate virtues. Angela Heywood, for example, praised women's greater ability for self-control and, in an attribution no doubt intended to jar and titillate the reader, branded men inferior on account of their lack of sexual temperance.[64] Men's refusal to accept women as human beings she identified, similarly, as a mark of men's incapacity: "man has not yet achieved himself to realize and meet a PERSON in woman."[65] In idealistic abstract terms, no doubt such male behavior is an incapacity. Yet that conceit failed to remark on the power and privilege over women that the supposed "incapacity" gave men.

This omission is characteristic of the cult of motherhood. Indeed, what made it a cult was its one-sided failure to recognize the privileges men received from women's exclusive responsibility for parenthood. The "motherhood" of the feminists' writings was not merely the biological process of gestation and birth but a package of social, economic, and cultural functions. Although many of the nineteenth-century feminists had done substantial analysis of the historical and anthropological origins of woman's social role, they nevertheless agreed with the biological determinist point of view that women's parental capacities had become implanted at the level of instinct, the famous "maternal instinct." That concept rested on the assumption that the

qualities that parenthood requires—capacities for tenderness, self-control and patience, tolerance for tedium and detail, emotional supportiveness, dependability and warmth—were not only instinctive but sex linked. The concept of the maternal instinct thus also involved a definition of the normal instinctual structure of the male that excluded these capacities, or included them only to an inferior degree; it also carried the implication that women who did not exercise these capacities, presumably through motherhood, remained unfulfilled, untrue to their destinies.

Belief in the maternal instinct reinforced the belief in the necessary spiritual connection for women between sex and reproduction and limited the development of birth control ideas. But the limits were set by the entire social context of women's lives, not by the intellectual timidity of their ideas. For women's "control over their own bodies" to lead to a rejection of motherhood as the *primary* vocation and measure of social worth required the existence of alternative vocations and sources of worthiness. The women's rights advocates of the 1870s and 1880s were fighting for those other opportunities, but a significant change had come only to a few privileged women, and most women faced essentially the same options that existed fifty years earlier. Thus voluntary motherhood in this period remained almost exclusively a tool for women to strengthen their positions within conventional marriages and families, not to reject them.

Notes

1. The word *feminist* must be underscored. Since the early nineteenth century, there had been developing a body of population control writings, which recommended the use of birth control techniques to curb nationwide or worldwide populations; usually called neo-Malthusians, these writers were not concerned with the control of births as a means by which

women could gain control over their own lives, except, very occasionally, as an auxiliary argument. And of course birth control practices date back to the most ancient societies on record.

2. There is no space here to compensate for the general lack of information about the Free Lovers. There is a fuller discussion of them in my *Woman's Body, Woman's Right: A Social History of Birth Control* (New York: Viking, Penguin, 1976). Some of the major Free Love writings include R. D. Chapman, *Free-love a Law of Nature* (New York: author, 1881); Tennessee Claflin, *The Ethics of Sexual Equality* (New York: Woodhull & Claflin, 1873); Tennessee Claflin, *Virtue, What It Is and What It Isn't; Seduction, What It Is and What It Is Not* (New York: Woodhull & Claflin, 1872); Ezra Heywood, *Cupid's Yokes; or, The Binding Force of Conjugal Life* (Princeton, Mass.: Cooperative Publishing Co., n.d., probably 1876); Ezra Heywood, *Uncivil Liberty: An Essay to Show the Injustice and Impolicy of Ruling Woman without Her Consent* (Princeton, Mass.: Cooperative Publishing Co., 1872); C. L. James, *The Future Relation of the Sexes* (St. Louis: author, 1872); Juliet Severance, *Marriage* (Chicago: M. Harman, 1901); Victoria Claflin Woodhull, *The Scare-Crows of Sexual Slavery* (New York: Woodhull & Claflin, 1874); Victoria Claflin Woodhull, *A Speech on the Principles of Social Freedom* (New York: Woodhull & Claflin, 1872); Victoria Claflin Woodhull, *Tried as by Fire; or, The True and the False Socially* (New York: Woodhull & Claflin, 1874).

3. Heywood, *Cupid's Yokes*, 20; Claflin, *The Ethics of Sexual Equality*, 9–10; *Woodhull & Claflin's Weekly*, 1870, 1, no. 6: 5.

4. "Contraception" will be used to refer to artificial devices used to prohibit conception during intercourse, while "birth control" will be used to mean anything, including abstinence, which limits pregnancy.

5. Heywood, *Cupid's Yokes*, 17–18.

6. Letter to her daughter Alice, 1874, in the Isabella Beecher Hooker Collection, Beecher Stowe Mss, Stowe–Day Library, Hartford, Conn. This reference was brought to my attention by Ellen Dubois.

7. Alice B. Stockham, M.D., *Karezza: Ethics of Marriage* (Chicago: Alice B. Stockham & Co., 1898), 84, 91–92.

8. Theodore Stanton and Harriot Stanton Blatch,

eds., *Elizabeth Cady Stanton as Revealed in Her Letters, Diary and Reminiscences* (New York: Harper & Bros., 1922), 2:210 (Diary, 9–6–1883).

9. Ben Barker-Benfield, "The Spermatic Economy: A Nineteenth Century View of Sexuality," *Feminist Studies*, 1, 1 (Summer 1972): 53.

10. J. J. Rousseau, *Emile* (New York: Columbia University Teachers College, 1967), 132. Rousseau was, after all, a chief author of the Victorian revision of the image of woman.

11. Dora Forster, *Sex Radicalism as Seen by an Emancipated Woman of the New Time* (Chicago: M. Harman, 1905), 40.

12. Norman E. Himes, *Medical History of Contraception* (New York: Gamut Press, 1963).

13. Heywood, *Cupid's Yokes*, 19–20, 16.

14. Ibid., 19–20; *Woodhull & Claflin's Weekly* 1, 18 (September 10, 1870): 5.

15. Heywood, *Cupid's Yokes*, 14–15.

16. Stockham, *Karezza*, 82–83, 53.

17. See for example, *Free Enquirer*, ed. Robert Owen and Frances Wright, May 22, 1830, 235–36.

18. Stockham, *Karezza*, 86.

19. Heywood, *Cupid's Yokes*, 19.

20. Charlotte Perkins Gilman, *Women and Economics* (New York: Harper Torchbooks, 1966), 38–39, 43–44, 42, 47–48, 209.

21. In England, for example, it was not until 1891 that the courts first held against a man who forcibly kidnaped and imprisoned his wife when she left him.

22. Ezra Heywood, *Free Speech: Report of Ezra H. Heywood's Defense before the United States Court, in Boston, April 10, 11, and 12, 1883* (Princeton, Mass.: Cooperative Publishing Co., n.d.), 16.

23. Quoted in Nelson Manfred Blake, *The Road to Reno: A History of Divorce in the United States* (New York: Macmillan, 1962), 108, from the *New York Tribune*, May 12, 1871, and July 20, 1871.

24. Letter of August 29, 1869, in Hooker Collection, Beecher Stowe Mss.

25. Isabella Beecher Hooker, *Womanhood: Its Sanctities and Fidelities* (Boston: Lee and Shepard, 1873), 26.

26. Elizabeth Cady Stanton Mss., 11, Library of Congress, undated. This reference was brought to my attention by Ellen Dubois.

27. See, for example, *Lucifer, The Light-Bearer,* ed. Moses Harman (Valley Falls, Kansas, 1894–1907) 18, 6 (October 1889): 3.

28. Victoria Woodhull, *The Scare-Crows,* 21. Her mention of the YMCA is a reference to the fact that Anthony Comstock, author and chief enforcer for the U.S. Post Office of the antiobscenity laws, had begun his career in the YMCA.

29. *The Word* (Princeton, Mass.) 20, 9 (March 1893): 2–3. Emphasis in original.

30. See, for example, the National Purity Congress of 1895, sponsored by the American Purity Alliance.

31. *Lucifer,* April 26, 1890, 1–2.

32. N.a., *Next Revolution; or, Woman's Emancipation from Sex Slavery* (Valley Falls, Kansas: Lucifer Publishing Co., 1890), 49.

33. Ibid., 8–9.

34. Linda Gordon et al., "Sexism in American Historical Writing," *Women's Studies* 1, 1 (Fall 1972).

35. *Lucifer* 15, 2 (September 1886): 3.

36. *The Word,* 1892–93, 20.

37. (Slenker) *Lucifer,* May 23, 1907; *Cyclopedia of American Biography,* 8: 488.

38. See for example *Lucifer* 18, 8 (December 1889): 3; 18, 6 (October 1889): 3; 18, 8 (December 1889): 3.

39. Heywood, *Free Speech,* 17, 16.

40. Ibid., 3–6. "Comstockism" also is a reference to Anthony Comstock (see note 28). Noting the irony that the syringe was called by Comstock's name, Heywood continued: "To name a really good thing 'Comstock' has a sly, sinister, wily look, indicating vicious purpose; in deference to its N.Y., venders, who gave that name, the Publishers of *The Word* inserted an advertisement . . . which will hereafter appear as 'the Vaginal Syringe'; for its intelligent, humane and worthy mission should no longer be libelled by forced association with the pious scamp who thinks Congress gives him legal right of way to and control over every American Woman's Womb." At this trial, Heywood's second, he was acquitted. At his first trial, in 1877, he had been convicted, sentenced to two years, and served six months; at his third, in 1890, he was sentenced to and served two years at hard labor, an ordeal which probably caused his death a year later.

41. *The Word* 10, 9 (March 1893): 2–3.

42. Ibid.

43. See for example Horatio Robinson Storer, M.D., *Why Not? A Book for Every Woman* (Boston: Lee and Shepard, 1868). Note that this was the prize essay in a contest run by the AMA in 1865 for the best anti-abortion tract.

44. Claflin, *Ethics;* Emanie Sachs, *The Terrible Siren, Victoria Woodhull, 1838–1927* (New York: Harper & Bros., 1928), 139.

45. Elizabeth Cady Stanton, Susan Anthony, Matilda Gage, eds., *History of Woman Suffrage,* 1: 597–598.

46. Heywood, *Cupid's Yokes,* 20; see also *American Journal of Eugenics,* ed. M. Harman, 1, 2 (September 1907); *Lucifer* (February 15, 1906; June 7, 1906; March 28, 1907; and May 11, 1905).

47. I deal with early feminists' ideas concerning eugenics in *Woman's Body, Woman's Right.* See note 2.

48. Elizabeth Cady Stanton to Martha Wright, June 19, 1871, Stanton Mss. See also Stanton, *Eight Years After: Reminiscences, 1815–1897* (New York: Schocken, 1971), 262, 297.

49. Stanton and Blatch, *Stanton as Revealed in Her Letters,* 132–33.

50. *Papers and Letters,* Association for the Advancement of Women, 1873. The AAW was a conservative group formed in opposition to the Stanton-Anthony tendency. Nevertheless, Chandler, a frequent contributor to Free Love journals, spoke here against undesired maternity and the identification of woman with her maternal function.

51. *Woodhull & Claflin's Weekly* 1, 20 (October 1, 1870): 10.

52. Woodhull, *Tried As by Fire,* 37; Lillian Harman, *The Regeneration of Society,* Speech before Manhattan Liberal Club, March 31, 1898 (Chicago: Light Bearer Library, 1900).

53. *Woodhull & Claflin's Weekly* 1, 20 (October 1, 1870): 10.

54. Ibid.

55. Harriot Stanton Blatch, "Voluntary Motherhood," *Transactions,* National Council of Women of 1891, ed. Rachel Foster Avery (Philadelphia: J. B. Lippincott, 1891), 280.

56. Rachel Campbell, *The Prodigal Daughter; or, The Price of Virtue* (Grass Valley, Calif., 1885), 3. An

essay read to the New England Free Love League, 1881.

57.   *Woodhull & Claflin's Weekly* 1, 14 (August 13, 1870): 4.

58.   In addition to the biography by Emanie N. Sachs, *"The Terrible Siren": Victoria Woodhull (1838–1927).* (New York: Harper Brothers, 1928), see also Johanna Johnston, *Mrs. Satan* (New York: G. P. Putnam's Sons, 1967), and M. M. Marberry, *Vicky: A Biography of Victoria C. Woodhull* (New York: Funk & Wagnalls, 1967).

59.   From an advertisement for her novel, *Perfect Motherhood: or, Mabel Raymond's Resolve* (New York: Murray Hill, 1890), in the *Next Revolution.*

60.   Helen Hamilton Gardner, *Pulpit, Pew, and Cradle* (New York: Truth Seeker Library, 1891), 22.

61.   Even the most outspoken of the Free Lovers had conventional, role-differentiated images of sexual relations. Here is Angela Heywood, for example: "Men must not emasculate themselves for the sake of 'virtue,' they must, they will, recognize manliness and the life element of manliness as the fountain source of good manners. Women and girls demand strong, well-bred, generative, vitalizing sex ability. Potency, virility, is the grand basic principle of man, and it holds him clean, sweet and elegant, to the delicacy of his counterpart," From *The Word* 14, 2 (June 1885): 3.

62.   Woodhull, *The Scare-Crows;* Charlotte Perkins Gilman, *Concerning Children* (Boston: Small, Maynard, 1900).

63.   See for example Blatch, "Voluntary Motherhood," 283–84.

64.   *The Word* 20, 8 (February 1893): 3.

65.   Ibid.

14 "About to Meet Her Maker": Women, Doctors, Dying Declarations, and the State's Investigation of Abortion, Chicago, 1867–1940

Leslie J. Reagan

Thougn+piece #5

In March 1916 Carolina Petrovitis, a Lithuanian immigrant to Chicago, married and the mother of three, was in terrible pain following her abortion. Her friends called in Dr. Maurice Kahn. The doctor asked her, "Who did it for you[?]" He "coaxed" her to answer, then told her, "If you won[']t tell me what was done to you I can't handle your case." When Petrovitis finally revealed that a midwife had performed an abortion, Dr. Kahn called for an ambulance, sent her to a hospital, told the hospital physician of the situation, and suggested he "communicate with the Coroner's office." Three police officers soon arrived to question Petrovitis. With the permission of the hospital physician, Sgt. William E. O'Connor "instructed" an intern to "tell her she is going to die." The sergeant and another officer accompanied the intern to the woman's bedside.

As the doctor told Petrovitis of her impending death, she "started to cry—her eyes watered." Sure that Petrovitis realized she was about to die, the police then collected from her a "dying declaration" in which she named the midwife who performed her abortion, told where and when it was done and the price paid, and described the instruments used. Later the police brought in the midwife and asked Petrovitis "if this was the woman," and she nodded yes. A third police officer drew up another dying statement "covering the facts." He read that statement back to Petrovitis as she lay in bed "in pain, vomiting," and she made her "mark" on the statement. Then she died.[1]

Carolina Petrovitis's experience in 1916 provides an example of the standard medical and investigative procedures used in criminal abortion cases. This account, drawn from the Cook County coroner's inquest into Petrovitis's death, illustrates this essay's major themes: the state's interest in obtaining dying declarations in order to prosecute abortionists; the intimate ques-

LESLIE J. REAGAN is Associate Professor of History, Medical Humanities and Social Sciences Program, and Women's Studies at the University of Illinois, Urbana-Champaign.

Reprinted from *The Journal of American History* 77 (March 1991): 1240–64, by permission of the publisher.

[handwritten margin note: State focused not medical as regulator of abortion]

tioning endured by women during official investigations into abortion; and the ways in which physicians and hospitals served the state in collecting evidence in criminal abortion cases.

[handwritten margin note: abortions were not always a crime]

Abortion had not always been a crime. During the eighteenth and early nineteenth centuries, early abortions were legal under common law. Abortions were illegal only after quickening, the point at which a pregnant woman could feel the movements of the fetus (at approximately sixteen weeks' gestation). In the 1840s and 1850s abortion became commercialized and was increasingly used by married, white, native-born, Protestant women of the middle and upper classes. In 1857 the newly organized American Medical Association (AMA) initiated the ultimately successful crusade to make abortion illegal. Regular physicians, such as those who formed the AMA, were motivated to organize for the criminalization of abortion in part by their desire to win professional power, control medical practice, and restrict their irregular competitors, including homeopaths, midwives, and others. Hostility toward feminists, immigrants, and Catholics fueled the medical campaign against abortion and the passage of criminal abortion laws by state legislatures.[2] In response to the physicians' campaign, in 1867 Illinois criminalized abortion and in 1871 outlawed the sale of abortifacients (drugs used to induce abortions) without a prescription. The new statutes permitted only therapeutic abortions performed when pregnancy threatened a woman's life. By the end of the century every state had restricted abortion.[3]

[handwritten margin note: abortion became illegal in response to physician campaign]

[handwritten margin note: try to ban abortion by laws]

The history of the nineteenth-century criminalization of abortion has been well studied, as has the movement to decriminalize it in the 1960s and 1970s. We know very little, however, about the practice or control of abortion while it was illegal in the United States.[4]

This is the first study to address a crucial question in the history of reproduction: How did the state enforce the criminal abortion laws? I examine the methods of enforcement in Chicago from 1867, when Illinois made abortion illegal, to 1940, when changing conditions of abortion during the depression brought changes in the control of abortion. This study is based on research in legal records from the city of Chicago, Cook County, and Illinois and in the national medical literature. Coroners' inquests have been especially rich sources. The evidence shows continuity in the patterns of control followed by government officials in Chicago, and across the nation, for over half a century. In that period, in cases involving abortion, the state prosecuted chiefly abortionists, most often after a woman had died, and prosecutors relied for evidence on dying declarations collected from women near death due to their illegal abortions. Furthermore, the state focused on regulating the use of abortion by working-class women.[5]

The history of abortion reveals the complexity of the medical profession's role in sexual regulation. Feminists, writing primarily in the 1970s, have tended to portray the medical profession as intent on controlling female sexuality. My analysis of the regulation of abortion shifts the focus away from the medical profession and to the state as the regulator of sexual and reproductive behavior. To obtain evidence against abortionists the state needed to have physicians reporting abortions and collecting dying declarations from their patients, which many doctors were reluctant to do. Without doctors' cooperation police and prosecutors could do little to enforce the criminal abortion laws. Illinois law did not require doctors to report evidence of abortions. But by threatening physicians with prosecution, officials successfully pulled doctors into a partnership with the state in the suppression of abortion. In cases of illegal abortion, doctors were caught in the middle between their responsibilities to their patients and the demands of government officials.[6]

My analysis of the actual workings of lower-

[handwritten margin note: Needed doctors to report to state about abortions gone wrong]

[handwritten note at top: women, being dead after abortion were used as evidence/punishment to stop abortion. while men were punished by jail.]

level government agencies in the control of abortion changes our understanding of legal processes and punishment. Most historians of crime and punishment have focused on police and prisons, while historians of women and the law have focused mainly on marriage and property rights, rather than on crime. Few have studied law in practice. This essay analyzes the experience of ordinary people caught in criminal investigations, the routine procedures of the legal system, rather than analyzing the volume of cases that reached the courts or judicial rulings.[7] In abortion cases the investigative procedures themselves constituted a form of punishment and control. Although women were not arrested, prosecuted, or incarcerated for having abortions, the state nonetheless punished working-class women for having illegal abortions through official investigations and public exposure of their abortions. Recognizing the impact of the criminal abortion laws on women requires looking closely at the details of women's experiences, especially the interactions between women and their doctors and between women and police and petty state officials. Our understanding of what punishment is needs to be refined and redefined, particularly in cases of women who violate sexual norms, to include more subtle methods of disciplining individuals. The penalties imposed on women for having illegal abortions were not fines or jail sentences but humiliating interrogations about sexual matters by male officials—often conducted at women's deathbeds. During investigations of abortion police, coroner's officers, and prosecutors followed standard procedures in order to achieve the larger end of putting abortionists out of business. No evidence suggests that officials consciously designed their investigative procedures to harass women, yet the procedures were punitive, and this punishment became a central aspect of the state's efforts to control abortion. For government officials the procedures were

[handwritten note: Abortion investigation was punishment for women getting abortions]

routine; for women subjected to them the procedures were frightening and shameful once-in-a-lifetime events. Moreover, media attention to abortion deaths warned all women that those who strayed from marriage and motherhood would suffer death and shameful publicity.

This essay also highlights the gendered character both of legally enforced standards of behavior and of punishment. The criminalization of abortion not only prohibited abortion but demanded conformity to gender norms, which required men and women to marry, women to bear children, and men to bear the financial responsibility of children. Although most women who had abortions were married, state officials focused on unwed women and their partners. Coroners' inquests into the abortion-related deaths of unwed women reveal the state's interest in forcing working-class men to marry the women they had impregnated. Historians of sexuality have given little attention to the regulation of male heterosexuality, concentrating instead on the sexual control of women and "deviants." Yet in the late nineteenth and early twentieth centuries, I was surprised to find, the state punished unmarried working-class men whose lovers died after an abortion. The sexual double standard certainly existed, but the state imposed penalties on men, in certain unusual situations, when they failed to carry out their paternal obligation to marry their pregnant lovers and head a "nuclear" family. Unmarried men implicated in abortion deaths were, like women, punished through embarrassing questions about their sexual behavior; in general, the state punished men in more conventionally recognized ways: arresting, jailing, and prosecuting them.[8]

[handwritten note in right margin: states growing intervention in medical practice and family members]

The state's attempt to control abortion reflected a turn-of-the-century trend toward growing intervention by the state in medical practice as well as in sexual and family matters. The regulation of doctors by the state in abortion cases coincided with expanding governmental control of medicine through licensing laws and medical practice acts. The state's interest in enforcing marriage when premarital sex led to pregnancy reveals another area where it took over the functions of the male patriarch. The punishment of unmarried men for the abortion-related deaths of unmarried women may also have reflected the influence of feminist critiques of male sexual irresponsibility.[9]

Despite legal prohibitions, each year thousands of women around the country had abortions. In 1904 Dr. Charles Sumner Bacon estimated that "six to ten thousand abortions are induced in Chicago every year." Both midwives and physicians performed abortions in the early twentieth century, and many women induced their own abortions at home. At drugstores women could buy abortifacients and instruments, such as rubber catheters, to induce abortions. Most women survived their abortions, and most abortions remained hidden from state authorities. Yet the number of deaths following illegal abortions was significant. In the late 1920s a Children's Bureau study documented that at least 11 percent of deaths related to pregnancy and childbearing followed illegal abortion.[10]

Working-class women's poverty—in both wealth and health care—made it more likely that they, rather than middle-class women, would reach official attention for having abortions. All forty-four Cook County coroner's inquests that I have examined recorded investigations into the abortions and deaths of white working-class women. Over half of the women were immigrants or daughters of immigrants. Working-class women may have had more abortions than did middle-class women. In addition, poor women, lacking funds, often used inexpensive, and often dangerous, self-induced measures and delayed calling in doctors if they had complications. By the time poor women sought medical attention, they had often reached a crit-

ical stage and, as a result, had come to the attention of officials. Affluent women avoided official investigations into their abortions because they had personal relations with private physicians, many of whom never collected dying statements, destroyed such statements, or falsified death certificates. If necessary, wealthier families might be able to pressure or pay physicians, coroners, the police, and the press to keep quiet about a woman's abortion-related death.[11]

I found only one case in which the Cook County coroner investigated the abortion-related death of a black woman, in 1916, and, unfortunately, there is no record of the coroner's inquest into it. Despite a tremendous increase in Chicago's black population after World War I, I did not find any cases involving black women in the 1920s or 1930s. The paucity of information on the abortion-related deaths of black women may be an artifact of bias in the sources or may reflect the relatively small size of Chicago's black population. Black women who called in doctors or entered hospitals after their abortions would presumably have been questioned as Petrovitis was. They might have found it particularly upsetting to be questioned by white police officers or coroner's staff. A black physician prosecuted for abortion complained of racist treatment by the coroner's office; black women dying due to abortions, or their family members, may have suffered similar incidents.[12]

From the late nineteenth century through the 1930s the state prosecuted abortionists primarily after a woman died. Popular tolerance of abortion tempered enforcement of the laws. Prosecutors discovered early the difficulty of winning convictions in criminal abortion cases. Juries nullified the law and regularly acquitted abortionists. As a result, prosecutors concentrated on cases where they had a "victim"—a woman who had died at the hands of a criminal abortionist. In 1903 attorney H. H. Hawkins re-

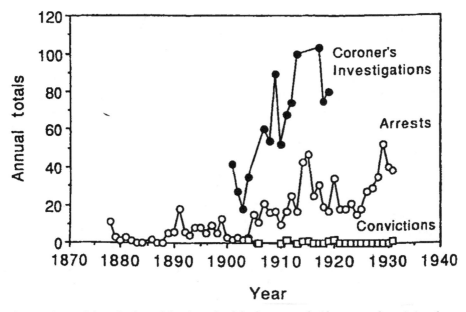

Fig. 14.1. Coroner's investigations of abortion-related deaths compared with arrests and convictions for abortions, Chicago, 1878–1931.

Sources: Chicago Department of Police, *Report of General Superintendent of Police of the City of Chicago,* 1878–1912 (Municipal Reference Library, Chicago, Ill.); Chicago Department of Police, *Annual Report,* 1913–1931, ibid.; Cook County Coroner, *Report,* 1907–1919, ibid.; "Chicago Medical Society. Regular Meeting, held Nov. 23, 1904," *Journal of the American Medical Association,* Dec. 17, 1904, p. 1890. Graph by D. W. Schneider.

from convictions of illegal abortion
to police arrest

viewed Colorado's record and concluded, "No one is prosecuted in Colorado for abortion except where death occurs. . . . The law only applies to the man who is so unskilful as to kill his patient." Thirty-seven out of the forty-three different abortion cases on which the Supreme Court of Illinois ruled between 1870 and 1940 involved a woman's death. Because prosecutors focused on abortionists responsible for abortion-related deaths, they relied for evidence on dying declarations, such as those obtained from Petrovitis, and coroner's inquests. In almost a third of the Illinois Supreme Court cases in which a woman had died because of an abortion, the opinions commented on dying declarations.[13]

Simply counting the convictions for abortion underestimates and obscures the state's significant effort to enforce the criminal abortion laws; analysis of the entire investigative process brings it to light. Police arrests for abortion and inquests into abortion deaths indicate a greater degree of interest in repressing abortion than suggested by the number of convictions (see Figure 14.1). Between 1902 and 1934 in Chicago the state's attorney's office prosecuted at most a handful of criminal abortion cases a year and never won more than one or two of them. In one ten-year period less than one-quarter of the prosecutions for murder by abortion resulted in a conviction. In contrast, police made at least ten arrests annually for abortion after 1905 and averaged twenty-five or twenty-six arrests annually during the 1910s and 1920s and almost forty

suspicion about the evidences collected to stop abortion

273

No care for dying patienc seen by the way they were interrigating the patience for a dying delaration on their death bed

rise in abortion arrests and inquests

arrests annually in the early 1930s. Police stepped up their arrests under political pressure. The coroner conducted even more inquests into abortion deaths every year. Between 1901 and 1919, the years for which figures are available, the Cook County coroner investigated an average of over 60 abortion deaths a year, from a low of 18 in 1903 to a high of 103 in 1917. Not all of those deaths followed criminal abortions. Some deaths followed "accidental," "self-induced," or "spontaneous" abortions or ones due to an "undetermined" cause, but because the coroner had to determine the cause of death, those deaths were investigated like criminal cases. Between 1905 and 1919 the coroner sent an average of twelve suspects a year to the grand jury. The level of legal action against abortion steadily increased, but the number of convictions did not change at all. While most of the rise in abortion arrests and inquests is probably explained by Chicago's growing population, the increase may also reflect intensified efforts by state officials to control abortion or an increase in police surveillance and control of the populace in general or of abortion specifically. Sorting out the role and relative importance of those factors in causing the rise in arrests and investigations is difficult, though perhaps possible with additional research.[14]

possible cause

Abortion investigations began, as in the Petrovitis case, when physicians or hospital staff members noticed "suspicious" cases and reported them to the police or the Cook County coroner. In the first stage of an investigation a woman was questioned by her doctor. She might be questioned again by police officers or special investigators sent from the coroner's office. Each interrogation was an attempt to obtain a legally valid dying declaration, in which the woman admitted her abortion and named her abortionist. A dying declaration not only led police to suspects; it also was crucial evidence that could be introduced at criminal trials. As one lawyer observed, it was almost "impossible" to obtain evidence of criminal abortion in any other way.[15]

The dying declaration was an unusual legal instrument that allowed the words of the dead to enter the courtroom. Legally, a dying declaration is an exception to the hearsay rule, which excludes the courtroom use of information that has been received secondhand. Common law allowed the admission of dying declarations as evidence in homicide cases, and states permitted this exception in abortion cases as well. Courts treated dying declarations as though given under oath, based on the common law assumption that a dying person would not lie since she was, as the coroner put it during the inquest on Petrovitis, "about to leave the worlds—to meet her maker." The exception allowed prosecutors to present the dying declaration in court as the dead woman's own accusation of the person who had killed her.[16]

If the woman died, the abortion investigation proceeded to a second stage: an autopsy performed by coroner's physicians and an official coroner's inquest into the woman's death. During the inquest the coroner or his deputy questioned witnesses and attempted to collect the facts in the case. There the police for the first time presented the dying statements they had collected, other information uncovered during their investigation, or individuals possessing information. Family members, lovers, friends, midwives, physicians, and hospital staff all testified at the inquests. A coroner's jury then deliberated on the proceedings and decided the cause of death. Although the legal purpose of an inquest was simply to determine the cause of death, the coroner in fact wielded significant power. The coroner's inquest was a highly important stage in the legal process since it generally determined whether anyone would be criminally prosecuted. The jury decided the guilt or innocence of various people involved in a case, and if the jury determined that the woman's

false sense of justice

Instead of going after all these illegal ibortionist, why not legalize abortion so it can be done in safe manner — state officials don't really care abort deaths they care about prosicuting those who have broken their laws death resulted from "murder by abortion," it ordered the police to hold the suspected abortionist and accomplices. The suspects remained in jail or out on bail until the case was concluded. Once the coroner's jury made its determination, prosecutors brought the case before the grand jury, which then indicted the suspects. Both prosecutors and the grand jury tended to follow the findings of the coroner's jury; if the coroner's jury failed to accuse anyone of criminal abortion, prosecutors generally dropped the case. Abortion cases did not come to trial exclusively after inquests into abortion-related deaths; some abortionists were caught during raids, and sometimes women testified in court against them, but most prosecutions for criminal abortion followed the death or injury of a woman.[17]

A visible and vocal minority of physicians actively assisted state officials in efforts to suppress abortion in Chicago. At various times members of the Chicago Medical Society's criminal abortion committee worked to remove abortion advertising from the pages of local newspapers, assisted coroner's investigations, and joined efforts to investigate and control midwives in order to end abortion in their city. The AMA, headquartered in Chicago, collected information on abortionists and abortifacients and shared it with local and federal officials investigating violations of the criminal law.[18]

Not all physicians wanted to help prosecutors bring abortionists to trial, however. Publicly, the leaders of the medical profession opposed abortion; privately, many physicians sympathized with women's need for abortions, performed abortions, or referred patients to midwives or physicians who performed them. Dr. Rudolph Holmes discovered during his work on the criminal abortion committee that physicians often declined to testify against abortionists. He reported that "so-called reputable members of our Chicago Medical Society regularly appear in court to support the testimony of some notorious abortionist." Furthermore, he concluded, "the public does not want, the profession does not want, the women in particular do not want any aggressive campaign against the crime of abortion." Dr. Charles H. Parkes, chairman of the Chicago Medical Society's criminal abortion committee in 1912, confessed that state authorities believed the society's members to be "apathetic in the extreme" regarding abortion. Some members were abortionists.[19]

The illegality of abortion made caring for patients who had had abortions not only a medical challenge but also a legal peril. In abortion cases physicians performed emergency curettements, repaired uterine tears and wounds, tried to stop hemorrhaging, and, most difficult in an age without antibiotics, fought infections. Once a woman had a widespread septic infection (characterized by chills and fever), it was very likely that she would die. If a woman died despite a doctor's efforts, he became a likely suspect in the criminal abortion case. According to the New York attorney Almuth C. Vandiver, police arrested physicians "simply because they were the last physician attending the patient and they had not made their report to the coroner."[20]

The state could not investigate abortion cases without medical cooperation; state officials won doctors' help by threatening them. Physicians learned that if they failed to report criminal abortion cases, the investigative process could be turned against them. At a 1900 meeting of the Illinois State Medical Society, Dr. O. B. Will of Peoria warned his associates of the "responsibilities and dangers" associated with abortion by relating his own "very annoying experience" when a patient died because of an abortion. He was indicted as an accessory to murder in an abortion case for "keeping the circumstances quiet, . . . not securing a dying statement from the patient, and . . . not informing the coroner." Will declared that he was not required to notify

if doc helped abortion patient and they die doc could be held responsible

witch hunt

doctors were for abortions, but ironically started the commiorialm of, [275] *[handwritten marginal note:] was away and every app along with you, you wanted doctor would refuse to attend to the woman if she didn't tell unmarried*

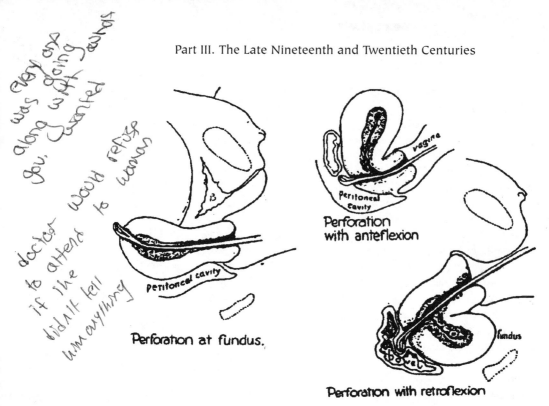

Perforation with anteflexion

Perforation at fundus.

Perforation with retroflexion

Fig. 14.2. Perforation of the uterus was a hazard of abortion: it could result from an induced abortion itself or from a physician's treatment following an abortion.

Illustration by Robert Latou Dickinson, in Frederick J. Taussig, *Abortion, Spontaneous and Induced: Medical and Social Aspects* (St. Louis, 1936), 231. *Courtesy C. V. Mosby Co.* *[handwritten:] The role of coroners*

the coroner and that the woman had refused to make a statement, but his story implied that co-operation with the authorities might help other doctors avoid similar notoriety. One doctor told his colleagues horror stories of Boston physicians who had been arrested, tried, and, though acquitted of abortion charges, nonetheless ousted by the Massachusetts Medical Society. Doctors who were associated with illegal abortion also risked losing their medical licenses. In Illinois a physician had to be convicted of abortion to lose his or her license, but some states revoked medical licenses without a trial. Physicians learned from tales like these that if they treated women for complications following abortions, they should report the cases to local officials or collect dying declarations themselves in order to avoid being arrested and prosecuted.[21]

Coroners' inquests into abortion deaths and the negative publicity coroners could cause helped enmesh doctors and hospitals in the enforcement system. At inquests into abortion-related deaths the Cook County coroner regularly reminded attending physicians of the unwritten "rule" (emanating from the law enforcers) requiring them to call the police or coroner whenever there was evidence that a woman had been "tampered with" and reprimanded those who failed to follow this policy. The fragile reputations of hospitals and physicians could be damaged if they were even named or associated with an abortion case in the newspapers. At a 1915 inquest into the death of a woman due to abortion, the coroner's jury suggested that Rhodes Avenue Hospital, which had cared for the woman, "be severely censured for lax methods in not complying with the rules re-

[handwritten: format of the death certificate to show unwed woman]

quired in notifying the proper officials . . . and the seeming indifference on the part of physicians and assistants . . . to ascertain[ing] who performed said abortion." The hospital's superintendent strongly objected to this censure and the notoriety the hospital subsequently received in Chicago's newspapers. Its superintendent wrote that the hospital had always cooperated with the coroner's office and that "the hospital was not on trial." Such publicized reproofs warned hospitals and physicians that if they failed to cooperate with state officials, their institutions and their individual careers could be hurt.[22]

By 1917, perhaps earlier, state authorities had persuaded Chicago's hospitals to pledge their cooperation in the investigation of abortion cases. The city's hospital superintendents reached an agreement with the coroner, chief of police, and state's attorney's office to notify the coroner's office when they saw patients who had had abortions. Furthermore, if it seemed that the woman would die before an official investigator arrived, hospitals agreed to collect the dying declaration themselves. The coroner even provided hospitals with a "blank form" for dying declarations. Although the official expectation that doctors would report abortion cases to the coroner was not codified, in the minds of both doctors and state officials, reporting abortions was, as one doctor described it in court, "compulsory."[23] *[handwritten: doctors being harrassed]*

A few New York physicians voiced the indignation that many doctors may have felt toward coroners and the treatment they received from them. Doctors who were the last attending physicians in abortion cases resented being pursued by the police and subjected to "disagreeable inquest[s]." New York physicians felt harassed by the city's coroners, who were, doctors complained, far too ready to arrest and investigate physicians in criminal abortion cases.[24]

One way to protect themselves from legal trouble and notoriety in abortion cases, physi-

cians learned, was to secure dying declarations. In 1912 the chairman of the Chicago Medical Society's criminal abortion committee, Dr. Parkes, reminded the society that "it is extremely easy for anyone to become criminally involved when connected with these cases, unless properly protected." Parkes presented to the medical society a model dying declaration drafted by State's Attorney John Wayman that would be legally admissible as evidence, that would "stand the supreme court test." Wayman advised the doctors to ask the dying woman the following questions: "Q. Do you believe that you are about to die? . . . Q. Have you any hope of recovery? . . . Q. Do you understand these questions fully? . . . Q. Are you able to give a clear account of the causes of your illness?"[25]

The state's attorney also provided a standardized format for the dying woman's answer. She should answer, "I am Miss ———. Believing that I am about to die, and having no hope of recovery, I make the following statement, while of sound mind and in full possession of my faculties." To be considered valid in court the statement had to establish that the woman believed she was near death. The woman was then expected to name her abortionist; to tell when, where, and how the abortion was done; and to name the man "responsible" for her pregnant "condition." Although most women who had abortions were married, the state's prosecutors focused on abortions by unwed women, and this formulaic dying declaration assumed that the dying woman would be unmarried.[26]

Physicians advised each other to deny medical care to a woman who had had an abortion until she made a statement. In 1902 the editors of the *Journal of the American Medical Association* endorsed this policy of (mis)treating abortion patients. The *Journal's* editorial quoted a physician who counseled his colleagues to "refuse all responsibility for the patient unless a confession exonerating him from any connection with the crime is given." Twenty years later Dr. Palmer

[handwritten: doctors mistreating female patients out of fear of getting in trouble with law]

Findley gave the same advice to obstetricians and gynecologists. "It is common experience," he reported, "that the patient will tell all she knows when made to realize her danger and a double purpose is attained—the physician in charge is protected and the guilty party is revealed."[27]

If a woman refused to give information, the smart doctor, according to these advisers, would walk out and refuse to attend her. And some physicians, like Dr. Kahn in the Petrovitis case, threatened to do just that. In 1916 a Chicago fireman called in Dr. G. P. Miller to attend his wife, who had been sick for three weeks following her abortion. Dr. Miller told her, "If I take this case . . . I want you to tell me the truth and who did it, who it was. Under the understanding that I was going to leave the house and have nothing to do with it, she told me the whole story." Physicians' refusals to treat abortion cases seem to have reached women's consciousness. In 1930 Mathilde Kleinschmidt, ill from her abortion, rejected her boyfriend's plan to call in a second doctor and insisted that he instead find the doctor who had performed the abortion. "Another doctor wont look at me," she explained. "He wont take the case."[28]

Fear of undeserved prosecution encouraged physicians to distrust their female patients. New York attorney Vandiver warned doctors, "Unscrupulous women and their accomplices have it within their power . . . to successfully blackmail the reputable practitioner, who omits the essential precautions for his protection." Dr. Henry Dawson Furniss told a story that encapsulated doctors' worst fears. He had "absolutely refused" to perform an abortion for a woman who later died from one. Under questioning, she got even with Furniss for spurning her plea by blaming him for her abortion.[29]

It seems that women rarely falsely accused physicians, and many, perhaps the majority, protected their abortionists by refusing to name

[handwritten left margin: lack of trust developed between physician and patient develop as a result of fear from prosecution]

them to doctors or policemen. The prosecuting attorney for St. Louis, Ernest F. Oakley, marveled at the loyalty of women who refused to reveal their abortionists' names. Dying declarations, he thought, were obtained in only "four out of ten cases." One New York woman, who was hospitalized following her abortion, told the doctors who pressed her to name her abortionist, "She was the only one who would help me, and I won't tell on her."[30] *[handwritten: women heath cares]*

Because the illegality of abortion compelled doctors to regard all miscarriages as suspect and to protect themselves against prosecution, women's health care suffered. Fearing prosecution, many physicians treated their female patients badly—rudely questioning them in attempts to gain dying declarations or delaying or refusing to provide needed medical care. For example, in 1915 when city, state, and federal officials began a "war" on abortion in Chicago following an abortion-related death, the *Chicago Tribune* discovered that "publicity has changed the attitude of hospital authorities in regard to the handling of abortion victims." One hospital superintendent would neither admit an abortion patient nor allow an operation "until he had received orders from the police." In this case the hospital allowed police officers to make medical and legal decisions as a result of a local antiabortion campaign. "It was not until detectives had assured the superintendent that an operation was necessary to save Mrs. Lapinski's life and that there would be no trouble for the hospital," the newspaper reported, "that Dr. G. M. Cushing . . . was permitted to take the sufferer to the operating room." In a less dramatic fashion doctors regularly avoided caring for abortion patients by sending them to hospitals instead. In 1929, when Dr. Julius Auerbach was called in to care for a woman who had had an abortion, he refused to examine her and sent her to the county hospital to avoid being "implicated" in the case.[31]

278

Took a stand against officials Gov. officials and state officials

Docs standing up against gov.

using medias to expose abortion dangers of abortion

While some frightened doctors threatened to deny medical care to women who had abortions and insisted that they make statements, others agreed with one physician who said that he refused "to act as a policeman" for the state. Dr. Parkes reported that Chicago officials "believe that the best hospitals now smother these cases and hinder in every way the work of investigation." Dr. William Robinson, a radical who advocated legalized abortion, scorned physicians who "badger[ed]" sick women to make dying statements. "The business of the doctor is to relieve pain, cure disease and save life," he declared, "not to act as a bloodhound [for] the state."[32] *Middle ground*

Other doctors found a middle ground between compliance with and rejection of the state's rules. Dr. Henry Kruse asked Edna Lamb about her abortion but did not inform her that she was about to die, omitting the explicit statement needed to make a dying declaration valid. He later explained, "We don't do that to patients because sometimes it is ver[y] discouraging and the result is bad." Other doctors questioned their patients about their abortions but only reported cases to authorities when women died. When women survived, the doctors destroyed their statements and kept the abortions confidential. In fact, the Cook County coroner accepted this practice as one that protected the interests of women who survived their abortions, shielded physicians from possible prosecution, and provided information to authorities in cases of death.[33]

Making a patient's medical history public undermined the private and personal relationships physicians had with their patients. Some physicians expressed their intention of maintaining patients' confidentiality, regardless of the wishes of authorities. Dr. Louis Frank of Louisville, Kentucky, commented on the issue in 1904, "If I was called in I would not give testimony compromising a young lady, and I would not put it

on record, no matter what the facts were, and I would not 'give away' a girl, but would attempt to protect her." Doctors Richard C. Norris and A. C. Morgan of Philadelphia strongly believed in the patient's right to confidentiality and proclaimed that medical ethics barred them from testifying against abortionists if such testimony violated their relationships with patients.[34]

The names of women who had had illegal abortions and the intimate details of their lives periodically hit the newspapers. Press coverage of abortion-related deaths warned all women of the dangers of abortion: death and publicity. Sometimes newspapers covered abortion stories on the front page and included photos; often abortion-related deaths and arrests of abortionists appeared in small announcements. The story of an unwed woman's seduction and abortion-related death made exciting copy and could dominate local newspapers for days. In 1916 Chicago and Denver newspapers published Ruth Merriweather's love letters to a Chicago medical student, who was on trial for his involvement in her abortion-related death. In 1918 the *Chicago Examiner* ran a series of "*tragedies*," excerpted from coroners' inquests, that told the stories of unwed women "*who were killed through illegal operations*." The articles in this series, and others like them, warned young women of the dangers of seduction and abortion and also warned rural fathers of the need to protect their daughters from the dangers of city life. The names and addresses of married women who had abortions often appeared in the press too, but their stories were not presented as seduction tales. Newspapers sometimes highlighted police officers' discovery of thousands of women's names in an abortionist's patient records. In doing so, they implicitly threatened women who had had abortions with the danger that they too could be named and exposed in the newspaper.[35]

Public exposure of a woman's abortion—

Media → Social Punishment for women + abortion

Press and media was working with gov. and state $ instead of against through the press or gossip—served as social punishment of women who had abortions and members of their families. A Chicago police officer recalled that when he questioned Mary Shelley, she "remark[ed] that she didn't want the statement in the newspapers." Some women whose abortions had been reported in the local press lived to face the shame of public exposure. Doctors observed that even when a woman died after an abortion, families did not want authorities to investigate because they wanted "to shield her reputation." Some families invited state investigation of abortions and pursued prosecution, yet even they may have resented publicity about the case. One mother whose daughter had died as a result of an illegal abortion cried at a public hearing that her whole family had "keenly felt the disgrace" of the crime. When Frances Collins died, police visited "all houses on both sides" of her home as well as "some ladies" in her old neighborhood and questioned them in hopes of finding a "woman confidant." The police failed to find any information, but they had informed the woman's entire community of her death by abortion and displayed the state's interest in controlling abortion.[36]

To the women whose abortions attracted the attention of medical and legal authorities, the demands of physicians and police for dying statements felt punitive. One woman described her hospital experience after an abortion as "very humiliating. The doctors put me through a regular jail examination." In their efforts to obtain dying declarations policemen and physicians, usually male, repeatedly questioned women about their private lives, their sexuality, and their abortions; they asked women when they last menstruated, when they went to the abortionist, and what he or she did. Were instruments introduced into "their privates"? If so, what did the instruments look like and how were they used? If the woman was unmarried, she was asked with whom she had been sexually

intimate and when, precisely the information that she may have hoped to conceal by having an abortion. Furthermore, as in the Petrovitis case, the police routinely brought the suspected abortionist to the bedside of the dying woman for her to identify and accuse. Hundreds of women who had abortions may have been questioned annually by physicians, police, or coroner's officers without their names ever entering official records because they survived their abortions.[37]

An investigation into a woman's abortion-related death was a shameful event for her relatives and friends because state officials required that they speak publicly about sexual matters that they ordinarily kept private and rarely discussed. At the Petrovitis inquest, police officer John A. Gallagher recalled that Petrovitis's sister had translated his questions, but when he reached the questions "about using instruments on her privates . . . the sister could not interpret anymore, didn't want to." To document a pregnancy and abortion the coroner asked questions about menstruation, sexual histories, and women's bodies. Family and friends often evaded such questions, but the coroner simply repeated his questions until he received an answer. Female witnesses, who may have discussed sexual topics only with other women, sometimes hesitated to speak before male officials, attorneys, and a jury of six men. At a 1917 inquest into an abortion death the dead woman's sister perjured herself during questioning. At a later trial she explained, "I knew everything but I could not answer all of them on account of all the men around. . . . Because there was so many men around I hated to talk about my poor sister more than I had to." The question she could not answer was, she explained, "about her body. . . . He asked me if the doctor had used any instrument, and I said no at that time." One immigrant woman commented during questioning by the coroner about her friend's abortion, "I am ashamed to tell." Despite her shame, she was forced to tell and to repeat her testimony at a criminal trial of the midwife-abortionist.[38]

The members of the coroner's staff understood that female family members often shared intimate knowledge about women's bodies and sexual behavior, and they tried to crack that female network to obtain information. The questioning of nineteen-year-old Julia McElroy at the 1928 inquest into the death of her sister Eunice McElroy is a vivid example of how the coroner's office expected sisters to have specific information, asked personal and shaming questions, and threatened those suspected of not cooperating. The deputy coroner questioned Julia intensively about the sisters' dating practices and Eunice's sexual behavior. He began with an offer to proceed in "a private chamber" because his questions might embarrass Julia, but he immediately denied the validity of her feelings in the legal arena. As he explained to Julia, "I may put some questions to you that you may think is embarrassing, but it is not. I am just merely questioning you because I am an officer of the law, the coroner." Julia denied knowing anything. The coroner established that the sisters shared a bed and asked, "When was the last time your sister had her last menstrual period? You tell the truth," he ordered. Julia told him that Eunice had menstruated over three months earlier and denied knowing of a pregnancy. "Well, you know they are supposed to come around and flow once a month unless a person is flowing irregularly? . . . you would know when your sister is unwell. She wears a napkin and you probably wear a napkin when you are not well on account of the odor, isn[']t that a fact?" Julia responded only with silence. The coroner continued to press her for information about Eunice's periods and sexual history. When Julia still maintained her ignorance a few days later, the deputy warned her, "If you don't tell a true story, you are going to get into a jam for a year." On the final day of the inquest, the deputy coroner told Julia that he knew that she had shielded the abortionist. With that, Julia revealed her knowledge and explained that she had been repeating the false story begun by her sister to protect her abortionist. The coroner concluded, "You are lucky you are telling the truth. I would sure send you to jail." The coroner eventually obtained from Julia McElroy evidence helpful in the prosecution of Eunice McElroy's abortionist, but he had also put Julia through a grueling experience, asked her graphic sexual questions, questioned her closely about her sister's and her own dating behavior, and threatened her with jail.[39]

Criminal abortion investigations reveal the importance of marriage—and especially of the lack of it—in the eyes of state officials. When police collected dying statements, they routinely asked about the woman's marital status, and at inquests into the deaths of unmarried women many of the coroner's questions focused on marital status. At inquests the coroner probed to discover whether the man had offered marriage. To a man, all claimed to have "promised to marry her." Perhaps men understood that this was the only way they could redeem themselves in the eyes of the law and the community. Yet there is evidence of genuine intention to marry. One man had bought a wedding ring; some even married after the abortion. In some cases the woman wanted to delay marriage; in others couples found marriage and children financially impossible. At the trial occasioned by the abortion-related death of his girl friend, William Cozzi testified "that he went to Dr. Rongetti to get rid of the baby because he could not afford it."[40]

Just as dying women endured intrusive questioning about their abortions, their unmarried lovers endured similar interrogations at inquests. For both unmarried women and men the official prying into their private sexual lives, and their own mortification, served as punishment for their illicit sexual behavior. The coro-

Men were interrogated as well

281

ner's questions to Marshall Hostetler about his sexual relationship with his sweetheart (who died in 1915) were not unusual at inquests into abortion deaths. The coroner asked Hostetler, "When did you become intimate with her? . . . have any relations with her? . . . When did that occur? . . . Had you been intimate with her before? . . . How many times? . . . Where did that occur?"[41]

During public investigations into abortion men too tried to avoid answering questions about sex. Charles Morehouse, for example, readily answered numerous questions about his girlfriend's abortion, how they borrowed money from an aunt, and how the family had tried to avoid an investigation. He also explained that the doctor had used "a spray." But when the coroner asked "Where?" Morehouse was silent. "A No response. Q What portion of the body? A Well, the privated parts."[42]

The "sweetheart of the dead girl" could be punished severely for having transgressed sexual norms. When an unwed woman died because of an abortion, her lover was automatically arrested, jailed, interrogated by the police and coroner and sometimes prosecuted as an accessory to the crime. Bob Berry's experience in 1931 was typical. When Alma Bromps died, policemen arrived at Berry's door, arrested him, and jailed him. The next morning he identified the body of his girlfriend and was questioned at the inquest into her death. He remained in jail for at least a week and ultimately became a witness for the state against the accused abortionist. Unmarried men involved in abortion deaths often spent at least one night in jail before the inquests. If they had no money to bail themselves out, they might spend several days or weeks, depending on the length of the inquest. Some spent months in jail waiting for their cases to come to trial. In 1917 Charles Morehouse spent four months in jail after the death of his girlfriend. The state prosecuted Patrick O'Connell,

a poor laborer, along with Dr. Adolph Buettner, for the abortion that led to Nellie Walsh's death. Although O'Connell was acquitted, it appears that he spent the nine months between Walsh's death and his criminal trial in the Cook County Jail. Other men were convicted and sentenced to prison for their part in an abortion.[43]

The actions of state officials toward unmarried men implicated in criminal abortion deaths reveal the state's stake in enforcing marriage in cases where an unwed woman became pregnant. The state punished young men for the moral offense of engaging in premarital intercourse and then failing to fulfill the implicit engagement by marrying the women they had made pregnant. Police routinely arrested and incarcerated unmarried men as accomplices in the crime of abortion, and the state's attorney sometimes prosecuted them. In contrast, husbands, who often had been just as involved as unmarried men in obtaining abortions, were very rarely arrested or prosecuted as accomplices when their wives died.[44] _Ironic_

Bastardy cases heard in Chicago also demonstrate the state's policy of coercing couples into marrying when pregnancy occurred. In bastardy cases the unwed woman brought the father to court to register paternity and to gain minimal financial support for the child. If the couple did not marry, the man could be fined up to $550 or sentenced to six months in jail. Of 163 bastardy cases studied by Hull House leader Louise DeKoven Bowen in 1914, a third ended in marriage. Fourteen couples "settled out of court" by marrying, while forty "married in court." A man who reconsidered his situation once in jail could gain his freedom if he decided to marry the woman and legitimate their child.[45]

The official response to unmarried men in abortion cases, as in bastardy cases, warned other young men of the dangerous consequences of avoiding marriage and children when pregnancy occurred. The newspaper story

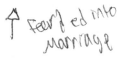
↑ forced into marriage

of an abortion-related death often told of the arrest, imprisonment, and interrogation of the "sweetheart of the dead girl," and young men probably traded detailed information about the events that transpired during abortion investigations. Newspaper coverage of abortions warned women that they could die and men that they could be thrown in jail—some may have concluded that it was better to marry.[46]

Jilted women could exploit the state's readiness to hold unmarried men accountable in illegal abortions as a weapon to strike back at their lovers. Alice Grimes of southern Illinois actively encouraged official investigation into her abortion for this reason. As she was dying in 1896, Grimes told her mother that her boyfriend, James Dunn, "ought to suffer some" as she had. When Grimes learned that her uncle had had Dunn arrested, she told her brother she was "glad of it."[47]

Conclusion

From the late nineteenth century through the 1930s the state concentrated on collecting dying declarations and on prosecuting abortionists when a woman had died, but the experiences of the 1930s changed the state's methods of abortion control. During the depression abortion increased just when advances in medicine were making it possible to save the lives of women who, in earlier decades, would have died from their abortion-related injuries and infections. The changes in the conditions and practice of abortion in the 1930s presaged changes in the investigation and control of abortion in the 1940s. Abortion control in the 1940s took two forms. First, hospitals took over much of the control of abortion through newly created therapeutic abortion committees; second, police and prosecutors stepped up raids on abortionists' offices. Rather than waiting for a death, the state's attorney's office sent police to raid the offices of suspected abortionists where they arrested abortionists and patients and collected

medical instruments and patient records. In criminal trials of accused abortionists the prosecution relied for evidence on testimony from women who had been patients of the abortionist, rather than on dying declarations. In the 1940s the system changed, but the process remained punitive for the women caught in it; criminal trials of abortionists required public exposure of women's sexual histories and abortions.[48]

The state's control of abortion was by no means entirely successful, but neither was it insignificant. Thousands of women regularly defied the law and had the abortions they needed. When questioned by doctors or police about their abortions, many defied their interrogators and refused to provide the information needed to prosecute abortionists. Yet the state punished women for having abortions, damaged the relationships between women and their doctors, and undermined women's health care. Officials focused on regulating the sexual behavior of working-class women and men and especially the unmarried. Investigations and inquests into abortions forcefully reminded all involved not only of the illegality of abortion but also of the power of the state to intervene in the private lives of ordinary people in order to prevent and punish violations of sexual codes that demanded marriage and maternity.

Medicine played a complicated role in the enforcement of the criminal abortion laws. Physicians both participated in the suppression of abortion and helped women obtain abortions. Organized medicine acted as part of the state in policing the practice of abortion by women *and* by physicians. While some physicians actively sought an alliance with the state in enforcing the criminal abortion laws, most physicians who cooperated with the state's investigations did so out of their very real fears of being arrested as suspects in abortion cases. By arming themselves with dying declarations naming others as

A

lesson of why abortion shouldn't be criminalized

lesson - learn from our mistakes

gendered punishment

the abortionists, doctors could avoid prosecution and help the state prosecute the real abortionists. The agreement in Chicago to notify the coroner of abortions became a "law," in a sense. Physicians who did not comply with the informal regulations were treated with suspicion by both their colleagues and state officials.[49]

This study of the investigation of abortion in Chicago calls attention to the social punishment inherent in the state's routine processes of investigation and illuminates the ways in which punishment has been gendered. The state may not have prosecuted women for having abortions, but it did punish women through persistent questioning by doctors and police and through public exposure of their abortions. The harassment of sick or dying women in the name of criminal investigation continued until the decriminalization of abortion.[50]

At inquests into abortion deaths the state reinforced the norms requiring men to marry the women they had made pregnant. Through arrests, incarceration, interrogation, and prosecution, unmarried women's lovers were punished for illegal abortions as well as for their illicit sexual behavior. The treatment of unmarried men in these cases reveals the implicit assumption of state authorities that the unwed women who had abortions had been forced to do so because their "sweethearts" had refused to marry. This underlying assumption ignored the evident agency of many women who sought abortions and delayed marriage. The punishment of unmarried men maintained age-old patriarchal standards that gave community support to fathers when they forced men to marry the women they had impregnated.

At a time when abortion is a political issue of national importance, the history of the enforcement of the criminal abortion laws should serve as a warning against recriminalizing abortion. If abortion is made illegal again, we can expect that the punitive procedures of the past will be revived. The antifeminist movement, which is pressuring the state to recriminalize abortion, will pressure the state to punish women if abortion is made illegal. We can expect that women will once again besiege doctors with requests for abortions and that the state will threaten to prosecute physicians who fail to report women who have, or perhaps even seek, abortions. As in the past, doctors are likely to be roped into assisting the state by interrogating and reporting women who have had abortions. Some women, like Carolina Petrovitis, will be injured or die from abortions induced by themselves or by inept practitioners. Many more may be interrogated, captured by police during raids of abortionists' offices, or publicly exposed. Today, as in the past, enforcement of any criminal abortion law will target the most powerless groups—poor and working-class women, women of color, and teen women—and their health care will be harmed the most. The history of illegal abortion is a history that should not be repeated.

Notes

1. Inquest on Carolina Petrovitis, March 21, 1916, case 234–3–1916, Medical Records Department (Cook County Medical Examiner's Office, Chicago, Ill.). For another example of a physician closely questioning a woman about an abortion, see Inquest on Matilda Olson, April 30, 1918, case 289–4–1918, Medical Records Department.

2. My summary of the history of abortion in the eighteenth and nineteenth centuries is based primarily on the pathbreaking study, James C. Mohr, *Abortion in America: The Origins and Evolution of National Policy* (New York, 1978). See also Linda Gordon, *Woman's Body, Woman's Right: A Social History of Birth Control in America* (New York, 1977), 49–61; Rosalind Pollack Petchesky, *Abortion and Woman's Choice: The State, Sexuality, and Reproductive Freedom* (New York, 1984), 67–100; Carroll Smith-Rosenberg, *Disorderly Conduct: Visions of Gender in Victorian*

REAGAN/"About to Meet Her Maker"

America (New York, 1985), 217–44; Michael Grossberg, *Governing the Hearth: Law and the Family in Nineteenth-Century America* (Chapel Hill, 1985), 155–95; Carl N. Degler, *At Odds: Women and the Family in America from the Revolution to the Present* (New York, 1980), 227–48; and Clifford Browder, *The Wickedest Woman in New York: Madame Restell, the Abortionist* (Hamden, 1988). On midwifery see Frances E. Kobrin, "The American Midwife Controversy: A Crisis of Professionalization," in *Women and Health in America: Historical Readings,* ed. Judith Walzer Leavitt (Madison, 1984), 318–26; Judy Barrett Litoff, *American Midwives: 1860 to the Present* (Westport, 1978); Jane Pacht Brickman, "Public Health, Midwives, and Nurses, 1880–1930," in *Nursing History: New Perspectives, New Possibilities,* ed. Ellen Condliffe Lagemann (New York, 1983), 65–88; Eugene R. Declerq, "The Nature and Style of Practice of Immigrant Midwives in Early Twentieth Century Massachusetts," *Journal of Social History* 19 (Fall 1985): 113–29; and Charlotte G. Borst, "Catching Babies: The Change from Midwife to Physician-Attended Childbirth in Wisconsin, 1870–1930" (Ph.D. diss., University of Wisconsin, Madison, 1989). On the role nativism and anti-Catholicism played in the campaign against abortion, see Mohr, *Abortion in America,* 90–91, 166–67, 180, 207–8. On the antifeminism of the leader of the campaign against abortion, Horatio Storer, see Mohr, *Abortion in America,* 168–69; and on his campaign against female physicians see Mary Roth Walsh, *"Doctors Wanted: No Women Need Apply": Sexual Barriers in the Medical Profession, 1835–1975* (New Haven, 1977), 109–13. On the antifeminist character of the anti-abortion movement see Mohr, *Abortion in America,* 94–95, 102–18, 188, 216; and Petchesky, *Abortion and Woman's Choice,* 82–83. Feminists challenged the sexual order by advocating voluntary motherhood, but they did not advocate contraceptives or abortions. Mohr, *Abortion in America,* 111–13; and Gordon, *Woman's Body, Woman's Right,* 108.

3. Mohr, *Abortion in America,* 205–260, 325. The Illinois statute, passed in 1867, read, "If any person shall, by means of any instrument or instruments, or any other means whatever, cause any pregnant woman to miscarry, or shall attempt to procure or produce such miscarriage, the person so offending

shall be deemed guilty of a high misdemeanor, and, upon conviction thereof, shall be confined in the penitentiary for a period not less than two nor more than ten years. 2. If any person shall, in the attempt to produce the miscarriage of a pregnant woman, thereby cause and produce the death of such woman, the person so offending shall be deemed guilty of murder, and shall be punished as the law requires for such offense. 3. The provisions of this act shall not apply to any person who procures or attempts to produce the miscarriage of any pregnant woman for *bona fide* medical or surgical purposes." 1867 Ill. Laws 89. The 1872 law read: "No druggist, dealer in medicines, or any other person in this state, shall sell to any person or persons any drug or medicine known or presumed to be ecbolic or abortifacient, except upon the written prescription of some well known and respectable practicing physician. . . . Any person or persons violating any of the provisions of this act, shall, upon conviction thereof, be punished by a fine of not less than fifty nor more than five hundred dollars, or by imprisonment in the county jail for not less than thirty days or more than six months, for each and every offense, or by both." 1872 Ill. Laws 369.

4. On illegal abortion see Petchesky, *Abortion and Woman's Choice,* 101–37; Kristin Luker, *Abortion and the Politics of Motherhood* (Berkeley, 1984), 40–65, 101–108; James Mohr, "Patterns of Abortion and the Response of American Physicians, 1790–1930," in *Women and Health in America,* ed. Leavitt, 117–23; James Mohr, "Iowa's Abortion Battles of the Late 1960s and Early 1970s: Long-Term Perspectives and Short-Term Analyses," *Annals of Iowa* 50 (Summer 1989): 67–70; and Linda Gordon, *Woman's Body, Woman's Right: A Social History of Birth Control in America* (New York, 1990), 406–11. On the movement to decriminalize abortion see Luker, *Abortion and the Politics of Motherhood,* 66–157; Eva R. Rubin, *Abortion, Politics, and the Courts: Roe v. Wade and Its Aftermath* (New York, 1987); Marian Faux, *Roe v. Wade: The Untold Story of the Landmark Supreme Court Decision That Made Abortion Legal* (New York, 1988); Angela Y. Davis, *Women, Race, and Class* (New York, 1981), 202–21; and Pauline B. Bart, "Seizing the Means of Reproduction: An Illegal Feminist Abortion Collective—How and Why it Worked," *Qualitative Sociology* 10 (Winter 1987): 339–57.

5. I have examined forty-four Cook County coroner's inquests into abortion deaths between 1907 and 1937, which can be found in the Cook County Medical Examiner's Office. Cook County coroner's inquests can only be located with a name and date of death; they cannot be located by topic. I found the names of women who had died following abortions through two sources: in published Illinois Supreme Court opinions on abortion and in the large collection of clippings from Chicago newspapers in the Abortionists Files in the Historical Health Fraud Collection (American Medical Association, Chicago, Ill.). Two of the inquests were found in transcripts of criminal abortion trials: Inquest on Nellie Walsh, 1907, in Transcript of *People v. Buettner*, 233 Ill. 272 (1908), Case Files, vault no. 30876, Supreme Court of Illinois, Record Series 901 (Illinois State Archives, Springfield, Ill.); and Inquest on Lena Benes, 1919, in Transcript of *People v. Heisler*, 300 Ill. 98 (1921), Case Files, vault no. 39077, Supreme Court of Illinois. I have also examined all of the Illinois Supreme Court opinions on abortion cases (45) from 1867 to 1940, twenty selected transcripts of criminal abortion trials in Illinois criminal and circuit courts that were appealed to the Illinois Supreme Court (held at the Illinois State Archives), and Chicago, Cook County, and Illinois government reports for the same period at the Municipal Reference Library, Chicago, Illinois. For studies showing that the state policed the working class more than the middle class, see Lawrence M. Friedman and Robert V. Percival, *The Roots of Justice: Crime and Punishment in Alameda County, California, 1870–1910* (Chapel Hill, 1981), 111–16, 310; and Sidney L. Harring, *Policing a Class Society: The Experience of American Cities, 1865–1915* (New Brunswick, 1983).

6. Ann Douglas Wood, "'The Fashionable Diseases': Women's Complaints and Their Treatment in Nineteenth-Century America," in *Women and Health in America*, ed. Leavitt, 222–38. For a critical response to Wood see Regina Markell Morantz, "The Perils of Feminist History," *Women and Health in America*, 239–45; Adrienne Rich, *Of Woman Born: Motherhood as Experience and Institution* (New York, 1976); Barbara Ehrenreich and Dierdre English, *For Her Own Good: 150 Years of the Experts' Advice to Women* (Garden City, 1978). For a different view see Judith Walzer

Leavitt, *Brought to Bed: Childbearing in America, 1750–1950* (New York, 1986).

7. On women and crime see D. Kelly Weisberg, ed., *Women and the Law: A Social Historical Perspective*, vol. I: *Women and the Criminal Law* (Cambridge, Mass., 1982). On the state's regulation of women's sexuality and sexual crimes, see Mary Ellen Odem, "Delinquent Daughters: The Sexual Regulation of Female Minors in the United States, 1880–1920" (Ph.D. diss., University of California, Berkeley, 1989). On the law in practice see Lawrence M. Friedman, *A History of American Law* (New York, 1985), 17, 584, 575; Marylynn Salmon, *Women and the Law of Property in Early America* (Chapel Hill, 1986), xi–xii; and James Willard Hurst, *The Growth of American Law: The Law Makers* (Boston, 1950).

8. Minnie C. T. Love, "Criminal Abortion," *Colorado Medicine* 1 (1903–1904): 57; Paul Titus, "A Statistical Study of a Series of Abortions Occurring in the Obstetrical Department of the Johns Hopkins Hospital," *American Journal of Obstetrics and Diseases of Women and Children* 65 (June 1912): 979; Raymond M. Spivy, "The Control and Treatment of Criminal Abortion," *Journal of the Missouri State Medical Association* 15 (Jan. 1918): 3; Frederick J. Taussig, *Abortion, Spontaneous and Induced: Medical and Social Aspects* (St. Louis, 1936), 391; Endre K. Brunner and Louis Newton, "Abortions in Relation to Viable Births in 10,609 Pregnancies," *American Journal of Obstetrics and Gynecology* 38 (July 1939): 88. In Minneapolis between 1927 and 1936, 57 percent of 109 women who died because of criminal abortion were married, according to Calvin Schmid, *Social Saga of Two Cities: An Ecological and Statistical Study of Social Trends in Minneapolis and St. Paul* (Minneapolis, 1937), 410–11. My thanks to Elizabeth Lockwood for sharing Schmid's work with me. Martha Vicinus, "Sexuality and Power: A Review of Current Work in the History of Sexuality," *Feminist Studies* 8 (Spring 1982): 133–56. On the control of male heterosexuality see Allan M. Brandt, *No Magic Bullet: A Social History of Venereal Disease in the United States Since 1880* (New York, 1987), 61–64, 66–70; G. J. Barker-Benfield, *The Horrors of the Half-Known Life: Male Attitudes Toward Women and Sexuality in Nineteenth-Century America* (New York, 1976); and

Odem, "Delinquent Daughters." For an overview of the history of sexuality and collections of recent work, see John D'Emilio and Estelle B. Freedman, *Intimate Matters: A History of Sexuality in America* (New York, 1988); and Kathy Peiss and Christina Simmons with Robert A. Padgug, eds., *Passion and Power: Sexuality in History* (Philadelphia, 1989). On prostitution see Ruth Rosen, *The Lost Sisterhood: Prostitution in America, 1900–1918* (Baltimore, 1982). On lesbians and gay men see Lillian Faderman, *Surpassing the Love of Men: Romantic Friendships and Love Between Women from the Renaissance to the Present* (New York, 1981); John D'Emilio, *Sexual Politics, Sexual Communities: The Making of a Homosexual Minority in the United States, 1940–1970* (Chicago, 1983); and Allan Berubé, *Coming Out Under Fire: The History of Gay Men and Women in World War Two* (New York, 1990).

9. Paul Starr, *The Social Transformation of Medicine: The Rise of a Sovereign Profession and the Making of a Vast Industry* (New York, 1982), 102–12, 118, 184–97; William G. Rothstein, *American Physicians in the Nineteenth Century: From Sects to Science* (Baltimore, 1972), 305–12; Richard Harrison Shryock, *Medical Licensing in America, 1650–1965* (Baltimore, 1967). Michael Grossberg found that the judiciary dominated nineteenth-century family law and claimed patriarchal authority over domestic relations. Grossberg, *Governing the Hearth*, 289–307. On the state's intervention in family violence and a critique of the social control model, see Linda Gordon, "Child Abuse, Gender, and the Myth of Family Independence: Thoughts on the History of Family Violence and Its Social Control, 1880–1920," *Review of Law and Social Change* 12 (1983–1984): 523–37. Nineteenth- and early twentieth-century feminists argued that men forced women to have abortions, either by forcing unwanted sex upon them or by refusing to marry their pregnant lovers. See for example, *Chicago Times*, Jan. 5, 1889, p. 5; and Smith-Rosenberg, *Disorderly Conduct*, 243–44. Gordon, *Woman's Body, Woman's Right* (1977), 95–115.

10. For the estimate by Dr. C. S. Bacon see "Chicago Medical Society. Regular Meeting, Held Nov. 23, 1904," *Journal of the American Medical Association* (hereafter *JAMA*) (Dec. 17, 1904): 1889. On self-induced abortions see Maximilian Herzog, "The Pathology of Criminal Abortion," *JAMA* (May 26, 1900): 1310; J. E. Lackner, "Serological Findings in 100 Cases, Bacteriological Findings in 50 Cases, and a Résumé of 679 Cases of Abortion at the Michael Reese Hospital," *Surgery, Gynecology, and Obstetrics* 20 (May 1915): 537; G. D. Royston, "A Statistical Study of the Causes of Abortion," *American Journal of Obstetrics and Diseases of Women and Children* 76 (Oct. 1917): 572–73; and Schmid, *Social Saga of Two Cities*, 411. On rubber catheters see Schmid, *Social Saga of Two Cities*; and Nelson M. Percy, "Partial Evisceration Through Vagina During Attempted Abortion," *Surgical Clinics of Chicago* 1 (1917): 979. On abortifacients and advertising see "Chicago Medical Society," 1891. The Children's Bureau study analyzed maternal deaths during 1927 and 1928 in fifteen states. Some abortion-related deaths could have been prevented, the study suggested, if physicians had provided better care. U.S. Department of Labor, Children's Bureau, *Maternal Mortality in Fifteen States* (Washington, 1934), 113, 115. In 1936 Dr. Frederick J. Taussig estimated that there were at least 681,000 abortions per year in the United States. Taussig revised his previous estimates of abortion mortality downward because he believed that the average abortionist's skills and equipment had improved. He estimated a mortality rate of 1.2 percent and 8,000 to 10,000 deaths due to abortions per year. Taussig, *Abortion*, 26–28.

11. On the number of working-class versus middle-class women having abortions, see Regine K. Stix and Dorothy G. Wiehl, "Abortion and the Public Health," *American Journal of Public Health* 28 (May 1938): 624; and Paul H. Gebhard et al., *Pregnancy, Birth, and Abortion* (New York, 1958), 109–10, 120, 139, 194–95, 198; Mohr, "Patterns of Abortion and the Response of American Physicians," 122. On delay in seeking medical treatment see James R. Reinberger and Percy B. Russell, "The Conservative Treatment of Abortion," *JAMA* (Nov. 7, 1936): 1530; and J. D. Dowling, "Points of Interest in a Survey of Maternal Mortality," *American Journal of Public Health* 27 (Aug. 1937): 804. On the high level of complications and fatalities associated with self-induced abortions, see Regine K. Stix, "A Study of Pregnancy Wastage," *Milbank Memorial Fund Quarterly* 13 (Oct. 1935): 362–63; Raymond E. Watkins, "A Five-Year Study of

Abortion," *American Journal of Obstetrics and Gynecology* 26 (Aug. 1933): 162; and Reinberger and Russell, "Conservative Treatment of Abortion," 1527. One doctor reported an abortion-related death as pneumonia. William J. Robinson, *The Law Against Abortion: Its Perniciousness Demonstrated and Its Repeal Demanded* (New York, 1933), 38–39. See also Children's Bureau, *Maternal Mortality in Fifteen States,* 103–4; and Ransom S. Hooker, *Maternal Mortality in New York City: A Study of All Puerperal Deaths, 1930–1932* (New York, 1933), 52. I have not found an incident of families bribing officials, but on the corruption of police and coroners see Mark H. Haller, "Historical Roots of Police Behavior: Chicago, 1890–1925," *Law and Society Review* 10 (Winter 1976): 306–7, 311, 316–17; and Julie Johnson, "The Politics of Death: The Philadelphia Coroner's Office, 1900–1956," paper delivered at the annual meeting of the American Association for the History of Medicine, Baltimore, May 1990, pp. 6, 7, 17 (in Leslie J. Reagan's possession).

12.   I requested the Inquest on Flossie Emerson, who died Feb. 28, 1916, but the Cook County Medical Examiner's Office has no record of her death. Emerson's abortion-related death was one of the cases for which Dr. Anna B. Schultz-Knighten was prosecuted, *People v. Schultz-Knighten,* 277 Ill. 238 (1917). Dr. Schultz-Knighten complained that the coroner's physician, Dr. Springer, had "sneered" at her and called her a "nigger." Abstract of Record, 117, *People v. Schultz-Knighten,* 277 Ill. 238 (1917), Case Files, Supreme Court of Illinois. Allan H. Spear, *Black Chicago: The Making of a Negro Ghetto* (Chicago, 1967), 140–46, 223. On black women's use of abortion see Jessie M. Rodrique, "The Black Community and the Birth Control Movement," in *Passion and Power,* ed. Peiss and Simmons with Padgug, 140–41; Titus, "A Statistical Study of a Series of Abortions Occurring in the Obstetrical Department of the Johns Hopkins Hospital," 973; Raymond Pearl, "Fertility and Contraception in New York and Chicago," *JAMA* (April 24, 1937): 1389; Brunner and Newton, "Abortions in Relation to Viable Births in 10,609 Pregnancies," 83; and Thordis Simonsen, ed., *You May Plow Here: The Narrative of Sara Brooks* (New York, 1986), 176, 177. My thanks to Susan Smith for bringing this book to my attention.

13.   "Symposium. Criminal Abortion. The Colorado Law on Abortion," *JAMA* (April 18, 1903): 1099; James Foster Scott, "Criminal Abortion," *American Journal of Obstetrics and Diseases of Women and Children* 33 (Jan. 1896): 77; Wilhelm Becker, "The Medical, Ethical, and Forensic Aspects of Fatal Criminal Abortion," *Wisconsin Medical Journal* 7 (April 1909): 633; Taussig, *Abortion,* 402; Gordon, *Woman's Body, Woman's Right* (1977), 57; Mohr, "Patterns of Abortion and the Response of American Physicians," 120–21; Roger Lane, *Violent Death in the City: Suicide, Accident, and Murder in Nineteenth-Century Philadelphia* (Cambridge, Mass., 1979), 93. Dying declarations may have been introduced or discussed in additional cases without having been addressed in the Illinois Supreme Court opinions. Furthermore, many dying declarations were collected but never used in court. For examples of Illinois Supreme Court discussion of dying declarations see *Dunn v. People,* 172 Ill. 582, 587–91 (1898); *People v. Huff,* 339 Ill. 328, 332–33 (1930); and *People v. Holmes,* 369 Ill. 624, 624–26 (1938). For nineteenth-century state supreme court opinions on the use of dying declarations in abortion cases from other states, see Grossberg, *Governing the Hearth,* 363–64.

14.   [Thomas E. Harris], "A Functional Study of Existing Abortion Laws," *Columbia Law Review* 35 (Jan. 1935): 91n17. The peak years for arrests for abortion reflect organized efforts against abortion. For example, the peak of abortion arrests in 1912 occurred at the same time as coordinated nationwide raids of people who used the mails to give information about, or to arrange the sale of, birth control or abortion instruments or services. "Fight Race Suicide in Raids All Over U.S.," *Chicago [News],* Nov. 20, 1912, Abortionists Files, Historical Health Fraud Collection; "Take Chicagoans in Federal War on Race Suicide," *Chicago Tribune,* Nov. 21, 1912, Abortionists Files. The peaks in 1914–1915 coincided with local and state investigations as well as newspaper exposés of abortions in Chicago. Chicago Department of Health, *Report,* 1911–1918, 1:1055–56 (Municipal Reference Library, Chicago, Ill.); resolution presented by Alderman Murray, June 1, 1915, Chicago, *Council Proceedings, 1916–1917,* 459, Municipal Reference Library. For numerous newspaper articles during these years see the Abortionists Files of the

Historical Health Fraud Collection. For examples of investigations of noncriminal abortions see Inquest on Milda Hoffmann, May 29, 1916, case 342–5–1916, Medical Records Department; and Inquest on Degma Felicelli, Oct. 11, 1916, case 224–10–1916, Medical Records Department. Cook County Coroner, *Biennial Report,* 1918–1919 (Municipal Reference Library).

15. William Durfor English, "Evidence—Dying Declaration—Preliminary Questions of Fact—Degree of Proof," *Boston University Law Review* 15 (April 1935): 382.

16. Ibid., 381–82; Inquest on Petrovitis, case 234–3–1916, Medical Records Department. One legal scholar described the dying declaration exception to the hearsay rule: "They are declarations made in extremity, when the party is at the point of death, and when every hope of this world is gone; when every motive to falsehood is silenced, and the mind is induced, by the most powerful considerations, to speak the truth. A situation so solemn and so awful is considered by the law as creating an obligation equal to that which is imposed by a positive oath in a court of justice." Simon Greenleaf, *A Treatise on the Law of Evidence,* vol. 1 (Boston, 1899), 245. About judicial rulings on this exception to the hearsay rule, see John Henry Wigmore, *A Treatise on the System of Evidence in Trials at Common Law, Including Statutes and Judicial Decisions of All Jurisdictions of the United States* (Boston, 1904), 1798–1819.

17. "Murder by abortion" was the standard phrase used in the coroner's jury's verdicts and grand jury indictments. See, for example, Inquest on Rosie Kawera, June 15, 1916, case 152–5–1916, Medical Records Department; Inquest on Augusta Bloom, March 4, 1916, case 176–3–1916, Medical Records Department; *People v. Dennis,* 246 Ill. 559, 560–61 (1910); and *People v. Heisler,* 300 Ill. at 100. John H. Wigmore, ed., *Illinois Crime Survey* (Chicago, 1929), 195, 596. For examples of raids see "Arrest Man as Quack Doctor," *Waukegan* [Illinois] *Sun,* Nov. 14, 1932, Abortionists Files, Historical Health Fraud Collection; and "Troopers Smash 'Stork' Ring," *Bridgeport* [Connecticut] *Herald,* Aug. 23, 1936, Abortionists Files. For examples of cases in which women who had abortions testified see *Baker v. People,* 105 Ill. 452 (1883); *People v. Patrick,* 277 Ill. 210 (1917); *People v.*

Pigatti, 314 Ill. 626 (1924); *People v. Wyherk,* 347 Ill. 28 (1931). In 1904 Dr. C. S. Bacon remarked, "There are practically no accusations nor indictments for abortion unless the mother becomes seriously ill or dies." "Chicago Medical Society," 1889. Roger Lane finds that in Philadelphia's circuit court most abortion cases did not follow the death of a woman but that women testified as a "result of the damage done." On the nineteenth century, see Lane, *Violent Death in the City,* 93. On the twentieth, see Roger Lane to Leslie J. Reagan, May 31, 1989 (in Reagan's possession).

18. Chicago Medical Society, Council Minutes, [Oct. 1905–July 1907], meetings of Oct. 9, 1906 and May 14, 1907, Chicago Medical Society Records (Archives and Manuscripts Department, Chicago Historical Society, Chicago, Ill). Chicago Medical Society, Council Minutes, [Oct. 1911–June 1912], meeting of Jan. 9, 1912, pp. 51, 55–56, Chicago Historical Society. On the Chicago Medical Society and midwives see Rudolph W. Holmes et al., "The Midwives of Chicago," *JAMA* (April 25, 1908): 1346–50; "Abortion Lairs Facing Cleanup by Authorities," *Chicago Herald,* May 30, 1915, Abortionists Files, Historical Health Fraud Collection; Mohr, "Patterns of Abortion and the Response of American Physicians," 119–20; Arthur J. Cramp, "The Bureau of Investigation of the American Medical Association," *American Journal of Police Science* 2 (July–Aug. 1931): 286–87. For numerous examples of letters with information about abortionists or abortifacients sent to the post office authorities, see Abortifacient Files, Historical Health Fraud Collection. Between 1934 and 1940 the bureau of investigation worked closely with California officials and medical societies in an investigation of a Pacific Coast abortion ring. Bureau of Investigation File, Historical Health Fraud Collection.

19. R. W. Holmes, commenting on Walter B. Dorsett, "Criminal Abortion in Its Broadest Sense," *JAMA* (Sept. 19, 1908): 960. Chicago Medical Society, Council Minutes, [Oct. 1911–June 1912], meeting of Jan. 9, 1912, pp. 53, 57, 59, Chicago Medical Society Records.

20. Taussig reported that mortality reached 60 to 70 percent when septicemia or peritonitis had occurred. White House Conference on Child Health and Protection, *Fetal, Newborn, and Maternal Morbidity and Mortality* (New York, 1933), 466–67; Al-

muth C. Vandiver, "The Legal Status of Criminal Abortion, with Especial Reference to the Duty and Protection of the Consultant," *American Journal of Obstetrics and Diseases of Women and Children* 61 (March 1910): 434–35, esp. 497.

21. O. B. Will, "The Medico-Legal Status of Abortion," *Illinois Medical Journal* 2 (1900–1901): 506, 508; Edward W. Pinkham, "The Treatment of Septic Abortion, with a Few Remarks on the Ethics of Criminal Abortion," *American Journal of Obstetrics and Diseases of Women and Children,* 61 (March 1910): 420. For the Illinois Medicine and Surgery Act of 1899 see *All the Laws of the State of Illinois Passed by the Forty-First General Assembly* (Chicago, 1899), 216. In Nebraska and New Jersey revocation of a physician's license for abortion did not require a criminal conviction. "Procedure Before State Board of Health and Revocation of License for Criminal Abortion," *JAMA* (Aug. 29, 1908): 788; "Revocation of License for 'Practice' of Criminal Abortion on Single Occasion," *JAMA* (June 24, 1922): 1988. Revocations of medical licenses in Washington and Wisconsin followed convictions for abortion. "Revocation of License for Conviction of Offenses Involving Moral Turpitude," *JAMA* (Feb. 10, 1917): 485; "Intent and Revocation of License in Abortion Case," *JAMA* (March 29, 1919): 957.

22. Inquest on Mary L. Kissell, Aug. 3, 1937, case 300–8–1937, Medical Records Department. See also Inquest on Edna M. Lamb, Feb. 19, 1917, case 43–3–1917, Medical Records Department; and Inquest on Anna P. Fazio, Feb. 14, 1929, case 217–2–1929, Medical Records Department. Superintendent of Rhodes Avenue Hospital to Peter M. Hoffman, March 17, 1916 in Inquest on Annie Marie Dimford, Sept. 30, 1915, case 75–11–1915, Medical Records Department. See also Inquest on Ellen Matson, Nov. 19, 1917, case 330–11–1917, Medical Records Department. D. S. J. Meyers and W. W. Richmond commenting on C. J. Aud, "In What Percent Is the Regular Profession Responsible for Criminal Abortions, and What Is the Remedy?" *Kentucky Medical Journal* 2 (Sept. 1904): 98, 99.

23. This "agreement" is discussed in Inquest on Matson, case 330–11–1917, Medical Records Department. Transcript of *People v. Zwienczak,* 338 Ill. 237

(1929), Case Files, vault no. 44701, Supreme Court of Illinois.

24. Vandiver, "Legal Status of Criminal Abortion," 496–501, esp. 500; "Criminal Abortion from the Practitioner's Viewpoint," *American Journal of Obstetrics and Diseases of Women and Children* 63 (June 1911): 1094–96.

25. Chicago Medical Society, Council Minutes, [Oct. 1911–June 1912], meeting of Jan. 9, 1912, pp. 53–54, 56–57, Chicago Medical Society Records. The New Orleans Parish Medical Society published a letter to be sent to every physician in New Orleans, which included a model dying declaration. N. F. Thiberge, "Report of Committee on Criminal Abortion," *New Orleans Medical and Surgical Journal* 70 (1917–1918): 802, 807–8. See also "Criminal Abortion from the Practitioner's Viewpoint," 1093.

26. Chicago Medical Society, Council Minutes, [Oct. 1911–June 1912], meeting of Jan. 9, 1912, p. 57, Chicago Medical Society Records. For judicial discussions on the validity of dying declarations and examples of dying declarations, see *Hagenow v. People,* 188 Ill. 545, 550–51, 553 (1901); *People v. Buettner,* 233 Ill. at 274–77; *People v. Cheney,* 368 Ill. 131, 132–35 (1938).

27. "Criminal Abortion," *JAMA* (Sept. 20, 1902): 706; Palmer Findley, "The Slaughter of the Innocents," *American Journal of Obstetrics and Gynecology* 3 (Jan. 1922): 37.

28. Inquest on Emily Projahn, Oct. 10, 1916, case 26–12–1916, Medical Records Department. Abortion convictions appealed to higher courts in Texas and Wisconsin revealed that dying declarations were obtained from women under threats by physicians to refuse medical care. "Dying Declarations Obtained in Abortion Case as Condition to Rendering Aid," *JAMA* (April 10, 1909): 1204; "Dying Declarations Made After Refusal of Physician to Treat Abortion Case Without History," *JAMA* (June 7, 1913): 1829–30. Inquest on Mathilde C. Kleinschmidt, Sept. 22, 1930, case 255–9–30, Medical Records Department.

29. Vandiver, "Legal Status of Criminal Abortion," 435. Dr. Henry Dawson Furniss commenting on "Criminal Abortion from the Practitioner's Viewpoint," 1096.

30. H. Wellington Yates and B. Connelly, "Treatment of Abortion," *American Journal of Obstetrics and Gynecology* 3 (Jan. 1922): 84–85. See also "Dying Girl Runaway Hides Name of Slayer," *Chicago Examiner*, March 8, 1918, Abortionists Files, Historical Health Fraud Collection; and "Slain Girl Dies Hiding Her Tragedy from Kin," *Chicago Examiner*, March 9, 1918, Abortionists Files; Robinson, *Law Against Abortion*, 106–7, 110. "Abortion 'Club' Exposed," *Birth Control Review* 4 (Nov. 1936): 5.

31. "End Murders by Abortions," *Chicago Tribune*, June 2, 1915, Abortionists Files, Historical Health Fraud Collection; Transcript of *People v. Heissler*, 338 Ill. 596 (1929), Case Files, vault no. 44783, Supreme Court of Illinois. See also Robinson, *Law Against Abortion*, 106.

32. "Symposium. Criminal Abortion," 1099. See also Will, "Medico-Legal Status of Abortion," 508; and "Criminal Abortion from the Practitioner's Viewpoint," 1095–96. Chicago Medical Society, Council Minutes, [Oct. 11–June 1912], meeting of Jan. 9, 1912, p. 55, Chicago Medical Society Records. For similar complaints from the coroner see Inquest on Fazio, case 217–2–1929, Medical Records Department; and Inquest on Kissell, case 300–8–1937, Medical Records Department. Robinson, *Law Against Abortion*, 105–11. On Robinson see Gordon, *Woman's Body, Woman's Right* (1977), 173–78.

33. Inquest on Lamb, case 43–3–1917, Medical Records Department; "Criminal Abortion from the Practitioner's Viewpoint," 1094.

34. Dr. Louis Frank commenting on Aud, "In What Percent," 100; "North Branch Philadelphia County Medical Society. Regular Meeting, held April 14, 1904," *JAMA* (May 21, 1904): 1375–76. Attorneys disagreed about whether or not physicians should act as informers in abortion cases. "Symposium. Criminal Abortion," 1097, 1098. Illinois law did not privilege communications between doctors and patients. "Chicago Medical Society," 1889.

35. "Girl's Letters Blame Dr. Mason in Death Case," *Chicago Tribune*, [April] 9, 1916, Abortionists Files, Historical Health Fraud Collection; "Voice from Grave Calls to Dr. Mason during Trial as His Fiancee's Betrayer," *Denver* [Colorado] *Post*, April 5, 1916, Abortionists Files. The death of Anna Johnson at the office of Dr. Eva Shaver, an abortionist, similarly filled Chicago's newspapers in late May and June 1915. Newspapers published photos of Johnson, her lover, and Dr. Shaver, as well as the couple's letters. For example, see "Who's Who in the 'Mystery House' Tragedy?" *Chicago Evening American*, May 31, 1915; and "Death of Girl Perils Schools for Abortions," *Chicago Tribune*, May 28, 1915; "Dying Girl Runaway Hides Name of Slayer," *Chicago Examiner*, March 8, 1918. See also "Girl Slain Here Gives Life to Hide Her Tragedy," *Chicago Examiner*, March 5, 1918; "Slain Girl Dies Holding Her Tragedy from Kin," *Chicago Examiner*, March 9, 1918; and "Little Jane's Tragedy Typical of Hundreds Who Disappear Here," *Chicago Examiner*, [1918], all of which can be found in Abortionists Files. On Progressive era fears of the sexual adventure and seduction of young women in the city, see the excellent study by Joanne J. Meyerowitz, *Women Adrift: Independent Wage Earners in Chicago, 1880–1930* (Chicago, 1988). On the stories of male seduction of young working women written by purity reformers, see Odem, "Delinquent Daughters," 16–74. On married women see "Mrs. Ruth Conn," *Chicago Herald*, Dec. 19, 1915, Abortionists Files, Historical Health Fraud Collection; "Aged Physician Is Jailed after Woman's Death," *Chicago Tribune*, June 22, 1929; "Death Arrest Bares List of 1,500 Women," *Chicago Examiner*, [1916], both in Abortionists Files.

36. Inquest on Mary Shelley, Oct. 30, 1915, case 352–10–1915, Medical Records Department; "Chicago Medical Society," 1889. Mamie Ethel Crowell's family tried to prevent an investigation into her abortion by lying to physicians and the coroner. Inquest on Mamie Ethel Crowell, April 16, 1930, case 305–4–30, Medical Records Department. For cases where family members wanted the state to investigate an abortion, see *People v. Hotz*, 261 Ill. 239 (1914); and Inquest on Frauciszka Gawlik, Feb. 19, 1916, case 27–3–1916, Medical Records Department. "A Maryland Abortionist Gets No Pardon," *JAMA* (Nov. 12, 1904): 1476; Inquest on Frances Collins, May 7, 1920, case 161–5–20, Medical Records Department.

37. Comments of "Esther E.," *Birth Control Review* 4 (Sept. 1920), 15. See also Robinson, *Law Against Abortion*, 106–7; *Hagenow v. People*, 188 Ill. at 551; *People v. Hagenow*, 236 Ill. 514, 526–27 (1908);

and *People v. Heissler,* 338 Ill. at 599. The coroner told Dr. Henry Kruse to "make it a rule" at his hospital to call police in abortion cases so that they could bring the suspect in for identification in the inquest on Edna M. Lamb. Inquest on Lamb, case 43–3–1917, Medical Records Department.

38.   Inquest on Petrovitis, case 234–3–1916, Medical Records Department. Transcript of *People v. Hobbs,* 297 Ill. 399 (1921), Case Files, vault no. 38773, Supreme Court of Illinois; Abstract of *People v. Heisler,* 300 Ill. 98 (1930), 38, Case Files, vault no. 44783, Supreme Court of Illinois.

39.   Inquest on Eunice McElroy, Nov. 14, 1928, case 486–11–28, Medical Records Department.

40.   Inquest on Dimford, case 75–11–1915, Inquest on Kissell, case 300–8–1937, Inquest on Esther Stark, June 12, 1917, case 65–6–1917, all in Medical Records Department; *People v. Carrico,* 310 Ill. 543, 547 (1924). For an example of a woman who did not want to marry see Inquest on Mary Colbert, March 25, 1933, case 7–4–1933, Medical Records Department. *People v. Rongetti,* 344 Ill. 278, 284 (1931).

41.   Inquest on Anna Johnson, May 27, 1915, case 77790, Medical Records Department.

42.   Inquest on Matson, case 330–11–1917, Medical Records Department.

43.   "Death Threat to Hostetler," *Chicago Tribune,* June 5, 1915, Abortionists Files, Historical Health Fraud Collection. The police and press often called the man in such a case "the sweetheart." See "Doctor Faces Manslaughter Charge in Girl's Death," *Chicago Tribune,* April 18, 1930; and "Woman Doctor Found Guilty in Death," *Chicago Tribune,* March 14, 1936, both in Abortionists Files. Inquest on Alma Bromps, April 27, 1931, case 35–5–1931, Medical Records Department; *People v. Ney,* 349 Ill. 172, 173–74 (1932). For similar cases see *Cochran v. People,* 175 Ill. 28 (1898); and *People v. Hobbs,* 297 Ill. 399 (1921). Inquest on Rose Siebenmann, April 16, 1920, case 266–4–20, Medical Records Department. Transcript of *People v. Hobbs;* Transcript of *People v. Buettner.* For convictions of boyfriends, see *Dunn v. People;* 172 Ill. 582 (1898); and *People v. Patrick,* 277 Ill. 210 (1917).

44.   I know of only one case where the husband was charged. Bertis Dougherty pleaded guilty to abortion and testified as a state witness against the abortionist in *People v. Schneider,* 370 Ill. 612, 613–14 (1939).

45.   In the 163 cases studied by Louise DeKoven Bowen, eight men were sent to jail for six months. It appears that sixteen men were ordered to make a financial settlement. Juvenile Protective Association of Chicago, *A Study of Bastardy Cases, Taken from the Court of Domestic Relations in Chicago* (Chicago, 1914), 18, 19, 22. *Notable American Women,* s. v. "Bowen, Louise DeKoven."

46.   On common perceptions of juvenile court proceedings regarding statutory rape, see Odem, "Delinquent Daughters," 75–136.

47.   Transcript of *Dunn v. People,* 172 Ill. 582 (1898), Case Files, vault no. 7876, Supreme Court of Illinois. Women may have pursued cases against their lovers in *Scott v. People,* 141 Ill. 195 (1892); and *People v. Patrick,* 277 Ill. 210 (1917). On women's use of state regulation for their own ends and the need to analyze the state and social control with attention to women and gender, see Linda Gordon, *Heroes of Their Own Lives: The Politics and History of Family Violence, Boston, 1880–1960* (New York, 1988), 289–99.

48.   *People v. Martin,* 382 Ill. 192 (1943); *People v. Stanko,* 402 Ill. 558 (1949); *People v. Smuk,* 12 Ill. 2d. 360 (1957).

49.   Inquest on Fazio, case 217–2–1929, Medical Records Department. Dr. Edward W. Pinkham suggested reporting abortion cases to the "prosecuting authorities" for "complete protection." He noted, "Good legal authority has declared that if the physician does not do this, but maintains secrecy in the treatment, he becomes *particeps criminis,* and subjects himself to a possible prosecution." Pinkham, "Treatment of Septic Abortion," 420.

50.   Jerome E. Bates and Edward Zawadzki, *Criminal Abortion: A Study in Medical Sociology* (Springfield, 1964), 61, 100, 103. See also Leslie J. Reagan, *When Abortion Was a Crime: Women, Medicine, and Law in the United States, 1867–1973* (Berkeley: University of California Press, 1977).

15 The Black Community and the Birth Control Movement

Jessie M. Rodrique

The decline in black fertility rates from the late nineteenth century to World War II has been well documented. In these years the growth rate of the black population was more than cut in half. By 1945 the average number of children per woman was 2.5, and childlessness, especially among urban blacks, had reached unprecedented proportions. Researchers who explain this phenomenon insist that contraception played a minimal role; they believe that blacks had no interest in the control of their own fertility. This belief also affects the interpretation of blacks' involvement in the birth control movement, which has been understood as having been thrust upon an unwilling black population.

This essay seeks to understand these two related issues differently. First, I maintain that black women were in fact interested in controlling their fertility and that the low birthrates reflect in part a conscious use of birth control. Second, by exploring the birth control movement among blacks at the grassroots level, I show that despite the racist ideology that operated at the national level, blacks were active and effective participants in the establishment of lo-

cal clinics and in the birth control debate, as they related birth control to issues of race and gender. Third, I show that despite black cooperation with white birth control groups, blacks maintained a degree of independence that allowed the organization for birth control in their communities to take a qualitatively different form.

Demographers in the post–World War I years accounted for the remarkable decline in black fertility in terms of biological factors. Fears of "dysgenic" population trends, coupled with low birthrates among native white Americans, underlay their investigations of black fertility. Population scholars ignored contraception as a factor in the birth decline even as late as 1938. Instead, they focused on the "health hypothesis," arguing that the drop in fertility resulted from general poor health, especially sterility caused by venereal disease. Although health conditions seem likely to have had some effect, there is no reason to exclude contraceptive use as an additional cause, especially when evidence of contraceptive knowledge and practice is abundant.[1]

In drawing their conclusions researchers also made many questionable and unfounded assumptions about the sexuality of blacks. In one large study of family limitation, for example, black women's lower contraceptive use was attributed to the belief that "the negro generally

JESSIE M. RODRIQUE received her Ph.D. in American History at the University of Massachusetts at Amherst.

Reprinted from *Passion and Power: Sexuality in History*, edited by Kathy Peiss and Christina Simmons, (Philadelphia: Temple University Press, 1989) 138–54, by permission of Temple University Press.

exercises less prudence and foresight than white people do in all sexual matters."[2] Nor is the entire black population represented in many of these studies. Typically, their sample consists of women whose economic status is defined as either poor or very poor and who are either illiterate or have had very little education. Population experts' ideological bias and research design have tended to foreclose the possibility of Afro-American agency and thus conscious use of contraception.[3]

Historians who have chronicled the birth control movement have focused largely on the activities and evolution of the major birth control organizations and leading birth control figures, usually at the national level. None has interpreted the interests of the movement as particularly beneficial to blacks. Linda Gordon, in her pathbreaking book *Woman's Body, Woman's Right,* focused on the 1939 "Negro Project," established by the Birth Control Federation of America (BCFA) as a conservative, elitist effort designed "to stabilize existing social relations." Gordon claims that the birth control movement in the South was removed from socially progressive politics and unconnected to any analysis of women's rights, civil rights, or poverty, exemplifying the movement's male domination and professionalization over the course of the twentieth century. Other historians concur, asserting that birth control was "genocidal" and "anathema" to black women's interests and that the movement degenerated into a campaign to "keep the unfit from reproducing themselves." Those who note its presence within the black community in a slightly more positive light qualify their statements by adding the disclaimer that support and information for its dissemination came only from the black elite and were not part of a grassroots movement.[4]

There is, however, an ample body of evidence that suggests the importance of birth control use among blacks. Contraceptive methods and customs among Africans as well as nineteenth-

century slaves have been well documented. For example, folklorists and others have discovered "alum water" as one of many birth control measures in early twentieth-century southern rural communities. The author of a study of two rural Georgia counties noted the use of birth control practices there and linked it to a growing race pride. In urban areas a "very common" and distinctive practice among blacks was to place Vaseline and quinine over the mouth of the uterus. It was widely available and could be purchased very cheaply in drugstores.[5]

The black press was also an abundant source of birth control information. The *Pittsburgh Courier,* for example, carried numerous mail-order advertisements for douche powder, suppositories, preventive antiseptics, and vaginal jellies that "destroyed foreign germs."[6] A particularly interesting mail-order ad was for a product called "Puf," a medicated douche powder and applicator that claimed to be a "new guaranteed method of administering marriage hygiene." It had a sketch of a calendar with the words "End Calendar Worries Now!" written across it and a similar sketch that read "Tear-Up Your Calendar, Do Not Worry, Use Puf." The instructions for its use indicate euphemistically that Puf should be used "first," meaning before intercourse, and that it was good for hours, leaving little doubt that this product was fully intended to be used as a birth control device.[7]

Advertisements for mail-order douches are significant because they appear to reflect a practice that was widespread and well documented among black women. Studies conducted in the mid-1930s overwhelmingly concluded that douching was the preferred method of contraception used by black couples. Yet contemporary researchers neglected to integrate this observation into their understanding of the decline in fertility because they insisted that douching was an "ineffective contraceptive." However ineffective the means, the desire for birth control in the black community was readily apparent, as George Schuyler, editor of the *National Negro News,* explained: "If anyone should doubt the desire on the part of Negro women and men to limit their families it is only necessary to note the large sale of preventive devices sold in every drug store in various Black Belts."[8]

Within the black community the practice of abortion was commonly cited by black leaders and professionals as contributing to the low birthrates. Throughout the 1920s and 1930s the black press reported many cases of abortions that had ended in women's deaths or the arrest of doctors who had performed them. Abortion was discussed in the *Pittsburgh Courier* in 1930 in a fictionalized series entitled "Bad Girl," which dealt with a range of attitudes toward childbearing among Harlem blacks. When Dot, the main character, discovers she is pregnant, she goes to a friend who works in a drugstore. The author writes: "Pat's wonderful remedy didn't help. Religiously Dot took it and each night when Eddie came home she sadly admitted that success had not crowned her efforts. 'All that rotten tasting stuff just to keep a little crib out of the bedroom.' After a week she was tired of medicine and of baths so hot that they burned her skin."[9] Next, she sought the advice of a friend who told her that she would have to have "an operation" and knew of a doctor who would do it for fifty dollars.

The *Baltimore Afro-American* observed that pencils, nails, and hat pins were the instruments commonly used for self-induced abortions, and the *Birth Control Review* wrote in 1936 that rural black women in Georgia drank turpentine for the same purpose. The use of turpentine as an abortifacient is significant because it is derived from evergreens, a source similar to rue and camphor, both of which were reported by a medical authority in 1860 to have been used with some success by southern slaves. Although statistics for abortions among black women are scarce, a 1938 medical study reported that 28 percent or 211 of 730 black women interviewed

said that they had had one or more abortions. A black doctor from Nashville in 1940 asserted in the *Baltimore Afro-American* that abortions among black women were deliberate, not only the result of syphilis and other diseases: "In the majority of cases it is used as a means of getting rid of unwanted children."[10]

These data, though somewhat impressionistic, indicate that a variety of contraceptive methods were available to blacks. Many were, and still are, discounted as ineffective "folk methods."[11] There was, however, a discernible consciousness that guided the decline in fertility. A discourse on birth control emerged in the years from 1915 to 1945. As blacks migrated within and out of the South to northern cities, they began to articulate the reasons for limiting fertility, and one begins to see how interconnected the issue of birth control was to many facets of black life. For women it was linked to changes in their status, gender roles within the family, attitudes toward motherhood and sexuality, and, at times, feminism. Birth control was also integral to issues of economics, health, race relations, and racial progress.

In these years blacks contributed to the "official" nationwide debate concerning birth control while also voicing their particular concerns. Frequent coverage was given to birth control in the black press. Newspapers championed the cause of birth control when doctors were arrested for performing abortions. They also carried editorials in favor of birth control, speeches of noted personalities who favored its use, and occasionally sensationalized stories on the desperate need for birth control. Often, the topic of birth control as well as explicit birth control information was transmitted orally through public lectures and debates. It was also explored in fiction, black periodicals, and several issues of the *Birth Control Review* dedicated to blacks.[12]

Economic themes emerged in the birth control discourse as it related to issues of black family survival. Contraceptive use was one of a few economic strategies available to blacks, providing a degree of control within the context of the family economy. Migrating families who left behind the economy of the rural South used birth control to "preserve their new economic independence," as did poor families who were "compelled" to limit their numbers of children. A 1935 study of Harlem reiterated this same point, adding that the low birthrates of urban blacks reflected a "deliberate limitation of families." Another strategy used by black couples for the same purpose was postponing marriage. Especially in the years of the depression birth control was seen as a way to improve general living conditions by allowing more opportunities for economic gain.[13]

Birth control was also linked to the changing status of black women and the role they were expected to play in the survival of the race. On this issue a degree of opposition to birth control surfaced. Some, most notably black nationalist leader Marcus Garvey, believed that the future of the black race was contingent upon increasing numbers and warned that birth control would lead to racial extinction. Both Garveyites and Catholic church officials insisted that birth control interfered with the "course of nature" and God's will.[14]

These issues were evident in an exchange between the journalist J. A. Rogers and Dean Kelly Miller of Howard University in 1925. Writing in the *Messenger*, Rogers took Miller to task for his statements concerning the emancipation of black women. Miller is quoted as saying that black women had strayed too far from children, kitchen, clothes, and the church. Miller, very aware that black women had been having fewer children, cautioned against race suicide. Using the "nature" argument of Garvey and the Catholic Church, he argued that the biological function of women was to bear and rear children. He stated, "The liberalization of women must always be kept within the boundary fixed by na-

ture." Rogers strongly disagreed with Miller, saying that the move of black women away from domesticity and childbearing was a positive sign. Rogers wrote, "I give the Negro woman credit if she endeavors to be something other than a mere breeding machine. Having children is by no means the sole reason for being."[15]

Other black leaders supported this progressive viewpoint. In his 1919 essay "The Damnation of Women," W. E. B. Du Bois wrote that "the future woman must have a life work and future independence. . . . She must have knowledge . . . and she must have the right of motherhood at her own discretion."[16] In a later essay he described those who would confine women to childbearing as "reactionary barbarians."[17] Dr. Charles Garvin, writing in 1932, believed that it was the "inalienable right of every married woman to use any physiologically sound precaution against reproduction she deems justifiable."[18]

Black women also expressed the need for contraception when they articulated their feelings about motherhood and sexuality. Black women's fiction and poetry in the years from 1916 to the early 1930s frequently depicted women who refused to bring children into a racist world and expressed their outrage at laws that prevented access to birth control information. Nella Larsen, for example, in her 1928 novella *Quicksand,* explored the debilitating physical and emotional problems resulting from excessive childbearing in a society that demanded that women's sexual expression be inextricably linked to marriage and procreation.[19]

Others spoke of the right not to have children in terms that were distinctly feminist. For example, a character in the *Courier* serial "Bad Girl" put it this way: "The hospitals are wide open to the woman who wants to have a baby, but to the woman who doesn't want one—that's a different thing. High prices, fresh doctors. It's a man's world, Dot. The woman who wants to keep her body from pain and her mind from worry is an object of contempt." [20] The changing status of women and its relation to childbearing were also addressed in Jessie Fauset's 1931 novel *The Chinaberry Tree*. Fauset's male characters asserted the need for large families and a "definite place" for women in the home. The female character, however, remained unconvinced by this opinion. She had "the modern girl's own clear ideas on birth control." [21]

Other writers stressed the need for birth control in terms of racial issues and its use to alleviate the oppressive circumstances of the black community. For example, Chandler Owen, editor of the *Messenger*, wrote a piece for the 1919 edition of the *Birth Control Review* entitled "Women and Children of the South." He advocated birth control because he believed that having fewer children would result in general improvements in material conditions. Observing that young black women in peonage camps were frequently raped and impregnated by their white overseers, Owen also linked involuntary maternity to racial crimes. [22]

Birth control for racial progress was advocated most frequently during the depression, and it helped to mobilize community support for clinics. Newell L. Sims of Oberlin College, for example, urged in his 1931 essay "A New Technique in Race Relations" that birth control for blacks would be a "step toward independence and greater power." In his opinion a controlled birthrate would free more resources for advancement. The black press hailed the essay as "revolutionary." [23] Other advocates insisted that all blacks, but especially poor blacks, become involved in the legislative process to legalize birth control. It was imperative that the poor be included in the movement because they were the ones most injured by its prohibition. One black newspaper, the *San Francisco Spokesman*, promoted a very direct and activist role for blacks on this issue. "To legalize birth control, you and

I should make expressed attitudes on this question a test of every candidate's fitness for legislative office," it argued in 1934. "And those who refuse or express a reactionary opinion should be flatly and uncompromisingly rejected." [24]

For many blacks birth control was not a panacea but one aspect of a larger political agenda. Unlike some members of the white community who myopically looked to birth control as a cure-all for the problems of blacks, most blacks instead described it as a program that would "modify one cause of their unfavorable situation." [25] They stressed that true improvement could come only through the "equalization of economic and social opportunities." [26] Sims summed up this position most eloquently in his 1932 essay "Hostages to the White Man." It was a viewpoint stressed well into the 1940s by numerous and leading members of the black community. He wrote:

> The negro in America is a suppressed class and as such must struggle for existence under every disadvantage and handicap. Although in three generations since slavery he has in many ways greatly improved his condition, his economic, social and political status still remain that of a dominated exploited minority. His problem is, therefore, just what it has been for three quarters of a century, i.e., how to better his position in the social order. Naturally in all his strivings he has found no panacea for his difficulties, for there is none. The remedies must be as numerous and varied as the problem is complex. Obviously he needs to employ every device that will advance his cause. I wish briefly to urge the merits of birth control as one means. [27]

Many also insisted that birth control be integrated into other health care provisions and not be treated as a separate problem. E. S. Jamison, for example, writing in the *Birth Control Review* in 1938, exhorted blacks to "present an organized front" so that birth control and other needed health services could be made available to them. Yet he too, like Sims, emphasized independence from the white community. He wrote that "the Negro must do for himself. Charity will not better his condition in the long run."[28]

Blacks also took an important stand against sterilization, especially in the 1930s. Scholars have not sufficiently recognized that blacks could endorse a program of birth control but reject the extreme views of eugenicists, whose programs for birth control and sterilization often did not distinguish between the two. The *Pittsburgh Courier,* for example, whose editorial policy clearly favored birth control, was also active in the antisterilization movement. It asserted in several editorials that blacks should oppose the sterilization programs being advanced by eugenicists and so-called scientists because they were being waged against the weak, the oppressed, and the disfranchised. Candidates for sterilization were likely to be those on relief, the unemployed, and the homeless, all victims of a vicious system of economic exploitation. Du Bois shared this viewpoint. In his column in the *Courier* in 1936 he wrote, "The thing we want to watch is the so-called eugenic sterilization." He added that the burden of such programs would "fall upon colored people and it behooves us to watch the law and the courts and stop the spread of the habit." The *San Francisco Spokesman* in 1934 called upon black club women to become active in the antisterilization movement.[29]

Participation in the birth control debate was only one aspect of the black community's involvement; black women and men also were active in the establishment of birth control clinics. From 1925 to 1945 clinics for blacks appeared nationwide, many of them at least partly directed and sponsored by local black commu-

nity organizations. Many of the organizations had a prior concern with health matters, creating an established network of social welfare centers, health councils, and agencies. Thus birth control services were often integrated into a community through familiar channels.[30]

In Harlem the black community showed an early and sustained interest in the debate over birth control, taking a vanguard role in agitation for birth control clinics. In 1918 the Women's Political Association of Harlem, calling upon black women to "assume the reins of leadership in the political, social and economic life of their people," announced that its lecture series would include birth control among the topics for discussion.[31] In March 1923 the Harlem Community Forum invited Margaret Sanger to speak at the Library Building in the Bronx, and in 1925 the Urban League asked the American Birth Control League to establish a clinic in the Columbus Hill section of the city.

Although this clinic proved unsuccessful, another, supported by the Urban League and the Birth Control Clinical Research Bureau, opened a Harlem branch in 1929. This particular clinic, affiliated with Margaret Sanger, had an advisory board of approximately fifteen members, including Harlem-based journalists, physicians, social workers, and ministers. There was apparently very little opposition to the work of this clinic, even among the clergy. One minister on the advisory board, William Lloyd Imes of the St. James Presbyterian Church, reported that he had held discussions on birth control at his church; at another meeting he announced that if a birth control pamphlet were printed, he would place it in the church vestibule. Another clergyman, the Reverend Shelton Hale Bishop, wrote to Sanger in 1931 that he believed birth control to be "one of the boons of the age to human welfare."[32] The Reverend Adam Clayton Powell of the Abyssinian Baptist Church both endorsed birth control and spoke at public meetings denouncing the "false modesty" surrounding questions of sex. Ignorance, he believed, led to unwanted pregnancies among young girls.[33]

Support for birth control clinics by black community organizations was also apparent in other locations throughout the country. Their activism took various forms. In Baltimore, for example, a white birth control clinic had begun to see blacks in 1928. In 1935 the black community began organizing, and by 1938 the Northwest Health Center was established, sponsored and staffed by blacks. The Baltimore Urban League played a key role in its initial organization, and the sponsoring committee of the clinic was composed of numerous members of Baltimore's black community, including ministers, physicians, nurses, social workers, teachers, housewives, and labor leaders.[34]

In Richmond, Fredericksburg, and Lynchburg, Virginia, local maternal welfare groups raised funds to pay for expenses and supplies for the birth control clinics at the Virginia Medical College and the Hampton Institute and publicized birth control services at city health departments. And in West Virginia the Maternal and Child Health Council, formed in 1938, was the first statewide birth control organization sponsored by blacks.[35]

Local clubs and women's organizations often took part in either sponsoring birth control clinics or bringing the topic to the attention of the local community. In New York these included the Inter-Racial Forum of Brooklyn, the Women's Business and Professional Club of Harlem, the Social Workers Club of Harlem, the Harlem branch of the National Organization of Colored Graduate Nurses, the Harlem YWCA, and the Harlem Economic Forum. In Oklahoma City fourteen black women's clubs sponsored a birth control clinic for black women, directed by two black physicians and one black club woman. The Mother's Health Association of the District of Columbia reported to the *Birth Control Re-*

view in 1938 that it was cooperating with black organizations that wanted to start a clinic of their own.[36]

Clinics in other cities were located in black community centers and churches. For example, the Kentucky Birth Control League in 1936 reported that one of the clinics in Louisville was located in the Episcopal Church for Colored People and was operated by a black staff. The Cincinnati Committee on Maternal Health reported in 1939 the opening of a second black clinic that would employ a black physician and nurse.[37]

Community centers and settlement houses were also part of the referral network directing blacks to birth control services. The Mother's Health Office in Boston received clients from the Urban League, the Robert Gould Shaw House, and the Harriet Tubman House. The Henry Street Settlement sent women to the Harlem clinic, and the Booker T. Washington Community Center in San Francisco directed black women to the birth control clinic in that city. In 1935 the Indiana Birth Control League reported that black clients were referred from the Flanner House Settlement for Colored People.[38]

In 1939 the Birth Control Federation of America (BCFA) established a Division of Negro Service and sponsored pilot clinics in Nashville, Tennessee, and Berkeley County, South Carolina. The division consisted of a national advisory council of thirty-five black leaders, a national sponsoring committee of five hundred members who coordinated state and local efforts, and administrative and field personnel. The project in Nashville was integrated into the public health services and located in the Bethlehem center, a black social service settlement, and the Fisk University Settlement House. Both clinics were under the direction of black doctors and nurses. The program was supplemented by nine black public health nurses who made home visits and performed general health services, including the distribution of birth control devices and information. The home visits served the large numbers of women who worked as domestics and could not attend the clinics during the day; five thousand home visits were made in Nashville in a two-year period. In South Carolina clinic sessions providing both medical care and birth control services were held eleven times each month at different locations in the country for rural women, 70 percent of whom were black.[39]

Simultaneously with the development of these two projects, the BCFA launched an educational campaign to inform and enlist the services of black health professionals, civic groups, and women's clubs. Although professional groups are often credited with being the sole source of birth control agitation, the minutes and newsletters of the Division of Negro Service reveal an enthusiastic desire among a broad cross-section of the black community to lend its support to birth control. In fact, black professional groups often worked closely with community groups and other "nonprofessionals" to make birth control information widely available. For example, the National Medical Association, an organization of black physicians, held public lectures on birth control in conjunction with local groups beginning in 1929, and when birth control was discussed at annual meetings the otherwise private sessions were opened up to social workers, nurses, and teachers. The National Association of Colored Graduate Nurses, under the direction of Mabel Staupers, was especially active in birth control work. Cooperation was offered by several state and local nursing, hospital, and dental associations. One nurse responded to Staupers's request for help with the distribution of birth control information by writing, "I shall pass the material out, we will discuss it in our meetings and I will distribute exhibits at pre-natal clinics at four health centers and through Negro Home Demonstration Clubs."[40]

The participation of Negro Home Demonstration Clubs in birth control work is significant because it is an entirely overlooked and potentially rich source for the grassroots spread of birth control information in the rural South. Home Demonstration Clubs grew out of the provisions of the Smith-Lever Cooperative Extension Act of 1914 and had, by the early 1920s, evolved into clubs whose programs stressed health and sanitation. The newsletter of the Division of Negro Service in 1941 reported that five rural state Negro agricultural and home demonstration agents offered full cooperation with the division. The newsletter included the response of H. C. Ray of Little Rock, Arkansas. He wrote, "We have more than 13,000 rural women working in home demonstration clubs . . . it is in this connection that I feel our organization might work hand in hand with you in bringing about some very definite and desirable results in your phase of community improvement work. We will be glad to distribute any literature." Also involved with rural birth control education were several tuberculosis associations and the Jeanes Teachers, educators funded by the Anna T. Jeanes Foundation for improving rural black schools.[41]

Other groups showed interest in the programs of the Division of Negro Service either by requesting speakers on birth control for their conventions or by distributing literature to their members. Similar activities were conducted by the Virginia Federation of Colored Women's Clubs, which represented four hundred women's clubs, the Negro Organization Society of Virginia, the National Negro Business League, the National Negro Housewives League, the Pullman Porters, the Elks, the Harlem Citizens City-Wide Committee, and the Social Action Committee of Boston's South End. In 1944, for example, the National Association for the Advancement of Colored People and a black boilermakers' union distributed Planned Parenthood clinic cards in their mailings to their California members. Twenty-one Urban Leagues in sixteen states as of 1943 actively cooperated with the BCFA in the display of exhibits, distribution of literature, promotion of local clinical service, and adult community education programs. These national and local black organizations advocated birth control as one aspect of a general program of health, education, and economic development in the late 1930s and early 1940s.[42]

Even in their cooperation with the BCFA, leading members of the black community stressed their own concerns and disagreements with the overall structure of the birth control movement. Their comments reveal important differences in orientation. At a meeting of the National Advisory Council of the Division of Negro Service in 1942, members of the council made it clear that birth control services and information must be distributed to the community *as* a community. Their goal was inclusion; members stated that they were disturbed at the emphasis on doctors and said that teachers, ministers, and other community members must be employed in birth control work. Even the black physicians on the council stressed the need for keeping midwives, volunteers, and especially women practitioners involved in the movement and suggested that mobile clinics traveling throughout the rural South distribute birth control information and other needed health services. This approach to birth control diverged significantly from the conservative strategy of the white BCFA leadership, which insisted that birth control services be dispensed by individual private physicians. Black physicians, it seems, were more sensitive to the general health needs of their population and more willing to experiment with the delivery of birth control services. They favored the integration of birth control into public health services, which many white physicians opposed.[43]

Others on the council stated that black women could be reached only through community or-

ganizations that they trusted, and they stressed again the necessity of not isolating birth control as a special interest to the neglect of other important health needs. Still others pointed to the need for birth control representatives who recognized social differences among urban blacks.

Clinicians also observed a difference in clinic attendance between white and black patrons. Black women, they noted, were much more likely to spread the word about birth control services and bring their relatives and friends to the clinics. Some rural women even thought of "joining" the clinic as they might join a community organization. A white woman, however, was more likely to keep the information to herself and attend the clinic alone. A statistician from the Census Bureau supported this observation when he speculated in 1931 that "grapevine dissemination" of birth control information contributed to low black birthrates. These reports are a testimony to the effectiveness of working-class black women's networks.[44]

Moreover, many local birth control groups were often able to maintain independence from the Planned Parenthood Federation of America (PPFA) even though they accepted and used PPFA's display and educational materials. This situation was evident at the Booker T. Washington community center in San Francisco. A representative from PPFA had sent this center materials and then did not hear from anyone for some time. Almost one year later the director of the Washington center wrote back to PPFA, informing the staff that birth control programs were flourishing in the center's area and that the group had used the federation's materials extensively at community centers and civic clubs and the local black sorority, Alpha Kappa Alpha, had accepted sponsorship of a mothers' health clinic. The PPFA representative described this situation as typical of many black groups. They would not respond to PPFA communications but would use PPFA materials and engage actively in their own form of community birth control work.[45]

In a speech delivered to PPFA in 1942 Dr. Dorothy Ferebee, a black physician and leader, said, "It is well for this organization to realize that the Negro at his present advanced stage of development is increasingly interested more in programs that are worked out with and by him than in those worked out for him."[46] This statement reveals a fundamental difference in the goals and strategies of the black and white communities. In the past scholars have interpreted the birth control movement as a racist and elitist set of programs imposed on the black population. Although this characterization may describe the intentions of the national white leadership, it is important to recognize that the black community had its own agenda in the creation of programs to include and reach wide segments of the black population.

As this essay demonstrates, black women used their knowledge of "folk" and other available methods to limit their childbearing. The dramatic fertility decline from 1880 to 1945 is evidence of their success. Moreover, the use of birth control was pivotal to many pressing issues within the black community. The right to control one's fertility emerged simultaneously with changing attitudes toward women in both the black and white communities that recognized their rights as individuals and not only their roles as mothers. And these changing attitudes contributed to the dialogue within the black community about the future of the family and strategies for black survival. Birth control also emerged as part of a growing race consciousness, as blacks saw it as one means of freeing themselves from the oppression and exploitation of white society through the improvement of their health and economic and social status. Birth control was also part of a growing process of politicization. Blacks sought to make it a legislative issue, they opposed the sterilization movement, and they took an active and often independent role in supporting their clinics, educating their communities, and tailoring programs to fit their needs. In their ideology and practice blacks were indeed a vital and assertive part of the larger birth control movement. What appears to some scholars of the birth control movement as the waning of the movement's original purposes during the 1920s and 1930s was within the black community a period of growing ferment and support for birth control. The history of the birth control movement, and the participation of black Americans in it, must be reexamined in this light.

Notes

1. Reynolds Farley, *Growth of the Black Population* (Chicago, 1970), 3, 75; Stanley Engerman, "Changes in Black Fertility, 1880–1940," in *Family and Population in Nineteenth-Century America,* ed. Tamara K. Hareven and Maris A. Vinovskis (Princeton, 1978), 126–53. For an excellent review of the demographic literature see Joseph McFalls and George Masnick, "Birth Control and the Fertility of the U.S. Black Population, 1880 to 1980," *Journal of Family History* 6 (1981): 89–106; Peter Uhlenberg, "Negro Fertility Patterns in the United States," *Berkeley Journal of Sociology* 11 (1966): 56; James Reed, *From Private Vice to Public Virtue* (New York, 1978), 197–210.

2. Raymond Pearl, "Contraception and Fertility in 2,000 Women," *Human Biology* 4 (1932): 395.

3. McFalls and Masnick, "Birth Control," 90.

4. Linda Gordon, *Woman's Body, Woman's Right* (New York, 1976), 332–35; Paula Gidding, *When and Where I Enter: The Impact of Black Women on Race and Sex in America* (New York, 1984), 183; Robert G. Weisbord, *Genocide? Birth Control and the Black American* (Westport, Conn., 1975); William G. Harris, "Family Planning, Sociopolitical Ideology and Black Americans: A Comparative Study of Leaders and a General Population Sample" (Ph.D. diss., University of Massachusetts, 1980), 69.

A brief chronology of early birth control organizations is as follows: the American Birth Control League was founded in 1921 and operated by Margaret Sanger until 1927. In 1923 Sanger had organized the Clinical Research Bureau and after 1927 controlled only that facility. In 1939 the Clinical Research Bureau and the American Birth Control League merged to form the Birth Control Federation of America. In 1942 the name was changed to the Planned Parenthood Federation of America (hereafter cited as ABCL, BCFA, and PPFA).

5.   For contraceptive use among Africans see Norman E. Himes, *Medical History of Contraception* (New York, 1936). For statements concerning birth control use among black Americans, see W. E. B. Du Bois, "Black Folks and Birth Control," *Birth Control Review* 16 (June 1932): 166–67 (hereafter cited as *BCR*); Herbert Gutman, *The Black Family in Slavery and Freedom, 1750–1925* (New York, 1976). Du Bois had first observed the trend toward a steadily decreasing birth rate in *The Philadelphia Negro: A Social Study* (Philadelphia, 1899). For folk methods see Elizabeth Rauh Bethel, *Promiseland: A Century of Life in a Negro Community* (Philadelphia, 1981), 156–57; Newbell Niles Puckett, *Folk Beliefs of the Southern Negro* (New York, 1926); Arthur Raper, *Preface to Peasantry: A Tale of Two Black Belt Counties* (Chapel Hill, 1936), 71; "Report of the Special Evening Medical Session of the First American Birth Control Conference" (1921), box 99, folder 1017, Margaret Sanger Papers, Sophia Smith Collection, Smith College, Northampton, Mass.

6.   *Pittsburgh Courier,* April 25, 1931, n.p. (hereafter cited as *Courier*).

7.   *Courier,* December 1, 1934, p. 7.

8.   McFalls and Masnick, "Birth Control," 103; George Schuyler, "Quantity or Quality," *BCR* 16 (June 1932): 165–66.

9.   See, for example, *Courier,* March 9, 1935, p. 2; and *San Francisco Spokesman,* March 1, 1934, p. 1 (hereafter cited as *Spokesman*); Vina Delmar, "Bad Girl," *Courier,* January 3, 1931, p. 2.

10.   *Baltimore Afro-American,* August 3, 1940, n.p. (hereafter cited as *Afro-American*); "A Clinic for Tobacco Road," *BCR* 3 (New Series) (January 1936): 6; Gutman, *Black Family,* 80–85; John Gaston, "A Review of 2,422 Cases of Contraception," *Texas State Journal of Medicine* 35 (September 1938): 365–68; *Afro-American,* August 3, 1940, n.p. On abortion see also "Birth Control: The Case for the State," *Reader's Digest* (November 1939): 26–29.

11.   McFalls and Masnick, "Birth Control," 103.

12.   "Magazine Publishes Negro Number on Birth Control," *Spokesman,* June 11, 1932, p. 3; "Birth Control Slayer Held without Bail," *Courier,* January 11, 1936, p. 4.

13.   Alice Dunbar Nelson, "Woman's Most Serious Problem," *Messenger* (March 1927): 73; Clyde Kiser, "Fertility of Harlem Negroes," *Milbank Memorial Fund Quarterly* 13 (1935): 273–85; Caroline Robinson, *Seventy Birth Control Clinics* (Baltimore, 1930), 246–51.

14.   Weisbord, *Genocide?* 43.

15.   J. A. Rogers, "The Critic," *Messenger* 7 (April 1925): 164–65.

16.   W. E. B. Du Bois, "The Damnation of Women," in *Darkwater: Voices from Within the Veil,* ed. Herbert Aptheker (1921: rpt. Millwood, N.Y., 1975), 164–65.

17.   W. E. B. Du Bois, "Birth," *Crisis* 24 (October 1922): 248–50.

18.   Charles H. Garvin. "The Negro's Doctor's Task," *BCR* 16 (November 1932): 269–70.

19.   For an excellent discussion of the theme of sexuality in black women's fiction, see the introduction to Nella Larsen, *Quicksand and Passing,* ed. Deborah E. McDowell (New Brunswick, 1986). See also Mary Burrill, "They That Sit in Darkness," and Angelina Grimké, "The Closing Door," *BCR* 3 (September 1919): 5–8, 10–14; Jessie Fauset, *The Chinaberry Tree* (New York, 1931); Angelina Grimké, *Rachel: A Play in Three Acts* (1920; rpt. College Park, Md., 1969); Georgia Douglas Johnson, *Bronze: A Book of Verse* (1922; rpt. Freeport, N.Y., 1971).

20.   Delmar, "Bad Girl," *Courier,* January 3, 1931, p. 2.

21.   Fauset, *Chinaberry Tree,* 131–32, 187.

22.   Chandler Owen, "Women and Children of the South," *BCR* 3 (September 1919): 9, 20.

23.   Quoted in *Courier,* March 28, 1931, p. 3, and *Norfolk Journal and Guide,* March 28, 1931, p. 1.

24.   J. A. Ghent, "Urges Legalization of Birth Control: Law Against Contraception Unjust to the Poor," *Spokesman,* July 9, 1932, p. 3; "The Case of Dr. Devaughn, or Anti-Birth Control on Trial," *Spokesman,* February 22, 1934, p. 6.

25.   W. G. Alexander, "Birth Control for the Negro: Fad or Necessity?" *Journal of the National Medical Association* 24 (August 1932): 39.

26.   Charles S. Johnson, "A Question of Negro Health," *BCR* 16 (June 1932): 167–69.

27. Newell L. Sims, "Hostages to the White Man," *BCR* 16 (July–August 1932): 214–15.

28. E. S. Jamison, "The Future of Negro Health," *BCR* 22 (May 1938): 94–95.

29. "Sterilization," *Courier*, March 30, 1935, p. 10; "The Sterilization Menace," *Courier*, January 18, 1936, p. 10; W. E. B. Du Bois, "Sterilization," *Courier*, June 27, 1936, p. 1; "Are Women Interested Only in Meet and Eat Kind of Club?" *Spokesman*, March 29, 1934, p. 4.

30. For examples of black social welfare organizations see William L. Pollard, *A Study of Black Self-Help* (San Francisco, 1978); Edyth L. Ross, *Black Heritage in Social Welfare, 1860–1930* (London, 1978); Lenwood G. Davis, "The Politics of Black Self-Help in the United States: A Historical Overview," in *Black Organizations: Issues on Survival Techniques*, ed. Lennox S. Yearwood (Lanham, Md., 1980). This statement is also based on extensive reading of the *Pittsburgh Courier, Norfolk Journal and Guide, Baltimore Afro-American, San Francisco Spokesman,* and *New York Age* for the 1920s and 1930s.

31. *Messenger* 7 (July 1918): 26.

32. "Report of Executive Secretary," March 1923, ser. I, box 4, Planned Parenthood Federation of America Papers, American Birth Control League Records, Sophia Smith Collection, Smith College, Northampton, Mass. (hereafter cited as PPFA Papers); Hannah Stone, "Report of the Clinical Research Department of the ABCL" (1925), ser. I, box 4, PPFA; "Urban League Real Asset, Clinic an Example of How It Assists," *Courier*, November 2, 1935, p. 1; William Lloyd Imes to Margaret Sanger, May 16, 1931, and November 23, 1932, box 122b, folders 1333 and 1336, Sanger Papers; Shelton Hale Bishop to Margaret Sanger, May 18, 1931, box 122b, folder 1333, Sanger Papers.

33. "Minutes of the First Meeting of 1932, Board of Managers, Harlem Branch," March 25, 1932, box 122b, folder 1336, Sanger Papers; "Companionate Marriage Discussed at Forum," *New York Age*, May 12, 1928, n.p.

34. E. S. Lewis and N. Louise Young, "Baltimore's Negro Maternal Health Center: How It Was Organized," *BCR* 22 (May 1938): 93–94.

35. "West Virginia," *BCR* 23 (October 1938): 121; "Birth Control for the Negro," Report of Hazel Moore

(1937), box 22, folder 10, Florence Rose Papers, Sophia Smith Collection, Smith College; "Negro Demonstration Project Possibilities," December 1, 1939, box 121, folder 1309, Sanger Papers.

36. For information on black organizations see box 122b, Sanger Papers, esp. March 25, 1932; "Minutes of the Regular Meeting of the Board of Directors of the ABCL," December 1922, ser. I, box 1, "Report of the Executive Secretary," November 11, 1930, ser. I, box 4, "ABCL Treasurer's Annual Reports for the Year 1936," ser. I, box 4, PPFA Papers; "Harlem Economic Forum Plans Fine Lecture Series," *Courier*, November 14, 1936, p. 9; "Birth Control Clinic Set Up for Negroes; Sponsored by Clubs," *Oklahoma City Times*, February 28, March 4, 1938; "Illinois Birth Control League," *BCR* 22 (March 1938): 64. By 1931 many black organizations in Pittsburgh supported the use of birth control; see "Pittsburgh Joins Nation-Wide League for Birth Control," *Courier*, February 21, 1931, p. 1.

37. "Annual Reports of the State Member Leagues for 1936, the Kentucky Birth Control League," ser. I, box 4, PPFA Papers; "Annual Report 1938–39, Cincinnati Committee on Maternal Health," box 119A, folder 1256, Sanger Papers.

38. "Mother's Health Office Referrals," January 5, 1933, Massachusetts Mother's Health Office, Central Administrative Records, box 35 and 36, Planned Parenthood League of Massachusetts, Sophia Smith Collection, Smith College; "PPFA Field Report for California, 1944," box 119, folder 1215, Sanger Papers; "Annual Meeting of the BCFA, Indiana Birth Control League, 1935," ser. I, box 4, PPFA Papers.

39. "Chart of the Special Negro Project Demonstration Project," box 22, folders 8 and 2, Rose Papers; John Overton and Ivah Uffelman, "A Birth Control Service Among Urban Negroes," *Human Fertility* 7 (August 1942): 97–101; E. Mae McCarroll, "A Condensed Report on the Two-Year Negro Demonstration Health Program of PPFA, Inc.," paper presented at the Annual Convention of the National Medical Association, Cleveland, August 17, 1942, box 22, folder 11, Rose Papers; Mabel K. Staupers, "Family Planning and Negro Health," *National News Bulletin of the National Association of Colored Graduate Nurses* 14 (May 1941): 1–10.

40. "Preliminary Annual Report, Division of Ne-

gro Service," January 7, 1942, box 121, folder 1309, Sanger Papers; "Doctors' Annual Meeting Marked by Fine Program; Local Committee Involved in Planning Meeting," *New York Age,* September 7, 1929, p. 8; "National Medical Association Meeting Held in Washington," *New York Age,* August 27, 1932, p. 4.

41. For information on the Smith-Lever Extension Act, see Alfred True, *A History of Agricultural Extension Work in the United States, 1785–1923* (Washington, D.C., 1928). Information on home demonstration clubs also appears in T. J. Woofter, Jr., "Organization of Rural Negroes for Public Health Work," *Proceedings of the National Conference of Social Work* (Chicago, 1923), 72–75; "Activities Report, Birth Control Negro Service," June 21–July 21, 1941, and "Progress Outline 1940–42," box 22, Rose Papers. For information on Jeanes Teachers see, for example, Ross, *Black Heritage,* 211.

42. Information on organizations is based on numerous reports and newsletters from the years 1940–42 in box 22, Rose Papers; see also "Newsletter from Division of Negro Service, December, 1941," box 121, folder 1309, and "PPFA Field Report for California, 1944," box 119, folder 1215, Sanger Papers.

43. "Activities Report, January 1, 1942–February 6, 1942" and "Progress Outline 1940–42," box 22, folder 4, Rose Papers; *Family Guardian* (Massachusetts Mother's Health Council) 5 (December 1939): 3, and 10 (July 1940): 3; "Minutes of the National Advisory Council Meeting, Division of Negro Service," December 11, 1942, box 121, folder 1310, Sanger Papers; Peter Murray, *BCR* 16 (July–August 1932): 216; M. O. Bousefield, *BCR* 22 (May 1938): 92. James Reed notes the opposition of the American Medical Association to alternative forms of health care systems in *From Private Vice to Public Virtue,* Part 4 and 254.

44. "Notes on the Mother's Clinic, Tucson, Arizona," box 119, folder 1212, Sanger Papers; "A Clinic for Tobacco Road," *BCR* 3 (New Series) (January 1936): 6–7; Leonore G. Guttmacher, "Securing Patients for a Rural Center," *BCR* 23 (November 1938): 130–31; "Chas. E. Hall [sic] Census Bureau Expert, Gives Figures for Ten States in Which Number of Children Under Five Shows Decrease," *New York Age,* November 7, 1931, p. 1.

45. "Activities Report, Birth Control Negro Service," November 21, 1942, box 22, Rose Papers.

46. "Project Reports," *Aframerican* (Summer and Fall 1942): 9–24.

16 Contraceptive Consumers: Gender and the Political Economy of Birth Control in the 1930s

Andrea Tone

In 1933 readers of *McCall's* probably noticed the following advertisement for Lysol feminine hygiene in the magazine's July issue:

> The most frequent eternal triangle:
> A HUSBAND ... A WIFE ... and her FEARS
> Fewer marriages would flounder around in a maze of misunderstanding and unhappiness if more wives knew and practiced regular marriage hygiene. Without it, some minor physical irregularity plants in a woman's mind the fear of a major crisis. Let so devastating a fear recur again and again, and the most gracious wife turns into a nerve-ridden, irritable travesty of herself.[1]

Hope for the vexed woman was at hand, however. In fact, it was as close as the neighborhood store. Women who invested their faith and dollars in Lysol, the ad promised, would find in its use the perfect panacea for their marital woes.

ANDREA TONE is Associate Professor of History in the School of History, Technology, and Society at Georgia Institute of Technology, Atlanta, Georgia.

Reprinted from *Journal of Social History* 29 (1996): 485–506, by permission of the *Journal of Social History*. Copyright 1996 by Peter N. Stearns, Carnegie Mellon University.

Feminine hygiene would contribute to "a woman's sense of fastidiousness" while freeing her from habitual fears of pregnancy. Used regularly, Lysol would ensure "health and harmony ... throughout her married life."[2]

The *McCall's* ad, one of hundreds of birth control ads published in women's magazines in the 1930s, reflects the rapid growth of the contraceptive industry in the United States during the Depression. Birth control has always been a matter of practical interest to women and men. By the early 1930s, despite long-standing legal restrictions and an overall decline in consumer purchasing power, it had also become a profitable industry. Capitalizing on Americans' desire to limit family size in an era of economic hardship, pharmaceutical firms, rubber manufacturers, mail-order houses, and fly-by-night peddlers launched a successful campaign to persuade women and men to eschew natural methods for commercial devices whose efficacy could be "scientifically proven." In 1938, with the industry's annual sales exceeding $250 million, *Fortune* pronounced birth control one of the most prosperous new businesses of the decade.[3]

Since the Depression the contraceptive industry's wealth and standing have steadily increased. Yet despite its meteoric rise, the contraceptive industry remains an unexplored chapter of American history. Studying the birth control movement chiefly as a medical or political phenomenon, historians have discounted the social significance of its commercialization. This historiographical lacuna can be explained in part by the belief that technological stagnation forestalled the emergence of a lucrative contraceptive industry prior to the mass marketing of oral contraceptives in 1960. In fact, the technological innovations of the 1960s and 1970s merely fortified the industry's already well-established position. Decades before the pill became a household word, the political economy of birth control in the United States had already been shaped.[4]

It was during the Depression that the structure of the modern contraceptive market emerged. Depression-era manufacturers were the first to create a mass market for contraceptives in the United States. Through successful advertising they heightened demand for commercial birth control while building a permanent consumer base that facilitated the industry's subsequent expansion. Significantly, this consumer constituency was almost exclusively female. Condoms, the most popular commercial contraceptive before the Depression, generated record sales in the 1930s. But it was profits from female contraceptives—sales of which outnumbered those of condoms five to one by the late 1930s—that fuelled the industry's prodigious growth.[5] Then, as now, women were the nation's leading contraceptive consumers.

An important feature distinguished the birth control market of the 1930s from that of today, however: its illegality. Federal and state laws dating from the 1870s proscribed the interstate distribution and sale of contraceptives. Although by the 1920s the scope of these restrictions had been modified by court interpretations permitting physicians to supply contraceptive information and devices in several states, the American

Medical Association's ban on medically dispensed contraceptive advice remained intact. Neither legal restrictions nor medical disapproval thwarted the industry's ascent, however. Instead, they merely pushed the industry underground, beyond regulatory reach.[6]

Contraceptive manufacturers in the 1930s exploited this vacuum to their advantage, retailing devices that were often useless and/or dangerous in a manner that kept the birth control business on the right side of the law. The industry thrived within a gray market characterized by the sale of contraceptives under legal euphemisms.[7] Manufacturers sold a wide array of items, including vaginal jellies, douche powders and liquids, suppositories, and foaming tablets as "feminine hygiene," an innocuous-sounding term coined by advertisers in the 1920s.[8] Publicly, manufacturers claimed that feminine hygiene products were sold solely to enhance vaginal cleanliness. Consumers, literally deconstructing advertising text, knew better. Obliquely encoded in feminine hygiene ads and product packaging were indicators of the product's *real* purpose; references to "protection," "security," or "dependability" earmarked purported contraceptive properties.[9]

Tragically, linguistic clues could not protect individuals from product adulteration or marketing fraud. Because neither the government nor the medical establishment condoned lay use of commercial contraceptives, consumers possessed no reliable information with which to evaluate the veracity of a product's claim. The bootleg status of the birth control racket left contraceptive consumers in a legal lurch. If an advertised product's implied claims to contraceptive attributes failed, they had no acceptable means of recourse.

Within this highly profitable and unfettered trade women became the market's most reliable and, by extension, most exploited customers. The rise of the birth control industry was an important episode in the advance of consumer society in interwar America. Mass production, a predominantly urban population, and innovations in consumer credit supplied the structural underpinning for the expansion of the consumer economy. The advertising industry, manufacturers, retailers, and political leaders provided a concomitant cultural ethos that celebrated the emancipating properties of consumption; the power to purchase was lauded as a desirable, deserved, and quintessentially American freedom.[10] Women became favored recipients of this self-congratulatory encomium. As Susan Porter Benson, Dana Frank, William Leach, Kathy Peiss, and Cynthia Wright have shown, gender has been central to how consumer economies and cultures have been configured.[11] In the 1920s, when advertising consultants agreed that purchases by women accounted for 80 percent of consumer spending (this, in an economy increasingly dependent on consumer sales), the gendered dimensions of consumption were readily apparent. Hoping to influence women's buying behavior, advertisers shrewdly cast women's time-worn role as consumers in a flattering light. Universally endorsing among themselves a psychological profile of the female shopper as mercurial and easily swayed by emotional appeals, advertisers attempted to convince women that consumption was an inherently empowering task. Advertising copy and images accentuated a common theme: that the freedom to choose between Maybelline and Elizabeth Arden lipsticks hallmarked women's newfound authority and liberation in the postsuffrage age.[12]

Depression-era manufacturers and retailers of birth control adopted the same consumption-liberation formula used to sell women lipstick, Hoover vacuum cleaners, and Chrysler cars to construct the first contraceptive mass market in the United States. Just as consumption was trumpeted as a characteristically female freedom, so too was reproduction portrayed as a distinctively female task. On this latter point women needed little convincing. By virtue of biology pregnancy was an exclusively female experience; by virtue of convention raising children in the 1930s was principally a female responsibility. Drawing upon and simultaneously reinforcing the prevailing gender system, the birth control industry reified the naturalness of women's twin roles as consumers and reproducers. Conjoining these functions, manufacturers and retailers urged women to use their purchasing "power" to assume full responsibility for pregnancy prevention. The industry's sales pitch struck a resonant chord with American women in the 1930s. At a time when the cost of raising children was rising and an unprecedented and increasing proportion of the laboring population was officially unemployed, controlling fertility assumed added urgency. With public birth control clinics few in number and privately prescribed diaphragms financially and medically out of reach to most women, access to easily acquired, affordable, and effective birth control became a widely shared goal. With advertisers' prodding, millions of women turned to the contraceptive market to achieve it.[13]

By the 1940s the commercialization of birth control had altered the contraceptive landscape of the nation. Fertility control in general, of course, was not a Depression-era invention; its long history has been well documented by historians. What birth control manufacturers succeeded in doing, however, was to increase the popularity of certain methods. By 1930 approximately 60 percent of white married women practiced some form of fertility control. As a spate of contemporary studies reveals, most of them depended on coitus interruptus (male withdrawal) for pregnancy prevention, with condom use—popular among immigrants and the working class—ranking a close second. By 1940 usage patterns had shifted appreciably. Notwithstanding the significant inroads made by doctor-prescribed diaphgrams and afford-ably priced condoms, the antiseptic douche, only one of many products bearing the feminine hygiene label, was the most popular birth con-trol method in the country. Strikingly, its com-mercial diffusion proved remarkably demo-cratic. The commercial douche was the favored contraceptive not only of middle-class women, who still made up a majority of birth control users, but of contracepting women of *all* eco-nomic classes. When the simultaneous popular-ity of other feminine hygiene products such as suppositories and jellies (used alone) is also considered, the trend toward feminine hygiene use becomes even more pronounced. By 1940 approximately one in three contracepting women depended on feminine hygiene as their primary birth control method. The success of contraceptive manufacturers' campaign was twofold: not only did it encourage more women to use birth control, but it also ensured that the single largest proportion of those who did used female-controlled, commercially acquired con-traceptives.[14]

The successful typecasting of women as con-traceptive consumers reveals the centrality of industry to the history of birth control in America. Manufacturers have not been impar-tial witnesses of contraceptive change; they have consistently tried to influence reproductive practice and guide consumer behavior. While we can reject, in the case of the 1930s, what Ro-land Marchand has dubbed "the hypodermic-needle theory" of advertising—the supposition that companies use advertising to create pre-viously nonexistent demands—we cannot ex-onerate companies' actions altogether.[15] Contra-ceptive manufacturers did not create the desire to control fertility, but they preyed on and com-pounded women's fears of pregnancy to reap higher profits. Printed ads and commissioned door-to-door sales representatives deliberately manipulated women's ignorance of the physiol-ogy of conception to hawk goods that were use-less as contraceptives and dangerous to women's health. Masquerading under the guise of medi-cal science, advertising promised women the lat-est advances in contraceptive technology. What women usually got instead were commercially prepared douches and suppositories less effec-tive than conventional methods of birth control. In addition, ads created new psychological anxi-eties by inflating the social significance of con-traception. If pregnancy signaled impending financial hardship, the means by which it was prevented—absent, of course, of the latest tech-niques and products—could destroy a marriage or ruin a family. Birth control advertising in the 1930s implicitly asked the obvious: Could a woman in the "modern age" afford *not* to buy the newest contraceptives?

When Congress enacted the Comstock Act in 1873, a new nadir in reproductive rights had arrived. The antiobscenity law, the result of the relentless campaigning of its namesake, purity crusader Anthony Comstock, proscribed, among other things, the private or public dis-semination of any

> book, pamphlet, paper, writing, advertise-ment, circular, print, picture, drawing, or other representation, figure, or image on or of paper or other material, or any cast, instrument, or other article of an immoral nature, or any drug or medicine, or any article whatever for the prevention of con-ception.[16]

Passed after minimal debate, the Comstock Act had long-term repercussions. Following Con-gress's lead, most states enacted so-called mini Comstock acts which criminalized the circula-tion of contraceptive devices and information within state lines.[17] Collectively, these restric-tions demarcated the legal boundaries of per-missable sexuality. Sexual intercourse rendered nonprocreative through the use of "unnatural"—that is, purchased—birth control was forbidden. Purity crusaders contended that if properly enforced, the Comstock and mini-Comstock acts would regulate birth control out of existence. Instead, they made birth control an increasingly dangerous, but no less popular, practice.[18]

By the time state and federal legislatures had begun to abandon their laissez-faire attitude toward birth control, a fledging contraceptive industry had already surfaced in the United States. Indeed, the two developments were integrally yoked: the initiative to regulate contraceptives arose out of the realization that there was a growing number to regulate. The nineteenth century witnessed the emergence of a contraceptive trade that sold for profit goods that had traditionally been prepared within the home. Douching powders and astringents, dissolving suppositories, and vaginal pessaries had supplemented male withdrawal and abstinence as mainstays of birth control practice in preindustrial America.[19] As the nineteenth century progressed, these conventional contraceptives became increasingly available from commercial vendors. Technologically upgraded versions of other standard contraceptives also entered the birth control trade. The vulcanization of rubber in the 1840s figured prominently in contraceptive commercialization, expanding birth control options even as it increased individuals' dependence on the market to acquire them. Vulcanization spurred the domestic manufacture of condoms, yielding American-made condoms that were cheaper than imported European condoms made from fish bladders or animal intestines.[20] The subsequent development of seamless condoms made of thinner latex, more appealing to users than earlier models, heightened condom demand. Vulcanization also facilitated the development of female contraceptives by supplying the requisite technology for the manufacture of rubber cervical caps and diaphragms. By the 1870s condoms, douching syringes, douching solutions, vaginal sponges, and cervical caps could be purchased from mail-order houses, wholesale drug supply houses, and pharmacies. Pessaries—traditionally used to support prolapsed uteruses but sold since the 1860s in closed-ring form as "womb veils"—could be obtained from sympathetic physicians. Thus when supporters of the Comstock Act decried the "nefarious and diabolical traffic" of "vile and immoral goods," they were identifying the inroads commercialized contraception had already made.[21]

After the Comstock restrictions were passed, birth control continued to be sold, marketed for its therapeutic or cosmetic, rather than its contraceptive, value. Significantly, however, commercial contraceptive use became more closely associated with economic privilege. The clandestine nature of the market prompted many reputable firms—especially rubber manufacturers—to cease production altogether. Those that remained charged exorbitant prices for what was now illegal merchandise. For many wage-earning and immigrant families the high price of contraceptives made them unaffordable. In addition, the suppression of birth control information reduced the availability of published material on commercial and noncommercial techniques, as descriptions previously featured openly in pamphlets, books, journals, broadsides, and newspaper medical columns became harder to find. In effect, contraceptive information, like contraceptives themselves, became a privileged luxury.[22]

Only in the 1930s were birth control manufacturers able to create a mass market characterized by widespread access to commercial contraceptives. This market developed in response to a combination of important events. The birth control movement of the 1910s and 1920s, spearheaded by Margaret Sanger, made birth control a household word (indeed, it was Sanger who introduced the term) and a topic of protracted debate and heated public discussion. Sanger insisted that women's sexual liberation and economic autonomy depended upon the availability of safe, inexpensive, and effective birth control. Sanger conducted speaking tours extolling the need for female contraception and published piercing indictments of "Comstockery" in her short-lived feminist newspaper, the *Woman Rebel*, the *International Socialist Review,* and privately published pamphlets. In October 1916 she opened in Brooklyn the first birth control clinic in the United States where she instructed neighborhood women on contraceptive techniques. The clinic's closure and Sanger's subsequent jail sentence only increased her notoriety. Sanger was not alone in her efforts to legitimize contraception, of course. The birth control movement was a collective struggle waged by hundreds of individuals and organizations, including IWW locals, women's Socialist groups, independent birth control leagues, and the liberal-minded National Birth Control League.[23] But Sanger's single-minded devotion to the birth control cause and her casual and frequent defiance of the law captured the media spotlight. In the 1910s it was Sanger, more than anyone else, who pushed contraception into the public arena and who, quite unintentionally, set the stage for the commercial exploitation that followed.

The momentum of the birth control movement in the 1910s persisted in the 1920s despite the absence of legislative reform. By the end of the 1920s state and federal legal restrictions on birth control remained operative and unchanged. Doctors-only bills which, had they been successful, would have permitted physicians to prescribe birth control, were introduced and defeated in New York, Connecticut, Pennsylvania, Massachusetts, New Jersey, and California. The Voluntary Parenthood League, suc-

cessor of the now-defunct National Birth Control League, pursued a different tactic. Fearful that a doctors-only strategy would make contraception a privilege of the elite, the League lobbied to have birth control struck from federal obscenity laws. It too was unsuccessful.[24]

Notwithstanding these legislative setbacks, significant advances were made. Capitalizing on a 1918 New York Court of Appeals ruling that exempted physicians from prosecution for prescribing contraception necessary to "cure or prevent disease," Sanger opened the first permanent public birth control clinic in the country in 1923. Within a year the clinic had supplied contraceptive information to 1,208 women.[25] By 1929 the number of medically supervised birth control clinics across the country had increased to twenty-eight, almost all of which were affiliated with Sanger's parent organization, the American Birth Control League.[26] Sanger was also responsible for facilitating the domestic manufacture of diaphragms and spermicidal jellies, clinics' contraception of choice. Frustrated by her inability to interest American manufacturers in the manufacture of female contraceptives, Sanger persuaded her second husband, J. Noah H. Slee, president of the Three-In-One Oil Company, to smuggle German-made Mensinga diaphragms and contraceptive jellies in oil drums across the border near the firm's Montreal plant. The smuggling system worked but not well: the method was unreliable and legally risky, the products acquired too few in number and vastly overpriced. In 1925, with Sanger's urging, Slee financed the Holland-Rantos Company which began manufacturing spring-type diaphragms and lactic acid jelly for Sanger's clinics.[27]

The cumulative effect of these activities—the sensationalist tactics, the organizational impetus, the failed legislative initiatives, and the expansion of public clinics—was to make Americans "birth control conscious." The popularization of the idea of birth control supplied the cultural backdrop to the economic birth control boom of the 1930s. In the absence of government approval and regulation the rising desire for contraceptives provided the perfect environment in which a bootleg trade could thrive. As journalist Elizabeth Garett explained in a 1932 article in the *New Republic,* "so long as contraception was wholly unknown and tabu [sic], saleswomen could not get very far with their prospects. But when 'birth control' became a familiar and at least partially respectable term, all that was needed to induce a woman to order contraceptive wares by mail, or to buy them from peddlers ... was skillful advertising."[28]

As the demand for birth control accelerated, the inability of existing institutions to satisfy it became apparent. By 1932 only 145 public clinics operated to service the contraceptive needs of the nation; twenty-seven states had no clinics at all. Each year in New York City birth control organizations received over ten thousand letters requesting contraceptive information; because of chronic understaffing, most went unanswered.[29] Many women, spurred on by public attention to birth control but unable to secure the assistance needed to make informed contraception choices, took contraception—and their lives—into their own hands. A Chicago physician noted in 1930 with alarm the growing number of doctors reporting the discovery of chewing gum, hair pins, needles, tallow candles, and pencils lodged in female patients' urinary bladders. The doctor blamed these desperate attempts to restrict fertility on the "wave of publicity concerning contraceptive methods that has spread over the country." Equally eager to control reproduction through self-administered means, other women turned to the burgeoning birth control market to purchase what they believed were safe and reliable contraceptives.[30]

That there was a commercial market to turn to was the result of liberalized legal restrictions that encouraged manufacturers to enter the birth control trade. The structure of the birth control industry of the early 1930s was markedly different from that which preceded it only a few years earlier. From 1925 to 1928 Holland-Rantos had enjoyed a monopoly on the manufacture of diaphragms and contraceptive jellies in the United States; other manufacturers expressed little interest in producing articles that might invoke government prosecution and whose market was confined to a handful of nonprofit clinics. A 1930 decision, *Youngs Rubber Corporation, Inc., v. C.I. Lee & Co., et al.,* lifted legal impediments to market entry. The *Youngs* case, in which the makers of Trojan condoms successfully sued a rival company for trademark infringement, forced the court to decide whether the contraceptive business was legal and thus legitimately entitled to trademark protection. The court ruled that insofar as birth control had "other lawful purposes" besides contraception, it could be legally advertised, distributed, and sold as a noncontraceptive device. The outcome of a dispute between rival condom manufacturers, the *Youngs* decision left its most critical mark on the female contraceptive market. Companies that had previously avoided the birth control business quickly grasped the commercial opportunities afforded by the court's ruling. Provided that no reference to a product's contraceptive features appeared in product advertising or on product packaging, female contraceptives could now be legally sold—not only to the small number of birth control clinics in states where physician-prescribed birth control was legal but to the consuming public nationwide. Manufacturers realized that the court's legal latitude would not affect the diaphragm market, monopolized, as it was, by the medical profession. Because diaphragms required a physician's fitting, the number of buyers, given financial and regional obstacles to this type of medical consultation, would remain proportionately small. Jellies, suppositories, and foaming tablets, on the other hand, possessed untapped mass-market potential. They could be used without prior medical screening. And because chemical compounds were cheaper to mass produce than rubber diaphragms, they could be sold at a price more women could afford.[31]

By 1938, only twelve years after Holland-Randos had launched the female contraceptive industry in the United States, at least four hundred other firms were competing in the lucrative market.[32] The $212 million industry acquired most of its profits from the sale of jellies, suppositories, tablets, and antiseptic douching solutions retailed over the counter as feminine hygiene and bought principally by women. Historians of birth control, attentive to the findings of medical studies in the 1930s, have rightly emphasized the rising popularity of diaphragms at this time, especially among urban middle-class women. Progressive physicians and public clinics consistently endorsed combined diaphragm and jelly use as the safest and most effective female-controlled contraception available. But as important as increased diaphragm use was to the medicalization of birth control, its surging popularity was incidental to the escalating profitability of the industry itself. The contraceptive industry thrived in the 1930s precisely because, while capitalizing on public discussions of birth control to which the medical community contributed, it operated outside customary medical channels. Manufacturers supplied women with something that clinics and private physicians did not: birth control that was conveniently located, discretely obtained, and, most important, affordably priced. While the going rate for a diaphragm and a companion tube of jelly ranged from four to six dollars, a dollar purchased a dozen suppositories, ten foaming tablets, or, most alluring of all, up to

three douching units, depending on the brand. Contraceptive manufacturers pledged, furthermore, that customer satisfaction would not be sacrificed on the altar of frugality. They reassured buyers that bargain-priced contraceptives were just as reliable as other methods. Without lay guides to help them identify the disjunction between advertising hyperbole and reality, women could hardly be faulted for taking the cheaper path. By the late 1930s purchases of diaphragms accounted for less than 1 percent of total contraceptive sales.[33]

Manufacturers' grandiose claims aside, not all contraceptives were created alike. The dangers and deficiencies of birth control products were well known in the health and hygiene community. Concerned pharmacists, physicians, and birth control advocates routinely reviewed and condemned commercial preparations. Experts agreed, for instance, that vaginal suppositories, among the most frequently used contraceptives, were also among the least reliable. Suppositories typically consisted of boric acid and/or quinine, ingredients not recognized as effective spermicides. Melting point variability posed an added problem. Suppositories, usually based in cocoa butter or gelatin, were supposed to dissolve at room temperature. In practice, weather extremes and corresponding fluctuations in vaginal temperature made suppositories' diffusion, homogeneity, and contraceptive attributes unpredictable. The "protection" given by foaming tablets was no better. Comprising an effervescent, moisture-activated mixture such as tartaric acid and sodium bicarbonate (which, when triggered, produced a protective foam), tablets often remained inert until *after* male ejaculation.[34]

But critics reserved their harshest comments for the most popular, affordable, and least reliable contraceptive of the day, the antiseptic douche. Noting the method's alarming failure rate—reported at the time to be as high as 70 percent—they condemned the technique as mechanically unsound and pharmacologically in-effectual. For one thing, the method's technique weakened its potential for success: by the time the solution was introduced, seminal fluid that had already penetrated the cervix and sur-rounding tissues was difficult to reach and ne-gate. In addition, the method's ineffectiveness was compounded by the benignity or toxicity of the solutions themselves. Scores of douching preparations, while advertised as modern medi-cal miracles, contained nothing more than wa-ter, cosmetic plant extracts, and table salt. On the other hand, many others, including the most popular brand, Lysol disinfectant, contained cresol (a distillate of coal and wood) or mercury chloride, either of which, when used in too high a concentration, caused severe inflammation, burning, and even death. Advertising down-played the importance of dilution by drawing attention to antiseptics' gentleness and versatil-ity; single ads praising Lysol's safety on "delicate female tissues" also encouraged the moneywise consumer to use the antiseptic as a gargle, nasal spray, or household cleaner. By the same token, the makers of PX, a less-known brand, sold a liquid disinfectant that ads claimed could be used interchangeably for "successful woman-hood" or athlete's foot.[35]

This strategy won sales, but it did so only by jeopardizing women's health. With even one-time douching a potentially deleterious act women, guided by the logical assumption that "more was better," strove to beat the pregnancy odds by increasing the frequency of their douch-ing and the concentration of the solution used. In one case a nineteen-year-old married woman relied on regular douching with dissolved mer-cury chloride tablets for birth control. Eager to avoid pregnancy, she doubled the dose and douched "several times daily." Her determina-tion landed her in a doctor's office where she was diagnosed with acute vaginal and cervical burns. In what must have seemed to her like a grave injustice, she also learned she was pregnant.[36]

Reports on douche-related deaths and injur-ies and the general ineffectiveness of popular commercial contraceptives were widely dis-cussed among concerned constituents of the health community. Sadly, however, these find-ings failed to prod the medical establishment as a united profession to take a resolute stand against the contraceptive scandal. Nor, regretta-bly, did blistering indictments of manufacturing fraud trickle down to the lay press where they might have enabled women to make informed contraceptive choices. The numerous women's magazines that published feminine hygiene ads—from *McCall's* to *Screen Romances* to the *Ladies' Home Journal*—were conspicuously si-lent about the safety and efficacy of the products they tacitly endorsed. The paucity of informa-tion impeded the development of informed consumerism. In advertising text and in many women's minds the euphemism "feminine hy-giene" continued to signify reliable contracep-tion. For unscrupulous manufacturers eager to profit from this identification, feminine hygiene continued to be a convenient term invoked to sell products devoid of contraceptive value.

Manufacturers absolved themselves of re-sponsibility by reminding critics that by the let-ter of the law, their products were not being sold as contraceptives. If women incurred injuries or became pregnant while using feminine hygiene for birth control, that was their fault, not manu-facturers'. Thus contraceptive firms whose profits depended on consumers' loose and lib-eral deconstruction of advertising text duplici-tously clung to a rigid literalist construction of language when defending their own integrity. The Norwich Pharmacal Company, for example, manufacturers of Norforms, the most popular brand of vaginal suppositories in the country, deployed precisely such an argument to justify its advertising policy. Norform suppositories were advertised exclusively as feminine hygiene, a term that the company's vice president, Webster Stofer, conceded had become synonymous with contraception in many women's minds. All the same, Stofer insisted, Norforms were not sold as birth control. Asked why the company did not then change its marketing slogan to avoid misunderstanding, Stofer expressed his regret that it was "too late" to advertise suppositories as anything else. "The term has become too closely associated with Norforms," Stofer contended. "And anyway, we have our own definition of it."[37]

Added to the growing list of groups unwilling to expose the hucksterism of the birth control bonanza was the federal government. Neither the Food and Drug Administration (FDA) nor the Federal Trade Commission (FTC) was in a strong position to rally to consumers' aid. The FDA, authorized to take action only against product mislabeling, was powerless to suppress birth control manufacturers' rhetorically veiled claims. The FTC, in turn, regulated advertising but only when one company's claims were so egregious as to constitute an unfair business practice. The subterfuge prevalent in all feminine hygiene marketing campaigns, as well as a unanimous desire on manufacturers' part to eschew protracted scrutiny, kept the FTC at bay. Sadly for the growing pool of female contraceptive consumers, without regulation and reliable standards for discriminating among products, the only way to discern a product's safety and efficacy was through trial and error.[38]

Clamoring for a larger share of the hygiene market, manufacturers did their utmost to ensure that their product would be one women would want to try. Aggressive advertising was instrumental to the industry's success. Appealing to women in the privacy of their homes, feminine hygiene companies blanketed middle-class women's magazines in the 1930s with advertisements, many of full-page size. Targeting

the magazines' predominantly married readership, advertisements were headlined by captions designed to inculcate and inflate apprehensions in readers' minds. Ads entitled "Calendar Fear," "Can a Married Woman Ever Feel Safe?" "Young Wives Are Often Secretly Terrified," and "The Fear That 'Blights' Romance and Ages Women Prematurely" relied on standard negative advertising techniques to heighten the stakes of pregnancy prevention.[39]

Ads conveyed the message that ineffective contraception led not only to unwanted pregnancies but also to illness, despair, and marital discord. Married women who ignored modern contraceptive methods were courting lifelong misery. "Almost before the honeymoon ends," one ad warned, "many a young bride is plagued by foreboding. She pictures the early departure of youth and charm . . . sacrificed on the altar of marriage responsibilities." Engulfed by fear, the newlywed's life only got worse—fear itself, women were told, engendered irreparable physical ailments. According to one douche advertisement, fear was a "dangerous toxin." "[It] dries up valuable secretions, increases the acidity of the stomach, and sometimes disturbs the bodily functions generally. So it is that FEAR greys the hair . . . etches lines in the face, and hastens the toll of old age."[40]

As if these physical penalties were not disconcerting enough, feminine hygiene ads insisted that a woman's apprehensions and their attendant woes could ruin the marriage itself. On this point the transcendent parable of ads was clear: the longevity of a marriage depended upon the right commercial contraception. "She was a lovely creature before she married," one ad began, "beautiful, healthy, and happy. But since her marriage she seems forever worried, nervous and irritable . . . always dreading what seems inevitable. Her husband, too, seems to share her secret worry. Frankly, they are no longer happy. Poor girl, she doesn't know that she's headed for the divorce court."[41] And as ads—whose sole purpose was to convince women, not men, to buy contraceptives—hastened to remind readers, women alone shouldered the blame for divorce. After all, why should a man be held accountable for distancing himself from a wife made ugly and cantankerous by her own anxieties? "Many marriage failures," one advertisement asserted authoritatively, "can be traced directly to disquieting wifely fears." "Recurring again and again," marriage anxieties were "capable of changing the most angelic nature, of making it nervous, suspicious, irritable." "I leave it to you," the ad concluded, "is it easy for even the kindliest husband to live with a wife like that?"[42]

Having divulged the ugly and myriad hazards of unwanted pregnancy while saddling women with the burden of its prevention, advertisements emphasized that peace of mind and marital happiness were conditions only the market could bestow. Readers of feminine hygiene ads, newly enlightened, returned to the world with the knowledge necessary to "remove many of their health anxieties, and give them that sense of well being, personal daintiness and mental poise so essential to wifely security." In the modern age the personal tragedies accompanying a woman's existence were easily avoided. "Days of depressing anxiety, a wedded life in which happiness is marred by fear and uncertainty—these need be yours no longer," one douche ad reassured. In the imagined world of contraceptive advertising, feminine hygiene was the commodity no modern woman could afford to be without. Fortunately, none had to. The path to unbridled happiness was only a store away.[43]

As advertisements reminded prospective customers, however, not all feminine hygiene products were the same. The contraceptive consumer had to be discriminating. Hoping both to increase general demand for hygiene products and

to inculcate brand loyalty, manufacturers presented their product as the one most frequently endorsed for its efficacy and safety by medical professionals. Dispelling consumer doubts by invoking the approval of the scientific community was not an advertising technique unique to contraceptive merchandising—the same strategy was used in the 1930s to sell women laxatives, breakfast cereal, and mouth wash. What was exceptional about contraceptive advertising, however, was that the experts endorsing feminine hygiene were not men. Rather, they were female physicians whose innate understanding of the female condition permitted them to share their birth control expertise "woman to woman."[44]

The Lehn and Fink corporation used this technique to make Lysol disinfectant douche the leading feminine hygiene product in the country.[45] In a series of full-page advertisements entitled "Frank Talks by Eminent Women Physicians," stern-looking European female gynecologists urged "smart-thinking" women to entrust their health only to doctor-recommended Lysol disinfectant douches. "It amazes me," wrote Dr. Madeleine Lion, "a widely recognized gynaecologist of Paris,"

in these modern days, to hear women confess their carelessness, their lack of positive information, in the so vital matter of feminine hygiene. They take almost anybody's word . . . a neighbor's, an afternoon bridge partner's . . . for the correct technique. . . . Surely in this question of correct marriage hygiene, the modern woman should accept only the facts of scientific research and medical experience. The woman who does demand such facts uses "Lysol" faithfully in her ritual of personal antisepsis.[46]

Another ad, part of the same series, underscored the point. "It is not safe to accept the counsels of the tea table," explained Dr. Auguste Popper, a female gynecologist from Vienna, "or the advice of a well-meaning, but uninformed friend." Only the advice of scientific experts could be trusted. While feminine hygiene "has alleviated woman's oldest fear," an Italian gynecologist advised readers in yet another Lysol douche ad, the greatest obstacle to realizing health and happiness lay in selecting the right hygiene merchandise: "Some are good, some are not." "My own preference is for 'Lysol,'" the gynecologist concluded, "in common with every other doctor I know."[47]

While insisting that women defer to medical opinion when choosing birth control, contraceptive ads simultaneously celebrated the tremendous "power" women wielded in the consumer market. The two claims were not antithetical; advertisements contended that women who heeded physicians' advice and purchased "scientific" birth control were intelligently harnessing the advances of modern medicine to promote their own liberation. Consistent with the consumer ethic of the day, birth control advertising successfully equated contraceptive consumption with female emancipation. An ad by the Zonite Products Corporation claimed that birth control was not only a matter of pragmatism but also a "protest against those burdens of life which are wholly woman's." When it came to as important an issue as birth control, Zonite explained, the modern woman was not interested in the "timid thoughts of a past generation"; her goal was "to find out and be sure." It was no surprise, the company boasted, that Zonite hygiene products were favored by "women of the independent, enlightened type all over the world."[48]

Contraceptive manufacturers' creation of a mass market in the 1930s depended not only upon effective advertising but also on the availability of advertised goods. Prospective customers needed quick, convenient, and multiple access to contraceptives. Manufacturers made sure that they had it. Flooding a wide array of commercial outlets with their merchandise, companies guaranteed that contraceptives became a commodity within everyone's reach. Here again gender was the crucial variable, determining product availability and sales venue. Condoms were sold in pharmacies but also in news stands, barber shops, cigar stores, and gas stations—locations where men were most likely to congregate. Women, on the other hand, were targeted in more conventional female settings: in stores and in the home.[49]

The department store became the leading distributor of female contraceptives in the 1930s. By the mid-1930s women could purchase feminine hygiene products at a number of national chains, including Woolworth, Kresge, McLellan, and W. T. Grant.[50] Already fashioned as a feminized space, department stores established sequestered "personal hygiene" departments where women could shop in a dignified and discreet manner for contraceptives and other products related to female reproduction such as sanitary napkins and tampons. Stores emphasized the exclusively female environment of the personal hygiene department as the department's finest feature. The self-contained department was not only separated from the rest of the store, where "uncontrollable factors . . . might make for . . . embarrassment," but it was staffed solely by saleswomen trained in the "delicate matter of giving confidential and intimate personal advice to their clients." As one store assured female readers in the local newspaper: "Our Personal Hygiene Department [has] Lady Attendants on Duty at all Times." Female clerks, furthermore, were instructed to respect the private nature of the department's transactions; sensitive, knowledgeable, and tactful, they were "understanding wom[en] with whom you may discuss your most personal and intimate problems."[51]

Contraceptive manufacturers actively promoted the creation of personal hygiene departments by emphasizing to store owners and managers the revenues their establishment would generate. Advertisements in retailing trade journals such as *Chain Store Age* recounted a plethora of feminine hygiene sales success stories; although the ads varied, their transcendent morale told the same good news: selling feminine hygiene guaranteed a higher volume of customers and sales. The Zonite Products Corporation warned retailers not to miss out on the hygiene bonanza. "Did you know that feminine hygiene sales are six times greater than combined dentifrice sales?" one Zonite ad queried. "You'll be amazed [at] the way your sales and profits ... will soar by simply establishing a feminine hygiene department. By this simple plan, many dealers have tripled volume almost overnight." Zonite offered free company consultations and sales training to encourage store managers to establish hygiene departments. Other firms with the same goal in mind sent complimentary counter displays, dispensing stands for "impulse sales and quick service," and window exhibits that could be strategically placed "where women predominate in numbers." An economic incentive undergirded company's promotional activities. The establishment of hygiene departments firmly committed stores to the long-term retailing of feminine hygiene products, while the dignified decorum of departments lent an air of credibility and legitimacy to the products themselves.[52]

Manufacturers reasoned that many prospective female customers would not buy feminine hygiene in a store. Many did not live close enough to one, while others, notwithstanding the store's discretion, might remain uncomfortable with the public nature of the exchange. To eliminate regional and psychological obstacles to birth control buying, companies sold feminine hygiene to women directly in their homes. Selling contraceptives by mail was one such method. Mail-order catalogues, including those distributed by Sears, Roebuck and Montgomery Ward, offered a full line of female contraceptives; each catalogue contained legally censored ads supplied by manufacturers. As a reward for bulk sales, mail-order houses received a discount from the companies whose products they sold. Other manufacturers bypassed jobbers and encouraged women to send their orders directly to the company. To eliminate the possibility of embarrassment ads typically promised that the order would be delivered in "plain wrapper."[53]

To create urban and working-class markets dozens of firms hired door-to-door sales representatives to canvass urban districts. All representatives were women, a deliberate attempt on manufacturers' part to profit from the prudish marketing scheme that tried to convince women that, as one company put it, "There are some problems so intimate that it is embarrassing to talk them over with a doctor."[54] At the Dilex Institute of Feminine Hygiene, for example, five separate female crews, each headed by a female crew manager, combed the streets of New York. The cornerstone of the company's marketing scheme was an aggressive sales pitch delivered by saleswomen dressed as nurses. As *Fortune* discovered in an undercover investigation, however, the Dilex canvassers had no medical background. In fact, the only qualification required for employment was previous door-to-door sales experience. Despite their lack of credentials, newly hired saleswomen were instructed to assume the role of the medical professional, a tactic the institute reasoned would gain customers' trust, respect, and dollars. "You say you're a nurse, see?" one new recruit was told, "That always gets you in." Canvassers walked from house to house delivering by memory the standard Dilex sales speech:

Good Morning. I am the Dilex Nurse, giving short talks on feminine hygiene. It

will take only three minutes. Thank you—
I will step in.

Undoubtedly you have heard of many different methods of feminine hygiene, but I have come to tell you of THE DILEX METHOD, which is so much more simple and absolutely sure and harmless, and which EVERY woman is so eager to learn about and have without delay.

At one time this was a very delicate subject to discuss, but today with all our modern ideas, we look at this vital subject as one of the most important of all time, and for that reason, we call to acquaint you with THIS GREAT SECRET, a most practical, convenient way.

The Dilex Method meets every protective and hygienic requirement. It is positive and safe and may be used with the utmost confidence. Each item has been given the most careful thought to fit the increasing strides in feminine hygiene.... ABSOLUTE FEMININE PROTECTION is assured.[55]

The saleswoman then attempted to peddle the company's top-of-the-line contraceptive kit. For seven dollars a woman could purchase jelly, a douching outfit, an antiseptic douche capsule, and—most alarming of all—a universal "one-size-fits-all" diaphragm. Poverty, women were told, was not an impediment to the personal happiness the company was selling: "luckily" for them, the Dilex kit was available on the installment plan.[56]

Contraceptive companies' tactics paid off. By 1940 the size of the female contraceptive market was three times that of the 1935 market.[57] The industry's unabated growth continued despite important changes in legal interpretation and medical attitudes in the late 1930s that might have reduced the industry's hold over American women. In 1936 the Supreme Court's *One Pack-*

age decision allowed physicians in every state to send and receive contraceptive devices and information. The following year the American Medical Association reversed its long-standing ban on contraception, endorsing the right of a physician to prescribe birth control. The Court's decision and the AMA's liberalized policy did not foster the immediate medicalization of birth control, a process that might have encouraged women to turn to the medical profession instead of the market for contraception. Indeed, in the short term these sweeping changes proved remarkably inconsequential to the state of the industry. Many Americans could not afford the luxury of a personal physician, and only a minority lived close enough to the 357 public birth control clinics operating in 1937 to avail themselves of clinic services. But of even more significance than medical barriers was manufacturers' enticing sales message. Companies' pledges to supply birth control that was affordable, immediate, and discretely sold—either anonymously or in a completely feminized setting—continued to strike a responsive chord with American women. In addition, manufacturers promised what no lay guide could dispute: that what was bought from the market was as effective as doctor-prescribed methods. Out of pragmatic necessity and personal preference, most women worried about pregnancy prevention continued to obtain birth control from the contraceptive market.[58]

The gradual expansion of government regulation of commercial contraceptives had a similarly negligible impact on the feminine hygiene sector of the industry. In 1938 the Federal Food, Drug, and Cosmetic Act enlarged the Food and Drug Administration's regulatory powers, authorizing the government agency to hold medical devices to some of the same standards as drugs. Paralleling the federal government's campaign to eliminate gonorrhoea and syphilis, the Food and Drug Administration announced that condoms, as articles sold to prevent venereal disease, would henceforth be monitored for product defects. Enforcing this policy, the FDA destroyed seventy-five batches of defective condoms in 1938 and 1939. FDA regulations encouraged manufacturers of condoms to adopt stricter quality control measures. They did little, however, to safeguard the health and safety of feminine hygiene consumers. Because manufacturers of feminine hygiene promoted their products as agents of vaginal cleanliness, their activities were protected by a linguistic legal safety net that made it nearly impossible for federal agencies to prosecute manufacturers for false advertising or technical flaws. Advertised and sold as useless goods, feminine hygiene products were beyond government reproach. Rather, ensconced in the interstice of advertising artifice and consumer expectation, the hygiene racket continued to thrive during the early 1940s, eclipsed only by the popularity of the condom and the diaphragm in the postwar era.[59]

In a 1936 letter to Harrison Reeves, a New York journalist studying the commercial aspects of contraception for the *American Mercury,* Margaret Sanger reflected on the state of the birth control business in America. Sanger was no stranger to the commercial scene. Since the early 1930s she had instructed her secretary to clip all commercial advertisements for birth control for Sanger's personal review. Sanger corresponded frequently with manufacturers eager to obtain her endorsement of new products. From direct involvement in the daily operation of public clinics that she had founded, to the multiple tests on birth control products conducted at her request by doctors at the Birth Control Clinical Research Bureau, Sanger was keenly aware of the perils and pitfalls of commercial contraception.[60]

And yet, in what amounted to more than a loyal defense of her husband's business activi-

ties, Sanger refused to vilify manufacturers for the commercial hucksterism, fraud, and misinformation that so many of them had spawned. As she explained to Reeves, "I do not feel as many do about manufacturing concerns.... They have not lagged behind like the medical profession but have gone ahead and answered [a] growing and urgent need."[61]

Sanger's observation, although anchored in what in hindsight appears to be misplaced charity, perspicaciously speaks to the expanding role of manufacturers in shaping contraceptive practices in the 1930s. The strictures and liberties of the law, the inertia of the medical profession, and the determination of American women to find affordable and effective female-controlled birth control, provided new economic opportunities that manufacturers eagerly seized. By the end of the 1930s manufacturers had created a lively and vigorous market that could be easily accessed in stores, by mail, and within the home. Drawing on a gendered culture that designated consumption and reproduction female roles, manufacturers implored women to "purchase" their happiness and security from the contraceptive market. Reinforcing Victorian sensibilities about female sexuality that self-servingly bolstered their marketing scheme, they feminized sites of birth control buying. Instead of visiting a male doctor or druggist, women were encouraged to acquire birth control in sex-segregated departments staffed by "discreet" female attendants, from visiting saleswomen, or through the mail upon the advertised recommendation of female physicians "who know."

The escalation of industry profits followed closely on the heels of the construction of the female contraceptive consumer; by World War II not only did sales of feminine hygiene products surpass those of condoms but more women depended on feminine hygiene for birth control than on any other method. Tragically, the very legal climate that permitted the birth control business to flourish in a bootleg state also encouraged it to peddle inferior goods. For too many women the freedom, pleasure, and security pledged by contraceptive manufacturers amounted to nothing more than empty promises.

The commercialization of birth control in the 1930s illuminates the important but overlooked role of industry in shaping birth control developments in the United States. Historians have typically framed birth control history as a tale of doctors, lawmakers, and women's rights activists. The events of the 1930s suggest that we need to recast this story to include the agency of a new set of actors, birth control manufacturers. The commercialization that manufacturers engendered at this time left an indelible imprint on the lives of ordinary women and men. It also revealed a world in which industry, gender, and reproduction were frequently and intimately intertwined.

Notes

The author wishes to thank John Tone, Michael Bellesiles, Peter Stearns, and Nancy Cott for their helpful comments and suggestions. Research for this article was funded by a President's Research Grant from Simon Fraser University.

1. *McCall's* 60 (July 1933): 85.

2. *McCall's* 60 (July 1933): 85.

3. "The Accident of Birth," *Fortune* (February 1938): 84.

4. The best treatments of the history of birth control in the United States have examined the rise of the contraceptive industry only peripherally. See, for instance, Linda Gordon's otherwise brilliant *Woman's Body, Woman's Right: A Social History of Birth Control in America* (New York, 1976); David M. Kennedy, *Birth Control in America: The Career of Margaret Sanger* (New Haven, 1970); James Reed, *From Private Vice to Public Virtue: The Birth Control Movement and American Society Since 1830* (New York, 1978); and Ellen Chesler, *Women of Valor: Margaret Sanger and the Birth Control Movement in America* (New York, 1992).

5. According to *Fortune*, sales from condoms accounted for $38 million of the industry's annual $250 million sales. See "The Accident of Birth," 84.

6. For a discussion of the relationship between the American Medical Association and the birth control movement, see J. M. Ray and F. G. Gosling, "American Physicians and Birth Control, 1936–1947," *Journal of Social History* 18 (1985): 399–408.

7. I have elected to use the term *gray market* to describe the sale of goods that, as they were marketed, were strictly legal but which, had they been packaged and labeled to reflect their intended application and purpose, would not have been. I am grateful to Michael Fellman for suggesting the suitability of the term to this study.

8. As one advertising leaflet put it, "Feminine hygiene is the 'nice' term . . . invented for the care and cleanliness of the vaginal tract from its outer opening to the cervix." Quoted in Rachel Lynn Palmer and Sara K. Greenberg, *Facts and Frauds in Woman's Hygiene: A Medical Guide Against Misleading Claims and Dangerous Products* (New York, 1938), 18.

9. Elizabeth H. Garrett, "Birth Control's Business Baby," *New Republic* (17 January 1934): 270; Dorothy Dunbar Bromley, "Birth Control and the Depression," *Harper's* (October 1934): 563; "The Accident of Birth," 110, 112.

10. For a discussion of the consolidation of American consumer society in the 1920s and 1930s, see Richard Wrightman Fox, "Epitaph for Middletown: Robert S. Lynd and the Analysis of Consumer Culture," in Richard Wrightman Fox and T. J. Jackson Lears, eds., *The Culture of Consumption: Critical Essays in American History, 1880–1980* (New York, 1983); Daniel J. Boorstin, *The Americans: The Democratic Experience* (New York, 1973), especially part two; Roland Marchand, *Advertising the American Dream: Making Way for Modernity, 1920–1940* (Berkeley, 1985); Stuart Ewen, *Captains of Consciousness: Advertising and the Social Roots of the Consumer Culture* (Toronto, 1976); Gary Cross, *Time and Money: The Making of Consumer Culture* (London, 1993), chapter six; Lizabeth Cohen, *Making a New Deal: Industrial Workers in Chicago, 1919–1939* (New York, 1990).

11. See Susan Porter Benson, *Counter Cultures: Saleswomen, Managers, and Customers in American Department Stores* (Urbana, 1986); Dana Frank, "Gender, Consumer Organizing, and the Seattle Labor Movement, 1919–1929," in Ava Baron, ed., *Work Engendered: Toward a New History of American Labor* (Ithaca, 1991); Kathy Peiss, "Making Faces: The Cosmetics Industry and the Cultural Construction of Gender, 1890–1930," *Genders* 7 (Spring 1990); William R. Leach, "Transformation in a Culture of Consumption: Women and Department Stores, 1890–1925," *Journal of American History* 71 (September 1984); Cynthia Wright, "Feminine Trifles of Vast Importance: Writing Gender into the History of Consumption," in Franca Iacovetta and Mariana Valverde, eds., *Gender Conflicts* (Toronto, 1992).

12. Christine Frederick, *Selling Mrs. Consumer* (New York, 1929), 43–44. For general discussions of women and American advertising in the 1920s, see Nancy F. Cott, *The Grounding of Modern Feminism* (New Haven, 1987), 170–74; Marchand, *Advertising the American Dream*, 66–69, 179–85; and Ewen, *Captains of Consciousness*, 159–76. Ruth Schwartz Cowan's insightful "The 'Industrial Revolution' in the Home: Household Technology and Social Change in the Twentieth Century," *Technology and Culture* 17 (1976) explores how household appliances were advertised to women in the 1920s. Advertisers in the 1920s consciously attempted to eradicate earlier, derisive perceptions of women's role as consumers by portraying female consumption as both psychologically fulfilling *and* economically functional. A sexual division of labor predating capitalist economic and social relations had designated the majority of household consumption women's work. The advance of industrial capitalism in the late eighteenth and early nineteenth centuries brought divisions between the household and the external market into sharper relief; in redefining "real" work as only that which possessed a tangible remunerative value, it amplified perceptions that women—largely excluded from wage labor—were dependent consumers rather than productive independent breadwinners.

13. A poll published in *Ladies' Home Journal* in 1938 found that 79 percent of American women surveyed favored birth control. The most frequent argument given in its favor was economic considerations.

See Henry F. Pringle, "About Birth Control," *Ladies' Home Journal* 55 (March 1938): 15.

14. See Gordon, *Woman's Body, Woman's Right*, 48–49, 62–64; Daniel Scott Smith, "Family Limitation, Sexual Control and Domestic Feminism in Victorian America," *Feminist Studies* 1 (Winter–Spring 1973): passim. Preclinic contraceptive methods were well catalogued by physicians. See Marie E. Kopp, *Birth Control in Practice: Analysis of Ten Thousand Case Histories of the Birth Control Clinical Research Bureau* (New York, 1934); Raymond Pearl, "Contraception and Fertility in 4945 Married Women: A Second Report on a Study in Family Limitation," *Human Biology* 6 (1934); Hannah M. Stone, "Maternal Health and Contraception: A Study of 2,000 Patients from the Maternal Health Center, Newark, N.J.," *Medical Journal and Record* (April 19, 1933 and May 3, 1933); "Feminine Hygiene Market," *Drug and Cosmetic Industry* 38 (May 1936): 647. The impact of the commercialization of contraception in the 1930s on contraceptive practice is documented by John Winchell Riley and Matilda White in "The Use of Various Methods of Contraception," *American Sociological Review* 5 (December 1940): 890–903 and discussed in Lee Rainwater, *And the Poor Get Children: Sex, Contraception, and Family Planning in the Working Class* (Chicago, 1960), 162.

15. Marchand, *Advertising the American Dream*, xx. Marchand himself rejects this theory, although he acknowledges "the power of frequently repeated media images and ideas to establish broad frames of reference, define the boundaries of public discussion, and determine relevant factors in a situation."

16. *Acts and Resolutions of the United States of America Passed at the Third Session of the Forty-Second Congress* (Washington, D.C., 1873), 234–35.

17. Twenty-four states enacted similar laws banning the circulation of contraceptives and contraceptive knowledge. The Connecticut legislature, heavily influenced by the state's Catholic constituency, earned the special distinction of criminalizing the very *use* of birth control (a prohibition not overturned until *Griswold v. Connecticut* almost a century later). Twenty-two other states also passed or strengthened existent obscenity laws, enabling prosecution of the purchase of contraceptives under the penumbra of the federal law. For a contemporary

analysis of legal restrictions see Mary Ware Dennett, *Birth Control Laws: Shall We Keep Them, Change Them, or Abolish Them* (New York, 1970 [1926]), passim.

18. For a general discussion of the impact of the Comstock Act, see Reed, *From Private Vice to Public Virtue*, ch. 3.

19. Gordon, *Woman's Body, Woman's Right*, 48–49, 64–71; Reed, *From Private Vice to Public Virtue*, 4–13; Norman Himes, *Medical History of Contraception* (Baltimore, 1936), ch. 11.

20. See Himes, *Medical History of Contraception*, 201–4; Vern L. Bullough, "A Brief Note on Rubber Technology and Contraception: The Diaphragm and the Condom," *Technology and Culture* 22 (January 1981): 104, 107–11; Reed, *From Private Vice to Public Virtue*, 13–15. According to Carl Degler, the price of condoms dropped by approximately 40 percent between 1847 and 1865. See Degler, *At Odds: Women and the Family in America from the Revolution to the Present* (New York, 1980), 219–20.

21. Michael A. LaSorte, "Nineteenth-Century Family Planning Practices," *Journal of Psychohistory* 4 (Fall 1976): 175–76; Bullough, "A Brief Note on Rubber Technology and Contraception," 105–6; Reed, *From Private Vice to Public Virtue*, 15–17; Gordon, *Woman's Body, Woman's Right*, 67–70.

22. Anthony Comstock, appointed U.S. Post Office inspector to enforce the Comstock Act, boasted that he had singlehandedly destroyed 160 tons of obscene literature between 1873 and 1915, bringing 3,760 "criminals" to "justice." See Margaret H. Sanger, "Comstockery in America," *International Socialist Review* 16 (July 1915): 46. In general, wealthy families were better situated than working-class families to circumvent legal obstacles—regardless of whether circumvention entailed retaining the services of a progressive physician or purchasing condoms from a private supplier. Indeed, statistics assembled by birth control clinics in the 1920s and 1930s on preclinic practices underscore the degree to which class background affected commercial contraceptive use. Clinic interviewers discovered that birth control in general, and commercial contraceptives in particular, were more likely to be used by women from professional and business background than by those with a working-class status. Many working-class women

purported to have no working knowledge of either the principles or the mechanics of contraception. See Degler, *At Odds*, 221–22 and Riley and White, "The Use of Various Methods of Contraception," 897–99. Also see Robert S. Lynd and Helen Merrell Lynd, *Middletown: A Study in Modern American Culture* (New York, 1929), 125, for similar observations.

23. Gordon, *Woman's Body, Woman's Right*, 231–32; Reed, *From Private Vice to Public Virtue*, ch. 9, passim; David Kennedy, *Birth Control in America*, 70–71.

24. Kennedy, *Birth Control in America*, 218–21.

25. Reed, *From Private Vice to Public Virtue*, 141.

26. "The Accident of Birth," 85; Bromley, "Birth Control and the Depression," 566.

27. "Accident of Birth," 108; Kennedy, *Birth Control in America*, 183; Reed, *From Private Vice to Public Virtue*, 114–15.

28. Garrett, "Birth Control's Business Baby," 269.

29. Bromley, "Birth Control and the Depression," 566; James Rorty, "What's Stopping Birth Control," *New Republic* (February 3, 1932): 313.

30. Dorrin F. Rudnick, "A New Type of Foreign Body in the Urinary Bladder," *Journal of the American Medical Association* 94 (May 17, 1930): 1565.

31. *Youngs Rubber Corporation, Inc. v. C. I. Lee & Co., et al.*, 45 *Federal Reporter*, 2nd Series 103; Morris L. Ernst, "How We Nullify," *Nation* (January 27, 1932): 114; Garrett, "Birth Control's Business Baby," 270. The effectiveness of contraceptive jellies when used alone was well documented. See "The Accident of Birth," 85.

32. "The Accident of Birth," 108.

33. Reed, *From Private Vice to Public Virtue*, 244–46; Ray and Gosling, "American Physicians and Birth Control, 1936–1947," passim; Riley and White, "The Use of Various Methods of Contraception," 896–900; "The Accident of Birth," 84; "Feminine Hygiene Products Face a New Marketing Era," *Drug and Cosmetic Industry* 37 (December 1935): 745; Harrison Reeves, "The Birth Control Industry," *American Mercury* 155 (November 1936): 287; "Birth Control Industry," *Drug and Cosmetic Industry* 46 (January 1940): 58; "Building Acceptances for Feminine Hygiene Products," *Drug and Cosmetic Industry* 38 (February 1936): 177.

34.   Robert L. Dickinson and Louise Stevens Bryant, *Control of Conception: An Illustrated Medical Manual* (Baltimore, 1931), 78–80; Dorothy Dunbar Bromley, *Birth Control: Its Use and Misuse* (New York, 1934), 99–100; Palmer and Greenberg, *Facts and Frauds in Woman's Hygiene,* 242–50.

35.   Dickinson and Bryant, *Control of Conception,* 39–45, 69–74; Bromley, *Birth Control,* 92–98; Palmer and Greenberg, *Facts and Frauds in Woman's Hygiene,* 12–15, 142–51; Lysol ad from pamphlet by Dr. Emil Klarmann, *Formula L-F: A New Antiseptic and Germicide* (Lehn & Fink Inc.) appended to letter from Lehn & Fink to Margaret Sanger, November 24, 1931, reel 29, Margaret Sanger Papers, Library of Congress; PX ad from Margaret Sanger Papers, box 232, folder "Commercial Advertisements, 1932–34," Library of Congress.

36.   "Effects of Corrosive Mercuric Chloride ('Bichloride') Douches," *Journal of the American Medical Association* 99 (August 6, 1932): 497.

37.   "The Accident of Birth," 110–12.

38.   Garrett, "Birth Control's Business Baby," 270–71; "The Accident of Birth," 110, 112; Bromley, "Birth Control and the Depression," 572; Reed, *From Private Vice to Public Virtue,* 114, Kennedy, *Birth Control in America,* 183; Palmer and Greenburg, *Facts and Frauds in Woman's Hygiene,* 21–24.

39.   For sample captions see Bromley, "Birth Control and the Depression"; the advertisement captioned "The Fear That Blights Romance and Ages Women Prematurely" is from *McCall's* 60 (October 1932): 102.

40.   "The Incompatible Marriage: Is it a Case for Doctor or Lawyer?" *McCall's* 60 (May 1933): 107; "The Fear that Blights Romance and Ages Women Prematurely," *McCall's* 60 (October 1932): 102.

41.   Advertisement cited in Garrett, "Birth Control's Business Baby," 271; "The Incompatible Marriage," *McCall's* 60 (May 1933): 107.

42.   "The Incompatible Marriage," *McCall's* 60 (May 1933): 107.

43.   "The Incompatible Marriage," *McCall's* 60 (May 1933): 107; Garrett, "Birth Control's Business Baby," 271; J. Rorty, "What's Stopping Birth Control?" *New Republic* 65 (January 28, 1931): 292–94.

44.   Mary P. Ryan, "Reproduction in America," *Journal of Interdisciplinary History* 10 (Autumn 1979): 330; Ray and Gosling, "Physicians and Birth Control," 405; Marchand, *Advertising the American Dream,* passim.

45.   "The Accident of Birth," 112.

46.   "The Serene Marriage . . . Should It be Jeopardized by Needless Fears?" *McCall's* 65 (December 1932): 87.

47.   "The Fear That Blights Romance and Ages Women Prematurely," *McCall's* 60 (October 1932): 64; "No Wonder Many Wives Fade Quickly with This Recurrent Fear," *McCall's* 60 (August 1933): 64.

48.   "Why Wasn't I Born a Man?" *McCall's* 60 (May 1933): 93; "Marriage Is No Gambling Matter: Better Find Out, Better Be Sure About It," *McCall's* 60 (March 1933): 107.

49.   Garrett, "Birth Control's Business Baby," 270; Reeves, "The Birth Control Industry," 286–87; Bromley, "Birth Control and the Depression," 570; Anne Rapport, "The Legal Aspects of Marketing Feminine Hygiene Products," *Drug and Cosmetic Industry* 38 (April 1936): 474; Himes, *Medical History of Contraception,* 202; "The Accident of Birth," 85.

50.   "The Accident of Birth," 112.

51.   "Feminine Hygiene in the Department Stores," *Drug and Cosmetic Industry* 40 (April 1937): 482; "Twelve Ways to More Sales in Feminine Hygiene Products," *Chain Store Age* (June 1941): 54.

52.   Zonite advertisements in *Chain Store Age* (January 1941): 5 and (March 1941): 66; "Twelve Ways to More Sales," 19; "Feminine Hygiene Products Face a New Marketing Era," *Drug and Cosmetic Industry* 37 (December 1935): 745–47; H. C. Naylor, "Behind the Scenes Promotion Builds Feminine Hygiene Sales," *Chain Store Age* (March, 1941): passim.

53.   Garrett, "Birth Control's Business Baby," 269; Reeves, "The Birth Control Industry," 287; Kennedy, *Birth Control in America,* 212.

54.   Ad quoted in Palmer and Greenburg, *Facts and Frauds in Woman's Hygiene,* 12.

55.   "The Accident of Birth," 114.

56.   "The Accident of Birth," 114; Bromley, *Birth Control: Its Use and Misuse,* 93; Dilex Institute to Mrs. M. Hoffman, 31 August 1931, reel 29, Margaret Sanger Papers, Library of Congress.

57.   "Birth Control Industry," 58.

58.   "The Accident of Birth," 108–14; "Birth Control Industry," 58; According to Mary Ryan, before

the pill became widely available, only 20 percent of American women consulted physicians about birth control. See Ryan, "Reproduction in American History," 330.

59. "The Accident of Birth," 108; Reed, *From Private Vice to Public Virtue,* 244–46; Rainwater and Weinstein, *And the Poor Get Children,* 149–62.

60. Sanger to Harrison Reeves, June 16, 1936, reel 29, Margaret Sanger Papers, Library of Congress. Sanger's careful monitoring of the commercial side of birth control is evidenced in reels 29 and 30 of the Margaret Sanger Papers, Library of Congress.

61. Sanger to Reeves, June 16, 1936, reel 29, Margaret Sanger Papers, Library of Congress.

Childbirth and Motherhood

Although American women and men became increasingly successful at limiting the size of their families in the nineteenth century—the number of children per married woman dropped from 7.04 in 1800 to 3.56 in 1900—childbirth remained a common female experience. When colonial women were "brought to bed," their women friends and relatives came to aid the midwife during labor and delivery and to help with domestic chores. This "social childbirth" continued to characterize American births throughout the nineteenth century, even as male physicians increasingly replaced female midwives as attendants in the birthing rooms of the urban elite. The change to physician attendants at childbirth occurred for most nonelite women, especially the poor and immigrant women, during the first half of the twentieth century, although certain small segments of America's women continued to give birth with female midwife attendants throughout the twentieth century. Even with individual male attendants, however, birth remained predominantly a female affair and an important part of women's domestic culture until the twentieth century, when it moved away from women's homes and into the hospital.

In this section Judith Walzer Leavitt describes women's responses to childbirth, especially their worries about dying or receiving permanent postpartum injury. For much of American history women walked under the "shadow of maternity" when pregnant, and many made choices about birth attendants and procedures based on such fears. Carolyn Leonard Carson describes how and why African-American women, having moved north in the great migration, changed their traditional birthing practices and instead went into the hospital to deliver their babies. The safety and promise of hospital-based obstetrics lured birthing women of all races and classes into the arms of the new science. Finally, Rickie Solinger offers an analysis of the two histories of single pregnancy in the post–World War II era, one for black women and one for white.

By 1938 half of American women delivered their babies in hospital delivery rooms, and by 1955, 95 percent of women rushed to the hospital when their labors began. Childbirth and motherhood had become heavily inflected by medicine, although broader social influences continued to be important. For a variety of reasons examined in this section American women gave up their women-centered childbirths for medicalized births, during which women left their families at the labor room door and, in the words of one, delivered their babies "alone among strangers." Similarly, as Rima Apple has explained elsewhere, women accepted medical advice about their mothering practices. Even though women participated in the transition from home to hospital, from the birthing women's point of view the changes have not all been positive. In many ways birthing women and mothers are less in control of their experiences today than were women in the past.

17 Under the Shadow of Maternity: American Women's Responses To Death and Debility Fears in Nineteenth-Century Childbirth

Judith Walzer Leavitt

In 1846 a young woman in Warren, Pennsylvania, gave birth to a son and soon after was taken with "sinking spells." Her female friends and relatives were there to help her; they took encouragement when she appeared better and consoled each other when she fell into a stupor. A woman who was with her during these days wrote to a mutual friend, describing the scene around Mary Ann Ditmer's bed.

> Oh my beloved Girl—You may imagine our sorrow, for you too must weep with us—How can I tell you; I cannot realize myself—Mary Ann will soon cease to be among the living—and numbered with the dead. . . . Elizabeth, it was such a scene that is hard to be described—L and I remained until the afternoon. Mrs. Mersel came and relieved us, also Mrs. Whalen

came. We took a few hours sleep and returned—She had requested us to remain as long as she lived—there was every indication of a speedy termination of her suffering—Mary had come over—Mrs. N remained to watch . . . all thought she was dying—She was very desirous of living till day light—She thought she might have some hope if she could stand it until morning—She retained her sense perfectly—She begged us to be active and not be discouraged that she might live yet—that life was so sweet—how she clung to it—Elizabeth I would wish you might be spared such a sight—we surrounded her dying bed—Each one diligent to keep life and animation in the form of one they so much loved and who at that very time was kept alive with stimulating medicines and wine—I cannot describe it for o my God the horrors of that night will ever remain in the minds of those who witnessed it—our hearts swelled at the sight.[1]

Mary Ann lingered a few days, during which time she bestowed rings and locks of hair upon

JUDITH WALZER LEAVITT is Ruth Bleier Professor of History of Medicine, History of Science and Women's Studies, and the Associate Dean for Faculty at the Medical School, University of Wisconsin–Madison.

Reprinted from *Feminist Studies* 12, 1 (Spring 1986): 129–54, by permission of the publisher, *Feminist Studies*, Inc., c/o Department of Women's Studies, University of Maryland, College Park, MD 20742.

her friends so they might remember her; she made her peace with God and provided for her child. Then she died. Her story, both in its recognition of the childbirth-related dangers to women's lives and in its revelations of women helping women, represents a reality visited upon countless numbers of American women in the eighteenth and nineteenth centuries, and it is a reality with supreme significance for understanding women's lives. Thesis

In this article I will examine the meaning of maternity's influence over women's lives by analyzing specifically the physical dangers of childbearing and women's responses to those childbirth-related risks. During most of American history women's anticipation of the possibility of dying or of being permanently injured during childbirth influenced their life expectations and experiences. But women's responses to their repeated and dangerous confinements suggest, instead of resignation to their difficulties, an active participation in shaping events in America's birthing rooms.

Feminists reject the notion that biology is destiny and deny that biological constraints themselves determine the course of women's lives. But feminist historians, looking back at women's experiences, are forced to acknowledge that biology has been a significant factor in setting the limits of life's choices for most women. Specifically, the "shadow of maternity," in the words of many nineteenth-century women,[2] provided the material conditions under which the culturally imposed social parameters of women's lives were defined. Most married women, and many unmarried women, had to face the physical and psychological effects of recurring pregnancies, childbirths, and postpartum recoveries, all of which took a toll on their time, energy, dreams, and bodies. The biological act of maternity, which carried with it severe risks, significantly marked women's lives as they made their way from birth to death. My exami-

nation of childbirth's biological domination over women's lives focuses on women's fear of the physical dangers of childbirth and examines the extent to which the fears reflected the reality of their life experiences, but I do not mean to suggest that these fears and dangers encompassed the totality of women's childbirth experiences, or that they alone determined the direction of life's choices. Childbirth had many meanings in women's lives, but this article will focus on the physical dimension of the birthing experience and on the female-centered birthing environment women developed to counteract the potentially deleterious nature of confinement.

My conclusions about biology's influence over women do not connote biological control, inevitableness, or women's passive acceptance of their fate. In contrast to the work of Edward Shorter,[3] which posits that biology held a stranglehold over women such that they could not escape until twentieth-century medical science came to their rescue, my work emphasizes the decisions women made within their confines to break the boundaries that biology seemed to set. My research reveals women in control of many aspects of their lives, including biological aspects, and I think women's culture developed through these biological bonds—even when those bonds were limiting their other activities. Women's biological functions might have led society to try to circumscribe women's nondomestic activities, but women themselves were not bound by their biological functions. Throughout American history women struggled to overcome their most significant biological limitations—as they perceived them—ofttimes successfully. The very processes of coping with and trying to change and expand available choices created in some cases improved conditions, but more important, the processes created a support system for women that opened wider worlds. Women used the strengths and help of

helping other women in their world.

other women to face their problems and in their unity developed coping mechanisms that were illustrative of what I think can be called a feminist impulse embedded within traditional women's experiences. Feminism did not develop when or because modern medicine saved women from their bodies (as Shorter posits), but, rather, feminist inclinations and the collective behavior they fostered developed out of the basic and shared experiences of women's bodies at times when those bodies seemed most confining and difficult. In this sense feminism, women grasping control and working together to overcome their commonly experienced burdens, can arise out of the very essence of biological femaleness and reside alongside the most traditional part of women's experiences.[4]

The shadow of maternity, under which so many women lived, has had numerous features, only some of which I will be able to discuss here. Underlying the shadow of maternity most significantly were high fertility rates. At the turn of the nineteenth century, American women bore an average of seven children before their fertile years ended. This implies considerably more than seven pregnancies, because many terminated before term. For many groups in the expanding American population, rates remained high throughout the nineteenth century.[5] Although control over fertility is not the subject of this article, we must view women's past life experiences within the context of limited choices over conception. Pregnancy, birth, and postpartum recovery occupied a significant portion of most women's adult lives, and the ensuing motherhood defined a major part of women's identity.

Take, for example, the life of Mary Vial Holyoke, who married into a prominent New England family in 1759. In 1760, after ten months of marriage, she gave birth to her first baby. Two years later her second was born. In 1765 she was again "brought to bed" of a child. Pregnant im-

mediately again, she bore another child in 1766. The following year she delivered her fifth and in one more year delivered her sixth. Free from pregnancy and childbirth in 1769, she gave birth again in 1770. During the next twelve years she bore five more children. The first twenty-three years of Mary Vial Holyoke's married life, the years of her youth and vigor, were spent pregnant or recovering from childbirth. Because only three of her twelve children lived to adulthood, she withstood, also, frequent tragedies. She devoted her body and her life to procreation throughout her reproductive years. Mary Holyoke had more pregnancies and suffered more child deaths than her average contemporary, and she presents a poignant example of the extreme physical trials women endured. Mary Holyoke had little choice in her frequent pregnancies: her life reveals how the biological capacity of women to bear children historically has translated into life's destiny for individual women.[6]

Mary Holyoke's experience became less common in nineteenth-century America, especially among white women, as fertility rates declined. By 1900 white women, showing the ability to cut their fertility in half over the century, averaged 3.56 children. Historians and demographers trying to understand this decline have suggested that as much as 75 percent of the dropping fertility rate can be explained by active fertility control, including abortion and birth control techniques. Some people seem to have succeeded in asserting partial control over the size of their families, but it is important to keep in mind that the fertility declines demographers have identified with nineteenth-century America apply only to white native-born women; immigrant and black women continued to have babies in larger numbers.[7]

Fertility rates explain only part of the impact of childbirth on women's lives. Maternity cast a shadow greater than its frequent repetition alone could have caused (Maternity, the creation of new life, carried with it the ever-present possibility of death.) The shadow" that followed women through life was the fear of the ultimate physical risk of bearing children. Young women perceived that their bodies, even when healthy and vigorous, could yield up a dead infant or could carry the seeds of their own destruction. As Cotton Mather had warned at the beginning of the eighteenth century, and as American women continued to believe, conception meant "your *Death* has Entered into you." Nine months' gestation could mean nine months to prepare for death. (A possible death sentence came with every pregnancy.[8])

Many women spent considerable time worrying and preparing as if they would not survive their confinements. During Nannie Stillwell Jackson's pregnancy in 1890 she wrote in her diary: "I have not felt well today am afraid I am going to be sick I went up to Fannies a little while late this evening & was talking to her, & I told her to see after Lizzie & Sue [other children] if I was to die & not to let me be buried here . . . & I want Lizzie & Sue to have *everything that is mine,* for no one has as good a rite [sic] to what I have as they have."[9] A pregnant Clara Clough Lenroot confided in her diary in 1891: "It occurs to me that *possibly* I may not live. . . . I wonder if I should die, and leave a little daughter *behind* me, they would name her 'Clara.' I should like to have them." Three days later she again was worrying. "If I shouldn't live I wonder what they will do with the baby! I should want Mama and Bertha [sister] to have the bringing up of it, but I should want Irvine [husband] to see it every day and love it so much, and I should want it taught to love him better than anyone else in the world." With the successful termination of the birth Clara's husband wrote in his wife's diary, "Dear Clara, 'mama and Bertha' won't have to take care of your baby, thank God." He continued, "Everything is all right, but

at what cost. My poor wife, how you have suffered, and you have been so brave. . . . I have seen the greatest suffering this day that I have ever known or ever imagined."[10]

When her sister Emma experienced a difficult pregnancy in 1872, Ellen Regal came to be with her and found her "so patient and resigned." Ellen wrote to their brother, "It is not strange that she should tremble and shrink at the thought of that Valley of the Shadow of Death which she must so soon enter." In the middle of the nineteenth century, Lizzie Cabot wrote to her sister when she was pregnant with her first child, "I have made my will and divided off all my little things and don't mean to leave undone what I ought to do if I can help it." Sarah Ripley Stearns, returning from church near her time of confinement, wrote in her diary: "Perhaps this is the last time I shall be permitted to join with my earthly friends."[11]

The extent to which these death fears spread to other family members beyond the parturients is evident in the diary of Albina Wight, whose sister living in another state was pregnant and near confinement. In 1870 Albina wrote, "I am so affraid [sic] she won't live through it." Three days later she continued to think of her sister, "I could not keep Tilda out of my thoughts. It has seemed like a funeral all day . . . I fear she is not living."[12]

I could go on with examples of how much some women feared that they might die, but I think the point is made. (Young, vigorous, healthy women who should have been anticipating a long life ahead instead faced the very real possibility that their pregnancies would bring their deaths, that in creating a life they would pay with their own.)

Women and family members were not the only ones who anticipated maternal death. Many physicians who attended parturients through the fearful hours of labor and delivery also brooded on mortality. In the 1870s Dr.

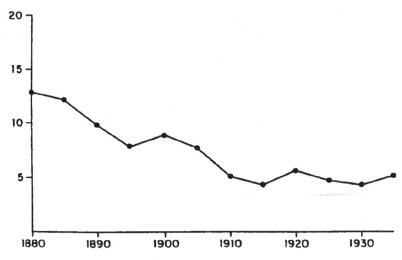

Fig. 17.1. Total Maternal Deaths per 1,000 Live Births
New York City 1880–1935

Source: Haven Emerson and Harriet Hughes, *Populations, Notifiable Diseases and Deaths Assembled for New York City, New York* (New York: Dehamer Institute of Public Health/College of Physicians and Surgeons, 1941).

James S. Bailey of Albany, New York, pondered the sometimes sudden and unexplained deaths of women following childbirth. He wrote: "To see a female, apparently in vigorous health until the period of accouchement, suddenly expire from some unforeseen accident, which is beyond the control of the attending physician, is well calculated to fill the mind with alarm and gloomy forebodings" and make it impossible "while attending a case of confinement, to banish the feeling of uncertainty and dread as to the result of cases which seemingly are terminating favorably."[13] Similarly, Dr. William Lusk, after relating the tragic story of a twenty-three-year-old "very beautiful young woman" who died after delivery, warned his fellow physicians that "the exhausted condition in which the woman is left after childbirth render[s] her an easy prey to the perils of the puerperal state."[14] Fort Wayne, Indiana, physician H. V. Sweringen wrote that "parturition under the most favor-able circumstances is attended with great risk."[15] Most physicians and families were well aware of the possibility of death: fears of confinement's dangers permeated society.

The extent to which social fears about maternal deaths reflected a reality of high death rates is almost impossible for historians to determine with any degree of confidence. Graphs of maternal deaths, such as fig. 17.1, created from New York City data, illustrate that deaths from causes associated with maternity seem to have been declining toward the last part of the nineteenth century. The statistics also show that the maternal death rate leveled out toward the end of the century and continued at a high level (and even increasing) in the twentieth century. But these figures do not tell the whole story. They record the deaths per live births that physicians or other attendants attributed to puerperal causes, but they remain silent on the deaths associated with childbirth, such as from tuberculosis, that physicians did not directly attribute to confinement-related causes. Officials might have recorded that women who died while pregnant or within a month after childbirth died from tuberculosis, which was exacerbated by maternity, when it would be accurate also to have attributed the deaths to maternity-related causes. Even the imperfectly reported maternal deaths and live births are available only for certain parts of the country for certain years in the nineteenth century.[16]

Perhaps more valuable to our understanding of the reality of maternal death is the observation that most women seemed to know or know of other women who had died in childbirth. One woman, for example, wrote that her friend "died as she has expected to" as a result of childbirth, as had six other of their childhood friends. Early in the twentieth century approximately one mother died for each 154 live births. If women delivered, let us estimate, an average of five live babies, these statistics can mean that over their reproductive years, one of every thirty women might be expected to die in childbirth. In another early twentieth-century calculation, one of every seventeen men claimed they had a mother or sister who had died as the immediate results of childbirth.[17] *affected men*

Despite the real decline in the number of births women endured, and a corresponding but smaller decline in the rate of maternal deaths, women remained fearful of maternity. Women might have been at risk for puerperal death fewer times during their lifetimes, but for them the fear of dying during childbirth continued to influence the possible parameters of their lives. Part of this, I believe, can be explained by the fact that maternal deaths continued at higher than acceptable (that is, preventable) levels. Physicians and women realized that by the turn of the twentieth century deaths from many infectious diseases, such as tuberculosis and diphtheria, were declining but that deaths from

childbirth-related causes remained high. Improvements in living conditions and in medical knowledge about disease transmission seemed to lead to improvements in mortality statistics, except for women in childbirth. Furthermore, maternal mortality statistics from other countries showed that women elsewhere fared better than women in the United States did. In 1910, when the United States recorded that one mother died for every 154 babies born alive, Sweden's record showed that one mother was lost for every 430 live births.[18] *Physical problems*

In the past the shadow of maternity extended beyond the possibility and fear of death. Women knew that if procreation did not kill them or their babies, it could maim them for life. Postpartum gynecological problems—some great enough to force women to bed for the rest of their lives, others causing milder disabilities—hounded the women who did not succumb to their labor and delivery. For some women the fears of future debility were more disturbing than fears of death. Vesicovaginal and rectovaginal fistulas (holes between the vagina and either the bladder or the rectum caused by the violence of childbirth or by instrument damage), which brought incontinence and constant irritation to sufferers; unsutured perineal tears of lesser degree, which may have caused significant daily discomforts; major infections; and general weakness and failure to return to prepregnant physical vigor threatened young women in the prime of life. Newly married women looking forward to life found themselves almost immediately faced with the prospect of permanent physical limitations that could follow their early and repeated confinements.

Chicago physician Henry Newman believed that the "normal process of reproduction [is] a formidable menace to the afterhealth of the parous woman."[19] Lacerations—tears in the perineal tissues—probably caused the greatest postpartum trouble for women. The worst of these,

Medical treatment for physical problems

Physical problems

the fistulas, which led to either urine or feces constantly leaking through the vaginal opening without the possibility of control, were, in the words of one sympathetic doctor, "the saddest of calamities, entailing . . . endless suffering upon the poor patient . . . death would be a welcome visitor."[20] Women who had to live with this condition sat sick and alone as long as they lived unless they were one of the beneficiaries of Dr. J. Marion Sims's operation in the second half of the century. Their incontinence made them unpleasant companions, and even their loved ones found it hard to keep them constant company.[21]

More frequent and less debilitating, but still causing major problems for many women, were tears in the vaginal wall or cervix that might have led to prolapsed uterus, uncomfortable sexual intercourse, or difficulties with future deliveries. One physician noted that "the widespread mutilation . . . is so common, indeed, that we scarcely find a normal perineum after childbirth."[22] Most perineal lacerations probably were minor and harmless, but if severe ones were not adequately repaired, women might suffer from significant postpartum discomfort.[23] Women complained most frequently of a prolapsed uterus. This displacement of the womb downward, sometimes even through the vaginal opening, often resulted from lacerations or postpartum relaxation and consequent elongation of the ligaments. One physician noted that fallen womb is often a temporary condition, but he also found it quite common: "Any woman subject to ill turns, lassitude, and general debility will tell you that not unfrequently upon these occasions she is sensible of a falling of the womb."[24] The condition caused misery for women. Albina Wight's sister Eliza, to give just one example from the 1870s, had a difficult delivery that was followed by prolapse. Six weeks following one of her sister's confinements, Albina recorded: "Eliza is sick yet can only walk

Uterus hanging down from vagina

across the room and that overdoes her. She has falling of the Womb. poor girl." Eliza's medical treatment by "a calomel doctor" who gave her "blue pills" did not help. Five months following the delivery she could only "walk a few steps at a time and cannot sit up all day." A second doctor predicted that "it will be a long time before she will get around again."[25]

The typical treatment for this common female ailment was the use of a pessary, a mechanical support for the uterus inserted into the vagina and left there as long as necessary (see fig. 17.2). Pessaries themselves often led to pelvic inflammations and pain for the women whose conditions they were designed to alleviate. In the opinion of one physician,

> I think it is indisputable that a pessary allowed to remain for a very short period will invariably produce irritation, and if continued longer, will produce almost as certainly, ulceration. I have removed many pessaries that have produced ulceration; one in particular, hollow and of silver gilt, was completely honey combed by corrosion, its interior filled with exuviae of the most horrible offensiveness, the vagina ulcerated through into the bladder, producing a vesico-vaginal fistula, and into the rectum, producing a rectovaginal fistula; the vagina in some portion obliterated by adhesive inflammation and numerous fistulae made through the labia and around the mons veneris for the exit of the various discharges.[26]

Uterine displacements puzzled physicians and pained women throughout the nineteenth century. A midcentury physician noted that anteflexion, an abnormal forward curvature of the uterus that seemed not amenable to pessary correction, "is the dread of almost every physician, and the constant, painful perplexity of many a patient." He told of his recent case:

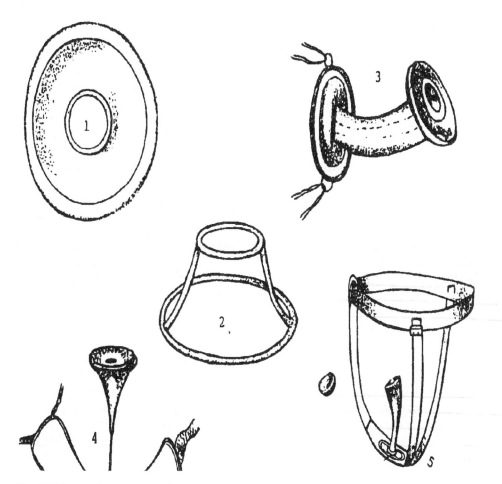

Fig. 17.2 Nineteenth-century pessaries
1. Smellie's ring pessary
2. Meig's double ring pessary, of gutta percha
3. Merriam's glass pessary to be fastened externally
4. O'Leary's hard rubber cap, with a screw to regulate its length, resting on a plate secured externally
5. Taft's ball and socket
6. Schaffer's spiral spring
7. Sims' pure, flexible Britannia pessary, capable of being bent in any desired form

Source: Augustus K. Gardner, "On the Use of Pessaries," *Transactions of the American Medical Association* 15 (1865): 109–22.

besides the fear of death; suffering

but

In the winter of 1863, I was consulted by a young lady from a distant part of the State, on account of a disease from which she had suffered for nearly four years. She had received the advice of many a physician of high and low degree—had worn the ring pessary—the globe pessary—the horseshoe pessary—the double S pessary and the intra-uterine stem pessary—and the common sponge. . . . The patient gave a history of frequent inflammation of the uterus and ovaries, and there appeared to be quite strong adhesions binding the womb in its assumed place. She had had too frequent menstruation—profuse and intolerably painful—frequent and painful micturition [urination].

The physician inserted his "modified" ring pessary and reported that "the patient went to her home after a few weeks entirely relieved from all bad symptoms."[27]

In the last half of the nineteenth century, physicians reported increased incidence of perineal and cervical lacerations and their accompanying gynecological problems, attributed by many observers to increased use of forceps and other interventions in physician-directed deliveries. If it is true that physicians' interventions caused problems for women in this period (and that is a question I explore elsewhere), it is also the case that physicians became increasingly adept at repairing the problems. The medical journals are filled with case studies of women whose badly managed deliveries had caused them problems, which could then be fixed by superior medical care. For example, an Iowa physician, Dr. Nicholas Hard, reported in 1850 a case he salvaged. A thirty-five-year-old woman "with her first child, having had the forceps applied at an improper time during her labour, suffered from inflammation of the vulva, vagina, and contiguous soft parts, and had a tardy convalescence. . . . The

vaginal orifice was perfectly closed . . . she suffered exceedingly from retained catamenial [menstrual] fluid." The doctor instrumentally reopened the vagina and happily reported that his patient now "walks to church, visits, and does house-work."[28]

Women who had already had children were more likely than first-time mothers to worry about the possible aftereffects of labor and delivery. They remembered how long it took them the first time to recover from the birth, they remembered how they had suffered, and they were particularly loath to repeat the ordeal. As one woman wrote about her second pregnancy: "I confess I had dreaded it with a dread that every mother must feel in repeating the experience of child-bearing. I could only think that another birth would mean another pitiful struggle of days' duration, followed by months of weakness, as it had been before."[29]

Apart from their concern about resulting death and physical debility, women feared pain and suffering during the confinement itself. They worried about how they would bear up under the pain and stress, how long the confinement might last, and whether trusted people would accompany them through the ordeal. The short hours between being a pregnant woman and becoming a mother seemed, in anticipation, to be interminably long, and they occupied the thoughts and defined the worries of multitudes of women. Women's descriptions of their confinement experiences foretold the horrors of the ordeal.

Josephine Preston Peabody wrote in her diary of the "most terrible day of [her] life," when she delivered her firstborn, the "almost inconceivable agony" she lived through during her "day-long battle with a thousand tortures and thunders and ruins." Her second confinement brought "great bodily suffering," and her third, "the nethermost hell of bodily pain and mental blankness. . . . The *will-to-live* had been massacred out of me, and I couldn't see why I *had* to." [30] Another woman remembered "stark terror was what I felt most." [31]

"Between oceans of pain," wrote one woman of her third birth in 1885, "there stretched continents of fear; fear of death and dread of suffering beyond bearing." [32] Surviving a childbirth did not allow women to forget its horrors. Lillie M. Jackson, recalling her 1905 confinement, wrote: "While carrying my baby, I was so miserable. . . . I went down to death's door to bring my son into the world, and I've never forgotten. Some folks say one forgets, and can have them right over again, but today I've not forgotten, and that baby is 36 years old." [33] Too many women shared with Hallie Nelson her feelings upon her first birth: "I began to look forward to the event with dread—if not actual horror." Even after Nelson's successful birth, she "did not forget those awful hours spent in labor." [34]

Regardless of the particular fear that women carried along with their swelling uteruses, the end result was similar for all of them. The prospect of often repeated motherhood promised hardship and anxiety. As Hannah Whitall Smith wrote in her diary in 1852:

> I am very unhappy now. The trial of my womanhood which to me is so very bitter has come upon me again. When my little Ellie is 2 years old she will have a little sister or brother. And this is the end of all my hopes, my pleasing anticipations, my returning youthful joyousness. Well, it is a woman's lot and I must try to become resigned and bear it in patience and *silence* and not make my home unhappy because I am so. But oh, how hard it is. [35]

Many women walked with Hannah Smith under the shadow of maternity, experiencing repeated and agonizing births in unrelenting succession with no relief throughout their fertile years. Many women suffered physical complications through their confinements that stayed with them the rest of their lives. For many women the physical hardships of childbearing determined the parameters of their lives and defined their social destiny. Although it is also true that childbearing and the ensuing motherhood held many happy times for women, it is the difficult part of the experience that created the boundaries within which most women had to construct their lives. The childbirth experience was, of course, heavily influenced by cultural and economic conditions, the particular time and place in which women lived, and their socioeconomic class or ethnic group. But much of the meaning of childbirth for women was determined not by the particulars of the event but by what women shared with each other by virtue of their common biological experience.

With all the horrors and dangers and worries, women could have easily given up hopes of improving their childbirth experiences or their hard domestic prospects. And no doubt some women did merely resign themselves to lives of invalidism and deprivation. But what comes through the written record much more strongly are the positive aspects of the experience that women chose to emphasize, the caring ways in which they tried to help each other, and the simple fact that women were able to change the childbirth experience for themselves in significant ways throughout the nineteenth century.

Let us examine first the cooperative nature of the labor and delivery experience. Throughout American history up until the twentieth century, when childbirth moved to the hospital, most women gave birth at home with the help of their female friends and relatives. Birth was a women's event, and women eagerly gave their aid when it was needed. Albina Wight, who was unable to attend her sister's confinement, wrote in her diary, "Poor poor girl how I pitty [sic] her. She says the two wimen [sic] that were there were as kind and good as Sisters could be. I am *unification* (handwritten note)

glad of that. . . . Oh how I do wish I could be with her."[36] When possible, sisters and cousins and mothers came to help the parturient through the ordeal of labor and delivery, and close friends and neighbors joined them around the birthing bed. One woman who described her 1866 confinement wrote: "A woman that was expecting had to take good care that she had plenty fixed to eat for her neighbors when they got there. There was no telling how long they was in for. There wasn't no paying these friends so you had to treat them good."[37] To this women's world, husbands, brothers, or fathers could gain only temporary entrance. In an 1836 account the new father was invited in to see his wife and new daughter and then: "But Mrs. Warren, who was absolute in this season of female despotism, interposed, and the happy father was compelled, with reluctant steps, to quit the spot."[38]

support during pregnancy (handwritten note)

Most crucial to the support networks women tried to gather around themselves during and following confinement were their own mothers. "If you could but be with me now, what wouldn't I give," wrote Anita McCormick Blaine to her mother, Nettie Fowler McCormick, in 1890 as she prepared for the birth of her first child. "Dearest mother mine—all would be complete if you were here." Nettie returned the sentiment. "Dearie," she wrote, "I wish I were there to thoroughly rub olive oil upon your hips, your groin muscles, your abdominal muscles all throughout—in short all the muscles that are to be called upon to yield, and be elastic at the proper time. See how reasonable it seems that they should be helped to yield, and to do their work if they are kneaded by the strong hand of mother, while olive oil is being freely applied."[39]

Some women could marshal only one or two women to help them, but many accounts list eight or ten women helpers in addition to the midwife or physician who might be attending. Nannie Jackson, in rural Arkansas in 1890, gathered six women to help when she delivered, and

her oldest daughter noted at the time that "Mama had a heap of company today."[40] Antebellum southerner Madge Preston gave birth to a child in 1849, which her husband, who waited in another room, reported this way: "At this birth were present Dr. J. H. Briscoe, Mrs. Margaret Carlon, Mrs. Connolly her friend—our servant Mary Miskel, and our Negroes Lucy and Betty. They inform me that Mrs. P. bore her protracted labor, difficulty, pain and anxiety, which endured forty eight hours, with calmness, courage, and fortitude."[41]

Women went to considerable sacrifice to help their birthing relatives and friends, interrupting their lives to travel long distances and frequently staying months before and after delivery to do the household chores. When relatives were not available, neighbors stayed for the labor and delivery and brought food and kept up with washing and other domestic duties. Christiana Holmes Tillson wrote in her journal of her second confinement in 1825: "I had made the acquaintance of Mrs. Townsend, who was with me and remained until John was a week old."[42] Ann Bolton recalled with enormous fondness and gratitude her good friend: "Of thirteen children which I brought into the world, she bore me company with ten of them."[43] Even in the relatively unusual event of a woman's being alone for her delivery, her friends rallied round afterward to make sure that she got the rest she needed.[44] Women who could not summon friends were described as in this diary entry: "[I] went thence to see Mrs. Ray who has been very ill in Childbed—Had a little girl which died in a few hours.—She is much to be pitied having no female relation or intimate friend to be with her."[45] Another woman pitied herself: "It seemed very gloomy when I found my time had come, to think that I was, as it were, destitute of earthly friends. No mother, no husband, and none of my particular friends that belong to the town; they happening to be out of town."[46]

The women's network that developed at least

[handwritten: womens network effects]

[handwritten: network made things bearable]

in part through the strong attachments formed across the childbirth bed had long-lasting effects on women's lives. When women had suffered the agonies of watching their friends die, when they had helped a friend recover from a difficult delivery, or when they had participated in a successful birthing, they developed a closeness that lasted a lifetime. Surviving life's traumas together made the crises bearable and produced important bonds that continued to sustain other parts of women's lives. "It was as if mothers were members of a sorority and the initiation was to become a mother," Marilyn Clohessy wrote.[47] Nannie Jackson's female support network offers one example of the importance of good friends. Her diary, only one year of which survives, is a litany of friends helping each other. Her best friend was Fannie, who lived one-half mile away; Nannie and Fannie visited each other daily and sometimes two, three, and four times a day. Once, during the eighth month of her third pregnancy, Nannie visited her friend three times in one evening and her husband got angry. But, Nannie noted, "I just talk to Fannie & tell her my troubles because it seems to help me to bear it better when she knows about it. I shall tell her whenever I feel like it."[48] Indeed, in this diary fragment there is evidence of significant rebellion against her husband's wishes and her strong reliance on her relationship with Fannie. When Nannie was confined, Fannie stayed over with her for four nights. But Fannie is only the most important in a long list of close friends. Nannie, who was white, visited daily with many other women, both white and black, cooking special things for them, sharing the limited family resources, helping them with sewing projects, sitting up with them when they were sick, helping out at births, arranging funerals. These women, whose economically limited lives left nothing for outside entertainment or expense, found rich resources within their own group. We see that the psychological dimension of the women's network played a significant role in making the difficult lives of these women bearable. Perhaps more significant to these women, however, was the very real support friends could provide during times of crisis. During labor and delivery, when a woman might not be able to stand up for herself, she could rely on her women friends to do her talking for her. Women influenced almost every aspect of birth procedures by having very important on-the-spot input into birth practices for most of the home birth period, and they made significant changes in birth procedures through their collective behavior.

Not until the last half of the eighteenth century did some urban women begin calling in physicians early in normal labor as the major attendants. Yet even then, and throughout the nineteenth century, most women continued to call their women friends and relatives to help them and relied on the advice of the women alongside or before the advice of the physician. Sometimes the women, who were called first, advised that no medical attendant be called. As one physician realized, "A certain amount of inconvenience is anticipated, and so long as this supposed limit is not passed, the patient contrives, with the advice of her female friends, to dispense with a medical attendant."[49] At other times the attending women suggested additional help. Many physicians attributed their obstetrics calls to midwives or to neighbor women who were already present at a progressing labor. Dr. John G. Meachem, struggling to establish himself after graduating from medical school, recorded a successful first case: "Mrs. Doolittle was present, and I always thought that she had a good deal to do with engineering this call. At least I gave her the credit."[50]

The collectivity of women gathered around the birthing bed made sure that birth attendants were responsive to their wishes. They made decisions about when and if to call physicians to births that midwives were attending; they gave or withheld permission for physicians' procedures; and they created the atmosphere of female support in a room that might have contained both women and men. (The result of women's seeming lack of control over biological imperatives was in fact increased control and the ability to influence events.)

Women friends were so common and so active in the birthing room that physicians tried to limit them to one or two in their efforts to gain some control over the birth process. Dr. Edward Henry Dixon, for example, wrote in his popular advice book that physicians "mildly, yet firmly exclud[e] from the room all who are not absolutely necessary as attendants."[51] Until birth moved to the hospital, however, physicians shared their authority with neighbor women, most of whom had had significant birth experience. There were times, of course, when the attending women encouraged physicians' interventions. One of Wisconsin's earliest physicians reported his first forceps operation "in the presence of all the old women of the neighbourhood." He was pleased to note that "all the relatives and friends expressed themselves quite satisfied with my exertions & skills."[52]

Often, however, physicians noted ruefully that they did not control events in the birthing rooms. One reported that "podalic version [turning the fetus before delivery] was again attempted but was forcibly interfered with by friends."[53] Another physician noted: "The officiousness of nurses and friends very often thwarts the best-directed measure of the physician, by an overweening desire to make the patient 'comfortable.' . . . All this should be strictly forbidden. Conversation should be prohibited the patient. . . . Nothing is more common than for the patient's friends to object to [bloodletting], urging as a reason, that 'she has lost blood enough.' Of this they are in no respect suitable judges."[54] Others wrote that they found it difficult to accomplish aseptic conditions because of the interference of the woman's friends.

In fact, the record shows that the parturients' friends and family made decisions about the use of forceps, anaesthesia, and other interventions usually considered to be within the control of the attending physician. Physicians went along with the women's demands or risked being removed from the case. One doctor related, for example, why he would not try to shave the pubic hair of his parturient patients. "In about three seconds after the doctor has made the first rake with his safety [razor], he will find himself on his back out in the yard with the imprint of a woman's bare foot emblazoned on his manly chest, the window sash around his neck and a revolving vision of all the stars in the firmament presented to him. Tell him not to try to shave 'em."[55] Dr. John Milton Duff of Pittsburgh unhappily recorded that "the obstetrician can not always control the general environment of his patient," and he worried about her "obstreperousness." This physician lamented the "sometimes dangerous interference of ignorant and superstitious neighbors and friends." His colleague, Dr. Joseph Price of Philadelphia, agreed that "very few households will permit the practice of well-organized and disciplined maternity work."[56]

(The power that the friends had did not necessarily result in better care for the parturient, but it does indicate a level of support that the birthing woman could count on.) Dr. E. L. Larkins of Terre Haute, Indiana, believed that pressures from these other birth attendants led physicians to poor practices. "The sympathy of attending friends, coupled with the usual impatience of the woman from her suffering, will too often incite even the physician, against his better judgment, to resort to means to hasten labor, resulting in disaster which time and patience would have avoided."[57] But from the birthing women's point of view this network of women supporting each other through this difficult ordeal assisted them in getting through a situation in which they felt so powerless.[58] One woman concluded that "the most important thing is not to be left alone and to know that someone is there who cares and will help you when the going gets rough."[59] Through their social network women were able to keep control over childbirth despite the presence and authority of male physicians.

I cannot quantify my findings to say what percentage of women found the kind of support that Nannie Jackson and others developed. Nor can I determine to what extent these friendship networks might have worked better among the middle classes. I can say that I have found them across class and ethnic lines, in the rural areas and in the cities, in the beginning of the nineteenth century and at its end. I am not positing a universal experience. I'm sure that there were countless women who underwent their severest suffering virtually alone or accompanied only by their husbands, with whom they may not have been able to share their deepest feelings. I understand that unremitting poverty took its toll on many suffering women who could not develop even the outlines of a support network. I think also that some rich women stood outside a meaningful network of friends, isolated perhaps by their status and background. But personal accounts of childbirth by women and birth attendants suggest that the birth experience was crucial in creating the social dimensions of most women's lives distinct from other socioeconomic factors. Anita McCormick Blaine, living an affluent life in Chicago at the end of the nineteenth century, shared in many respects the birth experience of Nannie Jackson, living in impoverished rural Arkansas at the same time. They both needed, sought, and got the help of their close women friends at the crucial time of their confinements. Although it is certainly true that Blaine had more advantages and more choice in the particulars of the birth experience, to both women the female context

in which they delivered their babies was crucial to a successful experience. The biological female experience of giving birth provided women with some of their worst moments and some of their best ones, and the good and the bad were experiences that all women could share with each other.

Despite the very real changes in the technical and physical experience of birth during the nineteenth century, women's perceptions of its dangers and methods of dealing with those dangers within a female-centered protective environment remained very much the same during the course of the time that birth remained in women's homes. When birth moved to the hospital for the majority of American women in the twentieth century, women lost their domestic power base and with it lost certain controls that they had traditionally held. This is the change, as I argue elsewhere, that took away women's traditional controls over childbirth and caused a basic transformation in womens' birth experiences.[60]

The silver lining in maternity's shadow that accompanied the crisis of childbirth during the home birth period—women actively helping each other shape their childbirth experiences—enabled women to find each other, to learn to give solace and support, and to receive them back in turn. It is in fact in the combination of shadow and light, of despair and of hope, that we can best view women's procreative experiences. Although the fears and dangers of childbirth followed women, the experience itself opened up new vistas and created practical and emotional bonds beyond the family that sustained them throughout the rest of their lives. Participating together in the function of procreation led women to share with each other some other aspects of their lives in close intimacy. This "female world of love and ritual,"[61] as Carroll Smith-Rosenberg so aptly called it, with its strong emotional and psychological supports

The good things about "making Shadow"

and its ability to produce real change in women's lives, was in large part created around women's shared biological moments, including repeated confinements and procreative death and debility fears. The valley of the shadow of birth gave women the essence of a good life at the same time it contributed to a strict definition of that life's boundaries. In uniting women, it ultimately provided the ability for women to stretch the boundaries of their world.

Notes

A grant from the University of Wisconsin Graduate School supported parts of the research for this paper. I would like to thank Evelyn Fine for her excellent research help. Nancy Schrom Dye, Susan Friedman, Ann Gordon, Gerda Lerner, Florencia Mallon, Ronald Numbers, Jane Schulenberg, Steve Stern, and Anne Stoler contributed valuable comments on earlier drafts of the paper, and for these I am very grateful. The editors and readers at *Feminist Studies* provided useful suggestions and aid. I presented a shorter version of the paper at the Sixth Berkshire Conference on the History of Women at Smith College on June 3, 1984.

1. Jane Savine (?) of Warren, Pennsylvania, to Elizabeth Gordon of Cleveland, Ohio, Feb. 26, 1846; see also letter of March 10, 1846, in Elizabeth Gordon Correspondence, Wisconsin State Historical Society Archives, Madison, Wisconsin.

2. See, for example, Augustin Caldwell, *The Rich Legacy: Memories of Hannah Tobey Farmer, Wife of Moses Gerrish Farmer* (Boston: Privately Printed, 1890), 97.

3. Edward Shorter, *A History of Women's Bodies* (New York: Basic Books, 1983).

4. I use the phrases "feminist impulse" and "feminist inclinations" to differentiate this analysis from any confusion with a publicly political feminism as manifested by those women who were active in the suffrage movement or in any public advocacy for women's rights and emancipation. In my interpretation of women's activity in the birthing rooms,

women consciously acted to keep childbirth within women's power, where it had traditionally been, when that power was being threatened by a medical profession growing in power and ability in the nineteenth and early twentieth centuries. By banding together to retain female traditions and values and to shape the events in their own birthing rooms, women acted in a way that acknowledged a specific women's agenda. I stop short of labeling the women as feminist, and refer instead to impulses and inclinations, because I think their actions were not consciously creating a new world so much as they were supporting an autonomous women's dimension within the existing one. For historians' debates on these issues see Ellen DuBois, Mari Jo Buhle, Temma Kaplan, Gerda Lerner, and Carroll Smith-Rosenberg, "Politics and Culture in Women's History: A Symposium," *Feminist Studies* 6 (Spring 1980): 28–64.

5. For the history of birth control and fertility patterns in America consult Linda Gordon, *Woman's Body, Woman's Right: A Social History of Birth Control in America* (New York: Viking, 1976); James Reed, *From Private Vice to Public Virtue: Birth Control in America* (New York: Basic Books, 1978); Robert V. Wells, *Revolutions in Americans' Lives: A Demographic Perspective on the History of Americans, Their Families, and Their Society* (Westport, Conn.: Greenwood, 1982). On the history of childbirth consult Janet C. Bogdan, "Care or Cure? Childbirth Practices in Nineteenth-Century America," *Feminist Studies* 4 (June 1978): 92–99; Nancy Schrom Dye, "History of Childbirth in America: Review Essay," *Signs* 6 (1980): 97–108; Catherine M. Scholten, "'On the Importance of the Obstetrick Art': Changing Customs of Childbirth in America, 1760–1825," *William and Mary Quarterly* 34 (July 1977): 426–45; Richard W. Wertz and Dorothy C. Wertz, *Lying-In: A History of Childbirth in America* (New York: Free Press, 1977); Jane B. Donegan, *Women and Men Midwives: Medicine, Morality, and Misogyny in Early America* (Westport, Conn.: Greenwood, 1978); Judy Barrett Litoff, *American Midwives, 1860 to the Present* (Westport, Conn.: Greenwood, 1978); and Judith Walzer Leavitt and Whitney Walton, "'Down to Death's Door': Women's Perceptions of Childbirth in America," in *Women and Health in America: Historical Readings,* ed. Judith

Walzer Leavitt (Madison: University of Wisconsin Press, 1984), 155–65.

6. *The Holyoke Diaries, 1709–1856*, Introduction and annotations by George Francis Dow (Salem, Mass.: Essex Institute, 1911).

7. Warren C. Sanderson, "Quantitative Aspects of Marriage, Fertility, and Family Limitation in Nineteenth-Century America: Another Application of the Coale Specifications," *Demography* 11 (August 1979): 339–58; Ansley J. Coale and Melvin Zelnik, *New Estimates of Fertility and Population in the United States: A Study of Annual White Births from 1855 to 1960 and of Completeness of Enumeration in the Censuses from 1880 to 1960* (Princeton: Princeton University Press, 1963). Significant variations even among white native-born women suggest that fertility control was exceedingly variable. Southern farm women, for example, continued to bear an average of almost six children at the end of the nineteenth century. In Philadelphia in 1880 German- and Irish-born women bore over seven children during their lives. See Stewart E. Tolnay, Stephen N. Graham, and Avery M. Guest, "Own-Child Estimates of U.S. White Fertility, 1886–99," *Historical Methods* 15 (Summer 1982): 127–38; Phillips Cutright and Edward Shorter, "The Effects of Health on the Completed Fertility of Nonwhite and White U.S. Women Born Between 1867 and 1935," *Journal of Social History* 13 (Winter 1979): 191–217; Michael R. Haines, "Fertility and Marriage in a Nineteenth-Century Industrial City: Philadelphia, 1850–1880," *Journal of Economic History* 40 (March 1980): 151–58.

8. Cotton Mather, ch. 53, "Retired Elizabeth: A Long tho' no very Hard, Chapter for A Woman whose Travail approaches with Remedies to Abate the Sorrows of Childbearing" (1710), in *The Angel of Bethesda* (Barre, Mass.: American Antiquarian Society, 1972), 235–48; quotations from 237.

9. Margaret Jones Bolsterli, ed., entry for July 1890, in *Vinegar Pie and Chicken Bread: A Woman's Diary of Life in the Rural South, 1890–1891* (Fayetteville: University of Arkansas Press, 1982), 38.

10. Clara Clough Lenroot, Journals and Diaries, pt. 1, 1891 to 1929, edited by her daughter, Katharine F. Lenroot, typescript (May 1969) in family hands. My thanks to Katherine Vila, who shared copies of this diary with my class at the University of Wisconsin–Madison, "Women and Health in America," during the spring semester 1983.

11. Ellen Regal to Isaac Demmon, May 13, 1872, Regal Family Collection, Michigan Historical Collection, Bentley Historical Library, University of Michigan; quoted in Carl N. Degler, *At Odds: Women and the Family in America from the Revolution to the Present* (New York: Oxford University Press, 1980), 60. Women also wrote of joyful anticipation and looked forward to surviving their pregnancies and enjoying their children. Fear seemed a common sentiment among those women who detailed their feelings. It should be noted here that many women chose not to talk at all about their feelings during pregnancy, and if these women differed from their more articulate sisters, their sentiments are lost. Also lost to us are the feelings of women who did not leave any written records of their lives.

12. William Wight Papers, Wisconsin State Historical Society; vols. 11–15 are the diaries of William's wife, Albina Wight, 1869–76. Quotations are from vol. 11, entries for April 14 and 17, 1870. See also entry for May 13, 1870.

13. James S. Bailey, "Cases Illustrating Some of the Causes of Death Occurring Soon after Childbirth," *New York State Medical Society Transactions* (1872): 121–29; quotation from 121. For more on physicians' experiences with sudden puerperal death, see James L. Taylor, "What Killed the Woman?" *Journal of the American Medical Association* 14 (June 14, 1890): 876–77; Fayette Dunlap, "Sudden Death in Labor and Childbed," *Journal of the American Medical Association* 9 (1887): 330–34; Edward W. Jenks, "The Causes of Sudden Death of Puerperal Women," *Transactions of the American Medical Association* 29 (1878): 373–91.

14. William Thompson Lusk, "On Sudden Death in Labor and Childbed," *Journal of the American Medical Association* 3 (1884): 427–31.

15. H. V. Sweringen, "Laceration of the Female Perineum," *Transactions of the Indiana State Medical Society* 32 (1882): 135. I am grateful to Ann Carmichael for this reference.

16. Haven Emerson and Harriet E. Hughes, *Populations, Notifiable Diseases, and Deaths, Assembled for*

New York City, New York (New York: DeLamar Institute of Public Health, College of Physicians and Surgeons, 1941) and *Supplement 1936–1953* (January 1955). I would like to thank Gretchen Condran and Morris Vogel, who brought these volumes to my attention, and Evelyn Fine, who deciphered the figures and drew the graph. The figures are for New York City and not meant to be representative of the rest of the country, although the trend shown in these figures was probably similar outside the city. Statistics gathered by the Metropolitan Board of Health in New York after 1866 were more complete than those gathered for other cities or states.

17. The use of the figure five births per married woman in the United States is not far off the mark, and may in fact be low, given that the 3.56 recorded average includes only white and mostly native-born women. The Northwestern Mutual Life Insurance Company recorded that of ten thousand applicants for life insurance, "one man in every 17.3 who applied for insurance had a mother or sister or both who died from the immediate effects of childbirth." The May 1920 *Crusader* noted: "It is believed that a considerable percentage of these deaths from childbirth were recorded on the death certificate as being due to tuberculosis, heart disease, etc., and that the applicant for insurance remembered the associated childbirth and not the cause of death given on the death certificate. Our present mortality records do not show the frequency with which childbirth is a contributing cause of death" (5). See also C. W. Earle's comments during a Chicago Medical-Legal Society discussion of J. H. Etheridge, "The Medico-Legal Aspect of Utterances Made in Medical Societies," *Journal of the American Medical Association* 10 (May 5, 1888): 570. The quotation is Anne Lesley, in Susan Inches Lesley, *Recollections of My Mother* (Boston: Press of George H. Ellis, 1889), 306.

18. I explore this issue in "'Science' Enters the Birthing Room: Obstetrics in America since the Eighteenth Century," *Journal of American History* 70 (September 1983): 281–304. See also Janet Bogdan, "The Mortality Experience of Nineteenth-Century New York City Women" (Paper presented at the Sixth Annual Berkshire Conference on the History of Women at Smith College, June 3, 1984). International com-

parisons of maternal mortality can be found in Grace L. Meigs, *Maternal Mortality from All Conditions Connected with Childbirth in the United States and Certain Other Countries,* Children's Bureau Publication no. 6 (Washington, D.C.: GPO, 1917). Women became aware of continuing high maternal deaths in part from their own experience and in part from articles appearing in popular health journals such as the *Crusader* and in women's journals. See, for example, S. Josephine Baker, "Why Do Our Mothers and Babies Die?" *Ladies' Home Journal* 39 (April 1922): 32, 174.

19. Henry Parker Newman, "Prolapse of the Female Pelvic Organs," *Journal of the American Medical Association* 21 (Sept. 2, 1893): 335.

20. S. D. Gross, "Lacerations of the Female Sexual Organs Consequent upon Parturition: Their Causes and Their Prevention," *Journal of the American Medical Association* 3 (1884): 337–38.

21. J. Marion Sims developed the surgical repair of the vesicovaginal fistula through his experiments on slave women in the 1840s; it gathered adherents in the second half of the nineteenth century and relieved many women of their suffering. See, for example, J. Marion Sims, *The Story of My Life* (New York: D. Appleton & Company, 1889); Seale Harris, *Woman's Surgeon: The Life Story of J. Marion Sims* (New York: Macmillan, 1950); Irwin H. Kaiser, "Reappraisals of J. Marion Sims," *American Journal of Obstetrics and Gynecology* 132 (Dec. 15, 1978): 878–84.

22. J. O. Malsbery, "Advice to the Prospective Mother: Assistance during Her Confinement and Care for a Few Days Following," *Journal of the American Medical Association* 28 (May 15, 1897): 932.

23. On postpartum lacerations, see, for example, the following articles in the *Journal of the American Medical Association:* Charles P. Noble, "The Causation of Diseases of Women," 21 (Sept. 16, 1893): 410–14; John C. Da Costa, "An Easy Method of Repairing the Perineum," 13 (Nov. 12, 1889): 645–47; Henry T. Byford, "The Production and Prevention of Perineal Lacerations during Labor, with Description of an Unrecognized Form," 6 (March 6, 1886): 253–57, 271; and H. V. Sweringen, "Laceration of the Female Perineum," 5 (Aug. 15, 1885): 173–77. See also the discussion of this last paper in the *Transactions of the Indi-*

ana State Medical Society, 258–264, during which Dr. Woolen of Indianapolis blamed physicians' interventions for the increased rate of perineal tears in women: "The frequent use of forceps is filling the country full of cases for our gynecologists, and it does seem to me that we are making a mistake" (263). Physicians, anxious to clear themselves of possible blame, more frequently named the baby's hard head rather than the hard forceps as the principal agent. Regardless of agent, however, the state of medical knowledge about repairing such damage remained at issue.

24. Augustus K. Gardner, "On the Use of Pessaries," Transactions of the American Medical Association 15 (1865): 110.

25. Albina Wight Diary, 11: entries of Aug. 20, Oct. 6, and 9, 1873; Jan. 18, 1874. Gerda Lerner noted that Angelina Grimke suffered from prolapsed uterus and other postpartum complications following the birth of her second child, and the biographies of numerous other famous and not famous women indicate the problem was a common one. Gerda Lerner, The Grimke Sisters from South Carolina: Pioneers for Woman's Rights and Abolition (New York: Houghton Mifflin, 1967), 288–92.

26. Gardner, "On the Use," 113.

27. Homer O. Hitchcock, "A Modified Ring Pessary for the Treatment and Cure of Anteflexion and Anteversion of the Uterus," Transactions of the American Medical Association 15 (1865): 103–106.

28. Reported in "Midwifery" section, Transactions of the American Medical Association 6 (1851): 361. I address the debate concerning physicians' blame for women's postpartum problems in "'Science' Enters the Birthing Room" and in Brought to Bed: Childbearing in America, 1750–1950 (New York: Oxford University Press, 1986). For some contemporary comment, see Hiram Corson, "On the Statistics of 3,036 Cases of Labor," Journal of the American Medical Association 7 (July 31, 1886): 138–39.

29. Agnes Just Reid, Letters of Long Ago (Caldwell, Idaho: Caxton Printers, 1936), 24.

30. Christina Hopkinson Baker, ed., Diary and Letters of Josephine Preston Peabody (Boston: Houghton Mifflin, 1925), 214–15, 226–29.

31. Elsa Rosenberg to the author in response to author's query in the New York Times Book Review,

July 30, 1983.

32. Elizabeth H. Emerson, Glimpses of a Life (Burlington, N.C.: J. S. Sargent, 1960), 4–5.

33. Lillie M. Jackson, Fanning the Embers (Boston: Christopher Publishing House, 1966), 90–91.

34. Hallie F. Nelson, South of the Cottonwood Tree (Broken Bow, Neb.: Purcells, 1977), 173.

35. This diary entry is all the more poignant because it influenced two generations in this family. Hannah Whitall Smith's niece, M. Carey Thomas, found the journal and was so moved by it that she copied it into her own diary in 1878. Marjorie Housepian Dobkin, ed., The Making of a Feminist: Early Journals and Letters of M. Carey Thomas (Kent, Ohio: Kent State University Press, 1979). Hannah Smith's diary entry was dated Dec. 20, 1852, and Thomas copied it on Sept. 1, 1878, p. 149.

36. Albina Wight Diary, 11 (April 14, 1870).

37. Malinda Jenkins, Gambler's Wife: The Life of Malinda Jenkins, as told in conversations to Jessie Lilienthal (Boston: Houghton Mifflin, 1933), 48.

38. Elizabeth Elton Smith, The Three Eras of Woman's Life (New York: Harper & Brothers, 1836), 85.

39. Anita McCormick Blaine to Nettie Fowler McCormick; Nettie Fowler McCormick to Anita McCormick Blaine, both in August 1890. McCormick Family Papers, series 1E, box 459; series 2B, box 46, Wisconsin State Historical Society.

40. Entries for Aug. 15, 16, 17, 1890, Bolsterli, ed., Vinegar Pie, 60–61.

41. William P. Preston to his daughter, May, on her fifteenth birthday, May 19, 1864, from the collection of the McKeldin Library Archives and Manuscripts, University of Maryland, College Park. I would like to thank Virginia Beauchamp for sending me this reference from her book, A Private War: Letters and Diaries of Madge Preston, 1862–1867 (New Brunswick, N.J.: Rutgers University Press, 1987).

42. Christiana Holmes Tillson, A Woman's Story of Pioneer Illinois, ed. Milo Milton Quaife (Chicago: Lakeside Press, 1919), 128.

43. The Life of Mrs. Robert Clay, Afterwards Mrs. Robert Bolton (Nee Ann Curtis), 1690–1738 (Philadelphia, 1928), 154.

44. Laura B. Gaye, Laugh on Friday, Weep on Sun-

day: One Woman's Reminiscence (Calabassas, Calif.: Loma Palaga Press, 1968), 55; May Harley, *Whither Shall I Go: A Story of an Itinerant Circuit Rider's Wife* (Southold, N.Y.: Academy Printing Services, 1975), 76.

45. Anna Maria Thornton, Diary of Mrs. William Thornton, 1800–1863, *Records of the Columbia Historical Society* 10 (1907): 100.

46. *Esther Burr's Journal* (Washington, D.C.: Howard University Print, 1903).

47. Marilyn Clohessy to the author, Sept. 9, 1983, in response to author's query in the *New York Times Book Review.*

48. Entry of June 27, 1890, Bolsterli, ed., *Vinegar Pie,* 35.

49. Fleetwood Churchill, *The Diseases of Females: Including Those of Pregnancy and Childbed,* 4th American ed. (Philadelphia: Lea & Blanchard, 1847), 340.

50. John G. Meachem, Sr., Papers 1823–1896, State Historical Society of Wisconsin, Autobiography, box 1, p. 6. See also Carl Binger, *Revolutionary Doctor: Benjamin Rush, 1746–1813* (New York: Norton, 1966), 77.

51. Edward Henry Dixon, *Woman and Her Diseases from the Cradle to the Grave,* 10th ed. (Philadelphia: G. G. Evans, 1860), 261.

52. Letter from Dr. Thomas Steel to his father, Dec. 12, 1844, in the Steel Collection at the State Historical Society of Wisconsin. I am grateful to Peter Harstad for calling my attention to Steel's obstetrical cases. For more of this kind of evidence see Leavitt, "'Science' Enters the Birthing Room."

53. Reported in James A. Harrar, *The Story of the Lying-In Hospital of the City of New York* (New York: Society of the Lying-In Hospital, 1938), 34. On the point of rebellion of women inside the hospital as well as outside, see Nancy Schrom Dye, "Scientific Obstetrics and Working-Class Women: The New York Midwifery Dispensary" (Paper delivered at the American Historical Association, San Francisco, December 1983).

54. Dixon, *Woman and Her Diseases,* 262.

55. S. H. Landrum, Altus, Oklahoma, letter to the editor, *Journal of the American Medical Association* 58 (1912): 576. I would like to thank Carolyn Hackler for calling this letter to my attention.

56. John Milton Duff, "Parturition as a Factor in Gynecologic Practice," *Journal of the American Medical Association* 35 (Aug. 25, 1900): 465. Price's comment came during the discussion of Duff's paper, 467.

57. E. L. Larkins, "Care and Repair of the Female Perineum," *Journal of the American Medical Association* 32 (Feb. 11, 1899): 284.

58. See "'Science' Enters the Birthing Room" for more examples of physicians' lack of control in home deliveries.

59. Clohessy to the author.

60. Discussed in "Alone Among Strangers: Childbirth Moves to the Hospital" (Paper delivered at the American Historical Association Annual Meeting, San Francisco, December 1983); and Leavitt, *Brought to Bed.*

61. Carroll Smith-Rosenberg, "The Female World of Love and Ritual: Relations Between Women in Nineteenth-Century America," *Signs* 1 (1975): 1–29.

18 And the Results Showed Promise . . . Physicians, Childbirth, and Southern Black Migrant Women, 1916–1930: Pittsburgh as a Case Study

Carolyn Leonard Carson

African-American women who migrated from the South to northern cities during the great migration period chose to shun familiar childbirth traditions and utilized available medical services due to the influence of the Urban League. Black northern urban women's childbirth experiences differed markedly from those of southern women although research suggests that changes which occurred in the North possibly already had begun in the South, but the transition to medical childbirth there took much longer. At the same time, and possibly as a result of the development of new medical services and the utilization of those services by black women, infant and maternal mortality rates showed a slight decline in the urban North. In addition, this research suggests that African-American

CAROLYN LEONARD CARSON is Visiting Lecturer of History and Coordinator of Urban Studies Program at University of Pittsburgh, Pittsburgh, Pennsylvania.

Reprinted by permission of Transaction Publishers. *Journal of American Ethnic History* 14 (1994). Copyright © 1994 by Transaction Publishers; all rights reserved.

women and white women may not necessarily have received the same medical care. This study emphasizes how historical inquiry can serve to inform current policy debate. The establishment and widespread utilization of medical services for childbirth suggested some promise of improved health in the twenties. This topic therefore deserves additional study to determine how these various factors may relate to current issues.

Between 1916 and 1930 approximately one-half million blacks, or 5 percent of the total southern black population, migrated from the South to northern cities, seeking a better way of life. They were lured by the anticipation of better educational opportunities for their children and sought higher wages and employment opportunities brought on, initially, by the high labor demand of wartime. Poverty, an oppressive sharecropping system, Jim Crow laws, a plague of boll weevils, and the sexual harassment of women were other factors that motivated them to head north.[1] Between 1900 and 1920 the black population of Pittsburgh increased by 85 percent.[2] By 1928, 7.7 percent of the population of Pittsburgh was African-American.[3] Blacks poured into the community and settled just at the time when immigration from Europe declined.

In spite of the horrible living conditions blacks encountered, medically speaking it was to the black woman's benefit that she arrived in the North when she did. Coincidentally, certain services for the poor expanded dramatically during the early years of migration.[4] Pittsburgh's black pregnant women were brought into the mainstream of medical care before they had a chance to settle into another system, unlike their European counterparts who had arrived earlier, before the advent of numerous programs to assist indigent parturient women. In other words, immigrant women, in order to benefit from new services, would have had to change the patterns of health care such as the use of midwives they had previously established when they arrived in the city and adapted to the American context. Blacks, however, were offered the new services immediately upon their arrival. When they settled in the city, the new programs were just beginning to develop. Northern African-American women, in contrast to southern black women, utilized medical services for obstetrical care. The Urban League, devoted to helping southern migrants adjust, played the key role in linking new arrivals with available health care facilities.

Two major trends were occurring simultaneously that explain the creation of new programs offering medical and obstetrical care to the indigent. Obstetricians were beginning to professionalize and raise their status in the medical community by improving their medical education. Second, social reformers were advocating services in an attempt to reduce the alarmingly high rates of infant and maternal mortality.

The northward migration of blacks occurred at the same time that childbirth practices were changing from a social to a medical phenom-enon.[5] Increasing numbers of women were choosing physicians over midwives as attendants and were selecting the hospital as the site for their confinements. In Pittsburgh in 1915, 12.9 percent of all births took place in the hospital, but by 1931 that figure had risen to 54.6 percent. During the same period the number of black women giving birth also grew. In 1915 only 3.6 percent of the infants born in the city were black. By 1931, 10.4 percent of all resident births were to black women.[6]

Judith Leavitt has argued that middle- and upper-class women, primarily those of urban areas, were the agents of change in moving childbirth from the home to the hospital.[7] This focus might be redefined, for agents of change were urban women more generally, including elements of the poorer classes and the black race. Residence, not class, may have been critical. Nancy Schrom Dye has suggested that poor and working-class women were central to the transformation of birth from a social to a medical phenomenon since they were the first to experience the new scientific obstetrics. It was those patients who were the first to restructure the doctor-patient relationship.[8] Dye has argued that the patterns of medical authority that evolved in those relationships served as the social basis for medical management of childbirth throughout the twentieth century.[9] It is now clear that African-American women, whether from the rural or urban South, played a key role in the process of medical childbirth as they migrated northward and utilized similar services.

It is not yet clear whether or not this Pittsburgh study is representative of a national pattern. However, so little work has been done in this area that a case study can at least raise important possibilities. Pittsburgh is representative of the urban northern areas to which southern blacks migrated so it is worth examining if only as a steppingstone to further research.

Migrant Families

Black families came from the Deep South as well as the border states, and Pittsburgh attracted married women more than single women because of the job market. Although the city offered industrial jobs to men, it had few job opportunities for black women outside of domestic service.[10] It has been estimated that 75 percent of the migrants were between eighteen and forty years of age.[11] This figure is not surprising since the incentive to migrate was often employment. The significance of this was that many women of childbearing age were arriving in Pittsburgh during this period. Of the entire black population in Allegheny County by 1930, 49.9 percent were between the ages of twenty and forty-four.[12] Further, 66.3 percent of the black population were born outside the state of Pennsylvania, and 62.1 percent of the total black population were born in the South.[13] Blacks were predominantly migrants.[14] By 1930, for every 100 black males, there were 97.1 females in the city.[15] The potential need for obstetrical services was obvious.

Better education as well as employment opportunities attracted migrants. They felt the school systems in the North would serve their children better than those in the South.[16] Urban blacks were much more likely to send their children to schools than were either the Poles or Italians.[17] In addition, the entire self-selective nature of the migration process meant that migrating women were likely to be better educated than black southern women who remained.[18] This information suggests that the newly arrived black women may have been much more receptive to the health education campaigns to which they were exposed.

Migrant families in Pittsburgh followed common patterns in living arrangements. Women generally married at a much younger age than white women or first- or second-generation foreign women. The average age of marriage for black women was before age twenty.[19] Newly married couples usually lived alone, apart from their parents, who encouraged their children to be self sufficient.[20] This emphasis on self sufficiency may have discouraged the supportive female networks so necessary to the social childbirth which Leavitt has described whereby women gathered together to support a friend or relative in labor.

It was common for black women to be employed. Men usually entered the industrial workforce at the bottom and had little success in moving up due to discriminatory attitudes of employers. Furthermore, the advent of labor-saving devices in the factories resulted in the decrease in the number of unskilled jobs available for men. Employers' labor demands tended to peak and fall, leading to obvious job insecurity.[21] Parents could not rely on income from older adult children who tended to live independently from their parents.[22] For these reasons mothers were forced to seek employment in order to support their families. Although his study was not quantitative, Abraham Epstein, surveying black migrants in 1918, formed the impression that "practically all of the mothers are doing some work outside the home." Another study suggested that approximately one-third of the black women were working in 1930, and of those, 90 percent were in domestic service.[23] Women lived separate and apart from one another, and it was found that women living next door to each other, though often having come from the same state or even the same city, hardly knew one another.[24] This family pattern further reduced the likelihood of establishing a significant female network or support system.

Black families were not part of a large cohesive group in Pittsburgh, as they were divided along class and geographic lines. Peter Gottlieb has identified three socioeconomic levels for blacks. The first group were the old elites, middle-class blacks who had resided in the city for several decades. The second group was comprised of economically stable skilled and unskilled workers, and the third group was a lower class of transients. The last two groups, the darker-skinned migrants from the South, often the Deep South, were primarily industrial laborers. According to Gottlieb, these three groups did not associate with one another.[25] The middle-class physicians and lawyers, for example, seldom offered aid to those less fortunate members of their own race.[26] Due to the hilly topography, the residential neighborhoods of the black community further discouraged any kind of black cohesiveness. None of the ghettos was all black, as blacks and immigrants frequently lived in the same areas. Further, blacks were not concentrated in one area of the city. This lack of cohesiveness within the race and within the family may have been one of the key factors that compelled black women to curtail the use of midwives and seek medical obstetrical care.[27]

Infant and Maternal Mortality

Infant and maternal mortality rates were higher among blacks than whites in the United States. In 1916 there were 185 black infant deaths per 1,000 live births, compared to 99 white infant deaths. By 1921 the death rate had declined but at approximately the same rate for both races so that the black infant death rate continued to exceed that for whites. The rate that year was 108 black infant deaths per 1,000 live births and 72 white infant deaths. The black infant death rate continued to exceed that for whites in the following decade. In the period 1933–1935 the rate per 1,000 live births in Pennsylvania was 84.3 for blacks and 51.4 for whites.[28]

It is, however, vital to note the differences between infant death rates among blacks in urban and rural areas. The urban infant death rate for black infants exceeded the rural rate in the United States as a whole but not within the northern states alone. In the northern states in 1933–1935 the rural death rate for black infants was 100.9, compared to an urban rate of 81.0. In Pennsylvania in 1933–1935 the urban infant death rate per 1,000 live births for blacks was 81.0, compared to a rural rate of 109.6. In the South the rates were reversed (rural: 80.2, urban: 109.3). These statistics offer a stimulus to research in order to determine what caused the distinctive pattern in the urban North.[29] Elizabeth Tandy, senior statistician for the Children's Bureau, suggested that the living conditions in the northern as well as the southern cities were characterized by crowded housing, "with inadequate facilities for care of their health and for recreation." She surmised that the high infant mortality rates in urban areas in the southern states were due to the lack of child-health activities in southern cities.[30] The existence of various health reform programs in the northern urban regions undoubtedly helps to explain this urban-rural difference.

In 1920 Pittsburgh lost more babies in proportion to its births than any other of the large American cities for which reliable records were available. Half of those deaths occurred within the first month of life, reflecting a national pattern (48 per 1,000 live births in 1920).[31] The overall infant death rate in Pittsburgh in 1920 was 110 per 1,000 live births, but for blacks the rate was 164. In 1926 the infant death rate per 1,000 live births was 81.5 for white residents and 111.8 for blacks. Although the rates continued to drop, as they did nationally, they still remained much higher among the black population than among the white population of Pittsburgh. In 1934 the white infant death rate was 51.5, and the rate for blacks was 77.8.[32]

Stillbirths in Pittsburgh, following the pattern of the national rates, were considerably higher for blacks, with a rate of 7.4 per 100 live births, and 3.5 for whites in 1923.[33] In 1931, although the stillbirth rates had declined, they were still considerably higher for the black population, at 6.1 per 100 live births compared to 3.4 for the white population.[34]

Maternal mortality rates were also higher among blacks than whites. During the period 1933–1935 in the United States the rate for black women was 96.1 per 10,000 live births compared to 54.6 for white women.[35] The statistics for Pittsburgh in 1935 show an even greater gap between maternal death rates for white and black mothers. The rate for white mothers was 37.0 per 10,000 live births compared to 91.0 for black mothers.[36] The most frequent causes of maternal mortality among blacks in the 1933–1935 period were puerperal sepsis and toxemia, both of which were recognized as preventable. Combined, those two causes accounted for 68 percent of the deaths of black mothers in the United States. The other deaths were due largely to puerperal hemorrhages and accidents of pregnancy or labor.[37]

During the migration period and shortly thereafter there were a number of theories regarding the causes of the high maternal and infant death rates among blacks. Members of the National Urban League suggested that the habits and conditions of the mother's living before childbirth, as well as her physical condition, contributed to these high death rates. "These deaths have been found to be most common among mothers who are forced to work and where proper medical care cannot be secured."[38] The Urban League in 1923 further contended that infant mortality, especially during the neonatal period, or first month of life, was largely caused by diseases classified as "developmental." These included congenital malformations, prematurity and congenital debility which resulted, it was thought, from the lowered physical condition and habits of the mother's living before

childbirth.[39] Rates of tuberculosis and syphilis, often the results of poor, overcrowded, unsanitary housing, were also higher among blacks, which suggests that black women were in poorer health than their white counterparts—which could have affected the outcome of their confinements.[40] Grace Abbott, director of the United States Department of Labor Children's Bureau, when addressing the Urban League in 1923, also noted the higher incidence of death among working women, whose husbands' wages were low and who lived in rear houses or houses in alleys.[41] These socioeconomic factors undoubtedly adversely affected the maternal and infant death rates among blacks.

Professionals of the period also recognized the prevalence of rickets among blacks.[42] Grace Abbot noted the disproportionate rate of rickets among black children. In the early twenties the Children's Bureau made a study of rickets in the Washington, D.C., area with the hope of finding a method to decrease the rate of the disease.[43] Dr. John Whitridge Williams, professor of obstetrics at the Johns Hopkins University, also noted the "prevalence of rickets among blacks," which led to the increased incidence of contracted pelvises among black mothers, a condition which had the potential of adversely affecting labor. An analysis of Baltimore women in 1911 revealed that 7.7 percent of all white patients had contracted pelvises compared to 33.2 percent of the African-American mothers.[44]

Current studies have noted the role of poor nutrition in the high maternal and infant death rates among blacks in the early part of this century. Phillips Cutright and Edward Shorter found that puerperal fever was the largest single cause of maternal death in childbirth, though there was also an increased incidence of deaths due to toxemia and other causes. They suggested that blacks' obstetrical mortality was due largely to infection and that this susceptibility to infection was probably due to a poor diet that had a chronic lack of protein. They also noted the increased incidence of rickets among blacks that led to contracted pelvises and prolonged labor, which increased the risk of infection.[45]

Other studies of the period suggest that the comparatively high death rates were due partially to the lack of adequate medical and nursing care. The Urban League recognized the high death rates among women who did not receive adequate medical care. "The most effective method by which the community can cut the high ratio of these losses is by providing care and instruction for the mother before her baby's birth and skilled attendance during her confinement."[46] Here was a link between the disproportionate health problems of maternal and infant mortality during the black migration period, the resultant attempted solutions, and changes in traditional health care patterns.

Medical and Social Reform

Progressive reformers had been taking an interest in infant health for over four decades prior to World War I, but at that time their focus began to change. After 1914 there was an increasing awareness of the value of improving the health of gestative and parturient women. Earlier attention had been given to reducing the neonatal death rate by dealing with postnatal threats to infant life by improving the domestic sanitary environment and the quality of infant care. In 1911 the New York Milk Committee conducted an experiment that suggested prenatal care reduced stillbirth and neonatal death rates. This information prompted welfare workers in other cities to institute prenatal care in their programs. In 1910 the American Association for the Study and Prevention of Infant Mortality (AASPIM) was founded by physicians and public health officials who were disturbed about the infant mortality rate.[47] They were partially responsible for the founding of the United States Children's Bureau in 1912, within the Depart-

ment of Labor, which worked to decrease infant and maternal mortality rates.

In 1917 the Children's Bureau conducted a landmark study entitled "Maternal Mortality from all Conditions Connected with Childbirth in the United States and Certain Other Countries." This study revealed that childbirth caused more deaths among women aged fifteen to forty-four than any diseases other than tuberculosis. Only two of fifteen countries surveyed had maternal death rates higher than the United States for 1900–1910.

Social reformers responded to these studies on a wide scale. Often with the aid of the federal Children's Bureau, cities and states in the United States established special bureaus to deal specifically with improving the health of infants and children. Most of these agencies, which included prenatal care in their services, were created between 1919 and 1921, the period when southern migration was growing and white immigration was on the decline.[48]

Physicians also responded to the high infant and maternal mortality rates. At the request of the AASPIM, John Whitridge Williams conducted a study in 1911 which revealed that medical schools were "inadequately equipped for teaching obstetrics properly." He referred to the degraded position of obstetrics in this country. He urged that people "be taught that a well conducted hospital is the ideal place for delivery, especially in the case of those with limited incomes."[49] As physicians sought to improve the training of obstetrics, they also sought a patient population from which they could learn. Crucial to their professionalization was the elimination of the practicing midwife for "she has charge of fifty percent of all obstetrics material of the country without contributing anything to our knowledge of the subject; a large percentage of the cases are indispensable to the proper training of physicians and nurses." Dr. Charles Ziegler, professor of obstetrics at the University of Pittsburgh and medical director of the Elizabeth Steel Magee Hospital, felt that students of medicine should be trained under careful supervision and by recent graduates who had themselves been trained in well-equipped and properly conducted maternity hospitals. Ziegler not only sought adequate obstetrical education but also argued that every woman had a right as a citizen and as a mother to proper care during and following childbirth. Ziegler's plan, which would allow for adequate training of students and proper care of patients, was implemented in Pittsburgh with admirable results.[50]

Influential Factors

Even with its limitations, the new medical experience entailed significant change, a real human transition. Black women arriving from the South had been familiar with midwives' services, and yet evidence suggests that they made no attempt to continue the tradition in the North. Their experience is unlike that of white immigrant women who appeared to be far more reluctant to accept medical care. Why did the change occur? The most elusive factor involves the African-American women's own attitudes, expectations, and motivations regarding medical care. There is no evidence to suggest why women sought medical care or how they felt about it. It is not clear whether they perceived physician care as better than the traditional care of a midwife or whether it was just more available.[51]

The fact remains that black women in large numbers began utilizing the services of nurses, white physicians, and integrated hospitals for maternity care fairly early in the migration period. Several organizations were influential in informing women of available services and encouraging women to take advantage of them. The major difficulty was not in getting women to use the services but in making them aware of their existence. "When the Negro mother is made acquainted with the existence of a prenatal clinic it is not difficult to get her to attend. The problem is to acquaint the ignorant Negro mother with the opportunity."[52]

Black women were brought into the medical system largely due to the efforts of the Urban League. There is no way to determine the numbers of women the league influenced, but its influence was spread throughout the city. All of the black neighborhoods were exposed to the health programs the Urban League offered. The Urban League had established a network between numerous hospitals, clinics, physicians, and social agencies which provided obstetrical care.

Blacks and whites first launched the league in 1915 as the Council for Social Service Among Negroes, which was organized within the Associated Charities to aid migrants seeking employment. On January 14, 1918, the council became the Urban League, part of a national organization whose chief activities were educational, promotional, or interpretive. Health was one of the major issues that it addressed. The Pittsburgh facility opened at 505 Wylie Avenue on March 1, 1918. The league's chief instruments were research, education, and publicity.[53]

Immediately following the organization of the Urban League its agents began their campaign for better health. They launched a Negro Health Education Campaign the last week of April 1918. The pastors of thirty-one churches cooperated by preaching on the seriousness of health conditions. Over twenty-thousand pieces of literature supplied by the government, the city, and the Metropolitan Life Insurance Company were distributed. In one week noted white and black doctors and educators conducted fifteen meetings to explain how to better preserve health. Two famous black physicians were brought in from New York and Chicago to speak at two mass meetings.[54]

The league later sponsored a program for the local celebration of National Negro Health Week. Booker T. Washington started this educational movement in 1914 through the National Negro Business League and soon thereafter won the cooperation of the National Urban League, the Surgeon General's office, and other organizations.[55] During this celebration in Pittsburgh, physicians and dentists gave at least fifty health lectures and talks.[56] Eugene Kinckle Jones cited this particular educational movement as "possibly the most effective educational movement for improving health among Negroes generally."[57]

Within the first few years of the Urban League's existence it had established a widespread health care network for poor blacks of which it was the center. Mutually beneficial relationships existed between several city hospitals, clinics, dispensaries, social agencies, and the Public Health Nursing Association. John T. Clark, executive director of the Urban League of Pittsburgh, stated that the cooperation of the Negro Medical Society, City Department of Public Health, Public Health Nursing Association, hospitals, physicians, and social workers in presenting health education campaigns caused growing attention to be paid to health problems in congested black sections. The result was that the proportionate black death rate gradually declined between 1918 and 1923, even though the population increased.[58] David McBride has suggested that black doctors and nurses in New York City played a key role in "effectively usurping working-class blacks' health beliefs and resistance to medical institutions."[59] In Pittsburgh it seems that they were also instrumental but only within the network of which the Urban League was the center.

Little is known of the African-American doctors and nurses who worked to improve the health of members of their own race. Clearly, many leaders associated with the league felt that black women may have been more receptive to black medical practitioners. Speakers brought in from other cities were often black. The Public Health Nursing Association hired two black nurses to work in the community, upon the recommendation of the Urban League.[60] The African-American physician and nurse "being accustomed to reckon with the personal idiosyncrasies of Negroes and to treat the patient as well as the disease is, therefore, best fitted to handle such cases, because those of the white race in quite a number of cases look upon Negroes as mere subjects for observation."[61] Although this may have been the ideal situation, and the Urban League certainly made every attempt to utilize black medical practitioners, there were limitations because of discriminatory policies in the local hospitals. Clearly, the Urban League's primary concern was that of providing medical care for black women, even in spite of those policies. The black practitioners were influential outside of the hospital setting by providing information and primary and preventive care. They advised patients to utilize hospitals and white physicians. Although little is known of these individuals and their activities, their roles should not be minimized.

A black migrant's first contact with the Urban League in Pittsburgh may have been with the Travelers Aid worker who met blacks at the railroad stations upon their arrival. During the last twenty days of May 1923, 1,165 migrants, 532 of whom were women, arrived in Union Station (the largest of five railroad stations in the city). Within six months the Travelers Aid worker rendered service to more than 1,100 black travelers.[62]

The Urban League had two departments which were particularly instrumental in establishing and utilizing this informal network of services. One of the first departments to be established within the league was the Department of Information and Advice. Individuals as well as organizations sought numerous types of assistance from this agency, which either responded directly or referred the problem elsewhere. Blacks sought advice concerning employment and housing, for example. Of importance here is the fact that individuals also sought health care advice. A man in April 1920 brought his sixteen-year-old pregnant daughter to the league, seeking a home for her until her confinement. She was referred to the Bethesda Home, a home for unwed mothers. Another man sought information regarding a hospital that treated diseases of the eyes and ears and was directed properly. One patient, ill in a boarding house, was unable to pay a private physician. The physician refused to treat him but did refer him to the Urban League, which obtained a bed for him at West Penn Hospital. According to the executive secretary's report of 1920, there were a number of requests from individuals needing help getting into hospitals.[63]

The Home Economics Department workers, employed by the Urban League as early as 1919, were key figures in acquainting black migrants with available medical facilities. John T. Clark described the workers as having "been valuable in overcoming their prejudices against hospitals . . . as well as in teaching economy and thrift in household management."[64] Workers visited the homes of migrants as early upon their arrival as possible and found that they were welcomed with open arms with very few exceptions. Cards were distributed which listed ten points of advice, "For the Negro in Pittsburgh." Point number five suggested that one "send for a doctor when sick, don't use patent medicines." Upon arrival, then, the migrant was introduced to an organization and to individuals who emphasized orthodox medical care.[65]

Home economics workers not only visited new migrants but other blacks in the poor communities as well and regularly made referrals to health care facilities, described as cooperating agencies, within the city. As many as 260 homes of poor blacks were visited in one month alone. Patients requiring medical care were referred to

the nurses at the settlement house, the University Maternity Dispensary, or the Public Health Nursing Association. The workers also "secured the services" of West Penn, Mercy, Homeopathic, Allegheny General, Saint Francis, Saint Margaret's, Passavant, and Magee Hospitals.[66] The migrant's health problems were varied, but maternity patients were included in those in need of care. One woman who had recently arrived from Virginia was pregnant, suffering from regular "nervous fits," and in need of immediate hospital care, which was secured for her at once.[67]

The relationship that the Urban League forged with the cooperating health agencies was reciprocal. Saint Francis and Mercy Hospital each requested that someone from the league visit several patients who had been receiving treatment at the hospitals. West Penn's social service department asked that follow-up care be provided for several "cases of unfortunate girls that had received attention at the hospital." Magee Hospital needed a place where a young mother and her fifteen-day-old infant could go to convalesce together as the hospital did not want them to be separated. The league found a Mrs. Washington on Wylie Avenue who agreed to take the mother and baby into her home. Magee Hospital and the Public Health Nursing Association each sought help in placing babies.[68]

Local hospitals, as part of the network that provided care, directly sought the assistance of the league in soliciting patients. The Western Pennsylvania Hospital wrote asking for the league's cooperation in promoting their outpatient obstetrical service. The service, available in several adjacent neighborhoods, provided standard prenatal care in the dispensary and a doctor and nurse in attendance at confinement in the patient's own home, as well as postnatal care of mother and baby. Only women unable to pay a private physician were eligible.[69]

The league divided the city into districts

where the home economics workers organized neighborhood meetings or clubs for African-American women, largely from the rural South. The purpose of the regular meetings, held in churches and settlement houses, was to acquaint each woman with proper nutrition, canning techniques, sewing skills, and other methods which would allow her to raise her family in a healthy environment with limited funds. Women also received instruction about good health habits and were encouraged to use public facilities such as clinics and dispensaries.[70] Clubs sometimes brought in speakers such as the Public Health nurses or Dr. Sarah Brown of Washington, D.C., who spoke on sex hygiene in February 1920.[71] Subcommittees were sometimes organized to study problems of child care or to report cases of sickness or need within the community to the home economics worker.[72]

The Urban League sponsored Better Baby Shows, sometimes organized by the neighborhood groups, on a regular basis. These shows capitalized on mothers' pride by encouraging them to show off their youngsters. The primary purpose, however, was to provide health screenings which were conducted by Jeanette Washington, the first black public health nurse in Pittsburgh, and Dr. Marie Kinner, a black physician. Within one year the attendance at these "clinics" tripled.[73] During the shows literature from the local health bureau and the Metropolitan Life Insurance Company was distributed. Dr. Kinner lectured to groups of women on baby hygiene and other subjects as well during the baby shows. Organizers showed moving pictures on proper health care. The result, according to a 1921 publicity letter, was that "hundreds of mothers [are] now registering their babies in various health centers and clinics."[74] As noted earlier, reformers recognized by this time the value of prenatal care in ensuring the health of a new infant. It is likely that this was also emphasized at the Better Baby Shows. Further, the

same individual providing assistance to the mother usually cared for infants during the first six postpartum weeks. Infant care and health of the pregnant mother were interrelated.

The Urban League's publication, *Opportunity*, also served to inform and educate African-American women. By 1930 northern urban black women under the age of forty-four were as literate as northern white women. The illiteracy rates for southern African-American women were five times higher than those of northern urban blacks.[75] It is therefore reasonable to assume that periodicals such as *Opportunity* may have been fairly influential for the southern migrants in Pittsburgh.

The Urban League was very much aware of the high rates of infant and maternal mortality among African-Americans and perceived that education regarding prenatal and infant care and medical obstetrical services would help to reduce the appalling rates. Furthermore, they published articles regarding these issues. Readers were informed that the handicaps of poverty, which were blamed for the high mortality rates, could be overcome by providing black mothers with information regarding her care during pregnancy and childbirth and care of her infant.[76] For example, diarrheal diseases accounted for 28.3 percent of infant deaths in the years 1916 to 1920 and were considered to be due to bad feeding habits and inadequate instruction of black mothers.[77] Maternal deaths were believed to be most common among mothers who had not secured "proper medical care."[78] In 1924, although black infant mortality rates were still high, they were decreasing. It was believed that they would steadily improve "as more Negroes migrate to the larger cities of the North and West where prenatal and obstetrical work is well organized and is open to all regardless of race."[79] Clearly, the Urban League was well aware of appallingly high infant and maternal mortality rates among African-Americans.

Furthermore, they saw education and medical care as a solution to the problem. Black women were not only directed by the Urban League workers but they may have read these materials that implied that medical care was superior to folk medicine, midwifery, and other traditional practices in coping with high death rates among infants and parturient women.

Settlement houses also played a role in bringing services to black women by making space available to other agencies. Although the Irene Kaufmann settlement, located in the Hill District, was primarily for Jews, "invaluable service has been rendered to Negroes through the personal service work of the Settlement. Negroes are constantly in attendance at the various clinics held there; they are clients of the social agencies quartered there."[80] The Kaufmann settlement also housed a substation for the Public Health Nursing Association. There were nine nurses employed there, including two black nurses, and two black student nurses.[81]

Churches played a vital role in educating blacks regarding proper health care. They provided space for Urban League meetings and activities and, as noted, the ministers assisted by preaching about health-related issues. Within the third and fifth wards there were forty-five churches, eighteen of which were storefronts; some of them reached less educated blacks to whom the Urban League by itself had limited access.[82]

Migration and the imposing network of medical and service agencies did not transform all health habits. Many Pittsburgh blacks continued to practice folk medicine, sometimes because they had neither the money nor the inclination to seek formal medical care. Jacqueline Jones suggests that the discrimination of northern urban hospitals influenced these practices. Pittsburgh blacks sought the help of spiritualists, crystal gazers, voodoo, or herb doctors. Many believed in amulets, assorted superstitions, or other conjuring devices that were designed to protect them from danger. Dream books sold for ten or twenty-five cents and sold by the thousands.[83] Yet two points are worth noting. The first is that these types of beliefs and the belief in the efficacy of scientific medical practice are not necessarily mutually exclusive. Women undoubtedly sought the help of medical practitioners while still retaining some of their superstitious beliefs. Dr. Paul Titus, director of the dispensary at West Penn Hospital, noted that blacks believed that intercourse at the beginning of labor would result in a light labor.[84] He cited numerous other superstitious practices followed by whites as well as blacks, even as they engaged a physician's care. Second, when medical care was available, the black women may have been willing to abandon some of their old practices. It is clear that blacks did indeed utilize available medical facilities, from the 1920s onward, in spite of the existence of assorted faith healers. It is possible, given the concentration of educational efforts if nothing else, that this evolution occurred particularly rapidly where childbirth was concerned.

From Southern Midwife to Northern Physician

When black women, who were frequently in poor health, arrived in the urban North, they faced inadequate living conditions and discriminatory practices in an unfamiliar environment. It would not have been surprising if they had sought familiar types of medical practitioners who had been prevalent in the South. In fact, however, urban black migrants utilized the care of hospitals, dispensaries, and clinics in order to have safe deliveries and healthy infants. They did so willingly upon the advice of the more experienced women with whom they had contact. There is no evidence to suggest that they utilized the services of midwives or other healers, although this remains possible in some cases. What is clear is that the maternity care they received was totally unlike that to which they were accustomed in the South.

It is well documented that southern black women had midwives attend them during delivery.[85] As late as 1940 midwives attended more than 75 percent of the births of black women in Mississippi, South Carolina, Arkansas, Georgia, Florida, Alabama, and Louisiana.[86] Several studies of midwives have treated immigrants and blacks as one group. Frances Kobrin has stated that primarily blacks and the foreign born employed midwives and that midwives themselves usually shared race with their customers.[87] Judy Barrett Litoff noted that immigrants and blacks were attracted to midwives because of lower fees and a wider variety of services provided. For example, a midwife functioned as birth attendant, nurse, housekeeper, cook, and baby sitter. Further, a study of blacks in Texas in 1924 revealed that blacks preferred midwives because of their accessibility. Immigrants and blacks alike, it was argued, opposed male attendants.[88] Against these generalizations, the evidence strongly suggests that southern black women who had migrated to the North did not use midwives on a large scale; an important change accompanied urban adaptation in this region.

It is very difficult to ascertain the actual feelings of the African-American mothers, in the North or South, who sought assistance with childbirth. It is not certain that they made choices or if they accepted whatever care was available. Evidence seems to suggest, however, that African-American women in certain regions of the South had little choice regarding birth attendants. Midwives were not necessarily preferred or chosen but may have been the only attendants obtainable. Studies have, in fact, pointed out that in some regions of the South

midwives were hired because doctors were virtually unavailable. Litoff noted that southern physicians usually lived in towns and cities, often quite far from black rural women's homes. Doctors, in fact, reluctantly supported the existence of midwives because the distances between homes of patients and doctors were too great. In Georgia in 1927 midwives attended 31.4 percent of all births. Doctors themselves felt it would have been impossible for them to perform all of those deliveries.[89] In addition, impoverished women had to consider fees charged by birth attendants. Poor southern women could not afford a doctor's care, and physicians were not interested in caring for them because of the low fee they would receive.[90] The assumption that southern African-American women preferred midwives is therefore somewhat debatable.

If southern African-American women could not afford physicians or if they lived in a region where the only physician was miles away, they had no choice but to create their own birth attendants. Other studies seem to suggest that midwives were established in communities where there was a need. Midwives clearly did not enter their professions for financial reward. They were seldom paid, or sometimes received animal products, vegetables, or some service in return for delivering an infant.[91] Sometimes they established themselves as midwives because they were "called by the Lord" or because of community pressure.[92] They learned their craft from observation, experience working with friends or relatives who were midwives, or occasionally from physicians.[93] Debra Anne Susie has suggested that a hospital birth was preferable to replacing a long-standing midwife who was no longer practicing.[94] Perhaps southern African-American women were beginning to desire physician-attended childbirths, if available, during this period but generally were unable to make that choice. Perhaps the tradition of midwife-attended childbirth among southern black women was not culturally defined but was in fact merely rooted in necessity.

Jacqueline Jones has stated that due to the discriminatory practices of northern urban hospitals and the fact that many migrants had neither the money nor the inclination to seek formal medical care, black women relied on elderly women skilled in the ways of healing, including granny midwives.[95] Evidence, however, suggests that if there were black midwives in Pittsburgh, or if immigrant midwives assisted the black women, few African-American mothers were assisted by them during childbirth. As previously noted, studies of southern midwives imply that where there was a need, black midwives were established. It seems reasonable to assume that if the southern black women who migrated to northern cities had wanted midwives, they would have pressured women into filling that role the same way they did in the South. Litoff claims that the preference of immigrants for midwives was reinforced by the fact that it was a long-established European tradition to have them serve as attendants at birth. When immigrant women arrived in America they continued to employ midwives.[96] It has long been known that southern black women practiced the same "tradition," and yet they did not continue the practice in the North, certainly not on a large scale. Some of the factors cited for the demise of the midwife include women's attitudes. As the fertility rates declined, women began to believe that delivery was rare and therefore necessitated increased concern and expense.[97] It is noteworthy that the fertility rates of northern urban African-American women were lower than those of southern women, black or white, urban or rural.[98] Black women in Pittsburgh were bombarded with information suggesting that medical care was the best obstetrical care. Perhaps black women too were becoming increasingly self-conscious about their own welfare and perceived that medical care was better. They were already well aware that midwives called physicians in situations which were dangerous to either the mother or child. The transition to physician-attended childbirth, in light of these new attitudes, is not so surprising. African-American women may have been agents of change along with their white native-born counterparts in the North.

General statistics, to be sure, suggest that white women in the United States went to the hospital for delivery sooner than black women. By 1935 only 17 percent of all black births occurred in hospitals, compared to 40 percent of all white births. Further, only 43 percent of black births were attended by physicians, compared to 94 percent of the white births.[99] Infants that were not delivered by physicians were delivered primarily by midwives, who, overall, attended the births of more than half of the black infants in the United States at that time.[100] Northern urban black women, however, had different experiences in childbirth than their southern counterparts. These women normally chose the hospital as a birthplace and chose a physician as the birth attendant, regardless of where the delivery occurred. They largely, if not completely, abandoned the southern midwifery tradition. Elizabeth Tandy's study in 1937 showed that 97.9 percent of all black births in northern cities were attended by physicians and that 61.8 percent of those occurred in the hospital. Findings for New York City were similar: 98.3 percent of black births occurring between 1917 and 1923 took place in the hospital or within the home under the care of hospital personnel.[101] This was in marked contrast to the rural districts of the South, where physicians attended only 20 percent of all the black births and only 0.6 percent occurred in hospitals. In southern cities, however, physicians attended

61.8 percent of all births, and 38.5 percent of the babies were born in hospitals.[102] This information suggests that African-American women in the South made a transition to medical childbirth when it was feasible. The transition moved much more quickly in the urban North. A rural-urban distinction existed in the South, but even the urban South differed from migrant behavior in the North.

The city of Pittsburgh exemplifies the northern cities cited in the other studies noted. Midwife use among whites and blacks dropped in Pittsburgh from 16.4 percent of all deliveries in 1920 to 1.6 percent by 1935.[103] Other statistics suggest that this was true for whites as well as African-Americans. In 1938, 77.2 percent of white births occurred in the hospitals, compared to 79.3 percent of black births. And while 0.9 percent of the white home births were attended by midwives, there is no evidence to suggest that midwives attended any of the black home births.[104] Another study done in 1936 of unmarried black mothers noted that of 159 mothers, 125, or 79 percent, delivered their babies in the hospital. Of the 11 women known to have delivered at home, 10 were attended by a physician and 1 had no medical care. The details regarding delivery were unknown for 15 of the women.[105] A black woman, born in 1903, who spent her entire life living in the Hill District and who also worked as a social service worker at Magee Women's Hospital, has no recollection at all of black midwives.[106]

Notable differences between the black southern migrant woman and her white immigrant counterpart have already been cited. Whereas immigrants, in a combination of tradition and necessity, might develop a pattern of childbirth based on the services of a midwife who was present in an already established community when they arrived, black women reaching the northern city were eager to shake off southern

ways and at precisely the time when other facilities became available and accessible.

Medical Maternity Care and Black Women

Black women in Pittsburgh had several options for prenatal care. In addition to six hospitals known to have accepted black women as patients in their outpatient obstetrical clinics, the Public Health Nursing Association (PHNA) provided prenatal care in the home.[107] This private agency, which was organized in July 1919, was supported by subscriptions from Red Cross, individuals, clubs, and societies, plus industrial affiliations, patient fees, and insurance payments.[108] Patients were referred by private physicians, the University Maternity Dispensary (UMD), hospital clinics, and other social agencies, including the Urban League. Individuals could also request nursing care by phoning or visiting one of the substations located at Saint Peter's Parish House in Oakland, the Kaufmann settlement on Center Avenue, or other stations located in Lawrenceville, Homewood, South Side, North Side, McKees Rocks, or Braddock.

Care was provided by a nurse in the home but under a physician's direction. Trained registered nurses visited patients' homes regularly (once a month for the first seven months and twice a month for the last two) to teach the hygiene of pregnancy, delivery preparation, and proper diet. Special emphasis was placed on early medical supervision. Patients were also instructed about the proper layette for the baby. Symptoms and complications were reported to the attending physician. Postpartum care was also provided for several weeks. The infant was referred to a PHNA Well Baby Conference Station after five weeks.

This service brought care to many of the indigent in Pittsburgh. During 1933 the association claimed to have made 20,480 home visits to 5,660 pregnant residents of the city. During that same year the City Bureau of Health recorded a total of 9,308 resident births.[109] Not all of the prenatal patients delivered in 1933, but the large percentage of patients receiving home prenatal care is noteworthy.

Most of the care provided was free to the patient. Patients who could afford to pay provided $1.00 per visit or $1.50 for mother and baby.[110] Patients with more limited funds paid on a sliding scale, but the majority of patients received care free of charge. In 1929, 55 percent of the patients did not pay anything, 3 percent paid a part of the fee, and only 2 percent paid the full fee. The other 40 percent were patients of service contracts.[111] The John Hancock Life Insurance and Metropolitan Life Insurance Companies, for example, had contracts with the association whereby they paid the organization directly for nursing care for their policy holders.[112]

Black southern migrant women were choosing physicians as birth attendants and the hospital as well as the home for a birthplace when they became residents of Pittsburgh. One of the services utilized by parturient black women was the Pittsburgh Maternity Dispensary (later known as the University Maternity Dispensary or UMD), a home birthing service established by Dr. Charles Ziegler in 1912.[113] Seven different clinic substations were located in the poor neighborhoods of the city by 1920 and served only women unable to pay a private physician. Several of them were housed in settlement houses, including the Kaufmann settlement and the Kingsley House. Women frequently referred each other to the UMD. Newspaper articles about settlement house activities also served to inform women of the availability of these services.[114] A salaried full-time obstetrician was in charge of the dispensary, which also employed graduate nurses and a social worker. The dispensary provided clinical experience for students of the University of Pittsburgh School of Medicine and nursing students from the Magee Hospital School of Nursing.[115]

Patients were urged to register with the dispensary as early in their pregnancy as possible so that pelvic measurements, routine examinations, a Wasserman test, and urinalysis could be performed. The importance of proper care during pregnancy was impressed upon each woman at that time. When a woman was in labor, a medical student, physician in charge, and a nurse, sometimes a student nurse, drove to the woman's home where the delivery occurred under sterile conditions. If the patient had complications she was taken to the hospital.[116] During the postpartum period a physician visited the patient five times. The nursing care was turned over to the Public Health Nursing Association under the supervision of the doctor in charge of the dispensary. Ten calls were made by a nurse, more if the physician deemed it necessary. At the end of five to six weeks a final examination of both mother and baby was made, and instructions were given to the mother regarding her personal care and that of her baby.[117] During the first six years of the Pittsburgh Maternity Dispensary (1912–1918), 16 percent (of 3,384 total home deliveries) of the mothers were African-American.[118]

Many southern black migrant women also chose to have their babies in the hospital. In 1921, 19 percent of the deliveries which occurred in the Elizabeth Steel Magee Hospital were to black mothers and, notably, 66.9 percent of those births were to southern born women. The 175 black births at the Magee Hospital represented 17 percent of all of the 1,026 black births which occurred in the city that year.[119] A local general hospital responsible for the second highest number of deliveries in the city delivered 41 black infants in that year.[120] Those two hospitals combined delivered 21 percent of all black babies born in Pittsburgh. At least six other hospitals were known to have accepted black maternity patients as well, so that the percentage of hospital births was almost certainly higher than 21 percent.[121] Yet in the same year only 27 percent of all births, white and black, occurred in city hospitals; black rates, in other words, already approximated the city average.[122] In 1932 there were 1,081 black resident births in the city and 521 black births at Magee Hospital alone.[123] These numbers are fairly significant when considering that there were other hospitals which accepted black maternity patients.

It is important to point out that black patients who chose to have hospital deliveries did so in spite of discriminatory policies and in spite of the existence of midwives (45 of whom were listed in the city directory in 1919).[124] Women who chose hospital delivery chose white doctors because black doctors were denied hospital privileges.[125] Hospitals also discriminated against the admission of black patients in terms of the types of beds that they made available. Institutions provided only ward accommodations for them, forbidding access to private rooms.[126] In some of the hospitals the wards were racially integrated but in at least two hospitals, separate wards were provided for black patients.[127] In other words, black women chose white doctors and discriminatory facilities over a midwife.

The care that black mothers received as patients in integrated hospitals was somewhat different from that of white private patients, which qualified the impact of the new medical settings. Records from the general hospital that delivered the second highest number of babies per year in Pittsburgh were examined during two different time periods, the summer of 1919 and the winter of 1920–1921.[128] Each sample comprised 130 patients. Although the samples

are small, some generalizations regarding the care of black women can be ventured. The records suggest some differences in the interference on the part of a physician during delivery and differences between white and black groups regarding the use of anaesthesia during delivery. By examining only patients whose care was supervised by one physician (the supervisor of the clinic population) the issue of variations in skill levels and judgments of physicians was alleviated, although not entirely eliminated because this was a teaching hospital and it is likely that residents and medical students were caring for the charity patients. It is reasonable to assume, then, that the white and black clinic patients should have received similar care. Three groups of patients were identified: private white patients were those admitted to private or semiprivate rooms, while white clinic patients and black patients were admitted to the ward.

The most notable trends during the eighteen-month span were the increase in the use of forceps for the white and black clinic patients and the increase in the use of anaesthesia for all three groups. It is important to note that the use of forceps almost always requires some form of anaesthesia and, in addition, the use of ether or nitrous oxide occasionally forces the physician to utilize forceps for delivery. It is not clear from the records which decision, to use forceps or anaesthesia, was made first. The differences in care which whites and blacks received, however, remained similar for both sample groups so the groups were combined for further analysis.

White private patients were far more apt to suffer interference from the physician than non-paying clinic patients, whether white or black (see Table 18.1). Thirty-four percent of all private patients had forceps-assisted deliveries, compared to 8 percent of the white clinic patients and 11 percent of the black patients. John Whitridge Williams, a leader in obstetrics and gynecology at the Johns Hopkins Hospital,

Table 18.1
Patients in sample who had
forceps delivery, by race

	Total	# Forceps deliveries	%
Private	115	39	34
White-clinic	100	8	8
Black	45	5	11
Total	260	52	

Source: Statistics compiled from hospital that requested anonymity.

Table 18.2
Uncomplicated patients in sample
who received anaesthesia, by race

	Uncomplicated patients	Anesthesia	%
Private	50	44	88
White-clinic	81	34	42
Black	37	8	22
Total	168	86	

Source: Statistics from unnamed hospital.

noted that black women suffered contracted pelvis several times more often than white women due to "the prevalence of rickets among blacks and the general physical degeneration which seems to overtake members of that race who live long in large cities. . . . That labor is not more disastrous to them is due to the fact that their children are smaller and have softer heads than white children." Williams also noted that in spite of pelvic deformities among blacks, operative delivery was required far more frequently among white women. According to him, the small size of black babies and their compressible heads compensated for the smaller size of the pelvis.[129] This does not explain, however, why operative delivery was less common among white clinic patients than among white paying patients. Rosemary Stevens has noted that forceps-assisted deliveries rose with the patient's income, signifying a higher level of technology for those who could afford it.[130] Joyce Antler and Daniel Fox also noted that the percentage of Caesarian sections performed in municipal hospitals was lower than in obstetric or voluntary hospitals, also signifying that paying patients were either demanding the use of new technology or that private physicians were more willing to offer it to them.[131]

What is more interesting still is the difference between the three patient groups regarding the administration of anaesthesia during labor and delivery (see Table 18.2), for here race as well

as wealth showed up clearly. Of all of the white private patients who did not have complications or who did not have operative deliveries (C-sections or forceps-assisted deliveries), 88 percent still received ether or nitrous oxide. Of the white nonpaying patients who were uncomplicated, 42 percent received anaesthesia, but only 22 percent of the uncomplicated black patients were given an anaesthetic agent. Stevens has suggested that patients with higher incomes enjoyed the benefits of medical technology, but that does not explain the difference between the two classes of indigent patients, especially when considering that some of the black ward patients may have been of the middle class and able to afford anaesthesia. It simply is not clear whether African-American patients were more reluctant to accept anaesthesia or whether physicians were less willing to offer it to them, but the differential treatment stands out whatever the explanation.[132]

Results of Medical Care

Many cities had evidence that increased prenatal services and medical care for confinement seemed to result in a lower infant and maternal mortality. Grace Abbott claimed that wherever communities had undertaken through well-baby clinics or child health centers to provide information to mothers, the death rate was promptly reduced. In 1924 she noted that the infant death rates in urban areas for the preceding five years decreased more rapidly than the rural rate because the organization of education work among urban mothers was so much easier with the same expenditure of time and money than among rural mothers.[133] Ziegler noted that at the New York Maternity Dispensary, intensive prenatal care was responsible for a 50 percent reduction in the number of stillbirths and infant deaths. Glenn Steele of the Children's Bureau cited studies of other cities that had similar results. In Cleveland in 1919 the death rate for infants under one month (in a district where the infant death rate had been much higher than the overall city level) dropped to 24.8 per 1,000; the city rate was 31.4. In Boston in 1920 the city death rate for infants under two weeks was 37 per 1,000. This dropped to 13 per 1,000 in an area where over 4,000 mothers received prenatal care by the district nursing association.[134]

The facilities which were available and utilized in Pittsburgh by poor black women also offered a solution, albeit of limited proportions, to the problem of high infant and maternal deaths. In Pittsburgh during the University Maternity Dispensary's first six years (3,384 confinements 1912–1918) the maternal death rate was 1.7 per 1,000 compared to the city's 1920 rate of 9.6 per 1,000.[135] Glenn Steele noted a reduction of infant mortality in 1920 in the third ward of the Hill District, which had a rate of 94 deaths per 1,000 births (rate for the entire city that year was 110). The wards on either side of the third ward, which were otherwise not dissimilar, had rates of 156 and 143. Steele attributed this to the existence in the third ward of two free maternity clinics, two city milk stations, a Public Health Nursing Association substation, and the only well-baby clinic in Pittsburgh, housed at the Kaufmann settlement. The rate in the fifth ward in 1920 was 105 infant deaths per 1,000 live births.[136] What is interesting is that the third and fifth wards, the Hill District, housed

45 percent of the city's black population by 1933.[137] In 1920 the fifth ward was 54 percent black.[138] Not only did these two wards have infant death rates lower than the 1920 national black rate of 164 per 1,000 but they had rates lower than the overall city rate. As a result there were some individuals at the time who perceived that the health of African-Americans was improving. Eugene Kinckle Jones, an executive officer of the National Urban League, delivered a speech before the National Conference of Social Work in May 1923 in which he cited various statistics regarding the health of urban blacks to "show conclusively" that the "Negro has actually improved in health."[139]

Conclusion

White immigrant women and black migrant women arrived in the urban North with different expectations and aspirations, which may help to explain the migrants' willingness to adapt to medical childbirth. African-Americans optimistically strived for social and economic improvement by pursuing specific occupational goals and by encouraging their children to attend school. They were also more likely than Poles or Italians to engage in small entrepreneurial opportunities.[140] Given this, it is not surprising that black women's obstetrical choices also differed from that of the white immigrants. Their apparently willing transition to medical childbirth provides another example of the distinctive attitudes of African-Americans compared to various white groups during the decades of the Great Migration. Medical childbirth may well have represented one aspect of the advancement in their lives that blacks so readily sought when they left the South.

Unfortunately, current mortality statistics suggest that all expectations were not met. This study therefore serves also to highlight the complexity of the historical dimension behind familiar current problems. Maternal and infant mortality rates in the 1980s remained at least twice as high for blacks as for whites. By 1986 infant mortality rates had been reduced to 8.9 per 1,000 live births for whites and 18.0 for blacks.[141] In 1986 the maternal mortality rate was 5.1 per 100,000 live births for white women and 16.6 for black women.[142] Infant mortality rates in Pittsburgh continued to be high as well. The black infant mortality rate in the city for 1989 was the highest it had been since 1972, three times above the white rate. The rate for whites was 9.9 per 1,000 live births compared to 34.8 per 1,000 black births.[143]

Numerous causes for the current high infant and maternal mortality rates and racial gap have been suggested. Adequate prenatal care seems to be the one factor which would reduce these causes, such as low birth weight, the incidence of which has decreased in whites but increased in blacks.[144] In addition, studies have suggested that maternal deaths could be prevented with adequate prenatal care.[145] Research has shown that blacks are not seeking adequate prenatal care on as wide a scale as whites, and socioeconomic factors have been cited as the cause.[146] A 1988 report stated that "outreach efforts are no match for the pervasive barriers faced by low income women in trying to secure adequate prenatal and maternity care services." These studies have identified the barriers to care as money and the fact that the system is not "user-friendly."[147]

Yet the ready historical explanation, that African-Americans long resisted medicalization for whatever reason, is clearly inaccurate. Despite discrimination, black women in the 1920s who utilized the new services, including available prenatal care, received some benefits in terms of reductions in infant and maternal mortality. To be sure, these services may not yet have attracted the majority of black women, and their impact should not be exaggerated. But some of the results can serve to inform current policy debate regarding the improvement of infant and maternal health, especially among blacks and minorities. They also emphasize the importance of continuing historical inquiry to trace the relationship between the initially prompt transition and current explanations for the serious mortality gap.

Urban black women during the migration period showed a remarkable ability to adapt to the new scientific obstetrics by utilizing available hospitals and physicians. As an increasing number of women, white as well as black, were drawn into mainstream medical care for obstetrics, an eventual reduction in the racial gap in mortality rates would have been expected. Within the past seventy years mortality rates have declined due to improvements in medical care and wider use of medical facilities for childbirth. The initial enthusiasm shown on the part of black women and the educational programs designed to assist them did not, however, generate the widespread use of medical care characteristic of white parturient women today. The experience of the 1920s suggested not only that black maternal and infant health would improve but that the racial gap would narrow further. Yet the anticipated trajectory was not realized, due possibly to racial differences in medical care, subsequent shifts in attitudes that might relate to medical discrimination, or broader characteristics of the black community and its relationship to the city at large. African-Americans were unable to overcome wider results of racial discrimination in cities like Pittsburgh. In addition to exploring possible medical reasons for the higher black rates, issues of discriminatory policies and other socioeconomic barriers to adequate care need to be addressed in greater depth, in linking, historically, the striking adaptation of the 1920s to the dilemma of the present day.

Notes

1. This time period was identified by Peter Gottlieb, *Making Their Own Way* (Chicago, 1987), 1–9. For this reason the study was confined to this period; however, some statistics cited from the mid- to late thirties are applicable to the migrants who arrived only several years earlier. Jacqueline Jones, *Labor of Love, Labor of Sorrow: Black Women, Work, and the Family from Slavery to the Present* (New York, 1985), 151, 153, 156, 157.

2. John Bodnar, Roger Simon, and Michael P. Weber, *Lives of Their Own: Blacks, Italians, and Poles in Pittsburgh, 1900–1960* (Urbana, Ill., 1982), 29.

3. Ira De.A.Reid, "Social Conditions of the Negro in the Hill District of Pittsburgh," *General Committee on the Hill Survey* (Pittsburgh, 1930), 20–21.

4. Laurence Glasco, "Double Burden: The Black Experience in Pittsburgh," in *City at the Point: Essays on the Social History of Pittsburgh,* ed. Samuel P. Hays (Pittsburgh, 1989), 79. Glasco states, for example, that "health conditions and crime rates reached scandalous levels." Crowding, high rates of tuberculosis, whooping cough, flu, scarlet fever, and VD were additional problems.

5. The term *social childbirth* was first used by Richard and Dorothy Wertz in *Lying-In: A History of Childbirth in America* (New Haven, Conn., 1977), ch. 1, according to Nancy Schrom Dye, "Modern Obstetrics and Working-Class Women: The New York Midwifery Dispensary, 1890–1920," *Journal of Social History* 20 (Spring 1987).

6. Statistics compiled from Report of the Department of Public Health, City of Pittsburgh. U.S. Department of Commerce, Bureau of the Census. *Birth Statistics for the Registration Area of the United States, 1915 and 1931* (Washington D.C., 1917, 1933).

7. Judith Walzer Leavitt, *Brought to Bed: Childbearing in America, 1750–1950* (New York, 1986), 8.

8. Nancy Schrom Dye, "Modern Obstetrics and Working-Class Women," 550.

9. Ibid., 560.

10. Jones, *Labor of Love,* 159. It is not that domestic service attracted married women but that it did not attract single women who preferred other types of work and therefore were attracted to other cities.

11. A. G. Moron and F. F. Stephan, "The Negro Population and Negro Families in Pittsburgh and Allegheny County," *Social Research Bulletin* 1 (April 20, 1933): 2; Abraham Epstein, *A Study in Social Economics: The Negro Migrant in Pittsburgh* (Pittsburgh, 1918), 18.

12. Philip Klein, *A Social Study of Pittsburgh, Community Problems, and Social Services of Allegheny County* (New York, 1938), 271.

13. Moron and Stephan, "Negro Population," 2.

14. Miriam Rosenbloom, "An Outline of the History of the Negro in the Pittsburgh Area" (M.S. thesis, University of Pittsburgh, 1945), 23.

15. Alonzo G. Moron, "Distribution of the Negro Population in Pittsburgh, 1910–1930" (M.A. thesis, University of Pittsburgh, 1933), 38.

16. Bodnar, Simon, and Weber, *Lives of Their Own,* 36.

17. Nora Faires, "Immigrants and Industry," in *City at the Point,* 11.

18. Jones, *Labor of Love,* 155.

19. Anne Rylance Smith, *Study of Girls in Pittsburgh* (Pittsburgh: Central Young Women's Christian Association of Pittsburgh, 1925), 82.

20. Jones, *Labor of Love,* 189.

21. Glasco, "Double Burden," 76–77; Bodnar, Simon, and Weber, *Lives of Their Own,* 117; Gottlieb, *Making Their Own Way,* 55–56, 89, 99.

22. Jones, *Labor of Love,* 189.

23. Bodnar, Simon, and Weber, *Lives of Their Own,* 92, 99, 101, 106, 108.

24. Abraham Epstein, *Study in Social Economics,* 61–62.

25. Gottlieb, *Making Their Own Way,* 185; Glasco, "Double Burden," 80.

26. Ralph L. Hill, "A View of the Hill—A Study of Experiences and Attitudes in the Hill District of Pittsburgh, Pa. from 1900–1973" (Ph.D. diss., University of Pittsburgh, 1973), 130.

27. Glasco, "Double Burden," 79.

28. Elizabeth C. Tandy, "Infant and Maternal Mortality Among Negroes," *Journal of Negro Education* 6 (1937): 329–34.

29. Ibid., 331.

30. Ibid., 332, 343.

31. U.S. Department of Labor, Children's Bureau, Glenn Steele, *Infant Mortality in Pittsburgh,* Bureau Publication 86 (Washington, D.C., 1921), 5.

32. Report of the City Department of Health, Pittsburgh, 1926–1935.

33. Reid, "Social Conditions of the Negro," 43.

34. U.S. Department of Commerce, Bureau of the Census, *Birth Statistics for the Registration Area of the United States, 1931* (Washington, D.C., 1932). Tandy, "Infant and Maternal," 329–34; The national rate in 1933–35 per 1,000 live births was 72 for blacks and 32 for whites.

35. Tandy, "Infant and Maternal," 337.

36. Report of the City Department of Health, Pittsburgh, 1926–1935.

37. Tandy, "Infant and Maternal," 337.

38. "Mortality of Negro Mothers," *Opportunity* 3 (April 1925): 99.

39. "Why Negro Babies Die," *Opportunity* 1 (July 1923): 195–96.

40. Tandy, "Infant and Maternal," 337.

41. Grace Abbott, "Methods by Which Children's Health May Be Improved," *Opportunity* 2 (January 1924): 10–11.

42. Rickets is a disease caused by a deficiency of Vitamin D in the diet which results in bone deformities.

43. Abbott, "Methods," 11.

44. John Whitridge Williams, *Obstetrics: A Text-Book for the Use of Students and Practitioners,* 5th ed. (New York, 1926), 783. It is recognized that contemporary medical literature addressed causes of high maternal and infant mortality rates, but of concern here are the opinions of those medical and social reformers who were influential in the development of programs to combat the problem.

45. Phillips Cutright and Edward Shorter, "The Effects of Health on the Completed Fertility of Nonwhite and White U.S. Women Born Between 1867 and 1935" *Journal of Social History* 13 (1979): 196. The authors also noted that a New York City study of the period found that 90 percent of black infants had rickets.

46. Steele, *Infant Mortality in Pittsburgh,* 16.

47. Judy Barrett Litoff, *American Midwives 1860 to the Present* (Westport, Conn., 1978), 57.

48. Ibid., 50–55.

49. J. Whitridge Williams, "The Midwife Problem and Medical Education in the United States," *Transactions of the American Association for Study and Prevention of Infant Mortality* 2 (1911), a condensed version of which was printed in *Journal of the American Medical Association* (*JAMA*) in January 1912.

50. Charles Ziegler, "The Elimination of the Midwife," *JAMA* 60 (January 4, 1913): 32–38. Opposition to midwives was based on other factors as well, including an anti-immigrant and anti-black sentiment prevalent during the period. Midwives were also viewed as an economic threat to the medical profession. See Litoff, *American Midwives,* 64–83 for a thorough discussion of midwifery opposition.

51. Records reflecting patient attitudes were sought in several hospitals and the Urban League but were not found.

52. *Negro Survey of Pennsylvania* (Harrisburg: Commonwealth of Pennsylvania Department of Welfare, 1927), 48.

53. Klein, *Social Study of Pittsburgh,* 403, 405; Antoinette Hutchings Westmoreland, "A Study of Requests for Specialized Services Directed to the Urban League of Pittsburgh" (M.A. thesis, University of Pittsburgh, 1938), 7; Urban League of Pittsburgh Records: FF235, report, 1918, Archives of Industrial Society, University of Pittsburgh (hereafter cited as AIS).

54. Urban League of Pittsburgh Records: FF235, Office files, report 1918, AIS.

55. Eugene Kinckle Jones, "The Negro's Struggle for Health," *Opportunity* 1 (June 1923): 4–8.

56. Klein, *Social Study of Pittsburgh,* 836.

57. Jones, "The Negro's Struggle for Health," 4–8.

58. John T. Clark, "The Migrant in Pittsburgh," *Opportunity* 1 (October 1923): 303–7.

59. David McBride, "'God Is the Doctor': Medicine and the Black Working-Class in New York City, 1900–1950," Presented at the Annual Conference of the American Historical Association, December, 1990, 3.

60. Clark, "Migrant in Pittsburgh," 304.

61. Charles H. Garvin, "Negro Health," *Opportunity* (November 1924): 342.

62. Clark, "Migrant in Pittsburgh," 303–4.

63. Records of the Urban League of Pittsburgh,

FF82, reports of Department of Information and Advice, April 1920, 13 February 1920, FF215, reports of the executive secretary, April 1920, AIS.

64. Clark, "Migrant in Pittsburgh," 305.

65. Urban League of Pittsburgh Records, FF80, Report of Home Economics Worker, February 1919, AIS.

66. Urban League of Pittsburgh Records, FF80, Home Economics Worker reports 1919–1923, AIS.

67. Urban League of Pittsburgh Records, FF80, Home Economics Reports, case work report, February 1925, AIS.

68. Urban League of Pittsburgh Records, FF80, monthly reports, April 1921, August 1921, August 1922; FF212, report, 30 April 1925, AIS.

69. Urban League of Pittsburgh Records, document, c. 1919: This service probably was the Pittsburgh Maternity Dispensary, which was taken over by West Penn Hospital when Magee Hospital closed temporarily in 1918 to care for soldiers suffering from influenza. Pittsburgh at this time had no hospitals that were specifically founded for the care of the African-American population.

70. Urban League of Pittsburgh Records, FF240, Brochure 7, 1, March 15 1924, AIS.

71. Urban League of Pittsburgh Records, FF215, reports of executive secretary, February 1920, AIS.

72. Urban League of Pittsburgh Records, FF80, report, February 1925, AIS.

73. Arthur J. Edmunds, *Daybreakers—The Story of the Urban League of Pittsburgh* (Pittsburgh, 1983), 61.

74. Urban League of Pittsburgh Records, Publicity letter, 1921, FF244; Bulletin 4, 2, March–May 1921, FF240, AIS.

75. Jones, *Labor of Love,* 193.

76. Frederick L. Hoffman, "The Negro Health Problem," *Opportunity* (April 1926): 120; Grace Abbott, "Methods by Which Children's Health May Be Improved," *Opportunity* (January 1924).

77. "Why Negro Babies Die," *Opportunity* (no date but c. 1921).

78. "Mortality of Negro Mothers," *Opportunity* (April 1925).

79. Charles H. Garvin, "Negro Health," *Opportunity* (November 1924): 341.

80. Reid, "Social Conditions of the Negro," 75.

81. Ibid., 111. This is somewhat curious as there were no nursing schools at that time in Pittsburgh that accepted black students, but they may have been affiliated with a school outside of the city. It was not unusual for nurses affiliated with a school to benefit from clinical experience elsewhere.

82. Ibid., 99; Alonzo G. Moron, "Distribution of the Negro Population in Pittsburgh 1910–1930" (M.A. thesis, University of Pittsburgh, 1933); Andrew Buni, *Robert L. Vann of the Pittsburgh Courier* (Pittsburgh, 1974), 56. Reid noted in his study a distinct apathy on the part of approximately half of the black population toward the church, whereas Moron described the black church as being more intimately part of the black community than the white church. These statements may not be as contradictory as they seem. Buni noted that many blacks, especially those who were better educated, looked elsewhere for answers to the rampant problems in their community, such as to the Urban League. Spiritual needs may not have been met, so perhaps the church's greatest function was in providing space and educational opportunities. Buni's comment also suggests a willingness on the part of the blacks to respond to the directives of the Urban League.

83. Jones, *Labor of Love,* 192; "The Negro in Pittsburgh" (Works Progress Administration Pennsylvania Ethnic Survey, 1938–1941), 59.

84. Paul Titus, "Obstetrical Superstitions," *Pennsylvania Medical Journal* (April 1918): 478–79.

85. Edward H. Beardsley, "Race as a Factor in Health," in *Women, Health, and Medicine in America,* ed. Rima Apple (New York, 1990), 125; Debra Anne Susie, *In the Way of Our Grandmothers: A Cultural View of Twentieth Century Midwifery in Florida* (Athens, Ga., 1988); Jane B. Donegan, "Safe Delivered, but by Whom? Midwives and Men-Midwives in Early America," in *Women and Health in America,* ed. Judith Walzer Leavitt (Madison, Wis., 1984), 313; Sharon Robinson, "A Historical Development of Midwifery in the Black Community: 1600–1940," *Journal of Nurse-Midwifery* 29 (July–August 1984): 247.

86. Robinson, "Historical Development of Midwifery," 247.

87. Frances E. Kobrin, "The American Midwife Controversy: A Crisis of Professionalization," in *Sick-*

ness and Health in America, ed. Judith Walzer Leavitt and Ronald L. Numbers (Madison, Wis., 1985), 197.

88. Litoff, *American Midwives,* 28–30.

89. Ibid., 30, 75, 105.

90. Ibid., 75, 105; Marie Campbell, *Folks Do Get Born* (New York, 1946), 8–9.

91. Campbell *Folks,* 25, 44–50; Susie, *In the Way,* 28–29.

92. Susie, *In the Way,* 12; Litoff, *American Midwives,* 32; Campbell, *Folks,* 7.

93. Litoff, *American Midwives,* 32.

94. Susie, *In the Way,* 17.

95. Jones, *Labor of Love,* 192.

96. Litoff, *American Midwives,* 29.

97. Kobrin, "American Midwife," 324–25.

98. Jones, *Labor of Love,* 189.

99. Tandy, "Infant and Maternal," 327. Although these statistics refer to a period later than the migration period, it is reasonable to assume that they refer to the same generation of women that arrived in the north 1916–1930.

100. Katharine F. Lenroot, "The Health-Education Program of the Children's Bureau. With Particular Reference to Negroes," *Journal of Negro Education,* 6 (1937): 509.

101. McBride, "God Is the Doctor," 13.

102. Tandy, "Infant and Maternal," 327.

103. City Bureau of Health records, 1935.

104. Ivan G. Hosack, *Public Health in Pittsburgh: Analysis-Progress-Recommendation, 1930–1933* (inclusive) with additional comments for 1939–1940 (Pittsburgh, 1941), 84–85.

105. *Services for Negro Unmarried Mothers in Allegheny County: A Study Sponsored by a Special Committee of the Child Welfare Division of the Federation of Social Agencies of Pittsburgh and Allegheny County* (Pittsburgh, 1938), 4.

106. Telephone interview with Mrs. Orlean Ricco, January 1991.

107. Reid, "Social Conditions of the Negro," 44–48; Marian H. Ewalt and Ira V. Hiscock, *The Appraisal of Public Health Activities in Pittsburgh, Pennsylvania, 1930 and 1933* (Pittsburgh, 1933), 50, 54.

108. Ewalt and Hiscock, *Appraisal of Public Health Activities in Pittsburgh,* 54; Margaret Chappell, "Public Health," c. 1925, School of Nursing Archives, Magee Hospital; Public Health Nursing Association brochure, 1923, 2–5, AIS.

109. Ewalt and Hiscock, *Appraisal of Public Health Activities in Pittsburgh,* 54; unpublished report of the City Bureau of Health, Pittsburgh 1935.

110. Public Health Nursing Association files, document, c. 1920, AIS.

111. Public Health Nursing Association files, brochure, 1932, AIS.

112. Public Health Nursing Association files, document, c. 1920, AIS.

113. When the University of Pittsburgh took over the service, it was known as the University Maternity Dispensary. The dispensaries were operational until 1957.

114. K. Emmerling, home delivery reports, 1925, School of Nursing Archives, Magee Hospital; *Pittsburgh Press,* March 15, 1929; *Sun-Telegraph,* November 11, 1928; *Chronicle Telegraph,* January 18, 1923.

115. *Pittsburgh Dispatch,* November 9, 1911; Barbara Paull, *A Century of Medical Excellence: The History of the University of Pittsburgh School of Medicine* (Pittsburgh, 1986), 108.

116. Paul Titus, "Dispensary Care of Obstetric Patients in Pittsburgh," *Weekly Bulletin* (Pittsburgh: Allegheny County Medical Society, June 1919).

117. Charles J. Barone, M.D., "Report of the Kingsley House Sub-Station of the University Maternity Dispensary," *Annual Report of the Kingsley Settlement House,* April 1919–April 1920.

118. Charles Edward Ziegler, "How Can We Best Solve the Midwifery Problem," *American Journal of Public Health* 12 (1922): 410.

119. Statistics compiled from birth records, Elizabeth Steel Magee Hospital; U.S. Department of Commerce, Bureau of the Census, *Birth Statistics for the Birth Registration Area of the United States,* 1921 (Washington, D.C., 1923), 143.

120. Statistics compiled from birth records of a local general hospital that requests that its name not be used in any publication.

121. Reid, "Social Conditions of the Negro," 44–48.

122. Report of the City Department of Health, Pittsburgh, 1926–1935.

123. Report of the City Department of Health, Pittsburgh, 1926–1935; Mabel Ammon Barron, "A

Study of Births at the Elizabeth Steel Magee Hospital, 1932–1944" (M.A. thesis, University of Pittsburgh, 1944), 11; comparisons of Census records with Department of Health records suggest that approximately 95 percent of black births within the city registration area were to resident mothers.

124.   The race of these women is unknown; however, most of the women's last names suggested southern and eastern European origins.

125.   "The Negro in Pittsburgh," Works Progress Administration, Pennsylvania Ethnic Survey (1938–1941), Microfilm F35, Reel 2.

126.   Epstein, *Study in Social Economics,* 58; Reid, "Social Conditions of the Negro," 12.

127.   Epstein, *Study in Social Economics,* 58; Reid, "Social Conditions of the Negro," 12; *Pittsburgh Press,* February 19, 1922.

128.   This large local general hospital, which recorded the second highest number of deliveries in the city per year at that time, agreed to open its records for research purposes with the stipulation that the hospital's name not be used in any publication.

129.   Williams, *Obstetrics,* 16, 783.

130.   Rosemary Stevens, *In Sickness and in Wealth: American Hospitals in the Twentieth Century* (New York, 1989), 174. The iatrogenesis factor and resultant implications for the patients who received anaesthesia or had forceps-assisted deliveries may be significant but is not relevant to this study.

131.   Joyce Antler and Daniel M. Fox, "The Movement Toward a Safe Maternity: Physician Accountability in New York City, 1915–1940," in *Sickness and Health in America,* 496.

132.   Although the sample is small, if similar results can be obtained in other cities, a more extensive study involving other aspects of obstetrical care would be warranted.

133.   Abbott, "Methods," 10.

134.   Steele, *Infant Mortality in Pittsburgh,* 14–16.

135.   Charles Edward Ziegler, "How Can We Best Solve the Midwifery Problem," 409.

136.   Steele, *Infant Mortality in Pittsburgh,* 11.

137.   A. G. Moron and F. F. Stephan, "The Negro Population and Negro Families," 4.

138.   Gottlieb, *Making Their Own Way,* 66.

139.   Eugene Kinckle Jones, "Negro's Struggle," 4–8.

140.   Bodnar, Simon, Weber, *Lives of Their Own,* 36, 42, 130, 264.

141.   "From the Assistant Secretary for Health," *JAMA* 262 (October 27, 1989): 2202.

142.   Hani, K. Atrash, Lisa M. Koonin, Herschel W. Lawson, Adele L. Franks, and Jack C. Smith, "Maternal Mortality in the United States, 1979–1986," *Obstetrics and Gynecology* 76 (December 1990): 1055.

143.   Roger Stuart and Mary Kane, "City's Black Infant Death Rate Jumps," *Pittsburgh Press,* March 26, 1991.

144.   "Infant Mortality Receiving Increasing Attention," *JAMA* 263 (May 16, 1990): 2604.

145.   Atrash, Koonin, Lawson, Franks, and Smith, "Maternal Mortality," 1057.

146.   "From the Assistant Secretary for Health," 2202. In 1987 it was reported that only 4 percent of white non-Hispanic mothers received inadequate prenatal care compared to 12 percent of blacks.

147.   Jody W. Zyke, "Maternal, Child Health Needs Noted by Two Major National Study Groups," *JAMA* 261 (March 24, 31, 1989): 1687.

19 Race and "Value": Black and White Illegitimate Babies, in the U.S.A., 1945–1965

Rickie Solinger

There are two histories of single pregnancy in the post–World War II era, one for black women and one for white. But for girls and women of both races, being single and pregnant revealed that either publicly or privately, their fertility could become a weapon used by others to keep them vulnerable, defenseless, and dependent, in danger without male protection. One aspect of single pregnancy which sharply and powerfully illustrates both the common vulnerability of unwed mothers and the racially distinct treatment they received is the matter of what an unmarried girl or woman would or could do with her illegitimate child.[1]

Throughout my study of unwed pregnancy in the period before the crucial Supreme Court decision of 1973 (*Roe v. Wade*),[2] racially distinct ideas about the "value" of the illegitimate baby surface again and again as central to an unmarried mother's fate. In short, after World War II the white bastard child was no longer the child nobody wanted. The black illegitimate baby became the child white politicians and taxpayers

RICKIE SOLINGER is an historian and independent scholar, Boulder, Colorado.

Reprinted by permission from *Gender & History* 4 (1992).

loved to hate. The central argument of this essay is that the "value" of illegitimate babies has been quite different in different historical eras and that in the United States during the midtwentieth century, the emergence of racially specific attitudes toward illegitimate babies, including ideas about what to do with them, fundamentally shaped the experiences of single mothers.

Social, cultural, and economic imperatives converged in the postwar era so as to sanction very narrow and rigid, but different, options for black and white unwed mothers, no matter what their personal preferences. Black single mothers were expected to keep their babies as most unwed mothers, black and white, had done throughout the history of the United States. Unmarried white mothers, for the first time in this country's history, were urged to put their babies up for adoption. These racially specific prescriptions exacerbated racism and racial antagonism in postwar America and have influenced the politics of female fertility into our own time.

During the Progressive era of the late nineteenth and early twentieth centuries, up through the 1930s, social commentators and social service professionals typically considered an illegitimate baby as a "child of sin," the product of a mentally deficient mother.[3] As such, this child was tainted and undesirable. The girl or woman, black or white, who gave birth to it was expected by family, by the community, and by the state to bring it up. Commentators assumed that others rarely wanted a child who stood to inherit the sinful character—the mental and moral weaknesses—of its parent. Before World War II state laws and institutional regulations supported this mandate, not so much because there were others vying for the babies but to ensure that the mothers would not abandon the infants. State legislators in Minnesota and elsewhere required mothers seeking care in maternity homes to breastfeed their babies for three months or more, long enough to establish unseverable bonds between infant and mother.[4]

Prewar experts stressed that the biology of illegitimacy stamped the baby permanently with marks of mental and moral deficiency and affirmed that moral conditions were embedded in and revealed by these biological events.[5] Likewise, the unwed mother's pregnancy both revealed her congenital and moral shortcomings and condemned her, through the illicit conception and birth, to carry the permanent stain of biological and moral ruin. The experience of pregnancy she underwent was tied to her moral status in a fixed, direct, and inexorable relationship. Equally important, her motherhood was immutable. While the deficiencies, the stain, and her ruination violated her biological integrity, as well as her social and moral standing in the community, the unwed mother's maternal relation to the child was not compromised. That was also fixed directly and inexorably by the facts of conception and birth.

These attitudes reflected in part the importance of bridal virginity and marital conception in mainstream American culture. They also reflected early twentieth-century ideas among moral and medical authorities regarding the strong link between physical, mental, and moral degeneracy and the degeneracy of sex. Until the

1940s illegitimacy usually carried one meaning; cultural, racial, or psychological determinants which admitted group or individual variability were not sought to explain its occurrence. In this prewar period social, religious, and educational leaders rarely called for the rehabilitation of unwed mothers or suggested that there were steps they could take to restore their marriageability and their place in the community. What was lost could not be regained; what was acquired could not be cast off. Consequently, most unwed mothers without family or kinship resources did not have choices to make in that era about the disposition of the bastard child.

After the war state-imposed breast-feeding regulations and institutional policies asserting the immutability of the white mother's relationship to her illegitimate baby became harder to sustain in the face of a complex and changing set of social conditions. First, the demographic facts of single pregnancy were changing. While birth control and abortion remained illegal and hard to obtain, more girls and women were participating in nonmarital, heterosexual intercourse; thus more of them became pregnant and carried babies to term.[6] As nonmarital sex and pregnancy became more common (and then very common during the later postwar period), it became increasingly difficult to sequester, punish, and insist on the permanent ruination of ever-larger numbers of girls and women. This was particularly the case since many of these single pregnant females were members of the growing proportion of the population that considered itself middle class. As a result it became increasingly difficult for parents and the new service professionals, themselves members of the middle class, to sanction treating "our daughters" as permanently ruined.

In addition, a strain of postwar optimism emerged that rejected the view that the individual white unwed mother was at the mercy of harmful environmental or other "forces" which had the power to determine her fate. The modern expert offered the alternative claim that illegitimacy reflected an emotional, psychological, not environmental or biological disorder and was, in general, a symptom of individual, treatable neuroses. Reliance on the psychological explanation redeemed both American society and the individual female. Moreover, by moving the governing imperative from the body (biology) to the mind (psychology), all of the fixed relationships previously defining white illegitimacy became mutable, indeterminate, even deniable.

Psychological explanations transformed the white unwed mother from a genetically tainted unfortunate into a maladjusted woman who could be cured. While there was no solvent that could remove the biological stain of illegitimacy, the neuroses that fostered illegitimacy could respond to treatment. The white out-of-wedlock child therefore was no longer a flawed byproduct of innate immorality and low intelligence. The child's innocence was restored and its adoptability established. At the same time psychologists argued that white unwed mothers, despite their deviant behavior, could be rehabilitated and that a successful cure rested in large measure on the relinquishment of the child.[7] The white unwed mother no longer had an immutable relationship to her baby.

In postwar America the social conditions of motherhood, along with notions about the psychological status of the unwed mother, became more important than biology in defining white motherhood. Specifically, for the first time it took more than a baby to make a white girl or woman into a mother. Without a preceding marriage a white female could not achieve true motherhood.[8] Accepting these new imperatives, social authorities insisted on the centrality of the male to female adult roles, thereby offsetting postwar concerns that women were aggressively undermining male prerogatives in the United States. Experts explained that the unwed mother

who came to terms with the baby's existence, symbolically or concretely, and relinquished the child enhanced her ability to "function [in the future] as a healthy wife and mother."[9]

Release from the biological imperative represented a major reform in the treatment of the many white unwed mothers who desperately desired a way out of trouble, a way to undo their life-changing mistake. The rising rate and number of white single pregnancies, particularly among unmarried middle-class women, would have created an ever-larger number of ruined girls and women if unwed mothers continued to have no option but to keep their illegitimate children. The opportunity to place an illegitimate child for adoption became, in a sense, an unplanned but fortuitous safety valve for thousands of white girls and women who became unwed mothers but—thanks to the sanctioning of adoption—could go on to become properly married wives and mothers soon thereafter. This arrangement could work only if there was a sizable population of white couples who wanted to adopt infants and who didn't mind if the babies had been born to unwed mothers. This condition was met in part because the postwar family imperative put new pressures on and suggested more intense pleasures to infertile couples who in the past would have remained childless. A social scientist in the mid-1950s referred to illegitimate babies as "the silver lining in a dark cloud" and a "blessing" to the many involuntarily childless couples trying to adopt a child.[10] In the early 1950s a leading social work theorist, using what was becoming a popular metaphor, worried about "the tendency growing out of the demand for babies to regard unmarried mothers as breeding machines ... [by people intent] upon securing babies for quick adoptions."[11]

Through adoption, then, the unwed mother could put the mistake—both the baby *qua* baby, and the proof of nonmarital sexual experience—behind her. Her parents were not stuck with a ruined daughter and a bastard grandchild for life. And the baby could be brought up in a normative family, by a couple prejudged to possess all the attributes and resources necessary for successful parenthood.

Some unmarried pregnant girls considered abortion the best way to efface their mistake, but the possibility in the mid-1950s of getting a safe, legal, hospital abortion was slim—in fact, slimmer than it had been in the prewar decades.[12] But if a girl or woman knew about hospital abortions, she might appeal to a hospital abortion committee, a (male) panel of the director of obstetrics/gynecology, and the chiefs of medicine, surgery, neuropsychiatry, and pediatrics. In hospitals, including Mt. Sinai in New York, which set up an abortion committee in 1952, the panel of doctors met once a week and considered cases of women who could bring letters from two specialists diagnosing them as psychologically impaired and unfit to be mothers.[13]

By the early 1950s doctors claimed that new procedures and medications had eliminated the need for almost all medically indicated abortions.[14] That left only psychiatric grounds, which might have seemed promising for girls and women desperate not to have a child.[15] After all, psychiatric explanations were in vogue, and white unwed mothers were categorically diagnosed as deeply neurotic, or worse. There was, however, a catch. These abortion committees had been set up to begin with because their very existence was meant to reduce requests for "therapeutic" abortions, which they did.[16] It was, in fact, a matter of pride and competition among hospitals to have the highest ratio of births to abortions on record.[17] But even though psychiatric illness was the only remaining acceptable basis for request, many doctors did not believe in these grounds. A professor of obstetrics in a large university hospital said, "We haven't done a therapeutic abortion for psychiatric reasons in ten years. . . . We don't recognize

psychiatric indications."[18] So an unwed pregnant girl or woman could be diagnosed and certified as disturbed, probably at considerable cost, but she couldn't convince the panel that she was sick enough. The committee may have in fact agreed with the outside specialists that the abortion petitioner was psychotic, but the panel often claimed the problem was temporary, with sanity recoverable upon delivery.[19]

The doctors were apparently not concerned with questions about when life begins. They were very concerned with what they took to be their responsibility to protect and preserve the links between femininity, maternity, and marriage. One doctor spoke for many of his colleagues when he complained of the "clever, scheming women, simply trying to hoodwink the psychiatrist and obstetrician" in their appeals for permission for abortions.[20] The mere request in fact was taken, according to another doctor, "as proof [of the petitioner's] inability and failure to live through the destiny of being a woman."[21] If such permission were granted, one claimed, the woman "will become an unpleasant person to live with and possibly lose her glamour as a wife. She will gradually lose conviction in playing a female role."[22] An angry committee member, refusing to grant permission to one woman, asserted, "Now that she has had her fun, she wants us to launder her dirty underwear. From my standpoint, she can sweat this one out."[23]

For many doctors, however, condemning the petitioner to sweat it out was not sufficient punishment. In the mid-1950s in Maryland a doctor would almost never agree to perform a therapeutic abortion unless he sterilized the woman at the same time.[24] The records of a large, midwestern general hospital showed that between 1941 and 1950, 75 percent of the abortions performed there were accompanied by sterilization.[25] The bottom line was that if you were single and pregnant (and without rich or influential parents who might, for example, make a significant philanthropic gesture to the hospital), your chances with the abortion committee were pretty bleak. Thousands of unhappily pregnant women each year got illegal abortions, but for thousands of others, financially, morally, or otherwise unable to arrange for the operation, adoption seemed their only choice.

Service agencies, however, found the task of implementing the adoption mandate complicated. Many who worked with white unwed mothers in maternity homes, adoption agencies, or public welfare offices in this period had to braid unmatched strands into a coherent plan. Agency workers were deeply uneasy about separating babies from the one individual who until recently had been historically and culturally designated as best suited, no matter what her marital status, to care for her own baby. In addition, the community response to out-of-wedlock pregnancy and maternity in the United States had historically been punitive.[26] Keeping mother and child together was simultaneously in the child's best interest and the earned wages of sin for the unwed mother. Until the postwar era most social workers had trained and practiced in this tradition.[27]

After World War II social workers struggled to discard the two most basic assumptions that had previously guided their work with white unwed mothers. These girls and women were no longer considered the best mothers for their babies. And they would no longer be expected to pay for their illicit sexual experience and illegitimate pregnancy by living as ruined women and outcast mothers of bastard children. Social workers were now to offer them a plan which would protect them from lasting stigma and rehabilitate them for normative female roles. The psychological literature supporting definitions of unwed mothers as not-mothers, the interest of many white couples in obtaining newborn babies, and postwar concepts of family helped social workers accept new ideas about the disposition of illegitimate white babies.

After the war in all parts of the country public agencies, national service organizations, and maternity homes allocated resources and developed techniques for separating mother and child. Services became so streamlined that in many maternity homes, such as the Florence Crittenton Home in Houston, "Babies [went] directly from the hospital to children's [adoption] agencies."[28] Indeed, public and private agencies were functioning in an environment in which the separation of single mother and child was becoming the norm. In Minnesota, for example, in 1925 there were 200 such separations; in 1949, one thousand; between 1949 and 1955, approximately seventeen hundred each year. Nationally, by 1955, 90,000 babies born out of wedlock were being placed for adoption, an 80 percent increase since 1944.[29]

To meet the demand agencies and individual operators not infrequently resorted to questionable tactics, including selling babies for profit. When the federal government undertook to investigate widespread coercive and profit-oriented adoption practices in the United States in the 1950s, the task was assigned to Senator Estes Kefauver's Subcommittee to Investigate Juvenile Delinquency. This committee was charged with redressing the problem of adoption for profit and assuring the "suitability of the home" for adoptable children, a criterion which could not, by definition, be met in homes headed by unmarried mothers.[30] While illegitimate pregnancy and babies had in the past been a private matter handled by family members, perhaps assisted by charity workers, by midcentury these issues had become public concerns and public business.

The Kefauver committee and the organizations and individuals it investigated defined white unmarried mothers out of their motherhood. If not by law, then de facto they were not parents. This judgment was in line with and supported various forms of state control over single pregnant girls and women, and those who might become pregnant, including, of course, the state's formal and informal proscriptions against birth control for unmarried girls and women, its denial of access to safe legal abortion, and its tolerance in many places of unsafe illegal abortions. The state determined what types of agencies and individuals an unmarried mother could deal with in planning for her child and either suggested strongly or legislated which ones were "morally wrong." These state prerogatives allowed some agencies and individuals to abuse and exploit childbearing single white women.

A very articulate eighteen-year-old unmarried mother from Minnesota wrote to her governor in August 1950, illustrating how some public agencies took direct action to separate white babies from their mothers even against the mother's will. She said that a welfare worker in her city told her she could not keep her baby, "that the baby should be brought up by both a mother and a father." Having gotten no satisfaction, she wrote in frustration and anger to President Truman:

> With tears in my eyes and sorrow in my heart I'm trying to defend the rights and privileges which every citizen in the United States is supposed to enjoy under our Constitution [but are] denied me and my baby. . . . The Welfare Department refuses to give me my baby without sufficient cause or explanation. . . . I have never done any wrong and just because I had a baby under such circumstances the welfare agency has no right to condemn me and to demand my child be placed [for adoption].[31]

A year earlier a young man living in Sterling, Colorado, wrote to the Children's Bureau about

a similar case. In this situation a young man and a young deaf woman had conceived a baby out of wedlock but planned to marry. When the man went to the Denver Welfare Department for assistance a few days before the baby was born, he found that the baby had already been targeted for adoption.[32]

This case in particular demonstrates a couple of key assumptions underlying the behavior of some agency workers in matters of out-of-wedlock adoptions. The young mother was deaf. As a handicapped person and an unmarried girl, her maternity, as well as her child, was considered illegitimate and could be rightfully terminated by the authorities. Physically defective women had curtailed rights as mothers, just as physically defective illegitimate babies had diminished opportunities to join the middle class. This case also suggests the very important notion that white babies were so valuable because in postwar America, they were born not only untainted but also *unclassed*. A poor "white trash" teenager could have a white baby in Appalachia; it could be adopted by an upper middle-class couple in Westport, Connecticut, and the baby would in that transaction become upper middle class also.

Courts also facilitated adoption abuse. A chief probation officer in the Richmond County, Alabama, Juvenile Court spent a great deal of her time finding and "freeing" white babies for adoption, using her position to legitimize these activities. One unwed mother told of her encounter with the officer, a Miss Hamilton. She said,

> Several hours after delivery [Miss Hamilton] informed me that my baby had been born dead. She told me that if I signed a paper she had, no one, my family or friends, would know about the situation, and that everything would be cleared up easily. She described the paper as being a

consent authorizing the burial of the child. . . . I signed the paper without really looking at it, as I was in a very distressed and confused condition at the time.[33]

This young woman went on to say that, "two years later I was shocked to receive in the mail adoption papers from the Welfare Department in California since I was under the impression that the child was deceased."

Illegalities and abuse existed in some mainstream institutions, but a great many of the worst abuses were committed by individual baby brokers—lawyers, doctors, and nonprofessionals cashing in on the misfortune of unwed mothers. In postwar consumerist America, institutions promoted services and attitudes to protect the out-of-wedlock child from market-driven deals and to see that it was well placed. On the other hand, these same institutions were themselves behaving in market-oriented ways as they promoted a specific, socially beneficial product: the two-parent/two-plus child family. This double message justified the baby brokers' commodity-like treatment of unwed mothers and their babies. Charlton G. Blair, a lawyer who handled between thirty and sixty adoptions a year in the late 1950s, justified his operation by denying he ever "paid one red cent" to a prospective mother of an illegitimate child to persuade her to part with the baby. But in suggesting why the adopting parents were willing to pay up to fifteen hundred dollars for a child, which included the lawyers $750 fee, Blair defined his sense of the transaction very clearly: "If they're willing to pay three thousand dollars for an automobile these days, I don't see why they can't pay this much for a child."[34]

A case which dramatically captures the plight of poor, white, unwed mothers was presented at the Kefauver hearings by Mary Grice, an investigative reporter for the *Wichita Beacon*. Grice testified about a woman, "Mrs. T.," who had been in the adoption business since 1951 or 1952. Mrs. T. warehoused unwed pregnant girls in the basement of her home. "She would have them on cots for prospective adoptive parents [who] would come in and she would take them downstairs, and she would point to the girls and say, 'Point out the girl that you want to be the mother of your child.'" Grice's investigation revealed that Mrs. T. kept on average seven unmarried mothers in her basement at a time and that she would oversee a number of the deliveries herself in the basement. According to Grice, between 150 and 164 adoptions each year of this sort were taking place in Sedgwick County, Kansas. Mrs. T. often collaborated with Grace Schauner, a Wichita abortionist. Unmarried pregnant girls and women would first see Schauner and if they decided not to have an abortion, they would be referred to Mrs. T. who would "care for them and sell their infants after birth."[35]

Mrs. T's girls were poor, so they did not have the information or other resources to resist baby market operators. Because they were female (specifically, white females), their socially mandated shame precluded self-protection and motherhood. Because they were white, their babies had "value." This combination of poverty, race, and gender—in a context which defined white unwed mothers as not-mothers and defined their babies as valuable—put some white unwed mothers in a position of extreme vulnerability.

Again, there is no question that for many white unwed mothers, the opportunity to place their babies independently meant that they could get exactly what they needed when they needed it: money to live on, shelter, medical care, and assurances about the placement of the baby, all with no questions asked. "These girls and women were often spared the delays, the layers of authority, the invasions of privacy, the permanent black mark engraved in the files of

377

the welfare department" and they were spared the pressure to reveal the father's name, all of which characterized the bureaucratic agency approach.[36] Their experience demonstrated how difficult it was for institutions to perform simultaneously as agents of social control and as sources of humanitarian assistance for the needy and vulnerable.

An intruder in the courtroom in Miami, Florida, where a section of the Kefauver hearings was held in November 1955, expressed the frustration of some girls and women who felt they had lost control over the disposition of their illegitimate children. This woman stood up, unbidden, and lectured the men before her in a loud voice. She said,

> Excuse me. I am not leaving no court. . . .
> You have to carry these children nine months and then you have them taken away by the Catholic Charities, and then they throw you out and drag you all over the street. . . . I'm no drunk, I'm no whore. . . . gave birth to two children and had them taken away from me. I don't sleep nights thinking about my children. What do you people care? Don't take my picture. You people have no feelings at all. That man [a judge testifying that there are plenty of services available for unwed mothers] is sitting there and lying—lying. These people just take other people's children away from them. All that he has said is a lie. My baby was born . . . and I haven't seen it since. . . . How would you like it? Year after year you have *to go to the people* . . . and ask them why you can't have your children.[37]

Clark Vincent, a sociologist who closely followed the treatment of white unwed mothers in this era, offered the following vision of a world in the near future where the state would have restrained authority to determine who is a mother:

> If the demand for adoptable infants continues to exceed the supply; if more definitive research . . . substantiates that the majority of unwed mothers who keep their children lack the potential for "good motherhood"; and if there continues to be an emphasis through laws and courts on the "rights of the child" superseding the "rights of the parents"; then it is quite probable that in the near future unwed mothers will be "punished" by having their children taken away from them at birth. Such a policy would not be enacted nor labeled overtly as "punishment." Rather it would be implemented under such pressures and labels as "scientific finding," "the best interests of the child," "rehabilitation goals for the unwed mother," and the "stability of family and society."[38]

In postwar America there was only one public intention for white unwed mothers and their babies: separate them. Toward black single mothers and their babies, however, there were three broadly different public attitudes.[39] One attitude, often held by middle-of-the-road politicians, social service administrators, and practitioners, maintained that blacks had babies out of wedlock because they were Negro, because they were ex-Africans and ex-slaves, irresponsible and amoral but baby-loving. According to this conception, the state and its institutions and agencies could essentially ignore breeding patterns since blacks would take care of their children themselves. And if blacks did not, they were responsible for their own mess. Adopting Daniel Moynihan's famous phrase from this period, I call this public attitude toward black illegitimacy *benign neglect*.

A second response to black mothers and babies was *punitive*. The conservative racist politicians who championed this position argued simply that the mothers were bad and should be punished. The babies were expendable because they were expensive and undesirable as citizens. Public policies could and should be used to punish black unmarried mothers and their children in the form of legislation enabling states to cut them off from welfare benefits and to sterilize or incarcerate "illegitimate mothers."[40]

I label the third attitude toward black unwed mothers *benevolent reformist*. Employees at the United States Children's Bureau and many in the social work community who took this position maintained that blacks who had children out of wedlock were just like whites who did the same. Both groups of females were equally disturbed and equally in need of help, particularly in need of social casework. Regarding the baby, benevolent reformers held that black unwed mothers should be accorded every opportunity to place the infant for adoption, just like whites.

Despite these different attitudes toward black women and their babies, proponents of all three shared a great deal. First, they shared the belief that the black illegitimate baby was the product of pathology. This was the case whether it was a pathology grounded in race, as it was for the benign neglecters and the punishers, or in gender, as it was for the benevolent reformers. Second, all commentators agreed that the baby's existence justified a negative moral judgment about the mother and the mother-and-baby dyad. The black illegitimate infant was proof of its mother's moral incapacities; its illegitimacy suggested its own probable tendencies toward depravity. Because of the eager market for white babies, this group was cleared of the charge of inherited moral taint while black babies were not. Indeed, proponents of each of the three perspectives agreed that the unwed black mother must, in almost every case, keep her baby. Where they differed was in explaining why this was so. The different answers reflected different strains of racism and carried quite different implications for public policies and practices regarding the black unmarried mother and her child.

The benign neglecters began to articulate their position at about the same time that the psychologists provided new explanations for white single pregnancy. In tandem these developments set black and white unwed mothers in different universes of cause and effect. According to these "experts," black and white single mothers were different from each other in several ways. When black single girls and women had intercourse, it was a sexual, not a psychological act, and black mothers had "natural affection" for their children, whatever their birth status. The white unwed mother had only neurotic feelings for her out of wedlock child. The "unrestrained sexuality" of black women and their capacity to love the resulting illegitimate children were perceived as inbred traits and unchangeable, part of black culture.

Thus by becoming mothers, even unwed mothers, black women were simply doing what came naturally. It was also important in this regard that the operative concept of "culture" excised considerations of environment. Environment was not a primary factor in shaping female sexual behavior or the mother's relationship to her illegitimate baby. These were determined by "culture," an essentially biological construct. Therefore, since professionals could have an impact on only the immediate situation—and could not penetrate or rearrange black "culture," it was futile to consider interfering. The absence of services for these women and their children was justified in this way. Issues regarding blacks and adoption were quickly dismissed by those who counseled neglect. Agencies claimed that blacks didn't want to part with their babies, and, just as important, black couples didn't want to adopt children.[41]

White policy makers and service providers often pointed to the black grandmothers—willing, able, loving, and present—to justify their contentions that the black family would take care of its own and that no additional services were necessary. Yet, when grandmothers rendered such service, policy makers labeled them "matriarchs" and blamed them for "faulty personality growth and for maladaptive functioning in children."[42] The mother was similarly placed in a double bind. She was denied services because she was black—an alleged cultural rather than a racial distinction—and then she was held responsible for the personal and social consequences.[43] The social service system was, in this way, excused from responsibility or obligation to black unwed mothers.

The punishers, both southern Dixiecrats and northern racists, drew in part on the "cultural" argument to target both the unwed mother and her baby. They held that black culture was inherited and that the baby would likely be as great a social liability as its mother. Moreover, they claimed that for a poor black woman to have a baby was an act of selfishness, as well as pathological, and deserved punishment.[44] Once the public came to believe that black illegitimacy was not an innocuous social fact but carried a direct and heavy cost to white taxpayers, many whites sanctioned their political representatives to target black unwed mothers and their babies for attack.[45]

The willingness to attack was expressed in part by a special set of tropes which drew on the language and concepts of the marketplace. The "value" assigned to the illegitimate child-as-commodity became useful in classifying the violation of the black unwed mother in a consumer society. Repeatedly, black unmarried mothers were construed as "women whose business is having illegitimate children."[46] This illicit "occu-

pation" was portrayed as violating basic consumerist principles, including good value in exchange for a good price, for a product which, in general, benefits society. Black unmarried mothers, in contrast, were said to offer bad value (black babies) at a high price (taxpayer-supported welfare grants) to the detriment of society, demographically and economically. The behavior of these women—most of whom did *not* receive Aid to Dependent Children grants for their illegitimate children[47]—was construed as meeting only the consumerist principle that everything can, potentially, be a commodity. These women were accused of treating their reproductive capacities and their children as commodities, with assigned monetary values. From this perspective black unmarried mothers were portrayed as "economic women," making calculated decisions for personal financial gain.[48]

The precise economic principle most grossly violated by these women was, according to many, that they got something (ADC) for nothing (another black baby); they were cheating the public with a bad sell. The fact that it was, overwhelmingly, a buyer's market for black babies "proved" the valuelessness of these children, despite their expense to the taxpaying public.[49] White babies entered a healthy seller's market, with up to ten couples competing for every one adoptable infant.[50]

Spokespeople for this point of view believed that black unmarried mothers should pay dearly for the bad bargain they foisted on society, especially on white taxpayers. But many felt that rather than paying for their sins, black women were being paid, by ADC grants, an exchange which encouraged additional sexual and fiscal irresponsibility.[51] Thus society was justified in punishing black unwed mothers. In addition, the black unmarried woman, allegedly willing to trade on her reproductive function, willing to use her body and her child so cheaply, earned the state's equal willingness to regard her childbearing capacity cheaply, and take it away, for example, by sterilization legislation.

The ironic truth was that ADC benefits were such inadequate support (and employment and child care opportunities so meager or nonexistent) that government policies had the effect of causing, not responding to, the economic calculations a woman made that might lead to pregnancy. The average welfare payment per child per month was $27.29 with monthly averages less than half that amount in most southern states.[52] The following encounter illustrates the relationship between illegitimacy and economics, from one woman's point of view. "When the case analyst visited the family, the little girl came in with a new dress and shoes. The mother explained that it was the last day of school and the child had begged for new clothes like the other children had. She got them, but the mother's comment was, 'hope that dress does not cause me another baby.'"[53] This mother's economic and sexual calculations were rooted in poverty and maternal concern, not in some desire to multiply inadequate stipends through additional pregnancies.

The public's interest in casting black unwed mothers and their babies as consumer violators was reflected in opinion polls which suggested that the American public wanted to withhold federal support, or food money, from illegitimate black babies.[54] Among dissenters were people who believed it was wrong "to deny food to children because of the sins of the parents."[55] Both groups, however, fell into a trap set by conservative politicians who found it politically profitable to associate black illegitimacy in their constituents' minds with the rising costs of public welfare grants. The ADC caseload increased in the postwar period for many reasons, including the basic increase in numbers of children and families and the increase in households headed by women because of divorce, separation, desertion, *and* illegitimacy. Between 1953

and 1959 the number of families headed by women rose 12.8 percent while the number of families rose only 8.3 percent.[56] While white sentiment was being whipped up to support punitive measures against black "subsidized immorality,"[57] only about 16 percent of nonwhite unwed mothers were receiving ADC grants.[58] Adoption, which was not an option for most blacks, was the most important factor in removing white children from would-be ADC families. Of unwed white mothers who kept their children, 30 percent, or nearly twice as large a percentage as blacks, were receiving Aid to Dependent Children grants in 1959.[59] Yet in the minds of large segments of the white public, black unwed mothers were being paid in welfare coin to have children. The "suitable home" laws which were originally designed, it was claimed, to protect the interests of children, were now instrumental in cutting off ADC payments. These were politicians who had no qualms about using "value-less" black illegitimate children as pawns in their attempt to squash black "disobedience" via morals charges.[60]

Led by Annie Lee Davis, a black social worker at the United States Children's Bureau, many members of the social work community worked unceasingly to convert benign neglecters and punishers into benevolent reformers. Davis was a committed integrationist. She was dedicated to convincing the white social service establishment that black unmarried mothers needed and deserved the same services as their white counterparts. In 1948 Davis addressed this message to her colleagues: "Within minority groups, unmarried mothers suffer guilt and shame as in the majority group." She added, "I know there are those who will challenge this statement," but she insisted, "In the process of adopting the cultural traits of the dominant group in America all groups are striving to be American."[61]

Davis insisted that white public officials and social workers must be brought to believe that black unwed mothers were psychologically and morally the equals of whites. Only then would blacks be eligible for the best available services. Ironically, Davis believed that a key element of proof was to establish that blacks were as interested in adoptive placements for their illegitimate babies as whites were urged to be. Her task was to convince her colleagues that lack of alternate options alone created the custom and the necessity that blacks kept their illegitimate children.

Benevolent reformers typically took the position that it was unacceptable and potentially racist to assume that blacks did not want every opportunity that whites had, including adoption. But it was extremely difficult for the reformers to suggest that some black single mothers wanted their children and others did not. It was not simply unwed mothers and their babies at issue, it was the Race. For the reformers, constructing an equivalency between black and white unmarried mothers was the most promising and practical route to social services and social justice.

But even if a black single mother did consider placing her child for adoption, she knew that the likelihood that the agency would expeditiously approve a couple as adoptive parents was slim.[62] While a white unwed mother could assume a rapid placement, the black one knew that her child would be forced, in part because of agency practices, to spend months in foster homes or institutions before placement, if that was ever achieved. For example, adoption agencies frequently rejected blacks who applied for babies, claiming they did not meet the agency's standards for adoptive parents. They also neglected to work with schools and hospitals in contact with black unwed mothers to improve referral services between these institutions and the agencies because they feared recruiting black babies when there might not be homes for them. The reformers had their integrationist vision,

but the institutions of society would not cooperate, even when some black unwed mothers did.

In fact, the evidence from postwar black communities suggests that the black unwed mother accepted responsibility for her baby as a matter of course,[63] even when she was sorry to have gotten pregnant.[64] A study in the mid-1960s cautioned the social work profession: "Social work wisdom is that Negroes keep because there is no place to give the baby up, but the study showed . . . that Negroes did not favor adoption, opportunities or their absence notwithstanding." Findings showed that the issue of disposition of the child was the only one that consistently yielded a difference between black and white respondents, no matter whether they were the unwed mother, her parents, or professional staff. In fact, the blacks revealed their determination to keep mother and child together and the whites their determination to effect separation, "no matter how [the investigator] varied the content of the questions."[65]

In the same period in Cincinnati several researchers captured the comments of the mothers themselves. Some girls and women focused on the needs of the baby. One typical respondent claimed, "An innocent child should not be denied his mother's love." Others focused on the strength of their own needs, "I'd grieve myself to death if I let my baby go." A few predicted they would have had nervous breakdowns if they hadn't been allowed to keep their babies. A representative outlook drew on the sanctified status of motherhood: "The Lord suffered for you to have a baby. He will suffer for you to get food for the baby." Still others expressed themselves in forward-looking, practical terms, "You were less apt to have regrets if you kept the baby than if you let him go."[66]

For many black unwed mothers the reasons to keep a baby were simply grounded in an immutable moral code of maternal responsibility. A young black woman said, "Giving a child away is not the sort of thing a good person would do," and a teenager asserted, "My parents wouldn't let me give up the baby for adoption."[67] A black woman in Philadelphia subscribed to this morality. She said, "I sure don't think much of giving babies up for adoption. The mother mightn't be able to give it the finest and best in the world, but she could find a way like I did. My mother had thirteen heads and it was during the Depression. . . . She didn't give us away."[68]

The central question for all of these black single mothers was how good a mother you were, not whether you were legally married.[69] The overriding stimulus in structuring the personal decisions of these girls and women was a "powerful drive toward family unity, even if the family is not the socially approved one of father, mother, and children."[70] In a study of thirty poor single black mothers, only two told the investigators that they would advise another woman in their situation to give the baby up, and both cited difficulties with the welfare office as their reason.[71] The author of the study referred to the "vehemence" of most black single mothers about their decision to keep the child.[72]

A research team in North Carolina investigating illegitimacy concluded in the early 1960s that one major difference between white and black unwed mothers was that the white girl generally felt that a "new maturity" had come with the experience of conceiving out of wedlock. The team claimed that this was not true for the black subjects and explained: "The white subculture demands learning from experience," so the white unwed mother must learn her lesson. The white girl "Has probably been encouraged to look within herself for the reasons for her mistake because the white subculture stresses individual responsibility for error."[73]

These observations capture a great deal of the intentionality underlying the white culture's treatment of unwed mothers under the adoption mandate. For these girls and women the "lesson" was twofold: no baby without a husband, and no one is to blame but yourself. Learning the lesson meant stepping on the road to real womanhood. The illegitimate child was an encumbrance or an obstacle to following this route. The ability to relinquish was constructed as the first most crucial step in the right direction.

Joyce Ladner, in her study of black women in the 1960s, dealt with the same issue—the relationship between illegitimacy and maturity. She suggested a strikingly different finding: "The adolescent Black girl who becomes pregnant out of wedlock changed her self-concept from one who was approaching maturity to one who had attained the status of womanhood. . . . Mothers were quick to say that their daughters had become grown, that they have 'done as much as I have done.'"[74] The road to maturity for black unwed mothers was unmediated. Maturity accompanied maternity, the baby's legal status notwithstanding.

Both black and white women in the postwar era were subject to a definition of maturity that depended on motherhood. The most pervasive public assumption about black and white unwed mothers, however, was that their nonmarital childbearing did not constitute maternity in the culturally sanctioned sense. The treatment of these girls and women reinforced the notion that legitimation of sexuality and maternity were the province of the state and the community and were not the rights of individual girls and women. In the case of white unwed mothers the community (including the mother and her family) with government support was encouraged to efface episodes of illicit sex and maternity. Outside of marriage neither the sex nor the resulting child had "reality" in the community or in the mother's life. They became simply momentary mental aberrations. In the case of black unwed mothers sexuality was brute biology and childbearing its hideous result. The state, with the support of public institutions, could deface the black single mother's dignity, diminish her resources, threaten her right to keep her child, and even threaten her reproductive capacity.

In both cases the policies and practices which structured the meanings of race and gender, sexuality and motherhood, for unwed mothers were tied to social issues—such as the postwar adoption market for white babies and the white taxpaying public's hostile identification of ADC as a program to support black unwed mothers and their unwanted babies—which used single pregnant women as resources and as scapegoats.

In the immediate pre–*Roe v. Wade* era the uses of race combined with the uses of gender, sexuality, and maternity in ways that dealt black and white unwed mothers quite different hands. According to social and cultural intentions, for the white unwed mother and her baby, relinquishment of the baby was meant to place all scent of taint behind them and thus restore good value to both. The black unwed mother and her child, triply devalued, had all their troubles before them.

Notes

1. Throughout this essay I have used such expressions as "unwed mother" and "illegitimate baby"—expressions heavily burdened with intended prejudicial meanings—in order to convey the flavor of public discussion of these matters and individuals in the postwar years. I trust it will be obvious that my use of period language does not reflect negative moral judgments on my part. My occasional use of the word *bastard* is a bit more complicated. The word is a technical term and also, of course, a profound derogation. It is rarely found in the record of public discussion of single pregnancy in these years, although sometimes a politician did not choose to bite his tongue in time. But as I uncovered the public and private fear and rage that structured the treatment of single pregnant women and their babies in this era, I sometimes felt

that bastard was the only word that expressed the meaning intended by my sources.

2. This essay is taken from a larger study, *Wake Up Little Susie: Single Pregnancy and Race Before Roe v Wade* (New York: Routledge, 1992).

3. See, for example, Charlotte Lowe, "Intelligence and Social Background of the Unmarried Mother," *Mental Hygiene* 4 (1927): 783–94; and Henry C. Schumacher, M.D., "The Unmarried Mother: A Socio-Psychiatric Viewpoint," *Mental Hygiene* 4 (1927): 775–82.

4. Maryland passed such a law in 1919 and Wisconsin in 1922. It was claimed that these laws would reduce high infant mortality rates, although they were never shown to do so. Maternity home residents were targeted since this group was considered most likely, in its search for secrecy, to abandon its babies. See Elza Virginia Dahlgren, "Attitudes of a Group of Unmarried Mothers toward the Minnesota Three Months Nursing Regulation and Its Application," M.A. thesis, University of Minnesota, 1940.

5. See, for example, Percy Kammerer, *The Unmarried Mother* (Boston: Little, Brown, 1918) and Schumacher, "The Unmarried Mother."

6. Even though many studies published in the postwar era claimed that rates of illicit coition were not rising, the fact that the illegitimacy rates and illegal abortion rates were higher than ever suggests otherwise. See, for example, Alfred C. Kinsey, Wardell B. Pomeroy, Clyde E. Martin, and Paul H. Gebhard, *Sexual Behavior in the Human Female* (Philadelphia: W.B. Saunders, 1953), ch. 8. Also see Phillips Cutright, "Illegitimacy in the United States: 1920–1968," in *Demographic and Social Aspects of Population Growth,* ed. Charles F. Westoff and Robert Parke, Jr. (Washington, D.C.: Commission on Population Growth and the American Future, 1969), 384.

7. See for example, Mary Lynch Crockett, "An Examination of Services to the Unmarried Mother in Relation to Age at Adoption Placement of the Baby," *Casework Papers, 1960* (New York: Columbia University, 1960), 75–85.

8. See Leontine Young, *Out of Wedlock* (New York: McGraw Hill, 1954), 216.

9. Janice P. Montague, "Acceptance or Denial— the Therapeutic Uses of the Mother/Baby Relation-

ship," paper presented at the Florence Crittenton Association of America Northeast Conference, 1964.

10. Winston Ehrmann, "Illegitimacy in Florida II: Social and Psychological Aspects of Illegitimacy," *Eugenics Quarterly* 3 (1956): 227.

11. Leontine Young, "Is Money Our Trouble?" paper presented at the National Conference on Social Work, 1953.

12. See, for example, Mary Calderone, ed., *Abortion in the United States* (New York: Harper and Brothers, 1958), 84; and Edwin M. Gold, Carl L. Erhardt, Harold Jacobziner, and Frieda G. Nelson, "Therapeutic Abortions in New York City: a Twenty Year Review," *American Journal of Public Health* 55 (1965): 964–72.

13. Calderone, *Abortion*, 92–93, 139; Alan Guttmacher, "Therapeutic Abortion: the Doctor's Dilemma," *Journal of Mt. Sinai Hospital* 21 (1954): 111; Lewis Savel, "Adjudication of Therapeutic Abortion and Sterilization," in *Therapeutic Abortion and Sterilization*, ed. Edmund W. Overstreet (New York: Harper and Row, 1964), 14–21.

14. Calderone, *Abortion*, 86–88.

15. See, for example, J. G. Moore and J. H. Randall, "Trends in Therapeutic Abortion: a Review of 137 Cases," *American Journal of Obstetrics and Gynecology* 63 (1952): 34.

16. Harry A. Pearce and Harold A. Ott, "Hospital Control of Sterilization and Therapeutic Abortion," *American Journal of Obstetrics and Gynecology* 60 (1950): 297; James M. Ingram, H. S. B. Treloar, G. Phillips Thomas, and Edward B. Rood, "Interruption of Pregnancy for Psychiatric Indications—a Suggested Method of Control," *Obstetrics and Gynecology* 29 (1967): 251–55.

17. See, for example, Charles C. Dahlberg, "Abortion," in *Sexual Behavior and the Law*, ed. Ralph Slovenko (Springfield, Ill.: Charles Thomas, 1965), 384.

18. Arthur Mandy, "Reflections of a Gynecologist," in *Therapeutic Abortion*, ed. Harold Rosen (New York: Julian Press, 1954), 291.

19. Gregory Zillboorg, "The Clinical Issues of Postpartum Psychopathology Reactions," *American Journal of Obstetrics and Gynecology* 73 (1957): 305; Roy J. Heffernon and William Lynch, "What Is the Status of Therapeutic Abortion in Modern Obstetrics?" *American Journal of Obstetrics and Gynecology* 66 (1953): 337.

20. Nicholson J. Eastman, "Obstetric Forward," in Rosen, *Therapeutic Abortion*, xx.

21. Theodore Lidz, "Reflections of a Psychiatrist," in Rosen, *Therapeutic Abortion*, 279.

22. Flanders Dunbar, "Abortion and the Abortion Habit," in Rosen, *Therapeutic Abortion*, 27.

23. Mandy, "Reflections," 289.

24. Manfred Guttmacher, "The Legal Status of Therapeutic Abortion," in Rosen, *Therapeutic Abortion*, 183. Also see Nanette Davis, *From Crime to Choice: The Transformation of Abortion in America* (Westport Conn.: Greenwood Press, 1985), 73; Johan W. Eliot, Robert E. Hall, J. Robert Willson, and Carolyn Hauser, "The Obstetrician's View," in *Abortion in a Changing World* 1, ed. Robert E. Hall (New York: Columbia University Press, 1970), 93; Kenneth R. Niswander, "Medical Abortion Practice in the United States," in *Abortion and the Law*, ed. David T. Smith (Cleveland: Press of Case Western Reserve University, 1967), 57.

25. David C. Wilson, "The Abortion Problem in the General Hospital," in Rosen, *Therapeutic Abortion*, 190–91. Also see Myra Loth and H. Hesseltine, "Therapeutic Abortion at the Chicago Lying-In Hospital," *American Journal of Obstetrics and Gynecology* 72 (1956): 304–311, which reported that 69.4 percent of their sample were sterilized along with abortion. Also relevant are Keith P. Russell, "Changing Indications for Therapeutic Abortion: Twenty Years Experience at Los Angeles County Hospital," *Journal of the American Medical Association*, January 10, 1953, 108–11, which reported an abortion-sterilization rate of 75.6 percent, and Lewis E. Savel, "Adjudication of Therapeutic Abortion and Sterilization," *Clinical Obstetrics and Gynecology* 7 (1964): 14–21.

26. See, for example, Michael W. Sedlak, "Young Women and the City: Adolescent Deviance and the Transformation of Educational Policy, 1870–1960," *History of Education Quarterly* 23 (1983): 1–28.

27. See Lillian Ripple, "Social Work Standards of Unmarried Parenthood as Affected by Contemporary Treatment Formulations" (Ph.D. diss., University of Chicago, 1953).

28. *Directory of Maternity Homes* (National Association on Services to Unmarried Parents, Cleveland, 1960).

29. U.S. Congress, Senate, Judiciary Committee, *Hearing Before the Subcommittee to Investigate Juvenile Delinquency, Interstate Adoption Practices,* July 15–16, 1955 (84th Congress, 1st sess. [Government Printing Office, Washington, D.C., 1955]), 200.

30. U.S. Congress, Senate, Judiciary Committee, *Hearings Before the Subcommittee to Investigate Juvenile Delinquency, Commercial Child Adoption Practices,* May 16, 1956 (84th Congress, 2nd sess. [Government Printing Office, Washington, D.C., 1956]), 6.

31. Duluth, Minnesota, to Governor Luther Youngdahl, August 2, 1950, and to President Truman, August 14, 1950, box 457, file 7-4-3-3-4, Record Group 102, National Archives (hereafter cited as N.A.).

32. Sterling, Colorado, to Mrs. Lenroot, November 21, 1949, box 457, file 7-4-3-3-7, Record Group 102, N.A.

33. U.S. Congress, *Hearings, Commercial Child Adoption Practices,* (May 16, 1956), 120.

34. *New York Times,* July 10, 1958.

35. U.S. Congress, Senate, Judiciary Committee, *Hearings Before the Subcommittee to Investigate Juvenile Delinquency, Interstate Adoption Practices,* Miami, Florida, November 14–15, 1955 (84th Congress, 1st sess. [Government Printing Office, Washington, D.C., 1956]), 54–56. Also see *Wichita Beacon,* July 31, and August 1–8, 1955.

36. U.S. Congress, *Hearings, Interstate Adoption Practices,* July 15–16, 1955, 206.

37. U.S. Congress, *Hearings, Interstate Adoption Practices,* Miami, Florida, November 14–15, 1955, 245.

38. Clark Vincent, "Unwed Mothers and the Adoption Market: Psychological and Familial Factors," *Journal of Marriage and Family Living* 22 (1960): 118.

39. See Solinger, *Wake Up Little Susie,* Chapter 7, for a fuller discussion of these three public policy perspectives.

40. See Winifred Bell, *Aid to Dependent Children* (New York: Columbia University, 1965), and Julius Paul, "The Return of Punitive Sterilization Proposal," *Law and Society Review* 3 (1968): 77–106.

41. See, for example, Andrew Billingsley and Jeanne Giovannoni, *Children of the Storm* (New York: Harcourt, Brace and Jovanovich, 1972), 142.

42. Patricia Garland, "Illegitimacy—a Special Minority-Group Problem in Urban Areas," *Child Welfare* 45 (1966): 84.

43. Ibid.

44. See, for example, the editorial, "It Merits Discussion," *Richmond News Leader,* March 22, 1957.

45. During the period considered here, black women in the South were among the first in the United States to receive publicly subsidized birth control, sterilization, and abortion services. See Thomas Shapiro, *Population Control Politics: Women, Sterilization, and Reproductive Choice* (Philadelphia: Temple University Press, 1985); Gerald C. Wright, "Racism and the Availability of Family Planning Services in the United States," *Social Forces* 56 (1978): 1087–98; and Martha C. Ward, *Poor Women, Powerful Men: America's Great Experiment in Family Planning* (Boulder, Colo.: Westview Press, 1986).

46. See, for example, *New York Times,* August 28, 1960.

47. See *Illegitimacy and Its Impact on the Aid to Dependent Children Program,* Bureau of Public Assistance, Social Security Administration, U.S. Department of Health, Education, and Welfare (Government Printing Office, Washington, D.C., 1960).

48. *Atlanta Constitution,* January 25, 1951. The *Constitution* reported that the Georgia state welfare director, making an argument for denying Aid to Dependent Children grants to mothers with more than one illegitimate child, noted that "Seventy percent of all mothers with more than one illegitimate child are Negro, . . . Some of them, finding themselves tied down to one child, are not adverse to adding others as a business proposition."

49. "A Study of Negro Adoptions," *Child Welfare* 38 (1959): 33, quoting David Fanshel, *A Study in Negro Adoptions* (New York: Child Welfare League of America, 1957); "In moving from white to Negro adoptions we are moving from what economists would call a 'seller's market' . . to a 'buyer's market.'"

50. See, for example, Lydia Hylton, "Trends in Adoption," *Child Welfare* 44 (1966): 377–86. In 1960 a government report claimed that in some communities, there were ten suitable applicants for every white infant, *Illegitimacy and Its Impact,* 28.

51. One of a number of readers responding irately to a *New York Times* editorial in support of giving welfare grants to unmarried mothers, wrote to the *Times,* "As for your great concern for those careless women who make a career of illicit pregnancy, they should either bear the expense or be put where they can no longer indulge their weaknesses." *New York Times,* July 7, 1961.

52. *New York Times,* August 9, 1959. In late 1958 average monthly family grants in the ADC program were $99.83 nationally but in the South ranged between $27.09 in Alabama and $67.73 in Texas. Bell, *Aid to Dependent Children,* 224.

53. Hazel McCalley, "The Community Looks at Illegitimacy," Florence Crittenton Association of America Papers, box 3, folder: FCAA Annual 11th, 1960–61, Social Welfare History Archives, University of Minnesota (hereafter cited as SWHA); see *Facts, Fallacies, and the Future—A Study of the ADC Program of Cook County, Illinois* (New York: Greenleigh Associates, 1960), 29, for a prominent, contemporary discussion concerning how small welfare grants to single mothers were directly responsible for increasing these women's financial and social dependence on men.

54. A Gallup Poll conducted in 1960 found that only "one in ten [respondents] favored giving aid to further children born to unwed parents who have already produced an out of wedlock child." *St. Louis Post Dispatch,* August 8, 1961.

55. *Milwaukee Journal,* August 9, 1961.

56. *Illegitimacy and Its Impact,* 30.

57. *Buffalo Currier Express,* December 5, 1957.

58. *Illegitimacy and Its Impact,* 36.

59. Ibid.

60. See "The Current Attack on ADC in Louisiana," September 16, 1960, Florence Crittenton Association of America Papers, box 3, folder: National Urban League, New York City, SWHA.

61. Annie Lee Davis, "Attitudes Toward Minority Groups: Their Effect on Services for Unmarried Mothers," paper presented at the National Conference on Social Work, 1948.

62. See Seaton W. Manning, "The Changing Negro Family: Implications for the Adoption of Children," *Child Welfare* 43 (1964): 480–85; Elizabeth Herzog and Rose Bernstein, "Why So Few Negro Adoptions?" *Children* 12 (1965): 14–15; Billingsley and Giovannoni, *Children of the Storm;* Fanshel, *A Study in Negro Adoption;* Trudy Bradley, "An Exploration of Caseworkers' Perceptions of Adoptive Applicants," *Child Welfare* 45 (1962): 433–43.

63. Elizabeth Tuttle, "Serving the Unmarried Mother Who Keeps Her Child," *Social Welfare* 43 (1962): 418.

64. See *Facts, Fallacies, and Future,* 19–20; 552 out of 619 mothers of illegitimate children in this study did not want another child but reported that they had no information about how to prevent conception.

65. Deborah Shapiro, "Attitudes, Values, and Unmarried Motherhood," in *Unmarried Parenthood: Clues to Agency and Community Action* (New York: National Council on Illegitimacy, 1967), 60.

66. Ellery Reed and Ruth Latimer, *A Study of Unmarried Mothers Who Kept Their Babies* (Cincinnati: Social Welfare Research, Inc., 1963), 72.

67. Shapiro, "Attitudes and Values," 61.

68. Renee Berg, "Utilizing the Strengths of Unwed Mothers in the AFDC Program," *Child Welfare* 43 (1964): 337.

69. Ibid. Also see Nicholas Lemann, *The Promised Land: The Great Black Migration and How It Changed America* (New York: Knopf, 1991), 59–108.

70. Renee M. Berg, "A Study of a Group of Unmarried Mothers Receiving ADC," Doctor of Social Work dissertation, University of Pennsylvania School of Social Work, 1962, 96.

71. Berg, "A Study," 93.

72. Ibid., 95.

73. Charles Bowerman, Donald Irish, and Hallowell Pope, *Unwed Motherhood: Personal and Social Consequences* (Chapel Hill: University of North Carolina Press, 1966), 261.

74. J. Ladner, *Tomorrow's Tomorrow: The Black Woman* (New York: Doubleday, 1971), 214–15.

Women and Mental Illness

Unlike the clearly gendered experiences of menstruation, abortion, and childbirth, both men and women suffer from mental illnesses. Yet the history of diagnosis and treatment of mental diseases also demonstrates a specifically female side. The two articles in this section demonstrate just how and why certain behaviors seem to be signs of a gendered pathology.

Elaine Abelson examines shoplifting as one such medical concern. Only at the turn of the twentieth century, in the context of a growing middle class and the new commercial milieu of the department store, did shoplifting acquire a new face, defined by gender roles, and a new name, *kleptomania*. The medical term not only evoked the image of a woman of means compulsively stealing merchandise but it simultaneously defended and excused the activities of these women.

Nancy Theriot concentrates on a more familiar notion of mental illness brought on by biological experiences, in this case childbirth. While puerperal insanity was a diagnosis that could apply only to postpartum women, Theriot discusses it as a cultural and medical construct changing over time. It diminished as a useful diagnosis when it no longer satisfied the needs of women to rebel or physicians to define their professionalism.

These articles address some overarching themes in the history of psychiatry. First, how much was this field, as others in medicine, used as an instrument of social control? That is, were diagnoses and treatments differentially offered to women as part of a larger social effort to control female behavior and funnel it toward acceptable domesticity? In short were specifically female-defined mental illnesses a mark of women's social oppression? Or, as historian Gerald Grob has suggested elsewhere, might social control have been a by-product of medical oversight rather than its primary goal? The historians here also offer a critique of the assumption that men and women are biologically and psychologically distinct and instead emphasize the importance of specific historical circumstance. While women and men may have suffered from different mental problems, the gendered worlds in which they lived may have created the differences from their historically rooted perceptions and experiences rather than from any inherent or necessary distinctions. As we have seen in other sections in this book, medicine and culture walk hand in hand in the responses to health and disease.

20 The Invention of Kleptomania

Elaine S. Abelson

Shoplifting is an ancient, though not honorable, art. Reports of criminal theft from shops and stalls appeared in Elizabethan England.[1] Moll Flanders was sent to Newgate Prison for shoplifting in midseventeenth-century London. "Light-fingered Sophie Lyons," a well-known shoplifter in nineteenth-century America, became a detective story heroine and wrote an autobiography that was syndicated by the Hearst newspaper chain.[2] But an entirely different kind of shoplifting appeared suddenly in the late nineteenth century and became the subject of medical concern and widespread popular interest. This shoplifting signaled a form of deviant behavior by a new group, the middle class, and its locale was that new commercial institution, the department store. Such shoplifting emerged from the intersection of new manufacturing capacity and new forms of merchandising in the context of burgeoning consumer capitalism. Moreover, it was linked to a rigid division of gender roles that assigned consumption activities to women, and, under the rubric *kleptomania,* it was used to define gender- as well as class-based notions of theft.[3]

ELAINE S. ABELSON is Associate Professor of History and Director of the Urban Studies Program at Eugene Lang College, the undergraduate division of The New School for Social Research, New York.

Reprinted from *Signs: Journal of Women in Culture and Society* 15 (1989):123–43, by permission of The University of Chicago. © 1989 by The University of Chicago. All rights reserved.

Kleptomania was a quasi-medical term that evoked the image of a woman of some means and indeterminate years who regularly took merchandise from large department stores without the formality of payment. The legal and moral innocence of this woman, as well as the compulsive nature of her actions, were taken for granted by professionals and the public alike.[4] Use of the kleptomania diagnosis to defend the actions of a select group of women suggests the distinctive role nineteenth-century doctors played in shaping and giving analytic visibility to gender-based definitions. It reflects as well the socially sanctioned privilege of the white middle class in the nineteenth century.[5]

Depictions of women in newspaper accounts detailing the arrests of middle-class shoplifters conformed to the cultural stereotype of the debilitated female. Mrs. Dora Landsberg, for instance, was described in the *New York Times* in January 1899 as a wealthy widow who was "suffering from kleptomania." Another woman, from a "most estimable family," was defended by her doctor, who said she was under his care for "general debility and her brain was affected."[6] The female kleptomaniac quickly became a stock character, a popular joke. Kleptomaniacs appeared in sketches by comedians Weber and Fields and Charlie Chaplin, as well as innumerable vaudeville acts, minor drawing room comedies, and popular songs with such catchy titles as "Mamie, Don't You Feel Ashamie."[7] Edwin Porter's 1905 silent film *The Kleptomaniac* was one of many movie depictions of this ubiquitous social phenomenon.[8] Kleptomania's powerful hold on the Victorian imagination can be traced to two parallel developments: the rising status and authority of medical science and the unique importance of the department store as an urban institution in the second half of the nineteenth century.[9] It is within this dual context that the figure of the kleptomaniac emerged.

In July 1887 the *American Journal of Insanity* published the annual proceedings of the Association of Medical Superintendents of American Institutions for the Insane. Among the papers delivered at that meeting was one with the provocative title, "Are Dipsomania, Kleptomania, Pyromania, etc., Valid Forms of Mental Disease?"[10] The author, Dr. Orpheus Everts, superintendent of the Cincinnati Sanitarium, answered his own question affirmatively but added a caveat. These manias exist, Everts said, but only in a dependent relationship to other symptoms. In respect to kleptomania he argued that a "natural desire to accumulate exaggerated by disease" constituted the reality of kleptomania, and he considered this combination of desire and disease to be a valid mental disorder.[11] The case of a thirty-nine-year-old widow with children and of "good society" illustrated his thesis. Admitted to the asylum as a hysteric with a history of kleptomania, this woman was diagnosed as suffering from "womb disease mania," which the doctor described as "larceny and eroticism with hysteria."[12]

With the hindsight of one hundred years we can look with a combination of horror and slight bemusement at this and similar late nineteenth-century medical diagnoses, but we must also keep in mind how these explanatory models, built on the science of the day, also reinforced established notions about class and gender. Although the medicalization of shoplifting implied a search for treatment and held out the

hope for a cure, there was a mixed message in the diagnosis of such behavior as disease. On the one hand, it suggested that older moral judgments of "bad" behavior were inappropriately simplistic, but, on the other hand, it defined the reproductive functions of women as inherently diseased. If manias could be traced to the womb, as Dr. Everts implied, the sexuality of women could be conflated with sickness and behavioral irregularities. Even as it became a socially and medically credible diagnosis, kleptomania reinscribed beliefs about female weakness.[13]

The well-documented and sensationally reported shoplifting case of Mrs. Ella Castle, which unfolded in 1896, provides a microcosm of these myriad and often conflicting strands of social, medical, and legal thought on kleptomania. On October 5, 1896, Mr. and Mrs. Walter Castle, wealthy and socially prominent American tourists, were arrested in London for stealing a sable muff from a fashionable West End establishment. Remanded to Holloway Prison, the couple spent a week in jail before their bail hearing and subsequently were forced to undergo the ordeal and spectacle of a public trial. American newspapers had a field day with the case; the *New York Times,* for instance, ran nineteen separate articles and editorials on the incident between the arrest on October 5 and the release of the Castles from English jurisdiction on November 13.[14]

Both Mr. and Mrs. Castle were brought to trial four weeks after their arrest, but her behavior and past history were the focus of the case. Although Mrs. Castle ultimately pleaded guilty to the shoplifting charge, in both the weeks before the trial and in court she was portrayed not as a thief but as a mentally unstable, physically ill woman who from the onset of puberty had exhibited mental "troubles incidental to female life."[15] Again and again press reports described her virtual physical collapse and spoke of her as subjected to a "disease which may have temporarily turned her mind."[16] The attending physician of Holloway Prison called her a woman of "highly nervous temperament and disposition."[17] The prestigious English legal journal, *Law Times,* seconded the diagnosis, referring to the "diseased condition of the moral nature" in this particular case.[18] An affidavit from the Castles' family doctor in California took note of "her excessive nervousness since the birth of her child," and testimonials from American friends recalled that her disordered state of mind had been a "pathetic secret . . . for many years."[19] English medical specialists summoned to examine Ella Castle were united in the opinion that she was neither mentally nor morally responsible for her crime.[20] Admitting his client's guilt before the trial, Mrs. Castle's English lawyer, Sir Edward Clarke, defended her as a pitiful kleptomaniac: "I have had her examined by some leading specialists in mental diseases, and have no doubt that the judge will admit that her symptoms are such as to warrant the defense of kleptomania. She suffers from pain in the head, from complete loss of memory and from other irregularities, which according to medical science, are known to be frequently associated with delusions. There is no reason in life why she should have taken these few trumpery bits of fur. . . . She has a well-to-do husband, who was willing to satisfy her every want."[21]

On November 7, the day of the trial, Mrs. Castle appeared in court sobbing, half-swooning, and supported by two uniformed nurses. Dr. William Chapman Grigg, specialist in women's diseases at Queen Charlotte's Lying-In Hospital, testified that "after repeated examinations of Mrs. Castle he had formed the opinion that the disease from which she was suffering was one of those which are almost always accompanied by great mental disturbance causing different manias, as kleptomania, religious mania, etc., in different women." Dr. George Henry Savage, lecturer on mental diseases of Guy's

Hospital and author of a widely read text, *Insanity and Allied Neuroses,* corroborated Grigg's diagnosis.[22]

It is significant (although not surprising) that Mr. Castle, who had been arrested and indicted *with* his wife on the shoplifting charge, was subsequently "exonerated from all responsibility for her pilferings."[23] At the bail hearing a week after the arrest, when the prosecution still considered this to be "a regular case of shoplifting," it was revealed that "a part of the property was found among the husband's clothes."[24] Three weeks later the prosecutor declined to offer any evidence against Walter Castle, arguing that even though the couple "occupied one room at the Hotel Cecil" and Mrs. Castle's trunks "contained a museum of articles" taken from various London stores, "in not a single instance was she detected in taking anything, so that she must have done her work so skillfully that even her husband, who was beside her was unable to see what she was doing."[25] A tea importer from a wealthy and prominent San Francisco family, Walter Castle provided his wife's spending money, and it is reasonable to wonder, as did Police Inspector Arrow who was in charge of the case, how Mr. Castle could have failed to notice such a vast accumulation of stolen merchandise.[26] The prosecution's decision to attribute the thefts solely to Mrs. Castle seems to have been prompted in part by pressure from American embassy officials, but it was facilitated by the opinions of medical experts. The availability of the kleptomania diagnosis, coupled with a deep-seated but unspoken assumption that "gentlemen don't act this way," made Mr. Castle's innocence easily rationalized.[27]

A repressive medical argument based on the belief that women were likely to be physically and mentally unstable, kleptomania was part of a complex pattern of psychological tension between Victorian men and women.[28] While Mr. Castle was transformed by both prosecutor and judge into the long-suffering, understanding husband, Mrs. Castle was neatly packaged as a mentally deranged woman who, in the throes of her illness, was a skillful shoplifter.[29] Her defense was grounded in the plea of kleptomania, based upon "her suffering from a woman's ailment which would account for her mania"; his rested on "his reputation for honor and integrity."[30]

Pleading guilty and convicted on seven counts of shoplifting, Ella Castle was deemed mentally and morally irresponsible for her actions and was speedily released by the British home secretary "on her husband's promise to take charge of her."[31] The couple sailed immediately for New York.

Upon her arrival in the United States Mrs. Castle went directly to Philadelphia to consult doctors at the Philadelphia Polyclinic Hospital. A team of senior physicians treated her, and their detailed report, "The Relations of Nervous Disorders in Women to Pelvic Disease," allows us to appraise mainstream medical and gynecological thinking about the relationship between women and kleptomania at the end of the nineteenth century. Dr. S. Weir Mitchell, who had been lecturing and writing about diseases of the nervous system in women for two decades, was called in on the case for "assistance and counsel," and he became the chief spokesman for the medical team. Citing his "long experience of many forms of neuroses associated with pelvic disease," Mitchell wrote,

I do not believe that Mrs. C. had any clear notion of the nature of her acts, or of their consequences, and I am of [the] opinion that very positive and long-neglected uterine and rectal disease had much to do with the disorder of mind from which she has suffered, and which is apt to be associated with hysterical conditions. . . . I think her hysterical, weak, and unbalanced, but not criminal. It is characteristic of her form of

mental disorder that she should show no other obvious signs of insanity than the overwhelming tendency which belongs to her form of monomania.[32]

Not surprisingly, Mitchell's diagnosis supported the conclusion of Dr. Grigg, one of the trio of specialists who had examined Mrs. Castle in London, who had testified at the trial and reported to the home secretary: "She is intensely neurotic. The condition of things—a disease of the upper portion of the uterus—is a very common accompaniment of various forms of mania in women, such as melancholia, religious mania, nymphomania, and I have seen it in several cases of kleptomania. It is invariably coupled with much mental disturbance. The condition I discovered is quite sufficient to account for any form of mental vagaries which are so well known to affect a certain class of women (neurotic) with disordered menstruation. Her bowel condition would aggravate this."[33] Dr. Solomon Solis-Cohen, the physician in charge of Mrs. Castle's medical care at the Polyclinic Hospital, identified her specific medical symptoms as disordered menstruation, hemorrhoids, and uterine irregularities. From physical problems such as these, the doctor reported, "various forms of mania in women," including kleptomania, commonly appear.[34]

What historical meaning does this drama, at once intensely personal and broadly social, hold for us? That Mrs. Castle was playing a culturally sanctioned role after the arrest seems obvious. Her exaggerated symptoms of hysteria accompanied by the appropriate props in the form of concealing black veils, smelling salts, and uniformed medical attendants may read to us like stage directions in a second-rate play, but they conformed to an appropriate behavioral norm for middle-class women in the late nineteenth century. Featured prominently in English and American newspapers for over a month, Mrs.

Castle exhibited every symptom a respectable kleptomaniac was supposed to possess: frequent nervous episodes, pains in the head, loss of memory, and menstrual problems. Further, she was married to a well-to-do merchant who was said to be "very generous" to his wife. Mr. Castle, in his turn, professed the proper ignorance of his wife's shoplifting and said the evidence came as a "frightful revelation," even though he was aware that "she had been subject at certain periods to mental delusions and loss of memory."[35]

Faced with the unsettling phenomenon of respectable women stealing merchandise from the dry-goods bazaars, physicians explained kleptomania in terms of feminine weakness and sexuality. The medical discourse permeated public discussion as well. In an attempt to understand what was deemed to be irrational behavior, doctors and the public alike embraced a view of women that limited them to biological dependency, to prescribed social roles, and to actions governed by the emotions. The medical and legal reactions to the Castle shoplifting incident illuminate what historian Carroll Smith-Rosenberg and others have similarly noted: the close association between popular ideology and medical "fact."[36]

We have little evidence of how women responded to this medical-sexual construct, which attempted to restrict their lives if not render them powerless. Mrs. Castle's voice and those of other women detained for shoplifting are not heard, but if sexual ideology mirrored social relations, the silence of these women is an element that functions as an integral part of what was said by doctors, husbands, and judges.[37]

In the earliest, French interpretation, kleptomania was characterized as the impulse of a diseased imagination and characterized by the absence of economic need. Kleptomania, the so-called thieving mania, was widely accepted in both Europe and the United States as the impairment of the individual's voluntary powers.

Although initially not explicitly gender specific, on both sides of the Atlantic the diagnosis was almost immediately associated with women, specifically with the female reproductive economy, which was understood to be the seat of the disorder.[38] Most shoplifters were women, but the association of shopping behavior with biological processes, so dramatically demonstrated in the Castle incident, was a cognitive leap that was deeply rooted in the intellectual assumptions of the Victorian period.

The Castle incident falls squarely within the ongoing nineteenth-century moral and medical debate about the relation of insanity to the female reproductive system. Over the course of three decades many formulations and criticisms jostled for hegemony at medical meetings and in the pages of the various internationally recognized medical journals. Despite the many differences evident in these debates, they had in common a construction of gender that showed little variation across national boundaries and cultures.[39] While many of the arguments hinged on "whether madness was at root an organic disease or a psychic disorder," the construct of the kleptomaniac invariably located the disease in the physical distinctions of female life.[40]

Menstrual disorders in particular were integral to medical explanations of apparently motiveless theft. What doctors designated "ovarian insanity" was transformed into the more specific discourse on kleptomania. In the widely read 1884 American edition of his textbook, *Clinical Lectures on Mental Diseases,* Scottish physician T. S. Clouston singled out disturbed menstruation as a "constant danger to the mental stability of some women." "It is often hard to determine," he explained, "whether disordered or suspended menstruation is a cause or a symptom."[41] When Solis-Cohen cited disordered menstruation as a contributing factor to Mrs. Castle's kleptomania, he based his diagnosis on this accepted medical paradigm.

Because doctors expected and often found

some pathology of the reproductive system, the localization of the disease, in this period of still uncertain and speculative gynecology, seemed beyond doubt.[42] Ella Castle may not have been totally representative of kleptomaniacs—she was wealthier and more socially prominent than most women detained for shoplifting—but physicians made the same connection between her mental and physical conditions as they did between those of other women labeled kleptomaniac, and the various surgical procedures she submitted to under the guise of alleviating these mental and physical disorders were well within the prevailing medical standards of the period.[43] Although we have to assume that the Polyclinic doctors found legitimate physical ailments, they operated, in any case, because she was a shoplifter. By treating her pelvic disease they hoped to cure her kleptomania.

Like hysteria, uterine disease became a diagnostic catchall.[44] As an explanatory model, it removed responsibility from the afflicted individual and made moral judgment of her behavior inappropriate. If kleptomania was under the control of biology, doctors reasoned, the kleptomaniac was physically defective but not evil. In a period where self control was the ideal, such dysfunctional behavior was easily labeled disease, "the symptom of which is crime."[45] Doctors labeled Mrs. Castle a kleptomaniac because in their view she had lost the powers of reason; she suffered from hysteria triggered by specific physiological malfunction. This understanding of her illness fit easily into a whole storehouse of popular assumptions and prejudices. Weir Mitchell's diagnosis of Ella Castle's shoplifting as the result of uterine disease was textbook perfect: sexuality was the root of female behavior.[46]

Beyond its implications of female irrationality, the diagnosis of kleptomania pointed to a social concern about middle-class conduct. Medical expertise, ever so compatible with popular notions of female character, allowed and even encouraged a court, representing a theoretically impartial law, to transform a criminal act into a physical symptom.[47] What, in another instance and with another couple, may well have been called criminal conduct was labeled disease. Dr. Arthur Conan Doyle captured the twin elements of the case in a letter to the *London Times* imploring that paper's intervention on behalf of Mrs. Castle: "If there is any doubt of moral responsibility," Doyle wrote, "the benefit of the doubt should certainly be given to one whose *sex* and *position* . . . give her a double claim to our consideration. It is in the consulting room and not to the cell that she should be sent."[48]

Doyle's letter illustrates that cultural assumptions about female sexuality went hand in hand with cultural assumptions about middle-class women. While neurologists and asylum superintendents were working out meanings and classifications of specific forms of mental disease, the popular understanding of kleptomania (aka shoplifting) was framed in terms consistent with commonly understood gender stereotypes and social concerns. The public discussion of the mania, not unlike the scientific, was in fact an extended commentary on women, on class, on role definitions, and on the dual questions about sickness and health.

A similar cultural discourse was taking place in France, where this "curious and frequent form of theft" had elicited a great deal of popular attention and was the object of several medical studies. French medical investigators, notably Paul Dubuisson, described in detail the department store environment at the turn of the twentieth century.[49] In an attempt to understand and systematize the relationship between illness and theft, Dubuisson sought to demonstrate the existence of "that special folly which seizes a woman the moment she crosses the threshold of a great department store."[50] But his arguments rationalized the traditional and defined narrowly the foundation of female activity. Within this highly restrictive model, women and department stores were seen in a symbiotic relationship in which the stores filled an elemental need for women. It was not that they needed to buy anything, Dubuisson wrote, but that they needed the atmosphere and the sight of "all those beautiful things."[51]

Although American doctors, if not merchants and lawyers, certainly were aware of the diagnosis of kleptomania by the French physician C. C. Marc, which had appeared in 1840, kleptomania as a form of middle-class shoplifting did not become an issue in the United States until the late 1870s.[52] There was no discussion of kleptomania, for example, in the uproar that followed Rowland Macy's arrest of the socially prominent New York feminist and philanthropist Mrs. Elizabeth B. Phelps in December 1870.[53] Charged with stealing a small package of candy in R. H. Macy and Company and summarily arrested, Elizabeth Phelps became a cause célèbre. *Lady* and *shoplifter* had not yet become synonymous terms, so the idea that Mrs. Phelps could stand accused of so tawdry a crime was inconceivable. The letter-writing public and the editors of various newspapers and journals were unanimous in their condemnation of the Macy employees who had precipitated the event and of the police who, it was charged, had dragged Mrs. Phelps away. The highly vocal critics of the incident never entertained the possibility that the lady might have been guilty as charged. In this first publicized incident of shoplifting in a dry-goods bazaar, class was the single issue.

By the late nineteenth century, shopping had become part of women's life and women's work. With a modicum of affluence and increasing amounts of free time, urban middle-class women integrated the dual roles of work and leisure into new patterns of behavior and became the shoppers so disparaged by contemporary journalists. In 1873 even the feminist *Woman's Journal* accepted the new role definition and advised its readers, "Next to mental improvement, shopping is now the business of life, and

a most bewildering and exhausting business it is. . . . Strong is the character demanded for wise shopping!"[54]

Increasingly understood as "the chief diversion of ladies," shopping became something which both men and women saw as being as innate and natural as any other female physiological function.[55] In the asymmetry of the Victorian world shopping became a woman's natural public sphere. By 1904 R. H. Macy and Company claimed it had 150,000 daily customers; estimates of the proportion of women among these customers ranged as high as 90 percent.[56] Although men were certainly not excluded from consumption choices, women were the primary consumers, and it was specifically to them that department stores directed their appeals.[57]

Free access within the stores was the critical appeal. Perhaps for the first time a woman could "circulate on her own, unattended, without interference from anyone and without rendering account to anyone." The freedom of the store environment was not often duplicated in other areas of a woman's life. Even child care often was provided. The Fair, a popular Chicago store, built an in-store park and playground that reportedly could accommodate two hundred children while their mothers shopped.[58]

Far more than simply a new and exciting shopping milieu, the Gilded Age dry-goods bazaar also proved to be an acceptable social location for the middle-class woman; it was a protected space in which she could eat lunch, have tea, meet her friends, rest, write a letter, and browse. The New York trade journal, the *Dry Goods Economist,* cited an unnamed store that assured women customers in 1902, "You may roam our floors unquestioned, without being urged unduly to buy . . . our place is to entertain you."[59] Many accounts of shoplifting describe women wandering aimlessly through the stores, killing time, going from counter to counter and floor to floor seemingly without any particular destination but fixated by the "new wilderness of goods."[60]

Merchants were acutely aware of this new phenomenon and obviously encouraged it; ultimately, they became dependent upon it. With newspaper advertising and window displays insufficient to fill the big stores on a day-to-day basis, crowds had to be manufactured, if not to view specific merchandise, then to become themselves a part of the aesthetics of the stores. Mullenmeister, the proprietor in Margrete Bohme's 1912 novel, *The Department Store,* speaks confidently of the crowds of the plainly curious he expects to lure into his monster new bazaar: "The best and most effective advertisement is *attractions,* which perhaps seem to have nothing to do with the business, but which draw great streams of people into, or even only through the house. Once we get them in, the buying can take care of itself."[61]

Although there was a never-ending barrage of journalistic ridicule and medical complaint about the dangers of novel reading for idle women, there was little opposition to the stimulation of desire within the "more controlled" dry-goods store.[62] In the toiletry departments the scent literally filled the air. "Keep the atomizer going," the *Dry Goods Economist* advised readers.[63] At the opening of Wanamaker's on Astor Place in 1896 the *New York Tribune* reporter was overcome by the display of silks in the great rotunda. "New weaves and designs of exquisite coloring and quality were on every counter, and the women who strayed into this spot were," like the reporter, "loathe to leave it."[64] Many observers noted what historian Rosalind Williams has called an "inescapable spectacle of mass consumption" in late nineteenth-century department stores.[65] Women did not have to buy, but they could not ignore the dazzling sensual impact of the environment.

Some women charged with shoplifting accused the stores of permitting too much free-

dom: they became "over excited" and over stimulated in the large stores; they could not refrain from handling things, and no one bothered them. Everything led to temptation, shoppers complained, the salespeople were either disinterested or too busy to be of real service, and there was a "deplorable liberty" to touch everything.[66]

The tension between traditional values, particularly the postponement of gratification, and the newer, more compelling gospel of consumption led many women to fear their own impulses.[67] In creating fantasy the stores encouraged an abandonment in consumers that produced behavior troubling in its implications for both moral and social restraint.

The temptation to possess found instant expression on the department store selling floor. In a world where seemingly everything could be bought, impulse buying became endemic. The press reported innumerable incidents of the excesses of women who sent merchandise home with no intention of keeping it or who ran up huge store bills despite their husbands' refusals to pay for the goods. The *Dry Goods Economist* related the tale of "a prominent citizen" who was sued by an unnamed store for fifteen hundred dollars' worth of toilet articles, soaps, and perfumes purchased by his wife. The man defended his action by denying his responsibility in the affair, saying his wife had become "possessed of a passion for such luxuries."[68] Yet in a very real sense women were only following the rules of the game; successful retailing assumed female susceptibility, and merchants relied on their ability to break down self-control and rational patterns of behavior. In succumbing to the lure of the merchandise, shoppers—both those who bought and those who stole—were doing what they were expected to do.

Observers of these women, whether social scientists, policemen, or trade spokesmen, seeking a way to explain and understand shoplifting, called them, almost unthinkingly, kleptomaniacs. Doctors and lawyers used the diagnosis of kleptomania to justify what they understood to be women's essential nature and allotted social role. Store managers, detectives, and the courts all worked within a framework that cast kleptomania as both a disease and as a type of antisocial behavior by women of a certain class. Casual references to kleptomania became part of the daily vocabulary in department stores, police stations, and courtrooms.[69]

Women in turn used the diagnosis in their own defense.[70] For many middle-class women kleptomania became an acceptable word, even a magical word, and a label that evoked understanding if not genuine sympathy. Women called themselves kleptomaniacs when they spoke of irresistible temptation, the physical inability to bypass an object, a counter, or a particularly attractive display. According to newspaper reports, one twenty-seven-year-old woman told court authorities that she was a kleptomaniac and "vainly sought to resist the temptation to steal." She added that "she had pains in her head at times during which she was subject to the disease." The report ended with the notation, "The police were inclined to believe Miss Degler's story."[71]

Stores lost merchandise from many sources—professional thieves, clerks, delivery men, and others—but only the middle-class female shoplifter was thought to be acting out of a medical disability. A large proportion of these women were excused either in the store or in the courtroom because they claimed to be kleptomaniacs or because they cited general malaise and physical debility without using the label. When Miss Degler complained of pains in her head at the very times she was subject to fits of kleptomania, she was, not unlike Mrs. Castle's lawyer, defining the disease for the layperson, conflating female illness and shoplifting.

More than gender was involved here. Kleptomania was a concept constructed upon cultural assumptions about gender, particularly about the irresponsibility and generative phases of women, but it was also a class concept. Only middle-class women seemed to suffer from it; stealing by members of other classes was simple theft. One Paris newspaper suggested, not altogether facetiously, that the definition of *thief* should be "a kind of kleptomania, but of the lower social class," while the definition of kleptomaniac should remain as it had always been, "a kind of thief, but of the better class."[72]

In 1882 Ellen Sardy, a "well-dressed middle-aged woman of respectable appearance," was caught shoplifting in the Sixth Avenue store of Simpson, Crawford, and Simpson. She took three pairs of stockings and five pieces of silk, of considerable value in the 1880s. Sardy was released when her lawyer admitted the larceny but "asked clemency on the grounds that the prisoner for some time past had been of weak mind and was not always responsible for her actions."[73] The Sardy case was probably not the first such plea in New York City, but it seems to be the first time the court accepted the excuse and the definition of the middle-class shopper. The release of the accused opened the way for a flood of such pleas.

Doctors and lawyers in the 1880s defended women like Ellen Sardy by saying little more than "she is a respectable, well-connected lady, but evidently a kleptomaniac."[74] A woman from Sandusky, Ohio, who was picked up for shoplifting in Ehrich's, appeared in court with a Dr. Augustine Daussé of West 22d Street, in New York City. The doctor claimed she was "a reputable woman, the wife of a judge . . . and at present under his care for special medical treatment." The woman contended she had no recollection of taking the merchandise as charged and "no object for taking them, as she is able to pay for everything she wants." For the judge the implication was clear: the lady was a kleptomaniac. The case was dismissed.[75]

The views of Chief Inspector Thomas Byrnes of the New York City Police Department also conformed to the popular medical understanding of kleptomania. In 1889, when two well-dressed women of "apparent respectability" were arrested for shoplifting in McCreery's and brought to New York's central police headquarters, Byrnes interviewed them. Three years earlier he had written *Professional Criminals of America*, and in it he had singled out the middle-class shoplifter, whom he called a kleptomaniac, for special notice. "These women," he wrote, "were so carried away by admiration of some trinket or knicknack as to risk home, honor, everything to secure it."[76] Now, three years later, Byrnes made a public statement defending the arrested women in their plea of kleptomania. "He believes," reported the *Times*, "they are kleptomaniacs and are no more legally responsible for these thefts than a lunatic would be for assault and battery."[77] Byrnes's understanding of kleptomania was by this time shared by a wide segment of the public.

The explanation offered in the case of a Mrs. Henry von Phul, a wealthy fifty-two-year-old woman from New Orleans, reflected this understanding. A spokesman told the judge that "the family has been subject to insanity, and Mrs. von Phul only left the asylum a few months ago. When in one of her fits of mental derangement, she has no control over her actions."[78] Although the term kleptomania was not used, the court declared Mrs. von Phul of unsound mind, her moral perversity in this instance taking the form of shoplifting. Neither this woman's behavior nor the court's interpretation was exceptional. Whether mentally healthy or truly insane, Mrs. von Phul was part of a socially determined view of women and madness.

Such a broad acceptance of the kleptomania diagnosis is plausible only if we recognize the centrality of the common cultural understanding of gender difference and the very real power of expectations. In the cultural climate of the late nineteenth century, the excessive behavior patterns of a few middle-class women were deemed representative of the constitution and fragility of the group. This determination hinged on a distortion of the female image and the assumption that women were totally ruled by their biology. For women and a middle class intent on preserving its privilege, disease thus became a defense. Weir Mitchell understood this perfectly when he wrote of Mrs. Castle's guilty plea in the English court: "She is now under a stigma from which it will be difficult to escape. . . . This [guilty plea] involves long explanations; the plea of insanity would have involved none."[79]

Instability was thought to be rooted in woman's nature; therefore it is easy to see why kleptomania found such ready acceptance. It served both as explanation and excuse. Had shoplifting, defined as a form of female delinquency, not been interpreted as illness, it would have to have been understood as crime, and the possibility that respectable, middle-class women could "sink into such a moral cesspool and forefeit the esteem and love of their best friends for a bottle of cosmetic" was unthinkable.[80]

Notes

1. A. V. Judges, ed., *The Elizabethan Underworld* (New York: Octagon, 1965), 170–71.

2. Sophie Lyons, *Autobiography of Sophie Lyons* (New York: Star Publishing, 1913).

3. There is a continuing debate among historians about the formation of the middle class during the nineteenth century. See, e.g., Stuart M. Blumin, "The Hypothesis of Middle-Class Formation in

Nineteenth-Century America: A Critique and Some Proposals," *American Historical Review* 90 (April 1985): 299–338; Arno J. Mayer, "The Lower Middle Class as Historical Problem," *Journal of Modern History* 47 (September 1975): 409–36; Karen Halttunen, *Confidence Men and Painted Women: A Study of Middle-Class Culture in America, 1830–1870* (New Haven, Conn.: Yale University Press, 1982); Daniel Horowitz, *The Morality of Spending: Attitudes Toward the Consumer Society in America, 1875–1940* (Baltimore: Johns Hopkins University Press, 1985). For a late nineteenth-century journalist's view of what it meant to be middle class, see George Ade, "The Advantages of Being Middle Class," in *Stories of the Streets and of the Town, from the Chicago Record, 1893–1900*, ed. Franklin J. Meine (Chicago: Caxton Club, 1941), 75–79. I am using *middle class* throughout this article as a designation for the growing number of urban residents. A group in the process of formation by the midnineteenth century, they were mostly native-born, Protestant, white-collar, business and professional men and their families. There were, however, hierarchies within each of these categories, and many people in economically precarious situations feared that social mobility went two ways. Not limited to occupation and income, the working definition of this new middle class has a significant cultural component that must include the home and its contents, residential location, levels and patterns of consumption, childrearing strategies, and leisure activities.

4. See, e.g., "Kleptomania as a Disease and Defense," *American Lawyer* 4 (December 1896): 533.

5. Carroll Smith-Rosenberg and Charles Rosenberg, "The Female Animal: Medical and Biological Views of Woman and Her Role in Nineteenth-Century America," *Journal of American History* 60 (September 1973): 338; Michel Foucault, *The History of Sexuality* (New York: Vintage, 1980), 1:44.

6. *New York Times* (January 14, 1899), p. 1, col. 4, and (January 14, 1896), p. 2, col. 7.

7. Felix Isman, *Weber and Fields: Their Tribulations, Triumphs, and Their Associates* (New York: Curtis, 1924), 280–81; Charles Chaplin, *The Floorwalker* (Mutual Film Corp., May 1916); Mark Melford, "Kleptomania," a farcical comedy in three acts, in *French's Acting Plays* 138 (London and New York: T. Henry French, 1888); Margaret Cameron, "The Kleptomaniac," a comedy in one act, October 30, 1901, reviewed in *New York Dramatic Mirror* (March 29, 1902, and April 15, 1905); Gus Edwards and Will D. Cobb, "Mamie, Don't You Feel Ashamie" (1901), in *Song Hits from the Turn of the Century, Complete Original Sheet Music for 62 Songs*, ed. Paul Charosh and Robert A. Fremont (New York: Dover, 1975), 162–66.

8. Edwin S. Porter, *The Kleptomaniac* (Edison studio, 1905, film).

9. Charles E. Rosenberg, *No Other Gods: On Science and American Social Thought* (Baltimore: Johns Hopkins University Press, 1976), 1–21.

10. Orpheus Everts, M.D., "Are Dipsomania, Kleptomania, Pyromania, etc., Valid Forms of Mental Disease?" *American Journal of Insanity* 44 (July 1887): 52–59.

11. Ibid., 56.

12. Ibid., 57.

13. Patricia O'Brien, "The Kleptomania Diagnosis: Bourgeois Women and Theft in Late Nineteenth-Century France," *Journal of Social History* 17 (Fall 1983): 65–77, esp. 71.

14. *New York Times* (October 10, 11, 14, 15, 18, 21, 23, and 31, November 3, 4, 7, 8 [three separate articles], 9, 10, 11 [two different articles], 13, 1896).

15. *San Francisco Chronicle* (November 7, 1896), p. 1, col. 7; *New York Times* (October 18, 1896), p. 1, col. 5, and (November 11, 1896), p. 9, col. 5.

16. *New York Times* (October 14, 1896), p. 9, col. 1.

17. Ibid.

18. "Kleptomania," *Law Times* 102 (November 14, 1896): 28; for a full discussion of this case and the issue of moral insanity, see "Kleptomania," *Atlantic Medical Weekly* 6 (December 26, 1896): 401–6.

19. *San Francisco Chronicle* (October 13, 1896), p. 1, col. 4, and (October 14, 1896), p. 1, col. 1; *New York Times* (October 18, 1896), p. 1, col. 5.

20. See, e.g., *New York Times* (October 23, 1896), p. 6, col. 7; and the remarks of Solomon Solis-Cohen, M.D., in S. Weir Mitchell, "The Relations of Nervous Disorders in Women to Pelvic Disease," *University Medical Magazine* (March 1897), 1–37, esp. 33.

21. *San Francisco Chronicle* (October 18, 1896), p. 18, col. 1.

22. *San Francisco Chronicle* (November 7, 1896), p. 1, col. 7; George Henry Savage, *Insanity and Allied Neuroses* (London: Cassell, 1884).

23. *New York Times* (November 7, 1896), p. 9, col. 5.

24. *San Francisco Chronicle* (October 14, 1896), p. 1, col. 1; *New York Times* (October 14, 1896), p. 9, col. 1.

25. *New York Times* (November 7, 1896), p. 9, col. 5.

26. *San Francisco Chronicle* (October 10, 1896), p. 1, col. 6. The *Chronicle* (October 14, 1896), p. 1, col. 1, published a partial list of the goods found in the Castle's trunks: "18 tortoise shell combs, 7 hand mirrors, 2 sable boas, 2 muffs, 2 neckties, 7 gold watches, 9 clocks, 17 valuable fans, 16 brooches, 7 tortoise shell eye glasses, 2 plated toast ranks [sic] marked Hotel Cecil, and a large number of smaller articles of less value."

27. Ibid. See also the statement by James Roosevelt, secretary of the U.S. embassy, shortly after the Castles' arrest, *San Francisco Chronicle* (October 10, 1896), p. 1, col. 6.

28. Carroll Smith-Rosenberg, *Disorderly Conduct: Visions of Gender in Victorian America* (New York: Oxford University Press, 1986), 3–52.

29. *San Francisco Chronicle* (November 7, 1896), p. 1, col. 2.

30. *San Francisco Chronicle* (October 31, 1896), p. 1, col. 7.

31. Up until the 1880s English law had been amenable to pleas of kleptomania, but the system of individual decisions in such cases had "worked so badly" and the defense was pleaded in so many cases that the law had been changed. After that it was no longer possible for a middle-class woman in England to be assured of acquittal by claiming irresistible impulse as a predisposing cause. A guilty plea and prison or insanity and a mandatory sentence to the asylum were the choices, and Sir Edward Clarke chose a guilty plea, confident that his client would be immediately released (*New York Times* [November 8, 1896], p. 17, col. 3).

32. Mitchell, "The Relations of Nervous Disorders," 35; see also S. Weir Mitchell, "Nervous Disorders (Especially Kleptomania) in Women and Pelvic Disease," *American Journal of Insanity* 53 (April 1897): 605–6.

33. Mitchell, "The Relations of Nervous Disorders," 34; *San Francisco Chronicle* (November 7, 1896), p. 1, col. 7.

34. Remarks of Solomon Solis-Cohen in Mitchell, "The Relations of Nervous Disorders," 34.

35. *San Francisco Chronicle* (October 11, 1896), p. 18, col. 1.

36. The work of Carroll Smith-Rosenberg has been fundamental to our understanding of the medicalization of deviant behaviors in the nineteenth century. In *Disorderly Conduct* she has again underscored the importance of social structure and male medical language in the determination of behavior as disease. Other historians, notably Joan Jacobs Brumberg, Nancy Tomes, and Regina Morantz-Sanchez, have moved in another direction; using clinical case records, they have begun to fashion a new social history of medicine. In my work I try to bridge the gap between the two approaches. See also Wendy Mitchinson, "Gynecological Operations on Insane Women: London, Ontario, 1895–1901," *Journal of Social History* 15 (Spring 1982): 467–84, esp. 467. Bert Hansen very kindly brought this article to my attention. See also O'Brien, "Kleptomania," 66–67.

37. Foucault, *History of Sexuality,* 27.

38. The literature revealing this association is voluminous. See, e.g., Raymond de Saussure, "The Influence of the Concept of Monomania on French Medico-Legal Psychiatry from 1825–1840," *Journal of the History of Medicine and Allied Sciences* 1 (July 1946): 365–96; Théodule A. Ribot, *The Disease of the Will,* trans. J. Fitzgerald (New York: Humbolt Library of Popular Science Literature, 1884).

39. For a sample of this literature see Henry Maudsley, M.D., "On Some of the Causes of Insanity," *British Medical Journal* 2 (November 24, 1866): 586–90; M. Legrand du Saulle, "The Physical Signs of Reasoning Madness," in the British *Journal of Psychological Medicine and Mental Pathology* 2 (1876): 317–23; Alice May Farnham, M.D., "Uterine Disease as a Factor in the Production of Insanity," in the American *Alienist and the Neurologist* 8 (October

1887): 532–47; Dr. L. G. Hanley, "Mental Aberration Consequent upon Pelvic Disease," Clinical Report, *Buffalo Medical Journal* 40–56 (March 1901): 672; A. T. Hobbs, M.D., "The Relation of Ovarian Disease to Insanity, and Its Treatment," *American Journal of Obstetrics* 43 (April 1901): 484–91; and John C. Doolittle, "The Relation of Pelvic Disease to Insanity," *Bulletin of Iowa Institutions* 3 (July 1901): 294–98.

40. Remarks of Solomon Solis-Cohen, M.D., quoting English physician, in Mitchell, "The Relations of Nervous Disorders," 33. See also John Young Brown, "Pelvic Disease in Its Relationship to Insanity in Women," *American Journal of Obstetrics and Diseases of Women and Children* 30 (September 1894): 360–64; C. C. Hersman, M.D., "Relation of Uterine Disease to Some of the Insanities," *Journal of the American Medical Association* 33 (September 16, 1899): 710.

41. T. S. Clouston, *Clinical Lectures on Mental Diseases* (Philadelphia: Henry C. Lea's Son, 1884), 171, 339.

42. Two of many smaller articles on the connection between reproductive pathology and moral insanity are Ernest H. Crosby, "The Legal Aspects of Partial Moral Mania," *Physician and Pharmacist* 12 (December 1879): 157–60; and George H. Roche, "Some Causes of Insanity in Women," *American Journal of Obstetrics and Diseases of Women and Children* 34 (December 1896): 801–6.

43. Mrs. Castle underwent the following treatment: "The sphincter ani [was] dilated, the fissures cauterized . . . the ulcers treated . . . and the hemorrhoids clamped and cauterized. The uterus was curetted and then the trachelorrhaphy performed by denudation of the cicatricial tissue and suturing with silkword gut." Remarks by Solomon Solis-Cohen, in Mitchell, "The Relations of Nervous Disorders," 31. For a sample of medical thinking and the practice of gynecology in this period, see W. B. Goldsmith, M.D., "A Case of Moral Insanity," *American Journal of Insanity* 40 (October 1883): 162–77; B. D. Evans, M.D., "Periodic Insanity, in Which the Exciting Cause Appears to Be the Menstrual Function—Report of a Typical Case," *Medical News* 62 (May 20, 1893): 538–40; Eugene G. Carpenter, M.D., "Pelvic Disease as a Factor of Cause in Insanity of Females and Surgery as a Factor of Cure," *Journal of the American*

Medical Association 35 (September 1, 1900): 545–51; Doolittle; and Hanley, 672.

44. For a discussion of uterine disease as a diagnostic catchall see Smith-Rosenberg and Rosenberg, "The Female Animal," 332–56; Barbara Sicherman, "The Uses of Diagnosis: Doctors, Patients, and Neurasthenia," *Journal of the History of Medicine* 32 (January 1977): 33–54, esp. 41.

45. For the connection between so-called dysfunctional behavior and crime, see H. Tristram Engelhardt, Jr., "The Disease of Masturbation: Values and the Concept of Disease," in *Sickness and Health in America: Readings in the History of Medicine and Public Health,* ed. Judith Walzer Leavitt and Ronald L. Numbers (Madison: University of Wisconsin Press, 1985), 18; "Moral Mania," *American Journal of Insanity* 27 (April 1871): 445; Ely Van de Warker, M.D., "The Relations of Women to Crime," *Popular Science Monthly* 8 (November 1895): 2; *New York Times* (December 26, 1905), p. 4, col. 5.

46. Medical literature on female disease was largely the source of these popular prejudices. See Mitchell, "Nervous Disorders (Especially Kleptomania) in Women and Pelvic Disease." For an earlier but essentially similar view see V. H. Taliaferro, M.D., "The Corset in Its Relations to Uterine Diseases," *Atlanta Medical and Surgical Journal* 10 (March 1873): 683.

47. *San Francisco Chronicle* (October 14, 1896), p. 1, col. 1, and (November 8, 1896), p. 18, col. 3.

48. *San Francisco Chronicle* (November 10, 1896), p. 1, col. 4 (my emphasis).

49. Paul Dubuisson, "Les voleuses des grands magasins," *Archives d'Anthropologie Criminelle* 16 (1901): 1–20, 341–70. (Note, *voleuse* is the feminine form of thief.) See also Roger Dupouy, "De la kleptomanie," *Journal de Psychologie Normal et Pathologique* Année 2 (1905), 404–26.

50. Dubuisson, "Les voleuses," 16.

51. Ibid., 343; Thomas Byrnes, *Professional Criminals of America* (New York: Cassell, 1886), 31–32; "Editorial," *New York Times* (December 26, 1905), p. 6, col. 3.

52. C. C. Marc, a French forensic specialist and physician to Louis-Philippe, was among the first to give the word *kleptomanie* scientific recognition. Building upon earlier interpretations, Marc defined

it as "a distinctive, irresistible tendency to steal" and thought social class and educational level of the kleptomaniac important, as well as certain biological determinants. For a fuller explication of Marc see O'Brien, "Kleptomania," 70; see also George L. Shattuck, "Kleptomania," *Atlantic Medical Weekly* 6 (December 26, 1896): 402–6. The *New York World* (March 31, 1872), p. 6, col. 3, referred to a "shoplifter's kleptomaneous propensities." This is the first citation that I could find in a New York newspaper specifically mentioning kleptomania. By the 1880s, however, middle-class shoplifters were regularly referred to as kleptomaniacs. See, e.g., *New York Times* (January 18, 1882), p. 8, col. 6; and *New York Tribune* (December 23, 1883), p. 6, col. 1. For earlier usages in medical literature see *American Journal of Insanity* 2 (January 1846): 275: "This prompts him to action by a kind of irresistible instinct, while he either retains the most perfect consciousness of its impropriety, and horror at the enormity of the conduct to which it would impel him, and with difficulty, restrains himself or gives way, as in desperation, to the impulse which urges him on. Examples—Cleptomania, or propensity to theft." John C. Bucknill ("Kleptomania," *American Journal of Insanity* 19 [October 1862]: 148–49) also cites Marc as the originator of the term and refers to its use in the *London Times* (April 1855). The *Oxford English Dictionary* traces the first public use of the term to the *New Monthly Magazine* 28, 15 (1830). For a popular interpretation of the term see the *New York World* (March 31, 1872), p. 6, col. 3.

53. New York newspapers were full of the details of the arrest of Mrs. Phelps. See, e.g., the *New York World* (December 25–28, 1870); the *New York Sun* (December 26–27, 1870); the *New York Daily Tribune* (December 26–29, 1870); *New York Commercial Advertiser* (December 27, 1870).

54. *Woman's Journal* (April 26, 1873), p. 135, col. 2.

55. *Dry Goods Economist* (February 6, 1892), 112; *Dry Goods Reporter* (June 7, 1902), 31; "Shopping," *Living Age* 251 (December 22, 1906): 758–60.

56. R. H. Macy & Co. (no signature) to Mr. Paul Kyle, Flushing, New York (July 6, 1904), R. H. Macy Collection, Baker Library, Harvard Graduate School of Business Administration, box 4, doc. 3375: "You are perfectly safe in making the assertion that about 150,000 persons visit the establishment of R. H. Macy & Co. daily." Estimates about the percentage of shoppers who were female vary, but for the generally recognized figure of about 90 percent, see *Dry Goods Economist* (June 13, 1896), 16 (February 27, 1897), 65 (February 4, 1899), 4; *Ladies' Home Journal* advertisement in *Dry Goods Economist* (April 3, 1897), 8; *Dry Goods Reporter* (January 21, 1899), 37.

57. Women make it clear in their diaries that their husbands were involved in consumer choices. See Clara Burton Pardee, *Diaries*, 1883–1938, New York Historical Society, New York, entries June 1883, November 1888; see also Joan M. Seidl, "Consumers' Choices: A Study of Household Furnishing, 1880–1920," *Minnesota History* 48 (Spring 1983): 183–97. Nevertheless, male customers were still very much the exception in nineteenth-century department stores. Clerks at Marshall Field & Co. called men who tagged along with their wives "Molly Husbands." See Lloyd Wendt and Herman Kogan, *Give the Lady What She Wants! The Story of Marshall Field & Company* (New York: Rand McNally, 1952), 277.

58. *Dry Goods Reporter* (July 5, 1902), 61; Dubuisson, "Les voleuses," 17; any history of the American department store mentions the free entry principle. See, e.g., Wendt and Kogan, 32, 34. Henry E. Resseguie ("Alexander Turney Stewart and the Development of the Department Store, 1823–1876," *Business History Review* 39 [Autumn 1965]: 310–11) notes that when Harry G. Selfridge, general manager at Marshall Field & Co., established his American-style department store with free entry for all in London in 1902, he was "severely censured by London magistrates for his alleged encouragement of shoplifting in doing so." See also *Dry Goods Economist* (November 5, 1892), 30–31: "British dry-goods stores have no goods openly displayed or on counters. . . . The bane of the American stores, the kleptomaniac, is scarcely known here and the force of floorwalkers and inspectors is correspondingly small."

59. *Dry Goods Economist* (March 15, 1902), 75.

60. See, e.g., *New York Herald* (October 10, 1895), 1. Ehrichs invited "ladies who are not prepared to purchase to come in and examine the goods," *Dry Goods Economist* (January 2, 1897), 37, and (September 11, 1897), 61.

61. Margarete Bohme, *The Department Store: A Novel of Today* (New York: Appleton, 1912), 110–11. For a description of one of these attractions, a huge silver statue of Justice originally displayed at the Chicago World's Fair and subsequently installed in Brooklyn's Abraham & Straus, see *Brooklyn Eagle* (November 5, 1893), p. 24, col. 6; for a historical discussion of this phenomenon see Hugh Dalziel Duncan, *Culture and Democracy* (Totowa, N.J.: Bedminster, 1965), 113.

62. For the effects of sentimental literature on women see Mary T. Bissell, M.D., "Emotions Versus Health in Women," *Popular Science Monthly* 32 (February 1888): 506; *Dry Goods Economist* (March 10, 1900), 21. See also Dee Garrison, "Immoral Fiction in the Late Victorian Library," in *Victorian America,* ed. Daniel Walker Howe (Philadelphia: University of Pennsylvania Press, 1976), 154–59.

63. *Dry Goods Economist* (August 3, 1893), 49.

64. *New York Daily Tribune* (November 17, 1896), p. 2, col. 1.

65. Rosalind Williams, *Dream Worlds: Mass Consumption in Late Nineteenth-Century France* (Berkeley and Los Angeles: University of California Press, 1982), 265.

66. Dubuisson, "Les voleuses," 342; *Dry Goods Economist* (March 15, 1902), 75; a twentieth-century English study of shoplifting suggested that in a few cases sexual excitement was a contributing factor (T. C. N. Gibbens and Joyce Prince, *Shoplifting* [London: Institute for the Study and Treatment of Delinquency, 1962], 72–73). See also Michael B. Miller, *The Bon Marché: Bourgeois Culture and the Department Store, 1869–1930* (Princeton, N.J.: Princeton University Press, 1981), 205–6.

67. *New York Times* (November 3, 1889), p. 10, col. 1, is one of the many articles that speaks of women's derangement as a result of the fatigue and anxiety that accompanied their shopping.

68. *Dry Goods Economist* (September 28, 1901), 53. This was one of a number of similar court cases. For an earlier view of women and overbuying see R. Heber Newton, *The Morals of Trade: Two Lectures* (New York: Whittaker, 1876), 76: "If women insist upon keeping up appearances, then husbands will very likely fail to pay or fail in paying. Cabinet ministers will not be the only men ruined by their wives."

69. In a *New York Times* article (April 11, 1895), p. 8, col. 1, two former police detectives noted "the cloak of kleptomania has been stretched out very thin, sometimes, in the test of compassionate friends and relatives." See also Benjamin P. Eldridge and William B. Watts, *Our Rival the Rascal* (Boston: Pemberton, 1897), 28–29.

70. *New York Times* (June 30, 1894), p. 1, col. 6, and (January 2, 1906), p. 15, col. 1; *Boston Globe* (December 8, 1897), p. 4, col. 4.

71. *New York Times* (June 30, 1894), p. 1, col. 6.

72. *L'Oeuvre* (June 7, 1925), cited in A. Anthéaume, *Le roman d'une épidémie parisienne, la kleptomanie?* (Paris: Librarie Octave Doin, 1925), 90. The basic organization of buying and selling in the department store is indicative of these understandings about class and role difference among women. See *Dry Goods Reporter* (February 1, 1902), 5; (February 6, 1904), 57; (August 15, 1903), 45.

73. *New York Times* (January 18, 1882), p. 8, col. 6.

74. See, e.g., the case of "A Well-dressed Female Shoplifter," *New York Times* (November 8, 1883), p. 8, col. 2.

75. *New York Times* (February 4, 1886), p. 2, col. 6.

76. Byrnes, *Professional Criminals,* 31–32. Byrnes's language followed French reasoning almost word for word; see, e.g., Dupouy, "Les voleuses," who describes those women whose ordinary sensibility and will are overcome by the desire for an object: "Reason is subsumed by desire . . . conscious will is diminished" (410–11).

77. *New York Times* (March 27, 1889), p. 2, col. 5.

78. *New York Times* (January 8, 1892), p. 10, col. 2. Belief in the persistence of inherited insanity was strong in the second half of the nineteenth century; see I. Ray, M.D., *Treatise on the Medical Jurisprudence of Insanity,* 5th ed. (Boston: Little, Brown, 1871), 173–74.

79. Mitchell, "The Relations of Nervous Disorders," 35.

80. *Woodhull and Claflins Weekly* (March 18, 1871), 4–5.

21 Diagnosing Unnatural Motherhood: Nineteenth-Century Physicians and "Puerperal Insanity"

Nancy Theriot

On December 16, 1878, Elizabeth S., age twenty-seven, was admitted to the Dayton Asylum for the Insane. The cause was "puerperal"; the form was "mania." About three weeks before admission she had given birth, and her insanity appeared a few hours after the child was born. When Elizabeth was admitted to the hospital, she was "very noisy and excited, clapping her hands and talking incessantly." She would sometimes tear her clothing and "expose her person." She had a poor appetite, did not sleep at night, and was in poor physical condition. Her physician "insisted" that she take plenty of milk and beef tea every day, gave her iron three times a day, and thirty-five grains of hydrate of chloral (a sedative) at bedtime. Under this treatment Elizabeth remained the same for almost two months, except that she rested at night. Near the end of February she began to improve. She started to "take an interest in things around her, was more neat in her dress; thought she ought

NANCY THERIOT is Associate Professor of History and Chairperson of Women's Studies at the University of Louisville, Louisville, Kentucky.

Reprinted from *American Studies* 30, 2 (Fall 1989): 69–88, by permission of the publisher.

to have something better to wear, and would help do the work." She continued to improve and was removed from the institution by her husband on June 19, 1879.[1]

The case of Elizabeth S. was one of hundreds reported by physicians in nineteenth-century medical journals. Elizabeth's was a case of puerperal mania, the most common type of puerperal insanity. Physicians also described two other forms of the disease which usually had melancholic symptoms: "insanity of pregnancy" and "insanity of lactation." Although doctors described puerperal insanity in various ways and although medical opinion about the nature of the malady changed over the course of the century, most physicians agreed that it was a very common ailment and that it was responsible for at least 10 percent of female asylum admissions. Yet by World War I the disease had all but disappeared. Except for "postpartum depression," the twentieth-century renaming of "insanity of lactation," puerperal insanity was cured by the world wars.

Like other nineteenth-century female diseases that have disappeared or been redefined in the twentieth century, puerperal insanity raises many questions about the relationship between the predominantly male medical profession and women patients. Was puerperal insanity an invention of men? Was it an expression of male physicians' ideas about proper womanly behavior, defining women's antimaternal feelings and activities as "insane"? Or was puerperal insanity only incidentally a gender issue; could it be understood as a professional struggle between male gynecologists and male alienists (nineteenth-century psychiatrists) over the treatment of insane women? Given the sexual politics involved when women's illness is named and treated by a male medical establishment, can physicians' accounts of puerperal insanity provide valid information about the meaning of the disease for women? If so, was puerperal insanity an indication of dissatisfaction with motherhood, disappointment with marriage, or anguish over abandonment or financial problems? In short, was puerperal insanity an expression of sexual ideology, medical professionalization struggles, or gender tension?

These questions cannot be answered adequately using either the traditional approaches to the history of insanity or the more critical approaches taken by historians interested in the history of women and madness. Both traditionalists and critics explain nineteenth-century insanity (or specific insanities) from one of three perspectives: that of the disease, the physician/medical institution, or the patient. Each vantage point is important but incomplete.

Although concentrating on the disease itself can provide information essential to interpretation, disease-focused studies deal with the disease either as an idea or as an essence gradually becoming known/named. Treating insanity or insanities as histories of ideas is interesting and useful, but this approach sidesteps questions of power.[2] Understanding how the idea of puerperal insanity changed over time and how it re-lated to other insanities is essential, for example, but this understanding does not begin to answer the questions posed earlier about gender and power. Similarly, it would be a mistake to see puerperal insanity as a "real" disease, misunderstood or misnamed by nineteenth-century physicians but understood and rightly differentiated by twentieth-century psychiatry.[3] This approach to insanity or insanities ends up begging all the questions of the meaning of insanity: Why was this set of symptoms seen in a particular way at this particular time? Why was this group of patients seen as "at risk"? Why was this disease named one way in 1850 and another way in 1910? Interpreting changing insanities as a change in medical nomenclature leaves all of the important questions not only unanswered but also unasked.

Interpreting the history of insanity from the perspective of physicians or medical institutions is more fruitful than the disease-centered approach because focusing on the medical establishment demands that insanity be situated within a specific socioeconomic setting. From this point of view the "reality" of the disease is questioned or ignored, as the historian concentrates on the role of professional and institutional politics or individual physicians in the creation of insanity. Perhaps the most well-known example of this approach is *Madness and Civilization* in which Michel Foucault argues that medical discourse on insanity helped to define *reason* by medicalizing and silencing an ever-increasing category of "unreason."[4] Similarly, many twentieth-century medical sociologists see insanity as a "label" applied by a powerful medical establishment to society's deviants.[5] Historians writing about nineteenth-century insanity have also noted the role of professional rivalries between alienists and neurologists in defining the nature of insanity, as well as the role of individual physicians (such as Charles Beard and S. Weir Mitchell) in discovering, classifying, and treating insanities.[6] What all of these approaches share is an emphasis on the power of organized medicine to define certain behavior as "insane" or "neurotic."

Many feminist historians and sociologists writing about women's insanity have concentrated on the power of physicians to categorize women's behavior as normal, neurotic, or insane and have pointed out how such categorizations both reflect and help maintain gender stereotypes and the imbalance of power between women and men.[7] While this perspective is superior to a disease-focused approach because it makes visible the sexual politics of medicine, there are problems with the physician-oriented interpretation. A major difficulty with concentrating on the medical establishment as the creator of insanity categories, or as the agent of "Society" in its quest to control deviants, is that patients/the public/women are seen as passive victims of medical definition. Reducing insanity to a behavior pattern defined as "sick" by a powerful profession tells us little about the meaning of that behavior in the lives of the patients.

Since Carroll Smith-Rosenberg's early article on hysteria, some feminist historians have interpreted women's insanity from the point of view of the patient, asking what the symptoms meant to the women afflicted. Like the physician-oriented perspective, concentrating on the meaning of the disease for the patient involves situating insanity in a particular cultural location. Smith-Rosenberg's study, and a later study of anorexia by Joan Jacobs Brumberg, interpret the illness within a specific family dynamic: woman as wife or daughter in a constricted or contradictory life pattern.[8] This patient orientation moves away from the "essence" of the disease and the politics of defining it, and instead asks why a woman might have behaved in a certain way. When trying to understand women's

insanity it is absolutely essential to focus on the meaning of the behavior within the context of women's lives, but there are at least two risks involved in relying solely on this perspective. Insane behavior might be misconstrued as heroic, as the only "sane" thing to do when confronted with a particular life situation. And, in concentrating on the family dynamics or the specific gender constraints of the patient, one might miss the medical dynamic and the process of defining/labeling behavior as insane or abnormal.

In order to understand the relationship between gender and insanity in general, and puerperal insanity in particular, we need a method of analysis that will encompass all three perspectives—that of the disease, the physician/medical establishment and the patient—and will describe the three in dynamic interrelationship. We need an interpretation that will be able to offer an explanation of both the meaning of symptoms in the lives of patients and the translation of symptoms into disease categories by medical professionals. What follows is an interpretation of puerperal insanity that divides the symptoms into "illness" and "disease" and sees both as social constructions.[9] The illness of puerperal insanity was a behavior pattern expressing dissatisfaction or even despair over the constraints of womanhood in a particular time, while the disease of puerperal insanity was a definition given by physicians to the illness symptoms, a definition which both legitimized the behavior pattern and played a role in medical specialization. As both illness and disease, puerperal insanity involved relationships: between the woman and her family, between the woman and her doctor, between the husband and the doctor, and between different medical specialists. Puerperal insanity can be interpreted as a socially constructed disease, reflecting both the gender constraints of the nineteenth century and the professional battles accompanying medical specialization. Male physicians and their female patients together created puerperal insanity, and that creation both reflected and contributed to sexual ideology and medical specialization.

Before elaborating this interpretation, a more thorough examination of puerperal insanity is in order. As mentioned earlier, most physicians believed puerperal insanity manifested itself differently in the three phases of the reproductive process. Milton Hardy, the medical superintendent of the Utah State Insane Asylum, defined puerperal insanity as a condition developing "during the time of and by the critical functions of gestation, parturition, or lactation, assuming maniacal or melancholic types in general" and characterized by "a rapid sequence of psychic and somatic symptoms which are characteristic not individually, but in their collective groupings."[10] Some physicians preferred to classify puerperal insanity as maniacal, melancholic, or depressive instead of dividing it according to reproductive phase, but in both groups there was consensus as to the type of insanity most associated with pregnancy, parturition, and lactation.

Insanity of pregnancy was thought to be the rarest of the three and usually involved melancholic (and suicidal) symptoms or depressive symptoms. Nineteenth-century physicians described patients as "melancholic" who appeared to be apathetic, hopeless, and prone to suicide, while "depressive" patients were those with "low spirits." In cases of insanity of pregnancy the symptoms sometimes lasted only a few weeks or months, but in other cases the patient was cured only by childbirth. Insanity of pregnancy was thought to occur most often with first pregnancies; however, some women who had developed symptoms once would develop symptoms in subsequent pregnancies. This form of puerperal insanity was rarely fatal.[11]

Lactation insanity was similar to gestation in-

sanity in its symptoms, melancholic and depressive, but was seen as more frequent. Lactation insanity differed from insanity of gestation and parturition in that it seemed to occur most often in women who had several children rather than in women going through their first pregnancies. In some cases of lactation insanity the melancholy ended in dementia and lifelong commitment to an asylum, but most cases recovered in under six months.[12]

Insanity of parturition was considered the most common type of puerperal insanity and was associated with maniacal symptoms. Usually, puerperal mania began within fourteen days of childbirth, but some cases started up to six weeks later. Like the insanity of pregnancy and lactation, puerperal mania was rarely fatal and usually lasted only a few months. Of the three forms of puerperal insanity, puerperal mania was the most baffling to medical writers in the nineteenth century. Indeed, most of the medical literature on puerperal insanity was a description of puerperal mania. Characteristic symptoms included incessant talking, sometimes coherent and sometimes not; an abnormal state of excitement, so that the patient would not sit or lie quietly; inability to sleep, with some patients having little or no sleep for weeks; refusal of food or medicine, so that many patients were fed by force; aversion to the child and/or the husband, sometimes expressed in homicidal attempts; a general meanness toward caretakers; and obscenity in language and sometimes behavior.[13]

Until the end of the century, when doctors began to express suspicion about puerperal insanity as a specific illness, there was widespread agreement about its frequency, duration, and prognosis. A physician writing in 1875 asserted that puerperal insanity was a "class of cases to be met with in the practice of nearly every physician," others cited asylum records indicating that the disease was responsible for "a very large proportion of the female admissions to hospitals," and still others claimed that puerperal insanity affected anywhere between 1 in 400 or 1 in 1,000 pregnant women.[14] Doctors also agreed that most cases of puerperal insanity lasted only a few months, with most recovering completely within six months.[15] Except for those cases with suicidal or homicidal tendencies, the prognosis was good for patients suffering from puerperal insanity, and doctors asserted that most cases could be, and were, treated at home.[16]

Treatment for puerperal insanity remained mostly the same over the course of the century, and the change reflected a more general change in medical therapeutics. In the first part of the century bleeding was considered the proper treatment, no matter if the symptoms were manic or melancholic. By midcentury that treatment was no longer recommended, and instead physicians were treating puerperal insanity patients with rest, food, a little purging, and sedation. Most physicians also recommended that patients be restrained or watched closely and that family and friends be kept away.[17]

One of the first explanations of puerperal insanity to occur to an historian sensitive to gender as a category of analysis is that the disease represented male physicians' definitions of proper womanly behavior.[18] To nineteenth-century men, a woman who rejected her child, neglected her household duties, expressed no care for her personal appearance, and frequently spoke in obscenities had to be "insane." Certainly there is much in the medical literature to support this explanation. Many physicians wrote in very sentimental terms about the mysterious beauty of motherhood being defiled by insanity. Dr. R. M. Wigginton wrote of the special horror of puerperal insanity: "The loving and affectionate mother, who has so recently had charge of her household, has suddenly been deprived of her reason; and instead of being able to throw around her family that halo of former

love, she is now a violent maniac, and feared by all."[19] Physicians commented on a woman's "letting herself go" or being "indifferent to cleanliness" as symptoms, and many listed willingness and ability to perform household tasks as evidence of a cure.[20]

By far the most shocking symptoms of puerperal insanity were women's indifference or hostility to children and/or husbands and women's tendency to obscene expressions. The first upset physicians' ideas about women's maternal and wifely devotion, while the second undermined doctors' assumptions about feminine purity. Allan McLane Hamilton described a patient who before her labor was "a loving and devoted wife, but shortly after lost all of her amiability, and treated her husband and mother with marked coldness, and sometimes with decided rudeness."[21] Even more difficult to explain than coldness was a woman's "thrusting the baby from the bed, disclaiming it altogether, striking her husband," a woman who looks at her baby "and then turns away," or a woman who "commenced to abuse it [the newborn child] by pinching it, sticking in pins, etc." So frequent was "hostility or aversion to husband and child" noted in cases of puerperal insanity that this was considered one of the defining characteristics of the disease, and physicians recommended that the woman not be left alone with her infant.[22]

If doctors were horrified at women's treatment of husbands and children, they were equally shocked at women's obscene words and behavior during an attack of puerperal insanity. "The astonishing familiarity of refined women with words and objects and practices of obscene and filthy character, displayed in the ravings of puerperal mania, gives a fearful suggestion of impressions which must have been made upon their minds at some period of life," wrote George Byrd Harrison, a Washington, D. C., physician. W. D. Haines of Cincinnati described a case in which the woman would repeat one word a dozen or so times "then break forth into a continuous flow of profanity. The subject of venery was discussed by her in a manner that astounded her friends and disgusted the attendants." Another doctor wrote of the typical puerperal mania patient's "tearing her clothes, swearing, or pouring out a stream of obscenity so foul that you wonder how in her heart of hearts such phrases ever found lodgment." An Atlanta physician expressed similar puzzlement: "It is odd that women who have been delicately brought up, and chastely educated, should have such rubbish in their minds." And still another physician described this symptom as "a disposition to mingle obscene words with broken sentences . . . modest women use words which in health are never permitted to issue from their lips, but in puerperal insanity this is so common an occurrence, and is done in so gross a manner, that it is very characteristic." W. G. Stearns, a Chicago physician, went so far as to note that in "all such cases [puerperal mania] there is a tendency to obscenity of language, indecent exposure, and lascivious conduct."[23]

Clearly, these physicians were shocked and dismayed by their patients' "indecent" behavior and use of language, as well as by their hostility toward husbands and infants, their neglect of household duties, and their refusal to pay attention to personal appearance. Even in their empirical reporting of patients' symptoms doctors revealed their disgust and horror over such unwomanly women. In naming the behavior "puerperal insanity," physicians were both reflecting and supporting nineteenth-century sexual ideology.

As authoritative spokesmen for the new scientific view of the nature of humanity, physicians were also helping to *create* sexual ideology in their explanations of puerperal insanity. Many doctors wrote of insanity as a logical by-product of women's reproductive function. George Rohe, a Maryland physician, asserted

that "women are especially subject to mental disturbances dependent upon their sexual nature at three different epochs of life: the period of puberty when the menstrual function is established, the childbearing period and the menopause."[24] Dr. Rohe regarded insanity as an ever-present danger to all women throughout their adult lives. Other doctors, however, wrote of pregnancy as a special challenge to women's mental balance, asserting that most women suffer mild forms of mental illness throughout their pregnancies. "In females of nervous temperament, the equilibrium of nerve force existing between these two organs [the brain and the uterus] is of the most delicate nature," wrote a Denver physician. He went on to say that "pregnancy is sufficient to produce insanity."[25] Probably the clearest statement along these lines was made by a professor of gynecology who wrote: "From the very inception of impregnation to the completion of gestation, some women are always insane, who are otherwise perfectly sane." He went on to say that others "manifest defective mental integrity in the form of whimsical longings for the gratification of a supposed depraved appetite."[26]

It would seem that nineteenth-century physicians' views of proper womanly behavior, along with their ideas about the power of the uterus to disrupt women's mental balance, influenced their perception and definition of puerperal insanity. It would be a mistake, however, to conclude that puerperal insanity was simply an indication that male doctors reflected their time or that the medical establishment influenced sexual ideology. Focusing too closely on the obvious ideological content of physicians' accounts of puerperal insanity, one might overlook that physicians' guesses about the nature of the disease were very much in keeping with nineteenth-century ideas about insanity in general and that many physicians offered what late twentieth-century people would call "sociological" explanations for women's behavior. Indeed, much of the medical discourse on puerperal insanity seems to have been influenced very little by male doctors' concepts of femininity but instead reflected the state of medical knowledge about insanity, on the one hand, and a jurisdictional dispute between alienists and gynecologists over the treatment of insane women, on the other.

For example, throughout the nineteenth century, physicians asserted that mental illness in general, not just women's mental illness, reflected a connection between mind and body; if the mind was unbalanced, a brain lesion was responsible, and the "exciting" cause of the brain lesion could be physical or emotional.[27] Indeed, this argument was one of the ways physicians convinced the public that mental illness was a medical problem. From the general assumption of a mind-body link as part of the nature of mental disease, it was logical to conclude that puerperal insanity was in some way caused by the physical state of pregnancy, parturition, or lactation. Doctors reasoned that the physical system was taxed by the reproductive process and that this added strain could be an "exciting" cause of insanity. A Pennsylvania physician wrote that "[t]here is no organ or portion of viscera which is not intimately connected with the brain through the sympathetic nervous system," and the Ohio physician who admitted Elizabeth S. to the Dayton Asylum noted more specifically about puerperal insanity: "The physical derangements attendant upon pregnancy, childbearing and nursing, are the principal causes of the insanity, which would be equally produced by any other physical suffering or constitutional disturbance of the same intensity."[28] Another indication of this line of reasoning was physicians' notation of any physical problem associated with labor as the probable cause of the insanity. If there was infection or a mild fever, if the labor was unusually long or difficult, if the woman re-

quired forceps, if her perineum was torn were seen as explanations for the puerperal insanity.[29]

Physicians also cited "heredity" as a primary cause of puerperal insanity, especially by the middle of the century. Like the mind-body theme, this too reflected a more general trend in medical ideas about the nature of mental illness. If there was insanity in a woman's family, regardless of how remote a relationship, this was considered a "predisposition" to mental unbalance. In such a case, pregnancy, childbirth, or lactation was seen as the stress that pushed the already unstable mind over the edge.[30]

Finally, many physicians argued that puerperal insanity was caused by situation, what the nineteenth-century writer called "moral" factors and what the late twentieth-century writer would call "sociological" factors. This too was in keeping with nineteenth-century theories about insanity in general. Just as financial problems or job stress were seen as possible causes of insanity in men, women were thought to develop puerperal insanity sometimes because of being abandoned or poorly treated by husbands, being pregnant and unmarried, being overburdened with too many children and household cares, or being emotionally drained because of grief or fear. In such cases physicians were very clear that the woman's insanity was brought on by her situation and that the puerperal state simply lowered the woman's strength so that she could no longer deal with the adverse environmental conditions. Kindness, rest, and reassurance were the best treatment.[31]

The mind-body connection and the possibility that physical or moral factors could be the "exciting cause" of puerperal insanity were both stressed throughout the century, but by the 1870s gynecologists began to emphasize the physical causes. The earliest proponent of this point of view, cited later as a man ahead of his time, was Horatio Storer. He argued as early as

1864 that most insanity in women is "reflex" insanity; that is, the primary cause of the insanity is a malfunction of the reproductive organs. For Storer and his post–Civil War followers this meant that women's insanity could be prevented, treated, and cured by medical and/or surgical means.[32] It also meant that a gynecologist should be consulted in any case of female insanity. Medical ideas about the nature, cause, and treatment of puerperal insanity were complicated by this professional struggle. Because it was in their best interest to link women's insanity with their reproductive organs, gynecologists "saw" a connection that other physicians saw less clearly. Furthermore, they wrote authoritatively, as the medical "experts" on women, and assumed disagreement was the result of ignorance. Charles Reed, professor at the Cincinnati College of Medicine, expressed surprise to hear any dissent from "the long-recognized doctrine of the genital origin of insanity in the female sex."[33] One Washington, D.C., physician claimed that puerperal insanity could be prevented only by good prenatal care.[34] These gynecologists directed their arguments to general practitioners and to alienists, who ran asylums. Many, though not all, of the gynecologists' articles about puerperal insanity or about women's insanity in general concluded that asylums should employ gynecologists—a clear expression of the professional struggle influencing medical perceptions of women's insanity.[35]

The medical discourse among gynecologists, alienists, and general practitioners about the nature of female insanity affected practice, which in turn affected discourse. From the mid-1870s to the 1890s gynecologists practiced their medical and increasingly surgical techniques on private patients and institutionalized women. Increasingly, diseases of the reproductive system were listed as the cause for the insane symptoms of women admitted to asylums.[36] More and more asylums employed gynecologists to examine female patients upon admission, and physicians found a variety of gynecological disorders among the women. Believing that there was a direct connection between these disorders and the women's insanity, the doctors administered medical and surgical cures. In the surgical category removal of the ovaries was the most popular operation, but more and less extreme operative procedures were also tried, such as hysterectomy and birth repair surgery.[37]

Some physicians reported patients being cured of insanity as a result of a gynecological procedure, and puerperal insanity was said to be especially responsive to physically oriented therapy. However, as gynecologists treated more insane women in and out of asylums, medical discourse reflected their growing disillusionment with surgical and medical treatment. Even those physicians who supported operative treatment reported disappointing cure rates.[38] By the 1890s there was lively debate over surgical treatment of insane women, with some physicians denouncing "mischievous operative interference" and others asserting that only physical (not mental) symptoms should prompt a surgical response. What made the debate different from the earlier one in which gynecologists successfully fought for the right to treat insane women was that the later debate was based on empirical studies. Having won access to asylum patients, gynecologists generated the numerical evidence against their own case. Two Minnesota physicians working at the state hospital at St. Peter found a large number of women asylum patients with serious pelvic disease in whom "there was not only no apparent relation between the pelvic disease and the mental disturbance, but there was no complaint or evidence of physical discomfort." They called this finding "the most unexpected result of our investigation."[39] Other physicians recorded the effect of surgery on women's insanity and found no significant link between operations and cures. Although they argued that gynecological problems could add to a woman's worry and discomfort and that all women (in and out of asylums) should have those problems treated, most gynecologists by the end of the century no longer claimed that women's diseased reproductive organs caused their insanity.[40]

If the empirical evidence, most of it gathered by gynecologists themselves, would not support a straight physiological explanation of women's insanity, how were physicians to account for puerperal insanity? Gradually, beginning in the 1890s, puerperal insanity was seen as a suspect category, and the emerging specialty of psychiatry emphasized the similarity between puerperal mania and any other mania, between the melancholy some women experienced during pregnancy or lactation and any other melancholy.[41] The particular physiological process was seen as less and less significant, and so the very term *puerperal* insanity was eventually dropped. Just as its appearance and growth was complicated by struggles of medical specialization, the disappearance of puerperal insanity from medical discourse was due to the empirical studies of one specialty and the reconceptualization of insanity that accompanied the rise of a new specialty (psychiatry).

Seen from this angle, puerperal insanity was not simply an expression or creation of sexual ideology by the medical profession. Certainly, gynecologists were able to convince other physicians of the physiological basis of women's insanity (and puerperal insanity) because the argument fit common ideas about woman's nature. Physicians "saw" mad women in a particular way because of generally held cultural ideas. That medical discourse was altered by empirical investigation at a time when most Americans, including feminists, believed in a biologically determined "woman's nature" indicates that gender was not a simple factor in the medical debate. Perhaps the most significant way gender

affected the medical construction of puerperal insanity was in the absence of women from the professional discourse until the late nineteenth century. There is no way to measure the impact of women's silence, but it is interesting to note that women physicians in the 1880s and 1890s were overrepresented in the group of doctors gathering evidence that separated women's insanity from their reproductive organs and eroded the assumptive framework for puerperal insanity as a specific illness.[42] It is safe to assume that the exclusion of women from medicine in the early and midnineteenth century affected the "scientific" view of women's mental (and physical) illness.

But what of the women who were diagnosed as having puerperal insanity? So far we have been concentrating on physicians and the ideological and professional issues influencing their conception of puerperal insanity. The medical discourse, however, also offers a way to understand the women who were patients. Most medical articles dealing with puerperal insanity included case studies, detailed descriptions of the situation, behavior, and treatment of the patients. Of course, what doctors selected as important information and what they recorded and did not record of patients' speech and behavior was subjective. Yet they were attempting "objective" observation. Although we cannot take case studies as the "complete picture" or as an entirely unbiased account, they reveal much about the possible meaning of puerperal insanity to the women who were so diagnosed, and they also provide a somewhat blurry snapshot of the doctor-patient dynamic.[43]

On the most literal and superficial level, case studies of puerperal insanity indicate that many women responded with melancholic or maniacal behavior to situations that they found unbearable. Illegitimacy, the fear that often accompanied first pregnancies, a traumatic birth experience or a stillborn infant, infection following birth, and extreme cruelty of husbands—were all cited in case studies, sometimes with the doctor attributing the insanity to the situation and other times not. One woman developed maniacal symptoms after her baby was delivered with forceps ("the head was extracted with considerable difficulty") and she suffered physical damage in this her first delivery. Another woman "frail and feeble" developed insane symptoms after her infant died a few days after birth. A woman whose symptoms included disclaiming her infant, striking her husband if he came near, and accusing people of trying to kill her was unimproved after five months in an asylum; her baby had died two months earlier and her husband, it turned out, had been continually abusive to her during her pregnancy.[44]

Other situational difficulties also appeared in case studies, such as women having many children in very few years and seemingly overburdened with work and responsibility. One woman, Mrs. S., who was thirty-five and had had five children, three of them within five years, developed "anxiety and slight confusion of ideas" during her last pregnancy. After the child was born she went into a "furious delirium . . . tried to leap from the window to avoid imaginary pursuers." A few days later she was no better; she said she "expects to be tortured soon, remonstrates bitterly." By the tenth day she was a little better: "Talks less and sleeps better. Tries to explain her sickness but cannot."[45] In another case a twenty-two-year-old woman was melancholic after the birth of her fourth child; her husband confined her and abandoned her once she was hospitalized.[46]

Case studies of puerperal insanity almost always included some physical or situational problem that late twentieth-century readers would see as cause enough for insane behavior, even when the physicians failed to note the connection. But while we may conclude that these women had good reason to act strangely for a

few months, the meaning of puerperal insanity is more complicated than this. The symptoms provide a clue to the meaning of the disease for women and also point to the doctor-patient relationship as a key factor in the waxing and waning of puerperal insanity.

Whether on a conscious or unconscious level, women who suffered from puerperal insanity were rebelling against the constraints of gender. The symptoms clearly indicate that rebellion. Case studies document that women refused to act in a maternal fashion by denying their infant nourishment or actively attempting to harm the child. Many women "did not recognize" the child, "ignored" its presence, or denied that that child belonged to them.[47] Similarly, women refused to act in a wifely fashion; they claimed not to know their husbands, expressed fear that the husband wanted to murder them, and sometimes struck out physically at their husbands.[48] Women were refusing the role of wife-mother, a role that most nineteenth-century Americans saw as the essence of "true womanhood."

Moreover, women suffering from puerperal insanity were not acting like women at all. They were "apathetic," "irritable," "gloomy," and "violent" instead of tuned in to the needs of those around them. In fact, these women required that others pay attention to them, in their constant talking and pacing the floor and in their refusal to care for themselves in the simplest ways, such as feeding themselves and keeping themselves clean. In a time when modesty was thought to be a defining characteristic of femininity, women with puerperal insanity "laughed immodestly," tore their clothing in the company of men, and used obscene language. Rebellion against cultural notions of "true womanhood" was the one thing tying together the various symptoms of puerperal insanity.

Physicians, new to the lying-in chamber, made these rebellious symptoms legitimate by defining them within a medical framework.

Doctors responded to women's behavior with a name: puerperal insanity. That naming was the result not only of the general ideas of the culture and the specific professionalization struggles of physicians but also was related to doctors' new relationship with women patients: as birth attendants. From the late eighteenth century, male physicians had begun to describe pregnancy and childbirth as a traumatic ordeal. Even doctors who did not think of birth as a sickness, but described it as a natural phenomenon, expressed a mixture of amazement, disgust, and respect at women's ability to undergo all the physiological changes associated with pregnancy, birth, and lactation. The assumptions of nineteenth-century physicians provided a framework both for their acceptance of women's strange behavior as a side-effect of reproduction and their definition of that behavior as, mostly temporary, insanity.

The medicalization of pregnancy, birth, and lactation provided a kind of permission for women to express rebellion and desperation in the particular symptoms of puerperal insanity. But if physicians and women patients both participated in the creation of puerperal insanity, the relationship was not a straightforward one. Women played out their rebellion against the male physician, and doctors translated that rebellion into an acceptable medical category. But doctors also "cured" the rebellion with their treatment and systematically silenced women in their case study reporting. In both cases women were unequal partners in the construction of the disease.

Treatment of puerperal insanity consisted of various levels of constraint and intrusion. In what late twentieth-century readers would judge the mildest, most humane treatment, women were confined to their rooms, denied the company of family and friends, and forced to rest by the admission of tranquilizers. If the woman refused to eat, which happened in an over-whelming majority of puerperal insanity cases, she was force-fed. Indeed, the element of force was characteristic of most treatment plans. One physician recorded force-feeding and threatening to cut the patient's hair if she continued to refuse food, and others noted that patients were confined to their rooms or their beds if their behavior did not change quickly enough.[49] In non-surgical cures force-feeding was the most intrusive aspect of the treatment, but surgical cures were penetrating in a more drastic sense. For the doctor these cures were restoring the unfortunate patient to her rightful and happy role. For the woman? Regardless of how women perceived the cures, and we will never know their perceptions, they certainly gave up their insane behavior, usually within a few months. If women were expressing rebellion in puerperal insanity symptoms and male physicians were defining that behavior as medically explainable and therefore legitimate, male physicians were also forcefully putting down the rebellion. In the social construction of puerperal insanity both parties were not equally powerful. A more interesting example of women's subordinate position in the relationship defining puerperal insanity is the judging and editing of women patients in the male-controlled medical discourse. The language physicians used to describe their women patients was often sympathetic but more often judgmental. One doctor described a woman before her insanity as having a "naturally obstinate and passionate disposition," and another wrote of a suicidal mother who tried to harm her four-month-old infant: "She should be hung."[50] More subtle than judgments of behavior were descriptions in case studies which substituted judgment for information. Physicians recorded "obstinate" and "indelicate" behavior and "immoderate" laughter. In some cases the physician's judgmental words were simply reflections of husbands' accounts of their wives' behavior, but that acceptance of the husbands' point of view was very much a part of the sexual politics involved in puerperal insanity. To many male physicians the women were to blame for their deviant unwomanly behavior, and physician case studies recorded the blame.

Although women patients and male physicians constructed puerperal insanity together, the clearest indication that men controlled the discourse was the near absence of women's words from the case studies. Over and over again physicians claimed that women suffering from puerperal insanity "talked incessantly," yet no attempt was made to record what the women talked about. Similarly, some women were said to complain of "imaginary wrongs," with no explanation of the content of those complaints. The most glaring omission in the case studies was physicians' refusal to record women's "obscene" language. An overwhelming majority of case studies referred to one or all of these speech acts, yet no content was provided.

If women were silenced partners in the construction of puerperal insanity, what can we conclude about the meaning of the disease for women? Although women's words were not reported, physicians' accounts of women's behavior and situations indicate that puerperal insanity was an unconscious act of rebellion against gender constraints for many women. The particular symptoms of puerperal insanity involved a denial of motherhood and a reversal of many "feminine" traits. Women presented these symptoms and acted out their rebellion; male physicians who for ideological and professional reasons were disposed to define women's behavior as "insanity" legitimized women's rebellion as illness. Yet part of the meaning of puerperal insanity for women must also have been the curing, the silencing. So many of the symptoms were aggressively, willfully expressive: the tireless pacing, the continuous talking, the laughter, the obscenity—all unlistened to, unrecorded. It is almost as if women usurped the power of

language only to find that it held no power at all. The woman cured of puerperal insanity surrendered these self-assertive symptoms and went back to being the "halo of love" in her family, without having been heard. There is no way of knowing whether she saw herself as victorious or defeated.

In spite of the sexual politics inherent in the doctor-patient relationship defining puerperal insanity, and in spite of women's silence in the case studies, women's symptoms were taken seriously enough to constitute a disease, at least until the turn of the twentieth century. What did it mean for women that puerperal insanity disappeared? Certainly, it can be argued that the constraints of gender were not as tight in the early twentieth as they had been in the nineteenth century. Women were having fewer children, childbirth was less dangerous and less painful, women had wider opportunities in terms of education and work, and women's marriage relationships were more companionate. If puerperal insanity was a rebellion against the constraints of nineteenth-century "true womanhood," women may have had less trouble with the twentieth-century variety and therefore ceased to manifest the symptoms of puerperal insanity.

Although changes in women's situation contributed to the demise of puerperal insanity, changes in the relationship between doctors and women patients also played a part. As we saw earlier, empirical studies and the rise of psychiatry altered medical perception of mental illness. Reliance on more "objective," "scientific" studies as the basis of medical discourse meant that there was less tolerance for puerperal insanity as a category. Regardless of how much or little women's situation had changed by the twentieth century, the symptoms of puerperal insanity were no longer a legitimate response to pregnancy, birth, or lactation in 1910, as they had been in 1870. Changing medical ideas, which had little to do with women patients, meant that physicians would no longer legitimize puerperal insanity as illness.

Elizabeth S. was admitted to the Dayton Asylum for the insane in 1878. Her illness was the product of several intertwined relationships: her own response to her marriage and motherhood; her physician's response to her story; and her story's resonance in the medical and cultural score of the nineteenth century. The interaction of these layers of relationship defined her condition as puerperal insanity. By the twentieth century, changes in all three layers made the disease obsolete. The creation and demise of puerperal insanity illustrates not only the social construction of illness but also the cultural embeddedness of medical categories.

Notes

Research for this article was made possible by a Summer Faculty Research Award (1987) from the University of Louisville Commission on Academic Excellence and by a Graduate Research Grant for Travel and Equipment (1987) from the University of Louisville, Graduate Programs and Research.

1. J. M. Carr, "Puerperal Insanity," *Cincinnati Lancet-Clinic* 7 (1881): 537–42.

2. I do not mean to imply that studies of the history of ideas about insanity are inappropriate. Such studies as Norman Dain, *Concepts of Insanity in the United States, 1789–1865* (New Brunswick, N.J. 1964); and Ellen Dwyer, "A Historical Perspective," in Cathy Spatz Widom, ed., *Sex Roles and Psychopathology* (Bloomington, Ind., 1984), 19–48, are absolutely essential to our understanding of the history of insanity. In her essay Dwyer goes beyond a history of ideas approach and attempts to test how ideas about gender and insanity related to actual practice. This relationship between ideas and practice deserves more attention. In another study Dwyer is particularly interested in comparing ideas and practice: Ellen Dwyer, "The Weaker Vessel: Legal Versus Social Reality in Mental Commitments in Nineteenth-Century New York," in

D. Kelly Weisberg, ed., *Women and the Law: A Social Historical Perspective* 2 (Cambridge, Mass., 1982), 85–106. A twentieth-century study that compares ideas and practice in that same volume is Robert T. Roth and Judith Lerner, "Sex-Based Discrimination in the Mental Institutionalization of Women," 107–39.

3. The approach to insanity taken by Mark S. Micale, "On the 'Disappearance' of Hysteria: A Medical and Historical Perspective," paper read at the annual meeting of the American Association for the History of Medicine, Birmingham, April 1989, I believe is not helpful. Micale explains the disappearance of hysteria as due to more specific medical definitions. The real question, however, is how/why definitions change.

4. Michel Foucault, *Madness and Civilization: A History of Insanity in the Age of Reason,* translated by Richard Howard (New York, 1965).

5. For discussions of the labeling theory see Roy Porter, *A Social History of Madness* (New York, 1987); Agnes Miles, *The Neurotic Woman: The Role of Gender in Psychiatric Illness* (New York, 1988); Elaine Showalter, "Victorian Women and Insanity," in Andrew T. Scull, ed., *Madhouses, Mad-Doctors, and Madmen: The Social History of Psychiatry in the Victorian Era* (Philadelphia, 1981), 313–31; Richard W. Fox, *So Far Disordered in Mind: Insanity in California, 1870–1930* (Berkeley, 1978); Nancy E. Waxler, "Culture and Mental Illness: A Social Labeling Perspective," *Journal of Nervous and Mental Disease* 159 (1974): 379–95.

6. See for example Bonnie Ellen Blustein, "'A Hollow Square of Psychological Science': American Neurologists and Psychiatrists in Conflict," in Scull, *Madhouses, Mad-doctors, and Madmen,* 241–70; and Barbara Sicherman, "The Paradox of Prudence: Mental Health in the Gilded Age," in the same volume, 218–40.

7. For example, Waxler, "Culture and Mental Illness"; Showalter, "Victorian Woman and Insanity"; and *The Female Malady: Women, Madness, and English Culture, 1830–1980* (New York, 1987); Miles, *The Neurotic Woman;* Dwyer, "The Weaker Vessel."

8. Carroll Smith-Rosenberg, "The Hysterical Woman: Sex Roles and Role Conflict in Nineteenth-Century America," *Social Research* 39 (1972): 562–83; Joan Jacobs Brumberg, *Fasting Girls: The Emergence*

of Anorexia Nervosa as a Modern Disease (Cambridge, Mass., 1988).

9. For an explanation of "illness" and "disease" see Arthur Kleinman, *Social Origins of Distress and Disease: Depression, Neurasthenia, and Pain in Modern China* (New Haven, Conn., 1986); Claudine Herzlich and Janine Pierret, *Illness and Self in Society,* translated by Elborg Forster (Baltimore, 1987); Arthur Kleinman, *Patients and Healers in the Context of Culture* (Berkeley, 1980); Peter Conrad and Rochelle Kern, eds., *The Sociology of Health and Illness: Critical Perspectives* (New York, 1981); Peter Wright and Andrew Treacher, eds., *The Problem of Medical Knowledge: Examining the Social Construction of Medicine* (Edinburgh, 1982), especially the essays: David Ingleby, "The Social Construction of Mental Illness," 123–41, and Karl Figlio, "How Does Illness Mediate Social Relations: Workmen's Compensation and Medico-Legal Practices, 1890–1940," 174–224; Bryan S. Turner, *The Body and Society: Explorations in Social Theory* (Oxford, 1984).

10. Milton H. Hardy, "Puerperal Insanity," *Western Medical Review* 3 (1898): 14.

11. W. H. B. Stoddart, "A Clinical Lecture on Insanity in Relation to the Childbearing State and the Puerperium," *Clinical Journal* 14 (1899): 242; Hardy, "Puerperal Insanity," 14; R. M. Wigginton, "Puerperal Insanity," *Transactions of the Wisconsin State Medical Society* 60 (1975): 40; L. R. Landfear, "Puerperal Insanity," *Cincinnati Lancet and Observer* 19 (1876): 54; Harry L. K. Shaw, "A Case of Insanity of Gestation," *Albany Medical Annals* 19 (1898): 459–62; Fleetwood Churchill, "On the Mental Disorders of Pregnancy and Childbed," *American Journal of Insanity* 7 (1850–1851): 297–317.

12. W. F. Menzies, "Puerperal Insanity: An Analysis of One Hundred and Forty Consecutive Cases," *American Journal of Insanity* 50 (1893–1894): 147–85; George H. Rohe, "Lactational Insanity," *Journal of the American Medical Association* 21 (1893): 325–27; W. G. Stearns, "The Psychiatric Aspects of Pregnancy," *Obstetrics* 3 (1901): 23–26, 32–36; Wigginton, "Puerperal Insanity," 41.

13. These symptoms were listed in numerous nineteenth-century articles. Since we will be dealing with the symptoms in more detail later, I will refrain from citing all of the articles here.

14. Landfear, "Puerperal Insanity," 54; Wigginton, "Puerperal Insanity," 41. See also R. E. Haughton, "Puerperal Mania," *Cincinnati Lancet and Observer* 9 (1866): 713; B. C. Hirst, "Six Cases of Puerperal Insanity," *Journal of the American Medical Association* 12 (1889): 29; F. C. Fernald, "Puerperal Insanity," *American Journal of Obstetrics* 20 (1887): 714; George Byrd Harrison, "Puerperal Insanity," *American Journal of Obstetrics* 30 (1894): 530. Arthur C. Jelly, "Puerperal Insanity," *Boston Medical and Surgical Journal* 144 (1901): 271, is the only medical writer I found claiming puerperal insanity was not common.

15. Hirst, "Six Cases of Puerperal Insanity," 29; Landfear, "Puerperal Insanity," 57; Anna Burnet, "Puerperal Insanity: Cause, Symptoms, and Treatment," *Woman's Medical Journal* 9 (1899): 269; "Abstracts," *American Journal of Obstetrics* 13 (1880): 641; T. W. Fisher, "Two Cases of Puerperal Insanity," *Boston Medical and Surgical Journal* 79 (1869): 233–34; Charles E. Ware, "A Case of Puerperal Mania," *American Journal of Medical Sciences* 26 (1853): 346; Churchill, "On the Mental Disorders of Pregnancy and Childbed," 309.

16. On the generally good prognosis expected of puerperal insanity patients see: Stearns, "The Psychiatric Aspects of Pregnancy," 24; William Mercer Sprigg, "Puerperal Insanity: Prognosis and Treatment," *American Journal of Obstetrics* 30 (1894): 537. On home versus hospital treatment see: Jelly, "Puerperal Insanity," 275; Wigginton, "Puerperal Insanity," 43; Landfear, "Puerperal Insanity," 59; W. W. Godding, "Puerperal Insanity," *Boston Medical and Surgical Journal* 91 (1874): 318–19.

17. On bleeding as a treatment for puerperal insanity, see: Sprigg, "Puerperal Insanity: Prognosis and Treatment," 540; Landfear, "Puerperal Insanity," 58; Churchill, "On the Mental Disorders of Pregnancy and Childbed," 315; J. A. Wright, "Puerperal Insanity," *Cincinnati Lancet-Clinic* 23 (1889): 651; Victor H. Coffman, "Puerperal Mania," *Nebraska State Medical Association Proceedings* 4 (1872): 18. On purging as a treatment see Haughton, "Puerperal Mania," 729; A. Bryant Clarke, "On the Treatment of Puerperal Mania by Veratrum Viride," *Boston Medical and Surgical Journal* 59 (1859): 237–39; J. MacDonald, "Observations on Puerperal Mania," *New York Medical Journal* 1 and 2 (1831): 279; Thomas Lightfood, "Pu-

erperal Mania; Its Nature and Treatment," *Medical Times and Gazette* 21 (1850): 274. Many physicians recommended that the patient be removed from family and friends; for example, see: J. A. Reagan, "Puerperal Insanity," *Charlotte Medical Journal* 14 (1899): 309; Fernaid, "Puerperal Insanity," 720; J. Thompson Dickson, "A Contribution to the Study of the So-Called Puerperal Insanity," *Journal of Mental Sciences* 16 (1870–1871): 390; C. S. May, "Puerperal Insanity, with Statistics Regarding Sixteen Cases," *Proceedings of the Connecticut State Medical Society* 85 (1877): 106; Carr, "Puerperal Insanity," 540; Wigginton, "Puerperal Insanity," 44; Landfear, "Puerperal Insanity," 59. Nearly every physician who wrote of treatment recommended rest, food, and some form of sedation. For examples not yet cited see: W. S. Armstrong, "Case of Puerperal Mania," *Atlanta Medical and Surgical Journal* 8 (1867): 419; Thomas H. Mayo, "Puerperal Mania," *Southern Medical Record* 4 (1874): 84; J. P. Reynolds, "Puerperal Mania," *Boston Medical and Surgical Journal* 72 (1865): 281; W. A. McPheeters, "Forceps; Puerperal Mania," *New Orleans Medical and Surgical Journal* 16 (1859): 660; Edward Kane, "Puerperal Insanity," *Medical Independent and Monthly Review of Medicine and Surgery* 2 (1856): 156; Horace Palmer, "A Case of Puerperal Mania," *Cincinnati Lancet and Observer* 2 (1859): 5; Horatio Storer, "Puerperal Mania; Recovery," *Boston Medical and Surgical Journal* 55 (1856–1857): 20.

18. See for example the explanation of Elaine Showalter in *The Female Malady,* especially 57–59, 71–72.

19. Wigginton, "Puerperal Insanity," 43.

20. For example, see: W. L. Worcester, "Is Puerperal Insanity a Distinct Clinical Form?" *American Journal of Insanity* 47 (1890–1891): 56; Edward J. III, "A Clinical Contribution to Gynecology," *American Journal of Obstetrics* 16 (1883): 264; Rohe, "Lactational Insanity," 325–26.

21. Allan McLane Hamilton, "Two Cases of Peculiar Mental Trouble Following the Puerperal State," *Boston Medical and Surgical Journal* 94 (1896): 680.

22. Lambert Ott, "Puerperal Mania," *Clinical News* 1 (1880): 337; W. P. Manton, "Puerperal Hysteria (Insanity?)," *Journal of the American Medical Association* 19 (1892): 61; W. I. Richardson, "Puerperal Septicaemia: Puerperal Mania," *Boston Medical and Surgical Journal* 102 (1880): 448. See also Carr, "Puerperal Insanity," 538; Fernald, "Puerperal Insanity," 717; W. D. Haines, "Insanity in the Puerperal State," *Cincinnati Lancet-Clinic* 23 (1889): 371; Worcester, "Is Puerperal Insanity a Distinct Clinical Form," 55; Burnet, "Puerperal Insanity: Cause, Symptoms and Treatment," 267–69. This is only a sampling of the articles listing these symptoms.

23. Harrison, "Puerperal Insanity," 532; Haines, "Insanity in the Puerperal State," 371; Godding, "Puerperal Insanity," 317; V. H. Taliaferro, "Puerperal Insanity," *Atlanta Medical and Surgical Journal* 15 (1877): 324; Carr, "Puerperal Insanity," 538; Stearns, "The Psychiatric Aspects of Pregnancy," 25. Almost every physician describing symptoms of puerperal insanity listed obscenity. For example see Stoddart, "A Clinical Lecture," 242; Reagan, "Puerperal Insanity," 308; Menzies, "Puerperal Insanity," 169; Rohe, "Lactational Insanity," 325; Worcester, "Is Puerperal Insanity a Distinct Clinical Form?" 56; May, "Puerperal Insanity," 106; Wigginton, "Puerperal Insanity," 43; MacDonald, "Observations on Puerperal Mania," 268–70; Clarke, "On the Treatment of Puerperal Mania by Veratrum Viride," 238.

24. George H. Rohe, "Some Causes of Insanity in Women," *American Journal of Obstetrics* 34 (1896): 802. See also John Young Brown, "Pelvic Disease in Its Relationship to Insanity in Women," *American Journal of Obstetrics* 30 (1894): 360; Montrose A. Pallen, "Some Suggestions with Regard to the Insanities of Females," *American Journal of Obstetrics* 10 (1877): 207. Carroll Smith-Rosenberg, "Puberty to Menopause: The Cycle of Femininity in Nineteenth-Century America," *Feminist Studies* 1 (1973): 58–72, is a discussion of this attitude.

25. P. V. Carlin, "Insanity of Pregnancy," *Denver Medical Times* 3 (1883–1884): 233. See also: Churchill, "On the Mental Disorders of Pregnancy and Childbed," 298; C. P. Lee, "Puerperal Mania," *Kansas Medical Index* 2 (1881): 200.

26. Pallen, "Some Suggestions with Regard to the Insanities of Females," 212.

27. For a clear explanation of this point, especially with reference to gender, see Dwyer, "A Historical Perspective."

28. C. C. Hersman, "The Relationship Between Uterine Disturbance and Some of the Insanities,"

Journal of the American Medical Association 33 (1899): 709; Carr, "Puerperal Insanity," 537. See also: Edward Jarvis, "Causes of Insanity," *Boston Medical and Surgical Journal* 45 (1851): 289–305; Reagan, "Puerperal Insanity," 309; Kane, "Puerperal Insanity," 147.

29. For example, see Fernald, "Puerperal Insanity," 716; Landfear, "Puerperal Insanity," 56; Jelly, "Puerperal Insanity," 271–72; G. H. Rohe, "The Influence of Parturient Lesions of the Uterus and Vagina, in the Causation of Puerperal Insanity," *Journal of the American Medical Association* 19 (1892): 59–60.

30. About heredity as a factor in puerperal insanity see Churchill, "On the Mental Disorders of Pregnancy and Childbed," 305; Fisher, "Two Cases of Puerperal Insanity," 233; Dickson, "A Contribution to the Study of the So-Called Puerperal Insanity," 382; Taliaferro, "Puerperal Insanity," 328; May, "Puerperal Insanity," 106; Landfear, "Puerperal Insanity," 55; Fernald, "Puerperal Insanity," 714; Jelly, "Puerperal Insanity," 272; Reagan, "Puerperal Insanity," 309.

31. This point of view was expressed in many of the articles. See, for example, Carr, "Puerperal Insanity," 537; Jelly, "Puerperal Insanity," 272; Wigginton, "Puerperal Insanity," 40; Fernald, "Puerperal Insanity," 715; Wright, "Puerperal Insanity," 648; Harrison, "Puerperal Insanity," 532; Menzies, "Puerperal Insanity: An Analysis of One Hundred and Forty Consecutive Cases," 162; Churchill, "On the Mental Disorders of Pregnancy and Childbed," 299–300; Kane, "Puerperal Insanity," 152.

32. Horatio Robinson Storer, "The Medical Management of Insane Women," *Boston Medical and Surgical Journal* 71 (1864): 210–18; Horatio R. Storer, "Cases Illustrative of Obstetric Disease—Deductions Concerning Insanity in Women," *Boston Medical and Surgical Journal* 70 (1864): 189–200.

33. Charles A. L. Reed, "The Gynecic Element in Psychiatry—with Suggestions for Asylum Reform," *Buffalo Medical and Surgical Journal* 28 (1888–1889): 571. See also Charles F. Folsom, "The Prevalence and Causes of Insanity; Commitments to Asylums," *Boston Medical and Surgical Journal* 103 (1880): 97–100; J. H. McIntyre, "Disease of the Uterus and Adnexa in Relation to Insanity," *Transactions of the Medical Association of Missouri* (1898), 191–95; Pallen, "Some Suggestions with Regard to the Insanities of Females," 207; W. P. Jones, "Insanity Dependent upon Physical

Disease," *Tennessee State Medical Society Transactions* 47 (1880): 97–104.

34. Fernald, "Puerperal Insanity," 719.

35. In addition to Reed, other gynecologists calling for specialists in insane asylums included Joseph Wiglesworth, "On Uterine Disease and Insanity," *Journal of Mental Sciences* 30 (1884–1885): 509–31; I. S. Stone, "Can the Gynecologist Aid the Alienist in Institutions for the Insane," *Journal of the American Medical Association* 16 (1891): 870–73; Ernest Hall, "The Gynecological Treatment of the Insane in Private Practice," *Pacific Medical Journal* 43 (1900): 241–56; Pallen, "Some Suggestions with Regard to the Insanities of Females."

36. Dwyer notes this trend of attributing women's insanity to problems with the female reproductive system in "The Weaker Vessel," "A Historical Perspective," and in her study of two New York asylums, *Homes for the Mad: Life Inside Two Nineteenth-Century Asylums* (New Brunswick, N. J., 1987).

37. Physicians who described medical and surgical attention to female asylum inmates include H. A. Tomlinson and Mary E. Bassett, "Association of Pelvic Diseases and Insanity in Women, and the Influence of Treatment on the Local Disease upon the Mental Condition," *Journal of the American Medical Association* 33 (1899): 827, 831; W. P. Manton, "The Frequency of Pelvic Disorders in Insane Women," *American Journal of Obstetrics* 39 (1899): 54–57; Eugene G. Carpenter, "Pelvic Disease as a Factor of Cause in Insanity of Females and Surgery as a Factory of Cure," *Journal of the American Medical Association* 35 (1900): 545–51; W. J. Williams, "Nervous and Mental Diseases in Relation to Gynecology," *Transactions of the Eighth Annual Meeting of the Western Surgical and Gynecological Association* (Omaha, December, 1898; published 1899), 49–57; W. O. Henry, "Insanity in Women Associated with Pelvic Diseases," *Annals of Gynecology and Pediatry* 14 (1900–1901): 312–20; W. P. Manton, "Postoperative Insanity, Especially in Women," *Annals of Gynecology and Pediatry* 10 (1896–1897): 714–19; Hall, "The Gynecological Treatment of the Insane in Private Practice."

38. See, for example, Henry, "Insanity in Women Associated with Pelvic Diseases"; Manton, "The Frequency of Pelvic Disorders in Insane Women"; Hall, "The Gynecological Treatment of the Insane in Pri-

vate Practice"; Rohe, "The Influence of Parturient Lesions of the Uterus and Vagina, in the Causation of Puerperal Insanity." I do not mean to imply that all doctors were treating puerperal insanity with surgery. "Rest and restoration" was probably the most popular therapy throughout the century. See Hardy, "Puerperal Insanity."

39. Tomlinson and Bassett, "Association of Pelvic Diseases and Insanity in Women," 827. See also Brown, "Pelvic Disease in Its Relationship to Insanity in Women"; Carpenter, "Pelvic Disease as a Factor of Cause in Insanity of Females and Surgery as a Factor of Cure"; and Williams, "Nervous and Mental Diseases in Relation to Gynecology" for examples of the argument that doctors should only resort to surgery when there is physical disease.

40. Stone, "Can the Gynecologist Aid the Alienist in Institutions for the Insane"; Manton, "Postoperative Insanity, Especially in Women"; Clara Barrus, "Gynecological Disorders and Their Relation to Insanity," *American Journal of Insanity* 51 (1894–1895): 475–91; Alice May Farnham, "Uterine Disease as a Factor in the Production of Insanity," *Alienist and Neurologist* 8 (1887): 532–47; Adolf Meyer, "On the Diseases of Women as a Cause of Insanity in the Light of Observations in Sixty-Nine Autopsies," *Transactions of the Illinois State Medical Society* (1895): 299–311; C. B. Burr, "The Relation of Gynaecology to Psychiatry," *Transactions of the Michigan Medical Society* 18 (1894): 458–64, 478–87. An excellent article about removal of the ovaries in the nineteenth century is Lawrence D. Longo, "The Rise and Fall of Battey's Operation: A Fashion in Surgery," *Bulletin of the History of Medicine* 53 (1979): 244–67.

41. See, for example, Helene Kuhlmann, "A Few Cases of Interest in Gynecology in Relation to Insanity," *State Hospital Bulletin* 1 (New York: 1896): 172–79; Mary D. Jones, "Insanity, Its Causes: Is there in Woman a Correlation of the Sexual Function with Insanity and Crime," *Medical Record* 58 (1900): 925–37; Edward B. Lane, "Puerperal Insanity," *Boston Medical and Surgical Journal* 144 (1901): 606–9; Manton, "Puerperal Hysteria."

42. For example, Jones, "Insanity, Its Causes: Is there in Woman a Correlation of the Sexual Function with Insanity and Crime"; Kuhlmann, "A Few Cases of Interest in Gynecology in Relation to Insanity"; Burnet, "Puerperal Insanity: Cause, Symptoms, and Treatment"; and the work of Mary Putnam Jacobi. An important study of women physicians working in asylums is Constance M. McGovern, "Doctors or Ladies? Women Physicians in Psychiatric Institutions, 1872–1900," *Bulletin of the History of Medicine* 55 (1981): 88–107.

43. A creative use of case studies to describe the doctor-patient relationship and the doctor-family relationship is Ellen Dwyer, "The Burden of Illness: Families and Epilepsy," paper read at the annual meeting of the American Association for the History of Medicine, Birmingham, April 1989.

44. McPheeters, "Forceps: Puerperal Mania"; Ware, "A Case of Puerperal Mania"; Lambert, "Puerperal Mania." Numerous case studies included information about cruelty, illness, illegitimacy, and stillbirth, although often doctors did not connect the situation to the symptoms. For example, W. H. Parish, "Puerperal Insanity," *Transactions of the Obstetrical Society of Philadelphia* 4–7 (1876–1879): 50–54.

45. Fisher, "Two Cases of Puerperal Insanity," 233.

46. Harrison, "Puerperal Insanity," 535.

47. Lee, "Puerperal Mania," 205; Lambert, "Puerperal Mania," 337; Denslow Lewis, "Clinical Lecture on Obstetrics and Gynecology—Mental and Nervous Derangements in Obstetric Practice," *Clinical Review* 11 (1899–1900): 181.

48. This theme of women denying their husbands and/or expressing fear or hatred of their husbands was common in the case studies.

49. Dickson, "A Contribution to the Study of the So-Called Puerperal Insanity," 383.

50. Storer, "Puerperal Mania; Recovery," 20; MacDonald, "Observations on Puerperal Mania," 270.

Health Care Providers: Midwives

Midwifery is the most traditional of women's healing occupations. For millennia women have attended other women as they labored to bring new life into the world. During the colonial period in America midwives, helped by other women neighbors and relatives, typically attended birthing women except those few whose emergency problems led them to call in physicians. Beginning in the 1760s, however, some male physicians entered the practice of normal obstetrics and began joining and ultimately replacing female midwives in the birthing rooms of those urban women who could afford to pay the doctor's higher fees.

Despite the attraction of male accoucheurs for some women, large numbers of American women, no doubt increasing with the rising immigration after 1840, continued to call midwives when their labors began. By 1910 midwives, many of whom were trained in Europe before they came to the United States, still delivered approximately half of America's babies. Charlotte Borst analyzes this period of midwifery through a case study of Wisconsin midwives. She describes the range of practicing midwives, including neighbor women and those who were apprentice trained or school educated, and demonstrates that changes within the profession as well as an expanding medical interest in obstetrics ultimately led in the twentieth century to the diminution of the practice of midwifery in the United States.

Borst emphasizes education, class, geography, gender, and ethnicity as important factors in the practice of midwifery and in its decline. Susan Smith uses the South, where lay midwifery had a long history, to examine the relationship between white public health nurses and African-American lay midwives in Mississippi. She concludes that black midwives were central figures in their communities and provided crucial liaison with public health departments. The work of the white nurses and the black midwives, crossing the boundaries of race and class, led to public health successes throughout the state.

Both historians address the extent and quality of midwife practice, the latter a particularly difficult subject for review given the extant records. While we know that many midwives remained relatively noninterventionist birth attendants, providing psychological and cultural support as much as physical aid, we do not know very much about the lengths to which midwives might have acted to help women suffering difficult labors. Podalic version (turning the fetus in utero) and manual stretching of the cervix were probably common midwife techniques, but to what extent did midwives also administer ergot, potentially a very dangerous drug, or cause perineal tears by trying to manually extract the baby before the cervix was fully dilated? Did midwives feel and respond to the same pressures from the birthing women and their friends to "do something" that physicians reported? Or did their shared gender protect them from this particular demand? As we come to learn more about the experiences and difficulties physicians had in America's birthing rooms, it will be easier to compare their experiences with those of midwives in order to understand more fully the varieties of American childbirth practices and to restore the midwife to her rightful place in the historical record.

22 The Training and Practice of Midwives: A Wisconsin Study

Charlotte G. Borst

On February 1, 1887, Caroline Kueny, thirty-five years old and the mother of five children, began her six-month course at the Milwaukee School of Midwifery. Attending daily lectures and sixteen confinements during the course of her instruction, Kueny was graduated after passing her final examinations to the satisfaction of the school's two physician examiners. She immediately registered with the Milwaukee Health Department and began attending births. Over the next fifteen years she worked as a midwife in Milwaukee and had several more children of her own.[1]

While Caroline Kueny was obtaining her midwifery skills at school, Dora Larson, living in the small rural village of Woodford, Wisconsin, became a midwife in a more traditional way. Thirty-seven years old and the mother of eight children, Larson learned midwifery techniques from assisting her midwife mother, by helping local physicians attending confinement cases, and from her own childbirth experiences.[2] She noted on her 1910 state license application that her midwife credentials included being the

mother of eight children and the grandmother of four.[3]

In 1910, when Kueny and Larson applied for state licenses from the Wisconsin Board of Medical Examiners, many other applicants had experience similar to theirs. Although the women who applied for licenses had become midwives in many ways—demonstrating the problem of defining the "American midwife"—patterns in the manner of their training and in the tempos of their practice are discernible. In addition, Kueny's and Larson's status as mothers and grandmothers typified most of Wisconsin's midwives: almost every midwife practicing in the state had children. Indeed, it seemed that motherhood was almost a prerequisite for the job.

Historical studies of American midwives generally have employed the term *midwife* to mean any nonphysician, usually female, attending a childbirth. The focus of most scholarly studies—the elimination of midwives by a hostile medical profession—explains why many historians have not analyzed the differences among these female practitioners.[4] By evaluating midwives only within the scope of the "midwife question" of the early twentieth century, historians have been forced to see these female practitioners through the eyes of the doctors engaged in the debate. That is, most of their

CHARLOTTE BORST is Associate Professor and Chair of History at St. Louis University, St. Louis, Missouri.

Reprinted from *Bulletin of the History of Medicine* 62 (1988): 606–27, by permission of the publisher. © 1988 by The Johns Hopkins University Press.

sources of information about midwives are articles from medical journals or papers delivered at conferences controlled by physicians.

Frances E. Kobrin's pioneering article suggested that midwifery was the victim of a crisis of professionalism within the medical profession over the role of the new specialty of obstetrics.[5] However, I will argue in this article that American midwifery also faced its *own* professional crisis. Like physicians and nurses, schooled midwives began to replace their untrained or apprentice-trained counterparts at the turn of the twentieth century. Unlike these other practitioners, however, midwives did not become professionalized as they became more educated; instead, they became increasingly subservient to physicians, their autonomy decreased, and, ultimately, they ceased functioning in the United States.

The evolution of midwifery training is not the sole explanation for midwives' professional problems at the beginning of the twentieth century. Most midwives, however they learned their trade, were married women who practiced within the limited confines of a discrete ethnic and class-based community. Thus any study of the demise of midwives in the United States must incorporate an analysis of the geographic, class, and ethnic dimensions of the midwives' practices, a consideration of their gender and marital status, and an examination of the results of changes in their training.

The best way to understand this crisis of professionalization is to examine the cultural context of these midwives' practice by focusing on a particular geographic location. This article examines midwifery in Wisconsin.

Wisconsin was fairly typical in its political stance toward midwives. Beginning informally in Milwaukee in the 1870s and culminating with the 1909 Midwife Registration Act administered by the State Medical Practice Board, physicians in the state helped to determine which midwives could deliver babies and the conditions under which they could work.[6] Thus Wisconsin did not ban these female attendants outright when there were still women who wished to practice, but neither did the state leave the practice of midwifery unregulated.[7]

Sociologists and historians have identified an occupation's movement toward professionalization with the "standardization allowed by a common and clearly defined basis of training. . . . [This training] is, in fact, the main support of a professional subculture."[8] Instruction, Magali Sarfatti Larson points out, historically proceeds from the relative informality of apprenticeships to formal education based on the standardization and the codification of knowledge.[9] While the rise of a profession is tied intimately to the development of formal training, a profession also needs a distinct body of knowledge to claim exclusively. To then achieve a monopoly of practice the profession needs to control both the "production of knowledge and the production of producers," preferably within one institution.[10] In summary, the professional practitioner must have mastered a body of knowledge unique to the field within a formal setting and then have the autonomy to decide when and under what circumstances to apply this knowledge.

By the first decades of the twentieth century, leaders in both medicine and nursing recognized clearly that better training fulfilled their professions' objective need for more knowledge and their subjective desire to improve the status of practitioners. As one historian has written, "The aim and end of reform . . . [was] professionalization."[11] Physician and nursing leaders understood professionalization to mean credentials based on education and licensing standards that would enable practitioners to assert the value to human life of their specialized knowledge and their special skills. At the same time professionalization was a movement to strengthen status and to consolidate authority, which allowed professionals to seek broad social prerogatives.[12]

This paper asks the same questions about midwives that medical historians have posed for physicians and nurses: Did midwife training follow the classic path toward professionalization—from casual learning to apprenticeships to school training? Were midwives able to claim an exclusive body of knowledge? Were these practitioners then able to control their own institutions and to determine who could practice midwifery?

From state and county records I identified 893 midwives practicing in Wisconsin between 1870 and 1920. Defined by the sources of their training, these practitioners may be divided into three groups: (1) the neighborwomen, usually native-born women who learned their techniques from giving birth themselves and from assisting at the deliveries of relatives and a few close friends; (2) the apprentice-trained midwives, typically native-born, who learned their craft from other older midwives or from physicians; (3) school-educated midwives, usually first- or second-generation immigrants, instructed in formal settings through both didactic and practical work. Though all three kinds of midwives could be found practicing in Wisconsin during much of the fifty-year period of my study, there was a distinct evolution from neighbor women to apprentice-trained practitioners to school-trained practitioners.

Neighbor Women

Distinguished from the other midwives by their completely informal training, neighbor women were the most "traditional" practitioners in my study.[13] Their experiences are exemplified by Susan Washburn, a sixty-year-old Millston, Wisconsin, woman who wrote in 1909: "I would state that my [midwifery] education has been limited; has more in careing [sic] for those that are going to be confined in my own neighborhood. I have followed this about 30 years with good results my experience has been much in midwifery: I never lost a case."[14]

Some neighbor women did not even consider themselves "midwives," for besides their own confinements, their birthing experiences were limited to a few births in their family or among their friends. A 1919 Children's Bureau report of rural problems illustrated how blurred the distinction was between merely helping out one's neighbor and assuming a title conveying authority. Reviewing maternity care in rural America, the report revealed that "many country women are cared for entirely by neighbors who may or may not have acquired some skill from experience." Some women, the report added, were seen "by midwives."[15]

Some neighbor women midwives developed more than the basic skills out of sheer necessity—many lived in isolated settlements where doctors or trained midwives were scarce and where even rudimentary skills were better than none. A study of childbearing conditions by the Children's Bureau in Marathon County, Wisconsin, portrayed the extreme isolation farm women faced in many rural areas. Only 5 percent of the county's roads were paved; the rest were dirt roads, impassable during much of the year. Families in this area, including childbearing women, were often forced to rely totally on nearby neighbors.[16]

The neighbor women's skills were probably the most heterogeneous of any of the three midwife groups. However, these women did share several characteristics: their training, which I have described as informal or self-taught, their age, and where they lived.

Thirty-nine of 396 women who registered with the State of Wisconsin after 1909 reported training like Susan Washburn's. These thirty-nine women shared two other characteristics: their mean age of fifty-four was older than the mean ages of other types of midwives,[17] and they were overwhelmingly rural residents—only one neighbor woman who registered with the state lived in Milwaukee, the state's most urban area.[18]

An analysis of the 588 women who registered with the Milwaukee Health Department between 1877 and 1907 reinforced the state license data correlating neighbor-women midwives with rural residence: over the thirty-year period of the Milwaukee registry, only twenty-one women reported having no training.[19]

Characterized by the female-centered networks within which they operated and by their self-acquired knowledge, neighbor-women midwives were also the most autonomous midwives—the conditions under which they practiced permitted them complete freedom. But even in isolated rural areas families turned to these practitioners only when they could not find a trained midwife. It is not surprising therefore that as frontier areas of Wisconsin became more settled and more accessible, trained midwives replaced neighbor women. By the beginning of the twentieth century, this family-based practitioner was found only in limited numbers and in the most rural parts of the state.

Apprentice-Trained Midwives

Throughout American history many midwives, like many physicians, received their training by apprenticing themselves to older and more experienced practitioners. Indeed, in the United States until the last decades of the nineteenth century, almost all trained midwives learned their trade in this way.

Anne Nowak of Lublin (in north-central Wisconsin) reported her training in terms that would be familiar to many medical historians: "I have not taken up the course in any school but have practiced with a professional midwife until I was fully-capable of doing the work alone."[20]

Education at the side of a preceptor midwife was not the only form of apprenticeship. Many nineteenth-century midwives did learn their craft from older, more experienced midwives, but many more female birth attendants apprenticed themselves with local physicians. Indeed, the most striking feature of apprentice-trained midwives in Milwaukee was that virtually all of them were trained by physicians.

Historians who have analyzed physician debates over the "midwife question" of the late nineteenth and early twentieth centuries have emphasized the hostility or at least the indifference most doctors felt toward these female practitioners. My analysis of Wisconsin data, however, indicates that midwives and physicians sometimes worked together in America's birthing rooms.[21]

Of the forty-nine apprentice-trained women who registered in 1909, thirty-five (or 71 percent) reported that their instruction came from physicians. Mary Greeley, a practitioner in western Wisconsin, was typical, declaring, "[I] have been practicing midwifery for 30 years. Usually under the direction of physicians of River Falls, Wis. and Ellsworth, Wis."[22]

The timing of this kind of instruction in the midwife's career varied. Some women received their initial training from physicians and then went out to deliver babies, as Ellen Stiefvator, practicing in northern Wisconsin, related: "I practiced with Dr. Rauls Gadsden [identified by her as an Alabama physician] 30 yrs ago and have practiced ever since: [the] last 18 years in Merrill, Wis."[23]

This cooperation between midwives and physicians in rural areas is not entirely surprising, as the 1919 Children's Bureau survey of childbirth attendants in northern Wisconsin pointed out. In one village, for example, the authors of the survey found that local doctors had ceded all of the obstetrical work to Mrs. M. (a local midwife). The doctors referred their maternity cases willingly, telling the survey authors, "We don't like that kind of work and have always more or less turned it over to [her]."[24]

A close interaction between midwives and doctors was not limited to rural maternity practice, however. Although historian Jane Pacht Brickman has concluded that "medical animus against the midwife focused on the urban practitioner," in Milwaukee many physicians trained midwives.[25] While some of these physicians were European-educated émigrés and may have been continuing a traditional association common to their ethnic groups, other Milwaukee doctors who worked closely with midwives had trained at American medical schools. Furthermore, most of these American physicians were well regarded; they were not marginal practitioners at the fringes of the Milwaukee medical community.[26]

Apprenticeships for these women lasted from several months to a year and ranged from what seemed to be private lectures to a combination of didactic and practical work. Frances Jahnz, for example, had "Studied with Dr. F. S. Wasielewski for (2) two months, [in 1901] then took up ten (10) practical cases of confinement in his presence according to his instructions."[27] Her entire course of instruction, she stated, lasted one year. Mary Browikowski had "received instructions in the art of midwifery for fully five months from Dr. F. M. Hinz, M.D."[28] Dr. Hinz, a homeopathic physician trained in Illinois, offered seven months of "private instruction" to another Milwaukee midwife.[29] Like Dr. Hinz, other Milwaukee physicians lectured to midwives, offering courses ranging in length from six to twelve months.[30] One physician continued his mentor relationship with his pupil midwife after training her. Mary Holub received "practical training under Dr. Andrew Munro, attended confinement cases with him and studied under

him. Have certificate from him and recommendation to practice. Attended cases with him for a period of seven years."[31]

Apprentice training, however, was not a major route to the profession for Milwaukee midwives. Overall, of the 588 women who registered with the Milwaukee Health Department between 1877 and 1907, only fifty-one (8.6 percent) reported an apprentice education. Analyzed year by year, the Health Department data also reveal the declining popularity of apprenticeships, from 15.7 percent of the registrants in the early years to 6.6 percent in the last years of the registry. The state licenses, covering a somewhat later period (1909–15), also demonstrate the decline in the twentieth century in apprentice training, especially for urban midwives: of 189 Milwaukee midwives who registered with the state, only twelve (6.3 percent) were apprentice educated.

School-Trained Midwives

Though many European school-educated midwives were among the large numbers of immigrants flooding Milwaukee and the rest of the state in the late nineteenth century, the most compelling explanation for the decline in apprentice training for midwives in Wisconsin lies with the establishment of two schools for midwives in Milwaukee.

Organized by physicians together with some local midwives, these schools offered Milwaukee and many other Wisconsin women a potentially more rationalized course of instruction in the theory and the practice of midwifery than they could obtain by themselves or in tandem with a physician or experienced midwife. Many Milwaukee midwives received their training at these schools: between 1890 and 1907 over half of the Milwaukee Health Department registrants reported having obtained a local school education.

But did the Milwaukee and other American midwifery schools organized at the end of the nineteenth century offer formal training that resulted from a "codification of knowledge" in midwifery? Did this standardized training become "the main support of a professional subculture" for midwives? A brief examination of these schools' curricula, personnel, and philosophy demonstrates that regardless of the new skills they offered their students, the schools were not, nor were they intended to be, a step in promoting the professionalization of midwifery. American midwifery schools developed from and found their roots in the immigrant community. It was not surprising therefore to find two schools based on the European model in Milwaukee. They were founded in the period of the heaviest immigration from Europe to America. Milwaukee in particular experienced an explosive growth in its population, and much of its growth was related to immigration from the Old World: between 1880 and 1910 the city was tied for first place among all American cities in its percentage of foreign stock. By 1910 the total foreign stock—total foreign-born plus natives with at least one foreign-born parent—was 78.6 percent of the total population. Of the foreign-born, almost half of them came from Germany.[32] Partly as a consequence of this immense immigrant population, the city had a large number of practicing midwives. As a *Milwaukee Sentinel* article pointed out in 1889: "No city in the country has as many practicing midwives as Milwaukee, it is said, at least in proportion to population. There are only a comparatively few more physicians than midwives, there being 171 of the former and 115 of the latter."[33]

European midwives and physicians emigrating to the United States brought their ideas about training with them. They were proud of their education and sometimes quite critical of unlicensed, unskilled American midwives. As

Agnes Pradzinska, a Milwaukee midwife, wrote to the *American Midwife*:

> I wish to express my concern that the so-called butcher-women be prohibited. I have already complained many times to the local health department, but I always receive the same answer: "We will look into the matter." But they are [still] looking into it. The births should all be registered in the state of Wisconsin, such female bunglers don't do it though. Among [female bunglers] is one, who calls herself a midwife and has hung out a shield with the inscription "midwife and birth assistant." But she can neither read nor write . . . [She] also knows fortune-telling, because when she performs a delivery, she tells the mother how many more children she will have; even better yet is that she also knows how many boys and girls she will have.[34]

The Milwaukee School of Midwifery, founded in 1879 by a German-born and -trained midwife, and the Wisconsin College of Midwifery, established in 1885 by a German immigrant woman trained in America, shared many of the characteristics of other American midwifery schools at the turn of the century. Established by immigrant midwives in conjunction with immigrant physicians and their sympathetic American colleagues, both schools drew their students from Milwaukee's foreign-born community. Like many American midwifery schools, the Milwaukee School of Midwifery offered its course in German and in English. Both schools, like their American and European counterparts, relied heavily on the prestige of local, usually young, physicians for some instruction and for the approval of the final examination.[35]

Finally, both institutions, established as profit-making ventures, encountered many of the problems that plagued proprietary medical schools of the era: poor facilities, a dearth of instructional materials, and coursework consisting of didactic lectures with limited clinical instruction. For example, a graduate of the Milwaukee School of Midwifery described her training as a course of "6 months—about 6 lectures per month . . . [I] assisted Precepteress Mrs. Wilhelmine Stein at her hospital as well as at [her] private confinement cases."[36] Mrs. Stein's hospital was in her home, having probably fewer than ten beds. The Wisconsin College of Midwifery offered a seemingly more rigorous course. Two graduates reported having taken a five-month course, with lectures in anatomy, biology, and physiology daily for two hours.[37] Students at this college gained practical experience by attending obstetrical clinics at least twice weekly, though many graduates reported that much of their entire four- to six-month term was spent working at the hospital. By 1912 the school's term had expanded to eight months, and students attended at least twenty-five confinements.[38]

These midwifery schools, despite their faults, probably offered more clinical instruction in obstetrics than many of the medical schools surveyed by J. Whitridge Williams in 1912.[39] However, both institutions suffered from a problem unique to the field of midwifery. Unlike nursing or medicine, midwifery could claim neither a distinct body of knowledge nor control over the production of its own practitioners. Milwaukee's two schools, like other midwifery schools in America and in Europe, needed physician support to establish and maintain respectability.

One way to maintain physician support was to downplay midwives' roles as teachers. For example, advertisements for the Wisconsin College of Midwifery in Milwaukee's city directory show how Mary Klaes's central role as founder and teacher was subsumed by her need to attract

doctors to her school. Listing the physician president's name first and in bold print, the text of the advertisement urged other doctors to bring their cases to the school's hospital, with the promise that "trained nurses" were "always in attendance."[40] By 1901 the school referred to itself as "St. Mary's Sanitarium," having the "best physicians in charge." Though midwives were still being trained in the hospital at that date, the city directory advertisement made no mention of the fact.[41]

The problem the Milwaukee schools experienced in holding on to their "best physicians" was not unusual and reflected a lack of autonomy by midwives at even the most prestigious institutions. The Bellevue Hospital School in New York City, for example, generally considered the best American midwifery school, denied any teaching role to midwives. Local doctors taught all of the courses.

The midwifery schools' curriculum reinforced and institutionalized their students' dependent role by emphasizing that good midwives should attend only normal uncomplicated births. A physician associated with the Bellevue School explained this philosophy:

> Not the least advantage of our primitive attempt to educate the midwife at the Bellevue School is the thorough teaching of each candidate for graduation her limitations. . . . One important fact is instilled into the brain of each midwife, and that is the knowledge of her own limitations— the knowledge of what not to do, and when to seek the aid of a practising physician.[42]

In another study by American physicians of midwifery schools in Scandinavia, the writer noted approvingly that the students "are taught to revere the physician, and they are distinctly shown their limitations."[43] Although Milwaukee physicians left no written records of their feelings about the midwifery schools they were associated with, it is reasonable to assume, given their standing in the medical community, that the physicians instructed their students the same way the instructors at the Bellevue and Scandinavian schools did theirs.

Midwifery in the United States in the late nineteenth century faced a "crisis of professionalism" that related directly to the natural evolution of the occupation. At each stage of the developmental process midwives' practice became less autonomous and more circumscribed. But a limitation on midwife practice that did not diminish as the period unfolded related directly to the nature of the practitioner herself: midwives in Wisconsin, as midwives everywhere else, were almost always mature married women who had children themselves. Most midwives therefore were limited by family responsibilities to practicing part time.

Like many other working women in their community, midwives left their homes in order to earn money to help support their families. Their workforce participation as married women placed them in a unique category of working women, however. Historians studying women's work have emphasized the rarity of married women participating in the workforce in the late nineteenth and early twentieth centuries. Leslie Woodcock Tentler found that only 5.6 percent of all married women were reported at work in 1900, though she admits that the census data probably underenumerated wives working part time in the home.[44] Married women who did work outside the home, she points out, chose jobs with relatively short or flexible hours or part-time employment, and they often tried to stay within the immediate neighborhood.[45]

Eugene Declercq and Neal Devitt have shown that early twentieth-century midwives were predominantly mature women of diverse ethnic backgrounds.[46] However, neither of these historians has considered how these women's demographic characteristics might have influenced the practice of midwifery. In my investigation of some of Wisconsin's midwives, I analyzed the effects of age, marital status, immigrant background, and social class on the occupation and professionalization of midwifery in America.

Women training for or practicing midwifery in Wisconsin were usually immigrant, working-class, married women. Midwives' age, marital status, ethnicity, and social class, and the number and the ages of their children, conceivably influenced the patterns of their practice. In order systematically to study midwives as working women, I linked practitioners who identified themselves as midwives to the descriptions of their families in the federal or state manuscript census schedules.[47] From four Wisconsin counties that were chosen to sample an urban-to-rural-to-frontier continuum in a fifty-year period (1870 to 1920), 893 different midwives were identified.[48] Of these practitioners, 398 were then linked to the 1870, 1880, 1900, or 1910 federal census, or to the 1905 state census (see Table 22.1). The census data were then analyzed to determine basic information about midwife families and about the practitioners themselves.[49]

To couple the data describing midwives with information about their patterns of practice, I sampled birth certificates in the same four counties every five years between 1870 and 1920.[50] The data sets contain 28,924 cases for all four counties and provided information on the patterns of practice of 893 midwives and 1,149 physicians.[51]

Neal Devitt's assertion that midwives were predominantly elderly women and that "advanced age" played a role in the demise of midwifery fails to acknowledge the fact that midwifery has always been an occupation for mature married women. As historian Jane Do-

Table 22.1
Distribution by county and by census
of 398 Wisconsin midwives

County of residence	Number	Year of census	Number
Dane	33	1870 federal	6
Milwaukee	301	1880 federal	77
Trempealeau	26	1900 federal	191
Price	38	1905 state	38
		1910 federal	86

Sources: U.S. Census Office, 9th, 10th and 12th Censuses, 1870, 1880, 1900, *Population Schedules: Wisconsin* (Washington, D.C.: Census Office, 1870, n.d., 1901); U.S. Bureau of the Census, 13th Census, 1910, *Population Schedules: Wisconsin* (Washington, D.C.: Government Printing Office, 1913); National Archives microfilm publications. Wisconsin Dept. of State, *Wisconsin State Census, 1905: Population Schedules,* Wisconsin Secretary of State's Office (Madison: State Historical Society of Wisconsin, n.d.). Microfilm.

negan relates, "The acceptable midwife was the respectable older woman who carried herself correctly."[52] In fact, young single women often were not allowed to become midwives. Dr. Marie Zakrzewska, the famous director of the New England Hospital for Women and Children, trained as a midwife in one of the most prestigious schools in Berlin. Even with influential support, however, her first application for admission was refused specifically because she was not married and because, at age eighteen, she was considered too young to be a midwife.[53]

Wisconsin's midwives, on the whole, were middle-aged but not elderly women. Two data sets provided information on the ages of these practitioners. The mean age of the women who registered with the state after 1909 was 48.1 years. Many women from this group were even younger—the numerous school-educated practitioners were a significantly more youthful group.[54] The census study, which identified practitioners earlier in their careers, confirmed the findings in the state file. In the census study, the 398 midwives had a mean age of 43.9, almost five years younger than the average age of the state licentiates.

The census study provided crucial information on marital status. This data emphasized the close relationship between the occupation of midwifery and the attendant's family situation. Of the 394 women whose marital status could be determined, 316 (80.2 percent) were married women living with their husbands. Another fifty-nine women (15.0 percent) were widowed, and eleven midwives were divorcées. Thus only eight women (2.0 percent) were single. In Wisconsin, at least, midwifery was most definitely practiced by married women.

Like the midwives in Declercq's Lawrence, Massachusetts, study, almost all of the midwives identified in Wisconsin were literate. Of the 359 midwives in the Census study for whom literacy could be ascertained, 343 (95.5 percent) were described as being able to read or write (not always in English). Only sixteen women in this sample were not literate.[55]

Midwifery, like the prevalent nineteenth-century custom of keeping boarders and lodgers, was ideally suited in many ways as an occupation for mature married women. Though midwives worked outside of their own homes, they delivered babies in other women's homes, remaining protected from the shop floor and the effects of working there, popularly accepted as deleterious. Like women who kept boarders, midwives could adjust their job schedule to meet the needs of their families, taking fewer cases if they needed the time at home, and taking more cases if they needed the extra money. Indeed, when viewed in the context of home-based wage earning, midwifery may have replaced keeping boarders. Only thirty-one midwife families in the Census study (7.8 percent) also had boarders present. Instead, most midwife families were nuclear—287 midwives (72.1 percent) lived solely with their husbands and children.[56]

Many Wisconsin midwives were responsible for their own minor children even as they were on call to deliver other women's babies. Of the midwives in the census study, 323 (81.2 percent) had at least one child living at home with them, and most women had more. The mean number of children was three. The range was quite narrow, however, as most practitioners (77 percent) had four or fewer children at home.[57] Married, divorced, and widowed women did not have significantly different numbers of children.[58]

These children were not preschoolers. Midwives, unlike other mothers working outside of the home, did not work until their children were of school age. Most working-class mothers who earned money outside of the home, on the other hand, *stopped* working when their oldest child reached the age of nine or ten. Thus the mean age of midwives' oldest children, sixteen, and the mean age of the youngest, 8.1, deviated considerably from Lynn Y. Weiner's findings for children of working mothers.[59]

To measure the effect of the number and the ages of midwives' children on the midwives' patterns of practice, data from the vital statistics were aggregated and merged with data from the census study. In a multivariant analysis of 274 midwives who were found in both the census study and in the birth records, no relationship was found between the number of babies a midwife delivered and the number of children she herself had, the age of her oldest child, or the age of her youngest child. A very weak relationship did exist between the size of a midwife's practice and her age.[60]

Like the age distribution of midwives' children, the distribution of midwives' husbands' occupations did not fit the usual pattern for married women workers. The census sample provided data on 314 husbands of midwives. Unlike the husbands of other married women wage earners, who were employed in very low status jobs, most husbands of midwives were skilled artisans (41 percent, n = 131). The next largest occupational groups were much smaller: 17.2 percent (54) of spouses of midwives

worked on farms or in the lumber trade, and 16.9 percent (53) worked as laborers. Most midwives therefore were the wives of men who earned a comfortable working-class income.[61]

The data on patterns of practice provide no direct evidence of a midwife's economic incentive. When data on the size and the pace of practice were linked to the midwives' demographic characteristics found in the census study, tests of significance showed no connection between the total number of births a midwife attended and her husband's occupation. Thus poorer practitioners were not more active than those who might be better off. Midwives married to day laborers were not significantly more active practitioners than midwives married to men having more secure occupations like carpentry or clerking.

While midwife families may not have been as desperately poor as the families of women working in factories, the status of their homes demonstrates their precarious financial position and provides some evidence of a midwife's economic incentive. Of the 320 midwife families whose housing status could be verified, 47 percent rented their homes (151 families), 29 percent held mortgages (92), and only 24 percent (77) owned their homes outright. As Stephan Thernstrom's work shows, home ownership in the nineteenth century was the result of a very slow and expensive process.[62] About half of the midwife families in my study were involved in the process of home ownership, a figure consistent with Thernstrom's findings for working-class families in Newburyport, Massachusetts.[63]

Although a simple analysis of the size of a midwife's practice showed no statistical connection with the status of her family's house,[64] a more complex examination of home ownership by midwife artisan and laborer families revealed one potential use of a midwife's income. One would perhaps expect artisan families predominantly to own their homes and laborers mostly to rent, but in actuality many artisan families

Table 22.2
Home ownership in artisan and laborer midwife families

Occupation	Renters	Mortgage holders	Owners
Artisan			
Number	61.0	34.0	10.0
Expected number	47.1	31.8	25.2
Laborer			
Number	13.0	13.0	13.0
Expected number	18.4	12.4	9.8

Sources: U.S. Census Office, 9th, 10th and 12th censuses, 1870, 1880, 1900, *Population Schedules: Wisconsin* (Washington, D.C.: Census Office, 1870, n.d., 1901); U.S. Bureau of the Census, 13th census, 1910, *Population Schedules: Wisconsin* (Washington, D.C.: Government Printing Office, 1913) National Archives microfilm publications. Wisconsin Dept. of State, *Wisconsin State Census, 1905: Population Schedules,* Wisconsin Secretary of State's Office (Madison: State Historical Society of Wisconsin, n.d.). Microfilm.

Note: A chi-square analysis of midwives' husbands' occupations by home status was significant at the 0.0001 level. The artisan and laborer groups provided most of the deviations from the expected numbers.

rented and not many of them owned houses outright. On the other hand, fewer laborer families rented and more of them had mortgaged or fully owned homes than would be expected (see Table 22.2). These surprising findings for artisan midwife families indicate that some of them lived near the edge of economic hardship. Thus for those families who did not own their homes, the practice of midwifery may have been an important secondary source of income or a supplement to be saved for a house.

While the Wisconsin data provided no clear-cut economic rationale for midwifery practice, they demonstrated strongly the relationship between midwives' immigrant status and their chosen occupation. Most of the midwives in the Wisconsin census sample were first-generation immigrants; all but ten had parents born in Europe. The majority of these women, not surprisingly, came from Germany (52 percent). Indeed, the German-speaking midwives, including those born in Germany, the Austro-Hungarian empire, and Switzerland, constituted 72 percent of all of the foreign-born women in the sample.

The next largest groups of foreign-born attendants were Poles (13.3 percent) and Scandinavians (12.7 percent). An analysis of the 274 midwives who appeared both in the census study and as birth attendants in the vital record study strengthened the evidence for a link between an immigrant background and the practice of midwifery. The 166 German and Swiss midwives delivered an average of twenty-eight babies each, both the Austrian and the Polish practitioners averaged twenty-two births, and the American-born midwives attended an average of only 11.8 births. Scandinavian-born birth attendants, living mostly in rural counties, were called to a mean of only 7.6 births, the smallest average number of deliveries of any group (see Table 22.3).

Historians of midwifery in the United States have long noted the large percentage of midwives who were immigrant women. Frances Ko-

Table 22.3
Distribution of births by ethnic group

Group	Mean number of deliveries	95 Percent confidence levels
German	28.01	23.11 to 32.91
Swiss	28.01	23.11 to 32.91
Scandinavian	7.61	2.37 to 12.85
Austrian-Hungarian	22.05	8.47 to 35.62
Polish-Russian	22.45	14.45 to 30.45
American	11.83	2.05 to 21.62

Sources: Milwaukee, Price, Trempealeau, Dane Counties [Wisconsin], *Births,* Vital Records Office, Wisconsin Dept. of Health and Human Services, Madison. Also U.S. Census Office, 9th, 10th and 12th censuses, 1870, 1880, 1900. *Population Schedules: Wisconsin* (Washington, D.C.: Census Office, 1870, n.d., 1901); U.S. Bureau of the Census, 13th census, 1910, *Population Schedules: Wisconsin* (Washington, D.C.: Government Printing Office, 1913), National Archives microfilm publications; Wisconsin Dept. of State, *Wisconsin State Census, 1905: Population Schedules,* Wisconsin Secretary of State's Office (Madison: State Historical Society of Wisconsin, n.d.). Microfilm.

Note: A one-way analysis of variance was very significant when attendants' total number of deliveries was tested with their own birthplace. The f-ratio was 2.86, with a probability of 0.0103. The German and Swiss groups also showed the least variability around the mean as measured by the 95 percent confidence levels about the mean. Other groups had much wider-ranging confidence levels.

brin describes turn-of-the-century midwives as being employed primarily by the foreign-born and their children.[65] Jane Pacht Brickman writes, "Wherever immigrants settled, the midwife flourished . . . The midwife, almost always foreign-born and living in the community, lay in the buffer that immigrant groups maintained against an already overwhelming cultural shock."[66] Almost none of the midwives Eugene Declercq found practicing in early twentieth-century Lawrence, Massachusetts, were born in the United States: "Like the population they served, they came predominantly from Southern and Eastern Europe."[67]

Contemporary observers of these birth attendants documented their prevalence and popularity in immigrant communities all over the United States. As Michael Davis noted in his study of immigrant health:

> The immigrant mother has rarely been accustomed to a man doctor at the time of confinement. She and her friends have used the midwife, who, in most European countries, is a woman of some standing, trained, and in many countries, carefully supervised. The midwife is the most important single element in the general question of the care of immigrant mothers.[68]

Grace Abbott's 1917 work on immigrant communities in Chicago also recounted immigrant women's preference for midwives at their lyings-in,[69] and Joseph B. DeLee, a virulent opponent of midwives, noted approvingly in 1915 that midwives seemed to be "slowly disappearing in America" and were only found in "crowded communities of foreigners."[70]

"Communities of foreigners" were not restricted to crowded urban areas, however. A 1924 study of Minnesota midwives found that only 13.7 percent of them were born in America; 86.3 percent (101 of 117 midwives found by the

survey) were foreign born. Twenty-five percent of the practitioners in this sample were from Scandinavian countries; 17.1 percent were German.[71] The Children's Bureau study of maternity conditions in Marathon County, Wisconsin, found that the large number of German and Polish settlers clung "to their foreign customs and habits of thought and to a certain extent their languages, making the district as a whole distinctly foreign in its atmosphere."[72] Midwife attendance at childbirth was one of the "foreign customs" these women clung to. Twenty-four midwives, all members of the German or Polish communities, practiced in the county, though only a few were reported to be earning their living as midwives. Most female attendants "went out" as an accommodation to their neighbors. Of all of the midwives in the region, only six practitioners delivered more than ten babies per year, and only two oversaw the births of more than thirty babies.[73]

As the Children's Bureau study suggests, midwives practiced not only in cities but also in rural areas where there were significant communities of immigrants. In my census study of Wisconsin midwives, American-born midwives were a distinct minority in each of the four counties. Tests of significance showed distinct ethnic groupings in each county, with Scandinavians predominantly located in the three rural areas and Polish and Russian midwives exclusively based in Milwaukee. Very few American-born midwives practiced anywhere in Wisconsin. In Milwaukee County only 7.6 percent of 301 birth attendants were born in America, and the numbers of American-born midwives in the other counties were so small that statistical analysis was not feasible.[74] In Milwaukee the overwhelming majority of practitioners, 185 (61.5 percent), had emigrated from Germany. In Dane County, German-born midwives formed the largest group (n = 15, 45.5 percent), but Scandinavian attendants were almost as numerous (n = 10, 30.3 percent). In the other two rural counties in the study Scandinavian women were the dominant ethnic group. In Trempealeau County 65.4 percent (n = 17) of the attendants were Scandinavian, and in Price County 34.2 percent (n = 13) were born in Norway, Sweden, or Denmark. Five midwives in Price County were natives of Finland.

Summarizing the practice of midwifery in the early twentieth century on the lower East Side of New York City, historian Elizabeth Ewen described it as an occupation that was "an acceptable means for a married woman to contribute to her family economy."[75] This article has demonstrated that midwifery certainly was an occupation practiced by mature married women. But the evidence from the analysis of Wisconsin's midwives also demonstrates that these women were not motivated solely by economic necessity. Though many Wisconsin practitioners undoubtedly contributed significant amounts of money to their family's economy, they practiced midwifery because it was an acceptable occupation in the *immigrant* community. Both historians and contemporary observers of European settlers in the United States found that these settlers had high regard for midwives. Midwives respected birth traditions of immigrant families and, as female birth attendants, they did not offend religious families' sense of moral propriety.[76]

Midwifery, unlike nursing or medicine, was practiced according to established customs. Women could adapt the size of their practice to the needs of their families and themselves. Unlike other wage-earning women, midwives did not have to punch a time clock or listen to a boss. However, despite having some very highly developed skills and the autonomy to set their own hours, midwives had more in common with housewives who kept boarders and lodgers than they did with nurses and female physicians. In these two branches of the healing arts, prac-

titioners worked full time, accepting people from many diverse classes and ethnic groups as patients. Midwives, despite their education, practiced part time, birthing babies in their own immigrant neighborhoods. While the trained midwife may have delivered more babies and attended births on a more regular basis than the neighbor woman, midwifery for both kinds of female practitioner was an occupation that was fitted in around the needs of the family. Given this priority within the context of women's lives, midwifery could not develop into a profession.

The demise of American midwives can best be understood through an analysis of the practitioners themselves. Many historians have emphasized the physician debates concerning midwives at the beginning of the twentieth century. This work on Wisconsin's midwives adds a new dimension to these authors' analyses, one that is important to complete the story: that dimension is an understanding of the process of change within this quintessentially female occupation. Midwives, I have argued, faced the same problems as nurses and physicians at the turn of the century: practitioners needed to build a professional ethos based on standardized education, licensure, and self-governance. Nursing and medical leaders of the early twentieth century gave direction to and helped shape their professions' goals. But there were no positions as leaders for midwives to assume. The presumptive leaders—trained midwives—relied on others to decide their qualifications and to license them for practice. In addition, the personal characteristics of the practitioners and the characteristics of their practice prevented the emergence of a united profession.

The demise of midwifery at the beginning of the twentieth century in this country was due to a number of complex and interrelated factors, including, but not limited to, physician opposition to these birth attendants. In an era when Americans increasingly insisted on professionals in many aspects of their lives, the evidence on Wisconsin's midwives shows that midwifery remained a traditional female occupation. Failing to achieve any professional goals or identity, and lacking any outside professional support, midwives faced the inevitable: their replacement in the birthing room by the self-consciously professional obstetrician.

Notes

Revised version of a paper read at the fifty-ninth annual meeting of the American Association for the History of Medicine, Rochester, New York, 1 May 1986. I wish to thank Judith Walzer Leavitt for her patient reading of and generous comments on several drafts of this paper. The Alice E. Smith Fellowship from the Wisconsin State Historical Society, and grants from the University of Wisconsin Graduate School and the Maurice Richardson Fund for the History of Medicine, helped defray the cost of the data entry and computer analysis.

1. Information about Kueny's schooling was found in her midwife license application, Midwife File, State Board of Medical Examiners, Archives Division, Wisconsin State Historical Society, Madison, Wisconsin (hereafter Midwife File). Information concerning her family life came from the 1900 federal manuscript census for Milwaukee.

2. License application of Dora Larson, Midwife File.

3. Ibid.

4. For example, Frances E. Kobrin carefully delineates the characteristics of the various physician groups who were debating the "midwife problem" at the beginning of the twentieth century. However, she only identifies the midwives as sharing "race, nationality, and language with their customers" (351). See Kobrin, "The American Midwife Controversy: a Crisis of Professionalization," *Bulletin of the History of Medicine* 40 (1966): 350–63.

Judy Barrett Litoff's sweeping study of American midwives also focuses primarily on the physician debates of the early twentieth century. While her evidence demonstrates that many varieties of midwives practiced in different areas of the country, she allows no regional distinctions for the "American midwife," grouping together such diverse women as southern black grannies, northern, native-born white neighbor-women, and immigrant midwives trained in the European hospital schools (Litoff, *American Midwives, 1860 to the Present* [Westport, Conn.: Greenwood Press, 1978], 53, 60). Despite the disparate nature of these groups of female practitioners, Litoff argues that there was no clear distinction between the official midwife and the neighbor (27). Litoff acknowledges the difficulties in drawing a composite picture of early twentieth-century midwives, but this problem relates directly to her sources. Because she did not use data directly focusing on the midwives themselves, such as midwife licenses or birth registration certificates, she could not provide detailed information about individual midwives and their practices.

5. Kobrin, "American Midwife Controversy," 350.

6. The Milwaukee Health Department began requiring midwife registration in the 1870s (see "Dr. Wight's Wisdom," *Milwaukee* (Wisconsin) *Sentinel*, July 20, 1878). In 1909 the state of Wisconsin began requiring all midwives to have licenses. Women already in practice were allowed to continue, regardless of their training. But new practitioners were required to have a diploma from a "reputable school of midwifery," defined as being "connected to a reputable hospital or sanitarium, [offering] a twelve month course in the science of midwifery and experience in twelve confinements." Candidates were also required to pass an exam relating to the clinical practice of midwifery (*Wisconsin Statutes*, 1909, ch. 528, s. 1435f-12).

7. In 1894 Massachusetts was the first state to forbid midwifery practice by refusing to license midwives and by recognizing attendance at childbirth as medical practice. Thus midwives delivering babies were subject to prosecution for practicing medicine without a license. Midwives continued to deliver babies in Massachusetts, however, often under the protection of a local physician who would sign the birth certificate. Contemporary critics of this approach to

midwife regulation noted this discrepancy. See J. M. Baldy, "Is the Midwife a Necessity?" *American Journal of Obstetrics* 73 (1916): 399–400. A recent historical study analyzed this phenomenon: Eugene Declercq, "The Nature and Style of Practice of Immigrant Midwives in Early Twentieth-Century Massachusetts," *Journal of Social History* 19 (1985): 113–29. Many southern states completely ignored their midwife practitioners in the early twentieth century. Midwives in these states, mainly black "grannies," did not have to hide their practice, but the state had no records about them.

8. Magali Sarfatti Larson, *The Rise of Professionalism: A Sociological Analysis* (Berkeley and Los Angeles: University of California Press, 1977), 45.

9. Ibid., 44–45. Also Paul Starr, *The Social Transformation of American Medicine* (New York: Basic Books, 1982), 81.

10. Larson, *Rise of Professionalism*, 17.

11. Barbara Melosh, *"The Physician's Hand": Work, Culture, and Conflict in American Nursing* (Philadelphia: Temple University Press, 1982), 4.

12. Ibid., 17. For a longer discussion see Eliot Freidson, *Profession of Medicine: A Study of the Sociology of Applied Knowledge* (New York: Dodd, Mead, 1970).

13. Judith Walzer Leavitt has defined *traditional* childbirth as a birth that remained within a female-centered network. The attendant was a midwife, trained or not (Leavitt, *Brought to Bed: Childbearing in America, 1750–1850* [Oxford and New York: Oxford University Press, 1986], 73–74). As I try to show in this article, the definition of who was a midwife in the nineteenth and early twentieth centuries did not remain static. However, despite this occupation's move toward school training, its practitioners and its philosophy toward birth remained tied to gender and cultural norms that were not part of a "modern" professional outlook. For the best interpretation of the historical meaning of modernization, see Thomas Bender, *Community and Social Change in America* (New Brunswick, N.J.: Rutgers University Press, 1978).

14. License application of Susan Washburn, Midwife File.

15. Elizabeth G. Fox, "Rural Problems," *Chil-dren's Bureau Conference*, Children's Bureau Publication 60 (Washington, D.C.: Government Printing Office, 1919), 187.

16. Florence B. Sherbon and Elizabeth Moore, *Maternity and Infant Care in Two Rural Counties in Wisconsin*, Children's Bureau Publication 46 (Washington, D.C.: Government Printing Office, 1919), 20–22. Many areas of this county were served by minor roads, which the study noted were impassable by automobile and difficult even for wagons. Many parts of the county had no mail delivery, which required that some people send twelve miles for their mail.

17. Compared with an overall mean age of forty-seven for all of the registrants and a mean age of forty-five for the school-trained group.

18. All of the licenses (396) were examined and coded for computer analysis. Using SPSS to analyze the data, t-tests to evaluate differences in mean age were run between the different groups. The t-test comparing the mean age of the school-trained versus no training was highly significant, with a t-value of (-4.56) and a significance level (2-tailed) $= .00001$. To investigate the link between region and type of midwife the fifty-two counties represented by the registrants were grouped into six regions. When these six regions were cross-tabulated with kind of training, the result was highly significant: $X^2 = -102.33$, with 15 df, the significance level $= .00001$. Particular individual cells contributed to this finding: the western and central regions had three times more than the expected number of midwives reporting "no training," and the City of Milwaukee, where twenty untrained midwives would be expected, reported only one.

19. "Physicians' Register," City of Milwaukee Health Department, Archives Section, Milwaukee Public Library. (The title of this registry is a misnomer, for the registry contains registration data on both physicians and midwives.) There were 658 entries between June 1877 and October 1907, but many of the women who registered in 1906 and 1907 had previously registered. The registry, then, provided information on 588 different women.

A more detailed check of the data suggests that the twenty-one women who reported "no training" did not represent a larger contingent of unregistered, infrequent practitioners, perhaps singled out for prosecution. Dividing the registry into five-year periods, I found one or two neighbor-women midwives in Milwaukee signing up with the health department every year.

The figures (from the "Physicians' Register") were as follows:

	1877–79	1880–89	1890–99	1900–1907
Total registered	46	171	227	144
"No training"	6	0	3	12

20. License application of Anne Nowak, Midwife File.

21. Leavitt finds that physicians called to a birth in the nineteenth century were forced to work within the limits imposed by "women's domain of the birthing room," which often included a midwife and the parturient woman's friends (*Brought to Bed*, 100).

22. License application of Mary Greeley, Midwife File.

23. License application of Ellen Stiefvator, Midwife File.

24. Sherbon and Moore, *Maternity and Infant Care*, 32–33.

25. Jane Pacht Brickman, "Public Health, Midwives, and Nurses, 1880–1930," in *Nursing History: New Perspectives, New Possibilities*, ed. Ellen Condliffe Lagemann (New York: Teachers College Press, 1983), 70.

26. License applications for both the City of Milwaukee and the State of Wisconsin required two physician references. Each woman's references were coded, and the names of physicians in selected counties were checked against their own license applications. (Physician Licenses, Board of Medical Examiners, Archives Division, Wisconsin State Historical Society). For physicians' activities within the medical community I consulted Louis F. Frank, *The Medical History of Milwaukee, 1834–1914* (Milwaukee: Germania Publishing, 1915).

27. License application of Frances Jahnz, Midwife File.

28. License application of Mary Browikowski, Midwife File.

29. Frank, *Medical History of Milwaukee*, 255. Hinz practiced in Milwaukee from 1875 to 1903.

30. State license applicants reported four other Milwaukee physicians who offered lectures. Some of these doctors were instrumental later in the several midwifery schools in Milwaukee (License applications of Kondaneygli Sythowski, Helena F. Mueller, Fredericka Wirth, and Mary Dudek, Midwife File).

31. License application of Mary Holub, Midwife File.

32. Roger Simon, "The Expansion of an Industrial City, Milwaukee: 1880–1910" (Ph.D. diss., University of Wisconsin–Madison, 1971), 82.

33. "Figures That Don't Lie," *Milwaukee* (Wisconsin) *Sentinel*, June 23, 1889.

34. Agnes Pradzinska, "Letter to the Editor" in the German-language section of the *American Midwife* 2, 6 (St. Louis, 1896): 40–41. My thanks to Rebecca Bohling for her expert help in translating this letter for me.

35. These schools exerted a statewide influence: the Milwaukee registry listed eighty-four graduates of these schools who gave addresses outside the city, perhaps implying that some students came to Milwaukee specifically for their education. The State License file also showed the wide influence of the schools on Wisconsin's midwives: of the 121 women in the state file who reported their school was based in Milwaukee, thirty gave addresses outside of the city.

36. License application of Sophia Kegel, Midwife File.

37. Licenses of Caroline Fuss and Maria Timm, Midwife File.

38. Two 1913 graduates of the school, Stefonia Staigwillo and Gertrude Turzynski, noted on their state license applications that they had attended the "full course" of the Wisconsin College of Midwifery. The full course was described as an eight-month term from September to May with hospital time and attendance at twenty-five cases (License applications, Midwife File).

39. J. Whitridge Williams, "Medical Education and the Midwife Problem in the United States," *JAMA* 6, 58 (January 1912): 1–7. Williams found the general attitude toward obstetric teaching to be "a very dark spot in our system of medical education" (1).

40. *Wright's Milwaukee City Directory* 26 (1893): 1178, advertisement under "Lying-In Hospitals." Midwives were never mentioned.

41. Ibid., 34 (1901): 1262, advertisement under "Hospitals."

42. J. Clifton Edgar, "Why the Midwife?" *American Journal of Obstetrics* 78 (1918): 249.

43. George W. Kosmak, "Results of supervised midwife practice in certain European countries." *JAMA* 10, 89 (December 1927): 2009.

44. Leslie Woodcock Tentler, *Wage-Earning Women: Industrial Work and Family Life in the United States, 1900–1930* (Oxford and New York: Oxford University Press, 1979), 137.

45. Ibid., 142.

46. Declercq, "Nature and Style," 118, and Neal Devitt, "The Statistical Case for Elimination of the Midwife: Fact Versus Prejudice, 1890–1935" (pt. 1), *Women and Health* 41 (1979): 82.

47. Practitioners were identified from the Milwaukee Health Department licenses, the licenses filed with the state of Wisconsin, advertisements in city directories, and the names of women who signed birth records as the attendant.

48. The four counties were Milwaukee, Dane, Trempealeau, and Price. Milwaukee, of course, was the urban center for my study. Madison, Wisconsin's capital city, is located in Dane County, which nevertheless was largely rural in the late nineteenth and early twentieth centuries. Trempealeau County was chosen both because it was the site of Merle Curti's famous study (*The Making of an American Community: A Case Study of Democracy in a Frontier County,* with the assistance of Robert Daniel et al. [Stanford, Calif.: Stanford University Press 1959]) and because, like Dane County, it was a fine example of a place with a settled rural population. Price County in northern Wisconsin, was a frontier county at the turn of the twentieth century. The major industry in this county was lumbering, but a few farmers had settled on the clear-cut land.

49. Within the four counties, I identified 893 different midwives in the fifty-year period of my study (1870–1920), and I was able to link 398 of these practitioners to the census (44.5 percent). I attempted to link practitioners to the census nearest to the date when they first began to practice, because I wanted to capture the full effects of family life on midwifery practice.

Note that the large number of practitioners linked to the 1900 census reflects the absence of the 1890 manuscript census schedules. The unavailability of this schedule (it was burned in a Washington, D.C., fire near the turn of the century), coupled with the high mobility of midwives' artisan families, was the reason more midwives were not found in the census.

50. To obtain an estimate of the frequency of practice and to maintain a reasonable sample size, every birth was coded for a given three-month period in Milwaukee County, every birth for six months in a sample year for Trempealeau County and Price County after 1905, and every birth for the entire year for Dane County and Price County between 1880 and 1905.

51. The birth data are contained in two data sets. Milwaukee, Price, and Trempealeau Counties were coded together in one set, with a total of 24,302 cases. Dane County data were coded alone, having a total of 4,622 cases. For each case, the following variables were coded: date of birth, including day, month, and year; place of the birth, including county, city, whether home or at a hospital; parents' home address; the occupation of the father; the birthplace of the mother and the father; the mother's parity; the attendant at the birth (if noted, otherwise a code for parents or other was included); the status of the attendant (midwife, physician, or other). Each attendant was given a unique five-digit number that reflected his or her county and the year he or she first appeared in any record of birth attendants. The data were analyzed using various statistical subroutines of SPSS.

52. Jane B. Donegan, "'Safe Delivered,' But by Whom? Midwives and Men Midwives in Early America," in *Women and Health in America: Historical Readings,* ed. Judith Walzer Leavitt (Madison: University of Wisconsin Press, 1984), 304.

53. Agnes C. Vietor, ed., *A Woman's Quest: The Life of Marie F. Zakrzewska, M.D.* (New York: D. Appleton, 1924), 39. Zakrzewska was eighteen years old when she first sought admission to the Midwifery School of Berlin. Finally, by the entreaties of a power-

ful physician friend to the king of Prussia she was admitted to the school (39–42).

54. The mean age of all of the respondents was 48.05, with a standard deviation of 12.37 (Midwife License File, State Board of Medical Examiners, Archives Division, Wisconsin State Historical Society). When age was cross-tabulated with kind of training, the chi-square was 43.88, highly significant at the .00001 level.

55. Although the number of illiterate midwives was small, I tested for significance for birthplace and county. When literacy was cross tabulated with county, the chi-square was significant (15.82, significant at .001). Half of the sixteen illiterate midwives lived in Milwaukee County, three lived in Dane County, four in Trempealeau, and one in Price. The Dane and Trempealeau figures account for the significance, as fewer illiterate women were expected. Birthplace was also significant, with a chi-square of 22.9, significant at 0.0008. The Scandinavian group had more illiterate women than expected. For thirty-nine midwives, literacy was either not reported or unknown.

56. The percentage of families with boarders is significantly low when midwife families are compared with families sharing similar ethnic and class backgrounds and age distributions. Modell and Hareven's study of New England families, for example, found that as high a proportion as 40 percent of immigrant, working-class families where the head of the household and his wife were in their forties had boarders. John Modell and Tamara Hareven, "Urbanization and the malleable household: an examination of boarding and lodging in American families," *Marriage and the Family* 35 (1973): 472. The implications of boarding for women's work in the home are discussed on 473.

57. Information on children was obtained for 387 women. The mean number of children was 2.99, with a standard deviation of 2.41. Number of children ranged from one to thirteen, 77 percent of midwives had four or fewer children, 48 percent had two or fewer.

58. An analysis of variance showed no significant differences between groups. The f-ratio was 99 and the f-probability equalled 37.

59. Lynn Y. Weiner, *From Working Girl to Working Mother: The Female Labor Force in the United States, 1820–1920* (Chapel Hill: University of North Carolina Press, 1985), 85. The median age of oldest children of all midwives was fifteen, with a modal (most frequent number) age of nine. However, the range of ages was quite broad, from one to fifty-four years.

60. In order to analyze the number of births attended by each practitioner or group of practitioners, the births file was sorted and aggregated by county, attendant number, and attendant status. The resulting file had information on 274 midwives who had also been found in the Census data. A new variable "NBABES" captured the total number of babies delivered by each midwife. Matching the midwives in this file to the census data by linking the attendant numbers, I performed a stepwise multiple regression, with number of babies as the dependent variable. The number of children, age of the oldest, age of the youngest, and the midwife's own age were entered as independent variables. Only age of the midwife had an F-value significant enough to enter into the regression equation (f = 7.101, significance = .0082). The resultant R^2 equalled 0.02194, showing age only very weakly explaining the total number of babies delivered.

61. To code for husbands' occupation, I used the categories defined by the federal census of 1940. The categories are roughly hierarchical, except for the two categories in the middle defining farming and mining. To approximate "white-collar" versus "blue-collar" occupations, I grouped the professional managerial, clerical, and sales groups together and the transport, artisan, laborer, and service categories together. Only thirty-nine midwife husbands fell into the "white-collar" group (12.4 percent), while 201 husbands (64.0 percent) were clearly "blue-collar." The farm husbands, all located in the three rural counties, were hard to categorize, as farm property values varied tremendously.

62. Amortizing mortgages, common to today's home buyers, were rare. Working-class families were required therefore to save a large sum of money to pay off the principal at the end of the loan period, even as they paid the interest in semiannual pay-

ments. See Stephan Thernstrom, *Poverty and Progress: Social Mobility in a Nineteenth-Century City* (New York: Atheneum, 1974), 121–22, 129–30.

63. Ibid., 120–21.

64. One-way ANOVAs were used to test whether the size of a midwife's practice was related to her socioeconomic status. For husbands' occupation, the f-ratio was 1.3700, with 7 df, and an f-probability of .2202. For home status, the f-ratio was 1.7649, with 2 df, and an f-probability of .1736.

65. Kobrin, "American Midwife Controversy," 350.

66. Brickman, "Public Health, Midwives, and Nurses," 70.

67. Declercq, "Nature and Style," 115. Table 22.2 indicates that all of the "major midwives" in Lawrence were born in Europe.

68. Michael M. Davis, "Birth Rates and Maternity Customs," in his *Immigrant Health and the Community* (New York: Harper and Brothers, 1921), 195.

69. Grace Abbott, *The Immigrant and the Community* (New York: Century, 1917). Abbott declared that "sometimes the traditions and the prejudices of our immigrant population must be consulted in determining an important health policy. An admirable illustration of this necessity may be found in the question of what should be our state and city policy with regard to the practice of midwifery" (145–46).

70. Joseph B. DeLee, "Progress Toward Ideal Ob-

stetrics," *Study of the Prevention of Infant Mortality* 6 (1915): 119.

71. B. C. Hartley and Ruth E. Boynton, "A Survey of the Midwife Situation in Minnesota," *Minnesota Medicine* 7 (1924): 439–46.

72. Sherbon and Moore, *Maternity and Infant Care,* 10.

73. Ibid., 30–32.

74. A chi-square analysis of county by birthplace was highly significant; the chi-square = 207.949, significant at 0.00001. The three rural counties had many more Scandinavian midwives than would be expected, and Milwaukee City had many fewer. (Milwaukee had only four Scandinavian midwives; 37.1 were expected.) Trempealeau and Price counties also had many fewer German birth attendants than would be expected. (Trempealeau had no German midwives; 13.9 were expected. Price had eight instead of 20.3.)

75. Elizabeth Ewen, *Immigrant Women in the Land of Dollars: Life and Culture on the Lower East Side 1890–1925* (New York: Monthly Review Press, 1985), 133.

76. Several historians have noted the almost outright ban on male doctors in immigrant women's birthing rooms. See Brickman, "Public Health, Midwives, and Nurses," 70; Kobrin, "American Midwife Controversy," 321; Ewen, *Immigrant Women in the Land of Dollars,* 70–71.

23 White Nurses, Black Midwives, and Public Health in Mississippi, 1920–1950

Susan L. Smith

Black lay midwives were important health workers well beyond their midwifery role in southern rural areas in the first half of the twentieth century. They were part of a black community health network, including ministers, teachers, and a few black nurses and doctors, that provided health education and health services to African-Americans. Lay midwives worked primarily with women and children, the key targets of public health work, and they also provided health education to the entire community. Midwives proved to be a vital link between poor African-Americans and health departments. Indeed, the success of official state and county health projects among African-Americans depended on the work of black midwives.[1]

In the past few decades the burgeoning interest among scholars in the study of lay midwives has deepened our understanding of the history of health care in the United States. Much of the literature has focused on native-born whites and European immigrants, although there is a growing body of material about African-American midwives.[2] This article examines the complex relationship between public health nurses and lay midwives in African-American communities in Mississippi from 1920 to 1950. Based on records of the Mississippi State Board of Health, including public health nurses' reports and correspondence from midwives, this essay places the activities of nurses and midwives at the center of the story of public health work. Nurses, most of whom were white, and midwives, most of whom were black, worked together to implement the modern public health care system in Mississippi and other southern states.

Ironically, state regulation of midwifery, which brought public health nurses in close contact with lay midwives, led to the creation of an unexpected cadre of public health workers for the state. Beginning in the 1920s Mississippi enacted state restrictions on midwifery practice as health officials identified midwives as a public health problem. Public health officials claimed that midwives did not maintain clean environments and that they used unscientific, and therefore unsafe, folk medicine. Yet health officers and public health nurses soon discovered

SUSAN L. SMITH is Associate Professor of History and Women's Studies at University of Alberta in Edmonton, Alberta, Canada.

Reprinted from *Nursing History Review: Official Journal of the American Association for the History of Nursing* 2, edited by Joan E. Lynaugh. Copyright 1994 by the University of Pennsylvania Press. Reprinted by permission of the publisher.

that midwives provided valuable assistance in implementing health policy in black communities. Even though nurses never treated midwives as their equals, they did frequently note midwife contributions to public health work.[3]

Examining the health work of midwives demonstrates the significant contributions made by poor black women to the implementation of state health programs. Midwives assisted nurses with health education, promotion of clinics and immunization programs, and efforts to encourage women to receive prenatal and postnatal medical examinations. Midwives and public health nurses developed a reciprocal relationship in which midwives aided the work of nurses, even as nurses provided them with training. Health reform in Mississippi therefore developed out of a dynamic relationship between public health professionals, midwives, and clients.[4]

The Midwives of Mississippi

Across the nation, lay midwives delivered half of all babies as late as 1910. Much of their practice was among European immigrant women and southern black women. By 1930, with immigration restrictions and the preference of urban women for childbirth attendance by physicians, midwives delivered only 15 percent of all births.[5] At that time 80 percent of all remaining midwives practiced in the South, where midwives delivered over one-fourth of all babies and over one-half of black babies. Even as midwifery declined in significance in northeastern and midwestern urban areas, the number of practicing midwives did not drop significantly in the South until after 1950.[6]

In Mississippi, where half the population was black, the vast majority of midwives were black women. Black midwives, called "granny" midwives by white health officials, delivered most of the black babies and even a few of the white babies in the state. In Mississippi in the late 1920s midwives delivered 8 percent of white babies and 80 percent of black babies, and although they delivered only 5 percent of white babies by the 1940s, they continued to deliver 80 percent of black babies.[7]

Scholars have demonstrated the important role black midwives played in their communities in cultural transmission and community leadership, as well as in health care. Midwives were highly respected, prestigious members of the communities they served, even if they were at the bottom of the medical hierarchy. Midwives were the female counterparts to preachers as the most influential members of the community. They were at the center of traditional healing networks in rural black communities of the South and served as advisers and spiritual leaders. Midwifery was only part of the labor they performed, for most midwives also cared for their own families, tended crops in the fields, and worked as domestic servants or teachers.[8]

Black women, and some white, preferred midwife over physician deliveries for economic and cultural reasons. Midwives were cheaper than doctors and would travel to remote places. Midwives provided comfort and support to pregnant women before, during, and after delivery. They even looked after the cooking and cleaning, in addition to caring for the mother and newborn. Many women preferred delivery by another woman, especially one with a reputation for skillful service. Finally, midwives treated birthing women among the rural poor with dignity rather than disregard.[9]

Because they believed that they did God's work, many midwives did not insist on payment, although a few dollars or payment in kind, such as a gift of pigs, was typical. One woman in Alabama gave her cow to the midwife for her delivery because her husband could not pay the fee.[10] In the 1920s midwife Bessie Sutton received about $1.50 for a delivery, and when she retired in 1962 the fee had increased to $20. She explained that she did not do the work for the money but because she loved people. She emphasized, "If I'd a stopped 'cause they didn't pay me, I'd a stopped a long time ago."[11]

Midwife Training Campaigns and Public Health Nurses

Midwives and public health nurses together implemented the development of a modern public health care system in Mississippi. As midwives came under state regulation beginning in the early 1920s they developed a working relationship with public health nurses who were responsible for midwife training campaigns to "modernize" midwifery. Intervention of the state into the previously unregulated practice of midwifery meant the creation of a division between state-sanctioned midwives and those midwives identified by the state as unfit to practice and therefore vulnerable to prosecution, such as uncooperative women and any men. Whether or not midwives abided by the new regulations, the state placed limitations on the health services midwives were entitled to perform during childbirth and in general.

In 1921 Laurie Jean Reid, a white nurse with the United States Public Health Service (USPHS), came to Mississippi to survey the midwives of the state. Reid traveled around the eighty-two counties of Mississippi tracking down the names of midwives and registering the women county by county. She identified over four thousand midwives in Mississippi, and the state located an additional one thousand midwives a few years later.[12] As she explained to the white Mississippi State Medical Association, her purpose was to address the fact that the United States had higher infant and maternal mortality rates than many European countries. Reid advised that these rates were caused by poor health care for pregnant women from "careless physicians and by illiterate and ignorant midwives."[13] Sidestepping the issue of physician care, she proposed that the state eliminate those midwives who were too old to be educated and register and train those who remained. Targeting midwives was an easier solution for public health officials dealing with an impoverished rural population than criticizing physician care or altering the economic and living conditions that contributed to ill health.[14]

Despite the fact that health officials blamed midwives for maternal and infant deaths, their safety record was not any worse than that of physicians. Some contemporary studies even showed that maternal mortality rates were lowest where the percentage of midwife-attended births was highest.[15] In 1923 Nurse Supervisor Lois Trabert of Mississippi's Bureau of Child Hygiene proclaimed: "I firmly believe that when we do get these midwives properly trained, in as far as that is possible, that they will do better and cleaner work than the average country doctor."[16] By the 1930s the Mississippi Board of Health employed 125 white public health nurses and 6 black public health nurses to train lay midwives.[17]

From 1910 to 1930 health reformers and medical professionals engaged in a debate over the future of midwives, with some obstetricians arguing for their immediate elimination and public health officials arguing that midwives were needed, even if only temporarily.[18] Most public health doctors and nurses in Mississippi supported the gradual, not immediate, elimination of midwives, believing that they were a "necessary evil" until there were enough hospitals and doctors to care for rural women.

The decision to regulate, and only rarely eliminate, lay midwives rested on the fact that southern rural areas had an inadequate number of practicing physicians to serve as the sole childbirth attendants. Midwives were available where physicians were not. Between 1920 and 1950 there were consistently twice as many midwives as physicians in Mississippi, with at least 2,000 to 4,000 black and several hundred white registered midwives, and only 1,000 to 1,800 practicing physicians, less than 75 of whom were black.[19] Other southern states reported equal or greater numbers of practicing midwives, including approximately 3,000 in Alabama, 4,000 to 9,000 in North Carolina, 5,000 to 6,000 in Georgia, and perhaps as many as 10,000 in Florida. All of these figures are rough estimates because health officials had no definite idea of the number of unregistered midwives who delivered babies.[20]

Southern state boards of health argued that training midwives lowered mortality rates. For example, Dr. Felix J. Underwood, the executive officer of the Mississippi State Board of Health from 1924 to 1958, pointed to the decline in state maternal and infant mortality rates. The total maternal mortality rates for the state dropped from 9.5 per 1,000 deliveries in 1921 to 4.4 in 1942, with black rates dropping from 12 in 1921 to 5.5 in 1942. A similar pattern held for infant mortality rates, which are the number of deaths per 1,000 live births in the first year of life. These rates dropped from 68 in 1921 to 47 in 1942 for the state as a whole, and from 85 to 54 for black infants in the corresponding years.[21]

Board of Health's Educational Programs

From 1920 to 1950 public health nurses implemented the board of health's educational programs, using midwife manuals that carried virtually the same message for three decades.[22] The lessons taught by the nurses emphasized limitations on midwives and the ultimate authority of doctors. Above all else, the two issues on which nurses focused were: (1) the importance of cleanliness, and (2) the need to call on a physi-

cian in case of complications during delivery. Before state training began, midwives wore whatever they chose and carried the tools of their trade in an assortment of bags, such as flour sacks. Nurses permitted midwives to continue carrying midwife bags, but they had to upgrade them from flour sacks to black leather bags. Nurses instructed the midwife to perform deliveries wearing a clean white dress, with a white mask and white paper cap, which nurses taught them to make to keep hair out of the way. The nurses also insisted that midwives keep their hands and nails clean.[23]

State boards of health in the North as well as the South followed a similar pattern of midwife instruction. Mississippi and other states required midwives to file birth certificates for each birth they attended, therefore requiring illiterate women to find someone else to fill out the forms. Nurses forbade midwives to use any folk medicine or herbal remedies in their childbirth work. In addition, in order to prevent infections nurses instructed midwives not to perform any digital examination of women during labor; only doctors were allowed to put their hands into the birth canal. If midwives followed this rule, they no doubt had difficulty identifying how much a woman's cervix had dilated and the progression of labor.

Some of the nurses' instructions conflicted with long-standing midwifery practices and therefore resulted in midwife resistance. For example, nurses insisted that midwives deliver women on the bed and not the floor. Mothers and midwives particularly resisted this requirement because many women felt the urge to walk around during labor and found it more comfortable to deliver in a squatting position, assisted by gravity. Furthermore, some midwives preferred to keep the bed clean and have all the mess on the floor. Alabama midwife Margaret Smith explained why she ignored the rule. As she pointed out, "When you in misery, if there is any way you can ease that misery, you gonna ease it."[24]

Nurses intended to ensure compliance with these rules by occasionally supervising deliveries and inspecting midwife bags. However, midwives were not always cooperative. Former county health officer Dr. W. E. Riecken, Jr., recalled that some midwives engaged in the practice of "a bag to show and a bag to go." Midwives used this strategy to circumvent the rules by keeping "a clean, properly organized bag for inspection and licensing and . . . another that was used for deliveries."[25] Nurse Elsie Davis mentioned in 1931 that some of the midwives with whom she worked refused to bring their bags for inspection, while others did not maintain their bags in sterile condition, offering the excuse that they just came off duty.[26] Nurses even inspected midwives' homes to see that they understood the principles of cleanliness. Nurses Abbie G. Hall and Caroline Bourg of Sharkey and Issaquena counties reported: "We try to always keep in mind our midwives and never miss an opportunity to stop at their homes and inspect bags whenever we are in the neighborhood."[27]

Midwives resented nurses' implications that they had not performed their work well before regulation by the state. As midwife Otha Bell Jones of Itta Bena explained in 1938, she had learned how to be a midwife twenty-three years earlier through apprenticing with midwife Nancy Wright. "And dear friends," she wrote the board of health, "I had nursed 36 womens before I got any permit and I has fill the sum of 15 books with 25 leaves each." Despite her long lists of women she had delivered, she wrote: "[I] never lose a woman in childbirth since I have started out."[28] Midwife Onnie Lee Logan attended her midwife training classes and received some instruction, but she emphasized that "two-thirds of what I know about deliverin', carin' for mother and baby, what to expect, what was happenin' and was goin' on, I didn't get it

from the class. God gave it to me. So many things I got from my own plain motherwit."[29] Yet, another midwife observed that even though she had a great deal of experience, she could still learn more. "Each time I attend my monthly meeting I learn something new, and I am happy to admit that because so many won't."[30]

Midwife Monthly Club Meetings

Beginning in the 1920s the state required midwives to have monthly club meetings at which they were to further their knowledge. Dressed in their white delivery outfits, the women were supposed to read from the state's midwife manual and discuss any delivery problems they encountered. Although public health nurses initially organized the local clubs, the meetings were run entirely by midwives. Each midwife club selected a leader or president and a secretary, and they were in charge of reporting the minutes of the meeting to the county public health nurse or, if there was none, to the state supervisor of midwives. The midwife leader ran inspection of the midwives' bags and fingernails and handed out silver nitrate eye drops to prevent blindness in newborns caused by passing through the birth canal of mothers with gonorrhea.[31]

Midwife meetings had definite religious overtones. Usually, midwives held the meetings in churches, one of the few public buildings controlled by African-Americans in rural communities. For example, the midwives of Hub, Mississippi, met monthly at Sweet Valley Baptist Church.[32] Midwife leaders also ran the meetings like a church service, complete with opening prayer and song, a reading from the Bible, short readings from "the Book," or midwife manual, and singing of the midwife songs. Midwife songs included "Protect the Mother and Baby," sung to the tune of "Mary Had a Little Lamb," which had hand motions to go along with the lyrics describing the importance of clean clothes, clean hands, clean midwives.[33] In 1937 John Lomax, a native of Holmes County and curator of folk songs at the Library of Congress, recorded a dozen midwives singing such songs at the Mississippi State Board of Health.[34]

Even though midwives seemed to have enjoyed the opportunity to socialize with each other, share stories, and sing together, they could not always attend every monthly meeting. Attendance was always down during times of heavy field labor and when weather made travel difficult, especially for women who had to walk several miles to the meetings. In 1923 Nurse Agnes B. Belset reported: "I have had but one midwife meeting as they begged off because of the cotton picking season."[35] Another nurse, Louise James, mentioned that her "greatest trouble in having a full attendance was caused by this season being cane grinding season and most all old women are used to skim cane syrup. Very often I had to go to the mill to get my entire class."[36] Despite the difficulty maintaining attendance, the number of midwife clubs increased over the years, growing from 290 clubs in the state in 1928 to 505 in 1942.[37]

Registration for Permits and Regulation of Midwife Practice

Midwives were careful about state rules regarding registration for permits and regulation of midwife practice. They took the issue of registration very seriously, identifying themselves in correspondence by their permit numbers, as requested by the state. Midwives were cautious because the state threatened to revoke the permits of women who did not follow the regulations. In 1923 Mississippi nurse Mae Reeves indicated that some nurses "found one midwife treating gynecological cases and children. We revoked her permit."[38] In 1924 a nurse in Washington County reported: "In follow up work for midwives found that one had delivered case without

cap and gown, and had also made [digital] examination. Her permit was revoked for one month and at the end of that month she is not to take a case unless I can be with her at time of delivery."[39] Punishment for performing digital exams and failing to wear regulation uniforms could be swift. When one nurse revoked the permit of a midwife at a conference for the renewal of permits, the nurse reported that it "was so upsetting to the midwife that long wailing sobs penetrated the air. The sympathy of the other 75 midwives brought on more sobbing—needless to say I wondered just what I'd do to restore order and continue with the meeting."[40]

Midwives also had their permits temporarily revoked if they tested positive for syphilis. Nurses kept watch over midwives' health, fearing that they would spread diseases to birthing women and infants. The nurse required the midwife to have a Wassermann test for syphilis and vaccine against typhoid fever and smallpox.[41] Nurses administered Wassermann tests before they renewed permits each year, and nurses forbade midwives who had syphilis from performing any deliveries. If the midwife tested positive for syphilis, she had to undergo treatment and a doctor had to indicate when it was safe for her to perform deliveries again. In 1928 Nurse McDaniel reported that "the one who had a positive Wassermann is taking treatment regularly and we hope we can give her a permit before long. She asks about her permit each time she comes in."[42]

In extreme cases health departments called on police power to force compliance with the rules. In 1931 Nurse Josie Strum of Clarke County chose to handle one midwife with extraordinary measures as a warning to all midwives. Strum indicated in her report: "Another midwife was discovered practicing midwifery without a permit, warnings did not seem to do any good so we had her arrested. She was convicted and fined. Now she is very anxious for a permit as well as several others in that same neighborhood."[43] In later years the state even used its regulatory power to punish midwives who engaged in civil rights activity. A former midwife living in Greenville in the 1960s claimed she lost her permit because she "demonstrated and sat in down at Jackson."[44]

The Midwife as Public Health Worker

Ironically, Mississippi's regulation of midwives, which limited the health services that midwives were permitted to provide, led to the creation of additional public health workers for the state. Gradually, health department officials, especially public health nurses, realized that midwives were assets in black community health work. In the early 1930s Mississippi's executive officer of the board of health indicated that supervision, which was "primarily designed to render least harmful to the public health the services of these midwives, has converted them into important allies in the cause of human well-being."[45]

Midwives seized on the resources provided by public health nurses. They became an important conduit for health education in their childbirth work and elsewhere. They provided health instruction to adults and children through the churches and schools. The activities of midwife Mollie Gilmore of Vicksburg during 1936 and 1937 illustrate the range of midwives' health work. Gilmore accompanied the public health nurse in her home visits to mothers, and she encouraged people to go to a local church for typhoid shots. "By my influence," the midwife reported, "Dr. Smith was able to protect 80 people from typhoid fever." She gave health lectures at church and "prayed to the congregation after service." In the fall of 1936 she "talked to 100 different people on prenatal care, and health and care of the baby." She observed, "During

that time I referred 5 expectant mothers to doctor." In addition, she assisted a doctor with the home deliveries of four white women, and she worked on the annual May Day child health program.[46]

Midwives provided communities with information about proper home birthing environments through model delivery room demonstrations. Beginning in the early 1930s the board of health required that twice a year midwives set up a model room in their homes and invite their clients and other people in the community to visit. The purpose was to force the midwife to demonstrate to the public the correct birthing room and to educate future mothers and community members about modern childbirth requirements. The delivery room was supposed to be a neatly organized, clean bedroom with newspapers spread on the floor, a paper pad placed on the bed with a drip sheet leading into a bucket, and adequate lighting and heating.[47]

These demonstrations were a clever way to ensure that midwives were familiar with state regulations, and they also provided publicity to white and black people about the state health board's work with midwives. Plantation owners, insurance agents, doctors, teachers, students, mothers, and tenant farmers all attended the delivery room demonstrations. A sizable number of midwives complied with the demonstration requirement. For example, in 1934 over one thousand three hundred midwives sent in reports of their model delivery rooms.[48]

Midwives performed important public health work when they encouraged women to receive prenatal and postnatal care, often bringing them to the health department or mothers' clinics themselves. Retired county health officer Dr. W. E. Riecken, Jr., stated that midwives often accompanied their clients to the county health department for checkups and "frequently assisted in the exam room."[49] Midwives also aided the work of nurses by notifying them when a woman had a baby. In 1931 nurses in Sharkey and Issaquena Counties reported that they "have tried to visit more lying-in cases this month but it is hard to know when they are confined. Quite a number of midwives report their cases to us right after delivery and in that way we have come in contact with more mothers during the lying-in period."[50] Midwives saw themselves as important liaisons between black communities and white health professionals. In one case in Mississippi a midwife made requests of the local public health nurse to tend to a sick woman. In 1939 midwife Estelle W. Christian notified Nurse Viola M. Jones that a woman had come to see her at the house where she had just delivered a baby. The middle-aged woman had sores on her arms, legs, and buttocks. The midwife explained to the woman that she could not visit with the new mother or baby because she might infect them. Then the woman told the midwife that the family across the road from her had the same symptoms. The midwife wrote the nurse:

> Now could you go out there at once to see about this. I made a lecture on the 4th Sunday in August at a church at Willows, Miss, to the people about their health. . . . I advised them if they had any sores or the least suspicion they were infected with that dreaded disease syphilis to please tell some one before it is too late. So she says she heard me talk that day and when she heard I was in the neighborhood she came to tell me about it. So I advised her to stay at home until I talked with you. So if you could come out here one day I could go to their homes with you.[51]

Midwives advocated for increased care from doctors and nurses for poor women.

Midwives assisted with venereal disease control work by encouraging pregnant women and other people in the community to have their blood tested. Venereal diseases, particularly syphilis, were a major public health concern throughout the period from 1920 to 1950. Midwives proved to be quite effective in reaching people. For example, in 1944 several nurses wondered why attendance at a venereal disease clinic was suddenly increasing until they discovered that "a leader of a midwife club who had received literature on syphilis had made talks at churches, schools and in the homes. This midwife was instrumental in sending in several young girls under sixteen years of age who had infectious syphilis."[52]

As a community leader, a midwife who endorsed public health projects contributed to community cooperation. Midwives assisted health department efforts to protect African-Americans against typhoid fever and diphtheria. They promoted this preventive work at their midwife club meetings, churches, and schools.[53] In 1931 Nurse Ethel B. Marsh of Adams County mentioned that her county sponsored an anti-diphtheria campaign among black infants and preschool children. She explained that "the midwives in the various sections of the county are assisting by informing parents of the various stations and dates on which toxin-antitoxin will be given."[54] Elsewhere in the state, Nurse Nell E. Austin of Forrest County indicated that

> the midwives are very helpful in getting the colored people immunized against typhoid fever. At the last midwives meeting they were asked to round up all the children in their neighborhoods and bring them to the health department and have them inoculated against typhoid fever. One midwife was in the office the very next morning with fifty-five children.[55]

Parents knew and trusted their local midwives and so more readily permitted their children to participate in health programs that midwives endorsed.

Midwives not only assisted with official health department programs; they also successfully organized their own. They ran May Day programs on child health and participated in the annual Negro Health Week programs.[56] In 1935 a midwife in Smith County reported that at their May Day program "a talk was given about the care of the teeth. A few weeks later the peddler who sells extracts, spices, and tooth paste asked the leader of the midwife club what in the world had caused so many of the people in the community to buy tooth paste."[57]

Midwives were extremely proud of their health care contributions and saw themselves as part of a great health crusade. A Simpson County midwife wrote to one nurse: "Yes the sisters is praying the time to hasten so they can meet you again for further instruction. They think it so grand that old as they are they can do this great work to help foster in this great battle field of deficiency."[58] One midwife even contacted the board of health, asking for official public recognition of the work performed by midwives. She pointed out that there was a Thanksgiving Day, Christmas Day, Mother's Day, and Father's Day. She said, "It look to me we could have a Midwife Day —i ben on this job sense September 1900 an my work been close to a 1,000, an i am workin in my 81 year."[59]

The Decline of Midwives

Even though individual nurses commented in their reports on the valuable public health work performed by lay midwives, the official policy of the board of health remained to ultimately eliminate the midwives. In 1948 Nurse Supervisor Lucy E. Massey instituted a retirement program, first suggested to her by county nurses, in an effort to accelerate the elimination of midwives.[60] The plan strongly encouraged the retirement of older midwives by informing them and their families that they were too old to renew their permits and then honoring the women with ceremonies, complete with badges stating "Retired Midwife." According to the new policy,

"The midwife must promise not to practice and must hand in her permit when she receives the badge."[61] From 1948 on, not only was there a policy of retiring midwives but all new midwives had to get a physician to testify that there was a need for their services in their community before they could receive a permit.

The retirement ceremonies probably never fooled the midwives forced out of practice, but they did provide an opportunity for the nurses and people in the community to honor the women. The public health nurse usually held the event at a local church where the midwife "would sit queen for the day."[62] Sometimes people would drop money in her lap, especially women the midwife had attended. In November 1948 midwives Mollie Merrill and Josephine Franklin were honored in a retirement ceremony at St. John Methodist Church in Forrest County. According to a newspaper report of the event, "The aged pair, clad in white uniforms and caps, and clutching small American flags, sat solemnly in pink and white decorated chairs of honor, placed near the altar." The white county nurse and the black midwife club leader each offered speeches of praise for the women. Those attending sang hymns as the members of the local midwife club led a procession, "robed in white uniforms and caps, and also carrying tiny American flags." Midwives sang, read Scripture, offered up prayers, and presented gifts to the retiring midwives.[63]

The number of Mississippi midwives registered with the state began to decline in the post–World War II era, in part because of state policy but also because of other factors such as urban migration, which resulted in fewer black midwives living on the surrounding plantations. Instead of thousands of registered midwives, by 1966 only 600 midwives had permits in the state, although they still delivered 10,000 to 12,000 babies that year. By 1975 there were only 220 registered midwives who delivered about

1,000 babies. In 1982 the state had a mere 13 registered lay midwives, by which point the state no longer issued permits or held training sessions.[64]

Conclusion

Black midwives have played an important but overlooked historical role as public health workers for rural African-Americans, performing unpaid public health work for the state. In Mississippi the community work of black midwives was vital to the success of state health programs because of their respected position and close relationship with community members. State regulation created an opportunity for midwives, long concerned with the health of their communities, to assist with state public health work. Because regulation put midwives into constant contact with public health nurses, midwives became vital assistants to nurses in implementing state health policy. From 1920 to 1950 midwives actually broadened their responsibilities for community health care, despite state regulation that set limitations on the activities of midwives. Indeed, the success of health reform in Mississippi depended on the work of women at the bottom of the medical heirarchy—black lay midwives.

Notes

The author would like to thank the anonymous reviewers and the following people for their support and their comments on earlier drafts of this material: Andrea Friedman, Maureen Galitski, Linda Gordon, Judith Walzer Leavitt, Gerda Lerner, Donald Macnab, Leslie Reagan, Leslie Schwalm, Mariamne Whatley, and Nancy Worcester. Special thanks to Anne Lipscomb at the Mississippi Department of Archives and History in Jackson and to Edna Roberts, retired public health nurse and former state supervisor of nursing for the Mississippi Department of Health.

This research was funded by the Woodrow Wilson National Fellowship Foundation Rural Policy Fellowship and the Women's Studies Research Grant, with additional financial assistance from the University of Wisconsin–Madison in the form of a KNAPP Women's History Fellowship Grant, the Graduate School Domestic Travel Fellowship, and the Maurice L. Richardson Fellowship from the History of Medicine Department at the University of Wisconsin Medical School.

1. This article is based on material from my dissertation, now book, *Sick and Tired of Being Sick and Tired: Black Women and the National Negro Health Movement, 1915–1950* (Philadelphia: University of Pennsylvania Press, 1995), ch. 5. Although not directly parallel, a similar argument has been made about the important public health role prostitutes could play in the current AIDS epidemic. Gloria Lockett, "Black Prostitutes and AIDS," in *The Black Women's Health Book: Speaking for Ourselves*, ed. Evelyn C. White (Seattle, Wash.: Seal Press, 1990), 189–92.

2. For an excellent historiographical essay on midwives see Judith Barrett Litoff, "Midwives and History," in *Women, Health, and Medicine in America*, ed. Rima D. Apple (New York: Garland, 1990), 443–58. On black midwives see Molly C. Dougherty, "Southern Midwifery and Organized Health Care: Systems in Conflict," *Medical Anthropology: Cross-Cultural Studies in Health and Illness* 6 (Spring 1982): 113–26; Sharon A. Robinson, "A Historical Development of Midwifery in the Black Community: 1600–1940," *Journal of Nurse-Midwifery* 29 (July–August 1984): 247–50; Linda Janet Holmes, "African-American Midwives in the South," in *The American Way of Birth*, ed. Pamela S. Eakins (Philadelphia, Pa.: Temple University Press, 1986), 273–91; and Debra Anne Susie, *In the Way of Our Grandmothers: A Cultural View of Twentieth-Century Midwifery in Florida* (Athens: University of Georgia Press, 1988).

3. One must be careful not to overdraw the respect that the board of health in general, and nurses in particular, held for midwives. For example, the nursing division of the health board had a habit of collecting lists of examples of midwife illiteracy drawn from midwife club reports. Typewritten sheet of quotes, no date, box 27, Record Group 51, Mississippi Department of Health, Mississippi State De-

partment of Archives and History (hereafter cited as MSDAH), Jackson.

4. Molly Ladd-Taylor, "Women's Health and Public Policy," in Apple, *Women, Health, and Medicine in America*, 398. On the relationship between professionals and clients, see Linda Gordon, *Heroes of Their Own Lives: The Politics and History of Family Violence* (New York: Viking, 1988), especially introduction.

5. The major works on the history of lay midwifery include Frances K. Kobrin, "The American Midwife Controversy: A Crisis of Professionalization," in *Women and Health in America*, ed. Judith Walzer Leavitt (Madison: University of Wisconsin Press, 1984), 318–26; Jane B. Donegan, *Women and Men Midwives: Medicine, Morality, and Misogyny in Early America* (Westport, Conn.: Greenwood Press, 1978); and Judy Barrett Litoff, *American Midwives, 1860 to the Present* (Westport, Conn.: Greenwood Press, 1978). See also Charlotte G. Borst, "Catching Babies: The Change from Midwife to Physician-Attended Childbirth in Wisconsin, 1870–1930" (PhD diss., University of Wisconsin–Madison, 1989).

6. "Percentage of Births Attended by Physicians and Midwives and Others, in Certain States, as Reported by State Bureaus of Child Hygiene for 1925," Central File, 1925–1928, box 274, Record Group 102, U.S. Children's Bureau, National Archives (hereafter cited as NA); and Elizabeth C. Tandy, "The Health Situation of Negro Mothers and Babies in the United States," Children's Bureau, March 1941, box 27, RG 51, MSDAH. Although my focus is on black midwives in the Southeast, specifically in Mississippi, midwives in the Southwest faced similar historical patterns.

7. See chart on white and black births in Mississippi, box 27, RG 51, MSDAH; and "The Relation of the Midwife to the State Board of Health," Mississippi State Board of Health, January 1, 1944, box 57, RG 51, MSDAH. For one midwife's accounting see "History of Della Falkner: Mid-Wife and Register of Her Patients," published in Holly Springs, February 26, 1937, and newsclipping, "State Midwife Tells All in Booklet," December 15, 1937, box 27, RG 51, MSDAH. In the ten years from 1926 to 1936 she delivered 120 black babies and thirteen white. See also Litoff, *American Midwives*, 27; Judy Barrett Litoff, *The American Midwife Debate: A Sourcebook on Its Modern*

Origins (New York: Greenwood Press, 1986), 4; and Tandy, "The Health Situation of Negro Mothers and Babies in the United States."

8. "The Relation of the Midwife to the State Board of Health," Mississippi State Board of Health, January 1, 1935, box 2, RG 51, MSDAH; Dougherty, 116–17; and Holmes, "African-American Midwives in the South," 278–82. Oral interviews and autobiographical accounts of midwives illustrate these points. See Onnie Lee Logan as told to Katherine Clark, *Motherwit: An Alabama Midwife's Story* (New York: E. P. Dutton, 1989); Susie, *In the Way*, ch. 1; and Linda Janet Holmes, "Thank You Jesus to Myself: The Life of a Traditional Black Midwife," in White, *The Black Women's Health Book*, 98–106. I also learned much about black midwives from Edna Roberts, a retired white public health nurse and former state supervisor of nursing in Mississippi. Edna Roberts, interview by author, tape recording, Mississippi, September 17, 1989.

9. Logan, *Motherwit*, 52; Litoff, *American Midwives*, 28; Litoff, *The American Midwife Debate*, 4; "Report on the Midwife Survey in Texas," Bureau of Child Hygiene, Texas State Board of Health, 1924, Central File, 1925–1928, box 275, RG 102, NA, and reprinted in *The American Midwife Debate*, 67–81.

10. Thordis Simonsen, ed., *You May Plow Here: The Narrative of Sara Brooks* (New York: Touchstone Books, 1986), 172.

11. Jack Bleich, "Midwife's Delivery Eighty-two Years Ago Began Tradition for Bessie Sutton," *Jackson Clarion Ledger*, September 17, 1978, and "Grannies: The Roots of Midwifery," *Jackson Clarion Ledger*, January 31, 1982, both in folder-midwives, Subject File, MSDAH; and Emily Herring Wilson, *Hope and Dignity: Older Black Women of the South* (Philadelphia, Pa.: Temple University Press, 1983), 43; Holmes, "African-American Midwives in the South," 280–81; and Logan, *Motherwit*, 103–4.

12. Dr. Felix J. Underwood, director of the Bureau of Child Welfare of the Mississippi State Board of Health, to Jesse O. Thomas, May 3, 1921, General Office File, box A6, Records of the National Urban League Southern Regional Division, Library of Congress.

13. Laurie Jean Reid, "The Plan of the Mississippi State Board of Health for the Supervision of

Midwives," 1921 speech, box 354, RG 51, MSDAH. See also Grace I. Meigs, *Maternal Mortality from All Conditions Connected with Childbirth in the United States and Certain Other Countries,* U.S. Department of Labor, Children's Bureau Publication, 19 (Washington, D.C.: Government Printing Office, 1917), reprinted in Litoff, *The American Midwife Debate,* 50–66; and Litoff, *American Midwives,* 50–51, and 55.

14. Financial support came from the Sheppard–Towner Maternity and Infancy Protection Act of 1921, which provided states with federal funding for maternal and child health programs. States applied for matching grants to the U.S. Children's Bureau, which administered the act. On Mississippi see Grace Abbott, Children's Bureau, to Governor Lee M. Russell, December 12, 1921, 1921–1924, Central File, box 249, RG 102, NA; and "Plan for the Division of Maternity and Infant Hygiene For 1922," Correspondence and Reports, 1917–1954, box 17, RG 102, NA. See also Molly Ladd-Taylor, *Raising A Baby the Government Way* (New Brunswick, N.J.: Rutgers University Press, 1986), 5–29.

15. Litoff, *The American Midwife Debate,* 5.

16. Lois Trabert, Bureau of Child Hygiene of Mississippi State Board of Health, to Dr. Anna E. Rude, Children's Bureau, April 2, 1923, Central File, 1921–1924, box 248, RG 102, NA.

17. Mary D. Osborne, "Public Health Nursing," September 17, 1938, box 317, RG 51, MSDAH.

18. Litoff, *American Midwives,* 80; and Litoff, *The American Midwife Debate.*

19. Mississippi State Board of Health, "Plan for the Division of Maternity and Infant Hygiene For 1922," 1917–1954, Correspondence and Reports, 1917–1954, box 17, RG 102, NA; Mississippi State Board of Health, "Study of Midwife Activities in Mississippi, July 1, 1921–June 30, 1929," June 30, 1929, box 36; Mississippi State Board of Health, "The Relation of the Midwife to the State Board of Health," July 1, 1937, box 356; and Mississippi State Board of Health, "The Relation of the Midwife to the State Board of Health," January 1, 1944, box 57, all in RG 51, MSDAH. See also Neil R. McMillen, *Dark Journey: Black Mississippians in the Age of Jim Crow* (Urbana: University of Illinois Press, 1989), 169.

20. Proceedings of the Negro Health Week Plan-

ning Meeting, Tuskegee, January 20, 1927, box 18, Thomas Monroe Campbell Papers, Tuskegee University. See also H. G. Perry, State Registrar of Vital Statistics, Alabama State Board of Health, to Jesse O. Thomas, April 25, 1921; and State Epidemiologist of North Carolina Board of Health to National Urban League, May 12, 1921, both letters in General Office File, box A6, National Urban League Southern Regional Division; Wilson, *Hope and Dignity,* 39; Susie, *In the Way,* 35; and Edward H. Beardsley, *A History of Neglect: Health Care for Blacks and Mill Workers in the Twentieth-Century South* (Knoxville: University of Tennessee Press, 1987), 39.

21. These figures should be taken merely as rough estimates. Roberts, interview by author; Dr. Felix J. Underwood, "Midwife Activities in Mississippi," [1932?], box 354; Mississippi State Board of Health, "The Relation of the Midwife to the State Board of Health," July 1, 1937, box 356; and infant and maternal mortality rates drawn from the U.S. Bureau of the Census, reported in "The Relation of the Midwife to the State Board of Health," by the Mississippi State Board of Health, April 27, 1938, box 36; U.S. Bureau of the Census, reported in "The Relation of the Midwife to the State Board of Health," by the Mississippi State Board of Health, January 1, 1944, box 57, all in RG 51, MSDAH.

22. The first midwife manual in Mississippi was published in 1922. Lois Trabert, Narrative Report of Work with Midwives, June 1921 to June 1922, Central File, 1921–1924, box 248, RG 102, NA; Mississippi State Board of Health, *Manual for Midwives,* see 1928, 1939, and 1952 in box 354, and Mississippi State Board of Health, "Midwife Supervision," [1938?], box 354, all in RG 51, MSDAH. Most southern states published midwife manuals, including Georgia in 1922 and Florida beginning in 1923. See *Lessons for Midwives,* Georgia State Board of Health, Child Hygiene Publication 17, Prenatal Series 3 [1922], reprinted in Litoff, *The American Midwife Debate,* 200–207; and Susie, *In the Way,* 241, endnote 24.

23. Dr. Felix J. Underwood, "The Development of Midwifery in Mississippi," read before the Southern Medical Association, 1925, box 36, RG 51, MSDAH; and photo entitled "A group of midwives in Madison County before any instructions," no date, box 27, RG 51, MSDAH.

24. Quoted in Holmes, "African-American Midwives in the South," 287. On midwives and birthing positions see Laurie Jean Reid, "The Plan of the Mississippi State Board of Health for the Supervision of Midwives," 1921, box 354, RG 51, MSDAH; Litoff, *The American Midwife Debate,* 4; and Holmes, "African-American Midwives in the South," 286. See also Dougherty, "Southern Midwifery," 117.

25. Dr. W. E. Riecken, Jr., July 17, 1990, written responses to questions posed by author. See also George Stoney, "All My Babies: Research," in *Film: Book 1,* ed. Robert Hughes (New York: Grove Press Inc., 1959), 79–96, especially 83.

26. Nurse Elsie Davis, Holly Springs, in "Nurses Narrative Report, Mississippi State Board of Health, Bureau of Child Hygiene, Division of Public Health Nursing, July 1931," box 36, RG 51, MSDAH.

27. Nurses Abbie G. Hall and Caroline Bourg, Sharkey and Issaquena Counties, "Nurses Narrative Report, Mississippi State Board of Health, Bureau of Child Hygiene, Division of Public Health Nursing, July 1931," box 36, RG 51, MSDAH.

28. Otha Bell Jones in Itta Bena to board of health, April 13, 1938, box 27, RG 51, MSDAH. It was not unusual for midwives to keep excellent records of the births they delivered. Della Falkner kept a list of the name of every birth she attended for ten years from 1926 to 1936. "History of Della Falkner: Midwife and Register of Her Patients." See also Wilson, *Hope and Dignity,* 39.

29. Logan, *Motherwit,* 90.

30. Rosie Bell Rollins of West Point to Lucy E. Massey, November 13, 1948, box 36, RG 51, MSDAH.

31. Silver nitrate was first used for newborn's eyes in the 1870s. Susie, *In the Way,* 4. See also Roberts, interview by author; Litoff, *American Midwives,* 101; and Litoff, *The American Midwife Debate,* 10. On midwife club reports from 1930s and 1940s see Melissa Ann Mobley and J. E. Lucas of Carlisle Midwife Club to board of health, [1941?], midwife report for Claiborne County, August 9, 1941, and Irene B. Brisco and Louise Ceal, Humphreys County midwife club report, September 5, 1942, all in box 27; and photos of Coahoma County midwife meeting, May 3, 1951, box 21, RG 51, MSDAH. See also Mary D. Osborne to Felix J. Underwood, Narrative and Statistical Report May 1924 of the Mississippi Bureau of Child

Welfare and Public Health Nursing, Central File, 1921–1924, box 249, RG 102, NA. For a discussion of midwife meetings in Florida see Susie, *In the Way,* 46.

32. Violor Dorsy and Earline Morris of Hub to board of health, box 36; photos of Holmes County midwife meeting, October 1938; and photos of Forrest County midwife meeting, no date, in box 27, RG 51, MSDAH. See also Dougherty, "Southern Midwifery," 116.

33. Lyrics to "Midwife Song: Protect the Mother and Baby" and "Song of the Midwives" sung to tune of "As We Go Marching On," box 27; and Robert Loftus, "Stork Loses Two Long-Time Forrest County Helpers," *Hattiesburg American,* November 29, 1948, box 36, RG 51, MSDAH.

34. John Lomax built a collection of black folk song recordings at the Library of Congress. *The Mississippi Doctor* 14 (June–May 1936–1937), and "Song's Recorder Here: Midwife Song Taken," newsclipping, March 9, 1937, box 36, RG 51, MSDAH. See also Dr. James Ferguson, who described a visit to a midwife meeting in October 1948 at which a phonographic recording of the entire meeting was made. James H. Ferguson, "Mississippi Midwives," *Journal of the History of Medicine* 5 (Winter 1950): 90–95, box 36, RG 51, MSDAH.

35. Nurse Agnes B. Belser and Nurse Inez Driskell, in Report from Mary D. Osborne to Dr. Felix J. Underwood, Dr. Anna E. Rude, and Dr. W. S. Leathers, September 1923, Central File, 1921–1924, box 248; and excerpts from nurses in "Narrative and Statistical Report, June 1926," Correspondence and Reports, 1917–1924, box 17, RG 102, NA.

36. Nurse Louise James, in Mississippi State Board of Health, Nursing Narrative and Statistical Report, November 1923, Central File, 1921–1924, box 249, RG 102, NA.

37. Mississippi State Board of Health, "The Relation of the Midwife to the State Board of Health," July 1, 1937, box 356; and Mississippi State Board of Health, "The Relation of the Midwife to the State Board of Health," January 1, 1944, box 57, RG 51, MSDAH.

38. Quote from Nurse Mae Reeves, in Mary D. Osborne report to Dr. Felix J. Underwood, Dr. Anna E. Rude, and Dr. W. S. Leathers, September 1923, Central File, 1921–1924, box 248, RG 102, NA.

39. Nurse report from Washington County, in Narrative and Statistical Report of the Mississippi Bureau of Child Welfare and Public Health Nursing, from Mary D. Osborne to Felix J. Underwood, May 1924, Central File, 1921–1924, box 249, RG 102, NA.

40. Zona C. Jelks, president of the Mississippi Public Health Association, "My Twenty-Five Years in Public Health in Mississippi," talk at the annual convention, November 29, 1967, box 317, RG 51, MSDAH.

41. Mississippi State Board of Health, "Midwife Supervision," [1938?], box 36, RG 51, MSDAH. Apparently, domestic servants were also required to be tested for syphilis. See Susan Tucker, *Telling Memories Among Southern Women: Domestic Workers and Their Employers in the Segregated South* (Baton Rouge: Louisiana State University Press, 1988), 177.

42. See Mississippi State Board of Health, *Manual for Midwives*, 1939, box 354, RG 51, MSDAH; and Nurse A. E. McDaniel in Tishomingso County, Mississippi State Board of Health, Bureau of Child Hygiene and Public Health Nursing, Public Health Nurses' Narrative Reports, July 1928, State Boards of Health, box 41, RG 90, United States Public Health Service (USPHS). Also see Mississippi State Board of Health, "Study of Midwife Activities in Mississippi, July 1, 1921–June 30, 1929," June 30, 1929, box 36; and Nurse Elois Conn of Amite County, newsclipping from *Southern Herald,* November 12, [1941?], box 27, RG 51, MSDAH.

43. Nurse Josie Strum, Clarke County, "Nurses Narrative Report, Mississippi State Board of Health, Bureau of Child Hygiene, Division of Public Health Nursing, July 1931," box 36, RG 51, MSDAH.

44. Quoted in "State Health Agency Calls Midwifery 'Dying Avocation,'" newsclipping in *Jackson Commercial Appeal,* [1966?], box 36, RG 51, MSDAH.

45. Dr. Felix J. Underwood, "Midwife Activities in Mississippi," [1932?], box 354; and Mississippi Department of Health, "Policies Regarding Midwife Supervision," June 1, 1948, box 36, RG 51, MSDAH.

46. "Report of Work of Mollie Gilmore, Midwife, for July 1936 to December 1937," box 27, RG 51, MSDAH.

47. Photo of a midwife and Mrs. Robley, possibly the nurse, showing a delivery room, no date, box 21; Photo entitled "Model Midwife Delivery Room" in Sharkey County, 1953, box 36; sketch of a model delivery room, no date, box 27; and Nurse Elois Conn, newsclipping from *Southern Herald,* all in RG 51, MSDAH.

48. Mississippi State Board of Health, "The Relation of the Midwife to the State Board of Health," January 1, 1935, box 2; newsclipping entitled "Midwives and Health Work," [1933?], box 36; and Mississippi State Board of Health, "Midwife Supervision," [1938?], box 36, all in RG 51, MSDAH. See also Roberts, interview by author.

49. Dr. W. E. Riecken, Jr., written responses to author's written questions, July 17, 1990; and Ferguson, "Mississippi Midwives," 89.

50. Nurses Abbie G. Hall and Caroline Bourg, Sharkey and Issaquena Counties, "Nurses Narrative Report, Mississippi State Board of Health, Bureau of Child Hygiene, Division of Public Health Nursing, July 1931," box 36, RG 51, MSDAH.

51. Midwife Estelle W. Christian to Nurse Viola M. Jones in Claiborne County, October 5, 1939, box 27, RG 51, MSDAH.

52. Mississippi State Board of Health, "The Relation of the Midwife to the State Board of Health," January 1, 1944, box 57, RG 51, MSDAH. See also Allan M. Brandt, *No Magic Bullet: A Social History of Venereal Disease in the United States Since 1880* (New York: Oxford University Press, 1987).

53. Letter from unknown midwife in Raleigh to board of health, no date, box 27; and Nurse Fannie Mae Howell of Holmes County, "Nurses Narrative Report, Mississippi State Board of Health, Bureau of Child Hygiene, Division of Public Health Nursing, July 1931," box 36, RG 51, MSDAH.

54. Nurse Ethel B. Marsh of Adams County, "Nurses Narrative Report, Mississippi State Board of Health, Bureau of Child Hygiene, Division of Public Health Nursing, July 1931," box 36, RG 51, MSDAH.

55. Nurse Nell E. Austin of Forrest County, "Nurses Narrative Report, Mississippi State Board of Health, Bureau of Child Hygiene, Division of Public Health Nursing, July 1931," box 36, RG 51, MSDAH.

56. Mississippi State Board of Health, "The Relation of the Midwife to the State Board of Health," January 1, 1935, box 2, RG 51, MSDAH. See also Nurse Mary L. Gregory, Narrative and Statistical Report, April 1924, Mississippi State Board of Health, Bureau

of Child Hygiene and Public Health Nursing, Central File, 1921–1924, box 249, RG 102, NA.

57. Midwife leader in Smith County, quoted in "The Relation of the Midwife to the State Board of Health," Mississippi State Board of Health, January 1, 1935, box 2, RG 51, MSDAH.

58. Midwife report from Simpson County, no date but 1930s or 1940s, box 27, RG 51, MSDAH.

59. Letter from unknown midwife to board of health, no date, box 36, RG 51, MSDAH.

60. Brooksie W. Peters for the nursing staff of the Lauderdale County Health Department to Lucy E. Massey, July 14, 1947, box 36; and Dr. Andrew Hedmeg, Jackson County Health Department, reporting on nurses' suggestions to Lucy E. Massey, September 8, 1947, box 36, RG 51, MSDAH. Florida's midwife retirement efforts also began in the 1940s. Susie, *In the Way*, 47.

61. Mississippi State Board of Health, "Policies Regarding Midwife Supervision," June 1, 1948, box 36; and memo from Lucy E. Massey to Dr. Felix J. Underwood, February 2, 1948, and memo from Dr. Felix J. Underwood to Massey, February 2, 1948, box 36, RG 51, MSDAH.

62. Roberts, interview by author.

63. Loftus, "Stork Loses Two Long-Time Forrest County Helpers."

64. By 1951 there were only about thirty white midwives. Dr. D. Galloway and Nurse Louise Holmes, Mississippi State Board of Health, to Dr. Lucille Marsh, U.S. Children's Bureau, December 3, 1951; "State Health Agency Calls Midwifery 'Dying Avocation'"; Note to Miss Ferguson on midwife figures as of December 1975, all in box 36, RG 51, MSDAH. See also "Grannies: The Roots of Midwifery"; Jelks, "My Twenty-Five Years in Public Health"; and Roberts, interview by author. A similar pattern held in Florida, with only a few registered midwives by the 1980s. Susie, *In the Way*, 55.

Health Care Providers: Nurses

Following the model established by Florence Nightingale in England, professional nursing began in this country after the middle of the nineteenth century. Before that time women, and some men too, had provided (often as patients) haphazard nursing services in the few hospitals that existed. Women traditionally had also provided nursing care to the families and friends within their homes. With the new training schools for nurses and the development of organized service, however, an identifiable occupation group emerged.

Many of the newly trained nurses found work in private homes, where they replaced family members as caretakers of the sick. Susan Reverby examines the sometimes anomalous position of these new professionals in private duty nursing, and she analyzes the tensions in their lives within the social and medical context of family health care. Often regarded as servants, white nurses worked hard to overcome the daily slights and establish their professional authority. Darlene Clark Hine demonstrates that trained African-American nurses, although perceived as inferior by their white colleagues, were held in high regard in their own communities, where they were seen as culturally sensitive, essential, and competent health care professionals.

One of the questions historians of nursing have often addressed is nursing autonomy and nurses' place in the hospital hierarchy. The hospital work environment that most twentieth-century nurses inhabit is very different from the home base of early private duty nursing. The divisions of labor within the hospital provide some new opportunities for nursing even within the stringent medical hierarchy. But if nursing remains subordinate to medicine, and—even more important—if the twentieth-century gender division between nursing and medicine continues, as historian Barbara Melosh has asked elsewhere, is nursing sufficiently autonomous to be described as a profession?

Emily K. Abel and Nancy Reifel examine a different part of the autonomy question by turning our attention to encounters public health nurses had with Native American clients on reservations during the 1930s. They examine the cultural meeting of two types of health practices and find that Sioux women accepted the trained nurses' practices and resources only when they met specific needs that Sioux defined as important. Abel and Reifel demonstrate that the Sioux women's needs and actions set a significant part of the context for nursing practices.

24 "Neither for the Drawing Room nor for the Kitchen": Private Duty Nursing in Boston, 1873–1920

Susan Reverby

"Neither for the drawing room nor for the kitchen" was how nursing leader Isabel Hampton Robb succinctly captured the ambiguous position of the private duty trained nurse working in a patient's home in 1900.[1] While this nurse's social standing in a household was uncertain, her particular position in the health care system was becoming clearer. At the turn of the century most Americans when ill, even seriously, took to bed at home, not in a hospital. If it appeared necessary and/or could be afforded, at the bedside stood a hired nurse. The omnipresent harried staff nurse employed by a hospital is a figure whose vintage dates only from the World War II years. Until then the majority of nurses we would now refer to as *registered nurses* worked in private duty as the employees of patients; and until the mid-1920s most of them did this work in patients' homes.[2] This article explores the conditions and practices of this historically critical form of caregiving and women's work.

To uncover the private duty world this essay draws upon nurses' written memoirs, letters, texts, journal articles, reports of nursing meetings, Census data, and a sample of 539 nurses taken from the records of the 4,550 nurses who registered for private duty work at the Boston Medical Library (BML) Directory for Nurses, the first major registry for nurses in this country. By the last quarter of the nineteenth century, directories or registries, run by hospitals, medical and nursing societies, or private businesses, were organized to keep lists of available nurses and to match them, for a fee, to the needs of inquiring physicians or patients. Between 1880 and 1914 the BML directory provided work for the majority of Boston's private duty nurses.[3]

The ubiquitous female ministering to the needs of the sick was not always a relative or an altruistic neighbor; women who were paid to nurse made their appearance in the colonies in the seventeenth century. Such "nurses" gained their label through self-proclamation coupled to some kind of experience of caring for the ill in their own families, domestic service, or hospital work.[4] In 1873 the first nursing schools were linked to hospitals and created to train women to nurse. By 1900 neither titles nor job functions clearly differentiated the "old-style" nurses from the new graduates. The label "nurse" was applied to a graduate of a two-or three-year training school, a nursing school dropout, an exper-

SUSAN REVERBY is Professor of Women's Studies at Wellesley College, Wellesley, Massachusetts.

ienced worker with no formal education, a person who had taken a few lessons at the YWCA, or an attendant who had worked in a hospital. Lack of uniformity characterized even those who received "training" as the education was frequently haphazard and unstandardized. A nursing leader lamented in 1893 that the title "'trained nurse' may mean then anything, everything, or next to nothing."[5] The nursing associations struggled to have licensing laws enacted to differentiate the trained or graduate nurse, as she was labeled, from her many untrained competitors. But even the strongest of these generally weak registration laws could not limit which "nurses" sought work in the private duty market.

Sheer numbers were part of the problem. There were fewer than 500 graduate nurses in the United States in 1890. The numbers rose 634 percent by 1900, another 136 percent between 1900 and 1910. In Boston the total number of nurses, both trained and nontrained, rose nearly 400 percent between 1880 and 1905.[6] By the turn of the century the supply of nurses began to outpace the demand for their services, particularly in the urban centers. The overcrowding in nursing was a function of the limited fields of employment open to the increasing numbers of women seeking work, exacerbated by the deliberate policies of the hospital training schools.

Many hospitals, once the idea of training women to nurse was institutionalized, began to staff their wards almost entirely with nursing students and untrained nursing workers. Control over nursing education, and who was admitted to the training, was often the subject of pitched battles between nursing superintendents and hospital administrators, trustees, and physicians. Under pressure from the hospitals to staff the wards as well as educate the students, nursing superintendents frequently sacrificed educational goals to the necessity of getting the work done. In many schools a two-track system

developed. A few students were encouraged upon graduation to become head nurses or nursing superintendents in the hospitals, while the majority were shunted into the increasingly crowded and undifferentiated private duty field.[7]

Private duty work quickly took on a peculiar ambiguous status in the nursing and medical world. A medical student completing his training went into private *practice;* the nursing student, however, went into private *duty*. The physician was expected to apply his skills in independent action. The nurse, even without the control of the hospital and medical hierarchy, was still supposed to be submissive to higher authority and morally obligated to her work. In private duty a nurse was working for a *doctor's* patients. Although employed by a family, she was primarily dependent on the physician to define what she did and to help her get work.

While devoted care to one private patient would seem to be the ultimate expression of a nurse's skill, the nursing superintendents expressed grave concern about the dangers of private duty and feared their students would find the work "exhausting." The exhaustion that worried them was both physical and spiritual, a loss of sheer strength and moral fiber. It was frequently assumed that a nurse could last only ten years in private duty, her collapse owing as much to the danger of "moral laxity" as to the physical labor.[8] Warnings were issued about the danger of nursing single men in hotels (no respectable woman should) or the tempting advances of a patient's unscrupulous husband (to be spurned at all cost), along with admonitions to get enough rest. These warnings reflected less a fear of the vanquishing of the nurse's physical virginity than of her loss of spiritual virginity— the collapse of her moral purity, her gentleness, humanity, sympathy, and tact because of the long hours and strain inherent in the work.

In private duty a nurse provided an array of

services from the purely domestic to skilled nursing care. As might any domestic servant, mother or wife, she often had to be an entertaining companion, an imaginative cook, and a competent laundress. She also had to monitor vital signs; insert catheters; prep and assist at operations and deliveries; provide cold packs, baths, and massages; even carry and decide when to give such medications as morphine. Alone with a patient twenty-four hours a day, often weeks at a time, she crossed the ambiguous line between nursing and medical care with regularity, if trepidation. Freed from the hierarchy and controls of the hospital setting, she could practice her best skills with relative autonomy or make terrible mistakes without supervision.[9]

The stress on the graduate nurse in private duty was due to the pressure to prove the role of the trained nurse; the ambiguity of her place in the household structure; the difficulty of the work; and the overcrowding in the field. In the 1880s and 1890s the trained nurse was still a new creation whose necessity had to be proven to both physicians and families. Although much of what she did seemingly could be done and was done by nontrained nurses, domestic servants, or female relatives, she was somehow through her personality, bearing, and character to present herself as a new and vital creation, necessary to patient survival, worthy of being paid a high wage. Many physicians were not convinced that this kind of nurse was necessary for their patients and perceived hospital-trained nurses as a threat. Physician preference, patient income, and the nature of the illness often determined what kind of nurse was hired and for how long.

Working in a patient's home at the time of the illness created a number of different stresses for the nurse and the family. In the patient's home, the nurse was an individual confronting a family social system. In the hospital the patient was the lone individual, subject to a set of defined rules and a structured hierarchy. A private duty nurse warned that hard work, with many patients, "under some circumstances [may] demand much less wear and tear on the nervous system than that consequent upon the supervision of her own solitary self while engaged in nursing one patient in the bosom of that patient's family."[10]

The nurse also had to learn to make do in the home without all the equipment, supplies, and paraphernalia which even then marked hospital care. The author of one of the many manuals on "how to be a good private duty nurse" recounted the story of an overzealous private duty nurse who "thought she was distinguishing herself by extreme neatness, used to put thirty-five sheets in the wash in a week. She defeated her own end, for the laundress, naturally thought this a folly, and smoothed out those that looked clean, without washing them."[11]

The most obvious contrast between hospital and home-based care was the lack of structure in private duty. A system and schedule for performing duties was the sine qua non of the hospital training schools. But as one private duty nurse cautioned, "Indeed a too loyal adherence to one certain system may prove a huge stumbling block in the way of success."[12] The very work rhythms of the home and hospital differed. In private duty a nurse's success depended on her ability to reset her work to the demands and whims of the patient and family, not in a premeasured routine set to a rigid schedule. Private duty nursing consisted of the performance of a series of tasks whose order was determined by the ups and downs of the patient's illness and the needs of the family, as much as a farmer's work depended upon the weather and seasons. The patient and the family, not the work as in the hospital, had to be the center of the nurse's attentions.

The difficulty of private duty work was compounded by the contrast between the class of patients in the hospitals and that of the families who hired private nurses. In the hospital the nurse confronted a patient population of primarily working-class men and women. In the home, however, most nurses were working for families whose class position was usually higher than their own. In a sample taken of the families who hired nurses through the BML Directory, more than 50 percent of the male family heads were either lawyers, owners of companies, or merchants. The others were skilled or white-collar workers and professionals. While the nurse could as easily be called to the home of a skilled bricklayer in a Dorchester triple-decker as to the bedside of a tea merchant in an elegant townhouse on Beacon Hill, few private duty nurses were hired to work in the tenements and boarding houses to care for the majority of Boston's populace.[13]

Once in a patient's home, there were few guidelines to govern social relations for either nurses or families. Nurses, as a character in a nursing novel explained, were "always afraid of being asked to do too much. They're always afraid of being treated like ordinary servants."[14] Nurses were told during training that there was nothing which was beneath their dignity to do, but once in a patient's home a nurse had to draw the line and decide for herself what was a reasonable demand. If a nurse washed out a baby's clothes in one home because there was no laundress, should she be expected to do so in another home where such a domestic servant was employed? Should a family be subject to a nurse's wrath because she was asked to eat at a second table or in a kitchen with the servants?

The unspoken rules of class conduct between employers and servants were continually violated by private duty nurses. Clashes were inevitable between a family's expectation of servantlike behavior and the nurse's need to assert her standing above that of servants and to establish her autonomy. Patients' objections to the nurses echoed those made to servant girls unfamiliar with the furnishings of a crowded bourgeois Victorian home. Nurses were faulted for their clumsiness with precious objects, lack of knowledge of how to handle exotic pieces of furniture, and willingness to use expensive items carelessly.

At a time of illness and stress small mistakes in social conduct by the nurse became magnified and compounded by the fears of death and disease which pervaded a household. "The families want to know 'what are the rules,'" a Philadelphia physician said. Yet, as one nursing superintendent noted, "there are no definite rules to be observed."[15] A nurse's inability to correctly judge the unwritten rules could be costly. In 1892 a Philadelphia nurse angrily reported:

> If by any chance a nurse gains the ill will of her first few patients, her career is ended. She is not told anything of this, simply waits in her boarding house until her last dollar is gone, . . . in suspense . . . and wondering why a "case don't come."[16]

The relationship of the nurse to the household's other servants was one of the problems of social conduct which caused the biggest difficulty. The dilemma centered on where the nurse would take her meals. Nurses often insisted (to make sure they were treated as ladies, not servants) that they be served their meals in the dining room with the family rather than in the kitchen with the servants. This demand was usually opposed by patients used to treating the nurse as a servant and uncomfortable about sharing their dinner conversations with a stranger from another class. To lessen the conflict on this question the BML Directory, for example, asked both trained and nontrained nurses on their application forms whether or not they would take their meals in the kitchen with the servants. In the sample from these records only 40 percent of the trained nurses were willing to eat with the servants, as opposed to 74 percent of those without formal nurses' training. Of those not willing to eat with the servants, 82 percent were trained nurses and less than 20 percent were the nontrained nurses.

Managing relationships to the family's servants required tact as the nurse had to be understanding but distant, above but not superior. A graduating nursing class at the Boston Training School for Nurses at the Massachusetts General Hospital was told by a physician: "Never assume an air of superiority when dealing with the servants; but on the other hand, never be too familiar with them. At best they recognize your superior position unwillingly, therefore do all you can to conciliate them."[17] The countless stories of the overbearing dictatorial nurse who left the household in an emotional and physical uproar suggest that finding a path to conciliation was not always easy. In desperation families often turned to the more expensive private room in the hospital or called for only an "old-style" nontrained nurse since in both situations the social relationships were clearer.

The letters and work records in the BML Directory suggest that the differences and difficulties in private duty existed between the trained and nontrained nurses as well. Class, age, and marital status often differentiated the two groups. Trained nurses, especially graduates of the larger and more elite schools, were likely to be middle to lower middle class in origin, while the nontrained nurses drew more heavily from women from the working class. However, graduates of the more numerous smaller training schools were more likely to be working class, compounding the difficulties of differentiating them on class grounds from the nontrained women. Except for large numbers of Irish and English Canadians from the Maritime Provinces, trained nurses in Boston were overwhelmingly native born. They were also much younger than the nontrained women: the graduates' average age was twenty-nine, while the nontrained average age was thirty-six. While the nature of nursing work demanded that its practitioners not be encumbered by family responsibilities, the nontrained nurses were more likely to be widowed or divorced, while 92 percent of the trained nurses had not yet married.

Few of the nurses of either type shared households with male relatives. As did other women workers, they crowded into the boardinghouse districts of Boston's South and West Ends. Many of the nontrained nurses were women already in Boston, who took up nursing as an extension of familial or domestic servant duties, while the graduates were more likely to be single women who had come to Boston or nearby towns to do their training and then stayed on in the city to work. Trained nurses, because they were younger and more often single, were twice as likely as the nontrained to leave Boston.[18]

Turnover figures had a morbid side as nursing was a dangerous occupation. Death reaped a much higher proportion of nurses than women of comparable ages in the Boston population. The BML Directory had a death rate of 16.36 per thousand, a rate greater than that for any adult age group, except those over fifty, in the general population. Even here, probably because of age and class, the nontrained nurses were twice as likely to die as the graduates.[19]

The work experiences and career patterns of the nurses varied considerably. The nontrained women had about two years more experience in nursing than the trained nurses, but both groups remained in the BML Directory on average for eight years. This figure does not measure their work commitment since dropping out of the Directory's records did not mean either giving up nursing or some other form of employment. Some nurses worked during their entire lifetimes, regardless of marital status or training. Others dropped out into related fields: to sell surgical equipment, to operate rest homes, or to work in training schools and hospitals. Some changed fields completely. Still others dropped out at marriage and never returned, while some came back to nursing when their children were older or widowhood necessitated employment.

While these nurses differed in age, morbidity, marital status, experience, and training, they competed in the same overcrowded labor market for the same jobs. It was certainly not clear to the physicians or patients who hired the nurses what skill really differentiated these women. Wage rates, specialities, and the seriousness of the patient's illness, however, were all factors in determining who was hired.

Wage rates clearly differentiated the nurses, as the graduates tended to charge the patients almost five to ten dollars a week more than the nongraduates. From the 1880s till the mid-1890s graduates received fifteen to eighteen dollars a week; by the late nineties they were commanding twenty to twenty-five dollars. But a graduate nurse just out of school and a graduate nurse with ten years' experience were paid the same. Neither trained nor nontrained nurses were rewarded for their experience with a higher wage. There was clearly a "customary" wage established for each group, although there was a slightly greater dispersion in the fees asked for by the nongraduates.[20]

Nurses were allowed in most registries to state their preferences on cases. Comments such as "only surgical cases" or "no obstetrical cases" cover the application forms. Trained nurses specialized more and were less willing to care for postpartum patients, presumably because such work almost always guaranteed they would be asked to do household labor. Nontrained nurses, in contrast, more willingly took what work they could get.[21]

The graduate nurses clearly felt their livelihood was always threatened by the nongraduates. There were frequent complaints that the nongraduates were receiving most of the work or that the directories did not apportion the work equitably.[22] But at least through the BML Directory it appears the nongraduates were the ones suffering discrimination. The graduates received consistently two or three times as much work as the nongraduates and about 20 percent more than they should have, given equal distribution.[23] Miss E. L. Blanchard, a nontrained nurse bitter over her lack of work, wrote to the physician in charge of the BML Directory in 1895: "I feel it is a . . . way of pushing one out, for opening the way for those younger, and . . . in the training school for a few months. . . . Thus . . . experience goes for nothing." Another nontrained nurse explained the dilemma by pointing out that her costs were too high for the poor and too low for the rich, who would, she believed, rather employ a graduate.[24]

Thus it appears that in the 1880s and early 1890s the trained and nontrained nurses in Boston were competing directly with one another. By the late 1890s the competition was caused by the enormous number of graduates entering the labor market. By 1915 in Massachusetts as a whole the trained nurses outnumbered the nontrained two to one.[25]

Competition then developed between the graduates of the different training schools. Physicians at the BML Directory told graduates of the smaller schools that their training was not equal to that of nurses from the more prestigious hospitals.[26] Graduates of the more elite schools demanded and received higher wages than those from the smaller schools and more often demanded a wage differential to nurse male and contagious patients. The graduates of the larger schools also received slightly more cases, although the reason for this may have been that they were better known by local physicians.

Class background as much as training differences was probably the reason for this division among the graduate nurses. Graduate nurses from Long Island Hospital, the site of the city's asylum, were told they probably would not receive cases if they charged the graduate nurses' going rate.[27] One of the physicians in charge of the BML Directory bluntly wrote:

> It is preposterous to put the Long Island Hospital nurses on the same plane as those of the Mass. and City Hospitals. Certainly our patrons would not accept it. When we consider the class from which they come, their lack of education, etc. . . . the answer would seem to be obvious. The Long Island nurses are worth say from $10–15 and I could not with any feeling of fairness send them to first-class families and serious cases with a supply of the others on hand.[28]

Nursing employment in private duty for every type of nurse was sporadic, seasonal, and uncertain. The average graduate received only 3.2 cases a year from the BML Directory, the nongraduates 2.3; this multiplies out to be a work time, on average, of thirteen and nine weeks a year, respectively. In contrast, a Cleveland nurse reported that in her eleven years of private duty work from 1895 to 1906 she worked thirty-three to thirty-five weeks each year. The only time she was employed fully was the fifteen months she spent on a "luxury" case as a companion, as much as a nurse, for one wealthy patient. A comparison to other women workers in Massachusetts in 1890, however, shows nurses and midwives as sixth in unemployment frequency out of a list of twenty-five other women's occupations.[29] A nurse, trained or not, did make more money *when* she was employed than most other working women.[30] But the wait between cases was often so long that the extra wage could not make up for the unemployment.

The expense of the nurse also limited her calls. As early as 1888 nurses were aware that

even what they called the "breadwinning middle class" could not afford their services. But because their own need for work made nurses feel they could not have an official sliding scale for patients, a nurse suggested that those who could not afford them should be "either taken to the hospital or put on the list of the visiting nurses."[31]

Nursing work was also seasonal, and the nurses, as much as any industrial workers, recognized that there were "slack" times. There was definitely a higher call for nurses in the midwinter months of January through March and a slowdown between May and July, caused perhaps as much by the exodus of the rich from the city during the summer as a difference in disease incidence. Similarly, the demand for nurses dropped during the economic depressions.[32]

One nurse echoed a common theme while reproaching the directory in 1895: "If I had depended entirely on the Directory for a living, of course I should have starved long ago."[33] But registering with a number of directories was still no guarantee of work. A nurse could and often did spend weeks at a time waiting at her boardinghouse for cases from any source. While the nursing leadership bemoaned the nurses' increasingly "mercenary spirit," a San Francisco nurse asserted that they could not afford to be "angels of mercy."[34] Worry and anxiety over where they would get their next case dominated the thinking of many nurses. Poignantly, a California nurse warned Boston nurses not to come to the West Coast and of the danger of "the gradual fading away of resources, courage, hope—too often self-respect—and sometimes suicide. The papers here suppress all that."[35]

One solution to the difficulty of finding work was for a nurse to "attach" herself to a physician and hope he would send her all his cases. Nurses had to make the rounds of physicians' offices to announce their presence and then wait to be called. Contacts made during training were critical to a nurse launching her career. If the nurse ventured to a new city, finding work took even more time. But nurses also competed for the patient dollar with the doctors since it was not just nurses at the turn of the century who were in oversupply.[36] A New York nurse complained in 1897 that a physician had her fired from a case and then began to make more visits to the patient himself.[37] There was the danger for physicians who had taken short apprenticeships or correspondence courses that a nurse with several years of hospital training might in fact know more scientific medicine.[38] The intensity of the economic competition makes more comprehensible the constant ideological stress by physicians on the need for the nurse to know her place and remain "loyal." But physicians held the master key which opened the employment door. An informal blacklist of sorts also circulated among physicians and directories of both dropouts and troublesome nurses.[39] Despite the fact that nurses were usually employed by the patients, in actuality they had to behave as if the patient's physician was their boss.[40]

Despite the uncertainty and limits on their autonomy, nurses fashioned a variety of *individual* means for surviving, reshaping, and enjoying the work. The stories of patients who took their nurses to Europe on trips, married them, left them large sums of money, or employed them for decades, however unusual and idiosyncratic, suggest that, as with servants and governesses, the step up in nursing could mean marriage or life as the family retainer. Nurses also found ways to create alliances in the household. Reexamining the figures on eating with the servants makes clear that nearly 40 percent of the graduate nurses and 74 percent of the nontrained nurses were willing to eat in the kitchen. Nurses quickly learned that a recalcitrant servant could make their lives miserable. Letters rebuking nurses for being too friendly

with the servants and gossiping with them about the patient's illness and the household life suggest that congenial working relationships frequently developed.

Nurses never lacked resourcefulness in asserting some kind of control of their work. Especially on long convalescent cases, they would sometimes leave patients to go home, to go to the theater, or to visit friends or family. Some nurses had other businesses and conducted their other work while on a case. If the job was really miserable, there were ways to be relieved. One nurse was disciplined for having another nurse call and lie for her, saying an aunt was ill and she was urgently needed at home. The nurse, when discovered, said she left the case because the patient was "poor" and conditions of work were terrible. Other nurses took to sleeping while on the job, or abandoning patients, or refusing to take cases they had agreed to.[41]

These attempts to control the work were perceived by the public and many nursing superintendents, however, as counter to devotion to duty. Nurses were, after all, expected to be more like mothers than workers, forever on call, cheerful, and devoted. In a 1904 editorial entitled "The Path of Duty," for example, the *American Journal of Nursing* chastised nurses for refusing to take cases: "Such failure to meet our highest obligations, such violation of our common standards of right and duty, cannot be too sternly censured. The women who permit themselves to conduct their professional work in this manner are in this, at least, wrong through and through."[42]

Different kinds of organized efforts were made to introduce some rationality into the private duty labor market and to distribute the work more evenly. Those in charge of the BML Directory, for example, tried to influence the wage rate, often to the detriment of all the nurses. With nurses of equal skill the Directory registrars would send out the nurse with the lowest fees.[43] Lavinia Dock, the outspoken socialist and feminist nursing leader, counseled nurses: "We must not undersell; that is treachery to fellow workers."[44] But Anna Maxwell, the nursing superintendent of Presbyterian Hospital in New York, wrote the Directory in Boston in 1899 to ask if there would be work if more graduate nurses lowered their wage rates.[45] In *principle* graduates tried to keep to an agreed-upon wage rate. But it was not uncommon for a nurse to lower her rates when necessary or, more often, to overcharge the patient on items such as carriage fares and laundry bills as well as on her fees.[46]

The Directory officials also attempted to rationalize the system by grading the nurses by experience and training. This was done on a haphazard basis since they had to rely upon the nurse's self-reporting and incomplete patient and physician recommendations for their assessments. Registry officials remained uncertain, however, about their legal right to discipline nurses. All they ultimately could do was to refuse to put the nurse on their lists, but she could go elsewhere.[47]

Much of the nursing concern over the lack of work focused upon the registries. "Most graduates," a nurse reported, "do not feel that they are fairly treated by the Directory, but are afraid to complain for fear that it will be visited upon them."[48] In an attempt to equally distribute employment, when a nurse reported off a case, her name was supposed to be placed at the bottom of the rotation list. But, nurses charged, the registries played favorites, did not always follow this system, and could not control whether or not a nurse received a "good" case which guaranteed employment for a length of time in a decent home.[49]

Aware of the discontent over the registries, the extent to which both physicians' groups and commercial agencies were profiting from such services, and the necessity to control distribution, some nurses began to advocate the organization of nursing-controlled, centralized, and officially sponsored registries in each city. In Boston, for example, the BML Directory was closed as graduate nurses began to register with the nurses' officially sponsored Suffolk County Nurses' Directory. One nurse commented, however, "It is not so much a share in the government of directories, as a share in the work given out by them that is asked by the majority of nurses. . . . Each one should have a share."[50] There was no guarantee, however, that a nursing-controlled registry would mean any more work. In fact, nurses admitted that the commercial agencies often allowed the more experienced trained nurses to charge more, thus making them more attractive than the official registries, which set one rate.[51] The official agencies often enrolled only graduates from the "better schools," did not provide nontrained nurses, and were thus less able than the commercial registries to meet the varied community demand for nurses. Hospital and alumnae registries often served only to provide the institution's own graduates with positions as private duty "specials" in their hospital.

Despite the overcrowding, private duty work continued to absorb the majority of nursing graduates because there was very little else they could do to remain in nursing. Some private duty experience was considered essential for every nurse, but status in nursing quickly accrued to those in more "executive" positions within hospitals or public health nursing agencies. Private duty nursing, seen even within nursing as often no more than domestic service or a mother's work, gave those who did it little status. Furthermore, because there was no supervision, women with the weakest skills could hide in private duty. With the growth of hospital-based care in the early 1900s and the decline in the debilitating sicknesses which required more long-term nursing care, the status decline of private duty nursing increased. By the late 1920s a major study of nursing's dilemmas could repeat the aphorism "Every nurse ought to do some private duty, but no good nurse ought to stay in the field more than a few years."[52]

Aware of the overcrowding as early as the 1890s, Lavinia Dock counseled nurses to specialize "by branching into auxiliary lines of work not strictly nursing, yet which can be better done by one having the training of nurses." Among Dock's suggestions were heading various departments in hospitals, becoming a dietician or pharmacist, directing nurseries, old people's homes, social service, settlement work, medicine and massage, or returning to nurse in small towns and the countryside. Her suggestions entailed the nurse's specializing *outside* of private duty work itself.[53] Lucy Drown, the superintendent of nurses at Boston City Hospital, was blunter in her advice. Sharing her concern over the lack of work with the physician in charge of the BML Directory, she wrote:

[there is] . . . less survival of the fittest [in nursing] than in some other walks of life. My advice to these young women would be that if they cannot make a place for themselves as nurses, to go back to the work they left when they came into the hospitals [for training], for they belong to the working class, and maintain themselves as teachers, stenographers, dressmakers, etc.[54]

But in fact, most stayed in nursing, waiting for cases, trying to find different kinds of nursing work, "making do" in an increasingly untenable form of employment.

The isolation of the work site, the severe competition for cases, and the acute divisions within the nursing ranks all worked to limit the forms of control over their labor nurses as a group could exercise. Individual nurses, sometimes with the help of others, found ways to make the work less difficult and to achieve a modicum of autonomy. But they could do little collectively to alleviate the competition and animosity between the trained and nontrained or to transform the economic structures which underlay and created the private duty system.

Sporadic efforts were made to find a more organized solution to the private duty dilemma. While waiting for a case or isolated in a patient's home, however, few nurses could attend regular meetings or put the time into the effort to sustain the few private duty leagues which briefly flourished. The professional nursing associations, dominated by nursing educators and focused on registration laws and educational reforms, often discounted the private duty nursing problems or hoped they would fade into oblivion. As women workers, providing services in the private employ of individual patients, nurses were similarly ignored by both male and female trade unionists.[55]

By the 1920s private duty was increasingly becoming a nursing backwater. Home-based private duty was relinquished to the nontrained nurses or the older graduates as the younger nurses sought employment in private duty as hospital "specials." There was a brief attempt to revitalize, rationalize, and reorganize private duty nursing at the end of the 1920s. But economics and hospital and nursing politics undermined these efforts. At the same time, secluded and many times embittered by their experiences, private duty nurses became increasingly conservative and isolated.[56]

The history of private duty nursing suggests some insights into the broader historical concern to comprehend how different groups of workers sought to gain individual and collective control over their labor. Historians of male artisans and skilled workers, in particular, have often romanticized this labor and described the "degradation of work" under monopoly capitalism in the twentieth century. As historians of women's work, most notably Susan Porter Benson and Barbara Melosh have argued, however, this "lament" for the lost artisan past is a very different tune when sung on higher notes. The nature of the sex-segregated and overcrowded labor market for women workers, the skills their jobs required, and the conditions under which they labored give their history and struggle a very different character from that of male workers.[57]

The private duty nurse cannot be equated with the independent craftsman or the skilled worker whose consciousness and struggles are the focus of much historical writing today.[58] Nor, however, was the private duty nurse in a position similar to women who labored next to one another in factories or department stores. The work site of private duty, the ideology of altruism and caring which pervaded this form of service work, the competition and lack of cases, in sum both the labor process and the social relations of production of this form of health care made the private duty nurse a particular kind of worker. She was indeed "neither for the drawing room nor for the kitchen" but also neither for the trade unions and professional associations nor for the powerful informal work groups.

It may well be that private duty nurses were sui generis. However, while private duty nursing was unique in its isolation and ambiguities, other women workers, workers of color, and the unskilled similarly faced high unemployment, competition for jobs, and lack of collective control over their work. Until we understand the political economy and work cultures of these less autonomous groups, we will have as partial an understanding of the American working class

as we have of the health workforce when we focus only on physicians.

Notes

This is a revised version of a paper presented at the Organization of American Historians Convention, April 15, 1978. Research for this article was supported by Grant Number 1 RO3 HS02879–01 from the National Center for Health Services Research, U.S. Department of Health and Human Services. As always, the trenchant comments of Diana Long Hall, David Rosner, Tim Sieber, and Lise Vogel were invaluable.

1. *Nursing Ethics: For Hospital and Private Use* (Cleveland: J. B. Savage, 1901), 32.

2. The exact date when Americans of all classes began to use the hospital cannot, of course, be set. For a careful analysis of this question see both Morris Vogel, *The Invention of the Modern Hospital* (Chicago: University of Chicago Press, 1979), and David Rosner, *A Once Charitable Enterprise* (New York: Cambridge University Press, 1982). For a discussion of the changes in private duty nursing in the 1920s, see Susan Reverby, *Ordered to Care: The Dilemma of America Nursing, 1860–1945* (New York: Cambridge University Press, 1987), and Barbara Melosh, *"The Physician's Hand": Work Culture and Conflict in American Nursing* (Philadelphia: Temple University Press, 1982), esp. ch. 3.

3. Boston Medical Library Directory for Nurses, A–V (J is missing), 1880–1914, Rare Books Room, Countway Medical Library, Harvard Medical School. The goodwill and support of Richard Wolfe, Carol Pine, and the entire Rare Books Room staff is gratefully acknowledged. The Directory claimed to represent the majority of Boston nurses (Boston Medical Library Association, *10th and 11th Annual Reports*, October 1887, 23). A comparison of the number of nurses in the Directory with the number in the Census for Boston suggests that, in any given year (with the exception of the closing years of the Directory), anywhere from one-third to three-quarters of Boston's nurses could be found in these volumes. The sample included 313 graduate nurses and 226 non-

graduates. The sample included only the female nurses, although the Directory did provide work for a small number of male nurses.

4. See Reverby, *Ordered to Care,* ch. 1.

5. Isabel Hampton Robb, "Educational Standards for Nurses," in *Nursing the Sick 1893,* by Isabel Hampton Robb et al. (New York: McGraw Hill, 1949), 5.

6. U.S. Bureau of the Census, *Historical Statistics of the United States* (Washington, D.C.: U.S. Government Printing Office, 1959), ser. B. 192–194; Carroll D. Wright, *The Social, Commercial, and Manufacturing Statistics of the City of Boston* (Boston: Rockwell and Churchill, 1892), 96–98; Secretary of the Commonwealth of Massachusettsets, *Census of the Commonwealth of Massachusetts* 2 (1905): *Occupations and Defective Social and Physical Conditions* (Boston: Wright and Potter, 1909), 138.

7. See Nancy Tomes, "'Little world of our own': The Pennsylvania Hospital Training School for Nurses, 1895–1907." The article was reprinted in the first edition of *Women and Health in America.* Originally it was published in *Journal of the History of Medicine and Allied Science* 33 (1978): 507–530; Jo Ann Ashley, *Hospitals, Paternalism, and the Role of the Nurse* (New York: Teachers College Press, 1976); Reverby, *Ordered to Care,* ch. 4, on the use of the student nurse in the hospital, and ch. 6 on a comparison of the career patterns of the graduates of different training schools.

8. Gertrude Harding, *The Higher Aspects of Nursing* (Philadelphia: W. B. Saunders and Co., 1919), 109; Sara E. Parsons, *Nursing Problems and Obligations* (Boston: Barrows, 1916), 115.

9. Katharine DeWitt, *Private Duty Nursing* (Philadelphia: J. P. Lippincott, 1913); Emily A. M. Stoney, *Practical Points in Nursing for Nurses in Private Practice* (Philadelphia: W. B. Saunders, 1897); Elinor Lason, "Characteristic Requisites for a Private Duty Nurse," *Report of the 18th Convention of the American Nurses Association,* June 1915, 65; E. B. M., Letter to the Editor of the *Transcript,* October 24, 1881, clipping in B, Boston Medical Library Directory for Nurses.

10. Annie E. Hutchinson, "Practical Nursing in Private Practice," *Trained Nurse and Hospital Review* 37 (August 1906): 83. "Don't imagine that you can discipline a patient in his own home as you would in a hospital ward. It can't be done," the *Trained Nurse* cautioned. "Don'ts for the Private Duty Nurse," 47 (October 1911): 202.

11. DeWitt, *Private Duty Nursing,* 70.

12. Hutchison, "Practical Nursing," 84.

13. A random sample was drawn of one hundred patients who hired nurses through the Boston Medical Library Directory between 1880 and 1914. The occupations of the male patients, or males in the family of the patient, were obtained by checking names and addresses in the *Boston City Directory* for the appropriate years. The class base of those who used the services of private duty nurses continued to be an issue in nursing and health care. For further discussion of this see Melosh, *The Physician's Hand,* ch. 3, and Susan Reverby, "'Something Besides Waiting': the Politics of Private Duty Nursing Reform in the Depression," in *Nursing History: New Perspectives, New Possibilities,* ed. Ellen Condliffe Lagemann (New York: Teachers College Press, 1983).

For quantitative evidence which suggests the changing class background of nurses, and the differences between the training schools, see Reverby, *Ordered to Care,* ch. 5; Jane Mottus, *New York Nightingales* (Ann Arbor: University Microfilms Books, 1981); and Janet Wilson James, "Isabel Hampton and the Professionalization of Nursing in the 1890s," in *The Therapeutic Revolution,* ed. Morris Vogel and Charles Rosenberg (Philadelphia: University of Pennsylvania Press, 1979): 201–44.

14. Brennan Gill, *The Trouble of One House* (Garden City, N.Y.: Doubleday, 1950), 142. This quotation was given to me by Barbara Melosh.

15. Dr. J. Madison Taylor, Letter to the Editor, *American Journal of Nursing* 4 (May 1904): 658. Stoney, *Practical Points in Nursing,* 24.

16. Philadelphia, Letter to the Editor, *Trained Nurse and Hospital Review* 8 (September 1892): 278.

17. William L. Richardson, *Address on the Duties and Conduct of Nurses in Private Nursing, June 18, 1886* (Boston: Press of George H. Ellis, 1886), 10. Richardson's address must have been very popular because it was printed and widely circulated.

18. The analysis of the nurses' class origins and

migration patterns is based upon samples drawn from both nursing school student records and the records of the Home for Aged Women in Boston which admitted nontrained women nurses. Other analysis is based upon the Boston Medical Library Directory sample compared with the Census and city directory evidence. The Directory does give addresses for about half the nurses. In vols. K, L, and O of the Boston Medical Library records, the hometowns for 220 nurses were given. These data were tabulated to determine where the graduates had gone to school before they came to Boston and to differentiate the nongraduates and graduates. For further discussion of class origins and migration patterns of nurses, see Reverby, *Ordered to Care*, ch. 1, 5, and 6.

The Boston Medical Library Directory does not give ethnicity data. The migration of women from the Maritimes into nursing may be a continuation of their earlier migration into domestic service in Boston; see Alan Brookes, "Migration from the Maritime Provinces of Canada to Boston, Mass., 1860–1900," master's thesis, University of Hull, 1974, 213–19. The role of Canadian nurses in the United States has not yet been examined, but these women were an important minority, especially in the leadership, from the 1870s through the 1930s.

Nurses were listed in the Directory either as "Miss," "Mrs." or "Mr." Marital status of the nurses given as "Mrs." was inferred by checking a subsample in the city directory and by using national Census data.

19.   This figure is based on all deaths *reported* to the Boston Medical Library Directory and is therefore an underestimate since not all deaths were reported. For all those who died *while registered* with the Directory, the average age at death for the men was only forty-one, for the women, thirty-seven. For comparative death rates for Massachusetts, see *Historical Statistics,* Bicentennial edition, ser. B. 201–213, 63.

20.   There is a statistically significant relationship between training and wages at the .00 level from 1880 till the close of the Directory in 1914, but no significant relationship between wages and years of nursing experience. On how the wages were set see Dr. Charles Putnam to Miss Gertrude Hamilton, August 7, 1913, Directory for Nurses, *Letterbook,* 1, February 11, 1891–January 7, 1914.

21.   This is not to suggest that the nontrained

nurses willingly did whatever the family wanted. One such nurse, for example, complained that families expected her to do the "spring housecleaning" along with the nursing. "A Graduate of the Training School of Life," Letter to the Editor, *Trained Nurse and Hospital Review* 51 (March 1916): 369.

22.   Richard Bradley, "Large Part of Hospital Work Performed in the Home," *Modern Hospital* 1 (November 1913): 227–31; Charlotte Aikens, "The Committee on Grading of Nurses," *Trained Nurse and Hospital Review* 37 (March 1914): 168; Lavinia L. Dock, "Directories for Nurses," *Report of the Second Annual Convention of the American Society of Superintendents of Training School for Nurses,* 1895, 59; Miss Hintze, Discussion on Dock paper, ibid., 60.

23.   There is a statistically significant relationship between training and total number of cases received throughout the Directory at the .01 level. Data on the number of nurses available to work, by type of training, were compared to the number, by type of training, who actually did the work, for the years 1889–93. The raw data for this latter calculation can be found in the *Letterbook.*

24.   Miss E. L. Blanchard to Dr. Brigham, February 11, 1895, Boston Medical Library Directory, E, 181; L. H., "What Shall She Do?" Letter to the Editor, *Trained Nurse and Hospital Review* 23 (January 1899): 42.

25.   Massachusetts Bureau of Statistics, *The Decennial Census, 1915* (Boston: Wright and Potter, 1918), 510.

26.   Dr. Putnam to Miss McBrien, January 20, 1894; Dr. Putnam to Dr. Brigham, September 10, 1894; Dr. Putnam to Miss Heintze, November 19, 1892, *Letterbook.*

27.   Alice N. Lincoln, secretary of the board, Long Island Hospital, to Dr. E. W. Taylor, October 23, 1899, *Long Island Hospital Letterbook,* 1, Long Island Hospital Collection, Rare Books Room, Countway Medical Library.

28.   Dr. Brigham to Dr. Putnam, November 3, 1899, Boston Medical Library Directory for Nurses, *Letterbook.*

29.   Case rates calculated from averages in the Boston Medical Library Directory sample. On the Cleveland nurse's experience see James H. Rodabaugh and Mary Jane Rodabaugh, *Nursing in Ohio* (Colum-

bus, Ohio: Ohio State Nurses' Association, 1951), 199–200. 1890 Massachusetts data were compiled from the 1890 Census and were given to me by Professor Alex Keyssar of Brandeis University from his ongoing study on unemployment in Massachusetts. Unemployment frequency was calculated by dividing the total unemployed by the total number in the occupation. Unfortunately, these numbers include both nontrained and trained nurses, as well as midwives.

30. In 1900, for example, the average weekly wage for domestic servants in Massachusetts was $3.61; in that year the nongraduates in the Directory were averaging $11.25; the graduates $20.93. For the domestic servant data see Stanley Lebergott, *Manpower in Economic Growth* (New York: McGraw Hill, 1964), 542. For comparisons to other women workers in Boston see Louise Marion Bosworth, *The Living Wage of Women Workers.* (New York: Longmans, Green, 1911), 33–39.

31. "The Social Side of Nursing," *Trained Nurse and Hospital Review* 2 (March 1888): 95–97; Letter to the Editor, *Trained Nurse and Hospital Review* 39 (October 1916): 369.

32. Daniel Gormon to Dr. Putnam, June 14, 1894; Elizabeth Bowness to Miss McBrien, December 9, 1903, *Letterbook.* The raw data on the number of nurses requested each month for 1891 through 1914 can be found in the *Letterbook.* This drop-off in demand was especially precipitous during the 1893 depression.

33. B. H. Giles to the Directory, March 10, 1895, vol. F, Boston Medical Library Directory, 176. For examples of letters which also reflect the nurses' bitterness at the lack of work, see Emilie Neale to Dr. Brigham, January 8, 1894; Dr. Putnam to Miss McBrien, February 24, 1900, *Letterbook.*

34. Stoney, *Practical Points,* 18; Henry Beates, Jr., M.D., *The Status of Nurses: A Sociologic Problem* (Philadelphia: National Board of Regents, 1909), 17.

35. Letter to the Editor, *Pacific Coast Nursing Journal* 3 (March 1914): 130.

36. Gerald Markowitz and David Rosner, "Doctors in Crisis: Medical Education and Medical Reform during the Progressive Era, 1895–1915," in *Health Care in America,* ed. Susan Reverby and David Rosner (Philadelphia: Temple University Press, 1979), 185–205.

37. "Report of the Monthly Meeting of the New York City Training School Alumnae, June 8, 1897," *Trained Nurse and Hospital Review* 19 (July 1897): 34.

38. "The Reasons and the Remedy—a Training School Symposium," *National Hospital Record* 11 (1907): 14–20; "Boston trained nurse in Chicago," *Boston Transcript,* October 24, 1881, clipping in B, Boston Medical Library Directory.

39. Memo from Dr. George Rowe, superintendent of Boston City Hospital, to the Directory, no date; numerous other letters in 1892, *Letterbook.*

40. "The members of the family regard themselves as the nurse's employers. She very often does not." Editorial, "The Patient's Family," *Trained Nurse and Hospital Review* 46 (March 1911): 166.

41. See letters and clippings laid in the volumes of the Boston Medical Library Directory for Nurses.

42. *American Journal of Nursing* 4 (October 1904): 2.

43. Dr. Charles Putnam to Dr. Edwin Brigham, January 31, 1894, *Letterbook.* Putnam's position was "it is but fair that a nurse who offers her services at a lower price and intrusts her engagements to us should either reap some benefit from that low price or else be informed that she will reap no such benefit." Putnam and Brigham were the physicians from the Boston Medical Library in charge of the Directory.

44. Quoted in A. S. Kavanogh, "The Indispensable Combination in Hospital Work," *Trained Nurse and Hospital Review* 37 (July 1914): 78.

45. Anna Maxwell to Miss C. C. McBrien, October 11, 1899, *Letterbook.* In his reply to Maxwell's letter Dr. Brigham was pessimistic about how much a lowering of the rates would help the overcrowding. Brigham to Maxwell, October 16, 1899, *Letterbook.*

46. A private duty nurse who secured her own cases did not of course have to keep to an agreed-upon wage rate (see Louise Darche, "The Proper Organization of Training Schools in America," *Nursing the Sick 1893,* 106). But it was considered a "breach of faith" with the Directory for a nurse to overcharge the patients when she had agreed upon one price (see Morton Prince to Miss C. C. McBrien, February 12, 1901, *Letterbook*). However, such overcharging was not uncommon (Florence LaFleur, Boston Medical Library Directory R, 66; L. M. B. Russell, I, 58; Annie Collimore to M. Adelaide Nutting, June 15, 1925; M.

Adelaide Nutting Papers, Teachers College, Columbia University, file X, folder 4, "Hours of work—domestic service").

47. Around 1904 the Directory officials began to ask patients and physicians to inform them when they felt the nurse was "superior or first class." On their uncertainty about the legality of their position, see F. Morison to Dr. F. Shattuck, February 20, 1884, *Letterbook*.

48. "A Graduate," Letter to the Editor, *Trained Nurse and Hospital Review* 15 (January 1895): 43.

49. On blacklists see memo from Dr. George H. M. Rowe, superintendent of Boston City Hospital, to the Directory, no date; numerous other letters in 1892, *Letterbook*.

50. Letter to the Editor, *Trained Nurse and Hospital Review* 23 (February 1904): 103.

51. "Private Duty Problems," *Trained Nurse and Hospital Review* 73 (July 1924): 57–58.

52. Committee on the Grading of Nursing Schools, *Nurses, Patients, and Pocketbooks* (New York: The Committee, 1928), 361.

53. "Overcrowding in the Nursing Profession," *Trained Nurse and Hospital Review* 21 (July 1898): 8–13.

54. Lucy Drown to Dr. Brigham, October 18, 1899, *Letterbook*.

55. Recent historical scholarship has begun to document and analyze the divisions and conflict within American nursing. For examples, see, in addition to my previous work cited, Melosh, *The Physician's Hand;* Susan Armeny, "Resolute Enthusiasts: the Effort to Professionalize American Nursing, 1893–1923," Ph.D. diss., University of Missouri, 1983; Lagemann, ed., *Nursing History.*

56. See Reverby, "'Something Besides Waiting.'"

57. The historiography on workers' control and capitalist rationality is growing rapidly. For the key works and critiques see David Montgomery, *Workers' Control in America* (New York: Cambridge University Press, 1979); Harry Braverman, *Labor and Monopoly Capital: The Degradation of Work in the Twentieth Century* (New York: Monthly Review Press, 1974); Susan Porter Benson, "The Clerking Sisterhood: Rationalization and the Work Culture of Saleswomen in American Department Stores," *Radical America* 12 (March–April 1978): 41–55; Melosh, *The Physician's Hand;* and James Green, "Culture, Politics, and Workers' Response to Industrialization in the U.S.," *Radical America* 16 (January–April 1982): 101–30.

58. For an example of an analysis of private duty work in this mode, see David Wagner, "The Proletarianization of Nursing in the United States, 1932–1946," *International Journal of Health Services* 10 (1980): 271–90.

25 "They Shall Mount Up with Wings as Eagles": Historical Images of Black Nurses, 1890–1950

Darlene Clark Hine

Although diverse and often contradictory images of nurses permeate American society, few writers have investigated or dissected the particular images of black nurses. Two general images of the professional black nurse prevailed during the first half of twentieth-century America. The black nurse was viewed, on the one hand, as an essential and competent provider of health care in black communities. She, more than most other health care personnel, was viewed as being completely responsive to the needs of black people. Within the largely segregated communities the black nurse represented an uncompromising voice speaking out for the best interests of blacks. On the other hand, the black nurse was perceived as an inferior member of the nursing profession when compared to her white counterparts. The image of the black nurse as an inferior professional was created and reinforced by discriminatory treatment. Accordingly, she was subjected to employment discrimination, educa-tional segregation, economic exploitation, pro-fessional exclusion, and social abuse. The real-life experiences of four black nurses, Mary Elizabeth Lancaster Carnegie, Frances Elliott Davis, Mabel Keaton Staupers, and Eunice Rivers (Laurie), enhance our understanding of the dual pro-cesses of image formation and transformation.

On an eventful day in 1942 Mary Elizabeth Lancaster Carnegie, a black nurse, reported for duty as a clinical instructor at the St. Philip Hos-pital, the separate Negro wing of the white-controlled Medical College of Virginia. Carnegie had earned her diploma in 1937 from the seg-regated Lincoln School of Nurses in New York City. After graduation she worked for several years as a general-duty nurse at the black Veter-ans Administration Hospital in Tuskegee, Ala-bama, before receiving a bachelor of arts degree in 1942 from the West Virginia State College. In spite of her previous nursing experience and unique academic preparation, Carnegie, recall-ing that unforgettable day at St. Philip, wrote: "Here began my first in-training lessons in what it means to be a Negro nurse in the South."[1]

In keeping with long-established patterns of racial etiquette, all the white administrators, physicians, and nurses at St. Philip addressed white nurses as "Miss" and black nurses as

DARLENE CLARK HINE is John A. Hannah Distinguished Professor of American History at Michigan State University, East Lansing, Michigan.

Reprinted from *Images of Nurses. Perspectives from His-tory, Art and Literature,* Anne Hudson Jones, ed. (Philadel-phia: University of Pennsylvania Press, 1988), by permission of the publisher.

"Nurse." This practice underscored the inferior status of black nurses and the low esteem in which they were held by white coworkers. Yet, as Carnegie noted, being called "Nurse so-and-so" was "a step up from being addressed by first names."[2] As if to compound their subordination, however, Carnegie observed: "Not only were Negro nurses addressed this way by the white nurses and doctors, they were instructed to address each other and refer to themselves in this manner."[3] Refusing to acquiesce to this social affront and professional slight, Carnegie declared to her black coworkers, "You can't control what someone else does, but you can control what you do."[4] Unmindful of the consequences and perhaps suspecting that her tenure at St. Philip would be brief, Carnegie admonished the black student nurses to "address themselves and each other as 'Miss.'"[5] She insisted that the black students extend this courtesy and manifest respect by addressing all their black patients as "Miss, Mrs., or Mr.," in spite of the fact that white nurses and doctors also addressed the Negro patients by their first names.

Carnegie's head-on collision with symbolic racism as reflected in the denial of appropriate titles to black women nurses was one small skirmish in a decades-long war for professional recognition and acceptance. Between 1893, when the first group of professionaly trained black nurses appeared, and 1951, the year of the dissolution of the National Association of Colored Graduate Nurses (NACGN), founded in 1908, black nurse leaders had struggled on every conceivable front to win equal pay, access to better quality training institutions, admission into advanced educational programs, broader employment opportunities, and individual membership in the American Nurses' Association. Progress toward these objectives and the eradication of the image of professional inferior was slow, as deeply entrenched negative white perceptions, attitudes, and actions toward black nurses halted advance.[6]

Actually, the many discriminatory practices of key American institutions, such as the United States military establishment and organized nursing bodies, contributed to the growth of negative images of black nurses and reinforced the already low esteem in which many whites held them. The American Red Cross's treatment of Frances Elliott Davis, a 1912 graduate of the black Freedmen's Hospital training school in Washington, D.C., is but one illustration of this point. Davis was the first black nurse to secure enrollment in the American Red Cross. At the end of World War I all nurses, with one exception, received identical pins indicating their enrollment and service in the Red Cross. However, the pin given to Davis was marked "1A," indicating that she was the first black nurse to be enrolled in the Red Cross. Thereafter, beginning with Frances Elliott Davis, from 1918 to 1949 all Negro nurses enrolled in the American Red Cross received special pins with the letter A inscribed.[7]

To be sure, Davis could have refused to accept the Red Cross pin with its discriminatory inscription. She had protested against segregation and other humiliating practices throughout her career. On her first Red Cross assignment in Jackson, Tennessee, Davis had pointedly objected to her white supervisor's introducing her to patients and coworkers as "Fannie." Moreover, she had successfully challenged the local custom requiring black patients and black health care personnel to enter the local hospital through the back door. Yet, when presented with the choice of accepting or rejecting the differently marked pin, Davis acquiesced. She swallowed her pride, accepted the pin, and consoled herself with the knowledge that she had, at least, opened a previously closed door through which other black nurses would enter. In short, to advance the professional interests of black nurses she chose to put aside personal considerations. As her biographer maintains, Davis ultimately could not "turn her back . . . simply because she had to face a certain amount of humiliation from white people who thought themselves superior."[8]

While Davis endured silently, the advent of World War II created a fortuitous array of circumstances that enabled some black nurses under the leadership of Mabel Keaton Staupers to protest loudly historical patterns of institutional racism and discrimination. Staupers, a 1917 graduate of the Freedmen's Hospital School of Nursing and executive secretary of the NACGN, took advantage of the war emergency and the increased demand for nurses to improve the status and image of black nurses as competent and valuable health care givers. Early on, War Department officials had declared that black nurses would not be called to serve in the Armed Forces Nurse Corps. As a result of pressures and protests from organized nursing groups skillfully orchestrated by Staupers, the army soon modified this policy and, in January 1941, announced that a quota of fifty-six black nurses would be recruited and assigned to the black military installations at Camp Livingston in Louisiana and Fort Bragg in North Carolina. Navy officials remained intransigent, and the Navy Nurse Corps continued to exclude black nurses throughout the war years.[9]

Staupers readily conceded that the army's quota of fifty-six represented an advance over World War I practices of total exclusion. Yet she remained determined to continue the assault on all such barriers. From 1941 to 1945 Staupers met repeatedly with white nursing groups, top military officials, First Lady Eleanor Roosevelt, and leaders of black civil rights organizations. She cultivated relations with editors of black newspapers, women's clubs, and white philanthropists, urging them to protest the imposition of quotas for black nurses.[10] Her strategic maneuvering eventually bore fruit.

The personal appeals of Eleanor Roosevelt and the protests of the National Nursing Council for War Service, combined with the acute nurse shortage toward the end of the war, eventually forced the army to increase the numbers of black nurses. By 1945 approximately 330 black nurses were serving in the Army Nurse Corps. Had black nurses been accepted in proportion to their numbers, as were white nurses, there would have been 1,520 of them in the Army and Navy Nurse Corps. There were at the time approximately eight thousand black graduate nurses active in nursing.[11] Unable to withstand the unrelenting pressure, the surgeon general of the navy declared on January 31, 1945: "There is no policy in the Navy which discriminates against the utilization of Negro Nurses."[12]

Although the American Red Cross, the United States Army and Navy, and some white nurses and doctors discriminated against black nurses, viewing them as inferior professionals, the black community's general reactions to and perceptions of black nurses were strikingly different. In part this divergence is explained by the insufficient health care available to large portions of the black population. The small numbers of black physicians, coupled with a growing trend to concentrate in urban areas, and the frequently insulting treatment meted out by white physicians, often meant that only a black nurse was available in most rural communities.[13] In many such areas rural black nurses, similar to Eunice Rivers (Laurie), a 1922 graduate of Tuskegee Institute's nursing program in Tuskegee, Alabama, played a pivotal role in the black health care delivery hierarchy in rural Alabama.

Rivers attributed the higher esteem that the black nurses found among blacks to the position they occupied as mediators between patients and physicians. Historian James H. Jones, in his provocative study of the Tuskegee syphilis experiment conducted by the U.S. Public Health Service (1932–1972), underscored the significant role played by Nurse Rivers. Her immense interpersonal skills won the trust and respect of hundreds of the male patients involved in the experiment. Jones describes Rivers as "a facilitator, bridging the many barriers that stemmed from the educational and cultural gap between the physicians and the subjects."[14] In one interview Rivers elaborated on her image of the role the black nurse played:

> the doctor saw the patient and he was gone and it was up to you to help that patient carry out his orders, do whatever the doctor suggested. The doctor said you do so and so. . . . First thing, the patient doesn't know how to do it. He doesn't know what his reaction is going to be. He doesn't want to be stuck. . . . So the nurse plays an important part there. She's closer to the patient. Patients would get to the point where if they're not sure, they're going to ask you. They get you in the middle.[15]

Rivers recalled, "A lot of my patients would not call a doctor until I had come to see them, to see how they were doing and see if they needed a doctor." She added, "I had an awful time training them to go ahead and get their own doctor."[16]

These capsule glimpses into the experiences of Carnegie, Davis, Staupers, and Rivers provoke more questions than answers concerning both the images of black women and the reality of their struggles as professional nurses. Why were black nurses denied the usual appellations denoting respect? Why were they given Red Cross pins inscribed with the letter *A*? Why did the U.S. Army and Navy first exclude black nurses altogether, as in World War I, and then in World War II establish quotas for recruiting them into the Armed Forces Nurse Corps? In light of the demeaning treatment accorded them in the larger society, why were black nurses held in such high esteem within the black community? What did the black community expect and receive from black nurses? Finally, how did discrimination, racism, and negative stereotypes

spur or impede the personal and professional development of the black nurse?

Traditional histories of nursing pay scant attention to the accomplishments and peculiar difficulties encountered by black nurses. Yet the struggle of black nurses for respect, recognition, acceptance, and status parallels and in many ways exemplifies the historical quest of all professional nurses. It is the difference that race made, giving rise to a store of derogatory images, which separates and distinguishes the black nurses' story from the larger history. Examining some of the historical images of black nurses and contrasting them with the objective reality provide deeper insights into the process of professionalization in nursing. Such an investigation enhances our understanding of the old images that American pioneer nursing leaders desired to destroy and the new ones that they substituted. In the years of the early professionalization, concerned nursing leaders initiated actions designed to limit the number of nurses, halt the proliferation of training schools, and recruit "women of the better classes" into the profession. From the outset the quest for image control and exclusiveness in nursing were essential components and characteristics of the overall professionalization process.[17]

Unfortunately, black women, and lower-class white women to some extent, were most affected by exclusionary tendencies and hence were sacrificed on the altar of nursing advancement. As nursing increasingly acquired the trappings of a profession, the restrictions and impediments placed in the paths of black women mushroomed. They were denied admittance into training schools, barred from membership in professional associations, refused listing in employment registries, and discouraged from aspiring to meet the higher requirements of nurse registration and licensing laws. Black women, because of their racial identity and slavery heritage, were seen as a permanently alien and inferior group that could not be assimilated. In the white mind the slavery-born images of the black woman as a defeminized beast of burden, a sexually promiscuous wanton, or a domineering mammy held sway long after the demise of the "peculiar institution." These negative images, mixed with entrenched racism, probably motivated those of the white nursing establishment seeking greater status and esteem to eschew association with or recognition of black women as professional peers.[18] Before I proceed, it will be useful to place these images of black women and nurses within the appropriate historical context.

In the latter part of the nineteenth century, the social conventions and normative attitudes of an industrialized society consigned women to the private sphere of the home. The ideology of "virtuous womanhood" sharply and oppressively defined their proper actions and behavior in very narrow and restricted terms. Women were considered to be repositories of moral sensibility, purity, refinement, and maternal affection in a male-dominated society. Thus woman's highest calling consistent with her biological destiny was deigned to be that of mother and nurturer. For many women growing adherence to the ideology of separate spheres occurred in tandem with constricting career opportunities in the public sphere of business and politics. Actually, the tension between theory and reality created a paradox, for as certain doors closed, other new female-stereotyped occupations and professions reserved for women opened.[19]

Of all such sex-segregated occupations, nursing was preeminent. The movement for formalized nursing training and practice provided an attractive alternative to middle-class, sphere-restricted women. Before the Civil War, nursing, as historian Janet Wilson James has pointed out, was a "low-paid, low-status job for laboring class women, who, over a twelve-hour day, attended to the physical needs of the patients while doing the heavy domestic work on the wards."[20] The opening of the first nursing schools in 1873 launched the movement to up-

grade nursing, to distance it from identification with domestic service, and to attract a "higher-class woman" into the profession. Traditionally considered a woman's job, nursing neither threatened nor challenged society's views of her traditional domestic functions. Rather, the substance of nursing and settings in which training was provided actually reinforced the image of the subordinate woman. Deemed less exacting and autonomous than medicine, nursing existed always under the control of the male-dominated medical and hospital professions.[21]

The late nineteenth- and early twentieth-century struggle to professionalize and upgrade the image of nursing concentrated in part on recruiting middle-class students while purging and excluding from the occupation the uneducated and untrained women of the lower socioeconomic classes. These efforts also coincided with the hardening of the color line in American society. As segregation pervaded the country, all southern and most northern nursing schools barred black women. Even in the most liberal northern institutions black women were subjected to restrictive quotas. The charter for the New England Hospital for Women and Children, for example, expressly stipulated that only *one* Negro and *one* Jewish student each year would be accepted. The first black trained nurse, Mary E. Mahoney, was graduated from the New England Hospital in 1879.[22]

If black women were to have access to professional nursing training, then it was incumbent on black leaders, in the name of racial self-help, to establish the corresponding institutions. Beginning in the early 1890s, black physician Daniel Hale Williams of Chicago and educator Booker T. Washington, founder of Tuskegee Institute, spearheaded a movement to found hospital nursing schools for black women. They solicited operating funds for these new institutions from their respective black communities and from private philanthropies. The rhetoric of the founders of the early black nursing schools reveals much about their own and the larger society's images of nurses. Washington and Williams espoused the new romantic and idealized image of the Florence Nightingale–type nurse. They merged this image with their views of what constituted the "proper" woman's role. Uppermost in their minds, however, was the belief that the black woman bore a large part of the responsibility for proving the humanity of black people to a skeptical white public. Accordingly, they portrayed the black nurse as a self-sacrificing, warm, and devoted mother figure and downplayed her as the efficient, autonomous, and assertive professional. Fund-raising campaigns for the hospitals and training schools employing this romantic image of the black nurse-mother proved most successful. Evoking this romantic image had other practical implications as well. Clearly, the black nurses trained in these black community-based and -supported institutions would forever owe primary allegiance to blacks and be responsible participants in the climb up the racial and social ladder.[23]

Williams, as founder of two black nursing schools, Provident Hospital and Training School in Chicago in 1891 and the Freedmen's Hospital nursing school in Washington, D.C., in 1894, frequently informed potential black supporters of nursing schools: "The servant class no longer furnishes the nurse."[24] Ironically, Williams was prone to invoke the ubiquitous "mammy" image to illustrate his claim that the black woman was a "natural" nurse possessed of a long heritage, in slavery and freedom, of caring for the sick of both races. He was perhaps unmindful that this continued association of black nurses with the mammy image would retard their advancement within the nursing profession. Williams insisted: "The young colored woman who chooses this calling enters the training school richly endowed by inheritance with woman's noblest attributes—fidelity, tenderness, sympathy."[25] After extolling all the nurturing qualities of black

women, Williams declared the black nurse an object lesson, who "teaches the people cleanliness, thrift, habits of industry, sanitary housekeeping, the proper care of themselves, and of their children. She teaches them how to prepare food, the selection of proper clothing for the sick and the well, and how to meet emergencies."[26] As Rivers would do thirty years later, Williams predicted that the trained black nurse would soon become a major force serving the black community in racial uplift work.[27]

Booker T. Washington, when launching the nursing program at Tuskegee Institute in 1892, linked it with the school's industrial education emphasis. He justified the new program on the grounds that nursing training would enable a black woman to have a career prior to marriage, one, however, that would also make her a better wife, mother, and homemaker. Moreover, he argued that should hard times befall the family, the trained nurse would always be able to help earn money. In describing the philosophical foundation on which the school was based, Washington reiterated tenets of the Victorian belief concerning woman's role and function: "A man can build the house but the woman must, for the most part, furnish the sort of culture and refinement that makes it a home," he said.[28] Washington wedded nursing training firmly to vocational work: "The course in child nurture and nursing has been established to complete the training in home building which is carried on as part of the industrial training of young women at Tuskegee."[29]

Williams's and Washington's images of the black nurse were shared for many years by most black hospital administrators and physicians responsible for their training. In a 1918 article John A. Kenney, a black physician named superintendent of the new and enlarged John A. Andrew Memorial Hospital and Nurse Training School of Tuskegee Institute, echoed Williams's and Washington's conviction that the black nurse was equal, if not superior, to white nurses. In an effort to persuade more black women to enter into the backbreaking endless toil euphemistically referred to as nursing training, he unabashedly lauded the black nurses' many womanly virtues of "devotion, endurance, sympathy, tactile delicacy, unselfishness, tact, resourcefulness, [and] willingness to undergo hardships."[30] The fact that he played an important role in training black women did not challenge Kenney to view the trained nurse as a serious professional rather than a mother-nurturer. Even when commenting on the good deeds performed by the black graduate nurses of Tuskegee Institute, Kenney interjected remembrances of his mother's unpaid nurturing activities. He wrote: "Regardless of the demands made upon her by the exacting duties of her own household, if there was a case of serious illness among her friends, white or colored, even miles away, she thought it her duty to go and care for them night after night, if necessary."[31]

The black nurse was entrapped in the vortex of the sexual and racial currents dominating black and white thought in late nineteenth-century America. To be sure, at this juncture much of the language describing black nurses and the "advantages" to be reaped by pursuing a nursing career were used with white women as well. Although reality and image frequently diverged, nevertheless the ideology of separate spheres severely reduced a woman's chances for challenging and remunerative work. Because of racial prejudice, most of the sex-segregated jobs were beyond the black woman's reach. The transformation of nursing into a skilled profession requiring formal training in a structured institutional setting further encumbered the black woman. With black women barred from the white nursing schools, black leaders, in order to provide access to the profession, proceeded to combine the ideology of woman's separate sphere with the doctrine of racial uplift.

Thus, because the black communities contributed so much to the start-up funds creating and sustaining the black nursing schools, the early generations of graduates were expected to repay the communities' investments. Hence, in addition to seeking professional acceptance and recognition, the black nurse bore the extra burden of providing health care for, and lifting up from the bottom of the American social scale, the entire black race. All future images in the black mind of the black woman nurse would be inextricably connected to her role within, and responsibility to, the black community.

By the 1920s the separate black training schools had produced approximately three thousand nurses. The burgeoning numbers caused white nurses, especially those engaged in private duty work, to fear increased economic competition, which only exacerbated the tenuous relations between the two groups. Meanwhile, economic exigencies and racism compelled many black nurses to act in ways that reinforced existing negative images of them in the minds of their white colleagues. For example, many black nurses worked for lower wages and longer hours than white nurses. Often, black nurses performed household and child care chores in addition to tending to sick members of a family. The fact that many white physicians spoke in glowing terms of the submissive and accommodating black nurse who adapted "well to the needs of the household" did not help matters.[32] Equally damaging were their expressions of delight with black private duty nurses, whom they perceived as being more "willing to render the small personal services only grudgingly performed by white nurses."[33] In a depressed job market, characterized by an oversupply of nurses and decreasing patient demands, black nurses were imagined as being the group least committed to advancing the profession and more willing to compromise on salary and working conditions.[34]

The scarcity of data precludes the development of a definitive analysis of white nurses' images of black nurses during the 1920s. Available, however, are a limited number of surveys and reports commissioned by philanthropic and nursing organizations, which canvassed white nurses, physicians, and hospital and public health officials for their personal evaluations and perceptions of the black nurse. In 1925 Ethel Johns, an Englishwoman trained in a Canadian hospital, conducted one of the most illuminating surveys and reports. Under the aegis of the Rockefeller Foundation Johns queried hundreds of white administrators and nurse superintendents in more than two dozen hospitals and visited scores of visiting nursing associations and municipal boards of health.

Johns's report, though frozen in time, does record the attitudes of the black nurses' professional colleagues while shedding light on the black nurses' social and economic status. The images white nurses and other health care personnel had of black nurses were informed by the widely held assumption of the poor quality of all black training schools. Moreover, the fact that blacks as a group occupied a subordinate position in America influenced the negative assessments of their leadership abilities. Most of the white superintendents of the twenty-three black hospitals Johns visited frankly admitted their displeasure with black nurses. They contended that black nurses exhibited "a marked tendency to concealment of what is going on in their respective wards. They will not report mistakes or accidents."[35] These claims, while perhaps accurate, did not take into consideration the forces that may have encouraged black nurses to cover for each other when a white supervisor was overhead. Few black nurses held supervisory positions in hospitals or sanitariums. Without dissecting the underlying reasons for the absence of black nurse administrators, Johns simply concluded that they were inherently lacking in leadership qualities: "My observations lead me to believe that the negro woman is temperamentally unsuited for the constant unremitting grind of a hospital superintendent's life. She finds it difficult to discipline her staff and yet to remain on friendly terms with them."[36]

While black nurse supervisors were scarce, the chances of black nurses for employment in the public health field during the 1920s were even more bleak. Visiting nursing association officials rationalized their aversion to hiring black public health nurses, insisting that they were of limited usefulness. Most agency leaders maintained that black nurses could work only with blacks, but white public health nurses could deal with both races. After all, these officials contended, black nurses were educationally deficient and ill prepared to assume the heavy responsibility of being a public health nurse. Supervisors of the nursing services for the municipal boards of health in New York, Chicago, and Philadelphia acknowledged that, as much as possible, they hired only white nurses because the employment of blacks "complicates the service and creates social friction."[37] Municipal boards of health supervisors in Birmingham, Baltimore, Atlanta, and Nashville regarded the black public health nurse as "admittedly inferior in intelligence to the white group."[38] They let it be known that, where employed, the black nurse was "paid substantially less than the white nurse . . . excluded from supervisory rank and . . . treated as a social inferior."[39]

Several directors of visiting nursing services did comment positively on the role of black nurses in public health nursing. Lillian Wald, founder of the Henry Street Settlement and a staunch friend of black nurses, employed 25 black and 150 white nurses, paid them equal salaries, and accorded them identical professional courtesies and recognition. Even here, however, black and white nurses were viewed and treated differently in two respects: black nurses were never sent to white homes, nor were they promoted to supervisory rank. Johns discovered similar conditions prevailing in Philadelphia, Chicago, and Saint Louis. Southern-based visiting nurse services employed black nurses; nevertheless, when they were hired they always received much lower salaries and were treated as social inferiors.[40]

Among the whites interviewed, Johns discerned nearly unanimous agreement that the black nurse was a professional inferior. According to most white nurses, supervisors, and administrators, the lower wages paid black nurses were entirely justified and simply confirmed alleged black shortcomings. The fact that black nurses were rarely promoted to or held supervisory and administrative positions reinforced white beliefs that blacks were incapable of leading. Only when she dealt with black patients did the black nurse stand a chance of being referred to as a competent and adept professional. Apparently, only as long as she remained in the black community, caring only for black patients, would the black nurse earn praise from her white counterparts and enjoy a better image. Johns well captured these mixed messages when she observed:

It is quite apparent that the negro nurse cannot be utilized successfully in public health work except among her own people. Even among them she has not the same authority as the white nurse although she has a better psychological approach. She has been very successful in overcoming their superstitious fears regarding immunization, vaccination and other preventive measures. The social and economic problems involved in case work are commonly too much for her but she can ferret out information and interpret domestic complications which would baffle a white nurse who lacks her intuitive understanding of racial characteristics.[41]

Fortunately, black nurses were highly regarded within black urban and rural communities. For thousands of poor blacks the nurse often meant the difference between living and dying. The black nurse allayed the fears and quelled the hostilities of those superstitious rural blacks who refused to seek medical care in hospitals and usually resorted to folk cures and questionable remedies. Frequently, the black nurse became the bridge connecting rural impoverished blacks mired in nineteenth-century notions of sickness with the twentieth-century reality of hospitals, physicians, scientific advances, and the germ theory of disease.[42]

Actually, the image of the race-serving, strong, and resourceful black nurse laboring among the poor and downtrodden is too one-dimensional. Eunice Rivers, to be specific, was described by her family, friends, colleagues, and black patients as a "born nurse." She had entered nursing out of a desire "to get closer to people who needed" her. Rivers attributed her success as a nurse to an innate ability to accept people on their own terms. She stated in an interview, "I go there and visit awhile until I know when to make some suggestions. . . . I don't ever go into any person's house, fussing with him about how he keeps his house, first." Rivers insisted, "I accepted them as they were and they accepted me."[43]

Rivers was more than a good country nurse. Her more than forty years' involvement in the Tuskegee syphilis experiment, in which treatment was deliberately withheld from patients, raises questions concerning relationships between the black nurses and the black community and between black nurses and white health care professionals. Indeed, it is fair to say that without her the white "government doctors" would not have been successful in engaging so many black males in such a detrimental and ethically bankrupt experiment. It was their unquestioning faith in Rivers as someone selflessly looking out for and protecting them that led the men to continue in the experiment for so many unrewarding years. Though they remained fun-

483

damentally suspicious of the "government doctors'" motives, they always tended to do what Rivers told them. According to historian James H. Jones, "More than any other person, [Rivers] made them believe that they were receiving medical care that was helping them."[44] They were not.

Nurse Rivers's motives for collaborating in this experiment and deliberately manipulating these black men are complex. It is possible that Rivers viewed the experiment as a way of ensuring for at least some blacks an unparalleled amount of medical attention. Jones offers several compelling explanations for Rivers's complicity: As a nurse, Rivers had been trained to follow orders and probably it simply did not occur to her to question a, or for that matter any, doctor's judgment. Moreover, she was incapable of judging the scientific merits of the study. For Rivers, a female in a male-dominated world, deference to male authority figures reinforced her ethical passivity. Finally, and perhaps most significantly, Rivers was black and the physicians who controlled the experiment were white. Years of conditioning and living in the South made it virtually impossible for Rivers to have rebelled against a white male government doctor, the ultimate authority figure in her world.[45] In this case the needs and interests of the black community of Tuskegee were not addressed and protected by the black nurse.

The image of the black nurse as self-sacrificing mother-nurturer, servant, and leader of the black community persisted with slight modifications. In the urban northern black communities the white uniform-clad black nurse with satchel at her side cut an imposing and impressive figure. One black nurse, Elizabeth Jones, recounted ways in which the black community viewed the black nurse as someone special. She describes, for example, her own approach one day to two children playing on a Harlem street. One child, as she walked toward them, ordered

his playmates, "'Get up, and let the lady pass.' While making a passage, all eyes were turned upon [her] with great intent. Suddenly, as if having solved a problem, one little voice chimed in, and said with much glee, 'Aw! She ain't a lady, she's a nurse!'"[46] Jones, musing about this reflection on the status of the black nurse in the black community, declared: "Not only is she a teacher, but she is looked upon by most of those with whom she comes in contact, as an example of the higher life."[47] For many blacks the black nurse became a symbol of white middle-class virtues and respectability. Residents of the black communities where the nurse visited, worked, and sometimes resided looked upon her with pride tinged with awe. The fact that she was there to tend to their needs and to help them solve their problems engendered feelings of possessiveness while intensifying desires to obey.

Throughout the brief history of professional nursing, black nurses struggled to create and sustain positive self-images while simultaneously pursuing an often frustrating quest for acceptance and recognition within their chosen occupation. Accommodation to their "place" and resistance to white efforts to devalue them developed as two forms of a single process by which black nurses accepted circumstances they could not change and vigorously fought individually and collectively for professional equality.

Throughout her long nursing career Red Cross nurse Davis demonstrated pride and a positive self-image. Her deep spiritual convictions and unyielding sense of responsibility to her race, combined with the support and love of her husband, sustained her quest for professional equality. These internal and external forces helped to deflect the psychological and moral aggression of a racist society. When gloom threatened to immobilize and depress her, she invariably reached for her Bible and reread her favorite passage from the Book of

Isaiah: "But they that wait upon the Lord shall renew their strength; they shall mount up with wings as eagles; they shall run, and not be weary; and they shall walk, and not faint" (40:31).[48]

Mabel Keaton Staupers was thirteen years old when her family moved from Barbados, West Indies, in 1903 to the Harlem community of New York City. She spent the early years of her nursing career in New York City and Washington, D.C.; in 1920, in cooperation with a couple of prominent black physicians, she organized the Booker T. Washington Sanatorium, the first facility in the Harlem area where black physicians could treat their patients. In 1921 she was awarded a working fellowship in Philadelphia and was later assigned to the chest department of the Jefferson Hospital Medical College in Philadelphia. The following year Staupers returned to New York, and under the auspices of the New York Tuberculosis and Health Association she made a survey of the health needs of the community, which eventually resulted in the organization of the Harlem Committee of the New York Tuberculosis and Health Association and her twelve-year stint as executive secretary of this body.[49]

In 1934 Staupers was appointed as the first nurse executive of the National Association of Colored Graduate Nurses (NACGN). Founded in 1908 by a group of black nurses in New York, the organization was virtually moribund by the time Staupers assumed the helm. Throughout the 1940s Staupers combined her struggle for the integration of black nurses into the Armed Forces Nurse Corps with the fight for full integration of black nurses into American nursing. In 1948 the American Nurses' Association (ANA) House of Delegates opened the doors to individual black membership, appointed a black woman nurse as assistant executive secretary in its national headquarters, and witnessed the election of black nurse Estelle Massey Riddle to the board of directors. In 1950 the NACGN membership elected Staupers president. The NACGN's board of directors charged Staupers with overseeing the dissolution of the organization. Her book, *No Time for Prejudice* (1961), details the history of the black nurses' victorious struggle to integrate into the American Nurses' Association.[50]

The fight for integration into state nurses' associations in the South continued even after the ANA openly accepted black members on an individual basis. Mary Elizabeth Lancaster Carnegie led the attack against the Florida State Nurses Association. After leaving St. Philip, Carnegie worked as assistant director of the Division of Nurse Education at Hampton Institute in Virginia. In 1945 she was named dean of the Division of Nursing Education at Florida A & M College. In 1951 Carnegie received a fellowship to earn a master's degree from Syracuse University in New York. She was assistant editor of the *American Journal of Nursing* from 1953 to 1956, then became associate editor, and in 1970 editor of *Nursing Outlook*. While pursuing her editorial career, Carnegie was a part-time student at New York University, where she received a doctoral degree in 1972. The next year she assumed the editorship of *Nursing Research*.[51]

Carnegie's somewhat exceptional career was unlike that of most black nurses; yet even her advancement was profoundly impeded by racist attitudes and discrimination. A clearly hostile larger social environment littered her path with innumerable reminders of her status as a Negro nurse. Carnegie, however, deliberately chose to resist those practices that assaulted her self-image. Carnegie's description of the black nurses' struggle for the right to participate in the meetings of the Florida State Nurses Association illuminates the strength of their determination to win professional recognition and acceptance and the ludicrous nature of the efforts of some white nurses to preserve segregation and subordination: "For many months, we played a game of 'musical chairs.' The white nurses would wait on the outside of the buildings for us to arrive and be seated; then they would proceed to sit on the opposite side of the meeting room. If we sat in the back, they would sit in the front, and vice versa."[52] Carnegie devised a clever scheme to end "the game." She and her fellow black nurses simply waited for the white nurses to arrive and would scatter throughout the room in order to ensure, at least, integration in the seating arrangements. When in 1950 she was elected to the board for a three-year term, she continued to press quietly for integration. After the 1950 meeting Carnegie observed that "for the first time, all Negro nurses attended all business and programme meetings, but were barred from the luncheon."[53] By the 1952 convention in Daytona Beach, Florida, she reported modest improvements: "There was integration in every respect but housing and the events that were strictly social."[54] (The luncheon meetings at the hotel included all members on an equal basis.)

By the 1950s it was evident that the self-image of the black nurse was formed in part by the twin realities of racism and sexism in American society. Black nurses recognized that their struggle for recognition, acceptance, and equality of opportunity within nursing was inextricably linked to overcoming this double-edged prejudice. Carnegie captured the relationship between the development of a positive self-image and struggle for unfettered access to professional opportunities when she asserted: "In the length and breadth of the United States of America, Negro nurses, many unknown to each other, have always fought for a common cause. . . . They were fighting on the same front in schools of nursing and in professional organizations in other states, and on other fronts—in the military, public health, hospital nursing service, industry, private duty, and the national or-

ganizations—throughout the country."[55] Actually, the dual processes of accommodation and resistance enabled black nurses to develop a collective self-consciousness and pride that allowed them to retain a viable sense of self-worth in the face of the oppression they endured. Central to their identity, however, was a strong conviction that, in the words of black nurse educator-administrator Gloria R. Smith, "black nurses were accountable to black people in a special way."[56] As late as 1971 Smith observed that the black nurses would always serve as "spokesmen who could articulate the needs of the black community for compatible care delivery systems as well as the dreams of black people for equal access to and mobility within the health care system."[57] In 1971 black nurses found it desirable again to organize in a separate body to continue the fight for full participation and equal access to opportunities in the profession. The new National Black Nurses' Association reminds us of the resiliency of negative images, racism, discrimination, and the will to overcome injustices of all kinds.

Notes

1. Mary Elizabeth Carnegie, "The Path We Tread," *International Nursing Review* 9 (September–October 1962): 26.

2. Ibid.

3. Ibid.

4. Ibid.

5. Ibid.

6. Joyce Ann Elmore, "Black Nurses: Their Service and Their Struggle," *American Journal of Nursing* 76 (March 1976): 435–37.

7. Jean Maddern Pitrone, *Trailblazer: Negro Nurse in the American Red Cross* (New York: Harcourt, Brace & World, 1969), 88.

8. Ibid., 69.

9. Darlene Clark Hine, "Mable K. Staupers and the Integration of Black Nurses into the Armed

Forces," in *Black Leaders of the Twentieth Century,* ed. John Hope Franklin and August Meier (Urbana: University of Illinois Press, 1982), 241–57.

10. Ibid., 254.

11. Philip A. Kalisch and Beatrice J. Kalisch, *The Advance of American Nursing* (Boston: Little, Brown, 1978), 567–68.

12. Quoted, ibid., 568.

13. Carter G. Woodson, *The Negro Professional Man in the Community* (New York: Negro Universities Press, [1934] 1969), 142. See ch. 10; the entire chapter, 133–48, is on black nurses.

14. James H. Jones, *Bad Blood: The Tuskegee Syphilis Experiment* (New York: Free Press, 1981), 6.

15. Interview with Eunice Rivers [Laurie], October 10, 1977, Schlesinger Library Black Women's Oral History Project, Radcliffe College, Cambridge, Massachusetts.

16. Ibid.

17. Janet Wilson James, "Isabel Hampton and the Professionalization of Nursing in the 1890s," in *The Therapeutic Revolution: Essays in the Social History of American Medicine,* ed. Morris J. Vogel and Charles E. Rosenberg (Philadelphia: University of Pennsylvania Press, 1979), 201–44; and Celia Davies, "Professionalizing Strategies as Time- and Culture-Bound: American and British Nursing, Circa 1893," in *Nursing History: New Perspectives, New Possibilities,* ed. Ellen Condliffe Lagemann (New York: Teachers College Press, 1983), 47–63.

18. Darlene Clark Hine, "From Hospital to College: Black Nurse Leaders and the Rise of Collegiate Nursing Schools," *Journal of Negro Education* 51 (Summer 1982): 224; George M. Fredrickson, *The Black Image in the White Mind: The Debate on Afro-American Character and Destiny, 1817–1914* (New York: Harper Torchbooks, 1971), 1–179ff; and Anna B. Coles, "The Howard University School of Nursing in Historical Perspective," *Journal of the National Medical Association* 61 (March 1969): 105–18.

19. Sheila Rothman, "Women's Special Sphere," in *Women and the Politics of Culture: Studies in the Sexual Economy,* ed. Michele Wender Zak and Patricia P. Moots (New York: Longman, 1983), 213–23; and Mary Beth Norton, "The Paradox of 'Women's Sphere,'" in *Women of America: A History,* ed. Carol Ruth Berkin and Mary Beth Norton (Boston: Houghton Mifflin, 1979), 139–49.

20. James, "Isabel Hampton," 205.

21. Ibid., 203–5.

22. Hine, "From Hospital to College," 224.

23. Daniel H. Williams, "The Need of Hospitals and Training Schools for the Colored People of the South," *National Hospital Record* 3 (April 1900): 3–7; and Booker T. Washington, "Training Colored Nurses at Tuskegee," *American Journal of Nursing* 2 (December 1910): 167–71.

24. Williams, "The Need," 5.

25. Ibid.

26. Ibid.

27. Ibid., 5–7.

28. Washington, "Training Colored Nurses," 171.

29. Ibid.

30. John A. Kenney, "Some Facts Concerning Negro Nurse Training Schools and Their Graduates," *Journal of the National Medical Association* 11 (April–June 1919): 53.

31. Ibid.

32. Ethel Johns, "A Study of the Present Status of the Negro Woman in Nursing, 1925," (Unpublished report, 43 pages of typescript plus 16 exhibits, Rockefeller Foundation Archives, Record Group 1.1, series 200, box 122, folder 1507, Rockefeller Archive Center, Pocantico Hills, North Tarrytown, New York 10591–1598), 27.

33. Ibid.

34. Estelle Massey Riddle, "Sources of Supply of Negro Health Personnel: Nurses," *Journal of Negro Education* 6 (Yearbook Issue 1937): 483–92; Johns, "A Study of the Present Status," 26–27; Darlene Clark Hine, "The Ethel Johns Report: Black Women in the Nursing Profession, 1925," *Journal of Negro History* 67 (Fall 1982): 212–28; and Donelda Hamlin, "Report on Informal Study of the Educational Facilities for Colored Nurses and Their Use in Hospital, Visiting, and Public Health Nursing," The Hospital Library and Service Bureau, 1924–25. A copy can be found in Rockefeller Archive Center.

35. Johns, "A Study of the Present Status," 25.

36. Ibid.

37. Ibid., 29.

38. Ibid., 30.

39. Ibid.

40. Ibid.

41. Ibid., 33.

42. Anna DeCosta Banks, "The Work of a Small Hospital and Training School in the South," *Eighth Annual Report of the Hampton Training School for Nurses and Dixie Hospital* (Hampton, Va.: Hampton Training School for Nurses and Dixie Hospital, 1898–1899): 23–28.

43. Interview with Eunice Rivers [Laurie].

44. Jones, *Bad Blood,* 160.

45. Ibid., 164–67.

46. Elizabeth Jones, "The Negro Woman in the Nursing Profession," *Messenger* 5 (July 1923): 764.

47. Ibid.

48. Pitrone, *Trailblazer,* 99–102; and Coles, "The Howard University School," 111.

49. W. Montague Cobb, "Mabel Keaton Staupers, R.N., 1890–," *Journal of the National Medical Association* 69 (March 1969): 198–99.

50. Mabel Keaton Staupers, "History of the National Association of Colored Graduate Nurses," *American Journal of Nursing* 51 (April 1951): 221–22; and Mabel Keaton Staupers, *No Time for Prejudice* (New York: Macmillan, 1961).

51. Hine, "From Hospital to College," 232.

52. Carnegie, "The Path," 32.

53. Ibid.

54. Ibid.

55. Ibid., 33.

56. Gloria R. Smith, "From Invisibility to Blackness: The Story of the National Black Nurses' Association," *Nursing Outlook* 23 (April 1975): 226.

57. Ibid.

26 Interactions Between Public Health Nurses and Clients on American Indian Reservations During the 1930s

Emily K. Abel and Nancy Reifel

One of the thorniest problems confronting contemporary social theorists is how to describe patterns of domination without casting subordinate groups solely as victims. A widely adopted solution is to focus on resistance, finding instances not just in large social movements but also in everyday practices.[1] An important form of resistance involves adopting the resources of the dominant but using them for purposes not originally intended. Marshall Sahlins demonstrates that the Hawaiians and Kwakiutl embraced many of the commodities the Europeans introduced between the mideighteenth and midnineteenth centuries while disregarding the interpretative grid in which they were embedded.[2] Nancy Fraser adds that this type of resistance frequently is available to clients of modern social service agencies. She cites a study of African-American pregnant teenage girls in a nonresidential facility in the late 1960s, who "made use of those aspects of the agency's program that they considered appropriate to their self-interpreted needs and ignored or side-stepped the others."[3] The girls were especially hostile to the agency's therapeutic approach, which focused on personal failure and cast any deviation from white middle-class norms in a negative light.[4]

This essay argues that some American Indians on reservations adopted a similar strategy in their encounters with public health nurses during the 1930s. Sioux people viewed the nurses as resources to be used strategically and selectively. Those who accepted the nurses' services did so because the services addressed specific needs the clients themselves defined as important. Most disregarded the health education program insofar as it assumed the superiority of Euro-American values.

Nurses hired by the Office of Indian Affairs (OIA) were responsible for dispensing the great bulk of government health services to American Indians on reservations.[5] Dubbed "field nurses"

EMILY K. ABEL is Professor of Health Services and Women's Studies at the University of California, Los Angeles.
NANCY REIFEL is an active duty member of the Commissioned Corps of the U.S. Public Health Service, Indian Health Service, Dental Branch and an Assistant Researcher in the American Indian Studies Center and School of Dentistry, University of California, Los Angeles.

Reprinted from *Social History of Medicine* 9(1996): 89–108, by permission of Oxford University Press.

to differentiate them from nurses employed by the Public Health Service, they replaced the field matrons, women without professional training who previously had been responsible for providing health education to American Indians.[6] In 1922 the OIA commissioned the American Red Cross to investigate the need for public health nurses on reservations. The report found extremely high morbidity and mortality rates and concluded that public health nurses could "assist in improving health conditions."[7] Two years later the first three nurses were hired.[8] The Merriam Survey conducted at the request of the secretary of the interior in 1928 recommended an enormous expansion of the field nursing service, arguing that "properly trained" nurses "could accomplish marvelous results" in "treating health habits for the prevention of illness and in raising living standards."[9]

Nevertheless, the staff grew slowly. Lack of funds prevented the Office of Indian Affairs from offering salaries that were competitive with those in other government agencies and private practice.[10] Poor working conditions also deterred qualified women from applying. The nurses had large case loads, were isolated from colleagues, had few resources at their disposal, lived in accommodations they considered substandard, and lacked opportunities for advancement.[11] As one field nurse later commented. "OIA Nursing was the step child of all Gov't Services."[12] The OIA was able to fill all openings only during the Depression, when other positions were scarce. The highpoint was in 1939, when the staff numbered 110. By 1955, when health services were transferred to the Public Health Service, fewer than 70 field nurses were employed.[13]

This essay relies on two different types of sources. The first are the accounts of the field nurses, including letters, memoirs, and, above all, their monthly and annual reports to Washington.[14] The reports consisted of both a two-page statistical section, in which the nurses enumerated the services they rendered, and a longer narrative section, in which they delineated their goals, chronicled their activities, and discussed individual cases. Like any historical document, these reports must be read with caution. Because the nurses had little direct supervision, the reports constituted the principal means of monitoring performance; the nurses had an obvious interest in presenting their work in the best possible light and expressing attitudes that mirrored those of officials. Nevertheless, the reports provide some insight into the assumptions and work of the nurses.

The second are oral histories of the field nurses' clients. In-depth semistructured interviews were conducted with residents of two Sioux reservations in South Dakota during August 1993. The twenty-three respondents included seventeen women and six men. All were at least seventy and thus old enough to recall having interacted with the field nurses during the 1930s.[15]

Again, caveats are necessary. The small sample size means that we have very few respondents in certain categories; moreover, because there are dramatic differences among Sioux people, we may draw questionable generalizations from our respondents.[16] Although we rely on reports of all the field nurses, we have interviews from members of only one tribe; nurses who worked in more than one region occasionally remarked that different tribes responded differently to their work. In addition, the interviews rely on recall of events occurring more than half a century earlier; we can assume that the intervening years affected many memories. Many respondents were children when they had contact with the field nurses and were asked to discuss actions taken by parents on their behalf. Although the respondents could identify fundamental family values and beliefs, they may have misconstrued some of their parents' actions. It is also possible that the age of the respondents influenced their perceptions. The nurses spent part of their work week in schools and thus directed many of their services toward children; many nurses asserted that members of older generations were far less receptive to their teachings. As a result of these limitations, we cannot assume that the interviews offer a more objective account than do the field nurses' reports. Nevertheless, our respondents powerfully challenge the nurses' interpretation of their interactions with American Indian people.

I

During the period when the field nurses worked on reservations, three different views about American Indians vied for acceptance among whites. The oldest held that American Indians, like other people of color, were inherently inferior, both mentally and physically, and thus doomed to extinction; neither education nor health care could prevent this.[17] Proponents of assimilation assumed that American Indian people would survive but that their culture would disappear. This doctrine underlay the establishment of boarding schools, which wrenched children out of their communities, and the Dawes Act of 1887, which allotted land to individual households. As Brian W. Dippie writes, the "values integral" to this act were "veneration of private property, individual initiative, and self-sufficiency—as well as white attitudes about everything from cleanliness to standards of beauty."[18] During the early twentieth century, cultural relativism began to challenge the belief that value judgments could be assigned to different societies and that American Indian people were tabulae rasae upon whom white middle-class culture could be inscribed.[19]

Public health nursing fit within the assimilationist approach. Historians of public health nurses have struggled to understand the partic-

ular blend of compassion and condescension these workers exhibited.[20] Both can be considered expressions of assimilationism. Rejecting the notion of the inherent weakness of disadvantaged groups, the nurses assumed that access to health care was essential, and many endured enormous personal sacrifices to deliver services themselves. But the nurses also were convinced that there was only one true culture, which was the preserve of dominant social groups. One of the key missions of public health nurses was thus to inculcate Euro-American attitudes and values.

Public health nurses themselves were overwhelmingly white, native born, and middle class. As one administrator wrote, "Public health nursing demands a young woman who has family background and ideals, so that almost unknowingly she inculcates."[21] Immigrants were believed to lack "the environment and training" necessary to fulfill the responsibilities of public health nurses.[22] Elinor Gregg, director of nursing in the Office of Indian Affairs between 1924 and 1937, later acknowledged that she refused to hire African-American women.[23]

The field nurses typically arrived on reservations knowing little or nothing about the people they were expected to serve. Although some sought opportunities to learn about indigenous cultures, others saw no reason to do so.[24] Moreover, the turnover rate was high.[25] The few nurses who remained in the OIA's employ for lengthy periods received frequent transfers.[26] Language differences contributed to the lack of understanding. Although all schools taught English, many older American Indians spoke only their own languages.[27] Most nurses were able to rely on patients' families or friends to translate, but some acknowledged that communication through interpreters was extremely restricted.[28] Only the rare nurse sought to learn an American Indian language.[29]

The field nurses' identification with scientific

medicine bolstered their sense of superiority. By 1930 such fearsome diseases as cholera, typhoid fever, and smallpox, which had attained epidemic proportions decades earlier, were virtually eliminated, and other common killers, including rickets, syphilis, and dysentery, had lost much of their menace. Although historians now debate the extent to which medical advances contributed to the decline in infectious diseases, many people were convinced that the credit belonged to medical science alone.[30] The growing faith in medicine both added a sense of urgency to the task of spreading its benefits to all sections of society and increased disdain for indigenous healing practices.[31]

Armed with scientific "truths," the nurses cavalierly dismissed American Indian health beliefs as arbitrary and bizarre. Like colonial health officers in nineteenth-century Africa and Asia,[32] the nurses frequently described the people they served as "ignorant," "primitive," "prejudiced," and "superstitious." Most nurses insisted that American Indians were capable of reason but had to be taught how to exercise it. Subscribing to a notion of reason transcending historical and social conditions, the nurses denied the possibility that American Indians could be active participants in the construction of meaning and knowledge. Ideal clients were "receptive," "amenable," and, above all, "cooperative." A basic assumption of the nurses was that American Indians would follow a linear progression from understanding the rules of health to the eradication of all traditional practices. This model left no room for ambiguity or syncretism.

II

Throughout the nineteenth century, Sioux people actively resisted encroachment by white entrepreneurs, farmers, and soldiers. Warfare between Sioux people and whites culminated in the Battle of Little Big Horn in 1876 and the Wounded Knee massacre of 1890. These events left the Sioux dependent on rations from the federal government. In order to survive, the Sioux made many accommodations, such as sending children to boarding schools, using the agricultural and ranching methods taught by government agents, and moving into log houses. Contrary to the assimilationist credo, however, these changes neither reflected nor produced wholesale repudiation of traditional values and beliefs. By selectively appropriating white practices, the Sioux were able to transform traditional culture without destroying it.[33] We will see that their response to the services provided by the field nurses followed that pattern.

Two major changes occurred during the 1930s. Grasshopper infestations and the economic depression caused new devastation, increasing reliance on government resources. The passage of the Indian Reorganization Act of 1934 provided at least the illusion of self-government; many Sioux claimed, however, that the tribal councils instituted in 1936 departed dramatically from traditional Sioux ways.[34]

Like many other American Indian people, the Sioux had a health care system based on the interrelationship between people and the environment. Because health was the result of harmony and balance, disease stemmed from some form of disequilibrium. The Sioux drew a distinction between diseases which originated in environmental disruptions and could be cured by healers trained in natural remedies and those which resulted from spiritual forces and required the intervention of spiritual healers.[35]

Various assaults had seriously weakened this system by 1930. Much knowledge was lost during epidemics because large numbers of healers died before teaching members of younger generations. The consolidation of reservations during the early twentieth century reduced access to traditional remedies. And the Europeans introduced lethal infectious diseases that were unfamiliar to indigenous healers. Although many Sioux people assumed that white practitioners were best equipped to provide some services, they did not embrace white medicine as a substitute for the declining Sioux health care system.[36]

III

Barbara Melosh notes that nursing manuals during the 1930s "emphasized health education as the special contribution of the public-health nurse, minimizing the traditional tasks of bedside nursing."[37] The field nurses on American Indian reservations engaged in a broad range of activities, including screening for such conditions as trachoma, tuberculosis, and sexually transmitted diseases, providing immunizations, delivering home care, and placing clients in institutions for sickness and childbirth. Most nurses insisted, however, that education was their primary focus.

By imparting information about the role of germs in disease causation, the nurses helped to democratize medical knowledge. But some components of health education also provided avenues for guiding American Indians toward what the nurses considered "right living." As the following section will note, a key ingredient of the health education program was teaching about the importance of cleanliness. Nurses on both Sioux reservations, for example, instructed women to boil contaminated water before cooking with it. But the concept of cleanliness shaded easily into neatness and order. Many nurses claimed to be promoting cleanliness when they encouraged American Indians to landscape their yards, decorate their homes with wallpaper and curtains, and adopt Euro-American dress. Under the rubric of diet the nurses not only taught about the nutritional content of different foods but also urged women to cook in the

Euro-American style. When instructing new mothers, the nurses provided information about the intestinal and respiratory diseases that caused a high proportion of infant deaths. In addition, the nurses advised women to give babies English names, feed them by the clock, and make them sleep alone.

Assuming a close relationship between personal habits and disease, the nurses were able to extend their purview as health educators still further. Some focused on what they considered the high rate of out-of-wedlock births. "If there is anything more prominent than another in my ambition for our people," wrote LaDora White about the Apache in Oklahoma, "it is an earnest desire to create higher standards of living—getting our unmarried mothers married, giving a heritage of a father's love and care to the children, a more stable home and mother."[38] Grace Olsen taught a Shawnee schoolgirl "that she must be polite and obedient and considerate of others at all times if she expected to get along."[39]

But if the nurses believed that American Indians had to conform to Euro-American standards of "right living" to promote health, the Sioux clients believed that the nurses had to adhere to Sioux standards of proper conduct to fight disease. According to the Sioux, the primary requirement of a healer was that he or she be a good person. Because healing powers could be used to inflict illness and death as easily as to cure, it was essential that practitioners be committed to the well-being of the community.[40] Eighteen of the twenty-three people interviewed for this study discussed the character of the nurses with whom they interacted. Asked what they recalled about the nurses, most respondents mentioned their dedication first. The following comments were typical from people who had made use of the nurses' services: "She's working for the people"; "She's pretty busy. Good worker. She helped people"; "I think she really worked. Worked hard. For the people."

One woman explained why she and her family avoided a particular nurse:

> When we're ready to leave Flandreau [a boarding school] she'll say . . . You're better off in school. Stay for the summer. She was that mean. She didn't want [the schoolchildren] to come home because their family are poor and they have outdoor toilet and no bathtub and like that. She talks like that. She's very rude, mean woman.

Because the family viewed the nurse as condescending and unsympathetic, they rebuffed her services.

IV

Although the nurses prided themselves on applying the new principles of scientific medicine, their reports were fraught with old assumptions about disease causation. Many wrote about the dangers not just of germs but also of "filth" and "odors."[41] The lingering association between dirt and disease meant that dirt floors were a special focus of concern. As the Merriam Survey had noted, such floors were "still the rule in primitive dwellings," causing "illness" as well as "discomfort."[42] Many nurses stressed the importance of laying down wood as quickly as possible.

The field nurses also resembled nineteenth-century sanitarians in their concern with environmental and social conditions. Some noted the futility of preaching the virtues of cleanliness to people who lacked adequate water supplies. Josephine Yanachek, a nurse in Fort Peck, Montana, wrote:

> For sometime I had been feeling that some of the lack of cleanliness might be due to the difficulty of getting water. There are as yet very few families who have wells, and those who have not must carry water either from the river or some neighbor's well. Frequently the neighbor is two and three miles distant. This condition is naturally not conducive to frequent tub bathing.[43]

Orpha Zoa Hall noted that when drought struck South Dakota in the 1930s, some Pine Ridge residents had to haul water from a source five miles away.[44] The only available drinking water was polluted on many reservations. The nurses took samples of water, which they sent to local doctors for testing, and they continually urged the OIA to dig wells. As the Red Cross wrote in its 1924 report, establishing an adequate water supply "depended upon the action of the government and not upon the efforts of the Indians."[45]

Nevertheless, the nurses also concentrated on individual change. Assuming that many American Indians were indifferent to their sanitary conditions, the nurses sought to "awaken a desire" for improvement.[46] After noting the lack of wells and privies in Pine Ridge, Orpha Zoa Hall commented, "The Indians see no need for these things, make no effort to get them and will have to learn appreciation of them after the need has been supplied."[47] Two years later she wrote that her "greatest thrill" was a visit from a group of Pine Ridge leaders, requesting that she use her influence to get wells.[48] The nurses also insisted that American Indian people knew nothing about cleanliness. The Red Cross asserted that it found "total ignorance of the most elementary principles of home and personal hygiene."[49] A prominent part of the health education program consisted of conducting classes in personal and domestic hygiene for girls and mothers, demonstrating the use of toothbrushes and paper handkerchiefs, examining schoolchildren for lice and "soiled hands and faces," and inspecting both the interior and exterior of homes during routine visits and sick calls.

Although the nurses viewed themselves as introducing concepts of cleanliness to American Indian tribes, the accounts of early colonial explorers and settlers describe high standards of personal hygiene among American Indian people. Some tribes bathed daily in streams regardless of the weather. Many also had some version of the sweat bath, which had spiritual as well as hygienic significance.[50] The nineteenth-century epidemics, migrations, and wars disrupted these traditional practices.[51] Nevertheless, by the 1930s most people had found ways to resume regular bathing rituals.

With a tone of voice indicating a great deal of pride in her mother's accomplishments, one Sioux woman we interviewed conveyed traditional respect for personal cleanliness in the following comment:

> My mom was very quiet. Just like a Sioux woman . . . She loves to wash. She would boil. She stood outside and boiled all the wet clothes or flour sacks. My mom was very quiet but boy she got ambitious. She kept us always clean. There was a boiler on the camp, our camp stove, outside. We had a fire going so that's where we boiled water.

One woman remembered a nurse teaching her about germs:

> Some [field nurses] was telling me. We was scared. In the house, you know, germs and stuff. Always tell me, just boil some water. Put two, three, oh two drops Lysol. That kills all the germs. Just boil that water. Now that's a field nurse that told me.

By the 1930s the new information about germs had spread widely; two of the women we interviewed indicated that their mothers learned about the relationship between sanitation and disease from sources other than the nurses.[52]

Our informants were not equally receptive to all advice about cleanliness. One man said, "When the nurse come over there, the field nurse, everybody's getting ready. Sweep the floor and get ready for the nurse to come." This quotation reminds us that some actions the nurses interpreted as evidence of the efficacy of their teachings may have been performed only in anticipation of their visits. Unlike daily bathing, sweeping the house floor was foreign to the Sioux. Many people who were adults in the 1930s had lived in tepees or tents as children. The few permanent houses had dirt floors. Because all homes were small, people routinely lived outdoors during the summer. Other comments of this informant suggest that the people in his community wanted to please the nurse, not because they considered themselves her inferiors but because Sioux customs demanded that visitors be made comfortable. Knowing that the nurse valued clean floors, the clients swept as a way to welcome her to their homes. Their action thus represented a continuation of traditional practices, not a break with them.

V

A recurrent complaint in the field nurses' reports was that American Indians had "preconceived ideas" about the nurses' responsibilities, expecting them to distribute large quantities of medications. Mollie Reebel, for example, wrote, "If I were given about fifty gallons each of Liniment and Cough Syrup, and an unlimited supply of Vaseline, would do a 'Land Office Business,' but following the policy outlined by the Office, have discouraged these."[53] The nurses gave several reasons for refusing to fulfill medication requests. American Indians often administered medicine inappropriately; large supplies of drugs enabled them to bypass the nurses' authority; some medications could be prescribed only by physicians; and American Indian preoccupation with drugs undermined the nurses' health promotion program. Mary Margaret Schorn, a nurse stationed in the Tulalip Agency, wrote: "These Washington Indians are great believers in drugs, pills, and liquid medicines for all aches and pains. It takes persistent tireless efforts to teach them that food, rest and personal hygiene have a great deal to do with minor illnesses."[54] It is also possible that the nurses wanted to distinguish themselves from OIA physicians, whose practice frequently consisted primarily of dispensing drugs.[55]

The nurses' language suggests that old notions of pauperism also influenced their stance. Like nineteenth-century social reformers who believed that strict control of charity was necessary to prevent the "demoralization" of the poor,[56] the field nurses condemned the "demanding" attitude of the clients and their "constant begging" and warned of the dangers of "indiscriminate giving." The implication was that access to pills not only undermined health but also retarded progress toward independence and self-reliance.

The elderly Sioux we interviewed seemed either indifferent to or oblivious of the nurses' condemnation of their behavior. Sioux people traditionally had relied on healers to provide medicine, and our informants had expected the nurses to follow that practice. From their perspective, requesting medications represented not a rejection of the nurses' program but rather an acceptance of the nurses into the community.

Our respondents fit the nurses' medicines into existing categories and used them the same way as traditional medicine. Asked what she remembered about the nurses, one woman responded, "I always got eyedrops . . . from the nurses." She continued by noting, "My grandma was good at that, too. She . . . boils some stuff and gives me eye drops for the eye." Dispensing cough syrup was the most commonly mentioned service the field nurses provided, perhaps because tuberculosis was prevalent and Sioux

people were accustomed to using a syrup remedy for the relief of coughs. A man described the nurses this way: "They carried a big bag on their saddle. And they go visit they take their bag inside and give you any kind of medicine that they have. Salve, Vaseline, maybe cough medicine." He described Sioux remedies: "They already know really what they're supposed to do. They're coughing they drink skunk oil. They use skunk oil, it's real bitter." The Sioux found the same extramedicinal uses for drugs, whether provided by nurses or natural healers. According to one of our respondents, older people used both kinds of cough syrup to flavor their pancakes. Viewing white medicines as the equivalent of Sioux ones, some people used them interchangeably, relying on whichever was more convenient. "I went both ways," said one woman, adding that she used the nurses' medicine in one instance because "it just happens that I was there when I had an ailment of some sort."

VI

The Sioux wanted pills because they resembled traditional medicine but accepted other services because they filled gaps in the indigenous healing system. Immunizations, for example, were a powerful new technology Sioux people lacked.[57] As part of the disease prevention program, the nurses provided immunizations against smallpox, diphtheria, and typhoid fever. For some of our respondents, receiving inoculations was not a matter of choice. Five, for example, had participated in mandatory school-based immunization programs.[58] One described her experience at a local day school this way:

They broke a glasses and put alcohol on here and they scratch it down like that. Ow, ow, ow, I said. That time they used the glasses. Those school kids when the doctors come everybody run away. Run to the field. When the doctors come the teachers locked the door and snap all the windows and everybody stay in the school. They cry and try to break the window.

Several respondents had received immunizations in their own homes or at the nurses' station, others when they had been brought to town for screening exams as part of a health survey clinic.[59] A woman who had been ten at the time of the survey related this incident: "There was a big tent put up, like a circus tent. And that's where we were all given shots. Everybody stand there and hovered like stand in line. And boy, when they stuck me why I'd really scream. I really embarrassed myself." Although many of our interviewees remembered the immunizations as traumatic fifty years later, they had submitted to the ordeal.

The comments of three respondents whose families typically used only Sioux medicines are especially illuminating. Although these respondents had welcomed the nurses during home visits and accepted rides and help with boarding school arrangements, they had relied exclusively on relatives who were healers to deliver medical care. All three, however, vividly described their smallpox vaccinations. One said:

We had our Indian doctors and my granddad was an Indian doctor. And he believed in herbs and everything for us so why when we got sick why my mama just took us to him. I don't remember us ever given any medication. She [the nurse] never gave medication. We depended on our granddad. . . . The only thing they allowed us I remember they vaccinated us. They took a needle and they scraped our [arms]. He did allow that.

VII

Tuberculosis was, in the words of the Merriam Commission, "without a doubt the most serious disease among Indians."[60] The tuberculosis death rate was seven times that of the rest of the population.[61] According to Sarah Smith, the disease was present "in almost every home" in Pine Ridge.[62]

The field nurses battled the disease in various ways. They established screening programs, encouraged "suspects" to receive X-rays, and in some cases brought them to be tested. In addition, the nurses made arrangements for clients diagnosed with the disease to be placed in sanatoria, encouraged them to enroll, and drove those who lacked other means of transportation. When tuberculosis patients remained at home, the nurses provided advice to them and their families.

The paucity of services hampered the nurses' work. Despite official statements of concern about the high prevalence of tuberculosis, the government never committed the funds necessary for effective control. A Rosebud nurse complained:

Clinic follow-up has been strenuously carried on, with most unsatisfactory results. Much time and effort has been used in arranging for and transporting X-ray recommendations to the hospital, where hours upon hours were spent in waiting for X-ray operation, etc. On several occasions an entire day was spent without accomplishing anything whatever.[63]

Because most state and county sanatoria refused to admit American Indians,[64] nurses typically could place tuberculous patients only in institutions established by the OIA. In 1934 the OIA operated fifteen sanatoria containing 1,315 beds.[65] Occupancy rates were extremely high.

Many field nurses convinced clients to enroll in institutions only to discover vacancies did not exist. "'No room at the inn,'" wrote Anna A. Perry, a Wisconsin nurse. "I wish I had a good stable somewhere for T.B. children of pre-school age."[66] The few places were often hundreds of miles away. Zelma Butcher reported driving three days to take Pine Ridge patients to a sanatorium in Albuquerque.[67] The poor quality of the roads, coupled with bad-weather conditions, made such trips extremely dangerous. After noting that tuberculous patients in Washington and Oregon had to travel two days by "day coach" before embarking on a drive by car to reach a Nevada sanatorium, the Merriam Commission commented, "Obviously such a trip is beyond all reason for a case of active tuberculosis."[68] Many of the government cars the nurses drove were open;[69] those that were closed lacked heat.[70] All were in poor condition and frequently broke down.

OIA sanatoria were seriously deficient. The 1935 Annual Report of the Phoenix Sanatorium stated, "The crying need of the entire institution . . . is an increase in the operating allotment. For the past fiscal year, the per diem cost per patient was less than $1.50." The majority of buildings were "in a most deplorable condition"; the equipment was "woefully inadequate."[71] Carrie Brilstra, a nurse at the sanatorium in Dulce, New Mexico, later wrote that she was shocked to discover that the institution had only four thermometers for eighty-eight patients and that X-ray films frequently were "unreadable."[72]

In explaining the ineffectiveness of their campaign to eradicate tuberculosis, the field nurses stressed the ignorance and negligence of the clients.[73] According to the nurses' reports, the disease spread not just because homes were small and cramped but also because patients failed to adhere to principles of basic hygiene. Janet Wallace, for example, claimed that although she "stressed the absolute necessity of many precau-

tions" to a Pine Ridge man, he refused to comply and thus was "exposing his whole family."[74] Family members and friends were also at fault. "Here is an example of where the Indians have no way of protecting themselves against their own people," wrote Charlotte Conrad, an Oklahoma nurse.

> They accept these patients as visitors in their homes, some times against their good judgment, endangering their health and the health of their children because it is customary among the Indians to accept all who care to come and visit them and to stay as long as they wish. These old customs and their rigid adherence to them, for fear of "losing face," with their own people, is a great drawback to progress in this phase of the work. . . . Until the Cheyenne and Arapaho will be able to take a more intelligent attitude toward this, progress will be slow.[75]

The nurses explained client hostility to sanatoria care in a similar way, focusing on personal fallibility rather than the deficiencies of available institutions. Because they were "superstitious," "backward," and "prejudiced," American Indians clung to traditional practices and believed even outrageous rumors about sanatoria. Such labels permitted the nurses to disregard the motives of their clients, treating them solely as obstacles on the path to progress. The nurses could then construct accounts in which all initiative came from them.[76]

Institutionalization occurred because the nurses "educated" their clients, "sold" them on the benefits of sanatoria, and "persuaded" them to enter. According to Gertrude Hosmer, for example, "It is somewhat like trying to raise the dead to persuade two patients with active Tuberculosis to go to a Sanatorium."[77] Margaret Mary Schorn wrote:

In many instances I made five to six trips to some families selling them the idea of hospitalization. All their prejudices, misinformation, suspicion etc was very hard to overcome. However, we won to the extent that we gained consent from every parent who had a tuberculous child. . . . I count this as a decided victory for the medical workers in the field, as the . . . tribe in the early spring had firmly made up their minds not to hospitalize their children.[78]

Arguing that American Indian parents could not be trusted to determine the best interests of their children, Schorn described herself as engaging in a campaign to wrest consent from the parents. Each child's entry into the sanatorium thus represented a personal triumph.

Three Sioux women we interviewed had suffered from tuberculosis as children; their comments help to restore agency to the field nurses' clients. All believed that their parents and grandparents had made their own decisions about institutionalization. They had not opposed sanatoria in all circumstances; however, just as they had viewed white health care solely as an adjunct to the Sioux system, so they had viewed institutional placement as an option to be considered only when home care proved inadequate. A woman who had lived with her grandmother said:

One time I was about seventeen years old I was really sick. At that time the people really have TB. And I had it too, at that time. So those tell me I should go to the government sanatorium. I didn't go, I stayed home. And I'd get up early in the morning to go outside and go for a walk and then come back. And I was all right. Miss Butcher, she was there. She give me clothes or stuff that I drink in the morning.

Despite the nurse's insistence on sanatorium placement, "Grandma didn't want me to go so I didn't. Stayed home." This girl had been relatively isolated from other family members and appeared to be in no immediate danger. The grandmother considered herself qualified to give care. Although she accepted the nurse's advice about home care, she rejected all entreaties to exile the girl.

The comments of the two other respondents suggest that even parents who complied with the nurses' recommendations about institutionalization did not simply surrender to the nurses' authority. One recalled:

At that time, you know, TB was going through the reservation. My mom was a good nurse. She kept me clean. There was a whole bundle of bandages there. Denver mud. That's what they treated me with. Part of the time I stayed at home. But there was a time that my dad took me to Rapid City on the train and left me over there. I really cried so bad. . . . There was another girl that was with me at that time. And she didn't go and she's dead. So my dad thought you better go for that so I went. And I survived.

This woman had been able to remain at home for a period because her mother was a "good nurse" who knew how to render care. Although the interviewee may have been unaware of the extent to which a field nurse influenced her father's decision, she at least perceived him as having been responsible for determining the timing of institutionalization.

The third woman reported yet another experience:

My mother . . . died. And then, after that . . . it just seemed like I didn't feel good at all. And that was when Mrs. Herm, the field nurse, took me into the clinic in

Rosebud and they run a series of tests on me and they said I had TB. So what did they do, they sent me to Iowa. It's on the Sac and Fox reservation. And the reason I was sent there was when my dad was in Carlisle [a boarding school] he befriended a man from there and became blood brothers, more binding than birth brothers. So anyway there he knew that he and his wife would give me good attention because he wrote him a letter explaining that my mother had died. So then I was in the hospital in there.

This woman's father did not have the resources to tend his daughter at home. Because he had a full-time job at the subagency, and his wife had died, he concluded that the best way to care for his daughter was to send her away. He did, however, take an active role in selecting a facility. Although patients from Rosebud and Pine Ridge did not normally go to the sanatorium at Sac and Fox, he insisted that his daughter be close to a man he trusted to look after her.[79]

VIII

Birthing women were another group the field nurses sought to institutionalize. During the 1920s and 1930s hospitals throughout the United States launched campaigns to enroll women for labor and delivery. Although no statistics existed to demonstrate the superior efficacy of these facilities,[80] hospital obstetrics grew rapidly; by 1940 more than 75 percent of urban women gave birth in hospitals.[81]

The field nurses faced special obstacles in encouraging institutional care. The number of beds in OIA hospitals rose from 2,261 in 1934 to 3,053 in 1940.[82] Nevertheless, the nurses frequently complained about the lack of vacancies. Travel time was an even more serious issue for birthing women than for victims of tuberculosis. "Being that hospitals are so far from here," wrote Bessie Holsworth in Rosebud in 1935, "it is very difficult to get many OBG patients admitted for confinements."[83] Some nurses reported distances of hundreds of miles. One strategy was to hospitalize pregnant women prior to confinement; however, such women frequently left while waiting to deliver. Several nurses who transported women in active labor described themselves as engaging in a "race with the stork."

Moreover, care in OIA hospitals, like that in sanatoria, was substandard. "There was a conglomeration of cases in the women's ward that shocked my nursing sensibilities," wrote Elinor Gregg about her visit to a Navajo hospital in the mid-1920s. A "mother and baby, the mother with puerperal insanity," were lying near patients suffering from various infectious diseases, including diphtheria, pneumonia, measles, typhoid fever, and meningitis.[84] Some hospitals were located in abandoned forts; administrators frequently warned of the dangers of fires.[85] Most lacked adequate sterilization and laboratory equipment. Understaffing was another problem. Most hospitals had only one doctor. If he quit unexpectedly or was temporarily absent, the nurses had to assume his responsibilities. When Janet Green delivered twins at Fort Wingate after the physician had left for the weekend, she "prayed hard that they would get no infection."[86]

As Rosemary Stevens notes, hospitals serving white middle-class clienteles during the 1920s advertised themselves as being "safe, scientific, and friendly place[s]."[87] Despite the dramatic differences between such facilities and those operated by the OIA, the nurses used similar language to describe the institutions they promoted. "I wish that all of the pre-natals would go to the hospital for their confinements," wrote Alice D. Divine. "It would be so much easier on them, the hospital has so much to offer, good doctors and nurses, clean beds and food."[88] Amy

Schreiber reported making several trips to a pregnant woman, telling her "what a wonderful opportunity it was to receive proper care."[89] The Rosebud hospital was "the safe haven for natal care," according to Edna M. Hardshaw.[90] The nurses' reports were filled with stories of mothers and infants whose lives were saved by hospitals; when institutionalization was rejected, the nurses warned, tragic consequences inevitably ensued.

The nurses insisted that institutional stays could offer moral as well as physical benefits. Remaining in the hospital approximately ten days after delivery, young mothers received instruction in infant care. As Adelia L. Eggestine, a Minnesota nurse, wrote. "With the mothers having the proper start, begun at the hospital, in feeding and caring for the baby, more progress is noted in their home care."[91] A Pine Ridge case showed what was possible. Visiting the home of a young woman who had delivered at the agency hospital, Janet Wallace found the mother "taking excellent care of her baby, bathing her daily, dressing her properly, and nursing her at regular intervals." In addition, the hospital had taught the mother to sleep in a bed rather than on the floor.[92] Hospital deliveries also were assumed to incline women to seek institutional care when sickness visited their households in the future.

Nevertheless, birthing women, like tuberculous patients, frequently shunned institutions.[93] Some women successfully hid pregnancies and failed to notify the nurses when labor began. Cecilia Severino complained that women in Leupp, Arizona, who acknowledged their pregnancies "simply will not come to the hospital for delivery." "I do not believe I have failed in my duty," she assured the OIA, "because I am particularly interested in this phase of my work as I think it very important, and I have done my utmost. But to no avail—they simply will not come in."[94]

Some field nurses attributed birthing women's hostility to hospital care to such factors as distance and unwillingness to leave older children. Most, however, viewed opposition to institutionalization as evidence of irrationality and backwardness, just as they had in the case of tuberculosis. "This problem of having the child born in the house has been a hard one to accomplish," wrote H. Louisa Harple, "for it is an Old Indian superstition and some will never change it."[95] Rosebud women who refused to go to the hospital were "the more primitive group," according to a nurse.[96] Hospital births were believed to demonstrate that women acknowledged the dominance of white medical knowledge.

The interviews offer a different perspective. The Sioux accepted white assistance at childbirth for the same reason they accepted other white medical care—to supplement the traditional healing system, not to supplant it. Most women who had access to midwives saw no reason to seek other birth attendants. Moreover, as one respondent explained, the Sioux feared all hospital care. "At that time, Lakota people, they're afraid to go hospitals. And they say that if some Lakota going to die they go to hospital and then they come back and die.... They really scared of the hospital." Just as nurses generalized from individual cases to conclude that hospital deliveries were safer, so one woman used her own experience as the basis for associating hospitals with poor outcomes: "You couldn't get the Indian women to go to the hospital. And the only one I ever had in the hospital, I lost. The rest I delivered at home."

Our respondents also criticized specific hospital practices. A woman whose own birth had been assisted by a nurse stated:

I don't think there were too many in my family that went to the hospital for their baby. They all had the baby at home. Hoisting their legs upon those things. That was totally against the Indian way of having a baby. . . . As I was growing up I do know a lot of things like people had gone to the hospital and they come back and the treatment they got there wasn't all that great. They didn't appreciate being treated that way. . . . So they were always reluctant to go to the hospital.

Another woman explained why home births were preferable: "When you're home you've got people that love you and you been with them and you're comfortable." Other respondents noted that certain cultural beliefs encouraged home births. For example, because babies were believed to assume the characteristics of the person who cleaned their mouths, it was imperative that friends and relatives rather than strangers assist at births.

Four women had had at least one baby in hospitals. Their comments suggest that situational factors rather than conversion to white medicine explained acceptance of institutional care. Two women had lived far from their families and had had little support from other women in the community. One said, "Knowing that I didn't have an older woman there to help me through my pregnancy [the field nurse] was always there. And so she was very helpful. My baby was born in the Pierre hospital."[97] The other two women went to the hospital because they encountered difficulties. As one stated, "I had trouble with one of them, one of my children. That hand came out, the rest of it was still inside. [The midwife] couldn't get her out. So they decided to rush me to Murdo." Although this woman did not substitute midwifery services for routine care, she sought additional help when special problems arose.

IX

By the 1930s American Indian people had succeeded in maintaining viable cultures through-out two centuries of catastrophic change by selectively incorporating white goods, services, and practices. Our interviews suggest, at least in a preliminary way, that some American Indians on reservations adopted a similar strategy in their encounters with field nurses. Although the field nurses approached their clients as passive consumers of Western knowledge and prede-fined services, Sioux people retained critical distance from the nurses' authority. Instead of accepting the nurses' normative judgments, the Sioux subjected the nurses to Sioux criteria of proper conduct. In addition, the Sioux challenged many recommendations, especially those about institutional placement. When the Sioux accepted the nurses' services, they did so on their own terms. They fit some services into existing categories and used others to fill gaps in the traditional health care system.

Although the field nurses hoped to under-mine traditional patterns of thought and behav-ior, the Sioux appear to have used those patterns to accomodate the nurses. When following the new instructions about eradicating germs, Sioux women built on their own hygienic practices. When sweeping their floors in preparation for the nurses' visits, families adhered to traditional norms about the proper respect due to visitors. And when choosing which nurses to trust, people employed customary criteria for as-sessing healers' competence. It is beyond the scope of this project to investigate how the nurses' work altered Sioux beliefs and behavior. Nevertheless, it is clear that change did not oc-cur in the straightforward way the nurses pre-dicted.

Despite the current preoccupation with resis-tance among scholars in a variety of disciplines, most historians have access only to the "texts of the dominant."[98] One solution is to "read against the grain,"[99] seeking evidence of subversion by members of subordinate groups. The interviews we conducted with elderly Sioux remind us of the limitations of that approach. Although the field nurses' reports contain numerous examples of client noncompliance, the reports provide no insight into the reasons why American Indians rejected or accepted specific services and the meanings American Indians assigned to the ones they chose to incorporate. For such information we must look to other types of sources.

Notes

1. See N. B. Dirks, G. Eley, and S. B. Ortner, eds., *Culture/Power/History: A Reader in Contemporary Social Theory* (Princeton, 1994).

2. M. Sahlins, "Cosmologies of Capitalism: The Trans-Pacific Sector of 'The World System,'" in Dirks, Eley, and Ortner, *Culture,* 412–56. See also N. B. Dirks, G. Eley, and S. B. Ortner, "Introduction," in *Culture,* 21.

3. N. Fraser *Unruly Practices: Power, Discourse, and Gender in Contemporary Social Theory* (Minneapolis, 1989), 180.

4. Fraser, *Unruly Practices,* 179–80.

5. For information on federal public health activity on American Indian reservations, see T. Benson, "Race, Health, and Power: The Federal Government and American Indian Health, 1909–1955" (unpublished Ph.D. thesis, Stanford University, 1993); D. Putney, "Fighting the Scourge: American Indian Morbidity and Federal Indian Policy, 1897–1928" (unpublished Ph.D. thesis, Marquette University, 1980); F. P. Prucha, *The Great Father: The United States Government and the American Indians,* vol. 2 (Lincoln and London, 1984), 841–64.

6. L. Emmerich, "'To Respect and Love and Seek the Ways of White Women': Field Matrons, the Office of Indian Affairs, and Civilization Policy, 1890–1938" (unpublished Ph.D. thesis, University of Maryland, 1987); R. M. Raup, *The Indian Health Program from 1800 to 1955* (Washington, D.C., 1959), 12.

7. F. Patterson Fox, "A Study of the Need for Public Health Nursing on Indian Reservations," in Senate Committee on Indian Affairs, *Survey of Conditions of Indians of the United States* (Washington, D.C., 1928–1943), 10,004.

8. M. A. Sandweiss, "Introduction," in *Denizens of the Desert, a Tale and Picture of Life Among the Navaho Indians: The Letters of Elizabeth W. Forster/Photographs by Laura Gilpin,* ed. M. A. Sandweiss (Albuquerque, 1988), 18; Raup, *Indian Health Program,* 12. Some state and local associations also employed field nurses (Raup, *Indian Health Program,* 12).

9. Brookings Institution, *The Problem of Indian Administration: Report of a Survey* (Baltimore, 1928), 24.

10. Brookings, *Indian Administration,* 189; Elinor D. Gregg, *The Indians and the Nurse* (Norman: University of Oklahoma Press, 1965), 89; Raup, *Indian Health Program,* 10.

11. Raup, *Indian Health Program,* 10.

12. Letter of Ruth Riss Seawright to Virginia Brown and Ida Bahl, January 20, 1978, Brown, Bahl, and Watson Collection, Special Collections and Archives, Cline Library, Northern Arizona University.

13. Raup, *Indian Health Program,* 12.

14. These reports are located in the records of the Bureau of Indian Affairs (Record Group 75, file E779) in the National Archives, Washington, D.C. Individual reports are identified by the nurse's name, the name of the agency where she was stationed, and the date.

15. The interviews lasted between one and two hours and were tape recorded and transcribed. Questions were designed to elicit information about experiences with the field nurses, attitudes toward the nurses' services, and use of traditional medicine. The primary interviewer was the second author, Nancy Reifel, a dentist employed by the Indian Health Service and a member of the Rosebud Sioux. Reifel was familiar with the population and geographical area and previously had conducted a community-based survey of elderly Sioux reservation residents at these locations. She was accompanied by Phoebe Little Thunder, a Pine Ridge Sioux resident of Rosebud Reservation, who located respondents and transcribed and translated the interviews. Although all but one of the interviews were conducted in English, many respondents used occasional Lakota words.

16. On the heterogeneity within American Indian societies see S. J. Kunitz, *Disease and Social Diversity: The European Impact on the Health of Non-Europeans* (New York, 1994).

17. For an excellent study of how similar views of

African-Americans affected public health programs in the 1930s, see A. M. Brandt, "Racism and Research: The Case of the Tuskegee Syphilis Study," in J. W. Leavitt and R. L. Numbers, eds., *Sickness and Health in America: Readings in the History of Medicine and Public Health* (Madison, 1985), 331–46.

18. B. W. Dippie, *The Vanishing American: White Attitudes and U.S. Indian Policy* (Middletown, 1982), 263.

19. See Dippie, *Vanishing American.*

20. See K. Buhler-Wilkerson, *False Dawn: The Rise and Decline of Public Health Nursing, 1900–1930* (New York, 1989); B. Melosh, *"The Physician's Hand": Work, Culture, and Conflict in American Nursing* (Philadelphia, 1982), 143–58.

21. Quoted in Melosh, *"Physician's Hand,"* 124.

22. Quoted in Melosh, *"Physician's Hand,"* 124.

23. Gregg, *Indians and Nurse,* 137.

24. "Questionnaire, Nursing Personnel Originals," Brown, Bahl, and Watson Collection, Special Collections and Archives, Cline Library, Northern Arizona University.

25. Gregg, *Indians and Nurse,* 89.

26. "Questionnaire, Nursing Personnel Originals."

27. Brookings, *Indian Administration,* 223.

28. See response of Grace Borgman, "Questionnaire, Nursing Personnel Originals."

29. "Questionnaire, Nursing Personnel Originals."

30. P. Starr, *The Social Transformation of American Medicine: The Rise of a Sovereign Profession and the Making of a Vast Industry* (New York, 1982), 134–40.

31. On the way the growing faith in scientific medicine affected popular concern with issues of access and distribution, see R. Stevens, *American Medicine and the Public Interest* (New Haven, 1971), 188.

32. See, for example, D. Arnold, *Colonizing the Body: State Medicine and Epidemic Disease in Nineteenth-Century India* (Berkeley, 1993); J. Comaroff, "The Diseased Heart of Africa: Medicine, Colonialism, and the Black Body," in S. Lindenbaum and M. Lock, eds., *Knowledge, Power, and Practice: The Anthropology of Medicine and Everyday Life* (Berkeley, 1993), 305–29; M. Harrison, *Public Health in British India: Anglo-Indian Preventive Medicine, 1859–1914*

(New York, 1994); M. Vaughan, *Curing Their Ills: Colonial Power and African Illness* (Stanford, 1991).

33. See R. Clow, "Northern Plains Indians," in *The Native North American Almanac,* ed. D. Champagne (Detroit, 1994), 276–85; E. S. Grobsmith, *Lakota of the Rosebud: A Contemporary Ethnography* (New York, 1981); M. N. Powers, *Oglala Women; Myth, Ritual, and Reality* (Chicago, 1986).

34. See T. Biolsi, *Organizing the Lakota: The Political Economy of the New Deal on the Pine Ridge and Rosebud Reservations* (Tucson, 1992); Powers, *Oglala Women,* 33.

35. See J. Adair, K. W. Deuschle, and C. R. Barnett, *The People's Health: Medicine and Anthropology in a Navajo Community* (Albuquerque, 1988), 3–11; S. J. Kunitz, *Disease Change and the Role of Medicine: The Navajo Experience* (Berkeley, 1983), 118–145; V. J. Vogel, *American Indian Medicine* (Norman, 1970), 13–35; A. G. Waxman, "Navajo Childbirth in Transition," *Medical Anthropology* 12 (1990): 187–206.

36. On changes in the traditional health care system of American Indian people, see J. R. Joe, "Traditional Indian Health Practices and Cultural Views," in *Almanac,* ed. Champagne; I. S. Kemnitzer, "Research in Health and Healing in the Plains," in *Anthropology on the Great Plains,* ed. W. R. Wood and M. Liberty (Lincoln, 1980), 272–83; Kunitz, *Disease Change,* pp. 118–23; G. C. Lang, "'Making Sense' about Diabetes: Dakota Narratives of Illness," *Medical Anthropology* 11 (1989): 305–27.

37. Melosh, *"Physician's Hand,"* 125.

38. "Monthly Report," Apache Agency, Oklahoma, March 1935.

39. "Monthly Report," Shawnee Agency, Oklahoma, August 1934.

40. See W. T. Corlett, *The Medicine-Man of the American Indian and His Cultural Background* (Springfield, IL, 1935), 105; Powers, *Oglala Women,* 65.

41. On the continuing impact of nineteenth-century ideas about disease causation on public health campaigns in the early twentieth century, see N. Rogers, *Dirt and Disease: Polio Before FDR* (New Brunswick, 1992).

42. Brookings, *Indian Administration,* 554.

43. "Monthly Report," Fort Peck, Montana, Feb. 3, 1935.

44. "Monthly Report," Pine Ridge, South Dakota, Aug. 1933.

45. Patterson with Fox, "Study," 32.

46. See H. Reinback, "Monthly Report," Tomah, Wisconsin, November 1934.

47. "Monthly Report," Pine Ridge, South Dakota, August 1993.

48. "Monthly Report," Pine Ridge, South Dakota, Sept. 1935.

49. Patterson with Fox, "Study," 66.

50. R. J. DeMallie, "Lakota Belief and Ritual in the Nineteenth Century," in *Sioux Indian Religion,* ed. R. J. DeMallie and D. R. Parks (Norman, 1987), 25–44; Vogel, *American Indian Medicine,* 253–61.

51. *Handbook of North American Indians,* 10: *Southwest* (Washington, D.C., 1983), 511.

52. It is possible that American Indian people were especially willing to incorporate the nurses' sanitary advice because it was compatible with their own principles of hygiene. See Gilbert Lewis, "Double Standards of Treatment Evaluation," in Lindenbaum and Lock, *Knowledge,* 208–209.

53. "Monthly Report," Western Navaho Agency, Arizona, June 1936.

54. "Monthly Report," Tulalip Agency, Washington, April 1933.

55. Brookings, *Indian Administration,* 234.

56. See E. K. Abel, "Middle-Class Culture for the Urban Poor: The Educational Thought of Samuel Barnett," *Social Service Review* 52 (Dec. 1978): 596–620.

57. Mark Harrison argues that vaccinations were among the more popular public health measures the British instituted in India between 1859 and 1914 (*Public Health,* 232–33). See also Vaughan, *Curing Their Ills,* 24.

58. In 1907 vaccinations became mandatory in all OIA schools. (Brookings, *Indian Administration,* 218).

59. This clinic was held in Pine Ridge in 1932.

60. Brookings, *Indian Administration,* 204.

61. Brookings, *Indian Administration,* 201.

62. "Monthly Report," Pine Ridge, South Dakota, August 1932.

63. Edna M. Hardsaw, "Monthly Report," Rosebud, South Dakota, August 1932.

64. Raup, *Indian Health Program,* 13–14.

65. Raup, *Indian Health Program,* 30.

66. "Monthly Report," Lac du Flambeau, Wisconsin, September 1933.

67. "Monthly Report," Pine Ridge, South Dakota, August 1934.

68. Brookings, *Indian Administration,* 290.

69. Brookings, *Indian Administration,* 223–34.

70. Letter of Ruth Riss Seawright to Virginia Brown and Ida Bahl, January 20, 1978, Brown, Bahl, and Watson Collection.

71. "Annual Report of the Phoenix Indian Sanatorium," July 1, 1935, Records of the Bureau of Indian Affairs, Record Group 75, National Archives, Pacific Southwest region, Laguna Niguel.

72. "Questionnaire, Nursing Personnel Originals."

73. OIA doctors and administrators also attributed the failure of the American Indian health program to patients' personal failings. See Benson, "Race, Health, and Power."

74. "Monthly Report," Pine Ridge, South Dakota, May 1934.

75. "Monthly Report," Cheyenne and Arapaho Agency, Oklahoma, February 1933.

76. According to L. Hunt, contemporary Mexican oncologists employ a similar strategy to reconcile the "gap between professional mandate and possible action." ("Practicing Oncology in Provincial Mexico: A Narrative Analysis," *Social Science and Medicine* 38, [1994]: 843–53.)

77. "Monthly Report," Elko, Nevada, April 1934.

78. "Monthly Report," Tulalip Agency, Washington, April 1933.

79. For a Sioux woman's account of institutionalization in a sanatorium during the 1940s and 1950s, see *Madonna Swan, A Lakota Woman's Story,* as told through M. St. Pierre (Norman, 1991).

80. In fact, historians now conclude that until the introduction of sulfa and antibiotics in the late 1930s and early 1940s, hospitalization probably increased rather than decreased maternal mortality rates. See N. S. Dye, "The Medicalization of Birth," in *The American Way of Birth,* ed. P. Eakins (Philadelphia, 1986), 21–46; I. Loudon, *Death in Childbirth,* (New York, 1992); J. W. Leavitt, *Brought to Bed: Childbearing in America, 1750–1950* (New York, 1986), 171–88.

81. Dye, "Medicalization," 41.

82. Raup, *Indian Health Program,* 30.

83. "Monthly Report," Rosebud, South Dakota, August 1935.

84. Gregg, *Indians and Nurse,* 103–4.

85. Brookings, *Indian Administration,* 277.

86. "Questionnaire, Nursing Personnel."

87. R. Stevens, *In Sickness and in Wealth: American Hospitals in the Twentieth Century* (New York, 1989), 110.

88. "Monthly Report," Cheyenne and Arapaho, Oklahoma, November 1935.

89. "Monthly Report," Tomah Field, Wisconsin, January 1934.

90. "Monthly Report," Rosebud, South Dakota, October 1932.

91. "Monthly Report," Ponemah, Minnesota, 1924.

92. "Monthly Report," Pine Ridge, South Dakota, August 1932.

93. Several studies conclude that at least 75 percent of Navajo deliveries occurred in homes in the late 1930s (see Kunitz, *Disease Change,* 91).

94. "Monthly Report," Leupp, Arizona, August 1935.

95. "Annual Report," Toledo, Iowa, July 11, 1934.

96. "Monthly Report," Edna M. Hardsaw, Rosebud, South Dakota, September 1934.

97. According to J. W. Leavitt, "Increased physical and psychological isolation" was an important factor motivating white women to give birth in hospitals during this period (*Brought to Bed,* 176).

98. See Dirks, Eley, and Ortner, "Introduction," 20.

99. See Dirks, Eley, and Ortner, "Introduction," 19.

Health Care Providers: Physicians

In the middle of the nineteenth century, when women sought medical education within male institutions, they met rejection more often than admission. Although Elizabeth Blackwell received her medical education and degree from the previously all-male Geneva Medical College in 1849, most women were frustrated when trying to follow in her footsteps. The women's response to the repeated rebuffs from the male schools was to open their own medical colleges: seventeen all-female medical schools were founded in the United States in the nineteenth century. Practical clinical experience was also scarce for women learning medicine, and in response to this need Dr. Marie Zakrzewska opened the New England Hospital for Women and Children, described here by Mary Roth Walsh. Walsh emphasizes the importance of the moral and financial support of the women's movement for this hospital as for women's medical education in general. She posits that the practice of the women doctors differed qualitatively from that offered at male-run institutions.

Because of the success of female medical education during the second half of the nineteenth century, many male schools began to admit women students, the largest breakthrough coming in 1893 when Johns Hopkins accepted $500,000 in contributions from the women's community along with its stipulation that women students be admitted equally with men. Sectarian medical schools more readily admitted women, yet coeducation remained controversial in all settings. Despite all the difficulties, women achieved significant successes in medicine. By 1910 more than nine thousand women practiced medicine in the United States, 6 percent of the total, and in some cities like Boston and Minneapolis almost 20 percent of practitioners were women.

Regina Morantz-Sanchez explores aspects of how women physicians fared in the medical world. She uses two careers to examine how representations of gender became embedded within medical professional ideology and concludes that "empathic expertise became an important component of women doctors' public and private image." By contrasting the medical interests and styles of two very different practitioners, Morantz-Sanchez expands ideas about the diversity of women's medical experiences.

Kimberly Jensen writes about the women physicians who sought commissions in the military medical corps during World War I. Approximately six thousand women made a claim for equal status and opportunity in 1918, hoping that war service might bring professional and civic equality back home.

For much of the twentieth century, women lost ground within the medical profession. All but one of the women's medical schools closed when male colleges in their areas admitted women students, and the remaining Woman's Medical College of Pennsylvania (which became coeducational in the 1970s) itself trained one-third of the women medical graduates during the first half of the twentieth century. Women found limited welcomes in the male schools and discovered that their minority status compromised their educational experiences. Only since the 1970s, under strict federal guidelines and with the full support of the new women's movement, have women in medicine again found as supportive an environment within medicine as they enjoyed at the height of their successes in the nineteenth century. Yet women at the end of the century experience a "glass ceiling" when trying to climb to the upper echelons of academic medicine and describe many medical specialties as unwelcoming.

27 Feminist Showplace

Mary Roth Walsh

Historians, by concentrating on the political side of feminism, have largely ignored the material and psychological support that the movement offered.[1] Nowhere is this better illustrated than in the case of the New England Hospital for Women and Children, established by Dr. Marie Zakrzewska.

Zakrzewska has attracted relatively little attention from historians, although she was one of the most influential female physicians of the nineteenth century. In many ways she played a greater role in developing careers for women in American medicine than the more famous Blackwell sisters.[2] Born in Berlin in 1829, her childhood experiences had a great deal to do with shaping her career. When Zakrzewska was ten, her father was dismissed from the Prussian army, forcing her mother to become a midwife and the family breadwinner. That same year Zakrzewska contracted an eye infection and was placed under the care of a physician who took a liking to her, allowed her to follow him on his hospital rounds, and loaned her medical books from his personal library. She quickly developed a keen interest in medicine and, as soon as she was old enough, began assisting in her moth-er's midwifery practice. At age twenty, after two years of petitioning the state authorities for a position in the government-sponsored midwifery school, Zakrzewska finally succeeded in gaining admission to the school at Charité Hospital, the largest hospital in Prussia. The fact that Zakrzewska was the youngest woman to have entered the school made her highly visible. In a short time her medical aptitude and outstanding performance as a student won the admiration of Dr. Joseph Schmidt, the director of the hospital. A French midwife, Madame La Chapelle, had won international fame in obstetrics; Schmidt, in a burst of patriotic pride, predicted that Prussia, as well as France, might boast of "a La Chapelle" and, before Zakrzewska ever had an opportunity to earn the title, dubbed her "La Chapelle the Second." The name stuck and, like a self-fulfilling prophecy, Zakrzewska became his star pupil.

In 1852, a year after Zakrzewska's graduation, Schmidt, although critically ill, was able to overcome strong internal opposition against the appointment of a woman as his successor. Zakrzewska became the chief midwife and professor in the hospital's school for midwives with responsibility for more than two hundred students, including men in the medical school. Unfortunately, the announcement of the appointment came only a few hours before Schmidt's death, and Zakrzewska was unprepared for the jealousy and hospital politics that followed. Schmidt's protective sponsorship for

At the time of her death, in February 1998, MARY ROTH WALSH was Professor of Psychology and American Studies at the University of Massachusetts at Lowell.

three years had eliminated the necessity of her learning strategies for coping with competition in the hospital. Within six months she found the job to be too burdensome and resigned, attempting for a time to establish a private midwifery practice. But an early taste of success in her field made it difficult for her to practice quietly and forget about assuming a larger role in medicine beyond that of an anonymous midwife. She recalled later: "My education and aspirations demanded more than this." Not surprisingly, in view of the recent tide of emigration, her search for expanded opportunities turned to America. She remembered the positive reaction of Schmidt to the news of the establishment of the Female Medical College in Philadelphia: "In America, women will now become physicians like the men: it shows that only in a republic can it be proved that science has no sex."[3]

Convinced that she would find greater freedom to practice medicine in the United States, she sailed with her younger sister to New York in 1853. There she contacted a family friend, a German-American doctor, who quickly dispelled her illusions about American medicine. Pointing out that female physicians were of the lowest rank, even below that of a good nurse, he offered Zakrzewska a position as his own nurse, which she politely but firmly refused, unwilling "to be patronized in this way."[4] Handicapped by her difficulty in learning English and unsuccessful in attracting midwifery cases, she was soon forced to turn to the establishment of a cottage industry in her tenement apartment in New York. She and her sister began the production of worsted materials, employing as many as thirty employees at one time.

The success of her business in no way diverted Zakrzewska from her goal of pursuing a medical degree, but her lack of contacts and difficulty in communicating made progress in this direction slow. Finally, a year after she had arrived in New York, she visited a Home for the Friendless and

was able to describe her frustrated attempts to learn more about the Female Medical College in Philadelphia. The woman in charge introduced her to Dr. Elizabeth Blackwell, who, in addition to her regular medical practice, had opened a one-room dispensary on the East Side of New York for poor women and children. Blackwell not only offered to tutor the young immigrant in English in return for her aid at the dispensary but also promised to help her gain admission to a medical college. Through Elizabeth Blackwell's efforts, Zakrzewska was accepted by the medical department of Cleveland Medical College (Ohio) from which Emily Blackwell had recently graduated. Up to this point, Zakrzewska had no sympathy for the woman's rights movement. The demands raised at one New York convention seemed so ridiculous that she found the caption in one newspaper, "The Hens Which Want to Crow," as "quite appropriate." However, when she received the news that ways and means had been found for her to attend medical school, she realized that she had been trying to crow as hard as any of the women without realizing it.[5]

A combination of Zakrzewska's neglect of her knitting business and changes in the fashions of the worsted industry had left her with little money for her education. Fortunately, Dr. Harriot Hunt had recently toured Ohio raising funds for female medical education in that state.[6] As a result of scholarship aid from this source, Zakrzewska's lecture fees were waived for an indefinite period, and since Elizabeth Blackwell supplied all the necessary medical textbooks, Zakrzewska had to pay only the twenty-dollar matriculation fee. Furthermore, Caroline Severance, a friend of Blackwell's and president of a ladies physiological society near Cleveland, agreed to pay for her board out of a fund established by the society to assist needy women medical students.[7] If one had attempted to construct a story revolving around the tangible benefit of sisterhood, one could hardly improve upon the example of Zakrzewska's early career.

It is impossible to exaggerate the importance of this web of feminist friendship. One of only four women out of two hundred students at the college, Zakrzewska encountered obstacles unknown to her male colleagues. It took several weeks for her, even with the aid of a society woman's sponsorship (in this case Caroline Severance), to locate a boardinghouse that would accept a female medical student. Even here, the other boarders would quickly leave when Zakrzewska and her roommate, another female medical student, entered the room.[8] Upon completion of her medical education, Zakrzewska returned to New York ready to hang out her shingle and begin private practice. Once again, however, she encountered resistance unknown to a male physician. The ordinarily simple task of securing an office elicited three types of negative response, all pointing out the stigma attached to being a female physician in the 1850s. One group of landlords could not accept the notion of a female physician and refused to rent space to her on the grounds that she was probably masking her real identity as a spiritualist or clairvoyant. The second group accepted her credentials but doubted that she would be able to support herself and pay the rent. The third group asked no questions but demanded such expensive rents that she could not afford to lease their quarters. After a month of fruitless searching, Elizabeth Blackwell allowed her to open an office in her back parlor.[9]

The burden of locating an office paled in comparison with the difficulty of finding patients to develop a medical practice. Rothstein has described the necessary elements for success in the nineteenth-century medical world: "Family background and wealth, social standing and friendship were paramount in gaining admission to a medical school, setting up a practice,

obtaining appointments to hospitals, medical schools, and elite medical societies, and attracting a wealthy clientele."[10] A female physician began her pursuit of a career severely stigmatized because of her sex, and this fact hindered her in every step she took in establishing herself as a professional. Elizabeth Blackwell had already experienced every possible discouragement in her five years in New York. She was barred from practice in the city hospitals and dispensaries, ignored by her medical colleagues, and was the target of anonymous hate mail. She had solved the problem of office space by buying her own home, and she had dealt with the loneliness by adopting a seven-year-old orphan in 1854. By the time Zakrzewska joined Blackwell, she was eagerly awaiting the return of her sister Emily who had been studying medicine for two years in Edinburgh.

In numbers there was strength as well as sociability. Both Zakrzewska and Elizabeth Blackwell were eager to reestablish the dispensary in which Zakrzewska had first assisted Blackwell and which failed for lack of funds. Zakrzewska, drawing on her business experience, drew up an operating budget for the proposed hospital. She then threw herself into fund-raising and traveled to Boston to meet with Harriot Hunt, Caroline Severance (who had moved to that city), and a number of other local women interested in woman's rights. She was able to extract a promise of $650 to be paid over a three-year period to the new hospital. This promise stimulated donors in New York to raise an additional $1,000, enough to open New York Infirmary for Women and Children on May 1, 1857; it was the first hospital staffed by women in the United States.[11]

For two years Zakrzewska served without salary as resident physician and general manager of the hospital, sharing with the Blackwell sisters the responsibilities connected with a growing institution. By 1859, with the infirmary firmly established, Zakrzewska felt that she had fulfilled her debt to the Blackwells. Moreover, her own financial situation had improved as a result of her private practice. In March of that year she accepted an offer from Samuel Gregory's New England Female Medical College to be professor of obstetrics and resident physician of the proposed hospital. At least three factors appear to have influenced her decision: the conviction that achievement would add to a woman's personal happiness; a desire for greater independence and leadership; and a conviction that the women of Boston were intent in "their desire to elevate the education of womankind in general and in medicine especially." In a letter to Harriot Hunt, Zakrzewska revealed how depressing she found working with the Blackwell sisters, especially after opening New York Infirmary. The two women had such gloomy outlooks on life that Zakrzewska found them bewildering: "These two women for instance have all right to be satisfied with their efforts as it resulted . . . but they won't acknowledge it either to each other or to themselves. . . . I feel sad that nothing can cheer them up . . . they do wrong not to reward their friends by showing them a pleased countenance." Zakrzewska's letter was filled with good wishes for all her friends in Boston, especially Hunt, whose pleasure in personal achievement was uninhibited and who had been hostess to Zakrzewska for several of her visits.[12]

Zakrzewska failed to realize her second objective in coming to Boston and resigned in 1862 after repeated clashes over school policy with Gregory. Yet, despite the problems connected with the Female Medical College, her experience there only reinforced her original convictions that a supportive body of women shared her belief in medical education for women: "I decided to work again on the old plan, namely to establish the education of female students on sound principles, that is to educate them in hospitals."[13] Hospital training was increasingly be-

coming recognized as an essential ingredient in a complete medical education,[14] yet the only American hospital open to women was the New York Infirmary. Fundamental to Zakrzewska's plans for a training hospital was her conviction that medical colleges such as Harvard would accept women students once pioneers like herself had demonstrated that women could perform a meaningful role in medicine.

Sound principles for Zakrzewska also meant uniting with the regular physicians and avoiding any association with the irregulars, particularly the homeopaths. In an effort to counter the conventional medical histories which dismiss the irregulars as quacks, recent historians have supplied us with a useful reinterpretation by focusing on the positive contributions of the irregulars. No doubt a thorough analysis of the irregulars will contribute to a new understanding of one particular strand of feminist ideology. Similarly, in order to understand the difficulties that have beset women seeking professional medical careers, we must study the pioneer women (such as Zakrzewska) with a view of understanding why they chose not to identify with the irregulars.

From both a political and a medicoscientific point of view Zakrzewska was convinced that women must follow the regular medical path. Formally trained as both a midwife and a physician, she had little patience with those whose remedies verged more on mind cure than body cure. Her experience with the male-dominated medical world of Europe, coupled with her initial experiences as a struggling immigrant, made her a realist about the significance of professional power. For her the only solution for medical women was to force men to deal with them as equals.

Accordingly, New England Hospital for Women and Children, a "sunny, airy house" at 60 Pleasant Street in Boston's South End, was rented for $600 and opened July 1, 1862. When it was incorporated the following March, its charter spelled out the two primary goals of the hospital: to furnish women with medical aid from competent physicians of their own sex and to provide educated women with an opportunity for practical study in medicine. Significantly, two-thirds of the first board of directors were the same women who had served on the board of lady managers of the New England Female Medical College.[15] There was no qualified woman surgeon available, so Zakrzewska was forced to employ a leading male gynecologist, Dr. Horatio Storer, in 1863; when he resigned three years later, he was replaced by a female specialist, making the hospital the first in New England to be entirely staffed by women physicians.

In the first few years these women physicians were a union of the weak, rather than a combination of the strong. For example, Zakrzewska recalled how she felt in 1863: "My coworkers were young and inexperienced, looking up to me for wisdom and instruction while the public in general watched with scrupulous zeal in order to stand ready for condemnation." Fortunately, there were a few male physicians who were willing to cross the sexual boundary. Samuel Cabot and Henry I. Bowditch, both Harvard educated, had told her they would "refuse all aid" so long as she remained at Gregory's school; they readily came to her assistance when she left.[16]

Cabot became the first consulting physician at the hospital, offering his advice in difficult cases. He also provided Zakrzewska with important psychological support. She recounted in a letter to Dr. Lucy Sewall that Cabot did not feel it necessary for her to call him for forceps deliveries. "You see," she wrote proudly, "he rightly supposes we use the forceps *skillfully.*" Bowditch had initially befriended Zakrzewska when she came to Boston in 1856 to solicit funds for New York Infirmary. She later recalled: "He remained the steadfast champion of medical women and

continued as consulting physician to the New England Hospital until his death in 1892." Another physician, Benjamin Cotting, was especially helpful in sending both rich and poor patients to the hospital. Both were welcome: one group contributed much-needed fees as private patients; students treated the others in the dispensary. Such gestures, infrequent as they were, were a welcome tonic to someone generally treated as a pariah by the city's medical establishment. As Zakrzewska noted: "Every slight word or act of endorsement, even though with reservations, was like a ray of hope that at last the dawn was breaking. . . . Such consolations helped to uphold me."[17]

Although Zakrzewska praised the early male consultants, except for Henry Bowditch they were for the most part fair-weather friends who withdrew support temporarily whenever there was too much pressure from the Boston medical establishment. Thus, when Zakrzewska took their advice and applied for membership in the Massachusetts Medical Society and was turned down because of her sex, Cabot and Cotting severed their ties with the hospital until the issue faded. Nevertheless, Zakrzewska felt their presence served to quiet those in the profession "who wanted to find fault but did not dare to do so openly so long as the two or three professional men stood as a moral force behind me."[18]

Finances were a problem of another dimension and, in the early years at least, were a major difficulty, even threatening the hospital's existence. The combined assets of the new institution in 1862 consisted of some $150 worth of hospital furniture brought from New England Female Medical College after Gregory evicted the lady managers and Zakrzewska. As the secretary of the board described the situation: "Our possessions were a few iron bedsteads, a few chairs, and bookcases, some straw, etc., our earnest purpose and our admirable Dr. Zakrzewska."[19] No gifts, however small, were refused, and the early annual reports are filled with lists of donations ranging from scissors and bandages to tea and cornstarch.

An important source of income during the hospital's first decade was an annual grant of $1,000 for maternity patients from the trustees of Boston Lying-In Hospital Corporation, which did not operate from 1856 to 1873 because it could not obtain patients.[20] In 1864, supported by a $5,000 grant from the Massachusetts legislature and matched by a number of donors, the hospital moved to larger quarters at 14 Warren Street. The state also offered additional assistance in the form of $1,000 each year for four years beginning in 1868. Fairs were also important sources of revenue; in 1871, $12,000 was realized in this fashion. Sizable donations from women during these formative years played a crucial role in the hospital's development. Zakrzewska's reputation inspired confidence as evidenced by a $2,000 bequest from the estate of Mrs. Robert G. Shaw that same year "to be used by Dr. Zakrzewska in aid of any Hospital or Infirmary . . . which may be under her superintendence in the City of Boston at the time of my decease." This money, coupled with other gifts (notably a $5,000 bequest from Miss Nabby Joy), enabled the hospital to move again in 1872 to what was to become its permanent location in the highlands section of Roxbury. By 1872 New England Hospital for Women and Children had become, in just ten years, one of the largest hospitals in Boston.[21]

The financial situation was further aided by the fact that a number of the staff, including Zakrzewska, donated their services. Driven by her desire to improve the cause of women in medicine, Dr. Lucy Sewall, for example, served as resident physician of the hospital for three years beginning in 1863 without salary or vacation.[22] Zakrzewska supported herself through her private practice and by renting rooms in her home in Roxbury to invalid boarders. Even in her personal practice, Zakrzewska sought to ex-

pand the role of women. Thus she regularly walked to night calls in any type of weather to prove "that a woman has not only the same (if not more) physical endurance as a man." Women physicians were handicapped by more than inclement weather in their pursuit of night calls. Zakrzewska always went with the messenger who called her. If he was unable to accompany her on her return home, she walked with the local policeman to the limit of his beat and traveled in similar fashion from one beat to another until she arrived home. This problem was solved in 1865 when she bought a horse and a secondhand buggy, which also enabled her to "uphold the professional etiquette and dignity of a woman physician on equality with men."[23]

Nevertheless, it is doubtful that Zakrzewska and her hospital could have succeeded without the support of the feminist movement. During its first fifty years the hospital had only three presidents: Lucy Goddard, Ednah Cheney, and Mrs. Helen F. Kimball. Ednah Cheney exemplifies the close bond between feminism and the hospital. One of the lady managers of the Female Medical College, Cheney was active at every stage of the hospital's development, beginning with her pledge along with three other women in 1862 to pay the first year's rent on the first building on Pleasant Street. A close friend and later neighbor of Zakrzewska, she served on the board of directors of the hospital for forty-eight years—including fifteen as president. In addition to her hospital activities, Cheney helped found the New England Women's Club, served on the executive committee of the New England Woman Suffrage Association, and actively campaigned for women's right to vote in school committee elections. Cheney was not unique, and an analysis of the early bequest lists shows that the hospital donors often left funds to suffrage associations. In 1887 over 80 percent of the donors to a $10,000 hospital fund were women.[24]

To view feminism as simply a struggle for

woman's rights and the vote is to ignore the support and companionship it offered those women who broke with their prescribed roles. It is clear that a female physician could not have functioned autonomously in nineteenth-century America. Zakrzewska's dependence on the woman's movement was total: she needed female supporters to help finance her education, to raise money, to promote the hospital, to help administer it, to serve as patients, and—probably most critically—to proffer their friendship during difficult times. It was Zakrzewska herself who had originally suggested an association of women which was translated into the New England Women's Club and which met initially in the home of Harriot Hunt. For many women it was the first time they came together not because of family, neighborhood, or church but as women. The club regularly supported the women doctors; for instance, it sponsored a "social levee" when Dr. Lucy Sewall went to Europe to study. The members also were active in running fairs to raise money for the hospital. In turn, female doctors regularly gave lectures to the women at the club and sponsored discussion groups.[25]

One of the most openly militant supporters of the woman physician was the *Woman's Journal,* which was edited by Lucy Stone and began publication in Boston in 1870. Zakrzewska had called for such a journal as early as 1862 when she and another woman doctor, Mary Breed, inserted a notice for a *Woman's Journal* in the *Liberator,* but at that date they were unable to secure enough support.[26] Lucy Stone's *Woman's Journal* championed the cause of women doctors and challenged the right of society to erect barriers to stand in their way. The editors encouraged reader response to this problem and published in full the letters of angry women doctors who felt blocked in their careers. An 1871 letter to the paper cautioned: "Let [men] not feel too sure that they alone hold the key that unlocks the door to medical science. They

bar and bolt the doors of their hospitals in Boston against all women medical students. They heap upon them undeserved ridicule. They hold up to the world their constitutional weaknesses in a manner to lead one to suppose that they possess no such weaknesses themselves. They scorn the very idea of holding a consultation with a woman physician. . . . True women physicians would be glad to have the men in the profession see the mistake they are making and become their friends as they ought, in this manner. They would be glad to see the city hospitals and dispensaries opened to women medical students. They blush for the city of Boston that this is not done." The writer concluded by warning: "But, aided or unaided, the day is not far distant when women will compel medical men to know that as physicians they are their equals, whether they have the magnanimity to acknowledge it or not."[27]

Each act of exclusion by the medical establishment brought the wrath of the journal down on the heads of the perpetrators. Pointing out that Boston's Free Hospital for Women was served entirely by male physicians, the editor noted: "This is a shame in a city where there are competent women physicians. It is a poor, empty and prating pretense, that of indelicacy of common study, by those men who clutch at and crowd for medical practice among women."[28] The journal also did a great deal in advertising the success of women in medicine. It was happy to report that a research article of Dr. Sara E. Brown, which had been refused by the *Boston Medical and Surgical Journal* "on account of her sex," was published by the *Archives of Ophthalmology and Otology* and reproduced in a number of international medical periodicals, thus gaining wider publicity than it would have received from the Boston publication.[29]

New England Hospital gained a great deal of free publicity in the pages of the *Woman's Journal.* Each year it published a lengthy report on the hospital's progress. Readers were reminded of their obligation to support this feminist project and urged to attend each fund-raising fair. Women doctors were especially hard hit by the medical strictures against advertising in the public press while at the same time their professional brethren refused to recognize their existence and include them in the directories put out by doctors themselves. For example, the *Medical Register of Boston* refused to list the names of the women physicians even after it began to include all manner of peripheral practitioners such as artificial limb makers, collectors, makers of optical instruments, vendors of patent medicines, and female nurses. In order to fight the prejudice confronting them, the women physicians had to advertise. Here the *Woman's Journal* was a valuable ally. It did all it could to right the balance by publishing testimonials about the competence of women physicians such as one from "M. W.," a schoolteacher, who described how she had been restored to health by Dr. Zakrzewska's "skill and kindness, a debt that words are feeble to portray." Similarly, the journal publicized each addition to the hospital staff. Thus Dr. Fanny Berlinerblau was introduced in a typical report that told the story of her difficulties in securing a medical education, extolled her "admirable scientific training," and informed the readers that she had abbreviated her name to Dr. Fanny Berlin "to suit the American tongue."[30] But the *Journal's* public relations efforts on behalf of the hospital could do only so much; in the final analysis the hospital's performance would be the ultimate arbiter of its fate.

The two primary objectives of the New England Hospital were to provide women with medical aid from doctors of their own sex and to contribute to the supply of competent women doctors by providing them with an opportunity for practical clinical experience. While specializing in obstetrics, gynecology, and pediatrics, the hospital also offered a full range of medical treatment, including surgery on a bed patient as

well as a dispensary basis. Zakrzewska's hospital filled an important void in Boston medicine. Boston City Hospital, which opened two years after Zakrzewska's institution, did not provide gynecological treatment until 1873 and then only on an outpatient basis. It did not create a gynecological department until 1892. Massachusetts General Hospital, which had been operating since 1822, did not provide obstetrical services until the twentieth century. At the time of its inception New England Hospital was unique in its provision for both obstetrical and gynecological treatment of patients. The only other hospital in the city to have specialized in obstetrics, Boston Lying-In Hospital, had closed its doors in 1856, "a white elephant of mastodonic proportions." Every effort had been made to attract patients, including a massive advertising campaign in eighty-five newspapers, but women, if they had any choice in the matter, avoided using Boston Lying-In.[31]

The unwillingness of such women reflected an accurate assessment of the dangers connected with most maternity hospitals. Puerperal disease, which frequently resulted in death, stemmed from the unsanitary techniques that were characteristic of midcentury hospitals. There were a few physicians early in the nineteenth century who had suspected the cause of the high mortality rates of women in childbirth. Dr. Oliver Wendell Holmes, for example, had attended a lecture in Paris in 1833 which suggested that doctors themselves may have played a role in communicating the disease. In 1843 he read a paper on his research findings to his Boston medical colleagues and published an article which demonstrated that the obstetrician, midwife, and nurse were active agents in transmitting the infection from one mother to another. He was promptly rebutted by Dr. Walter Channing, who delivered a paper on the noncontagious nature of the disease, although Channing later reversed his position.[32]

Zakrzewska's work in the large Charité Hospital in Berlin had given her far more opportunity than most American physicians to observe the unsanitary conditions which were conducive to spreading puerperal disease. Her experience confirmed Holmes's theory. She had observed that when the medical students appeared in the Berlin hospitals with their forceps, "untimely rupturing the membranes or by other meddlesome interference with nature," the cases of the disease soared. During her appointment as chief midwife in Berlin in 1852, not a single case of the disease occurred because of the precautions that she took in the administration of the hospital. Zakrzewska's scientific acumen and her experience with the advantages of cleanliness were enormous assets to the hospital in its early years when bacteriology and asepsis were still matters of debate. A number of leading Boston physicians, including Walter Channing, C.P. Putnam, Henry I. Bowditch, and Samuel Cabot, signed an 1864 circular attesting to the hospital's success in preventing various contagious fevers, so impressed were these men with Zakrzewska's leadership in this matter.[33]

By contrast, Boston Lying-In Hospital, reopened in 1873, was forced to close three times in the next thirteen years because of puerperal epidemics within its wards. In 1883, at the height of a puerperal disease epidemic in Boston, only one of the patients in New England Hospital died from the fever. In contrast, over five hundred women contracted the disease and fifty died from it at Boston Lying-In Hospital from 1878 to 1883. Whether women physicians offered medical care superior to their male counterparts remains speculative, especially in view of the dearth of evidence related to treatment and the difficulties connected with comparing different patient populations. One researcher, whose pioneering investigation of this question involved an examination of doctors' comments on patient records at four different nineteenth-century Boston hospitals, concluded that the male practitioners reflected a negative or even hostile attitude toward their female patients. Complaints that the maternity patients were too lazy to "work" in delivering their babies or that their infections were their own fault were quite common. On the other hand, she found that these remarks were absent from the patient records at New England Hospital.[34] While it is difficult to assess the effect of physicians' attitudes on their patients, it would be wrong to underestimate it.

One can make a strong case that a good deal of New England Hospital's success can be attributed to the fact that the physicians there were also women with special insights and sensitivity toward the medical problem of their own sex. In an age when medical techniques were generally undeveloped and often unsafe, the women physicians' restraint, coupled with compassion, may have done much to effect a healthier hospital environment. For example, unlike many male doctors, the women seem to have been more willing to let nature take its course in childbirth. Avoiding the temptation to demonstrate their virtuosity with scalpel and forceps, the female physicians also avoided the medical dangers these instruments caused to both mother and child.[35]

Much of the work of the hospital in the nineteenth century was given over to charity cases. A number of "Free Hospital Beds" were donated by friends of the hospital. The dispensary charged ten cents a visit and twenty-five cents to fill a prescription at the hospital pharmacy, but those who could not afford these modest fees were treated without charge. Zakrzewska, who supervised the dispensary during the early years, noted: "A crowd of women, some from towns miles distant, came every morning."[36] One of her difficulties was to persuade wealthy women not to use the dispensary but rather to visit staff members who maintained a private practice. In one letter she urged a friend of the hospital to recommend the services of two women physicians on the staff who were just starting their private practices: "Be sure to send them all the rich patients by telling [the patients] plainly that I don't want them."[37] In fact, charity cases became such an important part of the hospital case load that Lucy Sewall complained that many of the sick poor supposed the physicians were paid by the city, "and that they had a legal right to their services."[38]

The large number of charity and obstetrical patients in the early years led the hospital to establish the first social service department in an American hospital. Each patient was interviewed; for those women who had no family, places were found for them to board both prenatally and postnatally. Jobs were also found for those women who were the sole support of their children. The women who provided these counseling services were the same lady managers whom Samuel Gregory found so annoying at Female Medical College. At the hospital Zakrzewska used their desire to serve in a way which she described as mutually enriching: "It is thus the privilege of the [lady manager] to round off and finish the large charity done by the physicians, while she herself has her sympathies quickened and her experiences enlarged by intimate acquaintances with life flowing in different channels from her own."[39] The number of patients serviced by the hospital grew steadily throughout the century. During its first sixteen months the hospital treated 1,507 individuals; by the end of the century more than 19,000 patients annually passed through the dispensary doors. There were now specialized clinics for eye, ear, nose, and throat; maternity; and child health. The sophisticated turn-of-the-century hospital was a far cry from the single room that Zakrzewska supervised in 1862.[40]

By 1900 the hospital had made a great deal of progress. Most important, it had at last convinced its patients that women could be successful physicians. Far better than statistics in showing how far the hospital had come was Zakrzewska's encounter at the end of the century with an Irish immigrant, whose wife had been a charity case at the original, Pleasant Street location. He wanted to arrange an operation for one of his family, and he insisted on having one of the woman physicians at New England Hospital. When he noticed Zakrzewska's surprise, he explained: "Well, Doctor, when I came to this country with my wife, we were very poor and knew nothing. The good women of the Pleasant Street Dispensary attended to us and taught us to take care of ourselves. All our children were born under their care and they watched that we did right by them, all without any charge. Now that we can afford good pay, I am sure we want the same, for I swear by the woman doctors."[41]

The second objective of the hospital was to provide educated women with an opportunity for practical study in medicine. One of the most difficult obstacles to the advance of women in medicine during the latter half of the nineteenth century was the lack of adequate facilities for clinical instruction. When women sought to gain this practical experience to supplement their classroom education, they were rebuffed because of the alleged indecency of observing cases in the presence of men—despite the fact that very frequently the patients themselves were females. The existence of this deep-seated opposition to the participation of women in clinical situations is dramatically illustrated by the experience of women at Philadelphia's Pennsylvania Hospital in 1869. Some thirty female medical students were invited to clinical lectures, but the male students objected "with insolent and offensive language." During the last hour, despite the efforts of members of the faculty, the men showered the women with "missiles of paper, tinfoil, and tobacco quids." Thus ended the effort at coeducation at the Pennsylvania Hospital.[42] It was obvious that with the exclusion of women from existing hospitals, they needed their own institutions in order to obtain clinical instruction.

Zakrzewska's hospital anticipated a need that was only surfacing in 1862. Luckily, only a small number of women doctors applied to the hospital in the 1860s, for the staff was inadequate and the facilities severely limited. Of the twenty-seven interns in the first ten years of the hospital's existence, twelve were graduates of a medical college before coming to the hospital and the other fifteen were stimulated by their experience to complete the academic requirements for the degree shortly after leaving the hospital. While it was fairly commonplace for Boston hospitals to accept male "House Pupils" who had not yet earned their M.D. degree, what was unusual about New England Hospital interns was the distance they had to travel to finish a degree or to obtain advanced training. Five young women went to Europe for medical study, two of whom were pursuing postdoctoral training; seven went to the University of Michigan; four to the Woman's Medical College of Pennsylvania; one to Howard University. The reason, of course, was that in the early years there were few medical schools open to women students. Zakrzewska was never enthusiastic about Gregory's school when he was alive, and after the merger with Boston University she refused to recommend students to or accept applications from what was in its early years the city's only coeducational medical school, because of its irregular curriculum.[43]

The actual training that the hospital provided appears to have consisted of a minimum of formal instruction and a maximum of practical experience. The overtaxed staff had little time to devote exclusively to the students, and the education itself was on a learn-as-you-go basis. The hospital kept no intern records, and the only picture of the training is based on the variegated reactions of the students themselves. While some found all they had hoped for and others were disappointed, one is struck by how many influential women doctors passed through the hospital during its first decade.

Most notable of the hospital's "alumnae" of the 1860s was Dr. Mary Putnam Jacobi who, although twenty-one years old when she arrived, was a graduate of both New York College of Pharmacy and Woman's Medical College of Pennsylvania. She entered New England Hospital in 1864 but instead of the specialized training she expected, Jacobi found herself thrown into the work of the dispensary where close to two thousand women were treated during the year. No bed remained empty and many patients had to be treated in their homes. During one two-week period Jacobi counted eleven nights when Zakrzewska was called out on emergencies. Seeing so many patients suffering from such a variety of ailments day after day temporarily persuaded Jacobi that she was not cut out for a regular medical practice. Convinced that she was destined for a career in medical research, she went to Paris, where in 1868 she became the first woman to be admitted to the École de Médicine. After her return to America, she embarked on a career of research and medical school teaching which made her the leading woman physician in America in the late nineteenth century.[44]

Unlike Jacobi, Susan Dimock entered New England Hospital as the very first step in her medical career. Although only eighteen years old when she arrived in January 1866, her ability and prodigous capacity for work attracted the special attention of Zakrzewska and Sewall. With their encouragement she and another student at the hospital, Sophia Jex-Blake, applied to Harvard Medical School the following year and were both turned down. Despite their rejection, New England Hospital was able to temporarily arrange for a limited amount of clinical instruction for the two at Massachusetts General Hospital when the Harvard students were not at the hospital. The 1867 Annual Report of the New England Hospital proudly announced: "They have availed themselves of all the opportunities offered them."[45]

Meanwhile Zakrzewska and Sewall persuaded Dimock to apply to the University of Zurich, which had been accepting female medical students since 1864. Two Boston women paid Dimock's expenses with the only stipulation that she return to New England Hospital for three years and assist some other struggling women medical students in the future. After graduating with high honors from Zurich and then spending an additional year in Paris and Vienna, Dimock returned in 1872 to become the most skilled surgeon on the staff.[46]

Sophia Jex-Blake, Dimock's companion in the attempt to enter Harvard in 1867, first exhibited at the hospital the spirit that would eventually make her the spokeswoman for a woman's right to a medical education in her native Great Britain. Unwilling to accept Harvard's refusal, she embarked on a personal campaign while at the hospital to build up support to break this barrier. She arranged interviews with each member of the Harvard faculty and the Massachusetts General staff. Some, like Oliver Wendell Holmes, expressed a willingness to lecture to women "always provided that any special subject which seemed not adapted to an audience of both sexes, should be delivered to male students alone." A more representative response was recorded in Jex-Blake's diary: "Dr. A. 'not afraid of responsibility, of course'—only—he'd rather not admit us till other people do!"[47]

Georgia Sturtevant, an assistant nurse at the time the young women medical students were being permitted at Massachusetts General Hospital, noted in her memoirs that Dimock and Jex-Blake, "though championed by some of the most popular of the visiting staff, were really allowed this privilege under protest, and were under many restrictions, and were only allowed to visit in certain wards."[48] Jex-Blake, particularly, felt a constant sense of insecurity as one member of the staff was bitterly opposed to the presence of women and constantly searched for mistakes to bolster his prejudices. Such tensions took their toll as Jex-Blake wrote in her diary: "July 5th. Rest yesterday, but altogether weighed down yesterday and today with the fear and horror of this irritability which seems so fatally unconquerable." Dissatisfied with the situation at Massachusetts General and convinced that she had received enough practical experience, Jex-Blake went to New York where she was able to obtain private lessons in anatomy from the head demonstrator at Bellevue Hospital. A month later she left for home where she won fame as the leader of the movement to admit women to the medical profession in Great Britain and the founder of the London School of Medicine for Women.[49]

Elizabeth Mosher, whose only previous medical experience involved nursing her tubercular brother, entered New England Hospital in 1869. After a successful internship and a year of assisting Lucy Sewall in her private practice, Mosher left for the University of Michigan where she received her M.D. degree in 1875. She went on to a number of important medical posts which culminated in her appointment as the first dean of students and professor of physiology at the University of Michigan. She later credited her year at New England Hospital as the turning point in her career: "I believe I voice all of the women . . . when I say I feel I largely owe to the teaching, the spirit of devotion, and the high standard maintained by this hospital whatever of success I may have been able to achieve in medicine."[50]

With each year the hospital raised the quality of its training. By the mid-1870s the staff of the hospital began debating the problems of selecting from the large number of qualified applicants for the internship positions, and by 1879 they accepted, with reluctance, the solution of taking only those women who already possessed the M.D. degree.[51] Increasing numbers of women were also coming to the hospital for experience after having attended a liberal arts college as well as receiving the medical degree. This was the case with Dr. Minerva Walker, who attended Cornell University before receiving the M.D. from the Woman's Medical College of Pennsylvania in 1879.[52] The larger, more specialized staff of the last decades of the nineteenth century provided a far more intensive training period than had been possible in the 1860s. Dr. Kate Hurd-Mead, who interned at the hospital in the 1880s, described her experience there as highly structured: "Life was indeed serious to the young doctors under the watchful eye of resident and visiting physicians. If, in an unguarded moment, the intern was heard humming a little air or whistling softly at her work, or even if her shoes squeaked a trifle, she was taken to task by one of these dignified censors and questioned as to her reasons for studying medicine and for her unseemly deportment."[53]

By 1887, on the twenty-fifth anniversary of New England Hospital, Zakrzewska's original goal of a hospital run by women for women had been realized. From the board of directors to the delivery of health care to patients, women held full responsibility and authority, though, as the anniversary report pointed out, "the counsel and help of the other sex is gladly welcomed." This was especially true in regard to the consulting physicians, men who were selected not as mere status symbols but because "they have taken an active interest in the Hospital, and have been chosen for special eminence in some department."[54] As a result of the vision of Zakrzewska, an increasing number of trained doctors were being turned out to meet the rising patient demand. But the hospital was more than an institution where women absorbed the technical knowledge and skills of their profession. Equally

necessary in the sexually polarized world of the late nineteenth century was the psychic support and energy which they needed to enable them to practice a profession that did everything it could to discourage them. Herein lies the significance of the hospital. It was not only a showcase in which women physicians could prove themselves; it was also an island of feminist strength and sisterhood in a society only familiar with brotherhood.

Consequently, when a group of twelve Greater Boston women physicians, ten of whom had been associated with the hospital, gathered in 1878 to form the first female medical society in the United States, they named the organization the New England Hospital Medical Society. At a time when the Massachusetts Medical Society was closed to women, this separate group offered its members both a sense of colleagueship and a common voice. One of its first steps was to pressure the editors of the *Boston City Directory* to list its members under the heading of the society, a service which the directory had always rendered to the members of the Massachusetts Medical Society and any other local male medical group. The separate listing was important, for while the Massachusetts Medical Society members had been divorced from the other sectarian and irregular physicians, including patent medicine promoters, the women physicians had been indiscriminately lumped under the heading of "female physicians," which included phrenologists, magnetists, Christian Scientists, and electricians, as well as midwives and nurses. Zakrzewska was particularly eager to dissociate her hospital from any taint of homeopathy and sectarianism. Graduates of Boston University Medical School as well as other irregular schools were excluded from the New England Hospital Medical Society, and interns from such schools were likewise barred from the hospital. The women hoped that by not confusing the issue of women's competence in regular medicine with the sectarian controversies in the profession at large, they would advance the cause of women more directly.[55]

Thus New England Hospital fought for the causes of medical women and, indirectly, feminists on a variety of fronts. Its separatism was a means to an end; ironically, that end was the elimination of separatism and the movement of women into the mainstream of medicine. Whenever Zakrzewska spoke publicly about women physicians and particularly when she addressed her student interns, she always expressed the hope that the hospital would convince the medical profession of the ability of women physicians, "and shall thus force them to open Harvard College to such women as desire entrance there."[56] She believed that the presence of male consultants on the staff would demonstrate that men and women physicians could work side by side to the advantage of both sexes and society as a whole. In this spirit the New England Hospital Medical Society even invited male physicians to membership, an invitation that the men chose to ignore, though a few did toy with the idea briefly.[57]

Quite clearly, many female physicians believed that as women they brought a much-needed dimension to the practice of medicine. But it was equally obvious that these benefits would not accrue to the profession as long as women were isolated.

Notes

1. Books on political feminism include Eleanor Flexner, *Century of Struggle: The Woman's Rights Movement in the United States* (Cambridge, Mass., 1959); Andrew Sinclair, *The Better Half: The Emancipation of the American Woman* (New York, 1965); Robert Riegel, *American Feminism* (Lawrence, Kans., 1963); Aileen Kraditor, *The Ideas of the Woman Suffrage Movement, 1890–1920* (New York, 1965); William O'Neill, *Everyone Was Brave: The Rise and Fall of Feminism in America* (Chicago, 1969); Anne F. Scott and Andrew M. Scott, *One Half of the People: The Fight for Woman Suffrage* (Philadelphia, 1975).

2. Elizabeth Blackwell left America in 1869, spent most of her life in England, and died there in 1910. A well-documented study is Nancy Sahli, "Elizabeth Blackwell, M.D. (1821–1910): a Biography," Ph.D. diss., University of Pennsylvania, 1974. Emily Blackwell, who also died in 1910, was in many ways overshadowed by the fame of her older sister, though she devoted her life to the practice of medicine in America. Both women have been the subject of many popular articles and biographies. Zakrzewska's life, on the other hand, has gone relatively unnoticed except for two autobiographical memoirs cited in n. 3. Bibliographies of all three women are in Edward James, ed., *Notable American Women* (Cambridge, Mass., 1971).

3. Caroline Dall, ed., *A Practical Illustration of Woman's Right to Labor or a Letter from Marie Elizabeth Zakrzewska* (Boston, 1869), 60, 85; Agnes Vietor, *A Woman's Quest: The Life of Marie E. Zakrzewska* (New York, 1924), 84–85.

4. Dall, *Practical Illustration*, 105.

5. Vietor, *Woman's Quest*, 134.

6. Ibid., 485.

7. Ibid., 119–21.

8. Ibid., 131.

9. Ibid., 179–81.

10. William G. Rothstein, *American Physicians in the Nineteenth Century: From Sects to Science* (Baltimore, 1972), 206–7. Although there were important differences in the social stratification systems of England and America in the midnineteenth century, it is interesting to note how similar the plight of the beginning physician was in both countries—even without the added difficulty of sex discrimination. For an analysis of the problems of the male physician who started out alone in London, see M. Jeanne Peterson, "Kinship, Status, and Social Mobility in the Mid-Victorian Medical Profession," Ph.D. diss., University of California, Berkeley, June 1972, 153: "for all the growth in medical education and licensing and the advancement of medical science, the basis on which Victorian medical men built their careers was not primarily that of expertise. Family, friends, connections, and new variations of these traditional forms of social

relationships and social evaluation were the crux of a man's ability to establish himself in medical practice." Peterson uses statistical and biographical materials to substantiate this thesis.

11. Vietor, *Woman's Quest,* 211.

12. Ibid., 237–39, 149, 186, 192, 197; Marie Zakrzewska to Harriot Hunt, May 14, 1857, Caroline Dall Collection (Massachusetts Historical Society).

13. Vietor, *Woman's Quest,* 292.

14. Rosemary Stevens notes that the first use of the term *intern* in American hospital records was apparently in the Boston City Hospital Board of Trustees Report for 1865 (*American Medicine and the Public Interest* [New Haven, 1971], 116–17). See also "Background and Development of Residency Review and Conference Committees," *Journal of the American Medical Association* 165 (1957): 60–64. By 1904 the AMA Council on Medical Education found that as many as 50 percent of new medical graduates went on to hospital training; by 1914 it was estimated that 75 or 80 percent of graduates were taking an internship (see Stevens, 118). Rothstein claims that by 1865 about two-thirds of medical schools made arrangements for some hospital and clinical instruction of students—see 282.

15. Vietor, *Woman's Quest,* 486–87. See also the "Records of the Lady Managers of the New England Female Medical College" (Boston University Archives), and the first Annual Report of the New England Hospital (1864).

16. Vietor, *Woman's Quest,* 330, 256.

17. Ibid., 301, 336, 256, 332, 330–31.

18. Ibid., 277–78, 330.

19. Cited by Alice B. Crosby, *The Fiftieth Anniversary of the New England Hospital for Women and Children, October 29, 1912* (Boston, 1913), 17.

20. See the annual reports of the New England Hospital for Women and Children (AR-NEH) from 1863 to 1871; the 1871 report indicates final payment from the Lying-In Hospital Corporation, November 7, 1871.

21. AR-NEH (1871), 17; Vietor, *Woman's Quest,* 353. Francis H. Brown, M.D., *The Medical Register for the Cities of Boston, Cambridge, and Chelsea* (Boston, 1873) deliberately downplayed the importance of New England Hospital by omitting it from its otherwise comprehensive listing of Boston hospitals. The

judgment of the comparative size of New England Hospital is based on 1872 statistics given for the other Boston hospitals listed in the directory. Significantly, New England Hospital is the only "hospital" listed in the "Other Institutions and Societies" category of both directories.

22. Vietor, *Woman's Quest,* 348; *Woman's Journal,* February 22, 1890, 61.

23. Marie Zakrzewska to Paulina Pope, October 28, 1901, New England Hospital Papers (Sophia Smith Collection, Smith College).

24. Vietor, *Woman's Quest,* 335; Ednah Dow Cheney, *Transcript of the Memorial Meeting of the New England Women's Club* (Boston, 1905) (Schlesinger Archives, Radcliffe College); bequest lists appear in the annual reports of the New England Hospital.

25. Mrs. Walter A. Hall, Mrs. Joseph S. Leach, and Mrs. Frederick G. Smith, *Progress and Achievement: A History of the Massachusetts State Federation of Women's Clubs, 1893–1962* (Lexington, Mass., 1962), 16; Julia A. Sprague, *History of the New England Women's Club from 1868 to 1893* (Boston, 1894), 3; "Record Book of the Weekly Social Meetings, New England Women's Club, 1869–1871," and records of discussion groups for entire period of its history, New England Women's Club Collection, Schlesinger Archives, Radcliffe College.

26. *Liberator,* June 27, 1862.

27. *Woman's Journal,* July 29, 1871.

28. Ibid., November 8, 1879.

29. Ibid., December 12, 1874.

30. Ibid., April 14, 1877; May 26, 1883; December 3, 1887.

31. Frederick C. Irving, *Safe Deliverance* (Boston, 1942), 122–23; Frederic A. Washburn, *The Massachusetts General Hospital: Its Development, 1900–1935* (Boston, 1939), 364–65; Committee of the Hospital Staff, *A History of the Boston City Hospital from Its Foundation Until 1904* (Boston, 1906), 157–58.

32. Irving, *Safe Deliverance,* 145–59; Eleanor M. Tilton, *Amiable Autocrat: A Biography of Dr. Oliver Wendell Holmes* (New York, 1947), 169–76, 366, 409–10.

33. Marie E. Zakrzewska, "Report of One Hundred and Eighty-seven Cases of Midwifery in Private Practice," *Boston Medical and Surgical Journal* 121 (1889): 557–58; Marie E. Zakrzewska, "Report of the

attending physician," AR-NEH (1868), 9–21. An excellent source of information on maternity practices in the New England Hospital is Emma L. Call, "The Evolution of Modern Maternity Technic," *American Journal of Obstetrics and Diseases of Women and Children,* 58, 3 (1908): 392–404. Call, whose association with the hospital began in 1868, documents and analyzes the puerperal disease statistics of the hospital from 1862 to 1907. Her article is an invaluable source of information on the hospital procedures of this period, and it quotes from internal reports of the New England Hospital. For the earliest period there is a printed circular, 1864, with letter from John H. Stephenson endorsed by Drs. Horatio Storer, Walter Channing, C. P. Putnam, S. Cabot, and Henry Bowditch in New England Hospital Collection, Schlesinger Archives, Radcliffe College; Irving, *Safe Deliverance,* 143, and annual reports of Boston Lying-In Hospital.

34.   Laurie Crumpacker, "Female Patients in Four Boston Hospitals of the 1890s," Paper delivered at the Berkshire Conference on the History of Women, October 26, 1974; on file in Schlesinger Archives, Radcliffe College.

35.   Ibid. See also Virginia G. Drachman, "Women's Health through Case Records," Paper delivered at the Third Berkshire Conference on the History of Women, Bryn Mawr College, June 10, 1976.

36.   Alice B. Crosby, *The Story of New England Hospital for Women and Children Through Seventy-five Years, 1862–1937* (Boston, December 10, 1937), 4 and 13.

37.   Marie Zakrzewska to Caroline Dall, March 6, 1869; ibid., March 26, 1869, Caroline Dall Collection, Massachusetts Historical Society.

38.   Crosby, *Story of New England Hospital,* 11.

39.   Vietor, *Woman's Quest,* 497–98; Grace E. Rochford, M.D., "The New England Hospital for Women and Children," *Journal of the American Medical Women's Association* 5 (1950): 497; Felicia A. Banas, M.D., "The History of the New England Hospital," ibid. 10 (1955): 199; AR-NEH (1864), 4, and succeeding reports which annotate the social services rendered patients. A similar service was not established at Massachusetts General Hospital until 1905 and at Boston City Hospital until 1918. See Washburn, *Massachusetts General Hospital,* 570, and John J. Byrne, ed., *A History of the Boston City Hospital, 1905–1964* (Boston, 1964), 372.

40.   Crosby, *Story of New England Hospital,* 13; AR-NEH (1863), 11; AR-NEH (1900), 23.

41.   Vietor, *Woman's Quest,* 469; AR-NEH (1911), 10.

42.   *Evening Bulletin* (Philadelphia), November 15, 1869. Cited by Clara Marshall, M.D., *The Woman's Medical College of Pennsylvania: An Historical Outline* (Philadelphia, 1897), 20.

43.   AR-NEH (1863). Statistics on the early graduates, even names and dates, are extremely unreliable if one uses the compilations in the New England Hospital "fact sheets" published in the twentieth century, for example, one entitled "Former Interns of the New England Hospital for Women and Children," c. 1934, New England Hospital Collection, Sophia Smith Collection, Smith College. To insure an accurate portrait of the 1862–72 interns, I cross-checked information and verified it in a number of sources: New England Female Medical College graduate list; *Medical and Surgical Register of the United States* (Polk's) beginning with the first edition in 1886; *Woman's Medical College Graduates List;* and several listings in the AR-NEH to eliminate printing errors. There are eight women for whom information could not be obtained because they either died, left no forwarding addresses, or were of foreign birth and could not be traced in U.S. sources. The total of interns from 1862 to 1872 was twenty-seven.

44.   Jacobi was at New England Hospital for a few months during the summer of 1864. Neither her personal papers, her two-volume edited autobiography and articles, nor her later published works refer to her internship at New England Hospital. The reference to the busy schedule of Zakrzewska is taken from Rhoda Truax, *The Doctors Jacobi* (Boston, 1952), 36. Truax refers to private Putnam collections in writing her popular biography of Jacobi. See the complete bibliography on Jacobi in James, *Notable American Women.*

45.   AR-NEH (1867), 7; their letter of application to Harvard is in the Harvard Medical School Dean's Records, 1867, and in Chadwick Scrapbook (Harvard Countway Library Archives).

46. Dimock's correspondence with Samuel Cabot about her Zurich experience is in the New England Hospital Collection, Sophia Smith Collection, Smith College. *Notable American Women* contains a complete bibliography on Dimock. Dimock's reputation as a surgeon is demonstrated in a research article and obituary published simultaneously in "The Death of Dr. Dimock," *Medical Record* 10 (1875): 357–58.

47. Margaret Todd, *The Life of Sophia Jex-Blake* (New York, 1918), 192.

48. Sara E. Parsons, *History of the Massachusetts General Hospital Training School for Nurses* (Boston, 1922), 15. Sturtevant's memoirs are reproduced on 4–18 of Parsons; they originally appeared in "Personal recollections of hospital life before the days of training schools," *The Trained Nurse* (Boston, 1895).

49. Todd, *Life of Sophia Jex-Blake,* 201. See also Edythe Lutzker's pioneering studies: "Medical Education for Women in Great Britain," M.A. thesis, Columbia University, 1959; and *Women Gain a Place in Medicine* (New York, 1969).

50. *The Fiftieth Anniversary of the New England Hospital for Women and Children, October 29, 1912* (Boston, 1913), 11.

51. AR-NEH (1880), 13; Marie Zakrzewska, address to students (April 1, 1876), 3, New England Hospital Collection; Sophia Smith Collection, Smith College.

52. Frances Willard and Mary Livermore, eds., *American Women* (Buffalo, 1897), 2: 741.

53. Kate Campbell Hurd-Mead, *Medical Women of America* (New York, 1933), 34.

54. AR-NEH (1887), 9–11.

55. Margaret Noyes Kleinert, "Medical Women in New England: History of the New England Women's Medical Society," *Journal of the American Medical Women's Association* 11 (1856): 63–64, 67; "Memoirs of Dr. Emma Call, June, 1928," Schlesinger Archives, Radcliffe College; *New England Women's Medical Society Directory of Members, 1878–1928,* Sophia Smith Collection, Smith College, *Boston Directory* (1846–1910, annual editions).

56. Marie Zakrzewska, address to students (April 1, 1876), 7; Marie Zakrzewska, address to students (October 30, 1891), New England Hospital Collection, Sophia Smith Collection, Smith College.

57. Vietor, *Woman's Quest,* 336; *Woman's Journal,* January 6, 1872; for a discussion by a male physician (Dr. Derby) of whether or not the men should accept the women's invitation, see H. Derby to Dr. J. R. Chadwick, June 14, 1882, Chadwick Scrapbook.

28　The Gendering of Empathic Expertise: How Women Physicians Became More Empathic Than Men

Regina Morantz-Sanchez

Since the early 1980s feminist scholars have critically examined our culture's commonplace notion that women are more empathic than men. They have noted that caring labor is performed primarily by women and have asked why that has been so.[1] What can the historian offer to these deliberations? In particular, do we find anything in the historical record that can tell us about the development of a concept of empathic expertise in medicine? In answering this question in the affirmative, I intend to highlight aspects of the careers of two very different women physicians who achieved public distinction at the end of the nineteenth century. One, Elizabeth Blackwell, was a founder of the woman's medical movement in the United States and in England and spent much of her life formulating and disseminating her ideas regarding women physicians' role in society. The other, Mary Dixon-Jones, was a pioneer gynecological surgeon who practiced in Brooklyn and is the only woman I am aware of who gained entrée to the small transnational group of elite physicians attempting to shape the direction of gynecology.

The two women did not know each other. Had they met, I doubt whether they could have spent more than five minutes in the same room without coming to verbal blows. Whereas Blackwell thought deeply about the implications of the changes in medicine that were occurring because of the bacteriological revolution, worrying not just about women's role but about the future of patient care more generally, Dixon-Jones embraced those changes with single-minded enthusiasm; her raison d'être was to see to it that she was an integral part of them.

By locating my subjects within their particular social spaces and networks of communication, I will say something about the fate of what we now call "empathy" in the changing medical world of the nineteenth century. In addition, I hope to explore the ways in which representations of gender constituted an important element in the discourse of each of the professional communities of which these two women were a part, thereby coloring conceptions of professionalism and the obligations of physicians to patients.

Elizabeth Blackwell completed her medical training at midcentury, when the role of the

REGINA MORANTZ-SANCHEZ is Professor of History at the University of Michigan, Ann Arbor, Michigan.

Reprinted from *The Empathic Practitioner,* Ellen Singer Moore and Maureen A. Milligan, eds. (New Brunswick: Rutgers University Press, 1994), 40–58.

physician was shaped by a traditional system of belief and behavior that still explained sickness not as the specific affliction of a particular part of the body but as a condition affecting the entire organism. Therapy was consequently designed to treat the whole patient; the science of medicine lay with the doctor's ability to select the proper drug in the proper dose to bring about the proper physiological effect. To be sure, this task required a thorough knowledge of the therapeutic armamentarium, but it demanded "art" as well: the good practitioner was familiar with the patient's unique personal history and familial influences, all of which were assessed. Physicians treated patients in their own homes, a social context that emphasized the sacredness of personal ties with clients and the relevance of family history to clinical judgments.[2]

Though this system was labeled "scientific," Blackwell and her contemporaries understood the word *science* differently than we do today. For example, few physicians in the nineteenth century would have ignored the importance of intuitive or subjective factors in successful diagnosis and treatment. "The model of the body, health and disease," Charles Rosenberg has written, "was all inclusive, antireductionist, capable of incorporating every aspect of man's life in explaining his physical condition. Just as man's body interacted continuously with his environment, so did his mind with his body, his morals with his health. The realm of causation in medicine was not distinguishable from the realm of meaning in society generally."[3] Blackwell's colleague at the Woman's Medical College of Pennsylvania, Professor Henry Hartshorne, who held a professorship of hygiene similar to the one Blackwell had created for herself at the Woman's Medical College of the New York Infirmary, could have been speaking for her when he observed in an 1872 commencement address, "It is not always the most logical, but often the most discerning physician who succeeds best at the bedside. Medicine is, indeed, a science, but its practice is an art. Those who bring the quick eye, the receptive ear, and delicate touch, intensified, all of them, by a warm sympathetic temperament . . . may use the learning of laborious accumulators, often, better than they themselves could do."[4]

Blackwell's professional community consisted primarily of physicians and social reformers who held there to be a social, political, and moral component to sickness. The good physician addressed not only the health of the body but the health of the body politic. When advances in Parisian physiology during the first third of the century discredited much of traditional therapeutics, revealing the self-limiting quality of much disease, some practitioners, Blackwell included, responded by emphasizing the importance of preventive medicine. Many saw hygienic management as the best means of furthering clinical medicine, and some applied this logic by advocating public prevention as a way out of the excessive skepticism and therapeutic gloom of midnineteenth-century medical practice. Henry Bowditch of Harvard Medical School, for example, believed that the physician of the future would be concerned primarily with education, on both the individual and the state level.[5]

For others the dramatic bacteriological discoveries in the last decades of the nineteenth century led to a new paradigm of experimental science. Not only had researchers isolated pathogenic bacteria for numerous epidemic diseases but they offered a new ideology of science in medicine consisting of an acceptance of the germ theory, the isolation and identification of specific diseases, increasing specialization within medical practice, and a growing willingness to resort to evidence produced in the laboratory. While older practitioners continued to emphasize the importance of clinical observation and the inevitability of individual differences in treatment, laboratory enthusiasts argued that the chemical and physiological principles derived from experimentation must inform therapeutics. Patient idiosyncrasies and environmental differences were gradually stripped of their significance, while reductionist and universalistic criteria for treatment took their place. The experimental therapeutist focused less on the patient and more on the physiological process under investigation. The result was a competing definition of what constituted science in medicine and "a thoroughgoing rearrangement of the relationships among therapeutic practice, knowledge, and professional identity."[6]

Elizabeth Blackwell did not share the high hopes accompanying the new discoveries in the laboratory and remained suspicious of their usefulness. Others in her professional community also rejected the new medical materialism, clinging to traditional antireductionist approaches to patient care. Several historians indeed have demonstrated how and in what ways laboratory medicine threatened more traditional epistemological categories.[7] But what is especially intriguing about Blackwell's critique is that her arguments drew on the language of domesticity. Moreover, her thinking about medicine was deeply influenced by her conceptions of gender. Her writings about the good practitioner framed a discourse about gender that privileged empathic expertise over the new science of the laboratory and associated the one with women and the other with men.

At the core of the nineteenth-century ideology of domesticity was the concept of the moral mother. The female qualities of nurturing, sympathy, and moral superiority were depicted as naturally flowing from the experience of maternity. As the family was romanticized, women were increasingly depicted at its moral and spiritual center and assigned a pivotal place in the preservation of values intended to inform not only family life but the social institutions of society at large. In addition, women's elevated moral status was integrally connected to their

disinterestedness. "Only by giving up all self-interest and 'living for others,'" Joan Williams has observed, "could women achieve the purity that allowed them to establish moral reference points for their families and for society at large."[8]

Elizabeth Blackwell believed that motherhood, much like the practice of medicine itself, was a "remarkable specialty" because of the "spiritual principles" that underlay the ordinary tasks most mothers performed daily. These she called "the spiritual power of maternity," and they informed both her notions of moral responsibility and her formulations of what constituted good science. Indeed, for Blackwell this power had much in common with the psychologist Erik Erikson's idea of generativity, a concern for ensuring the healthy moral and physical growth of the next generation. Not only physicians but all mankind must learn to harness it. Moreover, the insights that could be derived from the social practice of mothering could not be measured or reproduced in the laboratory.[9] The microbe hunters posed three fundamental dangers to medicine as Blackwell understood it. First, their conception of disease etiology was reductionistic and materialistic. Although medicine deserved to be called scientific, the definition of science must not be forced within the narrow confines of bacteriology's deterministic model. Science is not, she insisted, "an accumulation of isolated facts, or of facts torn from their natural relations. . . . Science . . . demands the exercise of our various faculties as well as of our senses. . . . Scientific method requires that all the factors which concern the subject of research shall be duly considered. . . . [For example] the facts of affection, companionship, sympathy, justice . . . exercise a powerful influence over the physical organization of all living creatures."[10]

Blackwell's second objection was to the practice of vivisection, an experimental tool essential to laboratory physiology. It was not so much the plight of animals that concerned her but the process of detachment from the object of research that experiments on live animals inevitably encouraged. She believed such laboratory experiences would harden medical students and inure them to "that intelligent sympathy with suffering, which is a fundamental quality in the good physician." Soon, she predicted, they would be regarding the sick poor simply as "clinical material." In addition, Blackwell believed that laboratory research stimulated the increase in gynecological surgery, which rendered more and more women incapable of having children.[11]

These two fundamental objections inevitably led Blackwell to her third: the fear that a preoccupation with the laboratory would turn the profession away from an emphasis on clinical practice and severely threaten the doctor-patient relationship. Although research was indispensable to the physician's task, it must be focused on the patient, not on abstract physiological laws, and certainly not on the physiology of animals. Like other colleagues similarly contending with changes in medicine, Blackwell emphasized behavior over biomedical knowledge as the basis for professional identity. More important than long hours in the laboratory were a physician's skills in clinical observation and the ability to maintain "character" at the bedside. "It is not a brilliant theorizer that the sick person requires," she reminded her students, "but the experience gained by careful observation and sound commonsense, united to the kindly feeling and cheerfulness which make the very sight of the doctor a cordial to the sick." The "true" physician had two obligations to the patient: to cure disease and to relieve suffering through empathy, or what she and most Victorians called "sympathy."[12]

We need not invoke the etymology of the word *empathy* or consult its complex contemporary definitions to understand that Blackwell's notion of empathic expertise was essential to her concept of professionalism. By modeling the doctor-patient relationship on the interaction between mother and child, Blackwell was clearly gendering such behavior, though she was careful to assert that it was something that men could learn. She went even further in her elaboration of gender dualisms, however, when she labeled the new science "male." Indeed, she blamed bacteriology on the "male intellect" and warned her students against the tyranny of male authority in medicine. "It is not blind imitation of men, nor thoughtless acceptance of whatever may be taught by them that is required," she wrote. Women students, she regretted, were as yet too "accustomed to accept the government and instruction of men as final, and it hardly occurs to them to question it." They must be taught that "methods and conclusions formed by one-half the race only, must necessarily require revision as the other half of humanity rises into conscious responsibility."[13]

Blackwell's critique of bacteriology through the invocation of culturally available gender symbols represented a contestation of changing power relationships in medicine. The association of empathic behavior with femininity was something relatively new.[14] When we recall the comments of Henry Hartshorne cited at the beginning of this essay, we are reminded of an older concept of professional behavior that maintained a place for intuition and sympathy and stressed the therapeutic powers of moral and social concerns. Drawing on aspects of this older tradition, Blackwell also seems to have been reformulating it by valorizing a certain kind of clinical behavior and connecting it with women. What is implicit but not stated is that objectivity and professional disengagement—qualities intensely identified with the new version of scientific medicine—are male. Ironically, though her intention was to mount a critique of the changes in medicine, defining a

[handwritten margin note: Knowledge of germs and pain killers made operation safer]

particular form of behavior as female may have had exactly the opposite of her intended effect, because it linked interpersonal concern with a subordinate social group.[15]

Indeed, it seems that Blackwell was losing her audience. Her fault-finding with the new ideology of science reached a relatively small and circumscribed group of male and female practitioners, most of whom were losing ground in the face of rapid changes in the organization and practice of medical care. Although women physicians welcomed her ideas about the unique qualities they had to offer the profession, in part because the argument proved a still powerful justification for their occupational aspirations, more and more of them found her critique of the new science irrelevant to their experience.

It would be difficult to find a woman physician for whom Blackwell's discourse on medicine had less resonance than Mary Dixon-Jones. A graduate of the Woman's Medical College of Pennsylvania in 1873 at the age of forty-five, and only seven years younger than Blackwell, Jones's circuitous path to her profession was a familiar one for women physicians of that first generation. She began as a teacher, taught physiology at various female seminaries, and read medicine with a well-known Maryland physician, Dr. Thomas Bond. In the 1860s she received a sectarian medical degree from a hydropathic college in New York.

But like several women physicians in those early years who attended sectarian institutions because no regular medical school would accept them, Jones found herself drawn back into medical study later in her career, this time at an orthodox institution. She spent three years in the early 1870s matriculating in Philadelphia at the Woman's Medical College, displaying a particular interest in microscopy and pathology. In 1873 she passed a three-month preceptorship in New York with Mary Putnam Jacobi, a highly

respected and Paris-trained woman physician and medical professor, and later took courses at the New York Postgraduate Medical School.

It was probably during this period that Dixon-Jones came into contact with the new science of bacteriology. Moreover, as a student at the Woman's Medical College, she no doubt attended surgical clinics at Blockley Hospital and studied surgery with Professor Emmeline Horton Cleveland, another Paris-trained woman physician who was also dean of the school for two of the three years that Jones was there. A highly skilled technician and beloved by her students, Cleveland was the first woman surgeon to perform an ovariotomy in Philadelphia.[16]

One of the earliest by-products of the new experimental science occurred primarily in the operating room. While the gradual use of anaesthesia after midcentury had lessened the pain of surgery, Lister's adaptation of the germ theory in developing the principles of antisepsis had guaranteed the relatively safe surgical invasion of the body for a variety of hitherto incurable complaints. Many of these were gynecological; by far the largest proportion of abdominal operations between 1860 and the end of the 1890s was performed on women.

Jones must have watched these developments keenly, because in 1881 she became the chief medical officer of the Women's Dispensary and Hospital of the city of Brooklyn, a charitable organization whose Board of Lady Managers boasted some of the most prominent matrons in the city. Heightening discord with the hospital's board, however, probably over her increasing interest in ovariotomy, led to a severing of that professional relationship in January 1884. A few months later she established her own gynecological hospital, an institution that allowed her complete autonomy and flexibility in the medical decision-making process, since she dominated its board of trustees.

In the beginning the hospital was called the J. Marion Sims Hospital and Dispensary, though its name was soon changed to the Woman's Hospital of Brooklyn. The original name is revealing of Dixon-Jones's apparent desire to identify herself with the recently deceased pioneer gynecological surgeon of New York City, who had successfully presided over a revolution in gynecological surgery with the founding of the New York Woman's Hospital in the 1850s.[17] Much like many other would-be specialists in surgical gynecology, both in the United States and England, Dixon-Jones's relationship with a specialty hospital—in her case, the Woman's Hospital of Brooklyn—was crucial to her professional career. The rise of specialty hospitals like hers in Brooklyn was an important chapter in the development of gynecology in this period.[18]

Specialty hospitals provided a means for particularly ambitious practitioners to make a mark in their chosen field. Although the profession as a whole was skeptical of specialization, proponents justified their work by hailing the process of division of labor in medicine. By the end of the century, at least where women's hospitals were concerned, most practitioners had conceded the point of Dr. Charles Routh, a founding member of the British Gynaecological Society, who emphasized how important specialty hospitals were to the study of specific diseases. "Instead of three, four . . . there were . . . a hundred, or two hundred patients. The medical attendant could, therefore, reason on all of them. . . . No man could come to any positive conclusion as to the treatment of special diseases till he had many examples."[19]

As a woman, of course, Dixon-Jones's interest in surgery could not have been pursued with much success at any of the existing hospitals, except those few connected with a woman's medical school. For example, though the Board of Lady Managers at J. Marion Sim's New York

Hospital had stipulated that he appoint a woman assistant, Sims neglected to do so, and it was well into the twentieth century before a woman surgeon operated at the hospital he founded.

Jones's hospital was small, with perhaps only ten to fifteen beds, and seems to have been devoted almost exclusively to gynecological surgery. The medical staff consisted of Jones, her son Charles, a recent graduate of the New York College of Physicians and Surgeons, and another woman physician from the Woman's Medical College of Pennsylvania, Dr. Eliza J. Chapin-Minard. Both her son and Minard assisted Jones in her operations.

Once she had established her own hospital, Dixon-Jones both pursued her specialty and went about the task of meticulously constructing a professional identity in gynecological surgery and surgical pathology. Indeed, she was an aggressive self-promoter who instinctively moved to counteract the obvious barriers to advancement that presented themselves to an aspiring female surgeon. In the spring of 1884 she performed her first laparotomy, removing a diseased ovary and its appendages from a woman she diagnosed as a classic case of "hystero-epilepsy due to reflex irritation."[20] Four cases of ovariotomy followed the next year and seven the year after that. In 1886 she made an extended visit to Europe, studying and making herself known in various hospitals and visiting the clinics of some of the most renowned surgeons, including Lawson Tait, Theodore Bilroth, August Martin, Carl Schroeder, and Jules Péan. Upon her return the following year, she performed thirty-six ovariotomies and is credited with completing the first total hysterectomy for fibroid tumor ever attempted in the United States.[21]

Along with her surgical accomplishments, Dixon-Jones continued her interest in pathology, carefully studying microscopically tumors and tissue removed from the bodies of her patients. She developed an intimate acquaintance with Dr. Carl Heitzman, a Hungarian immigrant who was known as an expert microscopist and was a specialist in skin diseases.[22] He aided her in making slides and preparing specimens, and she accomplished much of her scientific work under his guidance. She became a member of the New York Pathological Society and frequently brought in specimens for discussion. Beginning in 1884, she commenced publishing pathological findings and clinical case reports in leading journals such as the *American Journal of Obstetrics,* the *Medical Record,* and the *British Gynaecological Journal,* thereby calling attention to herself both as a technical virtuoso in the operating room and as a careful scientist in the laboratory. During her lifetime she published more than thirty papers in gynecology and surgery. Indeed, the *Dictionary of American Medical Biography* credits her not only with being the first U.S. surgeon to perform a total hysterectomy for uterine myoma but also with describing and identifying two diseases—endothelioma, cancer of the lining of the uterus, and gyroma, a cancerous tumor of the ovary.[23]

Dixon-Jones successfully used her medical articles to create the sense among her readers that she was a member of a relatively small group of elite gynecological practitioners in the United States and Europe who were pioneering in operative approaches to women's diseases. Her first article, for example, entitled "A Case of Tait's Operation," was a rather audacious attempt to associate her work with the world-renowned ovariotomist from Birmingham, England, Lawson Tait. She continued to reference Tait over and over again in subsequent publications. She corresponded with him as well, printing part of one of his letters to her in a footnote to one article and telling her readers in another that her son Charles had served as his "first assistant" in 1886, when they made a grand European tour. She eventually attracted Tait's attention suffi-

ciently to prompt him to refer to her at length in one of his own publications.[24]

But it was not only Tait with whom she persistently identified in her published work. Her articles mention connections, conversations, and consultations with elite gynecological surgeons in New York, Boston, and Philadelphia and demonstrate familiarity with the work of most of the well-known ovariotomists who published in the leading journals. Her articles were characterized by incessant name dropping, coupled with continuous self-referencing to other of her publications and frequent claims to being "the first" to discover a particular cell formation or to try a certain procedure. In terms of her self-presentation, Dixon-Jones was a person who had succeeded in the world of gynecological surgery, someone who embraced and understood the new science.

Ironically, the aggressiveness with which she orchestrated her own success probably hastened her downfall. In April 1889 Dixon-Jones sent a note to the Brooklyn *Eagle* requesting that the paper print a feature story on her hospital in order to "draw public attention to its good works." The institution, she explained, drew its clientele primarily from the urban poor and depended on a combination of city charity funds and private donations. Mysteriously, that very day the newspaper received an anonymous communication accusing Jones of running a private enterprise with public funds. With the certainty of Greek tragedy Dixon-Jones's plan to promote herself and her hospital soon began to backfire.

A reporter assigned to investigate eventually crafted a series of unflattering articles about Dixon-Jones that touched off an avalanche of public criticism and resulted in two manslaughter charges and eight malpractice suits against her. The articles implied that she was an ambitious and self-promoting social climber, a knife-happy, irresponsible surgeon who forced unnecessary operations on innocent and unsuspecting women and used the specimens gleaned from them to advance her reputation in diagnosis and pathology. Although the first manslaughter case ended in acquittal and the rest of the charges were eventually dropped, the court battles took four years. In 1892 Dixon-Jones attempted to restore her reputation by charging the *Eagle* with libel.

Her lawyers sought $300,000 in damages, claiming that their client had been victimized by the newspaper, aided by certain disreputable members of Brooklyn's medical establishment. A legal spectacle of major proportions, the ensuing trial involved some of the most prestigious physicians in New York and Brooklyn. Medical journals and leading newspapers covered it daily. Testimony took almost two months; roughly three hundred witnesses were called, including former patients with babies in their arms. Jars full of specimens and surgical mannequins became common sights in the courtroom. When Jones lost the case, the state and city withdrew public funds from her hospital and its charter was revoked. Being deprived of her operating theater effectively ended Mary Dixon-Jones's surgical career. Relocating to New York City, she became an editor of the *Woman's Medical Journal* and passed the decade and a half before her death publishing articles on pathology, utilizing over and over again the specimens and slides collected from the hundred or so operations she had performed in the previous decade.

The trial testimony is a gold mine of complex and interrelated themes. We learn much about Dixon-Jones's status within the Brooklyn medical community, her relationship with a self-created group of elite gynecological surgeons in New York City and elsewhere, and the tensions over specialization seething within the profession at large. I concentrate here, however, on the ways in which representations of gender became embedded in the construction of new professional identities.

In retrospect it is clear that Dixon-Jones's behavior offended professional colleagues in Brooklyn from the very beginning. In the spring of 1884, for example, her application for membership in the King's County Medical Society was tabled on the grounds that "there is so much opposition to her name that it would be well to postpone action."[25] In contrast, her son Charles did become a member in good standing and remained so until 1892, despite his role as her surgical assistant. Dr. Landon Carter Gray testified that he had initially advocated Dixon-Jones's admission into the society. But he had been told by several colleagues that she had a poor reputation, though he "knew [of] no instance of unprofessional conduct on her part." Others who took the stand to respond to questions regarding her professional standing were equally vague.[26]

To complicate matters further, Dixon-Jones's medical detractors were not all male. Several women physicians, at least two of whom were members of the county medical society, voiced reservations regarding her medical reputation. Caroline S. Pease, an 1877 graduate of the Woman's Medical College of Pennsylvania, claimed that she had worked as Dixon-Jones's assistant briefly in 1886 and detected in her practice "a very marked discrimination" in favor of surgical cases.

Eliza Mosher and her partner Lucy Hall Brown also took the witness stand to express reservations about Dixon-Jones's character. In a letter to her friend and mentor Elizabeth Blackwell, written the week Dixon-Jones's first case report appeared in the *American Journal of Obstetrics*, Mosher reported, no doubt referring to Jones: "There are several regularly graduated women who are already members of the Kings Co. Med. Soc. There is one who, judging from her paper read before the Pathological Soc. not long since, is rather an able woman, but her manners are beyond description—we could not identify our selves with her safely and she is an element of evil because of her coarseness."[27]

One can only surmise what behavioral traits Mosher was referring to, but several witnesses and the newspaper itself drew a portrait of a strong-willed and outspoken woman who was not above tongue-lashing uncooperative colleagues and laypersons. A. J. C. Skene, for example, confessed that Dixon-Jones threatened him when he ceased consulting with her, warning him that "she had a tongue and would use it."[28] Even Kelly and Burrage, the authors of the *Dictionary of American Medical Biography,* published long after her death, remarked on Dixon-Jones's reputation for giving offense. They noted that Jones was "peculiar in person," "flashy and tawdry in appearance," and speculated that "lack of judgment and of intimate contact with the better members of the profession may have been responsible for a certain mental obliquity with which she is accredited."[29]

Expert medical testimony at the trial suggests that at least some of this professional hostility came from conservative Brooklyn physicians who not only disapproved of Jones's celebration of surgical solutions to pelvic disease but remained suspicious of the direction gynecology had moved in the last two decades. At the heart of the matter was tension over specialization, which continued to gain momentum in the second half of the nineteenth century. Suspicion of specialists in medicine had deep historical roots, given that specialization before this period had always been associated with quackery. But in the last decades of the century it became associated with the new ideology of science, which tended to deemphasize holistic approaches to disease in favor of localized pathological anatomy.

The work of the Paris school in the 1830s and 1840s, for example, proved crucial to the development of gynecological surgery, because researchers increasingly tended to break down the body into its component parts. New technology such as the stethoscope and the thermometer and new techniques such as auscultation, percussion, and palpation allowed practitioners to concentrate on specific organs or abnormal internal structures and develop new approaches to treatment. Ovariotomies could not have been attempted, Jane Sewall reminds us, "without a clear concept of local pathological anatomy–without believing that a woman with a grossly distended abdomen had ovarian lesions."[30]

When anaesthesia and antisepsis greatly increased the safety of surgery in the 1860s, the traditional criteria for operations—that surgery should be resorted to only in life-threatening situations—seemed no longer justifiable. Gynecological surgeons were among that segment of the physician population who became impatient with palliative treatment—primarily draining and tapping—which rarely offered permanent solutions to patients whose lives were blighted by chronic disease.

The men who created the specialty of surgical gynecology tended to be young and ambitious. They were bucking medical tradition in a number of ways, and their interest in women's diseases was stimulated substantially by the fact that surgical gynecology afforded them a place in the profession. Perhaps it is also worth pointing out that the gradual shift from art to science, from general practice to specialization, echoes in a very real sense the anxieties generated by the transition from craft traditions, where unique products were fashioned holistically by a skilled workman, to mass production, where uniform products were produced in a reductionist manner by a series of "specialists" created by the division of labor. Dixon-Jones's self-promotion, her making of herself into a professional commodity, is part of such a transformation.[31]

The successful ascendancy of surgical treatment led to a decided power shift in the medical hierarchy. The ambitious and entrepreneurial approach to professional disputes displayed by gynecological surgeons prompted more traditional physicians to lament the passing of an older, gentlemanly image. Suspicion of the new *case of controversy!*

professional style can be detected in the outcry against specialization as elitist and dehumanizing; indeed, similar accusations were hurled against laboratory science, and it is no coincidence that Elizabeth Blackwell spoke disparagingly of both in the same breath.

One hears echoes of these controversies in the Dixon-Jones trial testimony. Dixon-Jones's critics, all of whom practiced in Brooklyn, focused primarily on the uncertain validity of her therapeutics and on her questionable professional "character." The testimony of A. J. C. Skene was typical. Known to be a staunch conservative on the subject of ovariotomy, Skene was professor of gynecology at the Long Island College Hospital and a recognized authority on women's diseases. He acknowledged that he had known Dixon-Jones for fifteen or sixteen years, that she had operated on two of his patients against his recommendation, and that he no longer consulted with her. When Jones's lawyer tried to characterize Skene as a member of the "conservative" as opposed to the "radical" school of gynecology, Skene demurred, commenting that with good surgeons "surgery was never resorted to except in cases where life would be in danger if no operation should be performed."[32] Others questioned Dixon-Jones's pathological diagnoses, implying that she had invented them after the fact merely to justify her resort to the knife.[33]

In defense of Jones came an array of prominent surgeons from New York City and Philadelphia. A. M. Phelps, W. Gill Wylie, and H. Marion Sims each confirmed that they had done hundreds of laparotomies of the type performed by Jones and that they had consulted with her on numerous occasions. Wylie observed that the danger of laparotomy "had now become so slight that much less ceremony was observed than there used to be."[34]

What is particularly striking about this testimony is its subtext. One does not get the feeling from these statements that the speakers were

particularly close to Dixon-Jones or had an interest in promoting her. Yet these prominent practitioners, some of them at the pinnacle of their careers, traveled across the Brooklyn Bridge at considerable personal inconvenience to give evidence on behalf of a woman. In theory they were no more accepting of women physicians than any other of their male colleagues.[35] But they quickly grasped that it was not simply Dixon-Jones's surgical career that was on trial but theirs as well. In this instance gender antagonism played second fiddle to rivalries between newer and older views of medical professionalism.

Yet the gender themes in this extraordinary drama remain rich and complex. The trial offers us the spectacle of a woman physician accused of misusing professional power and expertise to manipulate and harm other women. While Dixon-Jones's lawyers employed her sex in her defense, urging that she embraced the "best in femininity" in her work, many detractors found her behavior particularly heinous because she was a woman. As we have seen, contemporary testimony suggests that Dixon-Jones was a classic example of what today would be labeled a "difficult woman"—a woman in authority who is outspoken and perhaps somewhat imperious. Not surprisingly, taking their cues from a number of former patients who spoke to the *Eagle,* several Brooklyn physicians, including Skene himself, accused Dixon-Jones of egregiously poor communication with patients and their families. Many of these, for example, knew only vaguely that there was to be an operation; others were simply told that the doctor would make them well.[36]

Because the concept of informed consent did not exist in this period, and there was a wide range of available opinion on how much information doctors should share with patients before surgery, I read this controversy also as a debate over Dixon-Jones's capacity for empathy—over her ability to treat patients as something other than "clinical material"—and not over whether she violated any formal rules of professional conduct. What is especially intriguing is that her supporters dismissed the idea of close physician-patient communication as either detrimental or unnecessary. A.M. Phelps, surgeon to City Hospital in New York, testified that he had performed over two hundred laparotomies and had sent many women patients to Jones in Brooklyn. He claimed that it was not usual for physicians to explain to ignorant patients the nature of imminent operations. He himself told them that "they must put themselves in his hands." Charity patients (the "sick poor" who Blackwell worried would be mistreated) were especially problematic because of their ignorance: Phelps believed that these women "might be frightened off the operating table" if they knew too much about what was going to happen to them. As already noted, W. Gill Wylie of Bellevue Hospital confirmed that because laparotomies now had become routine, much less attention was paid to "obtaining consents."[37]

In spite of the support of these colleagues, Dixon-Jones's experimental, active, and manipulative stance toward women's diseases heralded an image of medicine that her Brooklyn colleagues and much of the public were reluctant to accept. Given the prevailing image of the woman physician as nurturing and empathic, Dixon-Jones's being a woman may have actually heightened existing anxieties about the meaning of her various activities for the future of medical practice. Was she not performing aggressive scientific experiments on patients? Using pathological specimens in the name of science to advance her career? Refusing to inform patients properly of her intentions? Surely her conduct was inappropriate for a woman, but lurking below the surface was worry over how representative it was of the entire profession.

Indeed, Elizabeth Blackwell had warned of these developments in a letter to her longtime colleague Mary Putnam Jacobi, a professor at the Post graduate Medical School and the Woman's Medical College of the New York Infirmary, and the only woman physician to testify in Dixon-Jones's behalf. Worrying about the recent increase in gynecological surgery and connecting such activity with the horrors of vivisection, Blackwell proposed that Jacobi help her rally women physicians in the United States against unwarranted operative procedures. But Jacobi, like Dixon-Jones, was a woman physician who had embraced advances in technology and research. She gently suggested that Blackwell catch up on her medical reading and think more like a scientist.[38] Moreover, Jacobi added, it would be a terrible mistake to gender new approaches to the cure of disease; women also needed to keep abreast of these developments.[39]

conclusion

The careers of Elizabeth Blackwell and Mary Dixon-Jones aid us in exploring how representations of gender became embedded in the new ideology of medical professionalism that emerged at the end of the nineteenth century. Elizabeth Blackwell was merely the most eloquent spokesperson for a carefully crafted articulation of female professionalism which, using the language of domesticity, was supremely suspicious of the increasing tendency to treat human beings like objects and of the reductive, activist, and experimental approach of the new breed of ovariotomists. Empathic expertise became an important component of women doctors' public and private image, while motherhood emerged as a central trope of their discourse surrounding the physician-patient relationship. Although the ideology of female professionalism hearkened back to the doctor's traditional role in bedside care, the gendering of professional qualities like empathy was relatively new.

It follows from this that among Dixon-Jones's most virulent critics were other women physicians in Brooklyn who subscribed to a Blackwellian version of female professionalism. In contrast, Mary Dixon-Jones was either oblivious to the subtleties of these behavioral scripts or consciously rejected them. The medicine she practiced—diagnosing and excising diseased organs—evoked a materialist conception of the body that encouraged the practitioner to think, not in terms of the whole patient but about specific organs and localized infection. Moreover, Dixon-Jones's reference groups were exclusively male; she craved recognition by male colleagues as a surgical innovator of the first rank. Yet in playing the men's game she drew criticism from both sexes for failing to play it with the acceptable demeanor of a woman. In promoting herself she unwittingly exposed to the scrutiny of investigative journalism unresolved tensions regarding the medical procedures and behavior of the group of specialists to whom she so desperately wished to belong. Those professional tensions, and the gendered language invented in the nineteenth century to give them voice, remain very much with us.

Notes

I am indebted to Barbara Bair, Margaret Finnegan, Louise Newman, Ellen More, Anita Fellman, George Sanchez, Gerald Grob, and Tom Cole for helpful readings of this essay.

1. See Carol Gilligan, *In a Different Voice: Psychological Theory and Women's Development* (Cambridge, Mass.: Harvard University Press, 1982). See also Nel Noddings, *Caring: A Feminine Approach to Ethics and Moral Education* (Berkeley: University of California Press, 1984), and the essays in Janet Finch and Dulcie Groves, eds., *A Labour of Love: Women, Work, and Caring* (London: Routledge & Kegan Paul, 1983), for a sampling of this literature.

2. Charles Rosenberg, "The Therapeutic Revolution: Medicine, Meaning, and Social Change in Nineteenth-Century America," in M. Vogel and C. Rosenberg, eds., *The Therapeutic Revolution: Essays on the Social History of American Medicine* (Philadelphia: University of Pennsylvania Press, 1979), 3–25, 10–11.

3. Ibid., 10. See also John Harley Warner, *The Therapeutic Perspective: Medical Practice, Knowledge, and Identity in America, 1820–1885* (Cambridge, Mass.: Harvard University Press, 1986).

4. *Valedictory Address* (Philadelphia: Woman's Medical College of Pennsylvania, 1872), 1–23, esp. 6–7.

5. Warner, *The Therapeutic Perspective*, 235–43.

6. Warner, *The Therapeutic Perspective*, 258; Russell Maulitz, "'Physician Versus Bacteriologist': The Ideology of Science in Clinical Medicine," in *The Therapeutic Revolution*, ed. Vogel and Rosenberg, 91–107.

7. See especially Warner's enormously helpful and detailed volume, *The Therapeutic Perspective*.

8. Joan C. Williams, "Domesticity as the Dangerous Supplement of Liberalism," *Journal of Women's History* 2 (Winter 1991): 69–88, 71.

9. Erik Erikson, *Childhood and Society* (New York: W. W. Norton, 1950), 267; Blackwell, "The Influence of Women in the Profession of Medicine," in Blackwell, *Essays in Medical Sociology*, 2 vols. (1902; reprint New York: Arno Press, 1972), 1–32, 9–10.

10. Blackwell, "Scientific Method in Biology," in *Essays in Medical Sociology*, 87–150, 126–30.

11. Blackwell, "Influence of Women in the Profession of Medicine," 13; Blackwell, "Erroneous Method in Medical Education," in *Essays in Medical Sociology*, 3–46, 10–12.

12. Blackwell strongly supported clinical casework, postmortem and gross pathology, pathological chemistry, microscopic anatomy, and other types of patient-centered investigations. See her "Scientific Method in Biology," in *Essays in Medical Sociology*, 105. Sandra Holton has explored not only Blackwell's thought in this regard but that of other British physicians as well. See Sandra Stanley Holton, "'Christian Physiology': Science, Religion, and Morality in the Medicine of Elizabeth Blackwell," paper presented at the Pacific Coast Branch of the American Historical Association annual meeting, Kona, Hawaii, August 1991, and Sandra Holton, "State Pandering, Medical

Policing, and Prostitution: The Controversy with the Medical Profession Concerning the Contagious Diseases Legislation, 1864–1886," *Research in Law, Deviance, and Social Control* 9 (1988): 149–70.

13. Blackwell, "Why Hygienic Congresses Fail," 47–84, 57, 74–75; Blackwell, "Influence of Women in the Profession of Medicine," 12, 19–20, 27–29, in *Essays in Medical Sociology.*

14. As Londa Schiebinger and others have shown, the gendering of certain forms of cognitive thinking had been occurring in scientific discourse since the 1700s. See Schiebinger, *The Mind Has No Sex? Women in the Origins of Modern Science* (Cambridge, Mass.: Harvard University Press, 1989).

15. See Regina Morantz-Sanchez, "Feminist Theory and Historical Practice: Rereading Elizabeth Blackwell," *History and Theory* 31 (December 1992), for a more extensive analysis of Blackwell.

16. Gulielma Fell Alsop, *History of the Woman's Medical College of Pennsylvania* (Philadelphia: Lippincott, 1950), 109.

17. Deborah Kuhn McGregor, *Sexual Surgery and the Origins of Gynecology: J. Marion Sims, His Hospital, and His Patients* (New York: Garland Publishing, 1989).

18. See Charles Rosenberg, *The Care of Strangers* (New York: Basic Books, 1987), esp. chs. 7–8.

19. Quoted in Ornella Moscucci, *The Science of Woman* (New York: Cambridge University Press, 1990), 101.

20. Mary Dixon-Jones, "A Case of Tait's Operation," *American Journal of Obstetrics* 17 (November 1884): 1154–61, 1156. The diagnosis of reflex irritation referred to the commonly held view of many gynecologists that diseased reproductive organs could be manifested by a psychological response, in this case hysteria.

21. Mary Dixon-Jones, "Personal Experiences in Laparotomy," *Medical Record* 52 (August 1897): 182–92, 191.

22. See Howard Kelly and Walter Burrage, *American Medical Biographies* (Baltimore: Normon, Remington Co., 1920), 513.

23. Howard A. Kelly and Walter L. Burrage, *Dictionary of American Medical Biography* (Boston: Milford House, 1971), 677.

24. See Dixon-Jones, "Oophorectomy and Diseases of the Nervous System," *Woman's Medical Journal* 4 (January 1895): 1–11, 5; Dixon-Jones, "Removal of the Uterine Appendages—Recovery," *Medical Record* 27 (April 1885): 399–402, 400. For Tait's reference to one of Dixon-Jones's articles see "A Discussion of the General Principles Involved in the Operation of Removal of the Uterine Appendages," *New York Medical Journal* 44 (November 1886): 561–67. On Tait see Jane Sewall, "Bountiful Bodies: Spencer Wells, Lawson Tait, and the Birth of British Gynecology" (Ph.D. diss., Johns Hopkins University, 1991).

25. Council Minutes, April 9, May 14, 1884, Kings County Medical Society Archives. A spokesperson for the society told the *Eagle's* reporter in 1889 that her application had been rejected four times for "unprofessional conduct." Brooklyn *Eagle,* May 4, 1889. But there is no evidence of this in the minutes.

26. Brooklyn *Eagle,* February 10, 1892.

27. Pease to Dean Clara Marshall of the Woman's Medical College of Pennsylvania, January 18, 1892, Marshall MSS, Medical College of Pennsylvania; Brooklyn *Eagle,* February 9, 1892; Mosher to Blackwell, November 3, 1883, Mosher MSS, Bentley Library, University of Michigan.

28. Brooklyn *Eagle,* February 9, 1892. See also *Eagle's* comments regarding Jones's unruliness as a witness, her habit of making comments under her breath, and the testimony of Cornelia Plummer, February 4 and February 6.

29. Kelly and Burrage, *Dictionary of American Medical Biography,* 677. Howard A. Kelly was a distinguished surgeon at Johns Hopkins in the 1890s.

30. Sewall, "Bountiful Bodies," 44. This discussion is indebted to ch. 2 of Sewall's dissertation.

31. I am indebted to Barbara Bair for this insight.

32. Brooklyn *Eagle,* February 19, 1892.

33. Brooklyn *Eagle,* March 9, 1892. It is important to note that a similar controversy over too much gynecological surgery, which pitted followers of Lawson Tait against followers of Spencer Wells, occurred in Great Britain in 1886 in Liverpool. Dr. Francis Imlach was criticized by the senior surgeon at his hospital and the professor of midwifery at the Liverpool Medical Institution for "unsexing women" and not properly informing ovariotomy patients of the consequences. Imlach ultimately was denied reap-

pointment. See Moscucci, *The Science of Woman,* 160–64.

34. Brooklyn *Eagle,* March 8, 9, February 27, 1892.

35. Apropos of their feelings about Dixon-Jones, consider the following. When the Brooklyn *Eagle* first ran its series on Jones in 1889, the New York Pathological Society appointed a committee to investigate the accusations. The committee did a thorough job, soliciting corroborative letters from a number of people mentioned in the articles, clipping newspaper reports, corresponding with Jones, her son Charles, and the rest of the hospital's trustees. Although the committee concluded that there was not sufficient evidence to censure Dixon-Jones, a member in good standing, we find this curious note from the society's treasurer to the chairperson of the investigating com-

mittee: "Dear Doctor . . . Dr. Mary Dixon Jones' dues are *fully* paid up . . . you don't get Mary on the Hip in that way . . . women doctors are a nuisance." May 13, 1889, New York Pathological Society Minutes, New York Academy of Medicine.

36. Brooklyn *Eagle,* February 12, 13, 15, 19, 1892.

37. Brooklyn *Eagle,* March 8, 9, February 27, 1892. See Kenneth De Ville, *Medical Malpractice in Nineteenth-Century America* (New York: New York University Press, 1990); James Mohr, *Doctors and the Law* (New York: Oxford University Press, 1993).

38. Mary Putnam Jacobi to Elizabeth Blackwell, December 25, 1888. Blackwell Papers, Library of Congress.

39. For more on Jacobi, see Regina Morantz-Sanchez, *Sympathy and Science: Women Physicians in American Medicine* (New York: Oxford University Press, 1985), ch. 7.

29 Uncle Sam's Loyal Nieces: American Medical Women, Citizenship, and War Service in World War I

Kimberly Jensen

In the sad times coming before the war is over, our Uncle Sam will no doubt ask gladly for his loyal nieces, and grant them a place in his household equally honorable to that occupied by his nephews.
  Bulletin of the Woman's Medical College of Pennsylvania, December 1917

During World War I many American women physicians made interwoven claims for full citizenship and professional equality as they sought commissions in the military medical corps and other avenues of wartime medical service. As war was juxtaposed with the campaign for women's franchise in the United States, many medical women developed an ideology of full female citizenship which included the traditional citizen's obligation of military service. And as war placed the medical profession on the public stage, they made a claim for equal status and equal opportunity within the medical corps of the military,[1] believing that this would bring them professional equality during the conflict and in the postwar medical world. For many of the approximately six thousand medical women in the United States in 1918, full female citizenship, professional equality, and service and responsibility to the state all seemed to be part of one ideological whole. By voicing and acting on these claims, they recast traditional women's service into the politically significant sacrifice of fully participating citizens.

Both as individuals and through organized groups such as the American Women's Hospitals, women physicians struggled to claim a place in military medical service through petitions, resolutions, war service registration, contract practice, and medical service at the front. Their purposeful work was not only part of a

KIMBERLY JENSEN is Assistant Professor of History and Co-ordinator of the Gender Studies Program at Western Oregon University in Monmouth, Oregon.

Reprinted from *Bulletin of the History of Medicine* 67:670–90 by permission of the publisher. © 1993 by The Johns Hopkins University Press.

continuing struggle for equality within the profession but also part of a larger movement of women which sought to define the meaning of female citizenship. Their story is also a vivid illustration of the resiliency of the resistance to women's military service and the enduring boundaries of gender in the medical profession. At a time of unprecedented mobilization of government and civilian resources to win the war, women physicians believed that they could make a practical as well as an ideological case for equal acceptance as officers in the medical corps. Yet their campaign was vulnerable to the increasing bureaucratization of wartime service and to the power of government and wartime medical leaders to decide what was "right" for the war effort.

Women physicians made their wartime claims for equal opportunity in the context of a broader history of struggle for access to educational, professional, and organizational opportunity in the field of medicine, as well as the struggle for women's rights.[2] The second decade of the twentieth century was a time for optimism for many medical women. As the nation entered the war, women physicians could point to progress in educational opportunity and in occupational variety. In 1910 women physicians reached a peak of 6 percent of the medical profession. And a 1916 study showed that 1,313 American women physicians in active practice were specializing. Two-thirds of these specialists were in what were considered women's specialties (e.g., obstetrics and gynecology), and one-third were making inroads into fields considered to be male territory.[3] But there were still many reforms to be made, especially in increasing access to medical education for African-American women and other women of color, and in expanding internship and professional opportunities for all women.[4]

In addition, medical women constructed their case for a place within the military medical corps at the same time that male physicians were waging their own battle with the military for increased rank and authority. This wartime legislative campaign, led by medical leaders under the direction of the physician Franklin Martin of the General Medical Board of the Council of National Defense, and waged for almost the entire length of the war itself, was finally successful when the Owen-Dyer Bill was passed in both houses of Congress in July 1918. Officers in the medical corps were then given the same status as other officers and gained increased authority with respect to sanitary regulations and recommendations.[5] The confluence of wartime need and the medical profession's struggle for change made the construction of a "reform moment" possible. Would women physicians be part of such reforms? How might they construct their case for inclusion? And how would they confront the boundaries between women and military service with which other women interested in such service were grappling?

In June 1917, three months after the United States entered the war, the members of the Medical Women's National Association (MWNA) assembled at their second annual meeting in New York City. A purposeful audience unanimously approved a resolution by California women physicians to be sent to Secretary of War Newton D. Baker, calling for women to be accepted in the military medical corps on equal terms with men. In this atmosphere of enthusiasm for the possibilities of medical women's war service, the members present also supported the creation of a war service committee which they hoped would translate their desire for service and recognition into concrete plans and positive results. The New York surgeon Rosalie Slaughter Morton, who had visited and served at the front, was chosen to chair the committee. She gave it a new name, the "American Women's Hospitals" (AWH), linking the new group to the strength and accomplishment of the Scottish Women's

Hospitals already in service at the front. Morton was also asked to chair the Committee of Women Physicians as part of the General Medical Board of the Council of National Defense.[6]

Leaders of the AWH believed that they might prepare the way for women's acceptance in the medical corps by conducting a nationwide survey to determine the numbers of medical women who could serve and the specific skills these women possessed. Such a census would demonstrate concretely to officials in Washington, D.C., that medical women were ready for wartime service, and the AWH leaders believed that whoever controlled such information would be recognized as authoritative by military medical men and would be called on to set policy and oversee the wartime activities of women physicians. Under the auspices of the AWH and several affiliated groups, medical women throughout the country were sent registration blanks that were to be filled out with their preferences for wartime service and returned.[7] The registration was both a practical measure and a political one, in effect a petition to the government on behalf of medical women. In the summer and fall of 1917 and the early part of 1918 the completed registration blanks were returned to AWH headquarters and to the offices of the Committee of Women Physicians in Washington. They were so important that various AWH committees and the General Medical Board fought to control the physical evidence of medical women's desire for military and wartime service.[8]

This indication of women physicians' readiness for military service did not lead to the acceptance of women into the medical corps as officers, yet the individual act of registration and its cumulative effects had political meaning. Registration could demonstrate that women physicians were ready to act as full citizens by doing war service and that they wished to claim equality with their male colleagues. Some

women viewed registration as a strategic organizational step in preparation for full participation in the military medical corps. Rosalie Slaughter Morton also knew that some medical women "as a matter of principle and precedent" refused to register for war service unless they were first commissioned in the military medical corps.[9]

The results of the registration, with other information on 5,827 women licensed to practice medicine in the United States, were published by the AWH and the Committee of Women Physicians as the *Census of Women Physicians* in 1918.[10] According to the *Census,* almost one-third (1,816, or 31 percent) of the medical women in the country in 1917–1918, active and retired, signified through registration their willingness to provide medical service as part of the war effort. The significance of this percentage is amplified when we realize that this one-third comprised not simply those medical women who believed in the principle that women physicians should have the opportunity to participate in war service but those who were personally willing to register for such service. And it does not include those whose refusal to register was a protest against the lack of commissions for women physicians.

A commitment to medical professionalism was an important factor in the war registration of women physicians. Forty-one percent of the medical women who were members of the American Medical Association registered for war service; 59 percent of the members of the Medical Women's National Association registered; and 63 percent of those women who were members of both the AMA and the MWNA registered. The last of these figures was twice the percentage seen for the *Census* as a whole.[11] Only 24 percent of the women with no professional affiliation registered for war service. These figures indicate that those medical women who were most concerned with women's professionalization, those who were active in organizations and who perhaps believed that they had the most to gain in a bid for equality with male colleagues, registered for war service.

The war service registration rates of the medical women of the *Census* compare significantly with the war service numbers of their male colleagues. According to the army surgeon general's office there were 30,591 male physicians in the medical corps at the time of the Armistice in November 1918, which means that approximately 20 percent of the male physicians in the country were in military service in 1918.[12] Another contemporary listing, *The Physicians and Surgeons of Chicago* (1922), gives information on military and civilian war service for 1,507 Chicago medical men. Almost half of this group (734, or 49 percent) reported no war-related service, civilian or military. Twenty-six percent reported some form of civilian war service, and 25 percent reported military service in the world war.[13] These figures suggest that the percentage of women physicians who declared themselves willing to do war service compared favorably to the service rates of their male colleagues.

At the same time that the AWH's registration of women physicians got under way in the summer of 1917, officials in Washington received numerous letters and other inquiries regarding women and the medical corps, and many women applied for acceptance to the army's Medical Reserve Corps (MRC). Some physicians made personal visits to government and military leaders that summer, and prominent nonmedical women and men urged army surgeon general William Crawford Gorgas and Secretary of War Baker to admit women physicians to the corps on equal terms. At the same time medical men were developing their own campaign for increased status within the military. Pressure from many sides mounted. What would the military do about women physicians?

One official avenue for women's military service existed. Several women—including Anita Newcombe McGee, who later became the head of the Army Nurse Corps—had served as contract surgeons. Yet most women physicians opposed such work. For the Philadelphia surgeon Caroline Purnell contract work "would mean our ability to be under the cook, the head nurse, or others, and be ordered around." Following a trip west the Chicago physician Martha Whelpton reported that very "few of the Coast women, and few of the Colorado women also, will go as Contract Surgeons. They object with all their might." Chicago medical women believed that their "professional dignity" was at stake and "absolutely opposed" contract service.[14]

For these and other medical women across the country contract service in the military represented acceptance of inferior status based on gender. Physicians in the Civil War had been employed as contract surgeons and had performed "part-time work, the individual doctor so employed maintaining his own private practice at home and at the same time giving some hours of each day to his Army hospital duties." Since the Civil War male physicians had achieved increased rank and status within the medical department and were struggling for more, and the position of contract surgeon was weak by comparison, lacking rank, professional prestige, or the authority to command deference.[15] The surgeon general still had the power to appoint as many contract surgeons as might be needed in "emergencies," and "at places which did not justify the expense involved by the detail of a medical officer." Only two men served with the American Expeditionary Forces as contract surgeons; the other 887 men employed during the war years as contract surgeons served on the home front in part-time, limited capacities. The Chicago neurologist Peter Bassoe, for example, contracted with the army to teach a course in neurosurgery.[16]

Yet in response to pressure regarding women's medical service in World War I, the government articulated a renewed purpose for contract service. In August 1917 Acting Judge Advocate General Blanton Winship handed down his interpretation of the military regulations regarding contract surgeons. "The statute does not prescribe that contract surgeons shall be males," he wrote, "and, in the absence of such a limitation, I am clearly of the opinion that it is allowable by law to appoint female physicians as contract surgeons in the United States Army."[17] Many medical women saw this as nothing more than an attempt to create a separate and unequal category of military service for women physicians. The regulations dealing with service in the medical corps contained the same phraseology as did those outlining the qualifications for contract work. Yet that same month the judge advocate general interpreted these medical corps regulations as pertaining solely to men.[18] It seemed that whereas women would be accepted only as vulnerable "day laborers," men would be given the status, rank, and pay of officers in the medical corps.

Caroline Purnell, for one, did not believe that contract practice was a step along the road to equality in military service. "As a woman and as a physician and as a surgeon, I think our days for crawling are over," she said.

> I cannot see why women should demonstrate their patriotism in any different way from men. If the men respect themselves and demonstrate their patriotism according to their training and experience, why should not women do the same thing? Why should we have to have a different way when our ability is just the same? We would be more self-respecting if we should stand upon this.... Our brains are not in our sex.[19]

Apparently, the majority of medical women in the country agreed. In the fall of 1917 Caroline Towles, a Baltimore physician, was attempting to register medical women's opinions on such service. She wrote to the Medical Women's Club of Chicago that most women surveyed answered negatively to the question, "As the only manner of serving, would you consider contract practice if this form of service can be made less objectionable?"[20]

Some women held notions of patriotism and professionalism that allowed them to see contract service with the military in a more positive light. Such service would make it possible for them to use their professional skills to serve their country, the wounded, and the sick without delay. Some believed that if contract service was the place where women could push at the boundaries of military service, then they would join and push. And a few women found contract service to be the only way that they could circumvent military restrictions and still serve as members of the hospital units that they had joined for overseas work. (These units were formed in anticipation of the war and converted to base hospitals.) The physician Esther Pohl Lovejoy, who later compiled their history, summed it up: "They were without commissions," she wrote, but "they were on the job."[21]

Emma Wheat Gilmore, who in 1918 became chair of the Committee of Women Physicians of the General Medical Board of the Council of National Defense, was asked by the army's surgeon general to recommend women for contract service. Over the course of the war fifty-five medical women engaged in contract service with the U.S. Army—all, apparently, with her approval. They came from all parts of the country—19 of the fifty-five from the East, 23 from the Midwest, 7 from the South, 5 from the West, and 1 from Puerto Rico—and from urban areas such as Brooklyn, Boston, Memphis, and San Francisco, as well as smaller towns and rural communities ranging from Kalamazoo, Michigan, to Gilmore City, Iowa. They had been educated at both large and small medical institutions. And although more than half had graduated from medical school after 1908, they represented a range of age groups. The average length of service for the group was seven months, with two women serving only one month, and one serving longer than eighteen months. It appears that most of these women served as pathologists and anaesthetists and did laboratory work.[22]

Eleven women served as contract surgeons overseas. Frances Edith Haines of Chicago worked as an anaesthetist with a medical unit at Limoges, France.[23] The Cincinnati physician Elizabeth Van Cortlandt Hocker entered the army as a contract surgeon in May 1918 and served until August 1919. She was particularly proud that after the Armistice she was placed in charge of two hospital wards of forty-two beds each that were for women personnel of the U.S. Army at Savenay, France.[24] Forty-four women contract surgeons (80 percent of the total) performed their military duties while remaining in U.S. territory, some serving in administrative posts and others at military hospitals, many caring for convalescent soldiers.[25] Dolores M. Pinero had received her medical degree in 1913 and was practicing in Rio Piedras, Puerto Rico, when the war began. She became a contract surgeon with the army in October 1918 and was immediately assigned to the base hospital at San Juan, the only woman serving in the army in Puerto Rico. She did anaesthesiology and laboratory duty and helped to open a four-hundred-bed hospital in a school in Ponce during the influenza epidemic that fall.[26]

While these medical women accepted the terms of contract service, other women physicians wished to provide immediate war service overseas without the onus of contract work and volunteered with the Red Cross and other civilian organizations in Europe. They made individual contracts with these organizations, and their work ranged from service in private hospitals to service in civilian communities and in lo-

cations near the front. These women provided medical and surgical care and worked in public health with civilians and refugees. By November 1918 there were at least seventy-six American medical women serving abroad with various organizations. Forty-nine percent of them were from the East, 21 percent from the West, 17 percent from the Midwest, and 13 percent from the South. When we compare the average year of graduation of these women to that of all the women physicians listed in the 1918 *Census of Women Physicians*, we find that this overseas group was younger: for all medical women the average year of graduation was 1899, and for these seventy-six overseas women the average year was 1904. Sixty percent of the overseas women had graduated between 1902 and 1914.[27]

Voluntary service with organizations other than the U.S. military was more accessible to medical women of color, and most African-American women physicians worked within their communities to provide wartime health care. It appears that two African-American medical women served overseas. One of them was Mary L. Brown, who was commissioned by the Red Cross for service in France. A graduate of Howard Medical School who had received advanced training in Edinburgh, Brown had been living in France when she received her assignment in the spring of 1918.[28] The other, Harriet Rice, of Newport, Rhode Island, an 1891 graduate of the Women's Medical College of New York, was living in France with her brother at the outbreak of the war. Rice distinguished herself by her work in a French military hospital, where she served for most of the war. In August 1919 she received the Reconnaissance Francaise for her meritorious medical service with the French wounded.[29]

The physician Jessie W. Fisher wrote to her friend and colleague Kate Campbell Mead from Paris in May 1918 while with a group of medical women with the Red Cross who were "loafing around Paris waiting for our papers to different stations." Her detailed letter reveals a great deal about the way in which this type of war service raised questions about the broader issues of women's status and position in the war. Fisher wrote to her colleagues that there were "lots of good places to be filled in the civil work where you could do a lot of good and would be invaluable" and that the Red Cross would be willing to pay a good salary. Yet, she warned, "we are in a very anomolous [sic] position." Women physicians in Fisher's situation were not saluted, they had no standard uniform, and the authority accorded to them and the treatment they received depended on circumstance and on the inclinations of the men with whom they worked. "It is most embarrassing," Fisher wrote. She found the older men in the Red Cross and the army to be "as courteous as at home," but younger men were a problem. "I struck [slapped?]" a young medical officer "the other day who made me feel my lack of rank," she reported. "I think we are usually taken for army chauffeurs and of course the soldiers do not have the proper respect for us."[30]

Fisher's comments underscore the vulnerability of women involved in wartime service without official military status. In her relations with men both socially and professionally, she had much the same difficulties as did army nurses serving without definite rank. And her proposed solution was similar to theirs: she believed that medical women had to gain equal status as officers, both for professional integrity and also, especially, to enable them to control their social and professional relations with men. "My advice to you, and to all women M.D.s," she wrote, "is to *stay* in the U.S. and fight for commissions for the women." The demand for medical women was strong, she believed, and "if the women will refuse to come without commissions they will be compelled to give them to us." Her firsthand experiences with these issues gave authority to

her recommendations. And she was emphatic in her conclusion: "Now *stay at home* and fight tooth and toe nail for those commissions."[31] The message was clear. Unless they had officer status and rank, women physicians would be extremely vulnerable in both their working relationships and their social relationships with men in the military.

At least two groups of women physicians, and many others as individuals, sought to gain military commissions by direct application, taking the position that they were citizens who were eligible for service as stated in military regulations. Their strategies were similar to those used by Susan B. Anthony, Virginia Minor, and other late nineteenth-century suffrage activists who had presented themselves at the polls as citizens eligible to vote under the protection of the Fourteenth Amendment, which defined as citizens "all persons born or naturalized in the United States."[32] As Susan Anthony declared to audiences prior to her 1872 trial for attempting to vote, "In voting, I committed no crime, but simply exercised my 'citizen's right,' guaranteed to me and all United States citizens by the National Constitution, beyond the power of any state to deny."[33] Now these medical women, many of whom lived in states that by 1917 had granted women the right to vote, continued the cause of defining women's citizenship rights and obligations: it was as citizens that they challenged their exclusion from military medical service to the state.

In September 1917 the newly formed Colorado Medical Women's War Service League created a Committee on Recognition of Medical Women. Mary Elizabeth Bates, a prominent Denver gynecologist, officer of the Medical Women's National Association, suffrage activist, community organizer, and secretary of the league, was asked to chair the new committee and to "take up the question of the appointment of women physicians in the Medical Reserve

Corps of the U.S. Army."[34] The committee had two tasks: first, to identify and study the regulations governing the service of physicians in the Medical Department of the army; and second, to recommend and implement the actions necessary for women physicians to gain equal access to the army's Medical Reserve Corps.

Bates and her committee members began an investigation of the status of the regulations concerning service with the Medical Reserve Corps. Section 37 of the National Defense Act of June 3, 1916, restricted appointment in the Officers' Reserve Corps to "such citizens as, upon examination prescribed by the President, shall be found physically, mentally and morally qualified to hold such commissions."[35] No restriction based on gender was in the language of this act. The *Manual for the Medical Department* required an applicant for the medical corps to be

> between 22 and 30 years of age . . . a citizen of the United States, [who] must have a satisfactory general education, must be a graduate of a reputable medical school legally authorized to confer the degree of doctor of medicine, and must have had at least one year's hospital training, including practical experience in the practice of medicine, surgery and obstetrics.[36]

The requirements for service in the Medical Reserve Corps were similar: commissions were to be given "to such graduates of reputable schools of medicine, citizens of the United States [who shall be found] physically, mentally, and morally qualified to hold such commissions." Such officers were also required to be between twenty-two and forty-five years of age and to be qualified to practice medicine in their state or territory of residence.[37]

Sex was not an explicit category for acceptance in any of these regulations, and there were hundreds of women who could meet the profes-

sional and physical qualifications necessary for service. On the basis of their examination of military regulations, Bates and her committee concluded that "the word 'citizens' must include women, since women are citizens," and that it would not be "necessary to seek the enactment of a law to permit the appointment of women."[38] Unlike nurses, who needed to get new legislation passed to allow them as a professional group to gain officer status, medical women had to deal with the interpretation of laws that, in explicit language at least, were not gender specific and already granted officer status to their professional group.

Yet the interpretation of a law could be just as powerful as the absence or presence of a law. On August 30, 1917, Acting Judge Advocate General S. T. Ansell had written an official interpretation of military regulations regarding the service of women physicians in the Medical Reserve Corps. After quoting the National Defense Act of 1916, which stipulated the service of "citizens," Ansell referred to an opinion of the Massachusetts Supreme Court in a case dealing with the appointment of women as notaries. This decision in turn employed as precedent another Massachusetts court decision that raised the question of whether or not women could serve as justices of the peace:

> There is nothing in the Constitution which in terms prohibits women from being appointed to judicial offices, any more than from being appointed to military offices, or to executive civil offices, the tenure and mode of appointment of which are provided for in the Constitution. It was the nature of the office of justice of the peace, and the usage that always had prevailed in making appointments to that office, that led the justices to advise that it could not have been the intention of the Constitution that women should be

appointed justices of the peace. . . . In our opinion the same considerations apply to the office of notary public.[39]

Claiming these rulings as precedent, Ansell concluded, "For similar reasons it is the view of this office that it is not allowable by law to appoint female physicians to military office in the Medical section of the Officers' Reserve Corps of the Army."[40] In the body of his decision the acting judge advocate general stated four main reasons why, in his opinion, women physicians should not be commissioned as officers in the Medical Reserve Corps. Women physicians could not serve in the Medical Reserve Corps because they had not done so in the past; because soldiers were specified as "men" in other regulations not affecting the Medical Department; because, as women, they would not be physically capable; and because they were not to have the status of officers—a status that would allow them to command men.

To test the medical corps regulations that called for the appointment of "citizens," Mary Bates and her committee developed a plan that, Bates wrote, would "achieve the result desired with the minimum amount of trouble for the War Department."[41] They would select between six and twelve women physicians who would apply as a group for service with the Medical Reserve Corps. The medical women in this test case would come from states that granted women the right of suffrage, presumably because women in these states had already crossed an important conceptual boundary of citizenship. They would be professionally prepared and have all of the qualifications necessary to "make good" if appointed to the Medical Reserve Corps. And along with the standard documentation attesting to their education and accomplishments, these medical women would secure recommendations from "prominent and influential persons." Senators and representatives

from home states would be enlisted to present the applications together with Senator John Franklin Shafroth and Representative Edward Keating, both of Colorado, who already "enthusiastically endorsed the plan."[42]

In Colorado the Committee on Recognition of Medical Women selected eight women physicians to present themselves to the surgeon general of the army and the secretary of war for service in the Medical Reserve Corps.[43] By February 1918 the applications were in order, and the test case was put in motion. A group of eleven members of Congress, led by Shafroth, went to the office of Surgeon General Gorgas.[44] Anita Newcombe McGee, who had served as a contract surgeon in the Spanish-American War and had been appointed head of the Army Nurse Corps at its creation in 1901, accompanied the delegation. Because she was the woman physician most closely associated with actual experience in military medicine, McGee's presence was both symbolic and practical. In her account of the proceedings McGee reported that the application of Mary Bates was included with those of the other eight women, and all were presented to the surgeon general for his approval.[45]

By all accounts Gorgas was sympathetic with the movement. McGee reported that he told her he personally favored commissions for women physicians. But he was bound, he said, by the decision of the judge advocate general opposing the admission of women to the Medical Reserve Corps. He suggested that the next step for the group was to take the matter up with Secretary of War Baker. McGee believed that if Baker could be persuaded, then "the Judge Advocate General's Office would doubtless reverse its decision, and the Surgeon-General be given a free hand."[46]

On February 4 this same group met with the secretary of war, whose views on women in military service had already been publicly expressed. In a letter to the House and Senate military committees, which were considering a bill to commission women in the Signal Corps, Baker had written that he did "not approve of commissioning or enlisting women in the military service."[47] The group presented the women's applications to Baker, read the decision of the judge advocate general, and explained their position on the matter. Baker replied that "his main thought was to win the war, and . . . he did not think that commissioning women physicians would contribute to that end, nor did he want to make any unnecessary innovations now." Shafroth and McGee both argued that there was a need for the services of women physicians, McGee stating that the surgeon general wanted them especially for work as anaesthetists and pathologists. Baker terminated their conversation, saying that he believed that women physicians were not needed by the military but that he would consider the matter further.[48] The test case was at a standstill, with the applications shelved in the office of the secretary of war.

Other women made their claims for eligibility on the basis of citizenship locally. Four women physicians from Portland drove to the Vancouver, Washington, training facility for medical officers in the spring of 1918 to present themselves as eligible physicians desiring commissions as medical officers. Katherine Manion, Mae Cardwell, Mary MacLachlan, and Emily Balcom came representing many other women physicians of the area. "There is no word in the war department regulations that bars women," one of them told a reporter for the *Portland Journal* after their attempt, "and away we went to the [medical] officers' training camp." They presented themselves to an astonished major in charge of the camp, "ready and armed to take the examinations, don the uniforms and salute the privates." The four women stated their intentions, arguing that in their community women had full suffrage, that they were citizens, and that they were ready to meet the profes-

sional and other requirements necessary for acceptance into the Medical Reserve Corps. They brought with them the necessary documents. After checking these, the major told the group that he could not examine them because "it hasn't been done." After more discussion he asked them if they wouldn't like to go overseas as nurses. They replied firmly that they would not and asked the male physician-officer, "'Would *you?*'" The major finally said that he would telegraph Surgeon General Gorgas for an answer, and the negative reply came from Washington the next day.[49]

Medical women also mounted a campaign to bring the question of women's entrance into the medical corps to the formal attention of the American Medical Association, after which they hoped that the AMA would act to support commissions for women physicians. They were successful in achieving a voice—although it was a voice filtered through male allies—at the AMA's national convention in Chicago in June 1918. Here resolutions supporting commissions for women physicians were introduced and made their way to committees. Three resolutions on behalf of women physicians were introduced: one by H. G. Wetherill, a gynecologist and a Colorado colleague of Mary Bates; another, which specifically reflected the views of the Medical Association of the State of California, by George Kress, a California physician active in tuberculosis work and a professor of hygiene at the College of Medicine of the University of California; and a third by E. O. Smith, professor of urology at the University of Cincinnati.[50]

The language contained in these three AMA resolutions reveals a great deal about the way in which medical women and their allies constructed a case for women's service. In a broad sense, medical women employed the same two arguments that Aileen Kraditor has identified as the basic rationales in the call for woman suffrage: expediency and justice.[51] According to the AMA resolutions, it would be expedient for the government to commission women physicians: they were "fitted and equipped" to provide "valuable service"; they were graduates of medical schools and qualified to practice medicine; and "most if not all of them [had] signified their readiness" for service. In light of the demands of war all available skill should be utilized, and commissions would "further the utilization of women physicians in service."[52] The resolutions' call for women's entrance into the MRC was also based on the idea of justice. Women physicians in all fields—including surgery, one resolution emphasized—"render service as efficient and valuable as can be rendered by men." Medical women already serving overseas with such groups as the Red Cross had "performed invaluable services" and demonstrated their loyalty, employing, in the words of another resolution, "their skill and energy in our common cause." For these reasons the resolutions called upon the secretary of war to bring women physicians into the MRC "in full standing," with the same rank and pay as male medical officers.[53]

Medical women also addressed those objections to service which were based on women's supposed physical weakness or unsuitability for military service. American women physicians serving "in the war zone" with the Red Cross and other organizations, and women physicians from other nations, one resolution stated, "have demonstrated that it is possible for women to endure the hardships of life in the war zone and still do creditable work."[54] Here, as well as in the other aspects of their argument, they made the implicit claim that they were eligible to serve because they were citizens who were equal in their abilities to serve the state, even while facing the dangers of the war zone.

The resolutions were forwarded to the AMA's Reference Committee on Legislation and Political Action. This committee returned an opinion that appeared to be both supportive and cautious and drew the line for women's service at the boundary of "soldiering." The "very character of military service and women's natural limitation for such service must require wise discrimination in their employment in war work," the committee members wrote. Women physicians, in other words, were not to be frontline soldiers, owing to the "natural limitations" of their sex. However, they continued, the "principle of equal rank and pay for equal service is inherently just without regard to sex, and the committee feels this should be unhesitatingly approved by the House [of Delegates of the AMA]."[55] According to this AMA committee, women physicians could serve in the military without actually being soldiers, and therefore the committee avoided the issue of women's military service as combat service, a threshold most Americans were unwilling to cross. Yet by denying women the status of soldiers, the AMA also perpetuated a less-than-equal place for women in military medicine.

In her study of the relationship between male and female workers in the automobile and electrical industries during World War II, Ruth Milkman demonstrates that male union members supported female workers' claims for equal pay during wartime. In her analysis such support for women's "equality" by male workers was directly linked to the men's own self-interest, in that it maintained standards and pay levels for jobs that they believed would return to the hands of men following the war.[56] The AMA committee's decision to support the principle of "equal rank and pay for equal service" follows the same pattern. As we have seen, male physicians were fighting their own battles for increased status and authority in the military. They were interested in strengthening and maintaining the military status of their profession both during and after the war and could stand on such principles as "equal pay for equal work" without conceding much overall ground to women. This was especially true when the disclaimer regarding the "natural limitations" of women's military service was included in their resolution. "The men are not just ready to give us the ground yet," Caroline Purnell told her colleagues. "They got in on the ground floor first; they are going to get all out of it that they can for the sacrifices they have made."[57]

The resolutions were recommended to the AMA for action, but before that organization could respond, the war was over, taking with it many of the expediencies that had made the arguments for women's service possible. Like the applications of the women brought to Secretary of War Baker by the Colorado coalition, the resolutions, petitions, and other calls for legislative action were not formally acted upon before the end of the conflict, and yet they helped to set new definitions for what was possible. Officially buried, they nevertheless lived on in the memories of the women and men who had made their claims, and the influence of these early efforts was felt when another war broke out in Europe twenty years later.[58]

Full citizenship, service to the state, and professional opportunity during wartime involved complex issues for medical women. Some believed that as medical professionals they could ameliorate the effects of war by saving lives, thus placing themselves outside of the boundaries of responsibility for war. Perhaps this philosophy allowed pacifist medical women such as Alice Hamilton to register for war service. The physician Alice Wakefield believed that she "could not bear the war" if she were not working "purposefully every hour" in wartime medical service. Some believed that the war had a larger political purpose. "If we had not believed in the ideal of the war for the establishment of international peace," Rosalie Slaughter Morton recalled, "we could not have gone forward, planning, organizing, so strenuously."[59] And many medical

women believed, from sad experience, that it would take a catastrophe such as war to change the way they were viewed by their male colleagues and by others in American society. They did not welcome the war, but when it came they claimed the right to be civic and professional equals with men. Frances Van Gasken, of the Woman's Medical College of Pennsylvania, used the metaphor of Pandora's box to explain such views. Although the war let loose a host of disasters when it was opened, she believed that it also brought hope in the promise of equality and opportunity for medical women.[60]

From what we can see from the activities and reports of medical women themselves, and from their characteristics as revealed in the *Census of Women Physicians,* the campaign to permit women physicians to do war service represented a claim by many women physicians for full civic and professional participation. Almost one-third of medical women registered for war service. Along with the four Portland women, the members of the Colorado Medical Women's War Service League, the hundreds who sent in applications and inquiries, and the thousands who signed petitions to Washington officials, they seem to have agreed with Mary Bates that the real question—or perhaps the all-encompassing question—was "the rights of citizenship as applied to women."[61]

Notes

An earlier version of this paper was presented at the American Association for the History of Medicine Annual Meeting in Cleveland, Ohio, on May 3, 1991. The author also received helpful commentary on these issues at presentations to the History of Medicine Association at the University of Iowa and the Medicine and Society Group at the University of Iowa Hospitals and Clinics. The author wishes to thank Linda and Richard Kerber, Judith Leavitt, Ellen S.

More, Steven J. Peitzman, Janet Miller, and the staff of the Archives and Special Collection on Women and Medicine at the Medical College of Pennsylvania, and the editors and the two anonymous reviewers for the *Bulletin of the History of Medicine* for their assistance with this project.

1. The regulations for both the Medical Corps and the Medical Reserve Corps included the term *citizens* without explicit reference to gender. When most women talked about military service, they referred to service with the army's Medical Reserve Corps—in other words, service for the duration of the war and not career service with the army's regular Medical Corps. This was the type of service chosen by most of the male physicians who served in the war. Sometimes women used the terms interchangeably or referred to their right to service in both organizations. For the specific requirements see "Article I—The Medical Department, Its Organization, and Personnel," in "Manual for the Medical Department," reprinted in Charles H. Lynch, Frank Watkins Weed, and Loy McAfee, *The Surgeon General's Office*, vol. 1 of *The Medical Department of the United States Army in the World War* (Washington, D.C.: U.S. Government Printing Office, 1923), 762–67.

2. For background on these issues in medicine see Regina Morantz-Sanchez, *Sympathy and Science: Women Physicians in American Medicine* (New York: Oxford University Press, 1985); and Mary Roth Walsh, *"Doctors Wanted, No Women Need Apply": Sexual Barriers in the Medical Profession, 1835–1975* (New Haven: Yale University Press, 1977). For the context of the campaign for women's rights, see Nancy Cott, *The Grounding of Modern Feminism* (New Haven: Yale University Press, 1987).

3. Mary Sutton Macy, "The Field for Women of Today in Medicine," *Woman's Medical Journal*, (hereafter *WMJ*) 27 (1917): 49–58. These figures do not include those women physicians who were de facto specialists, for these were not traceable in official records. Undoubtedly, many more medical women specialized during these years.

4. The U.S. Census for 1910 reports that 84 percent of the women physicians in the country were white and born in the United States; 12 percent were white and born outside of the United States; 4 percent were African-American; and less than 1 percent were Native American, Asian, or of other racial/ethnic ori-

gin. U.S. Bureau of the Census, *Thirteenth Census of the United States Taken in the Year 1910*, 4, *Population 1910, Occupation Statistics* (Washington, D.C.: U.S. Government Printing Office, 1914), 428; see also Morantz-Sanchez, *Sympathy and Science*, 234n.2.

5. See "The Evolution of the Medical Department," in Lynch et al., *Surgeon General's Office;* Franklin H. Martin, *Fifty Years of Medicine and Surgery; An Autobiographical Sketch* (Chicago: Surgical Publishing, 1934), esp. 379–82. There was vigorous editorial comment in the medical journals. See, e.g., "Giving the Medical Officer the Rank to Which He Is Entitled," *Journal of the American Medical Association* (hereafter *JAMA*) 69 (1917): 292–94; and "Rank and Authority of Medical Officers," *Journal of the Iowa State Medical Society* 7 (1917): 352.

6. See the typescript report "American Women's Hospitals, Organized by War Service Committee of the Medical Women's National Association," box 1, folder 1, American Women's Hospitals Records, 1917–82, Accession 144, Archives and Special Collections on Women in Medicine, Medical College of Pennsylvania, Philadelphia, Pa. (hereafter AWH Records); "Origin of the American Women's Hospitals," box 1, folder 1, AWH Records; Bertha Van Hoosen, *Petticoat Surgeon* (Chicago: Pellegrini & Cudahy, 1947), 202; and Rosalie Slaughter Morton, *A Woman Surgeon: The Life and Work of Rosalie Slaughter Morton* (New York: Frederick A. Stokes, 1937), 270–94. For additional information on the origins of the MWNA see Ellen S. More, "The American Medical Women's Association and the Role of the Woman Physician, 1915–1990," *Journal of the American Medical Women's Association* 45 (September–October, 1990): 165–80. More also discusses Morton's leadership style and the activities of women physicians in the war in "'A Certain Restless Ambition': Women Physicians in World War I," *American Quarterly* 41 (1989): 636–60. See also Jayne C. DeFiore, "Rosalie Slaughter Morton: Founder of American Women's Hospitals?" *Collections: The Newsletter of the Archives and Special Collections on Women in Medicine, The Medical College of Pennsylvania* 16 (1987): 1, 2.

7. Information about registration may be found in various reports, especially in box 30, folders 292, 293, AWH Records.

8. See the controversy with the Women's Medi-

cal Society of Maryland, as reported in the minutes throughout box 30, folder 292, AWH Records; Morton, *Woman Surgeon,* 274–80; and *WMJ* 27 (1917): 186. See also Kimberly Jensen, "'The Battalion of Life': Medical Women, The American Women's Hospitals, and War Service in the First World War," *Collections,* 23 (1991): 1–4.

9. See Morton, *Woman Surgeon,* 283; and Emma Wheat Gilmore, "Report of Committee, Women Physicians, General Medical Board, Council National Defense," *WMJ* 29 (1919): 146–47.

10. The information that follows is taken from my analysis of the data base I created from *Census of Women Physicians* (New York: American Women's Hospitals, 1918).

11. From 1915 until 1918 the MWNA did not require members to hold AMA membership. But beginning in 1918 the leaders of the MWNA required AMA membership of all members to avoid the appearance of separatism. The *Census* therefore contains information on women who joined the MWNA from 1915 to 1918 without AMA membership.

12. Lynch et al., *Surgeon General's Office,* 138. The sixth edition of the *American Medical Directory* (Chicago: American Medical Association, 1918) contains the names of 159,144 physicians. I subtracted 5,827, the number of women physicians in the 1918 *Census of Women Physicians,* from this figure for a total of 153,317 male physicians in 1918.

13. *History of Medicine and Surgery and Physicians and Surgeons of Chicago* (Chicago: Biographical Publishing, 1922).

14. Caroline M. Purnell, "The Work of the American Women's Hospitals in Foreign Service," in *Transactions of the Forty-third Annual Meeting of the Alumnae Association of the Woman's Medical College of Pennsylvania* (Philadelphia: Alumnae Association of the Woman's Medical College of Pennsylvania, 1918), 97; Martha Whelpton to Rosalie Slaughter Morton, New York City, November 26, 1917, box 2, folder 14, AWH Records; and see *Bulletin of the Medical Women's Club of Chicago* (hereafter *Bull, MWCC*) 6 (1917): 5.

15. Lynch et al., *Surgeon General's Office,* 42.

16. Ibid., 151; Joseph H. Ford, *Administration, American Expeditionary Forces,* 1 of *The Medical Department of the United States Army in the World War* (Washington, D.C.: U.S. Government Printing Office,

1927), 102; see the entry for "Peter Bassoe" in *History of Medicine and Surgery and Physicians and Surgeons of Chicago,* 395.

17. U.S. Judge Advocate General's Department (Army), *Opinions of the Judge Advocate General of the Army,* vol. 1, *April 1, 1917, to December 31, 1917* (Washington, D.C.: U.S. Government Printing Office, 1919), 126; Blanton Winship, acting judge advocate general, to the surgeon general of the United States, August 13, 1917, quoted in Anita Newcombe McGee, "Can Women Physicians Serve in the Army?" *WMJ* 28 (1918): 26–28.

18. See my discussion of these interpretations in the text that follows.

19. Purnell, "Work of the American Women's Hospitals in Foreign Service," 97.

20. "Women in the Medical Reserve Corps," *Bull. MWCC* 6 (1917): 5.

21. Esther Pohl Lovejoy, *Women Doctors of the World* (New York: Macmillan, 1957), 303.

22. Lovejoy discusses Gilmore's role and other aspects of contract service in *Women Doctors of the World,* 302–4. I have gathered information for this profile on women contract surgeons from a five-page typewritten list dated November 13, 1919, entitled "Women Contract Surgeons, U.S. Army, Who Served During the War with Germany," box 17f, folder 142, AWH Records. At the end of the list is the typewritten notation "CHAIRMAN OF COMMITTEE OF WOMEN PHYSICIANS, Emma Wheat Gilmore, M.D." I have also used information from the *Census of Women Physicians* and individual sources listed in n. 23, n. 24, and n. 26 to augment the information in the 1919 list for this profile.

23. See "Haines, Frances Edith, Memoirs of War Service," Accession 103, Archives and Special Collections on Women in Medicine, Medical College of Pennsylvania.

24. Elizabeth Van Cortlandt Hocker, "The Personal Experience of a Contract Surgeon in the United States Army," *Medical Woman's Journal* 49 (1942): 9–11. In 1920, the *Woman's Medical Journal* (WMJ) changed its name to *Medical Woman's Journal* (hereafter abbreviated *MWJ*). See also "Women Contract Surgeons," 2.

25. Ford, *Administration, American Expeditionary Forces,* 102.

26. Information on Dolores Pinero may be

found in "Contract Surgeon: Dolores Mercedes Pinero, M.D.," *MWJ* 49 (1942): 310, 324; Lovejoy, *Women Doctors of the World*, 275, 303; "Women Contract Surgeons," 4.

27. The overseas women are listed in *WMJ* 28 (1918): 247. The figures come from the information on these women in my *Census of Women Physicians* data base.

28. Information on Mary Brown may be found in *Crisis* 16 (May 1919): 26.

29. Sara W. Brown reports on the achievements of Harriet Rice in "Colored Women Physicians," *Southern Workman* 52 (1923): 583.

30. The typescript copy of this letter from Jessie W. Fisher in Paris, dated May 12, 1918, is addressed to "Dear Lady," but on the top is written "Kate Campbell Mead, M.D." Evidently, copies were passed around the community of women physicians to spread Fisher's message. The typescript is in box 30, folder 295, AWH Records. Reports of sexual tension with younger officers are a prominent feature of her report.

31. Fisher to "Dear Lady," box 30, folder 295, AWH Records.

32. A number of cases of such direct action in relation to voting under a Fourteenth Amendment claim may be found in Elizabeth Cady Stanton, Susan B. Anthony, and Matilda Joslyn Gage, eds., *History of Woman Suffrage* 6 vols. (1881–1922; reprint, New York: Arno Press, 1969), 2:586–755.

33. Ibid., 2:630–31.

34. See Mary Elizabeth Bates, "A Most Interesting Report of Work of Colorado Medical Women's War Service League," *WMJ* 28 (1918): 39–40; quotation on 39. Information on the life and career of Mary Bates may be found in the following sources: "Mary Elizabeth Bates, M.D.," in *History of Colorado*, ed. Wilbur Fiske Stone, 2 (Chicago: S. J. Clarke, 1918), pp. 452–57; Doris Minney, "Mary Elizabeth Bates, M.D.," *MWJ* 55 (1948): 30–31; Lovejoy, *Women Doctors of the World*, 93; and Morantz-Sanchez, *Sympathy and Science*, 275, 301, 294–95.

35. Quoted in McGee, "Can Women Physicians Serve in the Army?" 26.

36. "Manual for the Medical Department," 763.

37. Ibid., 766. These regulations also illustrate the extent to which the professionalization of medicine had been encoded in the army.

38. Bates, "A Most Interesting Report of Work," 39.

39. The quotation appears in U.S. Judge Advocate General's Department, *Opinions of the Judge Advocate General*, 126–27.

40. U.S. Judge Advocate General's Department, *Opinions of the Judge Advocate General*, 127.

41. Bates, "A Most Interesting Report of Work," 39.

42. Ibid., 39.

43. They were Myra L. Everly, of Seattle, an 1893 graduate of Northwestern Medical College; Julia P. Larson, of San Francisco, an 1898 graduate of the Medical Department of the University of California at Berkeley; Mary McKay, of Macon, Georgia, an 1897 graduate of the Women's Medical College of Baltimore; Mabel A. Martin, of Livingston, New York, a 1912 graduate of Cornell Medical College; Regina M. Downey, of Beaver Falls, Pennsylvania, a 1914 graduate of the Woman's Medical College of Pennsylvania; Marion H. Rea-Lucks, of Philadelphia; M. Jean Gale, of Denver, an 1889 graduate of the Woman's Medical College of Pennsylvania; and Helen Craig, of Denver, a 1913 graduate of Rush Medical College of Chicago. All were from suffrage states except for McKay, of Georgia, and Downey and Rea-Lucks, of Pennsylvania, who, it appears, were chosen to round out the group's regional and institutional representation. I augmented the information on these women given by Bates in "A Most Interesting Report of Work" with information from the *Census of Women Physicians*. In the *Census* Rea-Lucks's name is given as Marion Reil Lucke.

44. The group included Senator McNary of Oregon; Senator Calder of New York; Senator Jones of Washington; Senator Phelan of California; Senator Knox of Pennsylvania; Senator Smith of Georgia; Representative Nolan of California; and Representatives Keating, Taylor, and Timberlake of Colorado, led by Senator Shafroth of Colorado. See Bates, "A Most Interesting Report of Work," 39.

45. McGee, "Can Women Physicians Serve in the Army?" 28.

46. "A Most Interesting Report of Work," 39, McGee, "Can Women Physicians Serve in the Army?" 28.

47. This is quoted in McGee, "Can Women Physicians Serve in the Army?" 28.

48. Ibid., 28.

49. This event is reported in the *Portland Journal* for May 5, 1918, and is reprinted in Bertha van Hoosen, "The American Press and Medical Women in War Work," *WMJ* 28 (1918): 155–56.

50. "Proceedings of the Chicago Session: Minutes of the Sixty-Ninth Annual Session of the American Medical Association, Held at Chicago, June 10–14, 1918," *JAMA* 70 (1918): 1855, 1858, contains an account of the resolutions and the action taken. Also see "Resolutions Pertaining to Women Physicians as Recommended to the A.M.A. by Legislative Committee," *WMJ*, 28 (1918): 138–39. For Wetherill see *Who Was Who in America* (Chicago: Marquis–Who's Who, 1968), 1: 999. For Kress see *Who Was Who in America*, 3: 490. For Smith see *Who Was Who in America* 1: 1139.

51. Aileen S. Kraditor, *The Ideas of the Woman Suffrage Movement, 1890–1920* (New York: Columbia University Press, 1965).

52. The resolutions may be found in "Proceedings of the Chicago Session," 1858.

53. Ibid.

54. Ibid. This was part of the Smith resolution.

55. "Report of Reference Committee on Legislation and Political Action," in "Proceedings of the Chicago Session," 1858.

56. Ruth Milkman, *Gender at Work: The Dynamics of Job Segregation by Sex During World War II* (Urbana: University of Illinois Press, 1987).

57. Purnell, "The Work of the American Women's Hospitals," 98.

58. Their actions provided precedent as another generation of medical women prepared for an assault on barriers to military service in World War II. The *Woman's Medical Journal* for 1941 and 1942 contains numerous reminiscences and biographies of women who served as contract surgeons and with medical units in France during World War I: tales of the past with hope for the present and future. And Emily Dunning Barringer, a member of the Executive Committee of the American Women's Hospitals during World War I, became the leader of the successful fight for commissions for women in the medical corps during the next "war to end all wars." See Emily Dunning Barringer, *Bowery to Bellevue: The Story of New York's First Woman Ambulance Surgeon* (New York: W. W. Norton, 1950).

59. For Hamilton's registration see *Census of Women Physicians,* 27. For the other views see Morton, *Woman Surgeon,* 276.

60. Van Gasken was speaking to students and faculty members in September 1917 at the opening of the first college session since the U.S. entry into the war. See Frances Van Gasken, "Introductory Address, Woman's Medical College of Pennsylvania," *Bulletin of the Woman's Medical College of Pennsylvania* 68 (December 1917): 3–5; quotation on 3–4.

61. Bates "A Most Interesting Report of Work," 40.

Women, Health Reform, and Public Health

Women have always been part of the informal practice of health care. Family caregiving formed a key aspect of how Americans took care of themselves. But as Emily K. Abel relates, family-centered care changed greatly over time. Using the example of one Kansas woman, who delivered extensive health care to her family at the turn of the twentieth century, Abel powerfully demonstrates the key roles women played in taking care of their families and how they changed over time with changes in medicine and medical institutions.

In a more organized fashion health reform flourished in the United States after the midnineteenth century, in part as a reaction against ineffective medical practice. Numerous Americans found existing heroic therapy—characterized typically by massive bleedings, and dosings with emetics and cathartics—painful and dangerous, and they sought to avoid it by teaching the public physiology and hygiene to help people stay healthy. Many middle-class women emphasized their nurturing abilities and became health reformers, expanding their domestic duties beyond their homes and traveling around the country giving lectures on hygiene and disease prevention. In the process they helped transform society's view of the role and abilities of women in general. Ronald L. Numbers and Rennie Schoepflin explore the parallel careers of two important religious leaders, Ellen G. White and Mary Baker Eddy, who responded to health reform in distinctive ways.

Health reform was a subject that in some ways transcended socioeconomic divisions in society. While many health reformers were middle- or upper-class men and women, the movement consciously made connections with the masses of Americans for whom sickness was a particularly grim burden. Many sectarian reactions to regular heroic medicine not only fostered participation as practitioners but authorized a degree of do-it-yourself medicine, which encouraged people to take control of their health. Such democratic appeals, especially during and after the Jacksonian period, in part accounted for the popularity and success of these reform groups. As Nancy Tomes illustrates, at the turn of the twentieth century, women reformers used the new germ theory of disease to help improve everyday home life. Women who combined teachings from the new home economics movement and the latest public health theories understood that American housewives could improve their family's health even while dusting, cooking, and redecorating their parlors, as well as when caring for their sick children.

Health reformers hit a particularly responsive chord among women. Whether consciously or not, women used their participation in the movement to move a little closer to full citizenship. Even the women who remained within their traditional role found gratifying the positive efforts to gain control of their family's health through understanding physiology and practicing better domestic hygiene. In a society in which men still had to be prodded to "remember the ladies," in Abigail Adams's words, health reform promised full participation and acceptance of women.

But public health and medical knowledge could be double-edged swords. While the new theories provided some weapons that women and families could use to their advantage, twentieth-century medicine and public health also—in a world in which many feared that families of European stock were diminishing and that women's roles should be bounded by the kitchen and the nursery—could add rationale for discriminatory actions. Judith Walzer Leavitt explores how gendered expectations for women's behavior influenced public health practices pertaining to healthy typhoid fever carriers. Lest we get complacent as the new century dawns, these articles together remind us how much medicine and public health remain embedded within political terrain and social ideology.

30 A "Terrible and Exhausting" Struggle: Family Caregiving During the Transformation of Medicine

Emily K. Abel

Despite the wealth of recent studies charting the metamorphosis of the formal health care system between 1890 and 1930, we know little about the changes wrought in informal care. Assuming that professional medicine absorbed family healing and nursing, most historical accounts follow care out of the home into physicians' offices and hospitals. James H. Cassedy justifies this approach by arguing that family care was shorn of medical significance: "The care and support provided by the family circle itself remained crucially important to the general well-being and morale of the seriously ill or dying person in America, but some of the family's traditional roles gradually diminished during this period. A major factor in bringing this about was the shifting of the locale of much medical care ... from the home to the hospital."[1] Ruth Schwartz Cowan expounds the substitution thesis even more unambiguously. After tracing the growth of the nursing profession, she asserts that "every hour of care" that nurses "offered to

EMILY K. ABEL is Professor of Health Services and Women's Studies at the University of California, Los Angeles.

Reprinted from *Journal of History of Medicine and Allied Sciences* 50 (1995): 478–506, by permission of Oxford University Press.

patients was an hour that would earlier, and under other circumstances, have been offered by a housewife."[2]

But family caregiving is not a timeless and static endeavor, changing only through the gradual loss of its medical component. Nor does an hour of professional care translate directly into an hour of relief for family members. Rather than assuming that families withdrew their services during the late nineteenth and early twentieth centuries, we should ask how the revolution in medical knowledge and practice transformed the content and meaning of the care that continued to be delivered at home.

Moreover, the rise of the health care industry was not the only factor affecting family care. This period also witnessed profound changes in transportation and communication. Although a few historians explore the impact of railroads, automobiles, and telephones on physician practice,[3] none investigates the way these developments affected the work of family caregivers. Changes in domestic technology are completely missing from the accounts of medical historians. Between 1890 and 1930 large corporations began to mass produce goods and services for private households.[4] Because electricity, gas, indoor plumbing, and store-bought foods helped to transform family care, they too demand our attention.

This essay discusses the ways a wide variety of changes molded women's caregiving responsibilities by examining the diary of Martha Farnsworth, a Kansas woman who delivered extensive care to two husbands, an infant daughter, and a niece. The growing interest in women's history since the mid-1970s has drawn unprecedented attention to women's personal writings; historians have uncovered a wide array of journals, diaries, and letters.[5] Most, however, are too fragmentary to enable us to place the authors' experiences in any meaningful context. Martha Farnsworth began keeping a diary in 1882, at the age of fourteen; by the time of her death in 1924 she had written over four thousand pages, an unusual length.[6] The early entries were short and matter-of-fact; as she grew older, her accounts became more detailed and introspective. Her diary has another advantage as well. Female diary and letter writers during the nineteenth and early twentieth centuries tended to be overwhelmingly eastern, affluent, and leisured. Martha, however, grew up in frontier Kansas, held a variety of paid jobs, and engaged in a constant struggle to make ends meet. Although we must be cautious in generalizing from her account, it provides insight into the experiences of a group of women too frequently ignored.

Martha was born on April 26, 1867, the oldest of three daughters of an Iowa farmer. Her mother died in childbirth in 1870. Two years later Martha's father remarried and moved the family to a village in rural Kansas near Winfield. Her subsequent childhood recollections included herding cattle while watching out for wolves.

At sixteen she moved to a neighbor's house, taking the first of a series of positions as a "hired girl."[7] In September 1887 she moved to Topeka and two years later married Johnny Shaw, a twenty-eight-year-old postman. Their troubles began almost immediately. On November 1, 1889, Martha wrote, "In bed most of day. Miscarriage: missed just one mo: feel awful bad." The next day she reported that Johnny verbally abused her. And, on November 7 she complained that Johnny arrived home drunk.[8] Above all, Johnny's illness profoundly shaped the course of the marriage. On November 23 Martha noted that Johnny "Came home at noon" with a fever; on November 27 she "went to the Doctor's in the evening for Medicine." Two days later she summoned another doctor, returning with Henry Roby, a prominent homeopath.[9] Although fetching medicine and doctors typically was men's work,[10] women took over when men were incapacitated. The task could be long and cumbersome, even in urban areas. In the late 1880s Topeka streets were unpaved and frequently muddy.[11] Martha's long skirts must have added to her difficulties. The absence of street lights meant that walking alone at night could be dangerous.[12]

When Roby returned on November 30, he pronounced Johnny improved. Roby enjoyed a substantial reputation in the community,[13] but Martha had little confidence in him. "I don't exactly like him," she wrote. "He seems to know all about him, without asking anything about his previous condition." It is unclear what "previous condition" Martha thought the doctor should have inquired about. Martha never mentioned a prior condition that may have troubled Johnny. Moreover, in the late nineteenth-century model of medical care, patients' prior health problems were considered less important than their habits and general lifestyles.[14] As we shall see, Martha later attributed Johnny's illness to his personal behavior.

On December 1 Roby's assistant, Charles F. Menninger, arrived. According to Martha, he "gave Johnny a thorough examination" and "we are *very* much pleased with him." Because Menninger remained the physician for Martha's family off and on for almost thirty years, he merits a thorough discussion. Although she preferred him to Roby, it is unlikely that he would have impressed her initially as very different from any of the doctors she had known as a child in rural Kansas. A recent graduate of Hahnemann Medical College, a small homeopathic medical school in Chicago, Menninger had been practicing less than a year when he and Martha met. He was well aware of the deficiencies of his education. Hahnemann had recently established a small laboratory and library, but it still did not teach new developments in biomedicine and provided little clinical experience.[15] Nor was Menninger entering a secure and highly paid occupation. Earning $40 a month, he relied on his wife's income from teaching to make ends meet. They lived in just two rooms in Roby's office building.[16]

This period of sickness introduced Martha not just to Menninger but also to the problems Johnny's poor health would create throughout their marriage. Illness appears to have aggravated his irritability and impatience. On December 8 she complained, "He got so very angry at me this eve and cursed me so hard, and was so mad he wouldn't take his medicine, because he thought I took more time than I needed, to eat a lunch." The following day she wrote, "He hardly lets me take time to breathe." Johnny's condition also interfered with his ability to fulfill the role of breadwinner. On January 6, 1890, Martha remarked that Johnny had returned to work after a four-week absence. But he continued to "lay off" some days. It may not be coincidental that Martha's diary also included the information that she had taken in two roomers. Although keeping boarders was a common way for women to contribute to the household economy during the nineteenth century,[17] the precariousness of Johnny's health may well have increased Martha's determination to find her own source of income.

With Johnny still sick in February Martha wrote that they were "studying his case and trying to learn what would be best to restore his health." Although Martha does not define the nature of their study, it is likely that they read the popular medical literature circulating during the period.[18] Martha and Johnny also critically evaluated the medical assistance they sought. When physicians offered advice, Martha felt competent to challenge it. "Doctors here . . . want to operate on Johnny for a Fistula," she wrote on February 12, "but I don't want them to: I am afraid it is Tuberculosis." This was a sensible diagnosis. Tuberculosis was a leading cause of death in the late nineteenth century, and Martha must have known several people who suffered from it.[19] She also may have been aware that even the most experienced doctors frequently missed its signs, especially in the early stages.[20]

On February 17 Martha announced a new decision: Johnny would "go to Chicago to be doctored." No overriding medical imperative compelled Johnny to travel to a distant city to seek care from strangers. Unlike Cook County or Michael Reese, the hospital he entered in Chicago was a fledgling institution with a local reputation. Founded in 1883 by the German-American community, the German Hospital opened its doors in 1884; in 1890 it had just sixty beds and served an annual patient population of 550.[21] Christ's Hospital, a comparable institution in Topeka, also admitted its first patients in 1884 and grew slowly during the remainder of the decade.[22] The explanation for Johnny's trip probably lay more with Menninger's background than with any special treatment available in Chicago. As a newcomer to Topeka and a member of a deviant medical sect, Menninger lacked ties to the local medical community. He did not gain admitting privileges at Christ's Hospital for another eight years.[23] If he remained outside the medical establishment, however, he was firmly rooted in the German-American community. The son of German immigrants, Menninger had grown up in Tell City, Indiana, among a large German population; he did not learn English until he entered school at the age of five.[24] His wife Flo also was the offspring of German immigrants and familiar with the language.[25] It is likely that nationality had influenced Menninger's choice of medical school. Samuel Hahnemann, the founder of homeopathy and namesake of the school, was a German doctor, and the sect enjoyed a large following among German immigrants throughout the nineteenth century.[26] Menninger apparently found it easier to admit a Kansas patient to a German institution in Chicago than to a hospital in Topeka.

Johnny left for Chicago on February 18. Eight days later Martha received a letter stating that he had undergone surgery; if she knew the nature of the operation, she did not mention it in her diary. Because Johnny noted that he "came very near dying," Martha decided to join him. Borrowing money from her roomers, she departed on March 13 and arrived in Chicago the following day.

As Joan E. Lynaugh points out, historians have tended "to telescope the story of the changing American hospital so that it seems as if hospitals were transformed directly from shelters for the chronically ill and homeless poor into highly technologic, medically dominated curative institutions." Lynaugh describes "a critical intermediate stage," which she dubs "the 'domestic era' of hospital development."[27] Martha's account suggests that the German Hospital was a "domestic era" institution. Although Johnny received an operation, surgery apparently had not replaced custodial care as the hospital's central function. Many patients Martha met on the ward were well enough to leave the premises during the day. Johnny too remained at the hospital for seven weeks, a lengthy stay by present-day standards.

The transitional nature of the institution shaped Martha's experience. Because German Hospital had acquired little of the bureaucratic rigidity characterizing contemporary medical institutions, it easily incorporated Martha's presence. The matron invited her to board at the hospital, sharing a room with a few convalescent women. When she developed a sore throat, she began to receive medical care. If Martha was treated more like a patient than a visitor, however, she was reminded of her caregiving obligations as Johnny's discharge approached. On March 28 she wrote, "They tell me I can take Johnny home now whenever I learn to *dress* the wound, made by operating, so I went into the Operating room with the Doctor and Nurse this morning but it looked so badly, I could not bear to touch him." Just as the hospital staff casually accepted her as a patient, so they allowed her access to space now considered the exclusive province of medical personnel.

Martha must have overcome her squeamishness because she and Johnny left the hospital on April 3. The improvement in Johnny's health proved ephemeral, however. In October Martha wrote, "Thro' the Summer, Johnny seemed to have regained his health, but now that Fall has come he is not real well and I feel he *never* will be well." The return of his cough confirmed her suspicion that he was "*going into Consumption.*" She attributed his condition to the personal habits she deplored; above all, she claimed, excessive drinking had undermined his health. This was not an unusual interpretation. Throughout the nineteenth century, individual behavior frequently was implicated in the etiology of tuberculosis.[28]

The remedy Martha and Johnny embraced also reflected popular assumptions. "We have *made up our minds* to *visit* my people in *Colorado*," she wrote on November 24, 1890. During the late nineteenth century, it was widely believed that the western climate had a beneficial effect on tuberculosis sufferers. As Billy M. Jones notes, "Faith in climatic treatment became so universal, that most patients accepted with unreserved approval a physician's charge to seek a more favorable treatment."[29] Known as "the World's Sanatorium," Colorado was an especially popular destination for invalids. In 1881 a railroad advertisement proclaimed that "nearly one-half of all the people now residing in Colorado were influenced in coming here, either directly or indirectly, by considerations of health."[30] Family ties may also have influenced Martha and Johnny's choice. Martha's parents and sister had moved to Colorado in 1887. Martha and Johnny could save money on room and board by staying with Martha's parents. Nevertheless, the train fare was a financial strain. The trip also exacted a psychological cost. *"Breaking up housekeeping and we had only just begun,"* she complained on the eve of their departure.

Most late nineteenth-century health seekers in the West initially penned optimistic accounts of the value of the air they found.[31] Martha wrote in a similar vein after her arrival in Colorado in December: "Johnny is much better than when we came. We brought ten dollars worth of medicine with us and I don't believe he will need any more." Martha was silent about why they decided to go back to Topeka in late March. She noted, however, that Menninger considered their return premature: "Met Dr. Menninger on street this morning," she wrote soon after their arrival. "He says, Johnny *is not yet strong enough,* to live here and must go back at once and stay at least three years." Although Martha took his advice seriously enough to record it in her diary, she and Johnny disregarded it for six months. On April 1, 1891, they moved into a new house in Topeka.

But when Johnny began to "cough considerably," he decided to heed Menninger's counsel. Menninger now added to that. Johnny should travel in a covered wagon and live on a farm where he could work in "open air." Martha beseeched Johnny to let her stay behind. She was in the middle of a difficult pregnancy and feared she would be unable to withstand the journey. The trip also would separate her again from her circle of friends and place her beyond the reach of medical help during the confinement.

Despite Martha's plea, she and Johnny left for Colorado in November. Shortly after moving into their new house, Baby Inez was born. In May 1892 a new worry arose. The baby began to cough and become fretful. At a time of high infant mortality, any sickness could presage disaster. Because infectious diseases were rampant and antibiotics nonexistent, infants and children frequently became ill and died.[32] In 1890, 58 percent of all burials in Leavenworth, Kansas, were for children under the age of five.[33] Martha tried to reassure herself that the baby suffered from nothing more serious than a cold but acknowledged that "sometimes it makes me heartsick." On May 17 she wrote, "I am convinced she is 'teething,' tho' many say 'too young.'" This was a reasonable, if terrifying, diagnosis. According to reigning medical beliefs, dentition was a dangerous passage, which could precipitate several fatal illnesses.[34]

Martha was contemptuous of customary healing practices. "How I wish there was a good Doctor in this part of the country," she wrote, "that I might consult him for my little 'joy-girl,' instead of using *all* the remedies, *all* the old 'grand-mothers' can think up, for *all* the 'diseases,' one ever heard or dreamed about." Although some feminists wax nostalgic about networks of women exchanging information about children's health care, Martha had no use for traditional sources of knowledge. A wide variety of unorthodox providers practiced in the Midwest in the early 1890s,[35] and on May 28 she decided to consult "Mrs. M. E. Martin, a sort of 'Home-made' Baby doctor." After a second visit in early June, Martha, however, grew skeptical:

"Mrs. Martin thinks Baby's trouble is her lungs and I'm just as sure it's her teeth."

The following day Martha resolved to return to Topeka "to get a good doctor." Arriving by train on June 11, she immediately summoned Dr. Menninger, who corroborated her diagnosis. "It *is* her teeth, as I suspected." In addition, Menninger stated, the baby suffered from "Brain-fever." Menninger continued to visit the baby regularly during the next few weeks. When Inez went into "spasms" on June 25, he "stayed some time" and returned the next day with medicine. Menninger had left Roby's practice in August 1890 but still had not acquired the rewards of professional status. Struggling to attract an adequate paying clientele, he remained dependent on his wife's income.[36]

Although Martha stayed with Johnny's mother and his sister Retta, she initially refused to allow them to help with the baby. "I do not undress, day or night and trust my child to *no one*," she wrote. "I only sleep with her in my arms." By the third week of June the stress of care had taken its toll. "Had to call Dr. Menninger to see *me* for I am all worn out with loss of sleep and rest, and care and anxiety and I have a very sore breast: in fact I am *sick*, all over, and had to give up awhile this afternoon and go lie down and sleep and leave my darling to the care of her Aunt Retta Shaw." As the baby worsened, Martha marshalled a broader network of support. On June 26 she reported, "Eva Herman and Mrs. Wm Baker helped me all day." The following day she commented, "Mrs. Pettit and Mrs. Ed Johnston were with me when baby died."

Although historians debate the extent to which early nineteenth-century patients were able to insulate themselves from the pain of infant death, they agree that the loss of offspring inflicted shattering grief by the end of the century.[37] Throughout the summer Martha poured out her anguish in the pages of her journal. On June 29 she wrote, "Oh! the emptiness of my arms, the loneliness of my heart. Oh! God, ease this terrible heart-ache."

In the fall Johnny's illness again preoccupied her. Because he had become too sick to farm, he decided to seek a cure in Los Angeles. Despite worries about the harmful effects of ocean fog, southern California coastal cities drew large numbers of tubercular patients during the late nineteenth century.[38] With a new move looming, Martha complained about her nomadic life. "This settling in a new home, and so soon 'tearing up' and going again takes the heart out of me."

On November 14, Martha wrote, "Johnny says he will divide with me what little we have and I can go my way and he will go his." She "could shout with very joy at the thought of freedom from such a life," but "duty" commanded her "to stay with him." Furthermore, she told him that "if he was to get sick, he would need me, and not every one would stand by him." She repeated this over and over. "No one wants me to go away with him," she wrote on November 28, "but if *he* insists on going, I feel it *my* duty to go with him; no one else would look after his welfare as I would." Such passages remind us of the role of moral beliefs in determining the nature of women's caregiving responsibilities. During the midnineteenth century, popular writers exalted women's special sphere. At the heart of the domestic code lay the belief that women were innately different from men. They were calmer, purer, and more loving. Several traits that are central to caregiving, such as patience and responsiveness to the needs of others, became part of the cultural definition of womanhood. In addition, prescriptive literature exhorted women both to safeguard the health of their families and strive to please others while subordinating their own needs.[39] Although we must be cautious about conflating prescription and behavior, it is significant that Martha ex-

women stuck as care givers

plained her motivation in terms of duty. To the extent that she accepted prevailing notions of female responsibility, she assumed that she should sacrifice to care for Johnny despite the poor quality of their relationship.

The move proved as disruptive as she had feared. Arriving in Los Angeles, she felt adrift in a new land. "Christmas Day, and what a mockery it seems to me," she wrote. "Strangers, in a strange land, among strangers, more than a thousand miles from home, is not conducive to a very happy Christmas." On New Year's Day 1893 she commented, "Johnny and I went for a walk after dinner, but how could it be pleasant, when you know not a soul to say 'Howdy-do' to."

Other members of the invalid community eventually helped to lessen her sense of loneliness. In March 1893 she wrote:

I have met some very wealthy people here and made some very pleasant friends. Mrs. Morey, whose hands are covered with diamonds—she wears *ten fine* diamond rings on one hand—has just about such a husband as I have: he is about the same size and looks much like Johnny: *he* also has consumption and is cross and cranky as can be. Well, *she* and *I* are *friends* and sympathise fully, with one another.

Martha's new job introduced her to another group of people. Johnny's deteriorating condition compelled her to reenter the labor force. In January 1893 she reported that she had been hired to serve meals at a boardinghouse: She was "thankful" for the job, although "pride made it embarrassing." Had they remained in Colorado, she noted, she would have been able to earn higher wages and avoid the indignity of waitressing. The demands of care also narrowed her options. Because she had to return home to tend Johnny in the middle of the day, she could not work as either a nurse or a clerk. But the flexi-

bility of her hours meant that she bore a double burden. On February 20 she described her daily routine: "To work at 6 A.M. Home again at 2 P.M. back to work at 4 P.M. and home at 7 o'clock for the night. 17 blocks to the [boarding house] walked four times a day, is 68 blocks, or more than five miles a day."

Nevertheless, work had its compensations. Martha found some companionship among her coworkers, including a Chinese dishwasher to whom she gave Bible lessons. Employment also offered relief from the stress of care. If her job was "hard," it also was "a blessing, in that I do not have to be so much with Johnny and run the risk of taking consumption, for he coughs dreadfully and the smell from his body is sickening: smells like his body was dead." The entry ended with the "wish" that she "did not have to sleep in the same room with him."

Four years earlier Martha had nursed Johnny without worrying about her own risk of infection. Now fear of contagion mingled with revulsion at his body, increasing the burden of care. Sheila M. Rothman notes that after Koch's discovery of the tubercle bacillus in 1882, "[e]ducated Americans slowly began to reckon with the fact that proximity to a person with tuberculosis could be dangerous." Kansas did not launch an extensive health education campaign about tuberculosis until after the turn of the century.[40] But in Los Angeles Martha associated with patients and their families and may well have learned about the communicable nature of the disease from them. Another family member later recalled her morning ritual in southern California in 1893: "I began to empty the cuspidors. Every room except mine had at least one cuspidor partly filled with water. . . . I knew cleaning the cuspidors was dangerous work."[41]

Martha's anxiety about Johnny's death grew throughout the spring and summer. "Johnny sick—in bed most of time," she wrote on March 31, 1893. "I am so afraid I will come home some evening and find him dead." On July 22 she wrote, "Death is so near my home, I shudder." Two days later she announced that she had "quit work" and was "home to stay, for Johnny had grown so much worse, it is not safe to leave him alone any more." On August 1 they received money from Johnny's mother and the following day boarded the train to Topeka. "I *return* home, with a heavy heart," she wrote, "for again, I am leaving many dear friends and most of all because I know, just ahead a little way, death is waiting to claim the one to whom I'm bound."

For the second time in fourteen months Martha arrived back in Topeka with a dying family member. Once again the burdens fell primarily on her. Although they stayed with Johnny's brother and had other relatives and friends around, Johnny insisted that she alone care for him. "Johnny slowly grows weaker and I am with him day and night," she wrote on August 16. "Have not undressed to go to bed, since I left Los Angeles. He won't let anyone else care for him, so day and night I sit by his bed-side, getting what sleep I can in a rocking-chair, and I never seem rested." The stifling summer heat must have added to her difficulties.[42] Moreover, some of her care involved hard physical labor. Because Johnny no longer could "raise himself up in bed" by the end of August, she had to "lift him so much and turn him." On October 5 she wrote, "Two months this afternoon, since we came home and not *once* in all that time, have I had my clothes off, only to change to others, nor have I been in bed. I sleep in a chair beside the bed, that I may minister to his every want. Several times I have lain down on the floor on a pallet, but he does not like to have me do so."

Observing the customary reticence of the period, Martha wrote nothing about the intimate aspects of care. She must have fed Johnny, washed blood, excrement, and sputum from his body, and changed his clothes and bedding. She

previously had recoiled from the sight of his surgical wound and the smell of his ailing body; the new information about contagion had added an element of fear. We therefore can assume that she found the tasks she rendered in the summer and fall of 1893 extremely distasteful. We also can only speculate about what the confrontation with physical frailty meant to her. Was she able to distance herself emotionally from Johnny's deterioration because she assumed that it stemmed partly from moral failure? To what extent did she dwell on the possibility that her own strength too could fail at any time?

Care of the soul was an important feature of nineteenth-century caregiving. Women sought to ensure that the dying were adequately penitent, "sensible" of their sins, and prepared to face death. Martha too assumed her responsibility for her husband's spiritual well-being. "I wish he was a Christian," she wrote in August. "It is hard to see him dying an unbeliever, an *Infidel*. I pray God to change his heart."[43]

Johnny's death on October 26, 1893, generated turbulent emotions. "*Tonight I am entirely alone and miserable,*" she wrote. "*Tonight I am a widow, I am free. My heart* would *cry out in very joy, because it is freed from a wretchedly miserable life,* and *my heart is breaking with pain, heartache and utter desolation.*" Although she claimed to "miss him more than anyone can know," she burned many of his possessions "that I may have no reminder of my unhappy life."

After Johnny's death Martha worked as a nurse in the home of neighbors. As Susan Reverby commented, "Marriage to a very poor man, divorce or abandonment, or widowhood were often preconditions for nursing. Widowhood, in particular, appears to have been an important, if cruel, pathway into nursing. . . . With no need for formal credentials, a woman could offer her experience of caring for a dying husband as her qualification to nurse."[44] On May 2, 1894, Martha married the son of the family, Fred Farnsworth, a postal carrier who had been Johnny's friend. This marriage was generally harmonious. Unlike Johnny, Fred did not demand wifely subservience, and Martha could retain her dignity and autonomy without a struggle. The major constraint on her life was responsibility for onerous domestic and farm labor. Fred's wage, like that of many working men, was insufficient to support the household. Instead of obtaining paid work, Martha raised poultry and sold eggs and milk. As late as 1922, when she was fifty-five, she still had sixty-six chickens. Her housework too was far more grueling than that of most middle-class urban women. Long after they had begun to send out laundry, she devoted much of one day to washing and another to ironing. She also baked the bread and canned the vegetables and fruit that better-off women bought in stores. Later, when the demands of care increased Martha's responsibilities, her burdens were staggering.

Like many other midwestern women in the early twentieth century, Martha delighted in the new goods and services that gradually lightened her tasks.[45] In March 1911 she noted that the Edison Company "sent men out . . . to 'wire' our house for Electric-lights." The installation of electricity made various appliances possible. In February 1912 she waxed lyrical about her new "suction sweeper," which she called "the finest thing *under* the sun." When indoor plumbing arrived in 1916, she pronounced herself "glad" to have it.

Martha also extolled the virtues of technological developments that did not liberate her from housework. After hearing her first "Graphaphone music" in 1903, she wrote, "I cannot *begin* to conceive, how the mind of man could *invent* such a *wonderful* machine." In 1916 she "fell completely in love with the Victrola" and decided to buy one for Fred for Christmas, despite its exorbitant cost. After seeing airplanes in 1919, she wrote that she was "crazy to take a Fly."

Martha's encounter with medical developments must be viewed within the context of her uniformly positive response to science and technology. It is reasonable to assume that her enthusiasm for consumer goods and services facilitated her acceptance of the innovations in diagnosis and treatment which she experienced as a caregiver.

There was, however, one technological advance Martha greeted with ambivalence. In November 1905 she wrote, "Men here stringing wires for a Telephone which I don't want but my neighbors want me to get, so as to visit over the Phone." As we will see, even after the telephone enabled her to obtain medical assistance in emergencies, she remained unconvinced of its advantages.

Two episodes during Martha's second marriage reveal the changes in care. On February 2, 1913, she wrote, "About 8 o'clock this evening Aunt Kate Van Orsdol Phoned from Silver Lake, that Freda had just been taken with appendicitis, and I told them to call a doctor." The town of Silver Lake was approximately eight miles from Topeka. The youngest child of Martha's sister May, Freda enjoyed a particularly close relationship with Martha. She had been born on the anniversary of the death of Martha's baby, and much of her early life had been spent with Fred and Martha. Two years after her father died in 1896, her mother had left Colorado and moved with her children into the Farnsworth home. Although May moved to her own house in Topeka in 1903 and returned to Colorado in 1908, Martha continued to feel responsible for Freda's well-being, sending her occasional money and advice. During the 1913 appendicitis attack Freda was again living with the Farnsworths.

Martha's relatives seem to have felt competent to diagnose appendicitis without consulting a physician. Moreover, the doctor they eventually called appears not to have recommended immediate surgery. When Martha phoned early on February 4, she learned that Freda "was doing alright, but wanted me." Martha therefore made arrangements to travel by train to Silver Lake and bring Freda back to Topeka. A phone call later that day abruptly altered her plans. Martha's relatives now informed her that "Dr. Dudley of Silver Lake said I must bring an Automobile and come and get her at once and bring her to Hospital for operation." Martha phoned the owner of a local garage, who drove her to Silver Lake. "What a hard race with death," she commented. "I hope I never have to take another." Arriving at her aunt's house, Martha

> found the poor, dear child suffering agony: we soon had her in the Automobile and Dr. Dudley too and began our terrible race—death kept close beside and it seemed at times, as if he *must* win. My heart bursting with grief, I held her close in my arms, across my body all the way, to make it easy, but she suffered terribly and when within four miles of Topeka, I saw the whiteness of death driving back the purple fever, I thought my heart would break. But at last we reached Stormont [Hospital], where everything was in readiness.

The telephone not only provided Martha with immediate news of family sickness but also saved essential minutes in the crisis. Martha phoned both the garage owner and the hospital, allowing the staff to have "everything in readiness." The transportation revolution enabled Freda to have emergency surgery in a hospital several miles away. But automobiles were still not commonplace. Although physicians were among the earliest car owners,[46] the Silver Lake

doctor apparently lacked one. Ambulances also were unavailable. Martha thus was compelled to hire a car in Topeka and travel to Silver Lake before the race to the hospital could begin.

When they reached the hospital, Martha watched the surgery, later giving a graphic description of the diseased organ: "I saw the appendix, a greatly enlarged, ulcerated and perforated rotten thing. Gangrene and also just a starting of peritonitis." Appendectomy was a powerful symbol of medical prowess. Before the development of antiseptic techniques surgery had been restricted to bodily extremities. According to David Rosner, "The successful removal of the appendix was perhaps the most dramatic testament [sic] to the ability of surgeons to operate and prevent infections."[47] Nevertheless, the marvels of biomedicine did not blind Martha to other forms of knowledge. Because Freda remained in danger, Martha called in Mrs. Goddard, a Christian Science practitioner. Christian Science was one of the alternative medical practices thriving in Kansas during the early twentieth century that aroused the greatest rage in the orthodox medical community.[48] Martha's reliance on Mrs. Goddard thus constituted a direct challenge to her doctors' authority.[49]

Martha's openness to inconsistent beliefs and practices may have been far more common than historians frequently assume. Paul Starr cites a survey conducted between 1928 and 1931, which found that "all the non-M.D. practitioners combined—osteopaths, chiropractors, Christian Scientists and other faith healers, midwives and chiropodists—took care of only 5.1 percent of all attended cases of illness." As a result, Starr concludes, physicians "had medical practice pretty much to themselves."[50] But this study appears to exaggerate the triumph of physicians. Recent research reminds us of the syncretistic nature of popular medical beliefs today and the broad array of medical practices that continue to attract large and loyal followings.[51] A 1990 study found that more than a third of adults in the United States use some form of unconventional therapy; many of those who visit regular doctors also consult unconventional practitioners.[52] We have no reason to assume that people in the early twentieth century were not as likely to subscribe simultaneously to a variety of beliefs.

After waiting to see Freda "come out of Anaesthetic," Martha returned home to eat and do her work. "I needed a moment of rest," she explained, "for I was exhausted from the long ride in the cold and with heart breaking grief." She was back at the hospital by 8 P.M. and stayed until almost 10 P.M. Arriving at the hospital at 8:30 the following morning, she "waited on Freda all day." Her report for February 7 was similar: "With Freda all day at Stormont Hospital. The Hospital is full, with only 16 nurses, so they neglect Freda's bathing and I had to make a *kick*." That evening Martha "went home by car as I was too tired to walk. I did not get to sit down five minutes during the day, but work over her constantly." On February 8 she wrote, "Another hard day at Stormont Hospital with Freda. I've eaten nothing but cold lunch, and little of that, since Freda's operation." Because the snow had melted, the walk home in the evening was not "so hard," but Martha herself was "so very tired and worn."

Martha's account lends some support to the familiar argument that the transfer of medical care to institutions relieved family members of critical obligations. When nursing Johnny, Martha had sat up through the night to watch for troublesome symptoms. Now she could leave Freda in the hands of institutional staff and return home to sleep. Nurses also relieved her of responsibility for at least some aspects of personal care. Nevertheless, Martha's diary suggests that change may not have been as rapid or absolute as previous accounts claim. In her history

of childbirth Judith Walzer Leavitt argues that when parturient women arrived at the hospital, they "found themselves abandoned to the impersonal routines of impersonal institutions."[53] But Martha not only tended Freda on the ward; she also accompanied Freda into the operating room and remained with her throughout the procedure.[54] It is true that Martha was an observer, not an assistant. Unlike previous generations of women, she had no responsibility for holding basins, supplying instruments, helping with sutures, or cleaning up afterward. In addition, although women did not have to request permission to watch the surgery performed in their own kitchens, Martha's admission to the hospital operating room occurred at the discretion of doctors.[55] Nevertheless, her experience raises questions about the extent to which family members relinquished jurisdiction over hospital patients in the early decades of the twentieth century.

If hospitalization released Martha from some tasks, it created others. For the first time she was responsible for supervising the work of paid caregivers. When the nurses' care fell short of her standards, she made a "kick." She also had to travel back and forth between home and hospital. Although she once returned home by car, she typically walked, often through snow. Because she could not intersperse caregiving with farm and household labor, her chores accumulated. And hospital care was expensive. Although we can assume that Johnny received free care in 1890,[56] Freda was a private-pay patient. When Martha subsequently had reason to chronicle the various sacrifices she had made on Freda's behalf, paying the hospital bill occupied a prominent place on the list.

Martha's caregiving obligations were not confined to Freda. Martha's sister May and the latter's daughter Zaidee moved into Martha's house soon after Freda entered the hospital, and on February 22 Zaidee required assistance. "I went as usual to Stormont Hospital to take care of Freda," Martha reported,

> but at 1:30 P.M. was called home to help Zaidee thro' confinement. . . . Well, I ran as much of way home, as I could thro' deep snow, and had to "get busy" at once, and the baby (a ten pound boy) came at 2:33 P.M. I got hold of the little fellow and pulled him away, let him lay 30 minutes, then cut the cord, rolled him in a blanket and laid him aside: not a soul but her mother and I with her. Dr. Jeffries was called but did not arrive for one hour and two minutes after the baby was born.

As long as birth remained at home, family members rather than hospital staff substituted when doctors were late.

Martha brought Freda home on February 25, three weeks after the surgery. During the early decades of the twentieth century, the average hospital stay dropped sharply.[57] Although Freda's stay was long by contemporary standards, it was much shorter than Johnny's seven-week hospitalization in 1890. Freda's discharge was a relief to Martha. "Another stormy day and it seems like Paradise, to be able to stay home," she wrote on February 27. "I *am* resting; at least it seems like rest, to be at home." But she also noted that her hands were "more than full"; her house was "a Hospital," with "Freda in one room and Zaidee in another and neither one able to go in to see the other." Freda continued to require skilled nursing care; the day she left the hospital her wound still contained "pus," and Martha was responsible for applying dressings.

Martha was unable to delegate any of this work to paid nurses. The first Kansas nursing schools were established in 1888.[58] The great majority of graduates worked as private duty nurses, providing care to patients in individual households.[59] But hiring a nurse was not an op-

tion for Martha. In at least some parts of the country nursing graduates earned an annual wage of approximately $950,[60] a sum far beyond the reach of the Farnsworths.

Martha's last major caregiving experience also incorporated both new and traditional elements. On November 14, 1915, she wrote, "My good Teddy went to see Dr. Menninger this afternoon, about his heel and finds he has Erysipelas and badly—caused by bruise from hard walking with heavy load of mail." Although Menninger still did not expect patients to make appointments, he increasingly saw them in his office rather than their homes.[61]

In other ways as well Menninger's practice had changed fundamentally during the twenty-two years since he last had ministered to a member of Martha's family. As medicine had consolidated into a regular and powerful profession between 1890 and 1915, he had gradually incorporated the precepts of biomedicine into his practice. His reputation steadily grew.[62] In 1898 he gained both admitting privileges at Christ's Hospital and acceptance into the Shawnee County Medical Society. Two years later he was elected president of the society.[63] His private practice also flourished, and by the second decade of the twentieth century he had become one of the most popular Topeka physicians.[64] It must have been reassuring to Martha to be able to rely on a physician who commanded enormous respect among both his professional colleagues and the lay public. We will see, however, that, although she regularly sought his assistance during Fred's long illness, she assumed that she alone made medical decisions.

Erysipelas was known to be a highly contagious disease in the early twentieth century,[65] but just as Martha initially had blamed her first husband's tuberculosis on his drinking, so now she viewed Fred's problem as an occupational injury. She may have been especially ready to do so because she often had complained about his working conditions, focusing on the length of his working day and the weight of his load. Unfortunately, Martha's diary does not tell us whether she took special precautions to safeguard her own health.

With Fred still out of work on November 27 and no indication of when he could return, money became a source of concern. In one respect, however, the Farnsworths were better protected against sickness than most working-class families; they had purchased "Insurance and sick allowance Policy of $10.00 per week." Fred's illness also underscored the critical role Martha played in the household economy. "We are not starving," she wrote, "for we have the cow and chickens and a cellar full of fruit, and vegetables."

On November 29 Fred's condition suddenly deteriorated. "Beautiful sunshine outdoors, but all darkness in our home," Martha wrote. "All last night and today *Death* has tried to come in and take my good Teddy: the struggle has been terrible and exhausting." (Menninger later explained that the cause of the problem was an embolism in the lungs.) Although Fred insisted he did not need a doctor, she "ran down stairs and phoned Dr. Menninger, who being out to Christ Hospital, had not far to come and was here in about ten minutes." Technological achievements had eliminated one onerous task. Instead of traipsing through dusty or muddy streets to fetch a doctor, Martha could pick up the telephone. And the time it took physicians to reach patients' homes had drastically declined. During the early years of his career Menninger had travelled around Topeka by a combination of street car, bicycle, and buggy.[66] Since purchasing a car in 1910,[67] however, at least some trips to the sick could be made in a matter of minutes.

After a "hurried examination" Menninger asked Martha to phone another doctor "to come at once." Although she "had trouble 'getting' Dr. Owen," she "finally succeeded." Martha also mo-

bilized nonmedical forms of assistance. Just as birthing women in the nineteenth and early twentieth centuries frequently marshalled female relatives and neighbors along with physicians,[68] so Martha phoned her "good friend Mrs. A. E. Jones." She arrived shortly after the doctors and spent the night at the Farnsworth house. The following day Martha wrote, "Mrs. Jones will go home tonight and Mrs. Wilcox come up. Any way *some* one will stay in house with me, so I won't be alone."

It was customary for serious illnesses to be treated at home during the second decade of the twentieth century.[69] Despite the proximity of Christ's Hospital, no one appears to have considered moving Fred there. One consequence was that doctors relied for assistance on Martha rather than a staff of subordinates. She helped Menninger "give Teddy five 'Hyperdermics' in the veins" and wrapped him in clothes she had heated with irons to ward off chill.

In the evening of December 5 "a most agonizing attack came on." This time it was more difficult to summon help. Menninger had hired a receptionist to answer his telephone soon after establishing his own practice,[70] but as long as he continued to make house calls, there was no guarantee he could be located in an emergency. Failing to reach Menninger by phone, Martha "raised a window called at the top of my voice for help and blew a whistle." Although her neighbors appeared to be at home, they did not respond. Martha therefore "gave up and worked over Fred for a time, then tried again and got Mrs. Wilcox who stayed at Phone until she got the Doctor." When Menninger finally came, "he could do nothing, as he had given the limit of medicine." As he was about to depart, Fred suffered another attack. "For five hours, he walked thro' the 'Valley of the Shadow' and we did not know whether he would come back it would have been easier to have seen him die than to sit by and see him in such agony and powerless to give him relief. At midnight he got easy, but at 3:30 A.M. he had another spell, which only lasted about an hour."

Martha's nursing responsibilities expanded as a result of this second crisis. On December 7 she noted that she "rubbed" Fred's leg and wrapped it in "hot Witch hazel woolens" every three hours and gave him medicine "every hour day and night." Although Fred improved during the next two weeks, he suffered yet another setback on December 20. Menninger diagnosed phlebitis, instructing Martha to "go back to the hot witch-hazel pack every three hours." "Well I could almost give up in abject despair," she confessed to her diary. "Will complications *never* cease to set in? We thought Fred would soon be up and now *this*—one of the most tedious and slowest of *disease:* it is dreadful."

During the long convalescent period that followed, Menninger continued to call regularly at the Farnsworth house. But Martha welcomed advice from other sources as well. On January 10, 1916 she reported:

> Mrs. Calvin came over and spent the evening with us and gave me a new recipe for Fleabitus (milk leg) which Dr. Weston of Chicago gave her (and which cured her after having it for four years) and she helped me put it on Fred. "Put cabbage thro' a meat grinder, making it very fine and put on a cloth and bind on Varicose vein, with five yards of bandaging, at bedtime: next morning take off and sponge limb with good rich buttermilk and bind on fresh cabbage—use about ¼ cup of buttermilk—the milk may be used quite cold."

Leavitt argues that "the price of the new science was a growing separation between expert and layperson.... The knowledge gap produced when medicine became increasingly technical put the uninformed in awe of medical science."[71]

But not all therapies required specialized training to be understood. Mrs. Calvin's treatment resembled the remedies contained in the domestic manuals circulating in the midnineteenth century; it relied on common household implements and food and was simple enough to be administered by family members.[72] Although the treatment may have originated with a physician, it could be disseminated through lay networks.

Martha failed to inform Menninger of the new therapy until five days after introducing it. "Dr. C. F. Menninger came out this morning and was fairly amazed, and greatly pleased, with Fred's improvement due to Cabbage poultice," she wrote on January 15. If Martha appreciated Menninger's stamp of approval, she felt enough confidence in her own judgment to disregard the modification he suggested: "He was thankful to know about the Cabbage treatment and said to continue its use: but he sees no merit in using the buttermilk—but as it is part of the treatment, we will continue *its* use also."

Although Fred's inability to work did not force Martha into the labor force, she again faced competing demands. She could integrate her regular round of chores with her nursing responsibilities because care took place at home. On December 15, for example, she noted that she "took the usual care" of Fred, doing her weekly washing "between times." Her workload, however, may have been as grueling as when she nursed Johnny in 1893. Electricity and assorted household appliances had lightened her tasks, but she still lacked indoor plumbing. Bathing Fred and washing his bedding and clothes must have involved considerable drudgery. The bitter Kansas winter added to her difficulties. "A howling Blizzard—15° below zero, 43 mile wind from North West, blowing sleet and snow," Martha wrote on January 12, 1916. "But I was out early, and cleaned the barn, fed the cow and chickens, carried water and coal and kept everything 'snug and warm,' took care of my good Teddy and did my usual house work." Although snow fell the following day and the temperature dropped still lower, she still "did *all* the usual rush of work."

Martha's complaints also fastened on her lack of support. Friends who initially rallied around seemed to lose interest as Fred's illness lingered. "I stay alone with Fred, all the time now," she wrote on December 13, "my neighbors all got tired quick." As the holidays approached, the absence of visits was especially painful. "We seldom see anyone any more, but the Doctor—everybody calls by phone and it's the Holiday season and everybody *rushed*." She expressed even greater bitterness on December 30. "We are all alone all the time now and being Holiday week, everyone so busy with his plans for the New Year, no one comes in." During the blizzard in the middle of January she wrote, "Not a neighbor came in to see how we are 'making it,' these bad days."

Martha assumed that the telephone was at least partly responsible for her plight. "All our Calls are by Phone and I almost wish I had no phone," she wrote at the end of December. Susan Strasser notes that "telephone conversation . . . constitutes a wholly different kind of communication from face-to-face conversation, in which participants read each other's faces, watch each other's hand gestures, and have an opportunity to really touch."[73] The telephone can weaken instrumental as well as emotional aspects of support. When communication is restricted to phone calls, undone chores remain invisible, and friends may be less likely to lend a hand.

We can interpret Martha's isolation in other ways as well. It is possible that she gave mixed messages, rebuffing offers of help she desperately wanted. The nature of Fred's disease may have been an additional factor. It was one thing to respond to Martha's pleas for help during

emergencies; it was a very different matter to provide sustained support over a four-month period. As caregiving has focused increasingly on chronic ailments, it has become a much lonelier endeavor.[74]

Finally, Martha may have based her expectations of support on the patterns of mutuality she had observed as a child in rural Kansas.[75] By 1915 various factors, including geographic mobility and urbanization, had weakened bonds of interdependence among women. Like other Kansas towns,[76] Topeka grew rapidly during the late nineteenth and early twentieth centuries, making many of Martha's neighbors relative newcomers. She herself had led a peripatetic existence during her first marriage, moving a total of six times. Her sister's departure in 1908 left her bereft of close relatives. And even her most loyal friend, Mrs. Jones, moved from Topeka shortly after Fred's recovery. The labor-saving devices that lightened Martha's daily burdens may have further attenuated her support networks. Instead of coming together to share household tasks, women increasingly performed their chores in isolation.[77] Although these changes affected women of all social classes, Martha's experience reminds us that the repercussions were especially serious for those with limited financial resources. Unable to hire servants or private duty nurses, Martha remained dependent on community ties for support. When these proved fragile, her responsibilities became overwhelming.

As Fred began to recover, Martha focused on her own health problems, which she attributed to the stresses of care. In February 1916 she reported that her nerves were "trying to go to pieces," her sleep was erratic, and she found herself "dropping things." On the nineteenth she wrote, "My nerves are all *unstrung* today. . . . I wish I could just let go of myself, and *scream*, and *scream*, and *scream*, and *scream*."

Fred's illness also imposed serious financial

burdens. Although he did not require technical equipment, the services of specialists, or hospital care, his medical bill was sizable. In April Martha commented that she paid Menninger $75 and the drugstore $17. A recent study by Joel D. Howell and Catherine G. McLaughlin of the medical care purchased by selected working-class urban families in 1917 helps us evaluate that expense. When Fred worked regularly, his annual income was $1,392, which placed the Farnsworths exactly at the median of the families Howell and McLaughlin studied. The $17 drugstore bill was only slightly higher than the mean medicine expense for households in the Howell and McLaughlin study that purchased any medicine during the year. The doctor's bill, however, was double the mean physician expense of $37.77 for families in the Howell and McLaughlin study that had paid doctors.[78] Moreover, Fred lost substantial income during his illness. Starr notes that workers' lost earnings as a result of sickness during the early twentieth century tended to be "two to four times greater than health care costs."[79] During Fred's four-month absence from work he forfeited a total of $580. The combination of wage losses and medical costs was $672, but Fred received just $205 from the insurance company. After settling their outstanding grocery bill, Martha reported that they had "nothing left."

Because Martha defined her job as maintaining as well as restoring health, Fred's working conditions remained a source of anxiety. In June 1916 she wrote: "A very hot day and Fred had so large a mail, he could not come home to dinner. One of the 'Subs' helped him with three blocks of his heaviest mail this morning, telling him to 'keep it secret' or it would anger the other 'Subs' who had all agreed to stand together, against Fred—not to help in any way and so if he was not strong enough, he could not work, would lose his job, and they would get it."

She was even madder at Fred's employers, writing in December 1919, "Deep snow and heavy mails and my poor Teddy had to work all day with no time to stop to get anything to eat. His foot is very sore and the P.O. authorities won't let him off to take care of it. I wish it were possible to make the Postmaster and the various Superintendents work as the Carriers do—that there could be a law passed, compeling *them* to go out and wade the deep snow, the same long hours without food, with the heavy loads on their backs, sick and hurt or injured—what a lesson in mercy it would teach these Government *task masters*." Although scientific advances may have increased her confidence in her ability to respond to medical crises, she remained powerless to exercise control over the hazards that threatened Fred at work.

In December 1923 Martha began to complain of illness, and on February 2, 1924, entered the hospital, where she died eleven days later.

It has become commonplace to note that the growth of social institutions, such as schools, prisons, and hospitals, removed critical functions from the home. Martha's diary tells a different story. Her obligations, rather than disappearing, changed form. As Laura Balbo points out, the growth of a vast health and social service system has increasingly required numbers of women to engage in "servicing work."[80] When her niece entered the hospital, much of the time Martha previously might have devoted to personal care now was spent mediating between Freda and formal health providers.

Many factors other than the expansion of the health care delivery system affected the amount and type of family responsibilities. The control of infectious diseases meant that women spent less time ministering to seriously ill and dying infants. Moreover, by the time of Martha's death in 1924, both malaria and typhoid fever were virtually eliminated in Kansas, and tuberculosis had lost much of its menace. The early decades of the twentieth century also saw the decline of the climate theory, which had kept Martha on the move during her marriage to Johnny Shaw. But not all changes alleviated family care. Had tuberculosis struck Martha's household in the 1920s, new concerns about ventilation and diet might well have augmented her tasks.[81] As chronic diseases replaced acute illnesses as the major cause of death, responsibilities for care shifted to the latter part of the life course.

In addition, Martha never relinquished control over medical decision making. Although she repudiated traditional practices, she felt fully able to assess doctors' competence and challenge their recommendations. Menninger's growing eminence did not dramatically transform their relationship. While relying on Menninger throughout Fred's illness in 1915, Martha also used other channels to acquire medical knowledge, and she made at least one important treatment decision without his approval. Nor did her encounter with one of the most dazzling achievements of modern medicine persuade her to surrender authority to physicians. Even after witnessing an appendectomy, she refused to view biomedicine as the only legitimate source of authority.

Issues that tend to escape the purview of medical historians helped to determine the nature of domestic care. Subscribing to dominant notions of womanhood, Martha assumed that her caregiving obligations transcended the particularities of her relationships. For four years she nursed Johnny Shaw, a man she neither liked nor respected. During her second marriage a variety of mass-produced goods and services eased some of her specific responsibilities. Telephones eliminated one task entirely. But here too the impact may have been complex and contradictory. Martha was convinced that the telephone weakened the support networks, which had relieved the caregiving burdens of previous generations of women.

Martha's diary also highlights the slow and erratic pace of change. Many late nineteenth-century features of caregiving remained intact well into the twentieth century. Because childbirth and serious illness frequently occurred at home, Martha continued to deliver skilled medical care. When surgery took place in hospitals, she accompanied the patient into the operating room and provided comfort throughout the procedure. Although cars and telephones greatly facilitated access to formal health care providers, doctors continued to spend many hours on the road and thus frequently were unavailable in emergencies.

In addition, change occurred unevenly, affecting various groups of women in different ways. The weakening of the bonds of kinship and community inflicted the greatest hardship on women who could not rely on personal servants or private duty nurses to replace the supportive networks their mothers had enjoyed. The rising cost of health care similarly had the most serious impact on low-income women. Fred's bill of $92 in 1915 represented an enormous outlay to a household getting by on $116 a month. Although Martha did not specify the amount she paid for Freda's hospital stay in 1913, she indicated that it imposed an excessive burden. And consumer goods and services reached Martha relatively late. She was still lugging pails of water inside to bathe Fred and wash his bedding and clothes long after more affluent women had indoor plumbing.

By challenging the persistent belief that family members lost caregiving responsibilities between 1890 and 1925, Martha's diary highlights the need to direct more attention to the world of family care. We should investigate how the phenomenal changes occurring during that period affected both the content and meaning of the care delivered by women in diverse social positions.

Notes

1. James H. Cassedy, *Medicine in America: A Short History* (Baltimore and London: Johns Hopkins University Press, 1991), 93.

2. Ruth Schwartz Cowan, *More Work for Mother: The Ironies of Household Technology from the Open Hearth to the Microwave* (New York: Basic Books, 1983), 73.

3. See, e.g. M. L. Berger, "The Influence of the Automobile on Rural Health Care, 1900–1929," *Journal of the History of Medicine and Allied Sciences* 28 (1973): 319–35; Helen Clapesattle, *The Doctors Mayo* (Minneapolis: University of Minnesota Press, 1941), 348–53; Guenter B. Risse, "From Horse and Buggy to Automobile and Telephone: Medical Practice in Wisconsin, 1848–1930," in *Wisconsin Medicine: Historical Perspectives,* ed. Ronald L. Numbers and Judith Walzer Leavitt (Madison: University of Wisconsin Press, 1981), 25–45. For a contemporary discussion of the impact of automobiles on medical practice, see "Satisfaction in Automobiling: A Symposium by Physicians on Their Experiences With Motor-Cars— How to Secure the Most in Comfort and Help at the Least Expense," *Journal of the American Medical Association* (hereafter *JAMA*) 58 (1912): 1049–80.

4. See Susan Strasser, *Never Done: A History of American Housework* (New York: Pantheon, 1982).

5. Two guides to women's writings are Andrea Hinding and Clarke Chambers, *Women's History Sources,* 2 vols. (New York: Bowker, 1979) and Joyce Goodfriend, *The Published Diaries and Letters of American Women: An Annotated Bibliography* (Boston: G. K. Hall, 1987).

6. The complete diary is located in the Kansas State Historical Society, Topeka, Kansas; for an excellent annotated and abbreviated version see Marlone Springer and Haskell Springer, *Plains Woman: The Diary of Martha Farnsworth, 1882–1922* (Bloomington and Indianapolis: Indiana University Press, 1988).

7. For the history of "hired girls" in the nine-teenth century see Faye E. Dudden, *Serving Women: Household Service in Nineteenth-Century America* (Middletown, Conn.: Wesleyan University Press, 1983).

8. On Kansas women and temperance see Glenda Riley, *The Female Frontier: A Comparative View of Women on the Prairie and the Plains* (Lawrence: University of Kansas Press, 1988), 178–79.

9. Walker Winslow, *The Menninger Story* (Garden City, N. Y.: Doubleday, 1956), 47.

10. Cowan, *More Work for Mother,* 81.

11. See Lucy Freeman, ed., *Karl Menninger, M.D., SPARKS* (New York: Thomas Y. Crowell, 1973), 2.

12. See Riley, *Female Frontier,* 94.

13. Winslow, *Menninger Story,* 62.

14. See Charles Rosenberg, "The Therapeutic Revolution: Medicine, Meaning, and Social Change in Nineteenth-Century America," in *The Therapeutic Revolution: Essays in the Social History of American Medicine,* ed. Morris J. Vogel and Charles E. Rosenberg (Philadelphia: University of Pennsylvania Press, 1979), 3–25.

15. Winslow, *Menninger Story,* 59–60.

16. Flo V. Menninger, *Days of My Life: Memories of a Kansas Mother and Teacher* (New York: Richard R. Smith, 1940), 238.

17. See Alice Kessler-Harris, *Out to Work: A History of Wage-Earning Women in the United States* (Oxford: Oxford University Press, 1982), 124–25.

18. On the popular medical advice literature see Anita Clair Fellman and Michael Fellman, *Making Sense of Self: Medical Advice Literature in Late Nineteenth-Century America* (Philadelphia: University of Pennsylvania Press, 1981); Guenter B. Risse, Ronald L. Numbers, and Judith Walzer Leavitt, eds., *Medicine Without Doctors: Home Health Care in American History* (New York: Science History Publications, 1977); James Harvey Young, *American Self-Dosage Medicines: An Historical Perspective* (Lawrence, Kansas: Coronado Press, 1974).

19. According to Joan E. Lynaugh, tuberculosis was the leading cause of death in nineteenth-century Kansas City. See *The Community Hospitals of Kansas City, Missouri, 1870–1915* (New York: Garland Publishing, 1989), 86.

20. See Barbara Bates, *Bargaining for Life: A So-*

cial History of Tuberculosis, 1876–1938 (Philadelphia: University of Pennsylvania Press, 1992), 17.

21. Information from Liana Overley, public relations coordinator, Grant Hospital, Chicago.

22. Thomas Neville Bonner, *Kansas Doctor: A Century of Pioneering* (Lawrence: University of Kansas Press, 1959), 91.

23. Winslow, *Menninger Story,* 90.

24. Ibid., 30–35.

25. Ibid., 45.

26. Paul Starr, *The Social Transformation of American Medicine: The Rise of a Sovereign Profession and the Making of a Vast Industry* (New York: Basic Books, 1982), 96–97.

27. Lynaugh, *Community Hospitals,* 22.

28. Bates, *Bargaining for Life.*

29. Billy M. Jones, *Health-Seekers in the Southwest, 1817–1900* (Norman: University of Oklahoma Press, 1967), 146. See also Sheila M. Rothman, *Living in the Shadow of Death: Tuberculosis and the Social Experience of Illness in American History* (New York: Basic Books, 1994), 131–75.

30. Cited in Jones, *Health-Seekers,* 96.

31. See Rothman, *Living in the Shadow of Death,* 161–67.

32. See Samuel H. Preston and Michael R. Haines, *Fatal Years: Child Mortality in Late Nineteenth-Century America* (Princeton: Princeton University Press, 1991).

33. Charles R. King, "Childhood Death: The Health Care of Children on the Kansas Frontier," *Kansas History* 14 (1991): 26.

34. See L. Emmett Holt, *The Diseases of Infancy and Childhood* (New York: D. Appleton, 1898), 243–45; Eustace Smith, *A Practical Treatise on Disease in Children* (New York: William Wood, 1894), 555–62.

35. Bonner, *Kansas Doctor,* 71–82.

36. Menninger, *Days of My Life,* 242.

37. See Linda A. Pollock, *Forgotten Children: Parent-Child Relations from 1500 to 1900* (Cambridge: Cambridge University Press, 1983).

38. John E. Baur, *The Health Seekers of Southern California, 1870–1900* (San Marino, Calif.: Huntington Library, 1959); Jones, *Health-Seekers,* 141; Rothman, *Living in the Shadow of Death,* 245–47.

39. See Mary P. Ryan, *The Empire of the Mother: American Writing about Domesticity, 1830–1860* (New York: Institute for Research on History and the Haworth Press, 1982).

40. Rothman, *Living in the Shadow of Death,* 182.

41. Lucy S. Mitchell, *Two Lives: The Story of Bishop Clair Mitchell and Myself* (New York: Simon and Schuster, 1953), 105.

42. According to Flo Menninger, *Days of My Life,* 87, the heat that summer was unusually oppressive.

43. For an example of a nineteenth-century wife who considered herself more successful in exerting a pious influence over her husband dying of tuberculosis, see S. S. Arpad, ed., *Sam Curd's Diary: The Diary of a True Woman* (Athens: Ohio University Press, 1984).

44. Susan Reverby, *Ordered to Care: The Dilemma of American Nursing,* 1850–1945 (Cambridge: Cambridge University Press, 1987), 15–16.

45. See Eleanor Arnold, ed., *Voices of American Homemakers* (Bloomington: Indiana University Press, 1985).

46. Starr, *Social Transformation,* 70.

47. David Rosner, *A Once Charitable Enterprise: Hospitals and Health Care in Brooklyn and New York, 1885–1915* (Princeton: Princeton University Press, 1982), 4.

48. Bonner, *Kansas Doctor,* 203.

49. See Cassedy, *Medicine,* 99.

50. Starr, *Social Transformation,* 127.

51. Ray Fitzpatrick, "Lay Concepts of Illness," in *Perspectives in Medical Sociology,* ed. Phil Brown (Belmont, Calif.: Wadsworth, 1989), 254–67; Raymond H. Murray and Arthur J. Rubel, "Physicians and Healers—Unwilling Partners in Health Care," *New England Journal of Medicine* 326 (1992): 51–61.

52. David M. Eisenberg et al., "Unconventional Medicine in the United States: Prevalence, Costs, and Patterns of Use," *New England Journal of Medicine* 328 (1993): 246–52.

53. Judith Walzer Leavitt, *Brought to Bed: Childbearing in America, 1750–1950* (New York: Oxford University Press, 1986), 189.

54. Lynaugh's study of Kansas City hospitals reveals that the presence of family members in operating theaters was not unusual. See Lynaugh, *Community Hospitals,* 82.

55. Permission was not always granted. Ac-

cording to Marilyn Ferris Motz, when Winnie Parker entered a Michigan hospital in 1897, her sister believed that she had a "right and duty" to accompany the patient into the operating room and was shocked to discover that the physicians would not allow this (*True Sisterhood: Michigan Women and Their Kin, 1890–1920* [Albany: State University of New York Press, 1983], 102).

56. See Rosemary Stevens, *In Sickness and in Wealth: American Hospitals in the Twentieth Century* (New York: Basic Books, 1989), 21.

57. Starr, *Social Transformation,* 157–58.

58. *Lamps on the Prairie: A History of Nursing in Kansas,* comp. Writers' Program of the Works Projects Administration in the State of Kansas (New York: Garland Publishing, 1984), 94.

59. Reverby, *Ordered to Care,* 95.

60. Ibid., 98.

61. Winslow, *Menninger Story,* 162. On the growing use of physicians' offices see Cowan, *More Work for Mother,* 84–85; Starr, *Social Transformation,* 75–76.

62. Winslow, *Menninger Story,* 67, 83.

63. Ibid., 90, 104.

64. Ibid., 111, 131.

65. Kenneth F. Kiple, *The Cambridge World History of Human Disease* (New York: Cambridge University Press, 1993), 720.

66. Menninger, *Days of My Life,* 245–46, 256.

67. Winslow, *Menninger Story,* 134.

68. Leavitt, *Brought to Bed.*

69. Charles Rosenberg, *The Care of Strangers: The Rise of America's Hospital System* (New York: Basic Books, 1987), 316.

70. Winslow, *Menninger Story,* 81; Menninger, *Days of My Life,* 246.

71. Leavitt, *Brought to Bed,* 174.

72. On midnineteenth-century domestic manuals see John B. Blake, "From Buchan to Fishbein: The Literature of Domestic Medicine," in Risse, Numbers, and Leavitt, eds., *Medicine Without Doctors: Home Health Care in American History,* 11–30.

73. Strasser, *Never Done,* 305.

74. See Emily K. Abel, *Who Cares for the Elderly? Public Policy and the Experiences of Adult Daughters* (New York: State University of New York Press, 1991).

75. Numerous commentators report that illnesses and deaths were communal events among nineteenth-century Midwest pioneers. Recalling his years of practice in frontier Kansas, one physician wrote, "When there was serious illness or accident . . . these rural settlers gave us town people lessons in applied Christianity. . . . These people were ready night or day to drive long distances to town to get groceries, fuel and medicines, or the doctor, as well as to do the work on the farm or in the home" (Samuel J. Crumbine, *Frontier Doctor: The Autobiography of a Pioneer on the Frontier of Public Health* [Philadelphia: Dorrance and Co., 1948], 49–50).

76. See Riley, *Female Frontier,* 93–94.

77. Strasser, *Never Done,* 235.

78. Joel D. Howell and Catherine G. McLaughlin, "Race, Income, and the Purchase of Medical Care by Selected 1917 Working-Class Urban Families," *Journal of the History of Medicine and Allied Sciences* 47 (1992): 439–61.

79. Starr, *Social Transformation,* 245.

80. Laura Balbo, "Crazy Quilts: Rethinking the Welfare State Debate from a Woman's Point of View," in *Woman and the State: The Shifting Boundaries of Public and Private,* ed. Anne Showstack Sassoon (London: Unwin Hyman, 1987), 45–71.

81. See Bates, *Bargaining for Life.*

31 Ministries of Healing: Mary Baker Eddy, Ellen G. White, and the Religion of Health

Ronald L. Numbers and Rennie B. Schoepflin

During the nineteenth century, medicine and religion appeared to be headed in divergent directions. Supernatural explanations of disease, even of epidemics and insanity, fell into disrepute, and the once common practice of combining physical and spiritual healing—called the "angelical conjunction" by the Puritan cleric-physician Cotton Mather—became anachronistic. But despite the secularization of medical theory and the professionalization of medical practice, religiomedical activities continued to flourish. Church-sponsored hospitals grew at an unprecedented rate, medical missionaries circled the globe, and faith healing experienced a late-century revival. Equally indicative of the continuing interaction between religion and medicine was the appearance of two new churches, Christian Science and Seventh-day Adventist, that actively integrated physical and spiritual concerns.[1] This essay focuses on the lives of the women who founded these sects, Mary Baker Eddy and Ellen G. White, and explores the ways in which their strikingly similar

personal experiences influenced their unique healing ministries.[2]

Religious and Medical Reform

The formative years of Mary Baker Eddy and Ellen G. White coincided with a period of intense cultural ferment, during which Jacksonian Democrats challenged the hegemony of elites, fire-and-brimstone revivalists awakened the churches, and unorthodox healers split medicine into competing sects. During the first half of the nineteenth century, the leading churches of the colonial period (Episcopalian, Presbyterian, and Congregational), which supported an educated clergy, gave way to evangelistic Methodists and Baptists, who prized personal experience over theological expertise. Within the various denominations charismatic men and women, each claiming "new light," broke off to form their own movements. It was a time, writes Sydney E. Ahlstrom, in which "farmers became theologians, offbeat village youths became bishops, odd girls became prophets."[3] Each decade seemed to feature a new religious attraction: Mormonism in the 1830s, Millerism in the 1840s, and spiritualism in the 1850s.

Spiritualism, a bane of both Eddy's and

RONALD L. NUMBERS is Hilldale and William Coleman Professor of the History of Science and Medicine at the University of Wisconsin–Madison.

RENNIE B. SCHOEPFLIN is Chair and Associate Professor of History at La Sierra University, Riverside, California.

White's, prospered in part because of its apparent connection with mesmerism (or hypnotism, as it would be called today). Although mesmerism, also known as animal magnetism, originated in Europe in the 1770s, it failed to attract much American attention until the 1830s, when a young French practitioner arrived in White's hometown of Portland, Maine, and began demonstrating his talents. Before long mesmeric displays became a favorite American pastime. "Animal Magnetism," recalled one observer, "soon became the fashion, in the principal towns and villages of the Eastern and Middle States. Old men and women, young men and maidens, boys and girls, of all classes and sizes were engaged in studying the mesmeric phenomena, and mesmerizing or being mesmerized."[4] In this way Americans gained a familiarity with trances, and in some instances, with spirit communication during these states. Thus when the notorious Fox sisters of Hydesville, New York, introduced the country to spirit rapping in the 1850s, they found a well-prepared audience. In fact, so many mesmerists embraced spiritualism that it became virtually impossible for the uninitiated to differentiate between the two. As R. Laurence Moore has noted, "Mesmerized persons, especially those who attributed their powers to the inspiration of guardian spirits, were indistinguishable in their actions from many of the later trance mediums of the spiritualist movement."[5] The same was true of Eddy and White.

The medical, like the religious, world of antebellum America produced numerous reformers.[6] Despite the nation's apparent vitality, sickness abounded, and Americans grew increasingly skeptical about the efficacy of regular physicians, who, in the absence of specific remedies, often bled, puked, and purged their patients. This unhappy state of affairs gave rise in the 1830s to a popular health crusade led by Sylvester Graham, an evangelist and temperance lecturer, who promised health to all who would obtain adequate rest and exercise, control their sexual passions, dress sensibly, avoid all stimulating and unnatural foods, and subsist "entirely on the products of the vegetable kingdom and pure water." He especially touted the benefits of homemade whole wheat bread (which bore little resemblance to the commercially made graham crackers of today). In the unlikely event of illness, Graham recommended letting nature take its own beneficent course. "ALL MEDICINE, AS SUCH, IS ITSELF AN EVIL," he declared.

Many Americans who shared Graham's antipathy toward traditional medicine nevertheless lacked his faith in the healing power of nature alone. Thus when sick they turned for therapeutic assistance to one of the many sectarian healers—most notably, botanics, homeopaths, and hydropaths—who arose in competition with regular physicians. Not surprisingly, health reformers displayed a particular fondness for the one medical sect that offered healing without drugs of any kind: hydropathy or the water cure.

Hydropathy originated with a European peasant, Vincent Priessnitz, who employed an array of water treatments—baths, packs, and wet bandages—to promote healing. When news of his methods reached the United States in the mid-1840s, it touched off a "water-cure craze" that continued unabated until the outbreak of the Civil War. Part of the popularity of hydropathy undoubtedly stemmed from the inadequacies of nineteenth-century medicine, but equally significant was its harmony with the democratic spirit of the times. "The water treatment of disease may fairly be said to originate with an untitled man," wrote one devotee. "This is the people's reform. It does not belong to M.D.'s of any school." Before long, however, enterprising hydropaths began opening schools to train professional hydropathic physicians, roughly one-fifth of whom were women.

Homeopathy, the invention of a German physician, Samuel Hahnemann, enjoyed even greater and longer lasting popularity. During the last decade of the eighteenth century, Hahnemann, dissatisfied with the heroics of orthodox medicine, began constructing an alternate system based in large part upon the healing power of nature and two fundamental principles: the law of similars and the law of infinitesimals. According to the former, diseases are cured by medicines having the property of producing in healthy persons symptoms similar to those of the disease. The latter held that medicines are more efficacious the smaller the dose, even as minute as one-millionth of a gram. Following its appearance in the United States in 1825, homeopathy rapidly grew into a major medical sect. By the outbreak of the Civil War there were nearly twenty-five hundred homeopathic physicians and a multitude of devoted patients. Its appeal is not difficult to understand. Instead of the harsh purgatives and emetics prescribed by regular physicians, homeopaths dispensed pleasant-tasting pills that produced no discomforting side-effects. In part because of its suitability for children, homeopathy won the loyalty of large numbers of American women, who constituted an estimated two-thirds of its patrons and who were among its most active propagators.

In a society that offered women few opportunities outside the home, that kept them out of the pulpit and in the pew, that allowed them to be patients but rarely physicians, medical and religious reform provided a possible way of escape. If women could not be theologians and evangelists, they could at least claim spiritual insight and, if particularly talented, perhaps found their own cult or sect. And if they could not enter regular medical schools and practice orthodox medicine, they could lecture on health reform and practice heterodox healing. Thus it is not coincidental that two of the most influential women in Victorian America found their social niche at the intersection of medical and religious reform.

Mary Baker Eddy

On a cold evening in February 1866 Mary Baker Patterson, the forty-four-year-old wife of an itinerant dentist, slipped on the icy streets of Lynn, Massachusetts. Friends carried the stricken woman to a nearby house and summoned a local homeopathic physician and surgeon, Dr. Alvin M. Cushing, to treat the semiconscious woman for severe head, neck, and back pains. By morning she had recovered sufficiently to endure the sleigh ride to her home in adjacent Swampscott, but when she showed no sign of improvement the following day, her friends feared the worst. Some time during the afternoon of February 6, 1866, while her friends attended Sunday services, Mrs. Patterson opened her Bible to read about the healing ministry of Jesus; "the healing Truth dawned upon my sense," she later recalled, "and the result was that I rose, dressed myself, and ever after was in better health than I had before enjoyed."[7] For the rest of her life she celebrated those moments of visionary insight as the birth pangs of a new age for Christianity.

Abigail Ambrose Baker gave birth to her sixth and last child, a girl named Mary, on July 16, 1821. For the first fourteen years of Mary's life the Baker family farmed a homestead near Concord in the township of Bow, New Hampshire; in 1836 they resettled near Sanbornton Bridge. From childhood on, religious anxiety and physical sickness plagued Mary and prevented her from obtaining more than a smattering of formal education. However, she did receive a little training from her oldest brother, Albert, who instructed her by mail and during visits home.

Later in life Mary fostered an image of her youthful self as a budding litterateur who wrote simple verse, short articles, and letters for newspapers.

Although Mary's parents were both long-standing Congregationalists, their theological orientations differed considerably. Abigail rooted her religious faith in a loving God who eagerly seeks to relieve the trials of his children, while her husband, Mark, measured true belief by adherence to theological tenets and strict moral behavior. Mary welcomed her mother's advice to take her troubles to God in prayer, which often provided relief from her mental and physical torment. During a period of about twelve months around her eighth birthday, Mary often heard voices calling her name. Her mother, initially confused, soon directed her to respond as Eli had instructed the child Samuel under similar circumstances: "Speak, Lord; for thy servant heareth" (I Samuel 3:9).[8] Such advice encouraged Mary to infuse spiritual meaning into the experiences of everyday life.

Mary often discussed religion with her father, who enjoyed engaging friends and acquaintances in religious debate. On one occasion, when she questioned the fairness of divine predestination, the discussion drove her into a fevered state that demanded the soothings of her mother and the attention of a doctor. She stumbled over the same doctrine just after her seventeenth birthday, while she was preparing for membership in her parents' church. Her father struggled to convince her of the fairness of the doctrine by emphasizing the evil of sin, the reality of eternal hell, and the justice of God, but this only made Mary ill, so the pastor accepted her despite her doubts.[9] Mary's inability to love a God who damns some persons to eternal hell later played an important role in the development of Christian Science theology.

Not all of Mary's sicknesses, however, stemmed from intense religious turmoil. Colds, fevers, chronic dyspepsia, lung and liver ailments, backache, and nervousness plagued her youth, and throughout her adult life she suffered from gastric attacks, severe depression, and short episodes of incoherent babbling or foaming at the mouth. In 1843 she married George Washington Glover, a successful builder in Charleston, South Carolina, who died seven months later, leaving her pregnant. This traumatic experience plunged her into a deep depression, and for several months she remained physically and mentally unable to care for her infant son, George. Upon temporarily recovering, she refused to be tied down by routine child care and sought instead to find fulfillment in a literary career. In 1851 young George moved in permanently with his nurse and her husband, thus freeing his mother from her maternal obligations. Desirous of a man's companionship, Mary married a ne'er-do-well dentist, Daniel Patterson, in the summer of 1853, but their marriage foundered when Mary's recurring maladies confined her to bed for long periods and her wandering husband intermittently deserted her. After 1866 they no longer lived together, and the marriage legally ended in divorce in 1873.

In her prolonged battle against physical and mental illness Mary tried practically every therapy available to the chronically ill of the nineteenth century. In 1837 she briefly adopted Graham's vegetarian regimen, but, although she returned to it off and on through the next twenty-five years, it brought little relief. In the 1840s she experimented with mesmerism and learned about the possibilities of mental healing. At times she consulted homeopathic physicians, and for a while in the 1850s she herself practiced homeopathy on neighbors and friends, in the process discovering the therapeutic power of positive thinking. In 1862 she visited the Vail Hydropathic Institute in Hill, New Hampshire, but the water treatments she received did little to improve her deteriorating health. Disillusioned with the water cure, she turned her hopes toward the healing hands of the famous mentalist of Portland, Maine, Phineas Parkhurst Quimby.

Quimby, an early convert to mesmerism, had carefully studied the principles of mesmeric healing and transformed them, believing that if a healer simply helped patients to create a positive mental attitude toward their illnesses, their bodies would heal themselves. The healer vicariously assumed the patients' symptoms, discussed their problems with them to instill a positive attitude, and occasionally manipulated their injured limbs or rubbed their heads. Mary Patterson flourished in Quimby's hands, and when she fell ill again in 1863 she returned to the Portland therapist and remained as his student during the following winter. With near adoration she served as his pupil, disciple, evangelist—and patient, whenever her maladies reoccurred. Several times she attempted to practice Quimbyism on her own, but she repeatedly encountered difficulty ridding herself of her patients' symptoms, vicariously acquired, and returned to Quimby for release. When she learned of his death on January 16, 1866, she felt a severe loss. Less than a month later she discovered "the healing Truth."

For the next four years Mary Patterson bounced from one boardinghouse to another as she struggled to write down the views of humans, God, sickness, and sin that she believed God had revealed to her. She continued to fight occasional relapses and the symptoms she acquired from her practice of Quimby-like healing and occasionally dabbled in spiritualism. Impressed that she could financially support herself by instructing others in her healing methods, she first advertised classes in the July 4, 1868, issue of the spiritualist journal *Banner of Light*. Her literary efforts resulted in notes for a commentary on Genesis and a pamphlet entitled *The Science of Man, By Which the Sick Are Healed*, which she first used in 1870 to instruct students in "Moral Science," an early form of her teachings that she later called Christian Science.[10] In the spring of 1870 she moved to Lynn, Massachusetts, established a partnership with Richard Kennedy, an early student whose healing practice helped support her teaching, and began to write the textbook of Christian Science, *Science and Health* (1875).

At the basis of Eddy's doctrine lay a radical idealism that denied the existence of anything but God and the ideas that generate from his being.[11] As she stated in the last edition of *Science and Health* to appear before her death, "God is incorporeal, divine, supreme, infinite Mind, Spirit, Soul, Principle, Life, Truth, Love."[12] God is one, and humans and the universe are his reflections (ideas). Since only God and synonyms for him, such as Mind, Life, Truth, and Good, exist, it follows that matter, death, error, and evil do not exist. On such grounds Eddy proclaimed a theodicy that erased all her anxieties about her father's doctrine of predestination and created a universe infused with the spirituality of her mother's proddings.

Drawing upon the language of religious contemporaries who believed that Christianity should be updated or reformed by scientific principles, Eddy called her religion Christian Science. She believed that her doctrines contained a kind of mathematical certainty that followed from their syllogistic form. To illustrate: God is All; God is Good; therefore evil (sin) does not exist. Or God is All; God is Spirit; therefore matter (sickness and death) does not exist. Or God is All; God is Truth; therefore error does not exist. Furthermore, Christian Science doctrine could be tested just like scientific knowledge. Eddy believed that healings and victories over sin and death demonstrated the truth of her claims for the Allness of God and the unreality of disease, sin, and death. By demonstrate she meant that one could test her teachings in the laboratory of life just as scientists test their theories by observing nature and experimenting in their laboratories. If one overcame sickness or sin by studying and practicing Eddy's teachings, then one had proven their truth.

To give demonstration a fair chance, however, the prospective practitioner of Christian Science needed to receive proper instruction, theoretically obtained by the careful study of *Science and Health* but preferably acquired by attending a course of instruction directed by Eddy or one of her appointed teachers. Since healing provided the ultimate proof of Eddy's teachings, she filled her writings with practical instructions, which changed over time to meet different situations, resulting in various sets of instructions and three basic types of practice. The first type of practice, commonly known as "audible treatment," closely followed the methods of Quimby. Since sickness was an erroneous belief, the practitioner corrected the belief of the patient through verbal argument. Using the second method, "mental argument," the practitioner aimed to correct the thought of the patient not by verbal instruction but by thought alone. The third and most esoteric level of treatment, "impersonal treatment," was based on Eddy's view that "if the *healer realizes* the truth, it will free his patient."[13] Therefore practitioners directed their attention toward correcting their own thoughts and argued with their own error-filled minds.[14]

The 1870s were filled with turmoil for Eddy as she met challenges to the uniqueness of her teachings. Tensions with Kennedy over her insistence that he reduce the amount of rubbing and manipulating in his practice led to a separation in 1872; that same year a former student, Wallace W. Wright, charged in a Lynn newspaper that Moral Science was nothing more than mesmerism. In the face of such criticism Mrs. Patterson ceased all manipulation and labeled it

a sign of mesmerism, a term she used to describe all forms of mental healing that did not coincide with her practices. Suspicious that certain students stole spiritual energy from her while others directed magnetic forces against her, she denounced such "malicious animal magnetism" and gathered her closest followers around her to ward off its evil influences. Buoyed by faithful supporters, and later encouraged by her new husband, Asa Gilbert Eddy, a converted sewing machine salesman whom she married in 1877, Eddy refused to allow anyone to wrest control of the movement from her hands.

From 1872 to 1875 Eddy industriously wrote and rewrote drafts of *Science and Health,* based on what she claimed to be divine insights gained through study of the Bible and early morning visions, which usually occurred at times of severe stress and anxiety and which provided supernatural answers to problems encountered in her work. Adamantly rejecting the claims of critics who found similarities between her writings and those of mental healers or idealist philosophers, Eddy asserted that she had "consulted no other authors and read no other book but the Bible for about three years," adding that "it was not myself, but the divine power of Truth and Love, infinitely above me, which dictated 'Science and Health with Key to the Scriptures.'"[15] Using an allegorical hermeneutic, which she generously applied to the Bible, she interpreted her visions and dreams in terms of her developing doctrines and metaphysical principles. At times when former students opposed her teachings or leadership, her visions featured apocalyptic images of cosmic struggle that revealed her human antagonists and identified her movement with the male child of Revelation 12, which the great red dragon sought to devour.[16] Although Eddy's basic views on sickness and health and her understanding of reality remained fairly stable, she believed that inspiration guided her continual revision of *Science and*

Health, and she adapted her beliefs to a growing movement and a changing world.

Eddy's attitudes toward nineteenth-century physicians verged on outright derision, as, for example, when she claimed that "when there were fewer doctors and less thought bestowed on sanitary subjects there were better constitutions and less disease."[17] In addition to denouncing the drugging of allopaths, the bathing of hydropaths, and the eating habits of Grahamites, she saved a special dose of invective for all mentalists and mind curists, whom she regarded as at best frauds and at worst criminals. Homeopathy, she believed, stood midway between the darkness of allopathy and the radiance of Christian Science, but since the truth of science had dawned, homeopathy too must be put aside. At times she acknowledged that many physicians possessed "great philanthropy of purpose," but she urged them to "make their endeavors more effectual by changing their basis of action" to the metaphysics of Christian Science.[18] Confronted by the concrete realities of medical licensing laws and public health ordinances requiring vaccination and quarantine, Eddy and her followers moderated their radicalism with some practical compromise. She instructed Christian Scientists to submit to vaccination and to report contagious cases to the proper authorities when the laws demanded, but her lieutenants continued to lobby to exclude religious healing from state regulation. Problems with metaphysical obstetrics, the application of Christian Science to childbirth, provide the best illustration of Eddy's adaptive ability.

In an attempt to add prestige to her movement and to garner the respect of the medical profession, Eddy chartered the Massachusetts Metaphysical College in 1881 and in 1882 published a prospectus that carried the names of cooperating physicians in the Boston area. Although she advertised herself as "Profesor of Obstetrics, Metaphysics, and Christian Science,"

she did not offer a course in obstetrics for five years; nevertheless, many women found the principles of Christian Science especially attractive during childbirth, when their beliefs removed or lessened pain.[19] Testimonies praising the methods of painless childbirth often appeared in the monthly *Christian Science Journal,* and Eddy capitalized on this application of her doctrines by instructing students during 1887 in the mental control of the errors of childbirth—belief in anatomy, physiology, physical intercourse, and pain. However, the tragic death three years later of a mother and her newborn during a mentally assisted childbirth led to an indictment of manslaughter against a Christian Science practitioner, Abby H. Corner. The case ended in acquittal, but in view of the harmful impact such incidents could have on her movement, Eddy instructed her followers to cooperate with physicians in cases of childbirth. Although she discouraged mixing the material with the spiritual as a general policy, she gave her blessing to a group of associates who enrolled in medical school with the purpose of uniting physical obstetrics with the metaphysical techniques of Christian Science, including pain control.[20]

Eddy best displayed her talents through the organization and administration of her movement. Whenever ambitious lieutenants or popular healers threatened to usurp her authority or infringe upon her popularity, she skillfully maneuvered them from positions of leadership or declared their beliefs unorthodox. When disaffection and dissent decimated the movement's leadership and decreased its membership by one-third during 1888, Eddy took the drastic measure of dissolving all of her existing church organizations. She withdrew from Boston, which had been the center of her activities for the past ten years, to Concord, New Hampshire, where she reflected, wrote, and reorganized her movement. From this distance she directed the construction of the Mother Church in Boston and formalized regulations for instruction in Christian Science. These two efforts centralized control of the church and placed it firmly within the hands of boards of directors appointed and controlled by Eddy. She devoted special attention to the publication of Christian Science literature, establishing the Christian Science Publishing Society in 1898 to spearhead a strong program of world evangelism. The society published numerous editions of Eddy's writings, as well as weekly, monthly, and quarterly journals; in 1908 it began publishing the *Christian Science Monitor*.

During this period of reorganization and consolidation, Christian Science membership exploded. Numbering only 8,724 in 1890, by 1906 the church had grown to 47,083, 72.4 percent of whom were women.[21] These years of growth coincided with the expansion of leadership roles for women in the movement. The Christian Science doctrine of a bisexual God, Eddy's own influence as a role model, and the career opportunities open to Christian Science healers had always attracted women, but before the mid-1880s they occupied few important positions of leadership in the movement. After the defection of many of Eddy's male leaders she came to recognize the talents of female practitioners, who outnumbered males five to one by the 1890s, and appointed them as journal editors, lecturers, and teachers.[22] However, most key positions of administration, such as board directorships, remained in the hands of men. Through the evangelistic efforts of public lectures, urban dispensaries, and reading rooms, Christian Scientists disseminated their doctrines and distributed their literature throughout America, and foreign evangelism soon transformed the church into a worldwide movement with major centers of activity in Germany and England.

Assailed by physical pains and mental anxie-ties that mirrored her fears for the future of her movement, Eddy struggled during her final years to direct the affairs of the church through subtle adjustments of the church manual and public defenses of her mental stability. Finally, her constitution, weakened by old age, succumbed to pneumonia, and she "passed on" in December 1910, leaving a movement best known for its striking physical healings to search for a new identity in a world increasingly devoted to the therapies of scientific medicine.

Ellen G. White

Ellen Gould Harmon was born November 26, 1827, in the village of Gorham, Maine, less than a hundred miles from where six-year-old Mary Baker lived in New Hampshire.[23] A few years after her birth her father, a self-employed hatter of modest means, moved his family to nearby Portland, where Ellen subsequently enrolled in school. After completing only three grades or so, she nearly lost her life when an angry schoolmate hit her in the face with a rock, knocking her to the ground unconscious. For three weeks Ellen lay in a stupor, while friends and relatives waited for her to die. She survived, but her injuries continued to plague her for years, and her facial disfigurement—initially so bad that her own father could scarcely recognize her—caused frequent embarrassment and prevented breathing through her nostrils. Frayed nerves rebelled at simple assignments such as reading and writing. Her hands shook so severely she could not write, and words appeared to be mere blurs on a page. Try as she might, she could not continue her studies. Thus frustrated, she resigned herself to the life of a semi-invalid, passing the time of day propped up in bed knitting or helping her father make hats.

In March 1840 a touring farmer-preacher named William Miller visited Portland and aroused the citizens with his prediction, based on biblical prophecies, that Christ would return to earth "about the year 1843." Ellen, who had been raised a Methodist, joined the crowds who turned out to hear Miller and became convinced that indeed the world would soon end—and that she would probably be lost. At times she sank into deep despair and spent sleepless nights bowed in prayer, "groaning and trembling with inexpressible anguish, and a hopelessness that passes all description." Sermons vividly depicting the red-hot flames of hell intensified her torment and pushed her to the verge of a mental breakdown.

While in this state of mind Ellen began having religious dreams. At her mother's urging she consulted her minister, who interpreted this development as a sign that God had chosen her for "some special work." Her mother agreed, on one occasion insisting that her daughter's fainting during an emotional prayer meeting was not a cause for concern but merely a manifestation of "the wondrous power of God." Reinforced by the views of her minister and mother, Ellen began speaking at public meetings and holding private prayer sessions for her teenaged friends.

As the months of 1843 slipped by, Millerite leaders began to suspect an error in their calculations; Christ's second coming, they finally decided, would actually occur on October 22, 1844. Ellen, by now convinced that she would be among the saints, eagerly awaited this date—only to have her hopes dashed when Christ failed to appear. Confused and disappointed, she sought divine guidance. One day in December, while praying with a few women friends, seventeen-year-old Ellen felt the "Holy Spirit" resting upon her in a new and dramatic way. Bathed in light, she seemed to be "rising higher and higher, far above the dark world." In this trancelike state, the first of her many visions, she received reassurance that her faith in Christ's return had not been misplaced. In a subsequent vision an angel guide explained that Christ could not come to earth until the Millerites began keeping Saturday instead of Sunday as the sabbath.

Ellen's visions followed no set pattern; she might be praying, addressing a congregation, or lying sick in bed, when suddenly she would be off on "a deep plunge in the glory." During these episodes, which lasted from a few minutes to several hours, she frequently described the colorful scenes—past and future, celestial and terrestrial—she was seeing. According to the testimony of numerous physicians and curiosity seekers, her vital functions slowed alarmingly, her heart beating sluggishly and respiration becoming imperceptible. Although she was able to move with complete freedom, strong men could not budge her limbs. Until the 1870s, when nighttime dreams gradually replaced daytime trances, Ellen averaged five to ten visions a year.

As both Ellen and skeptics recognized, her visionary experiences differed little from the trances of countless mesmerists, spiritualists, and religious enthusiasts. In fact, she herself for a time suspected that her visions might be only a mesmeric delusion. As divine punishment for questioning her gift, she was temporarily struck dumb and forced to communicate by means of a pencil and slate—the first time since her accident that she could write without shaking. Like Eddy, she greatly feared coming under the influence of unscrupulous mesmerists, whom she regarded as being "channels for Satan's electric currents."[24] When threatened by such a force, she requested an extra angel God had promised to send to protect her.

In one of her early visions Ellen received directions to travel among the scattered flock of Millerites, relating what she had seen and heard. Shy but ambitious, she worried that her new role as God's messenger might make her proud. But after an angel assured her that the Lord would preserve her humility, she determined to carry out his will. Only one obstacle stood in her way: the need for a traveling companion. At five feet, two inches, and barely eighty pounds, she was little more than skin and bones. Fatigue from long trips on steamboats and railway cars brought on dangerous fainting spells, during which she sometimes remained breathless for minutes. Obviously, she could not travel alone.

The solution to her problem appeared in the form of a twenty-three-year-old Millerite minister and erstwhile teacher, James White. Soon after meeting Ellen in 1845, James accepted her claim to be a latter-day prophet and volunteered to serve as her escort. Over the objections of Mrs. Harmon, who feared for her daughter's good reputation, the couple began contacting Millerites in New England. Because of James's conviction that marriage was inconsistent with belief in the imminent return of Christ, he and Ellen postponed marriage—until ugly rumors about them began to circulate. It was clear, said James one day, that "something had got to be done." So on August 30, 1846, they became husband and wife.

Married life for the Whites was far from glamorous. For years they worked as itinerant preachers, barely surviving on the meager contributions of their supporters. When their first child, born in 1847, was only one, they reluctantly left him with friends. The separation nearly broke Ellen's heart, but she vowed not to let her motherly affection keep her "from the path of duty." The arrival of a second son, in 1849, brought only a brief interruption to their nomadic life. He too was soon left with a kind Adventist sister in New York.

In 1852 the impoverished couple, worn out by years on the road, settled down to a semipermanent home in Rochester, New York, and collected their children about them. In 1854 Ellen gave birth to a third boy, and the following year she and James moved on to Battle Creek, Michigan, which remained their home until the 1880s. Here they established institutional headquarters

for their fledgling church, which by the early 1860s numbered thirty-five hundred members, who called themselves Seventh-day Adventists. These people believed in the imminent return of Christ, observed the seventh-day sabbath, and regarded Ellen White as a divinely inspired prophet. They also provided financial support for the Whites, allowing them for the first time to enjoy a relatively comfortable existence. During these years Ellen not only served as wife and mother to a growing family—a fourth baby boy arrived in 1860—but continued to fill speaking engagements and to publish her "testimonies" and other writings.

Her health, however, remained precarious. Sprinkled liberally throughout her various writings are complaints of lung, heart, and stomach disorders, frequent "fainting fits," paralytic attacks (at least five by her mid forties), pressure on the brain, and breathing difficulties. At least once a decade from her teens through her fifties she expected imminent death from disease. She frequently suffered from anxiety and depression and at times wanted to die. On one occasion her "mind wandered" for two weeks; on another it became "strangely confused." Although she commonly ascribed the illnesses of others to intemperate living, she tended to attribute her own mental and physical ailments to the machinations of Satan and his evil angels, who had made her and her husband "the special objects" of their attention and who had caused several near-fatal accidents.[25]

On the evening of June 5, 1863, Ellen White, now thirty-five, joined friends in rural Michigan for vespers. Lately, her mind had often turned to matters of health. Her sons had recently been threatened by diphtheria, James appeared to be on the verge of a physical and mental breakdown, and she herself was, in her own words, "weak and feeble, fainting once or twice a day." While praying, she went into a trance and began receiving instructions from heaven on preserving and restoring health. God's people, she learned, were to give up eating meat and other stimulating foods, shun alcohol and tobacco, and avoid drug-dispensing doctors. When sick, they were to rely solely on nature's remedies: fresh air, sunshine, rest, exercise, proper diet, and—above all—water. Adventist sisters were to abandon their fashionable floor-length dresses for "short" skirts and pantaloons, and all believers were to curb their sexual passions. As a result of this vision, health reform for Ellen White and her followers became a religious obligation, essential to salvation. It is, she declared, "as truly a sin to violate the laws of our being as it is to break the ten commandments."

The horrible consequences of self-abuse or masturbation especially impressed her. "Everywhere I looked," she later recalled in describing her vision, "I saw imbecility, dwarfed forms, crippled limbs, misshapen heads, and deformity of every description." Her first book on health reform consisted of *An Appeal to Mothers: The Great Cause of the Physical, Mental, and Moral Ruin of Many of the Children of Our Time* (1864), a work designed to strike fear in the most hardened of hearts. Graphically describing the effects of masturbation among girls, particularly vulnerable because they possessed less "vital force" than boys, she wrote: "The head often decays inwardly. Cancerous humor, which would lay [sic] dormant in the system their life-time, is inflamed, and commences its eating, destructive work. The mind is often utterly ruined, and insanity takes place."[26] For the first time she appreciated her childhood accident: it had allowed her to grow up in "blissful ignorance of the secret vices of the young." Only after her marriage to James, she insisted, had she learned about masturbation from "the private deathbed confessions of some females."

For several years most of White's sexual advice focused on self-abuse. In 1868, however, she received a second revelation on sex, in which she

was shown that even married persons were accountable to God "for the expenditure of vital energy, which weakens their hold on life and enervates the entire system." After this vision she began counseling Christian wives not to "gratify the animal propensities" of their husbands but to seek instead to divert their minds "from the gratification of lustful passions to high and spiritual themes by dwelling upon interesting spiritual subjects." Husbands who desired "excessive" sex she described as "worse than brutes" and "demons in human form." Although she never defined exactly what she meant by excessive, some evidence suggests that she frowned on having intercourse more than once a month.

From 1863 until her death in 1915 White, with varying degrees of zeal and success, proclaimed the gospel of health. Largely as a result of her crusade, many Adventists adopted a twice-a-day diet of fruits, vegetables, grains, and nuts and gave up tea, coffee, meat, butter, eggs, cheese, rich desserts, and "all exciting substances," which, she argued, not only caused disease but stimulated unholy sexual desires. Although she at first reported great progress in changing the eating habits of her followers, there soon appeared signs of "a universal backsliding." Fish and flesh reappeared on Adventist tables, and even among ministers vegetarianism became the exception rather than the rule. By the mid-1870s White herself was indulging her fondness for flesh foods, to the chagrin of disciples who remained true to her health reform message. It was not until the 1890s that she finally gained a permanent victory over meat and began leading her church back into the vegetarian fold.

Dress reform proved equally frustrating. During the 1850s a number of prominent feminists, including Elizabeth Cady Stanton, Susan B. Anthony, and Amelia Bloomer, briefly abandoned their corsets and long skirts for an outfit consisting of a short skirt over pantaloons. Although these women soon discarded the Bloomer costume, it remained popular attire at some water cures and among certain spiritualists. At first—even after her 1863 vision—White damned the reform dress and the political goals of its advocates. "Those who feel called out to join the movement in favor of women's rights and the so-called dress reform," she declared, "might as well sever all connection with the third angel's message," as she called her theology. Not only was it wrong for women to wear men's clothing, she explained, but the practice might lead nonmembers to confuse Adventists with spiritualists, a likelihood increased by White's claim that she too communicated with celestial beings.

Within two years of her vision, however, White changed her mind and began advocating a skirt-and-pants costume she claimed to have seen in her vision. She further confused her followers by first saying that God wanted skirts to clear the ground by "an inch or two," then declaring that they should reach "somewhat below the top of the boot," and finally settling on nine inches from the floor. Her version of the Bloomer—called a "woman-disfigurer" by her own niece—never caught on. It was, complained the prophetess in 1873, "treated by some with great indifference, and by others with contempt." Two years later God mercifully granted her permission to discard her pantaloons and end her divisive dress reform campaign.

During her seminal 1863 vision White, who had long held doctors in low esteem—in fact, she had once condemned the use of "earthly physicians," only to reverse herself after being accused of contributing to the death of a follower who failed to seek medical help—learned that God shared her suspicions about the medical profession. "I was shown that more deaths have been caused by drug-taking than from all other causes combined," she wrote. "If there was in the land one physician in the place of thou-

sands, a vast amount of premature mortality would be prevented." All drugs, vegetable as well as mineral, were proscribed; of the various medical sects, only hydropathy received divine sanction. This is not surprising, since White had discovered the water cure only a few months earlier and had employed it successfully to save two of her sons from possible diphtheria. Before the Adventists built their own water cure, White frequently gave hydropathic treatments to her neighbors in Battle Creek.

In 1864 and again the next year Ellen and James White visited one of the most successful water cure establishments in America, operated by Dr. James Caleb Jackson in Dansville, New York, and returned home to set up a similar operation in Battle Creek. Their Western Health Reform Institute, staffed by both male and female "doctors," experienced a rocky first decade. Then a young protégé of the Whites, Dr. John Harvey Kellogg, took over and turned the ailing institute into a world-famous sanitarium with branches from coast to coast. In his spare time he invented cornflakes and other health foods, from which his brother W. K. Kellogg made a fortune.

In 1907 a cabal of Adventist ministers excommunicated the imperialistic Kellogg, ostensibly for questioning the supernatural origin of White's visions. In departing the doctor took the Battle Creek Sanitarium and the church's only medical school, American Medical Missionary College, with him. To compensate for this loss White loyalists two years later converted a sanitarium in Loma Linda, California, into the College of Medical Evangelists (now Loma Linda University), which accepted both male and female students. According to White, it was "the Lord's plan" that women be treated by members of their own sex; ignoring such distinctions offended God and led to "much evil." Of the six M.D.'s on the original CME faculty, two were women: Julia A. White, professor of obstetrics and gynecology, and Lucinda A. Marsh, who taught pediatrics. Both had received their medical degrees from AMMC, which, like CME, taught a mixture of regular and hydropathic medicine.[27]

Over the years Ellen White wrote hundreds, perhaps thousands, of pages on health-related subjects. Although she repeatedly stressed the divine origin of her "testimonies" and denied acquaintance with earthly sources—"My views were written independent of books or of the opinion of others"—a comparison of her writings with those of other nineteenth-century health reformers reveals close parallels in both content and language, a problem discovered by Kellogg and other contemporaries.

For years Ellen relied on James to correct her grammar, polish her style, and publish her work. All that ended in 1881, when James died. Their marriage had not been without its trials. James was an impetuous man who easily took offense. He was excessively jealous of his wife's friendship with real or imagined rivals in the church hierarchy and during the last decades of his life suffered from long bouts of mental illness. At times he resented his wife's superior position and on occasion refused to sleep in the same house with her. Nevertheless, he was a man quick to forgive and to make amends, and, whatever his failings, Ellen loved and respected him and leaned on him in her hours of need. Without him, her career as a prophetess would probably have never gotten off the ground. Since the 1840s, publishing had been his passion— and the key to her success. In those early days it was he who insisted on printing her visions and on creating a strong church organization, over which for ten years he presided as president. Seventh-day Adventism would not have been the same without Ellen White; it would not have existed without James.

After her husband's death the grief-stricken widow sank into a yearlong depression. Then one night the Lord appeared to her in a dream and said: "LIVE. I have put My Spirit upon your son, W. C. White, that he may be your counselor. I have given him the spirit of wisdom, and a discerning, perceptive mind." Comforted by these words and the knowledge that her favorite son would remain by her side, she resumed her ministry with renewed zeal, spending two years in the mid-1880s in Europe and most of the 1890s in Australia and New Zealand. In 1900 she returned to the United States and settled on a comfortable farm in northern California, from which she continued to guide her growing church. Five years later she published her last major work on health, *The Ministry of Healing*.

On July 16, 1915, five months after a broken thighbone confined her to a wheelchair, Ellen White, aged eighty-seven, died. After a lifetime of illness and frequent brushes with death, she finally succumbed to chronic myocarditis, complicated by arteriosclerosis and asthenia resulting from her hip injury. In a fundamental way her life had been a paradox. Consumed with making preparations for the next world, she nevertheless devoted much of her energy toward improving life and health in this one. At the time of her death thirty-three Adventist sanitariums and countless treatment rooms spanned six continents. Although she had never received the national attention given to Mary Baker Eddy, whom she regarded as a satanically inspired spiritualist, over 136,000 devoted followers mourned her passing. By the late 1990s her church had grown to over eight million members, many of whom enjoyed healthier and longer lives for having adopted White's advice.

Parallel and Divergent Careers

The striking parallels between Eddy and White, from their births in New England in the 1820s to their creation of health-conscious churches, are more than historical curiosities; they illuminate the relationship between health and religion and the role of women in Victorian America. As youths both rebelled against orthodox theology, much as they would later reject orthodox medicine. Eddy's struggles against her "father's relentless theology" were so intense they occasionally precipitated spells of sickness, while White became so agitated by theological issues, she nearly went insane. As she herself surmised, "many inmates of insane asylums were brought there by experiences similar to my own."[28] Both women also suffered intensely from physical complaints, which restricted their activities and impelled them on a lifelong quest for health. Of course, nineteenth-century America abounded with physical invalids and religious rebels, but few turned their afflictions to such advantage as Eddy and White, who proudly displayed their sufferings as badges of divine calling.

Both women struggled to make their mark, to escape from the anonymity of domestic life and fulfill ambitious dreams. Eddy, who believed that her "mission was to write poetry," aspired to a literary career; White, who described herself as "naturally proud and ambitious," aimed "to become a scholar."[29] Above all, Eddy and White sought control: over sin, sickness, and society. For ambitious Victorian women moral reform offered one of the few avenues to power. Since the beginning of the century women had outnumbered men as church goers, and, as long as they submitted to male authority, ministers encouraged them to engage in religious education and moral uplift. This arrangement, writes Nancy F. Cott, "enabled them to rely on an authority beyond the world of men and provided a crucial support to those who stepped beyond accepted bounds."[30]

Both Eddy and White went out of their way to attribute their activities to divine selection rather than personal ambition, having been told since childhood that God had chosen them for a special mission. Both of their mothers, who no doubt viewed their daughters in the context of the biblical prophecy about the "last days"— "your sons and your daughters shall prophesy, and your young men shall see visions, and your old men shall dream dreams" (Acts 2:17) —attached religious significance to unusual dreams, hallucinations, and fainting spells.

Many nineteenth-century visionaries and mystics claimed divine inspiration for their teachings, but success depended as much upon organization as upon inspiration. Within Adventism, James White, who "discovered" Ellen and orchestrated her early career, was the organizational genius. While his wife provided spiritual guidance and theological validation, James ran the printing presses, built the institutions, and fought the political battles. Eddy, who seemed to prefer weaker men, tended to rely on her own formidable administrative skills, although she often delegated responsibilities to male subordinates.

An essential element behind Eddy's and White's achievement was their ability to combine pragmatism with dogmatism as the situation demanded. Both of them repeatedly adapted their teachings to meet the needs of the time. Eddy, for example, abandoned metaphysical midwifery when the costs became too great and, despite her conviction that Christian Science could cure all ills, allowed her followers to seek surgical assistance for such conditions as broken bones, reasoning that "in the present infancy of this Truth so new to the world, let us act consistent with its small foothold on the mind."[31] Over the years White lifted her ban on the use of physicians, discarded her reform dress, and moderated her dietary advice, especially with regard to the use of dairy products. Although such shifts led to criticism and charges of inconsistency, both women executed them with a deft timing that left their authority intact, if not untarnished. Against serious challenges to their authority, however, both Eddy and White stood firm. When evidence came to light of troublesome similarities between the views of Eddy and Quimby and between those of White and other health reformers, each insisted, with an intensity that betrayed her anxiety, that she depended solely on divine revelation.

Eddy and White attracted followers in part because their systems met real medical and religious needs but also because of the magnetic personalities they both possessed. Their styles, however, differed considerably. Eddy's piercing blue eyes, attractive countenance, and fashionable attire revealed a contagious self-assurance. White, a plain woman in both appearance and dress, felt less secure in public, but her personal piety inspired emulation and her clear forceful sermons could bring hardened sinners to their knees.

Such differences are important in understanding their respective teachings. Although both viewed illness in a moral light, as resulting from sin, the metaphysically inclined Eddy followed Quimby in identifying illness with wrong belief, while the practical White followed Graham and other health reformers in viewing sickness as resulting from wrong practices, violations of divinely ordained laws of physiology. Not surprisingly, both claimed divine endorsement for the type of therapy each had found most efficacious in her own life. Eddy benefited little from hydropathy and Grahamism but obtained at least occasional relief from homeopathy and Quimbyism, the influence of which is readily apparent in Christian Science. White's vision on the merits of health reform and hydropathy came within months after she had discovered the water cure and employed it to save the lives of her children.

Although both women might be called vi-

sionaries, visions played substantially different roles in their ministries. For White and her followers the visions validated her claim to divine inspiration. Often occurring in public and accompanied by dramatic physical signs, her visions induced a sense of awe and reverence and provided theological direction for the Adventist church. Eddy's visions, in contrast, occurred in private and served primarily a rhetorical function, guiding her writing and uplifting her spirits. Rather than looking to her visions for validation of her claims, Christian Scientists examined the internal logic of her doctrines and the evidence of healed bodies and changed lives.

Whatever their respective motivations and teachings, Eddy and White exerted a marked influence on nineteenth-century American culture: on religion, medicine, and women. They founded two of the largest churches of American origin, which ultimately touched millions of lives. The therapeutic success of Christian Science helped to promote a renaissance of faith healing in Protestant churches and provided physicians with compelling evidence of the relationship between mind and body. Adventist medical workers helped to transform sectarian hydropathy into the hydrotherapy of scientific medicine, and the health principles taught by White aided her followers in living healthier and longer lives. Through her influence on Kellogg she revolutionized the breakfast habits of a nation. Eddy and, to a lesser extent, White also expanded the opportunities for women interested in health-related careers. From the early days of Christian Science female practitioners far outnumbered males. Although White displayed little sympathy for feminism and discouraged Adventist sisters from following in her footsteps, she urged the creation of nursing schools and, by insisting on the immorality of men treating women, encouraged at least a few Adventist women to take up the practice of medicine, especially obstetrics. Few of their contemporaries—male or female—accomplished more.[32]

Notes

1. Ronald L. Numbers and Ronald C. Sawyer, "Medicine and Christianity in the Modern World," in *Health/Medicine and the Faith Traditions: An Inquiry into Religion and Medicine,* ed. Martin E. Marty and Kenneth L. Vaux (Philadelphia: Fortress Press, 1982), 133–60.

2. Born Ellen Gould Harmon, White used her married name, Mrs. White or Ellen G. White, for most of her professional career. Her followers often referred to her as Sister White. Eddy was known by many names over her lifetime. Born Mary Morse Baker, she became Mary Baker Glover after her first marriage in 1843. Usually called Mrs. Patterson or Mary M. Patterson after her second marriage in 1853, she returned to the use of Mary Baker Glover after a prolonged estrangement from her husband and her "discovery" in 1866. After her marriage to Asa Gilbert Eddy in 1877, she used Mary Baker Glover Eddy, Mary Baker Eddy, or simply Mrs. Eddy. Her followers often referred to her as Mother.

3. Sydney E. Ahlstrom, *A Religious History of the American People* (New Haven: Yale University Press, 1972), 475.

4. John D. Davies, *Phrenology—Fad and Science: A Nineteenth-Century American Crusade* (New Haven: Yale University Press, 1955), 126–27. See also Eric T. Carlson, "Charles Poyen Brings Mesmerism to America," *Journal of the History of Medicine and Allied Sciences* 15 (1960): 121–32; and Robert C. Fuller, *Mesmerism and the American Cure of Souls* (Philadelphia: University of Pennsylvania Press, 1982).

5. R. Laurence Moore, *In Search of White Crows: Spiritualism, Parapsychology, and American Culture* (New York: Oxford University Press, 1977), 9.

6. The paragraphs on medical reform that follow are extracted from Ronald L. Numbers, *Prophetess of Health: A Study of Ellen G. White* (New York: Harper and Row, 1976), ch. 3.

7. Mary Baker Eddy, *Miscellaneous Writings, 1883–1896* (Boston: The Trustees Under the Will of Mary Baker G. Eddy, 1924), 24. Eddy sometimes recalled Mark 3 and sometimes Matthew 9:2 as the passage of healing.

8. Mary Baker Eddy, *Retrospection and Introspection* (Boston: Allison V. Stewart, 1916), 8–9.

9. Eddy, *Retrospection,* 13; Robert Peel, *Mary Baker Eddy: The Years of Discovery* (Boston: Christian Science Publishing Society, 1966), 23, 50–51.

10. Eddy first called her teachings Moral Science, later Metaphysical Science or simply Metaphysics, and finally Christian Science.

11. Although Eddy believed God exhibited both masculine and feminine characteristics, except in the third edition of *Science and Health* (1881) she regularly used the masculine pronouns.

12. Mary Baker Eddy, *Science and Health with Key to the Scriptures* (Boston: First Church of Christ, Scientist, 1971), 465.

13. Mary Baker Eddy, *Rudimental Divine Science* (Boston: Allison V. Stewart, 1915), 13.

14. For this categorization of healing practices we acknowledge Charles S. Braden, *Christian Science Today: Power, Policy, Practice* (Dallas: Southern Methodist University Press, 1958), 336–49.

15. Mary Baker Eddy, *The First Church of Christ Scientist and Miscellany* (Boston: Allison V. Stewart, 1916), 114. *Science and Health* included *Key to the Scriptures,* a glossary containing allegorical meanings for key biblical words, after the sixth edition of 1883.

16. Robert Peel, *Mary Baker Eddy: The Years of Trial* (Boston: Christian Science Publishing Society, 1971), 25–28, 75, 135.

17. Mary Baker Eddy, *Science and Health* (Boston: Christian Science Publishing Company, 1875), 341.

18. Ibid., 365.

19. Peel, *Years of Trial,* 80–82, 111.

20. Ibid., 236–40.

21. Henry King Carroll, *The Religious Forces of the United States Enumerated, Classified, and Described* (New York: Charles Scribner's Sons, 1912); A. J. Lamme III, "Christian Science in the U.S.A., 1900–1910: a distributional study," Discussion Paper Series, Department of Geography, Syracuse University, 3, April 1975.

22. Stephen Gottschalk, *The Emergence of Christian Science in American Religious Life* (Berkeley: University of California Press, 1973), 244.

23. Except where otherwise indicated, the discussion of Ellen White is based on Numbers, *Prophetess of Health.*

24. *Early Writings of Ellen G. White* (Washington: Review and Herald Publishing Association, 1945), 21.

25. Ronald L. Numbers and Janet S. Numbers, "The Psychological World of Ellen White," *Spectrum* 14, 1(1983): 21–31. See also Numbers and Numbers, "Ellen White on the Mind and the Mind of Ellen White," in Ronald L. Numbers, *Prophetess of Health: Ellen G. White and the Origins of Seventh-Day Adventist Health Reform,* rev. ed. (Knoxville: University of Tennessee Press, 1992), 202-27.

26. Ellen G. White, *An Appeal to Mothers: The Great Cause of the Physical, Mental, and Moral Ruin of Many of the Children of Our Time* (Battle Creek, Mich.: SDA Publishing Association, 1864), 27.

27. College of Medical Evangelists, Calendar, 1909–1910.

28. Eddy, *Retrospection,* 13; Mrs. E. G. White, *Testimonies for the Church, with a Biographical Sketch of the Author,* 9 vols. (Mountain View, Calif.: Pacific Press Publishing Association, n.d.), 1:25.

29. Peel, *Years of Discovery,* 27; *Life Sketches of Ellen G. White* (Mountain View, Calif.: Pacific Press Publishing Association, 1915), 39; White, *Testimonies,* 1:13.

30. Nancy F. Cott, *The Bonds of Womanhood: "Woman's Sphere" in New England, 1780–1835* (New Haven, Conn.: Yale University Press, 1977), 140. On the role played by women in Victorian religious life, see also Ann Douglas, *The Feminization of American Culture* (New York: Alfred A. Knopf, 1977).

31. Eddy, *Science and Health,* 1875, 400.

32. A more complete development of the healing aspects of Christian Science appears in Rennie B. Schoepflin's *Lives on Trial: Christian Science Healers in Progressive America* (Baltimore: Johns Hopkins University Press, 1995).

32 Spreading the Germ Theory: Sanitary Science and Home Economics, 1880–1930

Nancy Tomes

In search of scientific principles to uplift the American home, the founders of the home economics movement at the turn of the century derived powerful insights from the new germ theory of disease, which gained widespread acceptance in the late 1800s. Teaching American women to master the invisible workings of the microbe in everyday life became a central element of the "science of controllable environment" envisioned by Ellen Richards and her compatriots. Bacteriology figured prominently in the early discipline's teachings about a wide range of topics, including home sanitation, interior decoration, and food preparation. Home economists stressed that the American housewife, whether dusting or canning, planning her parlor decor or nursing a sick child, could make an important contribution to the war against deadly diseases such as tuberculosis and typhoid. As Ellen Richards explained, "Sweeping

NANCY TOMES is Professor of History and Associate Dean for Faculty Affairs and Personnel, College of Arts and Sciences at the State University of New York, Stony Brook.

Reprinted from *Rethinking Women and Home Economics,* ed. Sarah and Virginia B. Vincenti, 34–54, by permission of the publisher, Cornell University Press. Copyright 1997 by Cornell University.

and cleaning and laundry work are all processes of sanitation and not mere drudgery imposed by tradition, as some people seem to think." What home economists variously termed *sanitary* or *bacteriological* cleanliness served as a prime example of the importance of truly scientific housekeeping.[1]

Perhaps no aspect of the early home economics movement has been more profoundly misunderstood than its promotion of these new standards of cleanliness. In the late 1960s, when the "second wave" of feminists began to criticize the ways scientific experts had fostered restrictive roles for women, home economists' teachings on the subject made an easy target. The most cogent of the new feminist critiques, Barbara Ehrenreich and Deirdre English's widely read book *For Her Own Good: 150 Years of the Experts' Advice to Women,* included a chapter tellingly titled "Microbes and the Manufacture of Housework," which argued that home economists employed pseudoscientific information about disease prevention to busy American women with a fruitless pursuit of cleanliness. "The scientific content of 'scientific cleaning' was extremely thin," Ehrenreich and English asserted. "The domestic scientists were right about the existence of germs, but neither they nor the actual scientists knew much about the transmission and destruction of germs."[2]

Contrary to this point of view, I will argue that the doctrine of scientific cleanliness taught by early home economists was not only consistent with the public health practice of the time but also represented far more than busywork. In an era when infectious diseases were the leading cause of death and municipal public health services were still limited and unreliable, the lessons of what S. Maria Elliott called "household bacteriology" had some real utility.[3] Home economists played a particularly important role in extending useful knowledge about evading microbial harms beyond the sanitarily privileged urban middle classes to recent immigrants, African-Americans, and farm families, who in the early 1900s had limited access to pure drinking water, efficient sewage systems, or safe food supplies. The significance of bacteriological cleanliness began to diminish only in the 1920s, as improving living standards and declining mortality rates from infectious diseases lessened emphasis on the domestic "battle with bacteria."[4]

Placing the home economics movement in the context of the prevailing disease environment and public health ideology of the time helps to explain the changing salience of domestic cleanliness to its educational mission. The waxing and waning of attention to disease prevention in the household is particularly interesting in light of the current resurgence of anxieties about the threat of infectious disease. For a variety of reasons, including the worldwide epidemic of AIDS, the increase in drug-resistant tuberculosis, and the general weakening of the public health infrastructure, Americans in the 1990s have become less complacent about their safety from the invisible world of microorganisms. Home economists' efforts to foster a sense

of mastery over the microbe earlier in this century suggest some intriguing parallels to contemporary confrontations with a new generation of "superbugs."[5]

In 1899, the year of the first Lake Placid conference on home economics, the United States was in the midst of a national crusade to control the spread of infectious diseases. Never before in American history had the public health movement been so focused on the dangers of such diseases or so confident of its ability to reduce their ravages by prevention alone. High rates of many communicable diseases, particularly tuberculosis; changes in the social order such as urbanization and immigration that made those diseases more visible and menacing; new forms of print culture which facilitated mass health education; and, last but not least, new scientific theories of contagion that made the prevention of disease seem more feasible all prompted Progressive-era campaigns against infectious disease.[6]

The public health crusades represented a response to a genuinely dangerous disease environment. Although the death rates from most infectious diseases began to drop in the late 1800s, for reasons that are still unclear, they still accounted for the majority of deaths in 1900. Tuberculosis led the list of fatal diseases, accounting for roughly 10 percent of all deaths; among people in the prime of life, aged twenty to forty, the death rate was even higher, an average of one in four people. Influenza, pneumonia, and typhoid added substantially to the death rolls. Rates of infant and child mortality from diseases such as diphtheria, scarlet fever, and nonspecific diarrheal infections remained high in large cities, even among affluent families.[7] The menace of infectious diseases was compounded by the paucity of means available to combat them. As of 1900, physicians could offer patients only immunizations against smallpox and rabies, plus a diphtheria antitoxin. The Progressive-era medicine chest held no "magic bullets," that is, drugs able to destroy the invading microbes without also killing their unlucky human hosts. The discovery of effective antimicrobial drugs remained a scientific ambition rather than a reality.[8]

Given the limits on medicine's curative resources, prevention offered the best method to reduce the toll of infectious diseases. Although researchers had not yet produced effective cures, they had developed a useful body of knowledge about how those diseases originated and spread so that they could be better avoided by conscious public and individual action. The late nineteenth-century public health movement's robust confidence in the power of prevention reflected growing certainty about the fundamental causes of disease.[9]

Beginning in the 1830s and 1840s, Anglo-American public health reformers had developed a "sanitary science" that linked organic chemical impurities or "ferments" in the air and water to the rising incidence of diseases such as cholera and typhoid fever. In the 1860s and 1870s a growing number of scientists, chief among them the French chemist Louis Pasteur, became convinced that the "ferments" of disease were in fact living microorganisms. The phrase "germ theory of disease" came into use around 1870 to denote the hypothesis that these organisms, variously called germs, bacteria, or microbes, were the cause of both human and animal diseases.

Many physicians, accustomed to explaining disease as a complex product of individual and environmental factors, initially found the germ theory of disease too simplistic. But from the late 1870s on, increasingly sophisticated laboratory work combining pure cultures, staining techniques, and animal experimentation gradually illuminated the complex ways that pathogenic microbes infected their human and animal

hosts. By 1900 the germ theory of disease and the experimental science of bacteriology that supported and elaborated it had been assimilated into the older discipline of sanitary science and widely accepted by the medical and public health establishments.[10]

Early bacteriological investigations confirmed that the soil, air, and water were all saturated with diverse microorganisms, including those responsible for human diseases. Experimenters found that most microbes could be killed with high heat or chemical disinfectants, but some species, such as the rare but deadly anthrax, had a spore form that could travel long distances and endure highly unfavorable environmental conditions, only to flourish again when given moisture and sufficient warmth. In whatever form, disease germs found many routes to reach susceptible hosts. Investigations in the 1870s and 1880s targeted fecal contamination of the air and water as the main sources of infection. After the German physician Robert Koch's startling announcement in 1882 that he had isolated the bacillus responsible for tuberculosis, a disease long thought to be hereditary, researchers looked more closely at the role of spitting, coughing, and sneezing in spreading respiratory ailments. Subsequent work in the 1890s implicated animal and insect vectors in diseases such as bubonic plague, typhoid, and malaria. Finally, bacteriologists became more aware of bacterial contamination of the food supply from tuberculous cows, dust, flies, and food handlers. Improper food storage or preservation was linked to food poisoning, particularly by the deadly *Clostridium botulinum* responsible for botulism.[11]

Armed with these new bacteriological insights, public health authorities agitated for the expansion of municipal and state departments of public health. From the 1880s on, these departments gradually improved the provision of sanitary services such as municipal water filtration, sewage systems, and garbage collection; they also acquired the legal powers to enforce an increasingly elaborate code of sanitary regulations concerning plumbing, food preparation, and other hygienic matters. At the same time public health leaders emphasized popular health education and voluntary sanitary reform of the household. Recognizing that existing sanitary laws were still limited in scope and difficult to enforce, late nineteenth-century health reformers stressed that individual citizens had to be educated not only to guard their homes against disease but also to lend political support to their local health departments.[12]

The late nineteenth-century concern about deficient domestic hygiene was by no means focused only on the poor and uneducated. Although public health authorities tended to assume, as did most white middle-class Americans, that working-class, immigrant, and nonwhite households were the worst sanitary offenders, they did not view their own class as exempt from hygienic ignorance. In their experience neither affluence nor education guaranteed an understanding of the simple but deadly errors of personal hygiene commonly found in American households. In light of the new scientific findings about how contagious diseases spread, public health reformers believed that all Americans stood in need of sanitary redemption.

Both sexes were included in the late nineteenth-century appeals for domestic sanitary uplift, but in practical terms the growing emphasis on "house diseases" imposed heavier, more constant burdens on women than men. In the gendered division of sanitary labor, men were apportioned responsibility for the external and structural spheres such as the construction of the house and installation of the plumbing system. Although these matters could be time consuming, they did not occupy a male home owner's constant attention and, if he had suffi-

cient financial means, could be contracted out to an architect, sanitary engineer, or plumber. The interior of the house was the woman's sanitary province. Public health authorities assumed that the housewife had a greater knowledge of domestic conditions and a stronger motivation to protect the family's health than her husband did. Thus the all-important daily minutia of sanitary cleanliness fell squarely in the female domain. Although middle- and upperclass women could hire domestic servants to do the heavy cleaning involved, they still had to maintain constant vigilance to make sure the work was done correctly or else suffer guilt over a loved one's illness.

The public health movement's equation of domestic hygiene with disease prevention offered the early leaders of the home economics movement a splendid field for action. Domestic disease prevention served as a perfect illustration of how science applied to the home could promote individual, familial, and social uplift. Here was a field in which women could naturally excel, as educators and researchers as well as wives and mothers. In addressing the deep anxieties about infectious disease common to their era, home economists found an issue made to order.

The career of Ellen Richards, the "mother of home economics," clearly illustrates the centrality of sanitary science to the emerging field. The first American woman to get a science degree, Richards trained as a chemist at the Massachusetts Institute of Technology in the early 1870s and became affiliated with its new laboratory for sanitary research in 1884, where she helped pioneer methodologies for measuring water pollution. Although she was trained as a sanitary chemist, not as a bacteriologist, Richards realized how the latter discipline could contribute to securing safe water, air, and food supplies. As early as 1887, in a household manual coauthored with Marion Talbot for the Association of College Alumnae, Richards wrote, "The general acceptance of the germ theory of disease makes it imperative for every housekeeper to guard against all accumulations of dust, since such accumulations may harbor dangerous germs." In making sanitary science an essential building block of home economics, she included not only the older chemical understanding of health and disease but also the newer insights of bacteriology.[13]

Among the slightly younger generation of women and men who joined Richards at the Lake Placid conferences in the early 1900s and helped found the American Home Economics Association in 1909—for example, S. Maria Elliott of Simmons College, Maurice Le Bosquet of the American School of Home Economics, Marion Talbot of the University of Chicago, and Martha Van Rensselaer of Cornell University—the appeal of sanitary science in general and bacteriology in particular was equally strong. The conception of the housewife as public health crusader nicely complemented the educational and political goals of early home economists.

In the first place, the threat of disease offered an effective way to challenge many women's unreflective attachment to traditional housekeeping methods. The Lake Placid conferees frequently complained that it was difficult to convince homemakers, especially those from the more affluent classes, to change their accustomed ways of cleaning or decorating in the name of some abstract ideal of science or utility. But if new housekeeping methods could be shown to safeguard their families against potentially fatal diseases, women would be more receptive to home economists' suggestions. As May Bolster noted in a talk at the 1902 Lake Placid conference, even the most complacent clubwomen realized the importance of municipal sanitation, a point that could be used to

interest them in the concept of scientific housekeeping.[14]

Likewise, the stress on domestic disease control fostered the larger reform agenda of home economics leaders. A basic education in bacteriology would not only make American women better wives and mothers but also more effective community leaders. Lessons about germs and disease mastered in the home could be applied to their dealings with storekeepers, milk dealers, and local politicians. Household bacteriology provided an excellent example of how domestic science could serve as the rationale for the broader forms of social housekeeping envisioned by Richards and other Progressive-minded home economists.[15]

Thus discussions of germ life and household bacteriology figured prominently in their early attempts to delineate the field's educational mission. In the various model curricula for domestic science courses drafted and debated by the Lake Placid participants, bacteriological perspectives were included at every level of teaching, starting from the basics of hygiene in the primary grades, through the practical work offered in manual education schools and the domestic science courses in the high schools, and culminating in the advanced courses in home economics at the college and graduate levels. Sanitary science and household bacteriology also became favored topics for educational programs aimed at adult homemakers, including clubwomen and farmwives.[16]

Home economists' conceptions of what they needed to teach regarding bacteriology and preventive hygiene varied depending on their audience's class and geographic location. For upper- and middle-class women living in large cities and towns, the worst dangers of infection seemed to lie in the exchanges between their homes and the rest of the community. Women of means could easily be instructed to make their domestic environments safer from microbial incursions by purchasing sanitary aids such as filtered water, good quality plumbing, certified milk, and iceboxes. Yet other common experiences, such as riding in railway cars, staying in hotels, purchasing factory-made clothing, or patronizing commercial laundries, exposed them and their families to the sanitary misdeeds of the less enlightened. Complete safety from disease required enlisting affluent women to take action beyond the household, such as supporting antispitting ordinances and white-label campaigns.[17]

Home economists realized that poor women, whether they lived in urban or rural areas, faced more fundamental difficulties in securing their homes from disease. In cities substandard housing, diet, and working conditions predisposed working-class families to infectious diseases, yet their tenement homes were the least likely to have the clean abundant water supply, sanitary toilets, well-ventilated rooms, and wholesome food deemed essential to resisting such illnesses. Rural women faced another distinctive set of sanitary hazards. Home economists recognized that in the early 1900s even relatively well-off farm households lacked the basics of safe sanitation. On many farms human and animal wastes polluted springs and outdoor wells, flies carried germs from manure piles to milk pails, and food storage and preservation were primitive.[18]

From city to countryside, then, the field of domestic sanitary science offered a mission made to order for the founding generation of home economists. Disease prevention was an issue bound to interest homemakers from every walk of life. Instructing women in simple effective ways to ward off infectious disease brought science into the home in the most meaningful way. Moreover, popularizing the rules of protective hygiene did not require highly specialized medical knowledge or equipment. Most meth-

ods of disease prevention seemed well within the reach of the average housewife. What she might lack in sanitary aids such as flush toilets or vacuum cleaners, she could, with the home economists' guidance, compensate for with ingenuity and hard work.

Although their intended audience was varied, the thrust of the home economists' turn-of-the-century sanitary gospel was remarkably uniform: dirt bred disease, therefore women of all classes had to keep themselves, their food, and their houses clean. In fact, much of the "science of controllable environment" hinged on the intelligent management of human wastes and other forms of organic dirt. But upon closer reading, home economists' conceptions of what dirt was dangerous and how bacteriology applied to the home were far more nuanced and complex than the simple equation of dirt with disease implies.

In their writings and lectures home economists attempted to give ordinary women a radical new understanding of their environment: that they shared the world with an invisible host of microbes; that far from being universally dangerous, many microorganisms performed useful functions; that a basic knowledge of bacteriology was essential to distinguish germ friends from foes; and that careful habits could furnish protection against the latter. As S. Maria Elliott told readers in her 1907 text on household bacteriology, "If these lessons point out dangers of which you were before unconscious, they also suggest ways of escape from those dangers." From the science of bacteriology modern society had learned that the "science of the infinitely small has become the infinitely important," she concluded.[19]

Early home economists believed that to appreciate the importance of sanitary cleanliness, women needed to understand both the germ theory of disease and the methods of bacteriology. To this end they taught not only rules for disease prevention but also simple scientific explanations for why observance of those rules was so important. Their insistence that ordinary housewives should learn even the rudiments of bacteriology was a novel idea in the early 1900s.[20]

Martha Van Rensselaer's experience in preparing the first bulletins for the Cornell Farmers' Wives' Reading Course provides a case in point. Although she lacked a college education and had no special training in sanitary science or bacteriology, Van Rensselaer nonetheless felt that these subjects should be an essential part of the educational program for farm women she began in the early 1900s. When she approached a Cornell bacteriologist with the request, "I would like to learn about the bacteriology of the dishcloth so that I may explain to farm women the importance of its cleanliness," he responded, "Oh, they do not need to learn about bacteria. Teach them to keep the dishcloth clean because it is nicer that way."[21]

Home economics educators emphatically rejected that approach in their early textbooks and circulars on the subjects of home hygiene and household bacteriology. While committed to making their explanations understandable to women with little or no formal education, they also tried to convey the complexity and fascination of bacteriology. For example, two of the standard books used in home economics coursework, H. W. Conn's *Bacteria, Yeasts, and Molds in the Home* and S. Maria Elliott's *Household Bacteriology*, stressed the importance of conducting simple experiments using petri dishes so as to understand the vitality of bacterial life in the household. They encouraged students and housewives to seek access to a microscope so that they could see the invisible world for themselves. Even Van Rensselaer's short extension bulletins suggested that farm women grow their own "dust gardens" and observe the

natural processes of decay and putrefaction that occurred in uncovered and unpreserved food items.[22]

Likewise, these early expositions of household bacteriology sought to convey a sense of the diversity of bacteriological processes in everyday life. Using drawings and photographs, they rendered the different classes of yeast, molds, and bacteria in substantial detail. Decrying the popular view that all microorganisms were bad, textbook authors emphasized the positive contributions that many species made to industry and agriculture. They contrasted the "bad" bacterial species that could spoil milk or even make it the agent of disease with the "good" species required for the manufacture of butter and cheese.

Such homely examples underlined the need for housewives to learn to discriminate between good and bad microbes and to keep the latter at bay. The exacting directions home economists gave about details of housecleaning, home decoration, food preparation, and the like grew out of this complex conception of the house's microbial environment. Three examples—the care of the toilet, the dangers of dust, and the hygiene of food—serve to illustrate the particularities of home economists' concerns about germs in the home.

In this era no public health precept was better established than the role of fecal contamination in spreading diseases such as cholera and typhoid. Because disease germs throve in human sewage, it was assumed that food, water, or air that came in contact with such wastes could become infected. Although by the early 1900s some public health experts had begun to question the disease-causing properties of sewer-contaminated air, or what was popularly referred to as "sewer gas," many health authorities continued to regard it as a potential source of danger.[23]

In line with prevailing public health opinion,

home economists considered sanitary toilets to be the foundation of modern household hygiene. Without a safe system for disposal of human sewage, a housewife's efforts to have a healthy home were doomed. For town and city dwellers whose homes had the benefit of modern water closets connected to municipal sewer systems, protection against disease required understanding and maintaining a complex plumbing technology. Texts on home hygiene routinely included long, detailed expositions of the air- and watertight system of pipes, drains, and traps needed to prevent germ-laden sewage vapors from fouling the interior air supply. While counseling women to rely on experts such as plumbers or sanitary engineers to install household sanitation systems, home economists warned that only the housewife could keep those systems in proper sanitary order.

For example, toilets required exacting daily maintenance. S. Maria Elliott outlined a three-step procedure, which included soaking, brush scrubbing, and cloth scouring the water closet with disinfectant soap powder. Elliott also recommended that women periodically check to make sure their plumbing was airtight by using the peppermint test, which involved pouring oil of peppermint into the waste pipes at their highest accessible point and then sniffing the air in every room to see if its telltale odor escaped.[24]

Early home economists recognized that many American women still lived in homes that lacked indoor toilets and sewer lines and thus faced an even more extreme sanitary danger. Elliott expressed their collective sentiments when she declared in 1910, "The ordinary privy is a menace to health as well as an odorous disgrace." Home economists urged that urban housing codes be written to force landlords to install and maintain sanitary toilets and that municipal health departments provide inspectors to enforce those codes. Until the landlord and the state had fulfilled these obligations, the vigilance of the individual woman living in the tenement counted for comparatively little. Once tenement buildings had been brought up to code, housewives needed instruction in how to keep their toilets functioning properly by being flushed after every use and kept clean and free of rubbish.[25]

Home economists required rural women to be even more active agents in the pursuit of the safe toilet. Early home economics extension work aimed at educating farmers' wives about the many sanitary dangers of the unimproved farmstead: the little "babbling brook" from which they fetched wash water might carry fecal wastes from upstream, the family's old privy could be polluting the well from which they drank, and the flies attracted to human and animal manure could contaminate their milk and food. Fortunately, in comparison to tenement dwellers, it was easier for the farm household to rectify such conditions by constructing an inexpensive "sanitary privy," that is, a sewage-tight, disinfecting, and fly-screened toilet, or installing an indoor toilet connected to a cesspool. Although external improvements such as sanitary privies and cesspools were seen as the primary responsibility of the male farmer, extension workers stressed that farm wives had to be actively involved in getting their husbands to undertake such projects and maintaining the sanitary facilities once they were completed.[26]

Home economists' teachings about "dust dangers" offer another good example of the way concerns about disease shaped household practices. Bacteriologists' ability to culture pathogenic organisms from house dust and other forms of common dirt seemed incontrovertible proof that dusty, dirty environments fostered the spread of disease. Home economists often cited the work of an eminent New York bacteriologist, T. Mitchell Prudden, whose 1890 book, *Dust and Its Dangers,* laid out, "in simple language, what the real danger is of acquiring serious disease—especially consumption—by means of dust-laden air." Thus when home economists advised women to keep their homes free of germ-infested dust and dirt, they had the force of experimental science behind them.[27]

Methods of safe dust removal figured prominently in home economists' directives concerning scientific housecleaning. The use of a wet mop, "dustless" broom, and oiled dust cloth maximized the capture of potentially germ-laden dust particles. Housewives were taught to let several hours elapse between dusting the house and preparing food to allow those particles that eluded them to settle. Home economists hailed the vacuum cleaner not only as an energy saver but also as a hygienic boon because it sucked up dangerous dust far more safely and thoroughly than traditional carpet sweepers.[28]

Cautions about dust and disease also figured prominently in home economists' efforts to do away with the overstuffed furniture, room-sized carpets, wallpaper, and extensive bric-a-brac favored in late nineteenth-century decorating schemes. For example, they promoted finished wood floors covered with small area rugs as far less hospitable to germ life than the huge, difficult-to-clean carpets that graced most parlors. Women were urged to take down their wallpaper, which home economists regarded as a germ breeding ground, and tint their walls with paint containing germicidal agents. In the kitchen easily washed surfaces such as enamel and linoleum provided the best stage for sanitary preparation.[29]

A final example of how a bacteriological perspective informed scientific housekeeping in the early 1900s can be seen in the area of food preservation and preparation. Home economists stressed that the safe preparation of meals required a clear understanding of microbial processes such as fermentation and decay. Their lesson plans taught the basic rules of sanitary food handling, beginning with the need to store food at precise temperatures to retard bacterial spoil-

age. Meats and homemade canned goods required lengthy cooking periods at high heat to destroy bacteria and their spores. Before and after cooking, dishes had to be kept carefully covered to prevent germ-laden dust and fly-specks from contaminating them. At every stage of food preparation, scrupulous cleanliness was necessary to kill germs; cooking and serving utensils had to be scalded in boiling water, dish towels must be boiled and dried before using, and the cook's hands and clothes must be washed with disinfectant soap.

Perhaps the most dramatic illustration of the importance of bacteriology to food preparation lay in the practice of home canning. Home economists advocated canning homegrown fruits and vegetables as an ideal way for farm women and girls to improve their diets and to make extra money to spend on home improvements. Teaching safe canning methods was a major focus of early extension work. Home canners learned what microorganisms were most commonly found on fruits and vegetables, why some substances such as vinegar and salt retarded bacterial growth, and what precise temperatures and lengths of cooking were necessary to kill the pesky spores. The elaborate protocols of canning summed up the need for absolutely clean hands and utensils, sterile food containers, and exacting observance of cooking procedures. The penalty for failing to observe these bacteriological precautions was obvious: the food would spoil. Over all these proceedings hovered the specter of food poisoning, particularly the dread of botulism, a deadly form of bacterial food contamination associated with improper canning.[30]

Even as early home economists popularized greater bacteriological awareness in the home, new trends in the public health movement began to diminish the emphasis on domestic hygiene in disease prevention. In 1913 a Minnesota public health official named Hibbert

Winslow Hill published a book entitled *The New Public Health,* which heralded an important shift in public health practice. At first glance Hill's book seemed only to reinforce the emphasis on domestic sanitation so central to home economics. "The infectious diseases in general radiate from and are kept going by women," he declared in the preface. "To teach women, girls, prospective mothers, that they may practice in their household and in turn teach their children to war on invisible germ-foes is one of the functions of public health bacteriology."[31]

Yet Hill's conception of the new public health represented a significant change from the home economists' program of sanitary cleanliness. The latest bacteriological research, Hill insisted, showed that "dirty people are no more subject to disease than clean." He noted that even a person of immaculate habits could catch smallpox when exposed to its highly contagious virus; conversely, a person living in horrific filth would never get typhoid fever unless the specific germ was introduced into his digestive system. Therefore isolating the sick was a more efficient preventive measure than ridding the environment of dirt. Exacting cleanliness might be sought for moral and aesthetic reasons, Hill conceded, but its pursuit was not the most scientific use of public health funds or energies. Rather, the state should concentrate on confining and regulating infected individuals and making sure their diseased discharges did not reach others through public water or food supplies.[32]

In the 1920s and 1930s, as the new public health became the guiding faith of the American public health movement, prevention of infectious diseases became more narrowly concerned with individual "case finding" and management, on the one hand, and improved municipal public health services, on the other.[33] The housewife's contribution to the control of infectious disease diminished accordingly; her duties came to consist primarily of obeying the health department's quarantine regulations and obtaining immunizations for her children.

The "battle with bacteria" also became less salient within the maturing discipline of home economics. A new generation of home economics leaders realized that consolidating the field's legitimacy required strengthening its base in the university, which in turn necessitated placing more emphasis on original research. In the late 1910s and early 1920s nutrition and textiles emerged as the most promising fields for home economics research and employment opportunities. In contrast, the technically sophisticated, medically dominated field of bacteriology offered few avenues for home economists to pursue independent research or to find jobs.[34]

As popularizers of public health information, home economists also faced increasing competition in the 1920s. After World War I the field of health education, especially with children, expanded rapidly. Local and state health departments, voluntary health organizations, and professional societies launched a dizzying variety of educational initiatives. The home economists' educational agenda overlapped significantly with those of other female-dominated fields, particularly nursing. School nurses and visiting nurses effectively dominated areas of instruction with children and adult women that home economists otherwise might have claimed. As the ideal of "positive health," as opposed to mere freedom from disease, came to dominate health education in the 1920s, the home economists found their skills in nutrition to be more in demand than their expertise in household sanitation.[35]

Meanwhile, mortality rates resulting from infectious diseases declined steadily. By the late 1920s heart disease and cancer had replaced tuberculosis, influenza, and pneumonia as the leading causes of death. As the number of hospitals and hospital insurance plans proliferated, those people who did contract serious infectious diseases were less likely to be cared for at home, diminishing the need for painstaking home nursing and sanitation procedures. Starting with the sulfanilamides in the mid-1930s, the discovery of effective antibiotics further reduced the threat of infectious diseases.[36]

Other long-term changes in the standard of living decreased the personal precautions individuals needed to take against their germ foes. Municipal water and sewer services gradually extended to all neighborhoods in American cities. Farmhouses lacking indoor plumbing or running water became increasingly rare except in the poorest parts of the country. Increased regulation of the dairy industry resulted in a safe, relatively cheap supply of pasteurized milk. Commercial processing techniques and "sanitary packaging" steadily decreased the likelihood of microbial contamination of foodstuffs. More homemakers acquired labor-saving appliances such as refrigerators, vacuum cleaners, and washing machines, which facilitated domestic cleanliness.[37]

The maxims of bacteriological cleanliness continued to be handed down from teacher to student, mother to daughter, but the fear of infectious diseases that had given them their original urgency gradually disappeared. The association between sanitary science and dread disease faded with each generation reared in greater safety from the killer diseases of the early 1900s, leaving behind only a vague conviction that cleanliness promoted health. By 1976, when Barbara Ehrenreich and Dierdre English conducted an informal survey of home economics professors, none could clearly articulate the relationship between cleanliness and good health.[38] Rather than read their confusion as evidence of the unscientific nature of early twentieth-century home economics, as Ehrenreich and English did, we might better see the atrophy of knowledge about domestic hygiene as a reflection of the lessening threat of infectious diseases over the last century.

To argue that early twentieth-century home economists provided American women with a bacteriologically informed perspective on household hygiene is not to claim that everything they taught was correct. In light of modern knowledge some of the practices they advocated did indeed protect families against infectious diseases, but others did not. For example, improvements in rural sanitation systems undoubtedly contributed to the decline in rates of typhoid, hookworm, and other diseases spread by human wastes. Thus the home economics movement's championship of the sanitary privy constituted a significant public health reform. The menace of microbial contamination of house dust, it turned out, was not so well founded. Scientists later discovered that the tubercle bacilli cultured from dust lacked infective power; the live bacilli expelled in droplet form by coughs and sneezes proved to be far more important to the spread of tuberculosis.[39]

The precepts early home economists taught have proved most enduring in the area of food preservation and handling. In an age when refrigeration was very expensive, when supplies of milk and water were liable to bacterial contamination, and when drinking glasses and utensils were frequently shared, we can only begin to guess what percentage of the gastrointestinal diseases so commonly suffered by Americans were the result of unsanitary food handling. To teach women that food needed to be stored at certain temperatures or that the surfaces and utensils used for preparing and serving food had to be kept scrupulously clean was neither busywork nor pseudoscientific. Such precautions remain the foundation of modern health departments' regulation of the food service industry.[40]

Through its educational outreach programs the early home economics movement played an important role in acquainting women of all classes and races with such precautions. Affluent women had access to this information through a variety of channels, including women's clubs, popular magazines, and private physicians, whereas poor women had many fewer such educational opportunities. By including immigrant and rural women in its extension programs, the home economics movement helped democratize the spread of potentially valuable hygienic information.

Historians and demographers cannot gauge precisely how much such campaigns actually contributed to the fall of mortality rates from infectious disease in the decades from 1880 to 1930. But enlisting the housewife's help in preventing fecal contamination of water supplies and ridding food and milk supplies of harmful microbial life may have made a significant contribution to that decline. Only from the insulated perspective of a society grown used to the protections guaranteed by safe water and food supplies, sewage systems, and the like could these efforts be dismissed as mindless makework.[41]

Unfortunately, the heightened sense of responsibility for preventing infectious diseases promoted by home economists and other health reformers probably brought in its wake a new intensity of guilt. They tended to overemphasize the efficacy of individual hygienic measures, thus leaving women vulnerable to shame and grief when their efforts at disease prevention failed. A poignant letter written by a farm woman who had completed a correspondence course on home hygiene given by the Cornell extension service suggests the hard contours of such responsibility: "Men, men, mud, mud, and my cellar," she wrote. "I wonder we are alive. Poor me. I know if everything had been kept properly my children would be alive and well."[42]

Limitations on income and resources severely constrained the ability of very poor women to implement the ideals of bacteriological cleanliness. Without running water, convenient-to-use cleaning supplies, or sanitary toilets, even superficial cleanliness could be obtained only with backbreaking labor. In this sense the home eco-

nomics movement's emphasis on domestic hygiene may have worked to reinforce oppressively personal solutions to the perceived threat of diseases. Here as in many other areas of American reform, individual adaptation and blaming the victim often proved easier than changing the inequities in income and housing which underlay the unequal distribution of illness.[43]

The legacy of these early twentieth-century efforts to spread the germ theory can still be seen in contemporary attitudes toward germs. Older Americans who remember the "bad old days" when siblings and friends died of typhoid or tuberculosis or polio, and who were first taught the rules of sanitary safety at an impressionable age, often have much stronger germ phobias than do Americans born in the postantibiotic era. Oral histories suggest that immigrant women raised in the 1920s and 1930s in very poor homes without the modern conveniences of toilets and refrigerators regarded exacting standards of housecleaning as a sacred duty to protect the family from disease. But their daughters and granddaughters proved harder and harder to convince that housecleaning is a prerequisite to good health. The oft-noted decline in the number of hours American women spend cleaning house may in part reflect the gradual weakening of the association between dirt and disease.[44]

The AIDs epidemic and growing concerns about environmental threats to health may reverse this more casual view of domestic hygiene. The appearance of deadly new viral diseases that have no cure, along with the resurgence of tuberculosis, a bacterial disease thought long under control, have reawakened many of the old fears associated with the spread of contagion. Widely publicized incidents of bacterial contamination of water and meat supplies have drawn attention to the vulnerability of existing hygienic protections. Confronted by a new generation of "superbugs," many Americans have begun to recall dimly remembered lessons about sterilizing dishes and cooking meats at high temperatures.

Rising concerns about new environmental threats to health have also focused attention on domestic hygiene. Worries about "sick" houses and buildings are couched in terms strikingly similar to earlier fears about "house diseases." Today people test their homes for radon gas instead of sewer gas and install water filters to remove chemicals, not bacteria, from their drinking water. Yet the impulse behind these individual efforts to protect the household against environmental dangers has many parallels to the early home economists' preoccupation with domestic disease prevention.

When we see these parallels between their concerns and our own, the history of the home economics movement serves to educate rather than to anger or to amuse. As we struggle in the late twentieth century to become more aware of the health consequences of our own interactions with the environment, to realize the limits of modern medicine in reducing our vulnerability to disease, and to appreciate the ways that race and class prejudice destroy health, the home economists' "battle with bacteria" provides an instructive historical precedent. In the wake of frightening new diseases, broken water mains, and tainted food, we may find ourselves turning to those old texts on household sanitation and bacteriology for more than historical insight.

Notes

This essay is drawn from a chapter in my book, *The Gospel of Germs: Men, Women, and the Microbe in American Life* (Cambridge: Harvard University Press, 1998). My research has been generously supported by the National Endowment for the Humanities (Grant No. RH-21055–92) and the National Library of Medicine (Grant No. R01 LM0579–01). Its contents are solely the responsibility of the author and do not necessarily represent the official views of either the NEH or the NLM. I want to thank Joan Jacobs Brumberg, Sally Gregory Kohlstedt, Emma Weigley, Sarah Stage, and Virginia Vincenti for their comments on earlier drafts.

1. Ellen H. Richards, *Sanitation in Daily Life,* 3d ed. (Boston: Whitcomb and Barrows, 1915), 3. For an overview of the early home economics movement see Sarah Stage, "From Domestic Science to Social Housekeeping: The Career of Ellen Richards," in *Power and Responsibility: Case Studies in American Leadership,* ed. David M. Kennedy and Michael E. Parrish (New York: Harcourt Brace Jovanovich, 1986), 211–28; and Emma S. Weigley, "It Might Have Been Euthenics: The Lake Placid Conferences and the Home Economics Movement," *American Quarterly* 26 (March 1974): 79–96.

2. Barbara Ehrenreich and Deirdre English, *For Her Own Good: 150 Years of the Experts' Advice to Women* (New York: Doubleday, 1978), 159. For a less critical assessment of the home economics movement, see Suellen Hoy, *Chasing Dirt: The American Pursuit of Cleanliness* (New York: Oxford University Press, 1995), esp. 153–56.

3. S. Maria Elliott, *Household Bacteriology* (Chicago: American School of Home Economics, 1907).

4. The phrase "battle with bacteria" appears in "Bacteriology of the Household," *Cornell Reading-Course for Farmers' Wives,* n.s. 1, 4 (February 1909): 89.

5. Laurie Garrett, *The Coming Plague: Newly Emerging Diseases in a World out of Balance* (New York: Farrar, Straus, and Giroux, 1994), illustrates the growing concern about so-called superbugs which include new viruses such as AIDS and known forms of bacteria that have become resistant to antibiotics.

6. For overviews of the nineteenth-century American public health movement, see John Duffy, *The Sanitarians: A History of American Public Health* (Urbana: University of Illinois Press, 1990), and George Rosen, *The History of Public Health* (New York: MD Publications, 1958).

7. On mortality rates, see Judith Walzer Leavitt and Ronald Numbers, "Sickness and Health in America: An Overview," in *Sickness and Health in America,* 2d ed. rev. (Madison: University of Wisconsin Press, 1985), 3–10; Michael Teller, *The Tuberculosis*

Movement (Westwood, Conn.: Greenwood Press, 1988), 3; Richard A. Meckel, *Save the Babies: American Public Health Reform and the Prevention of Infant Mortality, 1850–1929* (Baltimore: Johns Hopkins University Press, 1990); and Samuel Preston and Michael Haines, *Fatal Years: Child Mortality in Late Nineteenth-Century America* (Princeton: Princeton University Press, 1991).

8. For a short summary of the early history of immunology, see Rosen, *History of Public Health,* 304–10.

9. The best overview of the nineteenth-century public health movement is Duffy, *Sanitarians.*

10. For a concise summary of the arguments over disease theory, which I discussed in the two preceding paragraphs, see Lester King, *Transformations in American Medicine* (Baltimore: Johns Hopkins University Press, 1991), 142–81.

11. This rough chronology is based on my reading of the public health literature. For a general overview written toward the end of this explosion of knowledge, see Charles V. Chapin, *The Sources and Modes of Infection* (New York: Wiley, 1910).

12. On the expansion of state power in the provision of public health see Barbara Gutmann Rosenkrantz, *Public Health and the State: Changing Views in Massachusetts, 1842–1936* (Cambridge: Harvard University Press, 1972). On the importance of popular health education see Nancy Tomes, "The Private Side of Public Health: Sanitary Science, Domestic Hygiene, and the Germ Theory, 1870–1900," *Bulletin of the History of Medicine* 64 (1990): 528–30. The arguments in the next two paragraphs are also taken from this article. Two excellent works on England offer arguments about the "private side of public health" similar to my own: Anne Hardy, *The Epidemic Streets: Infectious Disease and the Rise of Preventive Medicine, 1856–1900* (Oxford: Clarendon Press, 1993), and Anamarie Adams, *Architecture in the Family Way: Doctors, Houses, and Women, 1870–1900* (Montreal: McGill-Queen's University Press, 1996).

13. Ellen H. Richards and Marion Talbot, eds., *Home Sanitation: A Manual for Housekeepers* (Boston: Ticknor, 1887), 9. On Richards's scientific training and career see Margaret Rossiter, *Women Scientists in America: Struggles and Strategies to 1940* (Baltimore: Johns Hopkins University Press, 1982), esp. 68–70.

On her contribution to the public health movement see George Rosen, "Ellen H. Richards (1842–1911), Sanitary Chemist and Pioneer of Professional Equality for Women in Health Science," *American Journal of Public Health* 64 (August 1974): 816–19; and Rosenkrantz, *Public Health and the State,* esp. 100.

The kind of chemical analysis Richards did in the laboratory was quite distinct from bacteriology. But though her conception of sanitary science and home sanitation remained heavily indebted to her chemical perspective, she seems quickly to have accepted the germ theory of disease and to have perceived the relevance of applied bacteriology to both home economics and health education.

14. May Bolster, "Standards of Living as Reflected Through Women's Clubs," Lake Placid Conference on Home Economics [hereafter LPC], *Proceedings of the Fourth Annual Conference,* 51–52.

15. For an excellent account of Richards's Progressive political philosophy, see Stage, "From Domestic Science to Social Housekeeping." Historians are only beginning to appreciate how central a role women played in Progressive-era sanitary reform in areas as diverse as tenement house reform, water and air pollution control, and food and drug regulation. See, for example, Suellen M. Hoy, "'Municipal Housekeeping': The Role of Women in Improving Urban Sanitation Practices, 1880–1917," in *Pollution and Reform in American Cities, 1870–1930,* ed. Martin V. Melosi (Austin: University of Texas Press, 1980), 173–98.

16. These generalizations are based on my reading of the Lake Placid Conference Proceedings, 1899–1909. See, for example, Mary Roberts Smith, "Report of Committee on Courses of Study in Home Economics in Colleges and Universities," LPC, *Proceedings of the Fourth Annual Conference September 16–20, 1902,* 17–21; Maria Parloa, "Suggestions for Home and Club Study," LPC, *Proceedings of Seventh Annual Conference,* 95–97.

17. Consumer groups began to organize white-label campaigns at the turn of the century to encourage women to buy goods made in factories that provided decent wages, hours, and sanitation for their workers. To get a sense of these concerns see, for example, "Railway Sanitation," *Proceedings of the Eighth Annual Conference,* September 15–22, 1906, 101–2;

and "Cleanliness in Markets, Hotels, and Restaurants," *Journal of Home Economics* 1 (June 1909): 289–90.

18. The tenor of home economists' concerns about the tenement dweller are nicely conveyed in Mabel Hyde Kittredge, "The Need of the Immigrant," *Journal of Home Economics* 5 (October 1913): 307–16. Their sanitary critique of the American farm is epitomized in Martha Van Rensselaer, "Suggestions on Home Sanitation," *Cornell Farmers Wives' Reading-Course,* ser. 3, 11 (November 1904).

19. Elliott, *Household Bacteriology,* viii, 32.

20. The aim of popular science education itself was not novel, but rather its extension to housewives with limited education. See John C. Burnham, *How Superstition Won and Science Lost: Popularizing Science and Health in the United States* (New Brunswick: Rutgers University Press, 1987).

21. This incident is reported by Van Rensselaer's colleague Flora Rose in "A Page of Modern Education, 1900–1940: Forty Years of Home Economics at Cornell University," in *A Growing College: Home Economics at Cornell University* (Ithaca: New York State College of Ecology, 1969), 22–23.

22. H. W. Conn, *Bacteria, Yeasts, and Molds in the Home* (Boston: Ginn, 1903); Elliott, *Household Bacteriology;* "Bacteriology of the Household," *Cornell Reading Course for Farmers' Wives,* N S, 1, 4 (February 1909). Maria Elliott helped Van Rensselaer prepare this bulletin for the Cornell series. For all their good intentions I suspect that home economists watered down the teaching of scientific principles when addressing their least educated audiences, such as immigrant women and sharecroppers' wives.

23. From the 1890s on, public health authorities more systematically trained in laboratory methods, chief among them Charles V. Chapin of Providence, Rhode Island, debunked the sewer gas threat. On Chapin's career see James V. Cassedy, *Charles V. Chapin and the Public Health Movement* (Cambridge: Harvard University Press, 1962).

24. S. Maria Elliott, *Household Hygiene* (Chicago: American School of Home Economics, 1910), 192–93 (care of the water closet), 137–38 (peppermint test).

25. Ibid., 120. On instructing tenement housewives on toilet care see Kittredge, "The Need of the Immigrant," esp. 307–8.

26. For a typical sanitary critique of the farm see Van Rensselaer, "Suggestions on Home Sanitation."

27. T. Mitchell Prudden, *Dust and Its Dangers* (New York: G. P. Putnam's Sons, 1890), iii. The dangers of dust were also popularized in the massive antituberculosis campaigns of the early 1900s.

28. See, for example, the instructions for dusting in Elliott, *Household Bacteriology,* 104–5.

29. For a characteristic critique of this sort see Claudia Q. Murphy, "Wall Sanitation," LPC, *Proceedings of the Eighth Annual Conference, September 15–22, 1906,* 47–50.

As was true of many turn-of-the-century sanitary measures, the call to simplify furnishings predated the popularization of the germ theory. For example, the midnineteenth-century domestic reformer Catharine Beecher praised the virtues of simple, easily cleaned furnishings. See Catharine Beecher, *Treatise on Domestic Economy* (New York: Marsh Capen Lyon and Webb, 1841).

30. Some representative examples of early canning directives are Maria S. Parloa, "Canning and Preserving," *Cornell Reading-Course for Farmers' Wives,* Ser. 4, 20 (February 1906); and Ola Powell, *Successful Canning and Preserving,* 3d ed. rev. (Philadelphia: J. B. Lippincott, 1917). Although cases of botulism were recorded as early as the eighteenth century, the Belgian researcher Emile Pierre Marie van Ermengen isolated the microorganism responsible for this form of food poisoning in 1895. See James Harvey Young, "Botulism and the Ripe Olive Scare of 1919–1920," *Bulletin of the History of Medicine* 50 (1976): 372–91.

31. Hibbert Winslow Hill, *The New Public Health* (Minneapolis: Press of the Journal Lancet, 1913), 5. Hill anticipated the book's arguments about women's role in disease control in "Teaching Bacteriology to Mothers," *Journal of Home Economics* 2 (December 1910): 635–40.

32. Ibid., 122.

33. The best single account of this transition is Rosenkrantz, *Public Health and the State.*

34. These generalizations are based on my reading of the *Journal of Home Economics* in the 1910s and 1920s. See, for example, Agnes Fay Morgan, "Physical and Biological Chemistry in the Service of Home Economics," *Journal of Home Economics* 13 (Decem-

ber 1921): 586–91. Other essays in this volume survey the interwar job opportunities home economists found in fields associated with nutrition and textiles; bacteriology is conspicuously absent.

35. In the 1920s articles in the *Journal of Home Economics* suggest that home economists were feeling the competition from other professions, particularly nursing, and that they saw nutrition as their best hope of aligning with the "new public health." See Martha Koehne, "The Health Education Program and the Home Economist," *Journal of Home Economics* 16 (July 1924): 373–80; Margaret Sawyer, "American Home Economics Association and the Health Program," *Journal of Home Economics* 16 (December 1924): 679–83.

36. On changing disease patterns see Leavitt and Numbers, "Sickness and Health in America," esp. the chart on page 8. On the expansion of hospitals and changing health care patterns, see Rosemary Stevens, *In Sickness and in Wealth: American Hospitals in the Twentieth Century* (New York: Basic Books, 1989). Although the sulfa drugs were discovered in the late 1930s and penicillin came into clinical use in the 1940s, widespread antibiotic use became common only in the 1950s.

37. On changing American living standards see Ruth Schwartz Cowan, *More Work for Mother: The Ironies of Household Technology from the Open Hearth to the Microwave* (New York: Basic Books, 1983); and Harvey Levenstein, *Revolution at the Table: The Transformation of the American Diet* (New York: Oxford University Press, 1988). The increasing difficulty middle- and upper-class women had in securing live-in domestic servants also contributed to their declining enthusiasm for rigorous housecleaning. See David Katzman, *Seven Days a Week* (New York: Oxford University Press, 1978).

38. Ehrenreich and English, *For Her Own Good*, 160–61.

39. Charles-Edward A. Winslow produced an exhaustive bacteriological study refuting the dangers of sewer gas in 1909. See the National Association of Master Plumbers of the United States, *Report of the Sanitary Committee, 1907–1908–1909* (Boston: The

Association, 1909), 39–95. On the importance of the sanitary privy in relation to the hookworm problem, see John Ettling, *The Germ of Laziness* (Cambridge: Harvard University Press, 1981). The point about the tubercle bacillus was explained to me by Barbara Bates.

40. The continuity in food-handling directives is strikingly evident in the course materials used in the Food Manager Certification Program for Suffolk County, New York, made available to me by Elizabeth Roberts Daily.

41. Two works that argue for the importance of change at the personal and household levels are Douglas C. Ewbank and Samuel H. Preston, "Personal Health Behavior and the Decline in Infant and Child Mortality: The United States, 1900–1930," in *What We Know About Health Transition: The Cultural, Social, and Behavioral Determinants of Health*, ed. John C. Caldwell et al. (Canberra, Australia: Australian National University Press, 1990), 116–49, and Hardy, *Epidemic Streets*.

42. "Reading Course Testimonials," Box 25, folder 16, Division of Rare and Manuscript Collections, Cornell University Library, Ithaca, N.Y. The author of the letter was not identified, nor was a specific date given, but it was written sometime between the early 1900s and the mid-1910s.

43. S. K. Kleinberg, *The Shadow of the Mills: Working-Class Families in Pittsburgh, 1870–1907* (Pittsburgh: University of Pittsburgh Press, 1989), provides an excellent overview of the poor health conditions common in working-class neighborhoods.

44. Nancy Tomes, "The Wages of Dirt Were Death," paper presented at the Annual Meeting of the Organization of American Historians, April 1991. On the decline in time spent housekeeping, see Molly O'Neill, "Drop the Mop, Bless the Mess: The Decline of Housekeeping," *New York Times*, April 11, 1993, I-1, 18. As O'Neill suggests, this decline reflects women's growing participation in the workforce and men's reluctance to take up the slack in domestic chores. But it has been facilitated by the absence of fear that the rising tide of domestic dirt would harm family members' health.

33 Gendered Expectations: Women and Early Twentieth-Century Public Health

Judith Walzer Leavitt

Two women stood before judges in the early twentieth century, both having been accused of transmitting typhoid fever as they carried out their work of cooking for other people. Ostensibly, they were in court because public health officials thought that they endangered the health of people around them. Indeed, both had been found to harbor typhoid bacilli and were capable of infecting others through food contaminated with the pathogenic bacteria from their urine or feces. In addition to the laboratory findings and the public health threat, another factor stands out as also prominent in determining the fate of these women before the bench: ideas about proper womanly behavior. The authorities concerned with these cases believed not just that the bacilli and food handling made the women dangerous but that women were more dangerous than men because cooking was a traditional and necessary female activity. One

JUDITH WALZER LEAVITT is Ruth Bleier Professor of History of Medicine, History of Science and Women's Studies, and the Associate Dean for Faculty at the Medical School, University of Wisconsin–Madison.

From *U.S. History as Women's History: New Feminist Essays,* edited by Linda K. Kerber, Alice Kessler-Harris, and Kathryn Kish Sklar. Copyright © 1995 by the University of North Carolina Press. Used by permission of the publisher.

woman was a New Yorker, Mary Mallon, known to the world as "Typhoid Mary." The other was Chicagoan Jennie Barmore, defended before the Illinois Supreme Court by Clarence Darrow. The stories of these two women can be told as episodes in early twentieth-century public health, whereby science, demonstrating new ways to control the spread of infectious disease, triumphs over sickness. Such a telling would mark the gendered words of the public health officials or the court (if it noticed them at all) as unimportant to the larger story of the healing potential and march of progress of medical science. But by centering the women as well as the science we can demonstrate that cultural gendered expectations about who women were and what they did formed important elements in determining the activities of early twentieth-century public health. The progress of science was not the only factor affecting public health actions. As women's historians have demonstrated over and over again in the last twenty-five years, we can enrich and make more complex our understanding of the past; indeed, we can often completely transform our understanding of the past, by including gender considerations as part of our analysis.[1]

In emphasizing the gendered significance of

the cases in this essay I do not in any way mean to diminish the importance of other elements. Elsewhere I have explored the Mary Mallon episode as an illustration of early bacteriological practices, and I am working on other aspects of the civil dilemma she posed for health officials who were trying to protect the public's health and at the same time not infringe on the rights of individual citizens.[2] Yet in studying the documents left to us, I have been struck by the prominence of the gendered language used to describe the experiences and fate of Mallon and Barmore; I am convinced that this language reveals an important determinant of the public officials' and the courts' actions. Public health officials employed gender stereotyping to support and uphold their activities and thereby to gain public acceptance of them. Both informal utterances and more formal legal discourse divulge some deeper meanings for some of the health officials' actions. Early twentieth-century public health texts provide insight into social views about women and how those views affected the treatment women received in medical and legal contexts.

The specific words chosen by the people who pushed Mallon and Barmore to seek legal redress reveal gendered meanings that historians cannot ignore. New York health officials portrayed Mary Mallon as "masculine" in her walk and her thinking. Chicago public health physicians insisted that Jennie Barmore, if freed, would immediately go and cook for her friends. Neither of these perceptions was a necessary part of the scientific indictment against these women (which could have been phrased in laboratory and scientific language alone), yet health officials used them nonetheless in framing their case. What impact did the gendered expectations embedded in these descriptions of behavior have on the outcome for the two women? Why did officials find it necessary to add gendered arguments to their indictment of Mallon and Barmore when establishing that their bodies were infectious? These questions must be explored in order to understand the full nature of public health activities and public policy during that period. Gendered subjective sentiments were woven into the fabric of public health even during the very years when the new science of bacteriology seemed to be pushing the field in the direction of greater objectivity.

Before examining the specific events in Mary Mallon's and Jennie Barmore's stories, it is necessary to explain the medical context in which they took place. The turn of the twentieth century was one of the most exciting and dramatic periods in all of medical history. At the end of the nineteenth century, medical scientists had come to accept germ theory, a new theory of disease causation brought to medicine through basic science research. Laboratory experiments established that microorganisms could cause disease, a conceptualization that radically altered previous views and seemed to make unnecessary the widescale urban cleanups and social welfare programs that had characterized activity under the older filth theory of disease.

Formerly, people's whole lives came under the purview of health officials who were trying to stem the tides of disease; in future, officials would need to concentrate only on the microbes. Because microbes could not distinguish between rich and poor, black and white, men and women, many people engaged in public health at the new century's debut believed that the new bacteriologically based public health would be more evenhanded than its predecessors. The reduction of public health science to the microbe would narrow public health activity and make it more equitable. While few cures had emerged from the study of microorganisms by the beginning of the century, bacteriology was brimming over with assurances for the future. The optimism led public health phy-

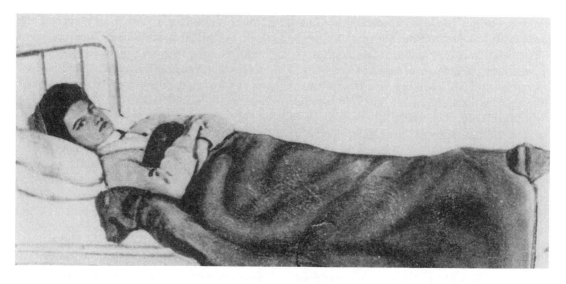

Fig. 33.1. Mary Mallon in bed at Willard Parker Hospital. The photograph was presumably taken when she was first admitted to the hospital in March 1907, but it did not appear in print until the time of the habeas corpus hearing in June 1909.
(*Courtesy UPI/Bettmann*)

sicians—those in the frontline confrontation with infectious disease in the cities where death rates soared along with the population—to search out answers from the new science and from the laboratory whenever possible.[3]

Typhoid fever had been one of the nineteenth century's major killers. A water- and food-borne systemic bacterial infection, typhoid caused sustained fever, headache, malaise, and gastrointestinal problems in its victims. Although many mild cases occurred, typhoid carried a case fatality rate of about 10 percent.[4] It struck most harshly those cities that sent untreated lake or river water through the pipes; thus it responded well to water filtration systems and sanitation efforts instituted during the last third of the nineteenth century or in the early years of the twentieth century.[5] Urban sanitation campaigns brought significant reductions in death rates from typhoid fever, but the disease did not disappear as a public health problem. Early

twentieth-century bacteriologists tried to understand how this bacterial infection continued to thrive in relatively clean city environments. In 1902 bacteriological studies in Europe finally led to the realization that typhoid, along with diphtheria and a few other diseases, could be transmitted by healthy people. Mary Mallon was the first healthy carrier to be carefully traced in North America. The excitement about her case, as well as some of the confusions surrounding it, must be seen within this context of the scientific breakthrough that she represented. She was the first of hundreds of New Yorkers whom the health department accused of sheltering typhoid bacilli in their gallbladders and, through their urine and feces via unwashed hands, transmitting the germs to susceptible and unsuspecting people and making them sick.

Mary Mallon provided the nation's first publicized test case of how to stop healthy people from transmitting typhoid fever to others; thus this essay concentrates on her story. Jennie Bar-mote's situation offered an important adjunct to Mallon's and was more significant as legal prece-dent. Although the sagas of these two women represent one extreme of what was possible in public efforts to protect the public health and welfare, and not the norm, they are instructive to study nonetheless, because they reveal the broad spectrum of factors that could and did influence public health policy. In this early twentieth-century period of medical break-throughs and scientific excitement, public health officials brought more than laboratory findings to bear in their decisions about how to best protect the public.

Mary Mallon was a peripatetic Irish-born cook living in New York City at the turn of the twen-tieth century. She hired herself out to wealthy families, finding most of her placements through an employment agency. Some of her as-signments lasted for years; often she stayed a few weeks or months before moving to a different family. Reputedly, she was a good cook; she did not have difficulty finding work. Her career pat-tern was completely ordinary, following the lim-ited opportunities available to single women of her ethnicity and class.[6] During the fine summer days of August 1906 she cooked meals in the household of a New York banker, Charles Henry Warren, at the family's rented summer residence in Oyster Bay, Long Island. Within weeks of her arrival six persons in the household of eleven were attacked by typhoid fever. Those stricken included Warren's wife and two daughters, two domestic workers, one white and one black, and the gardener. The initial investigation revealed that this particular household epidemic had not been caused by contaminated water or milk (frequent sources of typhoid infection), nor had the family consumed any infected clams from the bay. The owner of the house hired George Soper, a sanitary engineer trained at Columbia

University and known for his epidemiological work on typhoid fever, to investigate.[7]

Having read the latest literature out of Europe positing healthy people as carriers of typhoid fe-ver, and ruling out family members and other servants, Soper focused on trying to find the cook who had been employed by the family only three weeks before the outbreak and who left the family three weeks after it. By tracing the cook's job history before her arrival in Oyster Bay, with the help of her employment agency, he identi-fied eight families who had employed her; in seven of them typhoid fever had followed her stay. Soper became convinced that if Mary Mal-lon could be found, and her feces and urine tested, he could prove in the laboratory what his epidemiological study had already shown: that she, although healthy, had transmitted typhoid fever to those who unsuspectingly ate the food she prepared.

It took Soper until March 1907 to trace Mal-lon to her current employment in a home on New York's Park Avenue. He later described what happened: "I had my first talk with Mary in the kitchen of this house. . . . I was as diplo-matic as possible, but I had to say I suspected her of making people sick and that I wanted specimens of her urine, feces and blood. It did not take Mary long to react to this suggestion. She seized a carving fork and advanced in my direction. I passed rapidly down the long nar-row hall, through the tall iron gate, . . . and so to the sidewalk. I felt rather lucky to escape."[8] Unable to obtain Mallon's cooperation in the in-vestigation and thinking the epidemiological ev-idence sufficiently compelling, Soper reported his findings to Hermann M. Biggs, medical officer of the New York City Department of Health, with the recommendation that Mary Mallon be apprehended.

The health department did not find it any easier to get Mallon's cooperation. Possibly thinking that a woman could most easily approach another woman, Biggs sent Dr. S. Josephine Baker, a medical inspector in the department (later director of the Bureau of Child Hygiene), to collect specimens of urine, feces, and blood. Dr. Baker's first visit with Mallon in the East Side brownstone yielded nothing: she too was summarily dismissed. Interpreting this as a case of "blind, panicky distrust of doctors and all their works which crops up so often among the uneducated," Baker returned the next day, this time accompanied by three police officers and an ambulance. The young physician was again overpowered by the resistant cook: "Mary was on the lookout and peered out, a long kitchen fork in her hand like a rapier. As she lunged at me with the fork, I stepped back, recoiled on the policeman and so confused matters that, by the time we got through the door, Mary had disappeared. 'Disappear' is too matter-of-fact a word; she had completely vanished."[9]

Baker enlisted more police, and the search continued for five hours. The servants, showing what Baker recognized as "class solidarity," denied any knowledge of her whereabouts. Finally, one of the police officers saw a bit of calico showing in the doorway of the space under the outside steps, where they had not looked because of the dozens of filled ashcans heaped up in front of it (more evidence of her colleagues' support). Again Baker described the scene:

She came out fighting and swearing, both of which she could do with appalling efficiency and vigor. I made another effort to talk to her sensibly and asked her again to let me have the specimens, but it was of no use. By that time she was convinced that the law was wantonly persecuting her, when she had done nothing wrong. She knew she had never had typhoid fever; she was maniacal in her integrity. There was nothing I could do but take her with us.

The policemen lifted her into the ambulance and I literally sat on her all the way to the hospital; it was like being in a cage with an angry lion.[10]

Examination of Mallon's feces showed a high concentration of typhoid bacilli, proving in the laboratory that Soper's epidemiological study had been correct in its target: Mary Mallon was, in the press's words, a walking "human culture tube."[11] With active *salmonella typhi* bacilli in her feces, when she did not wash her hands thoroughly after using the bathroom, she transferred the germs to others for whom she cooked. She was held, against her will, in the Willard Parker Hospital and later moved to North Brother Island in the East River.

Although presumably proud of its discovery of the first healthy carrier in the country, the health department did not publicize the capture of Mary Mallon. Probably, it did not want to cause public panic. Moreover, the health officials were uncertain about the legal ramifications of their actions. Never before had the department faced the issue of locking up a healthy individual on the basis of laboratory reports. William Park, the health department bacteriologist, voiced the worry about the potentially long-term incarceration of healthy persons—the carrier state might last a lifetime—when he asked in his first paper on Mary Mallon: "Has the city a right to deprive her of her liberty for perhaps her whole life?"[12] Did police powers stretch that far?

While this question percolated, Mary Mallon remained in quarantine. Obviously bitter about her incarceration, she spent her days alone in a cottage on the grounds of Riverside Hospital, the isolation facility on North Brother Island. A hospital attendant brought her meals to the door and left; she saw no one. She wrote "violently threatening letters" to Hermann Biggs (none of which seems to have survived) and kept up with events by reading the daily newspaper.[13] Ultimately, the press discovered her whereabouts and publicized her story, although without using her name, which reporters were unable to learn from the health officials.[14] In the wake of public notice, and with a lawyer hired possibly through the Hearst newspaper, the *New York American,* which actively pursued the story, Mary Mallon decided to sue for release. On June 28, 1909, she and her lawyer filed a writ of habeas corpus with the New York Supreme Court, claiming that she was in perfect health and that she was being held "forcibly and without warrant or order of any character" and "that she is not in any way or any degree a menace to the community or any part thereof."[15] The *American* quoted Mallon as saying in court: "I never had typhoid in my life, and have always been healthy. Why should I be banished like a leper and compelled to live in solitary confinement with only a dog for a companion? ... I am an innocent human being. I have committed no crime and I am treated like an outcast—a criminal. It is unjust, outrageous, uncivilized. It seems incredible that in a Christian community a defenseless woman can be treated in this manner." The newspaper voiced sympathy with Mallon's plight.[16]

The court decision came on July 16, 1909: the judge ruled that indeed Mary Mallon was a menace to the public health and that she must remain in health department custody. Newspapers quoted the judge as saying, "While the court deeply sympathizes with this unfortunate woman, it must protect the community against a recurrence of spreading the disease."[17]

The one remaining letter in Mary Mallon's hand is from this period, probably written in July 1909 during the legal proceedings, and it is the only direct source we have for understanding her perspective on the events. She was angry, and she directed her emotion against William Park, who had presented her case before medical audiences. She insisted that she was being kept

a "prisoner without been sick nor needing medical treatment." The laboratory may have proved that her medical history included a case of typhoid fever, but she herself denied that she had ever been ill with the disease. She was aware of her situation and the stigma society attached to her. There had been, she said, a "visiting Doctor who came here in October he did take quite an interest in me he really thought I liked it here[,] that I did not care for my freedom." Mallon saw the situation clearly as one of lost liberty. She concluded her letter perceptively: "I have been in fact a peep show for Every body[;] even the Internes had to come to see me & ask about the facts already Known to the whole wide World[.] the Tuberculosis men would say there she is the Kidnapped woman[.] Dr. Parks [sic] has had me Illustrated in Chicago[.] I wonder how the Said Dr. Wm. H. Park would like to be insulted and put in the Journal & call him or his wife Typhoid William Park."[18]

After the court rejected her petition, Mary Mallon returned to North Brother Island, where she remained until a new health commissioner in 1910 decided to let her go. Ernst J. Lederle recognized that such a total isolation as Mallon had been subjected to was not medically indicated for typhoid fever carriers who were dangerous only when they cooked the food that others ate. He told the press, "She has been released because she has been shut up long enough to learn the precautions that she ought to take."[19] Lederle's compassion for Mallon was duly noted by the *American,* which quoted him as saying: "For Heaven's sake, can't the poor creature be given a chance to live? An opportunity to make her living, and have her past forgotten? She is to blame for nothing—and look at the life she led!"[20]

The health commissioner apparently helped Mary Mallon find employment in a laundry and did keep track of her for some time. But Mallon was not happy in another occupation and could not earn a good living away from the kitchen. She tried unsuccessfully to sue the city for damages and then faded from public view; in time, the health department also lost sight of Mary Mallon.

She did not hide for long, however. In early 1915 an outbreak of typhoid fever occurred at the Sloane Maternity Hospital in New York City. Twenty-five doctors, nurses, and hospital staff were stricken, and two died. A new cook, a Mrs. Brown, had been employed three months before the outbreak. Another dramatic arrest followed, as health department inspectors followed "a veiled woman," recognized by inspectors as Mary Mallon, to a Corona apartment, where, refused admission through the door, they pursued her with a ladder to the upper window. The health officers brought her back to the hospital, and Health Commissioner S. S. Goldwater promised that "she would never endanger the public health again."[21]

Indeed, she did not. This time Mary Mallon did not contest her quarantine, and she remained on North Brother Island for the rest of her life. Sometime in 1933 she suffered a stroke, and, paralyzed and unable to care for herself, she was hospitalized until her death on November 11, 1938. She had been in health department custody for a total of more than twenty-six years; at least forty-seven cases of typhoid fever had been traced to her, and three people had died. Nine mourners, three men, three women, and three girls, who would not identify themselves to reporters, paid her final tribute at her funeral in the Bronx. She was, it was estimated, seventy years old.[22]

The second arrest and proposed lifetime isolation of Mary Mallon evoked more newspaper copy and less sympathy than previous events. George Soper wrote sentiments that were echoed in the press: "Whatever rights she once possessed as the innocent victim of an infected condition ... were now lost. She was now a woman who could not claim innocence. She was known wilfully and deliberately to have taken desperate chances with human life.... She had abused her privilege; she had broken her parole. She was a dangerous character and must be treated accordingly."[23]

At the time of Mary Mallon's second incarceration, Soper declared to the *New York Times;* "Liberty is an impossible privilege to allow her."[24] At that moment, in 1915, after Mary Mallon had persisted in cooking with the knowledge of three years of incarceration behind her, Soper did not see any conflict or dilemma in his statement. Liberty was not a right of citizenship, it was a privilege to be earned, and Mallon had abused the privilege. She knowingly continued to infect unsuspecting others. Earlier she may have been an innocent offender (although Soper had supported her isolation from the first), in 1915 she became a guilty one. Soper's explanation seemed to fit previous legal decisions: a minority need not be allowed to define the health standards of a whole community in its disinclination to cooperate with health authorities.[25]

Previous police power quarantine rulings had applied to sick people for a delimited amount of time.[26] In this instance, however, with the penalty applied to a healthy person and for an indefinite period, the case for quarantine was considerably more shaky. When Mallon had been isolated the first time, in March 1907, she was not sick nor was she yet a repeat offender. Furthermore, it had not been demonstrated that total isolation was necessary for healthy carriers of typhoid fever. If such people could be instructed in strict personal hygiene and if they would not cook for anybody other than themselves, they could walk the city streets without endangering anyone. George Soper himself conceded in 1915 that while "proper precautions" ought to be taught to healthy carriers, "this does not mean isolation, nor anything drastic. There is only one way that the germs can be transmitted, and that is through contact with the waste products of the body."[27] In retrospect, it seems that the health authorities acted very quickly in taking away Mallon's liberty without exploring other options that may have been available.[28]

Mary Mallon herself provided some of the earliest answers to the medical questions about the carrier state. While she was in isolation between 1907 and 1909, physicians tried various drug regimens, hoping to kill off the *salmonella typhi* bacilli in her body. They offered to excise her gallbladder, the organ harboring the bacilli, a surgical procedure that many physicians advised against as too dangerous. They carefully observed the vicissitudes of Mallon's bacteria counts, monitoring them three times a week.[29] They were learning, and Mary Mallon's body was their laboratory.

Mallon's personal liberty became expendable in the name of science and as insurance against the future. Practical considerations, most explicitly the limits of the state's ability to handle large numbers of noncooperative carriers, created the need to encourage others to go along with health department policies. As the *Medical Record* put it: "It is evident that [healthy carriers] cannot all be segregated and kept prisoners. . . . It would be difficult to obtain popular sanction to such interference with the liberty of apparently healthy individuals, and even if the measure were recognized as justifiable the number of bacilli carriers would render it difficult of execution."[30] If for practical reasons the state could constrain only a few, officials needed convincing arguments that would resonate widely. They built a case against Mallon in their efforts to provide an example, presumably believing her to be expendable in the larger battle to contain the disease. The question for historians is why Mallon became the example and what she was thought to exemplify. Was she isolated for life because she was the first healthy carrier in the nation to be carefully traced? Or was it that

she personified other characteristics in the minds of the health officials?

There were many other healthy carriers found in New York City after 1907 who remained free, regulated by new department rules established just for them. Once apprised of the dangers that healthy people who handled food could pose, the health department in 1916 initiated a system of inspection of food handlers in the city.[31] In 1924, when a second healthy carrier went before New York courts, 150 healthy carriers were under health department observation; by the time of Mallon's death in 1938 there were 394 healthy carriers under surveillance in New York City. They were not permitted to handle food; Mallon was the only one held in long-term forced custody.[32]

Examination of the second healthy carrier court case in New York helps us to understand some of the dynamics at work in Mallon's situation. Under the systematic regulation of food handlers begun in 1916 and based on what was learned from Mallon, Alphonse Cotils had been identified as a healthy carrier of typhoid fever and had been denied a food handler's license. In March 1924 Cotils, a bakery and restaurant owner, was found preparing a strawberry shortcake in his restaurant despite his previous banishment from such an occupation. He defied health department rules, his physician said, because officials were "'annoying' him about working in his own bakery." Cotils knew that he was a typhoid carrier, and he knew that he was not allowed to prepare food for other people. But he, like Mary Mallon before him, refused to cooperate with the regulation and continued his work. When his case came before the court, the judge found Cotils guilty of violating the section of the sanitary code that prohibited food handling by typhoid carriers. Yet in his case the judge suspended sentence "after Cotils had promised to remain away from his restaurant and keep out of kitchens. He intends to conduct his business by telephone, he said." According to the *New York Times,* the judge reasoned: "I am thoroughly impressed with the extreme danger from these typhoid carriers, particularly when they are handling food. I could not legally sentence this man to jail on account of his health."[33] At the very moment the judge said that he could not legally imprison Cotils because he was not sick, a healthy Mary Mallon was securely held in her isolation on North Brother Island. Both had violated a previously imposed quarantine; only one was physically detained for it.

What were the reasons for the discrepancy? In which case, Mallon's or Cotils's, did the health department and the court more adequately protect the public health? Why did authorities see the need for denying personal liberty in one case and only restricting it in the other?

Both Alphonse Cotils and Mary Mallon were proven carriers of typhoid fever. Their work involved handling the food of others, and thus they endangered the health of those unsuspecting people whose food they prepared. Most important, they persisted in cooking, thus putting other people at risk. As the *New York Tribune* editorialized, those carriers who would not voluntarily give up their activities that spread typhoid were a menace to the public and needed to be locked up: "The plain and obvious fact is that we have no way of dealing with these unlucky persons except by keeping them where they cannot do harm to others. . . . when the danger has been so clearly demonstrated as in the case of 'typhoid Mary' there can be little doubt as to the right course to follow."[34] Previous public health experience taught that forcible incarceration could become part of normal procedures as a last resort in those cases deemed otherwise intractable. The question remains why health officials and the court deemed Mary Mallon intractable, and thus deserving of being denied her personal liberty, and why Alphonse Cotils, also a cook repeater whose case went to

court, although restricted, was allowed his liberty.

One acknowledged justification for the denial of liberty in Mary Mallon's case was her recalcitrance, her "perversity" as some called it. Soper claimed that if only Mallon had cooperated, her freedom could have been saved. Dr. Josephine Baker wrote too that Mallon "might have been a free woman all her life. It was her own bad behavior that inevitably led to her doom."[35] Because of the different and less restrictive treatment of other recalcitrant carriers, we know that other factors in addition to her refusal to cooperate must have existed. Indeed, we find in the health officials' words and deeds evidence with regard to class, ethnicity, and gender that bears closer scrutiny.

First, health officers gave much importance to Mary Mallon's social condition. Before her case, social differences had affected policy execution. People with no home or family had been particularly vulnerable to official control, because they did not have the social and physical supports to convince authorities that they could care for themselves and not endanger others. Alphonse Cotils owned his own restaurant; while we know comparatively little about his social situation, that fact alone put him in a class apart from Mallon. Health officials repeatedly called attention to Mallon's lifestyle. As Soper wrote:

I found that Mary was in the habit of going, when her work for the day was finished, to a rooming house on Third Avenue below Thirty-third Street, where she was spending the evenings with a disreputable looking man who had a room on the top floor and to whom she was taking food. His headquarters during the day was in a saloon on the corner. . . . He took me to see the room. I should not care to see another like it. It was a place of dirt and disorder. It was not improved by the presence of a large dog of which Mary was said to be very fond.[36]

These unsavory and disordered conditions—in addition, no doubt, to the unspoken judgment on out-of-wedlock cohabitation—became part of the justification for keeping careful watch over Mallon: it was obvious that she could not care properly for herself.

The rules adopted by the New York City Department of Health for typhoid carriers bore out this sentiment: "Typhoid fever carriers need not be retained in hospitals or institutions if not desired. They will be sent home *if home conditions are satisfactory.*"[37] People without homes and families, or, presumably, with unacceptable homes and families, thus became more vulnerable to detention because of the perception that they could not take care of themselves. Public health officials found it necessary to describe not just the bacillus-carrying nature of those deemed dangerous to the public health but also their social condition, even if that social condition was immaterial to describing the health dangers they posed.

Soper was concerned not just with Mallon's living conditions but with her personal hygiene as well, since she transmitted the typhoid bacilli in large part because she did not keep her hands clean. But the engineer epidemiologist did not limit himself to cleanliness in his observation: "She was careless in her personal habits, but so are most cooks."[38] Soper underscored his belief that Mallon needed to be under strict observation by placing her as part of a group, which he, and society, denigrated. The stigma attached to healthy carriers could be linked to the inferior status of domestic workers and together help justify Mallon's incarceration.[39]

Perhaps if Mary Mallon had had a home to shelter her, one authorities might have recognized as safe, and family to take care of her, she might have been released despite her refusal to

cooperate with health department guidelines. Perhaps if she had been a housewife and not a domestic laborer, no matter how hot her temper, health authorities may have found reason to liberate her. Certainly Cotils gained liberty, if we can read meaning into the judge's decision in his case, in part because he had some of these social options in his life: he could promise to carry out his business from his home on the telephone, for example. Perhaps if Mallon had not been Irish, with a stereotypical hot temper, she might have received less coercive responses to her actions. Although not often overtly a factor, her ethnic background lurked behind the scenes in many comments about her temper and lack of cooperation with health officials.[40] We can never know what would have happened if Mallon had been someone else, if the first healthy carrier followed so carefully in this country had been someone, even if uncooperative, who represented more respectable middle-class America. We can know that some class- and possibly ethnic-related perceptions were evident in the official and public thinking about Mallon's situation and that they affected attitudes about her and seem to have affected the treatment she received from the health authorities.[41]

The social class difference between Mary Mallon and the people she served also was a factor in how health officials regarded the dangers she posed. As a newspaper reported, "[A] well-known member of the Board of Health [revealed] that this human culture tube has worked for prominent families in this city and communicated the disease to some of its members."[42] Working-class domestics entered the private spheres of the rich; those who carried disease threatened the city's most powerful in their most vulnerable settings, the sanctity of their homes.

In addition to class and ethnicity, gender considerations played an important role in this story. Authorities used Mallon's case to illustrate the dangers all women, society's main food pre-parers, posed as potential typhoid fever carriers; they also used gendered language to single her out as unique and deserving of particularly harsh treatment.

Mary Mallon spread typhoid not just because she harbored the bacteria in her bowel movements but also because she was a cook. This predominantly female occupation made women especially dangerous. Health officers noted repeatedly that more women were identified as carriers because more women were in occupations, paid or not, that made it easy for them to spread germs to unsuspecting others.[43] Being a carrier was a gendered condition, one in part defined by sex-role expectations. As cooks, all women food handlers were potentially dangerous to the public health, whether they were employed outside the home or within it. In a fascinating passage, quoted in the *New York Times Sunday Magazine,* Soper blamed upper-class women, who hired cooks, for not being more careful to make sure they were not bringing danger into their homes. He blamed middle-class women, especially mothers, who went from the sick room to the kitchen to prepare the family's meals, thereby spreading typhoid fever. And he blamed working-class domestics who entered the homes of others as cooks, bringing their germs with them and spreading them to unsuspecting people. Women of all social classes, by virtue of their culturally defined tasks, were potentially dangerous.[44]

While all women thus became suspect for spreading typhoid fever, health officials could not for a moment have considered locking up all women identified as typhoid carriers. It was not sex alone or cooking per se that made Mallon a candidate for incarceration. Nonetheless, there are clear indications that cultural ideology about sex-appropriate behavior did affect her case. By contrasting Mary Mallon with a male working-class carrier followed by the health de-partment in these years, it is possible to see how social prejudices of the health officials affected their decisions about healthy carriers. In 1922 a New York City carrier, a man who had reportedly caused an outbreak of eighty-seven cases of typhoid fever and two deaths and whom the health department had been following, absconded from its purview and was found by New Jersey health authorities, who blamed him for still another outbreak that had resulted in thirty-five cases and three deaths. Rather than incarcerating this healthy carrier for repeated violations and breaking parole, the health officers instead added him to the list of carriers and concluded the case with the remark, "This carrier is now employed in this City as a laborer in building construction work and is required to report to us weekly."[45] From all indications this man was as dangerous to the public as was Mallon. In fact, he had already been identified with more typhoid fever cases and more deaths than Mallon. He had disobeyed the law in two states with repeated violations, certainly showing perversity and lack of respect, yet health officials allowed him his freedom to find construction work and continue at liberty. One reading of this incident, which unfortunately cannot be followed more closely with extant documentation, could conclude that as a male wage earner, he was viewed as a family breadwinner and therefore as necessary to the family economy. Despite his recalcitrance and his record as a menace to the public health, he was not locked up; Mary Mallon, in parallel circumstances, was denied her freedom and not retrained for a different job.[46]

There was precedent for finding other means of support for healthy typhoid carriers instead of isolating them. At the Pasteur Institute in Paris bacteriologist Ilya Metchnikoff had found employment in a library for a healthy carrier whose case interested him. In 1918 New York State began subsidizing the incomes of those carriers who were having difficulty finding adequate employment outside the food industry. The absence of a record of attempts to find funding or alternate employment for Mallon is notable.[47]

Another indication of how gendered expectations affected Mary Mallon can be seen in the language Soper, Baker, and others used to describe her. Soper noted that Mallon had "a somewhat determined mouth and jaw." He thought, "Nothing was so distinctive about her as her walk, unless it was her mind. The two had a peculiarity in common. . . . Mary walked more like a man than a woman and . . . her mind had a distinctly masculine character also."[48] These comments leap out of the page as we read them today because they are so obviously not connected to the health dangers that Mallon posed. Her walk or the character of her mind in no way influenced the bacilli lodged in her gallbladder. Yet Soper could not refrain from calling attention to them in his characterization of the stigmatized woman. The *Medical Record,* with similar disregard for relevance, described Mallon as "a perfect Amazon, weighing over 200 pounds." Josephine Baker emphasized Mallon's fierceness—she was like an angry lion, she was maniacal, she fought and swore.[49] Mallon was set apart in these accounts as different, deviant, unfeminine. The *World-Telegram* noticed too that she "was not imbued with that sweet reasonableness which would have allowed her to listen to the explanations of learned men about her particular case."[50]

None of these characteristics was relevant to Mallon's public health case. Her weight, her strength, her energy level, her lack of deference, and her degree of femininity were all extraneous to the health dangers she posed. Even more startling, the one early photograph I have been able to find of Mary Mallon, as well as line drawings in newspapers, indicate that the physical description of Mallon was not only immaterial, it was false. She looks to the camera's eye as a distinctly "feminine" woman. The robust, physically imposing, masculinized woman seems to have existed only in the eyes of the anxious beholders, in the vision of those who needed to see in her an aberrant "other" in order to justify their actions against her. The description of Mallon as deviant and masculine was part of building a comprehensive explanation to establish the new principle of healthy carriers as dangerous to the public health.

Health officials helped to justify Mallon's incarceration and bolster public support through this negative portrayal. They did not want the public to become too sympathetic to Mallon's plight, or people might not support the general policies instituted to regulate healthy carriers. Reflecting commonly held social values about proper class, ethnic, and female behavior, the portrayal of Mallon as a social pariah and an unnatural woman thus became as important a part of the case as notifying the public of her bacilli-carrying feces. Gender stereotyping bolstered the public health policy.

Mary Mallon's case thus functioned as a vehicle for public education. From the early newspaper accounts the public learned about the incarceration of this unfortunate, unlucky woman. When the public showed some sympathy with her plight, officials worked to recreate her image as a social undesirable, as a frightening and dangerous person, in order to discourage others from following her ways and to encourage public distance. The medical dangers could best be portrayed through building a specific individual case of social disorder. Chronic bacillus carrying itself was not sufficient. Health officials needed social as well as medical arguments to help shape public knowledge and discourse about healthy carriers. They constructed an image of a deviant and physically unappealing "Typhoid Mary" out of the story of Mary Mallon.

Health officials brought to bear the full strength of the state against Mary Mallon. In the name of protecting the public health of all citizens, they took away her liberty for more than twenty-six years, more than (as far as I can determine) any other healthy typhoid fever carrier in U.S. public health history. They made no concerted effort to try to teach her another skill, one that would not have endangered the public, until her productive years were almost over. Not until the later years of her incarceration did she work in a hospital laboratory. Mallon became, in reality and in symbol, as she characterized herself, a social "leper." She evoked some sympathy in the early years of her isolation, but she came to represent the worst in human behavior. Embodying the fullest extent of the law's ability to deny individual liberty, Mallon became an example for others, a reason to cooperate with health officials when they came knocking. Her name, her epithetical "Typhoid Mary," stood—and stands—for the ultimate stigma, someone to be feared, to be avoided, to be maligned.

By the beginning third of the twentieth century, health departments had forged for themselves wide powers to prevent the spread of infectious diseases by curtailing or denying the liberty of a few individuals, sick and healthy, and using them as examples to shape the thinking and behavior of the majority. They did not often explore the full reaches of their power, but the potential existed for significant denial of personal freedoms in the cause of protecting the public health. The role women played in forming the answers to the dilemmas posed in the new century bears witness to the limited and sex-defined expectations of the culture and illustrates how social prejudices affected public health history even during the period when scientific advances were most prominent.

In 1910, the year Mary Mallon was released from her initial quarantine, Jennie Barmore, then fifty-six years old, and her husband George opened a rooming house at 100 West 113th

Place in Chicago. They had grown children and a big mortgage. George Barmore had been invalided, and the family's sole means of support was the income collected from their lodgers for their rooms and the meals Jennie cooked for them.

In November 1919 the Barmores' economic security ended. An alert health department officer had noticed that during the previous few months at least five single men and women, who were receiving treatment in different hospitals around the city for typhoid fever, all gave 100 West 113th Place as their address. One subsequently died. In July 1919 David Barmore visited his parents and became ill soon after returning home. His physician also diagnosed typhoid fever, although David later denied any knowledge of this. Because of these cases a health department nurse visited the Barmore residence and informed the family of the sickness traced to them. Since Jennie Barmore had always prepared the food the roomers ate, the nurse took her to the hospital in November and December 1919 to obtain specimens of her feces and urine. Three specimens collected tested positive for typhoid.[51] As a result of these tests the health department, following its own guidelines for the management of healthy carriers, required Barmore to stop cooking for others and keeping boarders. Probably out of economic necessity and because she denied ever having typhoid fever, she did not comply. As a result of Jennie Barmore's lack of cooperation Dr. Herman Bundesen of the Chicago Health Department, on December 15, 1919, knocked on the door of the Barmore home and proceeded to take the sixty-five-year-old woman to Cook County Hospital as a suspected typhoid fever carrier.[52] At the hospital Barmore, "against her will," was "compelled" once again to submit excreta for examination. The health department ultimately released her to her home under a rigid house quarantine.

Barmore's house was placarded with a large red sign declaring that a typhoid fever carrier lived within:

DEPARTMENT OF HEALTH CITY OF CHICAGO
TYPHOID FEVER CARRIER
REMOVE NO MILK BOTTLES

All persons, not occupants of this apartment, are hereby notified of the presence of Typhoid Fever within, and are warned not to enter. The person having Typhoid Fever must not leave the apartment without permission of the Commissioner of Health, until this warning sign has been removed by the Department of Health. The Milkman must not enter the house. The family will set the vessels outside, into which the milk or cream may be poured by the Milkman, who MUST NOT HANDLE THE VESSELS. Milk or cream may be delivered in bottles, provided the bottles received are not returned to the Milkman, until this notice is removed by the Department of Health.

This warning card must not be concealed from public view, must not be mutilated or defaced and MUST REMAIN POSTED ON THESE PREMISES UNTIL REMOVED BY THE DEPARTMENT OF HEALTH.

PENALTY FOR VIOLATION OF THE RULES OF QUARANTINE; A fine of up to $200, or imprisonment not to exceed six months, or both.

By order of John Dill Robertson, M.D., Commissioner of Health[53]

In addition to having to submit to the sign on her house, a stigmatic symbol of governmental intrusion, Barmore was prohibited from keeping boarders unless they had been inoculated against typhoid fever. She was not to cook for anyone other than her husband, not even her visiting children. She was not to shop in any stores, go into crowds, use public toilets, or in any way endanger the health of the citizens of Chicago. She was, as one health department official later testified in court, "not allowed to leave her house and go out for a walk. . . . After she was quarantined she was not allowed to have any communication with the outside world at all."[54] This "modified" quarantine carried with it the constant threat of arrest and hospital confinement if its orders were violated.

Apparently right from the beginning of her house quarantine, Jennie Barmore resisted its terms, which her lawyers' brief later termed "arbitrary and absolute."[55] Repeatedly, according to the health officials (although this point was contested in court by the defense), she tore down the sign and continued to take in unvaccinated boarders. Herman Bundesen threatened to put her back in the hospital if she continued to disobey the quarantine. Indeed, he told the court that he would "cause her to be seized and hospitalized, and in so doing, he would not apply to any court for any writ or process or warrant of any kind for her arrest or detention."[56] Before Bundesen could rehospitalize Barmore, however, she filed suit for her release from the quarantine. The American Medical Liberty League provided financial support for the case, and the firm of Clarence Darrow represented Barmore. The writ stated: "Dr. Bundesen went to the home of your petitioner, Jennie Barmore . . . and confessedly, without any warrant, paper, document or process of law . . . forcibly took and seized the body of your petitioner without her consent and against her will, and took her into custody and detained her in the Cook County Hospital and other places and did then and there extract and take blood from her body and her bowel contents for examination."[57]

Darrow built his defense around three issues: that Jennie Barmore had been forcibly denied her liberty and her right of immunity from compulsory self-incrimination without due process, points that the health department never disputed; that she had never been sick with typhoid and had not been the cause of any epidemic or illness in any other person, a point answered by the use of laboratory evidence; and that the imposed quarantine was too rigid and oppressive in that it went beyond merely constraining her from handling food and serving food to others, an interpretation that engaged the court and respondents for most of the proceedings.

Judge Joseph Sabath heard the case in the Superior Court of Cook County, taking testimony from medical experts, health department officers, and nurses, and from Barmore and her son. On November 23, 1920, Sabath determined that Barmore was indeed a menace to the health of Chicago and could be held indefinitely:

> I am satisfied that to permit the relatrix free intercourse with the public, and particularly to continue in her business of preparing and serving food to others, would menace the health and safety of those with whom she would come in contact. . . . It is true that . . . her isolation is an apparent denial of her liberty, but . . . liberty of the individual may be restrained when the failure to do so would threaten the health and happiness of the community.[58]

The judge admitted that his decision would cause the Barmores "great hardship, loss and inconvenience," but he believed that "the present mental attitude of the relatrix is such that she clearly shows a lack of cooperation," and that it was therefore necessary to keep her under quarantine.[59]

Darrow appealed to the Illinois Supreme Court, where again the harshness of the terms of the quarantine were at issue. The health department insisted that "the relatrix in this case is not a prisoner. She has not been arrested in the meaning of the word 'arrested.'" Rather, the health officials insisted, the quarantine was imposed in order to protect the health of Chicagoans and only because of Barmore's refusal to obey orders. The printed testimony showed that there was no ambiguity about the meaning of Barmore's quarantine:

> BUNDESEN: These cases very seldom recover, your Honor.
>
> DARROW: Has she got to be taken care of all her life?
>
> BUNDESEN: Yes, sir; they are usually typhoid carriers all their lives.
>
> THE COURT: You don't contend that the city would have a right to detain her all her life, do you?
>
> BUNDESEN: . . . I will say that it is our experience with typhoid carriers that they very seldom recover.[60]

The defense tried to argue that Barmore, as the family breadwinner, should not be deprived of her ability to earn a living. One of Darrow's colleagues stated, "Her husband has been injured in an accident, and she wants to be free to take people in there and help take care of him, and with that sign on the house, she can't do it." By the end of the court proceedings, however, it was clear that the most the defense could argue was for leniency in the quarantine. Jennie Barmore showed her weariness and resignation. She told the court, "If my children want to come home, I want them to have the privilege of coming home, and very frequently." Even that was too much for the health department: "No; her children can go there if they don't eat food prepared by her, and if they do, we will take her back."[61] Judge J. Thompson filed the opinion on February 22, 1922: the Illinois Supreme Court remanded Barmore to the custody of the Chicago Department of Health, allowing the house quarantine to continue.

In March Darrow applied for a rehearing. The

application stated that the court's opinion assumed "that Jennie Barmore is an abnormal person" and that the medical theory upon which her quarantine was based was unproven. "If Jennie Barmore is to be quarantined, she is made a life prisoner not because of anything she has done, but because the medical men have evolved a theory which, so far, is not based upon any recorded facts."[62] Darrow rejected the laboratory evidence used to condemn Barmore as unimpressive. In light of the Mary Mallon precedent, not cited anywhere in this case, Darrow's last argument was even more interesting. The lawyer posited that the quarantine law was arbitrary in that it allowed health officials to "discriminate between individuals" who might be equally dangerous to the public's health:

> They may permit one carrier complete liberty to go about in his usual manner, transacting his ordinary business, while another may be confined under the strictest quarantine. Or one may be allowed his freedom upon merely giving his word that he would sterilize his discharges and another denied that privilege. If there is any reason for different treatment of typhoid carriers, the conditions under which the quarantine will be relaxed or modified should be set forth in the rule. . . . At present the Board assumes the right to arbitrarily discriminate between persons of the same class.[63]

The court did not allow the rehearing, and Barmore remained under virtual house arrest, as far as I can determine, for the rest of her life.[64]

Especially in light of Darrow's last argument about arbitrary discrimination between healthy carriers, it is important to scrutinize the arguments of the Chicago health officers. Much of their testimony centered on their perception that Barmore would not stop cooking if she were released from her quarantine. Despite this belief, they never discussed training her for an-other occupation, subsidizing her income, or helping to find alternatives to what was, at the time, her family's only support, even though all of these possibilities had been utilized with healthy carriers elsewhere in the country. Rather, they repeated the accusation that, no matter what, she would persist in her cooking. There is a clue to what undergirded the officials' belief that a rigid quarantine was necessary to constrain Jennie Barmore, and we see it in the discussion before the court about why she could not be released from her house quarantine:

> If relatrix is released from the present quarantine and isolation, she undoubtedly will have a big dinner party, inviting all her friends to partake of the food, which she has gone to the store for, purchased, brought home and prepared. While the relatrix is in the store handling food, she may there leave some deadly germs. While she is preparing the food, she may leave some deadly germs in the food. While she is serving the food, the same may happen. We believe the court can readily see that a great many persons could thus become afflicted with typhoid fever through no fault of their own.[65]

In their vision of how to prevent the spread of illness health officials could not see beyond traditional female domestic activity. The very job definitions of cooks, mothers, wives, and nurses implicated them as carriers of food- and waterborne disease and justified strong measures of control. Not only did sex roles define Barmore's activities; the authorities could recognize in her no ability to change her ways. Physicians argued that Barmore needed the strict quarantine by definition of her gender roles that would automatically come into play if she were released. She "undoubtedly will have a big dinner party," they claimed. Apparently, there was no way to stop such behavior in a woman; therefore it was necessary to keep her rigidly confined.

Accepted societal ideas about gender roles thus influenced public health policy. If women could not be expected to change their traditional duties, they would have to undergo the most severe penalties. Under this thinking it would not be so necessary to lock up a recalcitrant male healthy carrier, even one working in food preparation; such a person could change his job. But health officials seem to have had difficulty imagining women giving up their domestic duties, which were, supposedly, inherent in their being.

The distinction in public health treatment of men and women was underscored in Chicago's policy restraining the spread of venereal diseases. At the same time as Chicago health officials determined Barmore's fate, the same officers forcibly isolated only female sufferers of venereal disease even though they identified many male victims. They sequestered only women, "realizing that the application of the [isolation] rule [to men] may lead to interference in the business affairs of men involved and perhaps to trouble in their family affairs." This informal application of the law only to women remained policy in Chicago through these years, although women's groups in the city publicly objected to this particular form of the double standard. Health officials too declared that "adequate hospital facilities should be available for the retention and quarantine of males, the same as is provided for females." But regulation of female prostitutes continued to be the main focus of Chicago's venereal disease control program throughout the decade. When detained women filed writs of habeas corpus to try to gain their release from detention, as many did during the 1920s, they were all remanded to the health department's control.[66]

The stories of Mary Mallon and Jennie Barmore are instructive for historians because they open to scrutiny the underlying beliefs and social norms that, along with medical and scientific advances, influenced public health policy. Science was not impartially applied in the public arena in the early part of the century, and our reading of the triumph of science must be tempered by the knowledge that social prejudices entered into the determination of legal and medical decisions about the people's health.

Without denying the health dangers Mallon and Barmore posed, it is possible to understand that these two women received particularly harsh treatment at a time when most other healthy carriers retained their freedom. Although Clarence Darrow might not have recognized it as such, gendered expectations for women influenced their cases, and in his words they became the victims of "arbitrary discrimination." Mallon and Barmore both resisted the characterizations of themselves and their lives and fought back through the courts, and thus we know about them and can analyze their situations; we do not know how many other healthy women carriers were similarly affected by gendered stereotypes. Certainly, there were hundreds, nationwide, thousands, of healthy typhoid fever carriers, male and female, who did not suffer the stings of this particular public health arrow. Most of them no doubt cooperated with health officials enough to disappear from public view. There were other recalcitrant, perverse, uncooperative, and difficult healthy carriers who somehow managed to continue with their lives outside of isolation. If the arguments of this essay are transferable to their cases, most of them were male. But the argument here does not depend on that possibility. Historians can recognize that Mary Mallon and Jennie Barmore, regardless of what happened to anyone else, bore the brunt of health department authority, arbitrarily imposed. A lifetime of institutional isolation, to a healthy woman in the prime of life, and the lifetime house arrest of another older healthy woman were harsh penalties indeed, penalties informed by ideas of women's proper behavior and roles as much as by scientific evidence, penalties imposed in both cases before all options had been explored.

Notes

I want to thank many people for their help and encouragement in my Mary Mallon project, of which this essay is a part. John Duffy and Daniel Fox, both of whom have written on public health in New York, were most generous in discussing resources with me. Joel Howell and David Rosner offered helpful suggestions along the way. Joan Jacobs Brumberg, Susan Stanford Friedman, Linda Gordon, and John Harley Warner read drafts of earlier versions of the work and provided essential insights. R. Alta Charo, Hendrik Hartog, and Leslie Reagan gave needed advice with the legal sources. Research assistance by Dawn Corley and Sarah Leavitt was crucial to the completion of my first paper on the project, in time for its earliest presentation at Yale University in February 1990. Lewis Leavitt offered consultation on the medical issues and critical judgments throughout. I also wish to thank Lian Partlow, Sarah Pfatteicher, and Jennifer Munger for their help with research. I am grateful for all the suggestions and questions I received from my colleagues at the University of Wisconsin, Yale University, Harvard University, the University of Washington, and the Canadian Association for Medical History when I presented various versions of my work on Mary Mallon.

Since writing this essay, I have been fortunate to meet and to interview three people whose insights into Mary Mallon's situation have already proved very helpful to me. Even though I was unable to incorporate their wisdom into this chapter, I will be calling upon it in future writing. I would like here to acknowledge and to thank Ida Peters Hoffman, Dr. John Marr, and Emma Sherman for their gracious cooperation and support.

1. My favorite recent example of how adding women's perspectives to traditional history totally transforms it is Laurel Thatcher Ulrich, *A Midwife's Tale: The Life of Martha Ballard, Based on Her Diary, 1785–1812* (New York: Knopf, 1992), in which men's

medicine is shown to be at best only half of late eighteenth-century medicine. Ulrich's brilliant exposure of the full range of medical practice was built on the work of women's historians over the past twenty-five years, led so well by Gerda Lerner.

2. Judith Walzer Leavitt, "'Typhoid Mary' Strikes Back: Bacteriological Theory and Practice in Early Twentieth-Century Public Health," *Isis* 83 (December 1992): 608–29. My book addresses these and other factors that are part of the full story of Mallon's experiences. *Typhoid Mary: Captive to the Public's Health* (Boston: Beacon, 1996).

3. Many historians have described the narrowing of public health work in the early twentieth century. See, e.g., Paul Starr, *The Social Transformation of American Medicine* (New York: Basic Books, 1982), in which he writes: "The limitations on public health in the twentieth century were . . . profound. The early public health reformers of the nineteenth century, for all their moralism, were concerned with social welfare in a broad sense. Their twentieth-century successors adopted a more narrow and technical view of their calling" (p. 196). Recent historians, however, have altered this view and have called attention to the broad social concerns that continued to influence public health practices. For the latter see esp. Nancy Tomes, "The Private Side of Public Health: Sanitary Science, Domestic Hygiene, and the Germ Theory, 1870–1900," *Bulletin of the History of Medicine* 64 (1900): 509–39, and "The Wages of Dirt Were Death: Women and Domestic Hygiene, 1870–1930," paper presented at the Annual Meeting of the Organization of American Historians, Louisville, 1991; Allan M. Brandt, *No Magic Bullet: A Social History of Venereal Disease in the United States Since 1880* (New York: Oxford University Press, 1985); John Ettling, *The Germ of Laziness: Rockefeller Philanthropy and Public Health in the New South* (Cambridge: Harvard University Press, 1981); and my article in *Isis* (see n. 2).

4. Abram S. Beneson, ed., *Control of Communicable Diseases in Man*, 14th ed. (Washington, D.C.: American Public Health Association, 1985), 420–24.

5. See, e.g., George A. Johnson, "The Typhoid Toll," *Journal of the American Water Works Association* 3 (1916): 249–326, and Eric Ashby, "Reflections on the Costs and Benefits of Environmental Pollution," *Perspectives in Biology and Medicine* 23 (1979): 7–24.

6. On single working domestics see Susan Strasser, *Never Done: A History of American Housework* (New York: Pantheon Books, 1982), ch. 9; Barbara Mayer Wertheimer, *We Were There: The Story of Working Women in America* (New York: Pantheon Books, 1977); Rosalyn Baxandall, Linda Gordon, and Susan Reverby, comps. and eds., *America's Working Women* (New York: Vintage Books, 1976); David M. Katzman, *Seven Days a Week: Women and Domestic Service in Industrializing America* (New York: Oxford University Press, 1978); Faye E. Dudden, *Serving Women: Household Service in Nineteenth Century America* (Middletown, Conn.: Wesleyan University Press, 1983); Daniel E. Sutherland, *Americans and Their Servants: Domestic Service in the United States, 1800–1920* (Baton Rouge: Louisiana State University Press, 1981); and Phyllis Palmer, *Domesticity and Dirt: Housewives and Domestic Servants in the United States, 1920–1945* (Philadelphia: Temple University Press, 1989).

7. The first published account linking Mallon with typhoid fever was George A. Soper, "The Work of a Chronic Typhoid Germ Distributor," *Journal of the American Medical Association* (hereafter *JAMA*) 48 (1907): 2019–22. See also Soper's "The Curious Career of Typhoid Mary," *Bulletin of the New York Academy of Medicine* 15 (October 1939): 698–712, and his "Typhoid Mary," *Military Surgeon* 45 (July 1919): 1–15.

8. Soper, "Curious Career," 704.

9. S. Josephine Baker, *Fighting for Life* (New York: Macmillan, 1939), 73.

10. Ibid., 74–75.

11. See, e.g., "'Typhoid Mary' Has Reappeared: Human Culture Tube, Herself Immune, Spreads the Disease Wherever She Goes," *New York Times*, April 4, 1915, sec. 5, pp. 3–4.

12. William H. Park, "Typhoid Bacilli Carriers," *JAMA* 51 (1908): 981.

13. C.-E. A. Winslow, *The Life of Hermann M. Biggs, M.D., D.Sc. Ll.D.: Physician and Statesman of the Public Health* (Philadelphia: Lea and Fabiger, 1929), 199.

14. The first newspaper account of Mallon appeared in the *New York American* on April 2, 1907, fifteen days following her initial incarceration, and identified her as "Mary Ilverson"—knowing it was a

pseudonym. The 1907 and 1908 accounts, and the medical accounts from those years, refer to her as an Irish cook but do not name her. During the discussion of Park's paper (see n. 12), in June 1908, M.J. Rosenau used the term *typhoid Mary* (small *t*). George Whipple's 1908 book referred to her as Typhoid Mary, George C. Whipple, *Typhoid Fever: Its Causation, Transmission, and Prevention* (New York: John Wiley and Sons, 1908), 20. The major news coverage using her name did not begin until an *American* article of June 20, 1909.

15. The writ of habeas corpus and all accompanying documentation can be found in the county courthouse in New York City.

16. *New York American*, June 30, 1909, p. 3. See also *New York Times*, July 1, 1909, p. 8.

17. *New York Times*, July 17, 1909, p. 3. The decision was filed on July 22.

18. The undated letter is filed with the writ in the county courthouse (irregular spelling and capitalization in original). It had been addressed to the editor of the *American*, but this was crossed out, and George Francis O'Neil's name was penned in its place.

19. Quoted in *New York Times*, February 21, 1910, p. 18.

20. *New York American*, February 21, 1910, p. 6.

21. An account of the hospital outbreak, without naming Mallon, is M. L. Ogan, "Immunization in a Typhoid Outbreak in the Sloane Hospital for Women," *New York Medical Journal* 101 (March 27, 1915): 610–12. See *New York Times*, March 28, 1915, sec. 2, 1, and March 31, 1915, p. 8; *New York American*, March 28, 1915, p. 1; and *New York Tribune*, March 28, 1915, p. 7.

22. See the death and funeral accounts in all New York newspapers. Soper claims that the date of Mallon's stroke was Christmas 1932, and all commentators since have followed his lead. I am more convinced by two pieces of evidence suggesting the October 1933 date. First is a news article of December 1933 indicating that she suffered a stroke "two months ago," which would be in October, and the second is my informant Emma Goldberg Sherman, who says she was the person who found Mallon after her stroke. Sherman remembers the day as quite warm; she went looking for Mallon after she did not show up for work. She also believes that 1933, a date five

years before Mallon's death, is closer to the right date. She is unable to remember more specifically. She knows that she never worked on Christmas Day or other holidays. See "I Wonder What's Become of— Typhoid Mary," Sunday *Mirror* magazine sec., *New York Daily Mirror,* December 17, 1933, 19, and Emma Sherman to author, December 1993.

On the number of cases traced to Mallon, see my book. I have been able to account for forty-seven, with possibly two additional ones, whereas the official listing is fifty-three. Soper seems to have miscounted his own evidence.

23. Soper, "Typhoid Mary," 13. The *New York Tribune* editorialized similarly: "The sympathy which would naturally be granted Mary Mallon is largely modified for this reason: The chance was given to her five years ago to live in freedom, and . . . she deliberately elected to throw it away. . . . It is impossible to feel much commiseration for her" (March 29, 1915, p. 8).

24. Quoted in *New York Times,* April 4, 1915, sec. 5, p. 3.

25. The key case in making this argument was *Jacobson v. Massachusetts* (1905), 197 U.S. 11. I have written about this decision in "'Be Safe, Be Sure': New York City's Experience with Smallpox," in *Hives of Sickness,* edited by David Rosner (New Brunswick, N.J.: Rutgers University Press, 1995).

26. Some vaccination rules too were in effect only during epidemics.

27. Quoted in *New York Times,* April 4, 1915, sec. 5, 4.

28. The efficacy of quarantines through history needs more examination than it has thus far received, especially in light of recent suggestions about quarantining people infected with the HIV virus. For a good treatment of the issue of quarantines historically as relevant to AIDS, see David Musto, "Quarantine and the Problem of AIDS," *Milbank Quarterly* 64, supp. 1 (1986): 97–117. See also a longer discussion of the importance of isolation in this specific instance in my book.

29. I examine fully the medical aspects of Mallon's case in "'Typhoid Mary' Strikes Back."

30. *Medical Record* 71 (June 1, 1907): 924.

31. New York City Department of Health, *Annual Report,* 1916, 56.

32. *New York Times,* March 14, 1924, p. 19.

33. Ibid.; *New York Times,* March 15, 1924, p. 13 (quotations).

34. *New York Tribune,* March 29, 1915, p. 8.

35. Baker, *Fighting for Life,* 75. A California study of healthy typhoid fever carriers revealed that in that state a full 25 percent of identified carriers did not cooperate with authorities. See M. Dorthy Beck and Arthur C. Hollister, *Typhoid Fever Cases and Carriers: An Analysis of Records of the California State Department of Public Health from 1910 Through 1959* (Berkeley: State of California Department of Public Health, 1962).

36. Soper, "Curious Career," 704–5.

37. New York City Department of Health, *Annual Report* (1916): 56 (emphasis added).

38. Soper, "Typhoid Mary," 11.

39. On the status of single working domestics see Strasser, *Never Done;* Wertheimer, *We Were There;* Baxandall, Gordon, and Reverby, *America's Working Women;* Katzman, *Seven Days;* Dudden, *Serving Women;* Sutherland, *Americans and Their Servants;* and Palmer, *Domesticity and Dirt.*

40. See, for example, "Typhoid Mary," *Scientific American* 112 (May 8, 1915): 428. See my book for more discussion of ethnicity as a factor in Mary Mallon's case.

41. Educational level also seems to have been of interest to health officers. Josephine Baker, in her effort to explain why Mallon would not cooperate during her arrest, had described her as "uneducated," yet Mallon's clear handwriting in the letter cited earlier refutes that claim. Religion may also have been a factor working against Mallon. These points need to be examined elsewhere.

42. *New York American,* April 2, 1907, p. 2.

43. A 1916 Minnesota study found that 25 of 30 identified carriers were women. The investigators noted that because women cooked for their families and friends, as well as took many low-paying jobs in food-handling occupations, it was natural to find more women than men among healthy carriers. Men were less likely to be discovered, and also less likely to become public health hazards, because their daily tasks and their occupations did not center upon food preparation. A. J. Chesley, H. A. Burns, W. P. Greene, and E. M. Wade, "Three Years' Experience in the

Search for Typhoid Carriers in Minnesota," *JAMA* 68 (1917): 1882–85 (this finding, p. 1884). A Boston study similarly concluded that "we have a preponderance of women on our carrier list, this is largely because they handle food more frequently and are therefore more frequently discovered in connection with outbreaks. George H. Bigelow and Gaylord W. Anderson, "Cures of Typhoid Carriers," *JAMA* 101 (1933): 348–52. It was not until the 1940s that studies began to document more women carriers in the population at large, not just among those found in food-handling jobs transmitting the disease. In a New York State study published in 1943 the investigators concluded that "the rate of development of the carrier state at all ages is almost twice as high for females as for males." The most striking sex difference found in that study occurred in the 40–49 age group, in which 16 percent of female cases and only 3.5 percent of male cases resulted in the chronic carrier state. Wendell R. Ames and Morton Robins, "Age and Sex as Factors in the Development of the Typhoid Carrier State and a Method for Estimating Carrier Prevalence," *American Journal of Public Health* 33 (1943): 223. Medical science in the 1990s acknowledged similar differentials. I want to thank Dennis Maki, head of the Section of Infectious Diseases, Department of Medicine, University of Wisconsin, and Herbert Dupont, chief of Infectious Diseases, University of Texas, Houston, for generously consulting with me on this issue.

44. Soper is quoted in *New York Times,* April 4, 1915, sec. 5, 3.

45. New York City Department of Health, *Annual Report,* 1922, p. 92. See also *New York Times,* October 13, 1922, and January 21, 1923.

46. It was common practice to allow bread-winners greater latitude in public health restrictions. See, e.g., a national study of laws and regulations controlling infectious diseases in which researchers noted that "exceptions in favor of bread-winners . . . may be made by local health authorities." J. W. Kerr and A. A. Moll, "Communicable Diseases: An Analysis of the Laws and Regulations for the Control Thereof in the United States," *Public Health Bulletin 62,* July 1913 (Washington, D.C.: GPO, 1914), 66–67.

The health department did ultimately retrain Mallon, years after her second incarceration, when she began to work in the hospital laboratory. This oc-

curred probably in the mid-1920s and did not result in her release.

47. See, e.g., *New York Times,* March 30, 1913, and Herman F. Senftner and Frank E. Coughlin, "Typhoid Carriers in New York State with Special Reference to Gall Bladder Operations," *American Journal of Hygiene* 17 (1933): 711–23.

48. Soper, "Curious Career," 698.

49. *Medical Record* 71 (May 18, 1907): 818.

50. *New York World-Telegram,* November 12, 1938, p. 26.

51. *The People ex. rel. Jennie Barmore v. John Dill Robertson et al. Respondents* (1922), 302 Ill. 422. All of the papers in this case are in the Archives of the Supreme Court of Illinois, Springfield, RG 901, microfilm roll 30–1807. The final opinion can be found in *Reports and Cases at Law and in Chancery,* Supreme Court of Illinois, vol. 302 (Bloomington, Ill., 1922), 422–36. At Iroquois Hospital Ruth Moore, nurse, and Leonea Letourneau, clerk, collected the specimens and delivered them to the laboratory. Their testimony is recorded on 177–80.

52. The Chicago newspapers reported Barmore's story, although not often or fully. See, e.g., *Chicago Tribune,* November 24, 1920, sec. 2, 1; June 19, 1921, p. 14; and *Chicago Herald and Examiner,* November 24, 1920, p. 2. The health department recognized her extremely briefly in its publications. The legal record is most complete. The writ, return, traverse to the return, stipulations, briefs and arguments, typescript of all testimony, and opinions in the case are located in the Archives of the Supreme Court of Illinois, Springfield (hereafter cited as Illinois Supreme Court Archives). It is primarily through these records that I have reconstructed Barmore's story and can examine its meanings.

53. The placard was introduced as part of the stipulation (that part not disputed by either side) on December 9, 1921, p. 7, Illinois Supreme Court Archives.

54. Bundesen testimony in the printed Abstract of Record, 55–56, Illinois Supreme Court Archives.

55. Brief and argument for relatrix, p. 65, Illinois Supreme Court Archives.

56. Bundesen testimony is in the printed brief and argument for the relatrix, p. 24, Illinois Supreme Court Archives.

57. The case centered on due process, although

during the examination of expert witnesses, Darrow tried (and to some extent succeeded) to make the physicians sound inept. His challenge to science in this case stands in stark contrast to his defense of science in the Scopes trial a few years later.

58. Judge Sabath's opinion, 210, 214, Illinois Supreme Court Archives.

59. Ibid., 214.

60. Typescript of witness testimony, 198.

61. Quotations in this paragraph are from the court discussion about when the briefs would be filed and appear on pp. 200–201 of the printed testimony.

62. Application for Rehearing, 1, 4.

63. Ibid., 4, 9.

64. Darrow's autobiography does not mention this case, nor do various biographies of him. See Clarence Darrow, *The Story of My Life* (New York: Scribner's, 1932); Kevin Tierney, *Darrow: A Biography* (New York: Thomas Y. Crowell, 1979); Geoffrey Cowan, *The People Versus Clarence Darrow* (New York: Random House, 1993).

65. Brief and argument for respondents, 57, Illinois Supreme Court Archives.

66. *Chicago Tribune,* April 15, 1922, sec. 2, 17; *Report of the Department of Health of the City of Chicago for the Years 1919, 1920, and 1921* (Chicago, 1923), 189. In 1923 twelve women filed for release, in 1924 fifteen; in 1925 five women used the courts to attempt to get their release: "in every instance the defendants were remanded and ordered detained by the Commissioner of Health until such time as, in his opinion, they were no longer a menace to the public." *Report of the Department of Health of the City of Chicago for the Years 1923, 1924, and 1925* (Chicago, 1926), 32. I am grateful to Bonnie Ellen Blustein for helping me locate these reports and to Patricia Spain Ward for alerting me to health department activity in this area and for sharing her views about Herman Bundesen.

Twentieth-Century Medicalization and Women

The final two articles in this collection address how one of the big trends of the twentieth century affected women. Medicalization, the increasing authority of medicine over the lives of Americans, has characterized much of the change described in this book and elsewhere in the history of medicine. To compare the eighteenth and twentieth centuries is to notice that physicians and medical institutions intrude much more broadly into peoples' lives now than they did then.

Judith Walzer Leavitt examines the topic of childbirth, an area in which physicians and hospitals significantly increased their involvement. Before 1880, birthing women and their female attendants dominated decisions concerning labor and delivery. Between about 1880 and 1920, however, the medical profession gained control over obstetrical care, and birth increasingly took place within hospitals. Using the specific example of the debate about performing craniotomies (the surgical mutilation of the fetal head to permit vaginal extraction) on live fetuses, Leavitt analyzes how technological innovations, changing medical theory, moral and ethical considerations, and professional interests interacted to make physicians more powerful arbiters in America's birthing rooms.

Rickie Solinger takes up the medicalization theme and looks at hospital abortion committees in the post–World War II period. Her review of the discussions about abortion within the medical community reveals how broad cultural attitudes toward women and motherhood affected hospital policy and practices. Solinger believes that physicians in the decades preceding *Roe v. Wade* were less interested in the question of when life begins than they were in playing a key moral role in helping preserve social institutions and encourage women to serve the family. Physicians used their science to expand their social influence.

Some of the most stimulating and controversial research in the field of medical history at the end of the twentieth century concerns this question of medicalization, including its extent and its cause. While readers may find themselves joining sides in this debate after reading these articles, one hopes that they will in the process come to understand the complexity of writing history, be fascinated by the challenge of interpreting changes over time, and be forced to think about how we know what we think we know about the past.

34 The Growth of Medical Authority: Technology and Morals in Turn-of-the-Century Obstetrics

Judith Walzer Leavitt

At the end of the twentieth century, physicians individually and collectively speak with power, and their voices carry unique influence in the sick room. In hospital deliveries (approximately 98 percent of all births in the United States) the obstetrician is in control, making most decisions about labor and delivery interventions, while the other participants—the nurses, the laboring woman, and her husband or helpers—receive and accept medical wisdom. One hundred years ago this was not so: physicians around the birthing bed added their voices to others with equal or greater power and struggled to establish their authority. This paper explores the period from about 1880 to 1920 in which obstetrical authority became consolidated. This change in medical decision making affected the cohesiveness of women's friendship and self-help networks, which had dominated traditional childbirth practices. It was mediated in large part by the development of new medical

technology and, perhaps more important, by the perception of the effectiveness of this technology by both medical professionals and the laity. Using the example of the controversy over craniotomy on the live fetus (surgical mutilation of the head to permit vaginal extraction) I examine how and why physicians became more powerful arbiters in America's birthing rooms, as well as the larger issue of the growth—and limits—of obstetrical authority.[1]

For much of American history childbirth was almost exclusively a women's event (Leavitt 1986a, 1986b; Scholten 1985; Wertz and Wertz 1979). When a woman went into labor, she summoned her women friends and relatives to aid her; they took care of her and carried out her usual domestic chores. In turn, the parturient was expected to reciprocate during the confinements of her friends. This women's network around labor and delivery provided emotional support at the same time that it helped the parturient make important decisions about how to solicit interventions if the labor was not proceeding easily. Beginning at the end of the eighteenth and throughout the nineteenth century, women who could afford medical attendants increasingly incorporated technical aids in their deliveries. Physicians would be called to attend

JUDITH WALZER LEAVITT is Ruth Bleier Professor of History of Medicine, History of Science and Women's Studies, and the Associate Dean for Faculty at the Medical School, University of Wisconsin–Madison.

Reprinted from *Medical Anthropology Quarterly* 1, 3 (Sept. 1987), by permission of the American Anthropological Association. Not for further reproduction.

laboring women—still in the company of friends and relatives—to intervene with such procedures as internal version, forceps operations, or, by the middle of the nineteenth century, anaesthesia. In emergency situations physicians extricated dead fetuses or, under dire circumstances, performed a craniotomy on a live fetus in order to save the birthing woman's life.

Even in those emergency cases in which physicians were called to perform heroic acts, decision making in the birthing room remained a collective undertaking in this period. For example, in 1885 a pregnant Catholic woman, known to have a narrow pelvis, went into labor in the company of her husband, her clergyman,[2] and her physician. The physician soon realized that a successful vaginal delivery was impossible. Before he could recommend any course of action, however, the priest spoke in favor of a Caesarean section and rejected the alternative of a craniotomy. The parturient and her husband strongly desired that a craniotomy should be performed (despite the inevitable mortality of the fetus) in order to save the woman's life. The physician, Dr. Green of Dorchester, Massachusetts, himself uneasy with the alternatives, was reluctant to proceed given the divided opinion. He consulted with a colleague who specialized in abdominal surgery, and the noted surgeon, "for good [medical] reasons," declined to operate, recommending also that a craniotomy be undertaken. Green finally performed a craniotomy, and the woman recovered. However, the physician felt uncomfortable with his participation in the destruction of a live fetus and said he "was not sure that he should proceed in like manner again under precisely similar circumstances" (Dixon 1885:279).

When in 1901 Dr. Joseph DeLee of Chicago faced a similar situation, the outcome for the parturient was quite different. He was called to attend a woman with a pelvis deformed by childhood rickets who had been in labor for sixty hours. He advised against a Caesarean section because the duration of her labor at this point made the procedure a threat to her life. However, the priest in attendance insisted on a Caesarean section, overruling the woman's initial preference for a craniotomy. With priest and parturient ultimately in agreement, DeLee, feeling he had no other choice, performed the Caesarean section. The woman subsequently died without rallying from the operation. DeLee concluded, "This case made a strong impression on my mind, for I am certain craniotomy would have saved the patient" (DeLee 1901:463).

In both these examples the physicians in attendance felt pressured to carry out decisions that others around the birthing bed made. These cases were not unusual and followed contemporary patterns of decision making. What was new around the turn of the century was the nature of the choices available, the increasingly technical aspect of obstetrical procedures, and the growing discomfort of physicians, like Green and DeLee, about performing procedures that they themselves did not choose. Questions about the place of medical expertise became increasingly insistent as obstetrics incorporated more surgical procedures.

Technological Developments of the Period

At the end of the nineteenth century, physicians developed sophistication in understanding how germs spread infection, the major killer of parturient women. As a result of bacteriological findings medicine had new proposals to make for managing labor and delivery more hygienically. Physicians offered protection against infection by using strict sterile procedures: cleaning the birthing room thoroughly, restricting the activity of untrained attendants, using antiseptics, boiling instruments, washing hands, using rubber gloves, and shaving the perineum. Although a high degree of cleanliness was frequently difficult or impossible to achieve in the homes of birthing women, physicians worked hard to maintain such routines.

For many doctors, however, the teachings of bacteriology merely brought new frustrations into their obstetric work. Frequently, parturient women and their friends, feeling a significant and bothersome distance between their traditional knowledge base and the new medicine, rallied in resistance to the new techniques. As one physician told his colleagues, shaving pubic hair was especially likely to lead to trouble:

In about three seconds after the doctor has made the first rake with his safety [razor], he will find himself on his back out in the yard with the imprint of a woman's bare foot emblazoned on his manly chest, the window sash around his neck and a revolving vision of all the stars in the fermament [sic] presented to him. Tell him not to try to shave 'em. [Landrum 1912:577]

Another physician wrote that his foreign-born patients particularly objected to aseptic procedures:

Elaborate preparations for asepsis are extremely alarming [to them]. They absolutely refuse to be handled by an obstetrician wearing rubber gloves. A sterile gown reminds them of an autopsy. Cleansing the field of operation makes them angry, as they feel it implies uncleanliness on their part. If the physician persists in his preparations, the chances are that he will be discharged, and another called who is not so exacting. Under such conditions, I have seen and had many cases where through no fault of mine, the asepsis was as bad as it could possibly be. [Albert 1916:701]

Thus physicians felt increasingly thwarted as their reasons for intervention expanded, but their ability to carry out procedures of choice remained limited by tradition. For physicians were still invited guests within women's homes, and women—surrounded by their friends and in their own domain—felt comfortable challenging medical advice, asserting their own wishes, and helping to arbitrate the procedures used.

In addition to introducing sanitary procedures at the bedside, medicine also began to produce some new solutions for difficult and hazardous deliveries. These new solutions ultimately became the weapon with which physicians fought and gained control over decision making in the birthing room.

One cause of these hazardous deliveries was cephalopelvic disproportion—a mismatching of the head size of the fetus and the pelvic shape of the woman. Only a small percentage of birthing women suffered from abnormal pelves. Estimates about the prevalence in the United States varied widely. Dr. Edward Davis of Philadelphia concluded from the data available in 1900 that "among childbearing women in the United States of the white and negro races, 25 percent have pelves smaller than the average," and four-fifths of these women could be expected to deliver spontaneously (Davis 1900:15; Reynolds 1901b). Most physician-attended deliveries did not involve cephalopelvic disproportion or any other major complication, but the specter of childbirth-related difficulties was larger than reality. The hazards that birthing women faced during their confinements, as well as the heroic acts that physicians sometimes performed to save women from their fates, remained important images in America's birthing rooms. The power that such images of danger and salvation held over all birth participants greatly affected the evolution of birth patterns, as we shall see.

In cases of cephalopelvic disproportion an operation of some kind was imperative, or the parturient and her fetus would die. With the operation life for one or the other, or perhaps both, was possible. At the end of the nineteenth century the available procedures (in the order in which physicians would have considered them) included a high forceps operation, or internal version with forceps application on the aftercoming head; symphysiotomy, the surgical separation of the pubic bones; pubiotomy (also called hebotomy), the cutting of the pubic bone to increase the conjugate diameters; Caesarean section, the delivery of the fetus through an incision in the abdominal wall; or craniotomy, the reduction by various operations of the size of the fetal head so that it would fit through the pelvic opening.[3]

The first option, a high or floating forceps operation, was fairly commonly attempted in the nineteenth century in those cases in which physicians judged the pelvis adequate to deliver but in which the fetus remained high in the pelvis after a long labor. It was the most conservative option—that is, the least invasive one—and would have been the procedure of choice in many cases had it not been difficult to perform. Its success rate in adequately proportioned pelves was not high, and in small pelves delivery with high forceps could not be accomplished even in the most skilled hands. Yet birthing women and their helpers encouraged an attempt at such an intervention when labor was protracted and the woman seemed unequal to the task of coping with the difficulties. The use of high forceps frequently involved tearing perineal tissues, adding to a woman's postpartum recovery difficulties, but this seemed a small price to pay for a successful delivery in a complicated labor situation. Often the procedure was completely unsuccessful, however, leaving the fetus unengaged and the mother in a weakened condition.

It would then have been necessary to progress to a more difficult intervention (H. Williams 1879). The physician might have tried to turn the fetus in utero and deliver it feet first. This procedure, podalic version, was most commonly attempted in adequate pelves. In women whose pelves were small, flat, or misshapen, such deliveries were extremely difficult, however, and physicians reported a high degree of failure with attempted version/forceps delivery, necessitating the progression to a more invasive intervention (Brodhead 1907; Davis 1907). In the cases of version and forceps-assisted deliveries, parturient women and their helpers did not provide very much technical aid, but they did actively participate in making the decision for or against the procedures. That is, they made the decision to call physicians to perform such operations, or they encouraged the interventions when the physician was already present (Leavitt 1986a).

If forceps and version did not work, or if they seemed inadequate to the task because of the degree of cephalopelvic disproportion, symphysiotomy provided the next alternative for many physicians. The procedure involved inserting instruments through the vagina to sever the cartilage connecting the pelvic bones. One danger associated with the procedure was that postpartum healing might not approximate the original pelvic juxtaposition, and women risked difficulties walking and complications in subsequent deliveries. The separation of the pubic bones could be as much as two centimeters without risk to the ligaments, and this increased the potential for a live delivery. If the pelvis was so small or misshapen that the increased diameter would still not allow vaginal delivery, symphysiotomy was not indicated (Allen 1906; Ayers 1902; Boyd 1896; Broomall 1893; Coe 1896; M. L. Harris 1892; R. P. Harris 1883; Noble 1983; Zinke 1895).[4] Decisions to use this difficult procedure also rested on accurate pelvimetry,[5] itself a new and problematic technique. The use of such highly technical procedures distanced women from the decision-making process and helped physicians dominate decisions about whether or not to proceed with the operation.

In those cases where the size of the pelvis could not be increased enough by symphysiotomy, pubiotomy (sawing through the pubic bone) offered hope, because a larger opening could be produced by this procedure. Its advocates claimed that postpartum difficulties were minimal, compared to symphysiotomy, because the healing of the bone could be more precise and injury to the soft parts less likely. Significantly, debates about whether symphysiotomy or pubiotomy provided the best course were carried out among consulting physicians rather than between medical advisers and family members (Bill 1906; Fry 1907; J. W. Williams 1910).

Instead of symphysiotomy or pubiotomy, physicians increasingly considered the Caesarean section, a procedure that had altered most radically in this period. Although an ancient operation, the Caesarean section had for most of history been performed only on dead or dying women. Successful Caesarean sections, in which both mother and child lived, were reported throughout the nineteenth century in the United States, but it was only after the development of anaesthesia in the 1840s and antiseptic and aseptic techniques late in the century that the procedure became established (Cianfrani 1960). Caesarean section involved incising the abdominal wall and uterus, extracting the baby and placenta, allowing the unsutured uterus to contract, and stitching the abdominal wall. Various improvements in this method around the turn of the century brought Caesarean section into the list of relatively safe options available for women with contracted pelves.[6] However, it required specialist assistance and a surgically sterile environment and was best performed in hospitals. Physicians who performed this operation reported its greatest success early in labor before the parturient was exhausted by a long labor or infected from numerous vaginal proce-

dures. The medical profession, acknowledging the continuing danger to women, widely debated the relative merits of this procedure; as success rates improved, it gradually gained adherents (Allen 1909; Davis 1913; Garrigues 1886; Grandin 1890; Hirst 1916; Lusk 1880; J. W. Williams 1917; Zinke 1903).

The only other procedure that promised any hope for the survival of birthing women with cephalopelvic disproportion was craniotomy, also an ancient procedure that had been improved in the nineteenth century with the use of anaesthesia and the introduction of new implements (Cianfrani 1960). This destructive operation, which involved perforating the fetal head, evacuating the cranial contents, crushing the skull, and extracting the fetus (sometimes in pieces) became the focus of the most intense obstetrical debates in this period. Physicians performed craniotomies most often to extract dead fetuses that could not be delivered otherwise. However, they also perforated live fetuses in those dire cases in which rapid sacrifice of the fetus was an emergency last resort to save the mother's life. Although craniotomy was a difficult and repulsive operation, birthing women and their companions had a history of participating in the decision to do it. It was in fact the traditional means of saving women's lives in otherwise impossible labors (Bacon 1900; Busey 1884, 1889; Dixon 1885; Edgar 1893; Gushee 1907; Jaggard 1884; Myers 1895; Peterson 1915; Rosenberg 1892; Trawick 1904; Voorhees 1902; Wathen 1889).

The Ensuing Medical Debate

Technical criteria for selecting between craniotomy and Caesarean section occupied physicians' attention for years and caused heated arguments at medical society meetings. The debate was complex. In part, these criteria involved a judgment about the attending physician's expertise:

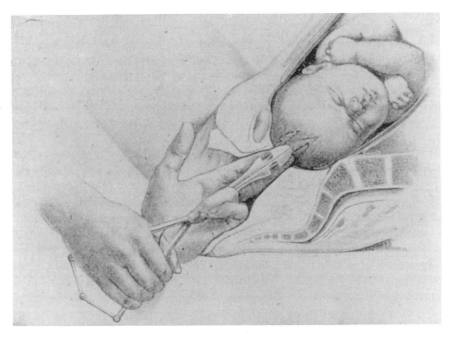

Fig. 34.1. Craniotomy: method of perforating fetal head.
Source: J. W. Williams 1920:503. Reprinted by permission of Appleton-Century-Crofts.

Which operation could he or she perform most skillfully? In addition, craniotomy was gory; no physician liked doing it, although many insisted there were times when it was necessary. One such physician acknowledged "that the performance of a Cesarean section gives far more pleasure than a craniotomy" (Rosenberg 1892:327). The decision between craniotomy and Caesarean section also involved a careful assessment of the woman's status in labor. Both procedures were safer when performed early, while the woman still had her strength and before she was already at risk for infection. Seeing every craniotomy as "a confession of a partial defeat at least," and realizing that many doctors failed to perform one "because the operation is in dispute," physician Reuben Peterson nonetheless suggested to his fellow practitioners in 1915 that when they believed that craniotomy could save

the mother's life, they should make the decision and deliver quickly.

> It is decidedly unscientific and cowardly on the part of the obstetrician to delay removal until after the death of the child, in order to escape the odium of perforating the living child. . . . Sentiment should be thrown aside and the child perforated and removed in the interests of the mother. [Peterson 1915:319, 321]

As with the caesarean section, the decision to perform a craniotomy had to be made early and decisively to give the mother the best chances for survival (DeLee 1901).

Increasingly around the turn of the century, however, physicians voiced the opinion that modern surgery was making craniotomy obsolete and that physicians who resorted to it sacrificed the fetus unnecessarily. The new surgery—especially the Caesarean section but also aseptic symphysiotomy and pubiotomy—offered life to both mother and fetus, and all parties deemed such operations preferable, if done safely.

A lengthy debate spanning three meetings of the Medical Society of the District of Columbia in 1887 illustrates the dilemmas raised by craniotomy and the passions evoked within the profession itself. Dr. Joseph Taber Johnson, president of the society, argued for open-minded decision making in each case, saying any outright rejection of craniotomy on moral grounds was unscientific. He admitted that he preferred to perform a Caesarean section (a "clean surgical procedure") rather than a craniotomy (a "horrid and detestable operation") but that there were many instances in which craniotomy was arguably safer for the mother and should be performed regardless of physicians' feelings (Johnson 1887a:171).

Johnson's paper aroused strong emotions among his colleagues. The medical society was sharply divided on the question of whether or not moral considerations should invoke absolute restrictions on craniotomies. The debate is instructive. Dr. Samuel Busey, who became a national spokesperson on the primacy of moral considerations, thought craniotomy so evil that it should not be taught in medical schools, "for just so long as teachers teach to kill at will the killing will be done." Dr. King responded, "The killing could hardly be said to be done at will." Busey retorted, "It was most assuredly done at will. The operator fully considered the mother before proceeding, and then voluntarily and at will killed the child." Dr. Acker then queried how Busey proposed to "get around the objections of the friends and relatives to the Caesarean section?" Busey stood firm that "their objections did not relieve the doctor of his responsibilities—that because they will not permit the right thing to be done, it does not make

it right to do a wrong act" (Johnson 1887a:191).

While these debates focused on the procedures of craniotomy and Caesarean section, decisions about symphysiotomy and pubiotomy were also at issue. Physicians, trying to decide how to act under conditions that made delivery of a live fetus impossible without surgery, had a hard time determining a course of action (Barnes 1890; Haggard 1889; Montgomery and Veer 1889; Rohe 1889). To be successful all possible operations required skill and experience, qualities that were not evenly distributed in the medical profession at the time. Many procedures were unsuccessful, many dangerous situations were made more dangerous, and women's and babies' lives were sometimes risked unnecessarily or lost because of procedures carried out by inept hands. Thus not only did physicians worry about which procedure to advocate but some were also deeply concerned about whether or not they could effect successful outcomes at all (Leavitt 1986a).

New Patterns of Decision Making

The debates within the medical community and the difficulty of making decisions under pressure of labor in cases of cephalopelvic disproportion put physicians in the position of remaining susceptible to outside influences. Yet we can see and begin to understand how in this process the medical profession tried to turn the traditional mode of collective decision making into one that recognized more firmly their own expertise.

New delivery techniques held significant attraction for women who felt their lives endangered by their efforts to give birth. Since the middle of the eighteenth century, women's fears of death in childbirth and postpartum debility, as I have argued elsewhere (Leavitt 1986a, 1986b), propelled them to incorporate medical interventions. Thus forceps, opium, anaesthetics, and other drugs and procedures gained their place in obstetric practices as women themselves recognized their value and potential and called in physician-experts to execute them. Bacteriology and the surgical procedures that became possible under its auspices added to the traditional list of interventions. However, the new medicine was considerably more complex than the old, and the interventions became more technical, complex, and difficult to carry out. The understanding of nonexperts about the different indications for, for example, symphysiotomy and pubiotomy, or their ability to recognize when in labor such procedures might be helpful, was necessarily incomplete. Birthing women and their companions, who had mastered earlier techniques, lost touch with the nuances of the new ones.

Although their knowledge and ability to control decisions might have diminished, birthing women continued to play an active role in choosing birth options. Many factors influenced a woman's point of view. Primary among them, no doubt, were her chances of survival. But she also took into account her previous childbearing history: How many living children did she already have? Would the procedure under consideration allow her future children? If the woman had already given birth to live children, her anxiety about herself would likely take precedence; if she had repeatedly come to term only to deliver dead children in the past, she might be highly motivated to take the additional risk to herself to be able to give birth to a viable child this time. No generalized rules for women of a certain age, class, or pelvic measurement could meet the very particular needs each birthing woman brought to any specific delivery. Women continued to try to keep labor and delivery decisions individualized to their own needs and fought medical efforts to generalize (Leavitt 1986a).

The husbands of parturient women in labor

trouble played an increasing role as solutions became more technical. While earlier birth accounts by physicians and family members, as well as contemporaneous accounts of noncomplicated labors, did not often refer to the husband's direct role in labor and delivery decision making, turn-of-the-century accounts of life-threatening deliveries increasingly named the husband as an important voice of authority. Dr. Barton Cooke Hirst, for example, explained that he "felt it necessary to put the case [of cephalo-pelvic disproportion] frankly and truly before the person most nearly concerned—namely, the husband. . . . I have never had a husband consent to a Cesarean section" (Barker 1892:137). Another physician bemoaned his lack of control in terms of the dual role played by the husband and the parturient:

> We cannot elect which operation we will perform in a great many cases. We state the case to the patient and to the husband, and they will almost always elect craniotomy, and you cannot get permission to perform the other operation. It is not always in your power to act as you might wish to. [McColl 1891:272]

When considering abdominal surgery, surgeons appear to have sought approval from the husbands of parturients as frequently as they did from the women themselves. Sometimes the physician, other consultants, and the husband would confer about whether or not to do a certain procedure, weigh the possible outcomes, and debate whether or not to inform the woman of their deliberations. Reports of such conferences, perhaps in a corner of the room or in the next room, starkly reveal a shift from women's to men's decision making.[7]

Dr. Edward P. Davis of Philadelphia reported a case of a breech presentation in a woman he attended in 1902 in which he expected a difficult delivery (1902:177–78). "The patient's husband was warned of the complications existing and the dangers of the case fully stated. . . . The husband was informed that he was at liberty to have consultation either before or during labor." Davis wanted to perform a Caesarean section. "Section was proposed to the husband and immediately accepted." The physician happily gave credit for the successful outcome to the husband: "Fortunately the husband appreciated the situation and by his prompt and efficient help made a successful termination of the case possible." Other husbands were not so cooperative. For example, Dr. A. McDiarmid reported a case in which he wanted to perform a Caesarean section, "but it was only on his sternly threatening to abandon the case, and with the agonizing cries and pleadings of his wife ringing in his ears that her husband reluctantly consented" (McDiarmid 1899:766–67).

As women, husbands, friends, and relatives continued to participate in communal obstetric decision making in this turn-of-the-century period, they frequently held discussions with religious advisers in cases when life and death issues were under debate. The Catholic Church was especially vocal and adamant about craniotomy.[8] In the language of the church this procedure involved "the destruction of an innocent life," and priests at the bedside therefore argued in favor of other forms of delivery. The case related by Dr. P. Pineo (1857:377–78) of Queechy, Vermont, illustrates the influence of the church:

> Had been in labor 48 hours; pains severe; head impacted in the pelvis, and no advancement. . . . I wished to apply the forceps, but the friends were not willing I should do so until the *priest came.* The priest did not arrive until the woman had been in labor 72 hours. He then consented that the forceps should be used, and they were applied by myself and another physi-

cian, whom I called in as counsel, but no impression could be made upon the condition of the labor. The woman was evidently sinking; and I told the priest that she would die if not delivered immediately, and that craniotomy was the only proper course to pursue. He wished to know if the child was dead I thought it was not. He refused then to allow the operation done. "*What!*" said he, "*shall we take the life of the innocent unborn, that never sinned, to save the guilty? No indade* [sic]." Several hours more elapsed, when I threatened to leave the house if I was not permitted to do my duty. With some representation to the priest that the child was probably dead, the operation was finally permitted. . . . The patient was exceedingly prostrated. [emphasis in original]

The distraught physician concluded, "I have no doubt the mother could have been saved if the prejudice exhibited had not prevented my operating at the proper time."

Other physicians seemed to find accommodation to church doctrines easier. Dr. Ralph Pomeroy (1908:514) reported the case of a Catholic woman he attended in 1908 in which he performed a pubiotomy:

> It was voted [the electors not specified] that the extraction of the unmoulded head would be much facilitated and the child's chances of living would be greatly enhanced by a preliminary enlargement of the pelvic girdle. The recognized exceptional interest of the tenets of the Catholic Church in the life of the child was also given weight in the decision.

Coming to a contrary judgment, however, a Nashville, Tennessee, physician decided that "religious belief of the patient should bear no weight in reaching a conclusion as to the plan of procedure" (Trawick 1904:212).

The church went to some lengths to make its position known in medical circles. In various medical journals, including the widely read *Medical Record* and *Journal of the American Medical Association,* Catholic theologians insisted that craniotomy in all cases was sinful and that saving the life of the mother, "however good and desirable, can never justify or excuse the use of means which are unlawful in themselves. Now, the killing of a human being, even though that being is within a few moments of death from other causes, is wrong in itself, and has all the guilt of homicide" (Sabetti 1885:606–607; see also Hund 1895; Monaco 1885). In fact, the Reverend A. J. Schulte told obstetricians in 1917, "Better that a million mothers die than that one innocent creature be killed" (Schulte 1917:886). The church's position was absolute and clear: Catholic physicians who disobeyed the teaching were told they committed homicide, and all physicians recognized that their ability to determine birth options for Catholic women who chose to follow church teachings was severely circumscribed.

As physicians felt its impact, the language of the church became echoed in the medical literature. Dr. T. Ridgway Barker of Philadelphia wrote, "Has the unborn infant no rights? Is it a matter of no importance whether we destroy that life locked up in its mother's womb? . . . Is there any moral difference in murder within the womb from that without?" (1892:134). Dr. Samuel Busey agreed: "Custom and usage may excuse, and civil and criminal law may acquit the accused, but neither of these avenues affords escape from the moral responsibility of intentional and deliberate killing" (1889:55).[9]

The interactions between church and family in the decision-making process provided particular dilemmas for physicians. Barton Hirst related one of his experiences with a Roman Catholic family in 1892:

> I explained to the husband that embryotomy [craniotomy] or Caesarean section was necessary. He refused absolutely to permit Caesarean section and the matter was referred to the priest who was summoned. He would not permit craniotomy, but would not assume the responsibility of ordering Caesarean section. In this dilemma I suggested version, knowing that the child would die during the attempt to extract the head. I did the operation, but could not extract the head. In the course of five or ten minutes, I was able to announce that the child was dead, and craniotomy was then allowed. I should much prefer Caesarean section in these cases, but I scarcely expect in the near future that a larger number will consent to it. The husband, *if told the truth* [about the dangers to the woman of Caesarean section], will demand craniotomy, and I think moreover that every one of us would do the same under similar circumstances. I know I should, and I feel confident that every gentleman here to-night would, too, if his own wife were in question. [Barker 1892:137, emphasis in original]

Although Catholics were in a minority nationwide, the church succeeded in injecting its moral position into the medical debate, because it insisted upon defining the problem in terms that had application beyond its specific religious teachings. By accusing physicians of "killing" and "murdering," the church placed physicians in a position where they could not ignore the impact of such powerful and thought-provoking language on their patients.

The law did not help physicians resolve their dilemma either. As they interpreted it, both the performance of a craniotomy and the refusal to do so could be legally protected. According to one medical analyst:

> When the obstetric attendant, in the presence of conditions justifying a truly formidable operation which has for its purpose the saving of the mother's life at the cost of the foetal life, proceeds to destroy the living child within the womb, does he violate any law in the criminal code? The answer is in the negative. So long as the child is within the womb or indeed within the maternal passages, it is regarded by the law as a part of the mother's body . . . it is not a "person" and cannot die by violence within the meaning of the law. [Draper 1890:53]

On the other hand, "A physician is not compelled to make a destructive operation against his own judgment or conscience," wrote Dr. C. S. Bacon of Chicago (1906:1983; Glenn 1911). As they themselves interpreted the law, physicians could choose either craniotomy or Caesarean section and be within their rights.

Not given clear legal guidance, physicians pursued the controversy intraprofessionally and medicalized the issues to distance both clergy and laypeople from decision making. In fact, it was through the church's direct challenge that physicians first seemed to understand their need to establish clearer lines of authority. It was when clergy stepped in and tried to dictate medical procedures that physicians faced their own impotence and devised tactics to overcome it.

The Beginnings of Physician Dominance in Decision Making

Thus by the turn of the twentieth century, physicians increasingly debated among themselves which alternatives suited particular labor situations, sometimes consciously trying to leave the birthing woman and her attendants out of the discussion. A general practitioner in attendance at a difficult labor would call in an obstetrical specialist, and the two would consult, for example, about whether or not to perform a Caesarean section, where to do it, and who to call for assistance. Physicians' case reports of such situations indicate that these consultations were difficult on many levels. The initial medical advisers first had to come to terms with their own limitations and admit the need for help. Then an assessment had to be made and agreed upon about the precise nature of the problem: Was the pelvis adequate for the size of the fetal head or not? Was the parturient strong enough to withstand major surgery? Was she already septic from earlier attempts at delivery? Did the fetus have a chance of survival? Could the woman be transported to a hospital? If the consultants could agree on answers to these questions, they then had to determine which procedure offered the best hope and to weigh and balance the mother's health and fetal chances (Coakley 1896; Johnson 1887b; McCoy 1899).

Physicians involved in such situations realized that answers to some of the concerns they faced were not solely "scientific" but that there were social and ethical considerations as well. As much as they might have liked to make decisions on the basis of medical assessments alone, they could rarely accomplish this. Indeed, justification for both Caesarean section and craniotomy could be found in the medical literature. At such difficult and decisive moments physicians had—in the absence of technical answers—to consult their consciences, their own religious and ethical beliefs, their particular experiences with the proposed interventions, their own skill and that of their helpers, and the environment under which the procedure would take place. Thus doctors had to rely upon extramedical factors to help them in obstetrical decision making.

Let us examine more closely the position of the medical experts in this period. In favor of their autonomy and authority was their exclusive control of the technical information necessary to both decision making and execution of certain procedures. Some of the objective information, including pelvic measurements and estimation of the size of the fetus, although open to interpretation, remained in the hands of medical observers alone. Without a technical evaluation of the nature of the problem and the identification of the possible solutions, there could be no informed decision making. Physicians thus set the stage for everyone else's understanding of the situation. To the extent that doctors withheld information, the ability of the other parties to understand the nature of the problem and thus participate in determining its solution was limited. In fact, doctors' ability to set the parameters of the discussion gave them extensive authority over the choices to be made, although such control over information did not necessarily lead to their ultimate power in decision making.

Rather than gaining confidence from their monopoly over information, however, physicians bemoaned the necessity of paying attention to all the other, extramedical factors that they felt they could not control. For amid all the changes that childbirth had undergone by the beginning of the twentieth century, much remained fairly constant (Leavitt 1986a). Often physicians were not called to attend a laboring woman unless her other attendants deemed the situation dangerous. In such cases, as one physician realized, "We must take things as they come, we cannot have them always as we want them" (Johnson 1887a:272). Most women, certainly all private patients, were confined at home in an environment unconducive to surgery. In cases when surgical intervention was indicated, parturients had to be convinced to move to hospitals; otherwise, physicians were faced with the

necessity of operating under what they considered to be substandard conditions. In addition, unless physicians brought along trained nurses with specific obstetrical skills, they were forced to rely for help on people who might not have relevant training. Ethical or religious considerations, as we have seen, might also circumscribe the choices available in any given case. When these factors were added to the fact that most doctors were called to attend or consult on women whom they had not previously known or examined, the lack of control from their point of view was overwhelming. It no longer seemed so important that they were the only ones in the room who could adequately measure a pelvis or define a medical problem: they did not feel in control of the situation. Even for those women whom physicians succeeded in moving to the hospital, precious time might have been lost in the transport, which itself limited what physicians could accomplish (Poucher 1913).

The nature of medical authority in this predominantly home-birth period, then, was circumscribed by significant nonmedical factors. At the same time as women's authority over birth procedures, environment, and companions was withering in face of the increasingly technical nature of the medical debate, medicine itself remained shackled. Women were losing their traditional birthing-room powers, and physicians were taking up only part of the slack. Who or what stepped into the breach?

We have already identified the newcomers as church and husband, both more active birthing-room participants in this period. Physicians were quick to turn the husband's role to their own advantage. Traditionally excluded from childbirth—often relegated to the next room until after delivery—husbands did not have the experiential knowledge on which to judge medical interventions. Because of the dire nature of the cases of cephalopelvic disproportion, they perhaps accepted medical authority rather easily. They had no choice but to seek medical assessment of the problems, for they were frightened for their wives' lives. Since men were accustomed to a certain amount of authority in their own homes, they posed a possible threat to the assumption of medical authority, and physicians had therefore to tread carefully. However, by almost conspiratorily including husbands in decision making, they won their loyalty and enlisted them in their own cause.[10]

Physicians felt the intrusion of clerical authority more strongly, and they voiced most anxiety about the church's teachings, which tied their hands so irrevocably. The church's claimed power to vote as a "proxy" for that other presence in the labor and delivery room, the fetus, especially upset traditional birthing-room balances and threatened physicians most significantly. Traditional obstetrics had explicitly valued the life of the mother above that of the fetus. If one had to be sacrificed to save the other, the teaching uniformly was to save the life of the mother. She was already a member of society, with responsibility and social connection, with love and commitment. She could bear other children, and she might have already borne children who needed her care. The most desired outcome would, of course, be to save the lives of both mother and baby, but under conditions that did not permit both to live, mother and physician agreed that all procedures should be directed to save her life, even at the expense of her offspring. Yet the church emphasized fetal life above that of the mother; the former was pure and innocent and had not sinned, while the latter was already guilty of sin. Physicians, in order to consolidate their own authority, learned to use this argument to their own ends. But they did it in a medical, and not a moral, context.

The new surgical procedures that provided the potential for saving the lives of both mother

and fetus, even though they put the mother's life in greater jeopardy, helped move the physicians in the direction of increasing medical authority. These are the words of Dr. George I. McKelway, speaking to an American Medical Association meeting in 1892:

> I do not regard the physician as primarily the conservator of individual life. His duty is to all human life and to its conservation, and his obligation is as great, other things being equal, to life incarnated in one body under his care as to it in another. I grant that things are not equal as between the mother and her unborn child, but they are very much nearer so than is argued by the habit and thought of many obstetricians. [McKelway 1892:7]

If symphysiotomy, pubiotomy, or Caesarean section were successful, the lives of both mother and child could be rescued. But who was to decide whether or not the potential life of the fetus justified the increased risk to the mother that these procedures entailed? Were physicians to carry out decisions made by others, even when they could not agree with them? One physician, for example, noted that he preferred to perform a Caesarean section over a craniotomy in most cases but that "in case this elective operation were not elected by those who had the right to vote," he might be compelled to perforate the head of a live fetus (Johnson 1887a:170).[11]

The stressful conditions of the birthing room in cases of cephalopelvic disproportion present a context for understanding decision making and power balances in turn-of-the-century obstetrics. All the involved parties would presumably have chosen to save the lives of both mother and baby, if this were possible. However, observers agreed that the parturient and her friends most frequently advocated craniotomy over Caesarean section in order to save the woman's life with the least risk to her health and future childbearing capabilities. The husband, as a rule, also spoke of the necessity of saving his wife's life first. The priest, in the case of Catholic families, spoke for the "innocent" fetus over against the "guilty" mother. In the midst of this debate obstetricians tried to establish medicine's authority to cast the "swing" vote—indeed, to make medicine's vote the most important one.

Medical discourse in this turn-of-the-century period reflected the difficult position that obstetricians experienced. Writing advice to each other in their medical journals, physicians grappled with the issue of how to weigh the relative value of lives, of how to choose interventions under imperfect conditions. Dr. T. E. MacArdle of Washington, D.C., concluded, "I contend that when the child is alive, we hold its proxy, and must cast its vote in favor of Caesarean section" (Johnson 1887a:271). Dr. E. E. Montgomery of Philadelphia understood that "the importance of the unborn individual has obtained a higher appreciation" within the medical community as a result of improved obstetric technique (1892:756–766; see also Grandin 1890).

But more than the specific choice of treatment, the process of deciding became all important; physicians focused on the time when they alone would be able to make the important obstetric decisions. "The choice of the patient must for the present be considered," wrote Fort Wayne, Indiana, physician William Myers, but he felt the time was fast approaching when the lines of decision making could more decisively emphasize the medical (Myers 1895:484). "The consent of the patient, obtained without direct or indirect coercion, is essential [to the performance of a cesarean section]," wrote W. W. Jaggard; but, he anticipated, "The time will probably come when under certain circumstances— ex. gr. in hospital practice—the woman shall not be permitted to elect as freely as she must be allowed to do at present" (Jaggard 1888:98).

The medical literature in this period began to claim authority with new language. Medicine's answer to the dilemma was to suggest that physicians take more authority in childbirth than they had in the past and that physicians turn this dilemma—stated in moral, ethical, religious, social, economic, and even political terms— into a more clearly and uniquely medical one. "We must strike directly at the root of the evil, which declares that 'it is the mother's right to save her life, even at the sacrifice of her child,'" wrote Dr. Samuel C. Busey of Washington, D.C. "We must in the interest of a broader humanity and a far wider field of usefulness, accept the progress of science, and offer chances to two lives rather than take the one which cannot assure the safety of the other" (1889:53).

Physicians told one another to eschew the advice of all other participants—birthing women, husbands, friends, clergy; to delimit the debate solely on the issue of medical safety; to remove their patients to hospitals, where medical considerations could be paramount; and to establish science as the primary focus. Dr. George McKelway discussed the importance of gaining medical control, while casting other birthing participants in a pejorative light:

It would seem hardly necessary for me to emphasize the fact that the obstetrician alone must be the judge of what is to be done. The preference of the ignorant or prejudiced parents or of the friends, the disposition on their part to save the life of one being from any added risk at the expense of the other being, or any sentimental considerations on their part of any nature, for or against either life at the expense of the other, should not weigh at all in the obstetrician's mind or decide his course. The responsibility is his for his conduct in the case and he cannot answer to his own conscience if he does that which he believes to be inadvisable or wrong, simply because someone else prefers that he should.

This physician believed that medically "trained intelligence and judgment" should triumph over the "sentiment and prejudices of other people" (McKelway 1892:8). "This is not a body of religionists," exclaimed Dr. J. F. Hartigan to his medical colleagues. "We meet to discuss medical subjects on a purely scientific basis" (Johnson 1887a:324). Obstetricians, for the first time in their history, tried to take command on their own terms, rather than respond to the wishes of others.

The physician's judgment should bow to no interference, claimed Dr. Egbert Grandin of the New York Academy of Medicine: "Theologians cannot decide it for obstetricians." Grandin also denied women the right to decide for themselves between craniotomy and Caesarean section, and he insisted that the medical voice should be and increasingly was paramount in determining birthing room events (Grandin 1891:641–42). The proposed hierarchy of control would allow medicine to reign with greater authority, an authority stated in scientific language and based at least in part on fetal rights.

It almost seemed as if the life the woman had nurtured for nine months—the life she had valued and hoped for, the life she wanted—was to be used against her. When physicians began defining their own terms for decision making, they continued informing and getting consent from the woman and her family, but by responding decreasingly to the woman's point of view, they began to see the parturient as the antagonist to the fetus. It was her life (or health) versus the life of the fetus. She could not be viewed as objective in such a decision; therefore she should be excluded from it. "Such a right [of decision making] cannot, however, be conceded to a woman in labor who is responsible for the existence of her child and the danger of both," wrote Busey, "since by it she imposes upon an innocent operator the act of killing, that her prospect of life may be slightly improved" (1889:67). It was through this definition of the problem that a hierarchy began to evolve with medicine at the top and the previously dominant partner, the birthing woman, deposed to a distinctly lesser position.

Physicians were increasingly successful at creating the hierarchical inequality specifically because of the dilemmas posed by such life-threatening situations as cephalopelvic disproportion and the nature of the treatment options available in this period. The new technology made it possible to consider fetal life as a viable option and provided physicians with the opportunity to wrest decision-making power away from its traditional place within the family and establish it in their own domain. It was in these high-risk cases requiring surgical intervention that physicians found their first commanding voice in the birthing room; they later learned to use their new authority in all obstetric cases.

The traditional cooperative pattern of decision making was superseded when the surgical options, especially the option of Caesarean section, increased the chances of life to the fetus. That it also put the parturient at greater physical risk was minimized in the new debate; the woman's voice grew weaker as the fetus became viable. As Busey put it:

> The child is entitled to life at the increased risk of the mother. Humanity demands it, and science clearly points to it as the line of duty. . . . Sentimentality, family ties, and other circumstances, may embarrass individual instances, but such considerations cannot affect the question in its scientific aspects. [Johnson 1887a:275]

Physicians came to understand that adoption of the new surgical procedures, the indications for which could be defined in medical and scientific terms, made their decision making easier at the same time as they could be said to speak for both mother and infant.

Craniotomy came to be associated with the old ways, Caesarean section with the new. "The reason that [craniotomy] is done so frequently is that the practice of obstetrics is so largely empirical and so little scientific in this country," wrote Dr. Charles P. Noble, calling for more rationally based decisions in the field (Barker 1892:136). His Philadelphia colleague, Dr. W. S. Stewart, went so far as to suggest that if families refused Caesarean section when physicians felt it was the operation of choice, "it is the duty of the physician to have nothing more to do with the case" (Barker 1892:138).

Because Caesarean section, symphysiotomy, and pubiotomy put women at greater risk than craniotomy, physicians had to defend the value of the fetus's life above the mother's. This could not be accomplished using a medical argument alone. Thus, at the same moment in history when they tried to make objective science paramount in their attempts to take control away from the biased parents and clergy, physicians were forced to couch their own arguments in social and moral terms. One physician, William Wathen, of Louisville, Kentucky, argued that the traditional belief that "the mother is of much more immediate use to the family, society, and state than the unborn child," was pagan in origin and should not influence modern Christians. In any case, he wrote, the women for whom the craniotomy was most frequently considered were

the women [who] have usually had no living children, nor are they capable of having any; so their existence is necessary only so far as they are able to contribute to the immediate interest and welfare of husband, society, and state, and at death their usefulness is ended. . . . [The fetus,

on the other hand] may become a useful member of society and state, and produce children that may continue to multiply. [Wathen 1889:1237]

Physicians justified their actions on medical grounds and in social terms, making social value comparisons between the birthing woman and the fetus. In so doing they revealed their own biases about women's role as childbearers. Dr. Egbert Grandin of New York in 1890 reported doing a Caesarean section on a twenty-eight-year-old single woman, pregnant for the first time. The woman died thirteen days postpartum. Grandin concluded from this case:

Notwithstanding the unfavorable issue in this case, . . . I do not regret having performed it. . . . She was always moody and melancholic, conditions which were aggravated by the fact that she was without friends and home. . . . The child is at this writing [approximately six weeks old] in excellent health, and were I anxious to apologize for the operation, I might question if of the two lives the more valuable had not been saved. [1890:393]

Grandin defended his medical actions in social terms—the woman's marital status, her lack of social network, and her melancholia made her life less valuable than the potential of her child.

The medical profession was not monolithic in its views of women's childbearing role, however. Just as the culture in which they lived, physicians voiced multiple views of women's social roles. As some in the profession refused to perform craniotomies because they valued unborn life more highly than imperfect womanhood, other physicians continued to advocate craniotomies, arguing the woman's social worth ahead of the fetus. Dr. J. R. Brownell, for example, said:

Shall we only consider the life of a woman of value because she can bring a live and viable child into the world? Is she only to be considered as a child-bearing machine? If so, a man is justifiable in putting away the wife who cannot bear him children, and seeking until he finds one who can. He who acts under such "conscientious discharge of duty" presumes beyond his prerogative in constituting himself jury, judge and executioner. I would ask if it may not be as much the fault of the child's head in being too large, as of the pelvic diameters in being too small? [Johnson 1887a:325]

J. Whitridge Williams tried to strike a social, as well as medical, balance in his attempt to insert a relativity to women's roles as reproducers:

To my mind, the answer [to the question about whether to perform a craniotomy or a Caesarean section] depends upon the social status of the woman, her desire for a living child, and particularly upon whether she is the mother of several children or is pregnant for the first time. In the former event, I hold that low amputation of the uterus is a justifiable procedure, as the patient has already done her duty to the State, and the possibility of further childbearing may be regarded as a matter of relative indifference. In the latter event, on the other hand, I feel strongly that such interference is highly reprehensible, and that decapitation is preferable to forever abolishing the reproductive function of a young woman. [1917:194]

Thus physicians claimed their right to make medical decisions using the whole range of considerations that had characterized collective decision making in the past. Decrying the "prejudices" of parents in judging difficult cephalopelvic disproportion cases objectively, as well as the inappropriateness of the clergy's medical

decisions based only on moral grounds, physicians tried to establish their own authority on the basis of science and presumably rationality. Yet medical rhetoric revealed that the very same social prejudices that had influenced collective decision making continued to help determine medical decision making. While claiming to make decisions on scientific grounds and thus increasing their monopoly over birthing decisions, physicians took power away from birth's nonmedical participants. Actually, however, the physicians relied just as heavily on their own social ideas as they gained medical authority. They made decisions on both social and medical grounds, with the social ones of continuing, if less acknowledged, importance. Their actions allowed them to begin to supersede traditional collective decision making with a more unitary professional one, even while they used all the traditional modes of integrating medical with nonmedical considerations.

Although medical authority expanded in the early decades of the twentieth century, it remained far from complete in those years. Physicians struggled to establish the power of their words and their medicine; sometimes they succeeded, other times they failed. They could not be sure of the reaction their suggestions would get in home birthing rooms, and they remained insecure about their authority until childbirth moved into the hospital. The debates and activities of the years examined here increased medical authority and helped move birth into the physician's domain. In this period the growth of medical authority during emergency situations laid the groundwork for later medical decision making. By the 1930s, when for the first time in history childbirth occurred more often in hospitals than at home, physicians were ready to assume more exclusive control over obstetric decisions (Leavitt 1986a). The establishment of physicians' authority over difficult home-based deliveries during this turn-of-the-century period was a necessary precursor to their ultimate assumption of hospital-based authority in obstetrical matters.

Notes

I am grateful to Linda Gordon and Lewis Leavitt for their comments on earlier drafts of this paper and to Charlotte Borst and Jennifer Langdon for their research assistance. Carolyn Hackler made a heroic contribution in processing the many drafts of the paper.

Correspondence may be addressed to the author at the Department of the History of Medicine, University of Wisconsin Medical School, 1300 University Avenue, Madison, WI 53706.

1. The data for this study were collected from articles, case reports, letters to the editor, editorials, and discussions of papers published in the major national and regional medical journals, as well as from obstetric textbooks, statistical studies, and, wherever possible, women's birth accounts. The study builds upon the extensive unpublished archival sources used in Leavitt (1986a).

The realignments in birth practices occurred in a period of great changes within medicine. In the late nineteenth century, the discovery of the role of microorganisms in causing human disease led to great optimism for an increased effectiveness of medical therapy. The potential of the practical applications of the new science infused both medical practitioners and the laity with a great desire to incorporate the new theoretical breakthroughs into the practice of medicine (Pernick 1985; Starr 1982; Vogel and Rosenberg 1979; Warner 1986).

2. It is not clear from the historical record when clergy became included in the decision-making group or whether such advisers were sought only in life-threatening situations. Most family and medical birth accounts of normal deliveries do not mention the presence of a clergyman; many of the medical case reports describing cephalopelvic disproportion do indicate the presence of such a person.

3. Of these procedures, symphysiotomy and pubiotomy have been abandoned almost universally today, and craniotomies are performed "exceedingly rare[ly]" (Pritchard, MacDonald, and Gant 1985:851).

4. An example of a "successful" symphysiotomy is the following, reported by M. L. Harris of Chicago in 1892:

The patient, aged 35 years, had been previously delivered of four dead children, two of them by craniotomy. Being unable to deliver with the forceps, with a conjugata vera under seven centimetres, Dr. Banga concluded to perform symphysiotomy, in the hope of securing a living child. A free incision was made down to the symphysis, which was divided with a scalpel from before backward, the pubes separating about one centimetre. The nurses who held the limbs were now told to separate the thighs, whereupon there was a sudden tearing of the soft parts, followed by a most profuse hemorrhage which was controlled with considerable difficulty. There seemed to be no spurting vessels, but a pouring forth of blood from everywhere. Mass ligatures, artery clamps, and solid tamponade, with counter-pressure made by introducing the hand into the vagina, finally succeeded in controlling the hemorrhage. It was one of the most profuse hemorrhages I have ever seen, and in the hands of a less experienced and cool-headed operator than Dr. Banga might have resulted fatally. There was a laceration of the urethra, within the formation of a fistula, which was subsequently closed, and the patient made a good recovery, happy in the possession of a living, healthy child. [M. L. Harris 1892:378]

5. All of the techniques under consideration relied on pelvimetry to determine pelvic capacity, which was difficult to determine with any degree of accuracy in this period. Although X-ray pelvimetry was available just at the turn of the century, it was not in common use and certainly not available to physicians delivering babies at home. Indications for the various procedures were set in centimeters: a normal pelvis measuring its conjugate diameter (the diameter that represents the shortest diameter through which the head must pass in descending through the pelvis;

specifically, the distance from the promontory of the sacrum to the upper edge of the pubic symphysis) at 11 cm; between 7 and 9 cm, symphysiotomy or pubiotomy would have the best success; between 3 and 7 cm, craniotomy could extract a fetal head successfully; below 2.5 cm, craniotomy was not indicated, and Caesarean section the only way to extract the fetus. But even the advocates of the various procedures that depended on accurate measurement admitted that "the difficulty of measuring a pelvis" limited physicians' abilities to determine which procedure might be indicated (M. L. Harris 1892:387).

6. In 1876 the Italian surgeon Eduardo Porro developed a new technique, which called for amputating the body of the uterus, thus removing the major source of infection. This operation proved much safer for birthing women, but it prevented future pregnancies (R. P. Harris 1878, 1880; Lusk 1880). In 1882 German physician Max Sanger reported an "improved" Caesarean section, in which the upper part of the uterus was opened and sutured separately from the abdominal wall. In 1880, with the Porro operation in use, physicians still reported as high as 50 percent maternal mortality with the Caesarean operation. Robert P. Harris, for example, reported a record of saving 47.5 percent of women in eighty Caesarean operations in the United States in 1878 (1878:625). Physicians' reluctance to advise and women's reluctance to undergo such a procedure rested on these figures. (See also R. P. Harris 1880; Lusk 1880.) By the turn of the twentieth century, physicians reported maternal mortality rates for Caesarean section of from 5 percent to 30 percent. J. Whitridge Williams calculated in 1917 that maternal mortality nationwide still averaged 10 percent. In Michigan Alexander MacKenzie Campbell of Grand Rapids surveyed the problem and concluded that maternal mortality from Caesarean sections in the state fell from 25.2 percent in 1913 to 10.9 percent in 1920 (Campbell 1920). The operation was getting safer, but it was clear that the training and abilities of individual operators, coupled with the time in labor of the operation, were crucial to predicting success (Campbell 1920; DeLee 1901; Reynolds 1901a; J. W. Williams 1917).

Comparisons between mortality rates for Caesarean section and other procedures, including symphysiotomy, pubiotomy, and craniotomy, were also be-

coming available. J. Whitridge Williams concluded in 1910 that "when pubiotomy is properly performed under suitable indications upon uninfected women, the maternal mortality should not exceed 2 percent, while approximately 95 percent of the children should be saved" (1910:734). Joseph DeLee insisted that maternal mortality from craniotomy was zero if performed at the appropriate time and with skill (1901).

Much of the debate about what obstetrical operation to perform centered around the issue of statistics and how to interpret their meaning. Most of the articles on the subject in medical journals quoted statistics gathered from other physicians' work as well as those from the individual's own practice. Physicians at the time understood the difficulty of comparing cases when data gathering was not systematic. How do you compare maternal mortality from Caesarean section, for example, when the operation was performed at different points in labor and under very different conditions in both women's homes and local hospitals? Statistics before and after Porro or Sanger were also mixed together, without any possibility of distinguishing technique or skill of operator. There is no way the historian can resolve the issue of relative safety given the state of the data.

7. Of course, female physicians found themselves in these kinds of situations as well, and the gender difference was not absolute; it was predominant, however, as most physicians and of course all husbands were men.

8. The connection between Catholic interest in craniotomy and in abortion in this period seems obvious but cannot be developed in the context of this paper. In fact, it may be that craniotomy became crucial to the Catholic Church only in the shadow of its larger interest in abortion in this period during which antiabortion legislation was passed in most states. On abortion in this period consult Gordon 1976; Mohr 1978; Petchesky 1984. On the role of priests in the delivery room see Hund 1895; Pineo 1857; Sabetti 1885; Schulte 1917; Whelan 1905.

9. For an example of how these issues reached beyond the practice of obstetrics into, for example, the establishment and staffing of medical schools, consult Weistrop (1987). It is not clear from the record whether or not Catholics were the only denomi-

nation worried about this issue, though the medical accounts suggest that this was the case.

10. The role of the husband needs more historical investigation. Most birth accounts of the eighteenth and nineteenth centuries do not mention his presence in the birthing room at all, although there is some evidence that some women preferred their husbands to other attendants (Suitor 1981). What is striking in the research on responses to emergency cephalopelvic disproportion is that husbands frequently appeared in the picture and were often consulted by the physicians.

11. The voting metaphor appeared frequently in physicians' writings about their role in obstetrical decision making. An editorial in *Medical News*, for example, noted that women generally chose craniotomy "and in such a matter as this the woman herself must where possible be given the deciding vote" (Editor 1884:364). Dr. Frank Stahl, commenting on a paper by Joseph DeLee, indicated his preference for Caesarean sections but admitted:

> If the family will not permit it, then the choice of operation lies with the family; then you are only the physician who must endeavor to overcome some obstruction to nature. You are not the elector. I hope the concensus [sic] of opinion throughout the world will be that we should never perform craniotomy upon the normal living child, unless compelled to do so by the family. [DeLee 1896:349]

References

Albert, Lionel L. 1916. Letter to the Editor. *Boston Medical and Surgical Journal* 174:701.

Allen, L. M. 1906. Symphyseotomy with the Report of Five Operations, and a Brief Consideration of Its Advantages and Disadvantages. *American Journal of Obstetrics and Diseases of Women and Children* 54:204–14.

Allen, L. M. 1909. A Plea for the More Frequent Performance of Cesarean Section. *American Journal of Obstetrics and Diseases of Women and Children* 59:189–202.

Ayers, Edward A. 1902. Symphysiotomy: Practical Deductions from an Experience in Thirteen Cases without a Death from the Operation. *Journal of the American Medical Association* 38:645–48.

Bacon, C. S. 1900. The Mutilating Operations in Obstetrics. *American Journal of Obstetrics and Diseases of Women and Children* 42:62–67.

Bacon, C. S. 1906. The Legal Responsibility of the Physician for the Unborn Child. *Journal of the American Medical Association* 46:1981–84.

Barker, T. Ridgway. 1892. When Is Embryotomy Justifiable? *Proceedings of the Philadelphia County Medical Society* 13:132–39.

Barnes, Robert. 1890. Embryotomy Versus Cesarean Section. *American Journal of Obstetrics and Diseases of Women and Children* 23:303–306.

Bill, Arthur H. 1906. Hebotomy (Pubiotomy): An Operation for the Enlargement of Contracted Pelves. *Surgery, Gynecology, and Obstetrics* 3:42–50.

Boyd, G. M. 1896. Symphyseotomy. *American Gynecological and Obstetrical Journal* 9:194–97, 340–45.

Brodhead, George L. 1907. Forceps Version and Craniotomy. *Journal of Surgery, Gynecology, and Obstetrics* 29:145–56.

Broomall, Anna E. 1893. Three Cases of Symphysiotomy, with One Death from Sepsis. *American Journal of Obstetrics and Diseases of Women and Children* 28:305–12.

Busey, Samuel C. 1884. Craniotomy upon the Living Fetus Is Not Justifiable. *American Journal of Obstetrics and Diseases of Women and Children* 17:176–93.

Busey, Samuel C. 1889. The Wrong of Craniotomy upon the Living Fetus. *American Journal of Obstetrics and Diseases of Women and Children* 22:51–69.

Campbell, Alexander MacKenzie. 1920. The Present Status of Abdominal Caesarean Section in Michigan. *Journal of the Michigan Society of Medical Sciences*. December: 556–58.

Cianfrani, Theodore. 1960. *A Short History of Obstetrics and Gynecology*. Springfield, IL: Charles C. Thomas.

Coakley, J. W. 1896. Cesarean Section Twice in One Person. *Journal of the American Medical Association* 26:720–22.

Coe, Henry C. 1896. Symphyseotomy. *American Gynecology and Obstetrics Journal* 8:355–61.

Davis, Edward P. 1900. The Frequency and Mortality of Abnormal Pelves. *American Journal of Obstetrics and Diseases of Women and Children* 41:11–15.

Davis, Edward P. 1902. Six Cesarean Sections. *American Journal of Obstetrics and Diseases of Women and Children* 45:169–79.

Davis, Edward P. 1907. The Delivery of Debilitated Women, with Especial Reference to the Interest of the Child. *Journal of the American Medical Association* 49:1334–37.

Davis, Edward P. 1913. The Present Status of Cesarean Section. *The American Journal of Obstetrics and Diseases of Women and Children* 68:12–20.

DeLee, Joseph B. 1896. Report of Six Cases of Craniotomy. *Chicago Medical Recorder* 11:318–30, 345–52.

DeLee, Joseph B. 1901. Three Cases of Cesarean Section, and a Consideration of the Indication for Craniotomy. *American Journal of Obstetrics and Diseases of Women and Children* 44:454–75.

Dixon, Robert B. 1885. Is Craniotomy upon the Living Fetus Ever Justifiable? *Boston Medical and Surgical Journal* 113:265–68, 278–81.

Draper, F. W. 1890. The Legal Relations of Obstetrics. *Boston Medical and Surgical Journal* 122:49–54.

Edgar, J. Clifton. 1893. Embryotomy, Its Prognosis, and Limitations. *American Journal of Obstetrics and Diseases of Women and Children* 27:496–509.

Editor. 1884. Craniotomy upon the Living Foetus Is Not Justifiable. *Medical News* March 29: 364.

Fry, Henry D. 1907. Pubiotomy in America, with a Report of Two Cases. *American Journal of Obstetrics* 55:834–40.

Garrigues, Henry J. 1886. The Improved Cesarean Section. *American Journal of Obstetrics and Diseases of Women and Children* 19:1009–1022.

Glenn, W. Frank. 1911. Is a Foetus a Person? *Southern Practitioner* 33:117–20.

Gordon, Linda. 1976. *Woman's Body, Woman's Right.* New York: Viking Books.

Grandin, Egbert H. 1890. The Caesarean Section from the Standpoint of Relative Indication. *Transactions of the American Gynecology Society* 15:382–400.

Grandin, Egbert H. 1891. Is Embryotomy of the Living Foetus Justifiable? *Medical Record* 39:641–42.

Gushee, E. S. 1907. Some Observations on Craniotomy. *Bulletin of the Lying-In Hospital* 4:12–15.

Haggard, W. D. 1889. The Improved Caesarean Section vs. Craniotomy. *Southern Practitioner* 11:505–10.

Harris, M. L. 1892. Symphysiotomy. *Transactions of the Chicago Gynecological Society* 2:375–88.

Harris, Robert P. 1878. Remarks on the Cesarean Operation. *American Journal of Obstetrics and Diseases of Women and Children* 11:620–26.

Harris, Robert P. 1880. The Porro Modification of the Caesarean Operation. *American Journal of Medical Science* 79:335–62.

Harris, Robert P. 1883. The Revival of Symphysiotomy in Italy. *American Journal of the Medical Sciences* January: 17–32.

Hirst, John Cooke. 1916. Cesarean Section as the Operation of Choice in Difficult Labor Cases. *American Journal of Obstetrics and Diseases of Women and Children* 74:784–92.

Hund, John. 1895. The Catholic Church and Obstetrical Science. *Medical Record* 47:283–84.

Jaggard, W. W. 1884. Is Craniotomy upon the Living Fetus a Justifiable Operation? *American Journal of Obstetrics and Diseases of Women and Children* 17:1131–41.

Jaggard, W. W. 1888. Conservative Caesarean Section. *Journal of the American Medical Association* 11:63–65, 97–102.

Johnson, Joseph Taber. 1887a. Can the Caesarean Section Be Safely Substituted for Craniotomy in the United States at the Present Time? *Journal of the American Medical Association* 8:169–74, 189–91, 270–75, 324–27.

Johnson, Joseph Taber. 1887b. Caesarean Section: Death on the Tenth Day. *Journal of the American Medical Association* 9:696–99.

Landrum, S. H. 1912. Letter to the Editor. *Journal of the American Medical Association* 58:577.

Leavitt, Judith Walzer. 1986a. *Brought to Bed: Childbearing in America, 1750–1950.* New York: Oxford University Press.

Leavitt, Judith Walzer. 1986b. Under the Shadow of Maternity: American Women's Responses to Death and Debility Fears in Nineteenth-Century Childbirth. *Feminist Studies* 12: 129–54.

Lusk, William T. 1880. The Prognosis of Caesarean Operations. *American Journal of Obstetrics and Diseases of Women and Children* 13:18–23.

McColl, Hugh. 1891. Is Craniotomy upon the Living Child Justifiable? *Transactions of the Michigan State Medical Society* 15:268–71, 271–73.

McCoy, J. C. 1899. Embryotomy and Two Cesarian Sections in the Same Patient. *Journal of the American Medical Association* 339:224–25.

McDiarmid, A. 1899. Cesarean Operation, with Report of Two Cases. *Journal of the American Medical Association* 32:766–67.

McKelway, George I. 1892. Delivery Through the Abdominal Wall vs. Craniotomy in Otherwise Impossible Births. *Journal of the American Medical Association* 19:5–9.

Mohr, James. 1978. *Abortion in America.* New York: Oxford University Press.

Monaco, Rev. Cardinal. 1885. The Ethics of Craniotomy. *Medical Record* 28:492.

Montgomery, E. E. 1892. Some Mooted Points in Obstetrics and Gynecology—President's Address. *Journal of the American Medical Association* 18:765–66.

Montgomery, E. E., and A. Vander Veer. 1889. The Cesarean Section. *American Association of Obstetricians and Gynecologists* 2:366–75.

Myers, William H. 1895. The Limitations of Craniotomy. *American Gynecology and Obstetrics Journal* 7:479–86.

Noble, Charles P. 1893. Symphysiotomy Versus Its Substitutes, with the Report of a Case of Symphysiotomy. *Medical News* 62:176–81.

Pernick, Martin S. 1985. *A Calculus of Suffering: Pain, Professionalism, and Anesthesia in Nineteenth-Century America.* New York: Columbia University Press.

Petchesky, Rosalind Pollack. 1984. *Abortion and Woman's Choice: The State, Sexuality, and Reproductive Freedom.* New York: Longman.

Peterson, Ruben. 1915. Under What Circumstances Is Craniotomy on the Living Child Justifiable? *Transactions of the Clinical Society of the University of Michigan* 14:319–24.

Pineo, P. 1857. Craniotomy. *Boston Medical and Surgical Journal* 56:376–78.

Pomeroy, Ralph H. 1908. Report of a Case of Impacted Breech Presentation Treated by Hebotomy. *American Journal of Obstetrics and Diseases of Women and Children* 57:511–15, 565–72.

Poucher, John Wilson. 1913. The Advantage of Cesarean Section over Other Procedures in Border-Line Cases. *American Journal of Obstetrics and Diseases of Women and Children* 68:1143–49.

Pritchard, Jack A., Paul C. MacDonald, and Norman F. Gant. 1985. *Williams Obstetrics: A Textbook for the Use of Students and Practitioners.* New York: Appleton-Century-Crofts.

Reynolds, Edward. 1901a. Circumstances Which Render the Elective Section Justifiable in the Interest of the Child Alone. *Transactions of the American Gynecology Society* 26:277–82.

Reynolds, Edward. 1901b. The Major Obstetrical Operations. *Journal of the American Medical Association* 36:415–20.

Rohe, George H. 1889. Is Craniotomy Justifiable on Living Children? *Transactions of the American Association of Obstetricians and Gynecologists* 2:351–55.

Rosenberg, Julius. 1892. The Indications for Craniotomy upon the Living Child, and the Contra-Indications to Cesarean Section. *American Journal of Obstetrics and Diseases of Women and Children* 26:319–30.

Sabetti, A. 1885. The Roman Catholic View of Craniotomy. *Medical Record* 28:606–7.

Scholten, Catherine M. 1985. *Childbearing in American Society,* 1650–1850. New York: New York University Press.

Schulte, A. J. 1917. The Rights of the Unborn Child as Viewed by the Catholic Church. *American Journal of Obstetrics and Diseases of Women and Children* 75:886–90.

Starr, Paul. 1982. *The Social Transformation of American Medicine: The Rise of a Sovereign Profession and the Making of a Vast Industry.* New York: Basic Books.

Suitor, Jill. 1981. Husband's Participation in Childbirth: Nineteenth-Century Phenomenon. *Journal of Family History* 6:278–93.

Trawick, George C. 1904. Should Embryotomy Be Done on the Living Foetus? *Southern Practitioner* 26:211–17.

Vogel, Morris, and Charles E. Rosenberg, eds. 1979. *The Therapeutic Revolution: Essays in the Social History of American Medicine.* Philadelphia: University of Pennsylvania Press.

Voorhees, James D. 1902. Craniotomy. *American Journal of Obstetrics and Diseases of Women and Children* 46:765–77.

Warner, John Harley. 1986. *The Therapeutive Perspective: Medical Practice, Knowledge, and Identity in America, 1820–1885.* Cambridge: Harvard University Press.

Wathen, William H. 1889. Is Craniotomy upon the Living Child Justifiable? *American Journal of Obstetrics and Diseases of Women and Children* 22:1233–41.

Weistrop, Leonard. 1987. *HVO: The Life & Letters of Dr. Henry Vining Ogden, 1857–1931.* Milwaukee: Milwaukee Academy of Medicine.

Wertz, Richard, and Dorothy Wertz. 1979. *Lying-In: A History of Childbirth in America.* New York: Schocken Books.

Whelan, W. P. 1905. Cesarean Section for Placenta Praevia. *Journal of the American Medical Association* 44:1868.

Williams, Harold. 1879. A Comparison Between the Cesarean Section and the High Forceps Operation. *American Journal of Obstetrics and Gynecology* 12:23–31.

Williams, J. Whitridge. 1910. Is Pubiotomy a Justifiable Operation? *American Journal of Obstetrics and Diseases of Women and Children* 61:721–52.

Williams, J. Whitridge. 1917. The Abuse of Caesarean Section. *Surgery, Gynecology, and Obstetrics* 25:194–201.

Williams, J. Whitridge. 1920. *Obstetrics: A Text-Book for the Use of Students and Practitioners.* 4th edition. New York: D. Appleton.

Zinke, E. Gustav. 1895. Symphyseotomy Versus Embryotomy upon the Living Fetus. *Ohio Medical Journal* 6:73–83.

Zinke, E. Gustav. 1903. The Limitations of Caesarean Section. *American Journal of Obstetrics* 48:604–15.

35 "A Complete Disaster": Abortion and the Politics of Hospital Abortion Committees, 1950–1970

Rickie Solinger

In the late 1950s the obstetrical staffs of twenty-six hospitals in Los Angeles and the San Francisco Bay area, responding to a questionnaire, evaluated a number of hypothetical abortion requests. Among them were these three cases:

> Mrs. C. is a thirty-eight-year-old woman who has had six children in the past ten years. At this time, she is two months pregnant and severely depressed. After an unhappy childhood and marriage, Mrs. C. sees herself as a failed mother, wife, and housekeeper.
>
> Each of her recent pregnancies made Mrs. C. tired and depressed. During her last pregnancy one year ago, she spent most of the time in bed, vomiting a great deal, unable to eat; twice she was hospitalized for dehydration and weight loss. Following delivery, Mrs. C. was chronically depressed and listless, with multiple physical complaints. Presently, she complaints of being tired, of not caring and says she

wants to rest and sleep most of the time. She states that she can't eat and that she vomits when she tries. She appears emaciated and hollow-eyed. Although she seems fairly well in contact with reality, she claims to be unable to face the prospect of a seventh pregnancy.

> Mrs. A. is thirty-two. She has three healthy children, ages four, six, and seven. She is now seven weeks' pregnant. There is conclusive evidence that she had an attack of rubella two weeks ago.
>
> Miss C. is a fifteen-year-old daughter of a minister. Eight weeks ago she was raped by an escapee from a state institution for mental defectives and became pregnant. As a result, Miss C. is experiencing serious emotional distress.[1]

Naturally, both the lawyers who devised the questionnaire and the physicians who responded to it were aware of Section 274 of the California Penal Code which "proscribe[d] as a felony the performance of an abortion upon a woman 'unless the same is necessary to preserve her life.'"[2] The lawyers and medical doctors were equally aware that in none of the three cases described did the pregnancy directly endanger the life of the petitioner. A physician or a hospital

Rickie Solinger is an historian and independent scholar, Boulder, Colorado.

Reprinted from *Feminist Studies* 19, 2 (Summer 1993): 241–68, by permission of the publisher, Feminist Studies, Inc., c/o Department of Women's Studies, University of Maryland, College Park, MD 20742.

agreeing to terminate any of these pregnancies would do so in violation of California law. The completed questionnaires were returned to professors Packer and Gampell of Stanford Law School and the results later published in the *Stanford Law Review*. The results indicated that almost one-half of the reporting obstetrical staffs were willing to break the law in cases where psychological or eugenic indications for abortion were present or in the case of rape. A greater number, however, were unwilling. But even many of the unwilling physicians believed that almost any woman could arrange a legal therapeutic abortion for herself if she shopped around among hospitals in Los Angeles or San Francisco. The Packer-Gampell questionnaire and numerous articles written by physicians and published in mainstream medical journals in the 1950s and 1960s reveal a profession deeply divided, embarrassed, angry, and frustrated over the issue.

This essay reviews discussion within the medical community in the postwar years concerning contraindications to pregnancy and the circumstances, if any, justifying therapeutic abortion. Such discussions reflect broader cultural attitudes toward women, mothers, babies, and pregnancy in the postwar era. They also illuminate the turmoil within the profession over these issues and the uneasy, insecure, but sometimes enduring, resolutions physicians devised to quell internal dissension and reinforce medical authority in the two decades immediately preceding *Roe v. Wade.*

Specifically, the essay argues that having been pushed into a defensive posture by the combination of medical advances, the specter of legal liability, and the emergence of women taking a new degree of initiative, physicians quickly transformed their uncomfortable defensiveness into an offensive posture toward women. To do so they adapted a legalistic tribunal method which tightened the association between two powerful professions—legal-izing medicine and medical-izing the law, at once.

In addition, many influential physicians in this era drastically redefined pregnancy in a direction prochoice advocates must still confront today. In the postwar decades medical and psychiatric discourse uncoupled the woman and the fetus while at the same time binding women in ever-tighter traces to their pregnancies. These experts claimed that medical-technological advances removed all physical impediments to pregnancy. The advances could also reveal the fetus-as-homunculus. Pregnant females in turn became carriers and agents of protective custody. Many medical commentators in these years came to cast pregnancy *first* as a process of fulfillment and realization for the fetus. Still important, but now secondarily so, pregnancy was viewed as an essential expression of female identity and destiny.[3]

This new approach to pregnancy, fetuses, and pregnant women (although subordinated for a time to a discourse of women's choice) has clearly provided the "scientific" underpinnings of the antichoice movement today. The postwar tribunal method of enforcing this perspective could be the state strategy of tomorrow. As we observe the dignity and protection of fertile women (embodied in *Roe v. Wade*) threatened in the early 1990s, it is well to consider the arguments and processes that experts used in our recent past to eliminate access to these rights. It is also worthwhile to consider that the strategies discussed here—as pervasively and powerfully promoted as they were—proved highly vulnerable to the grassroots counterclaim for choice.

Legal Abortions
Before *Roe v. Wade*

Dissension over abortion within the medical community was not a long-standing intra-professional problem. The post–Civil War state

[handwritten margin notes: "when abortions were legal"; "legal abortions allowed because it can be life-threaten"]

laws against abortion, which turned back the traditional right of girls and women to abort in the first trimester of pregnancy, stipulated that abortions were permissible only in cases where, due to a medical condition, the pregnant woman's life was in danger. These new late nineteenth-century laws granted the determination to licensed physicians only. Through the late 1940s legal abortions were performed often and routinely in most hospitals across the country. Medically approved contraindications to pregnancy included cardiovascular conditions (rheumatic heart, hypertensive-cardiovascular, coronary artery, and congenital heart disease); kidney dysfunction (chronic nephritis, hydronephrosis, polycystic kidneys, single kidney, renal stones, and pyelonephristis); neurologic diseases (epilepsy, multiple sclerosis, myasthenia gravis, and Ménière's disease); toxemia; respiratory disease (tuberculosis, bronchiectasis); uterine disease (cancer of the cervix, fibroids); orthopedic problems; and blood diseases such as leukemia, ulcerative colitis, diabetes, premature separation of the placenta, otosclerosis, bowel obstruction, lupus, and thyrotoxicosis. Physicians occasionally performed abortions on women suffering from severe psychiatric disorders.[4]

With such an extensive list of contraindications to pregnancy, abortion ratios at some hospitals were high in various decades before 1950, especially in comparison to what they would soon be, for example: 1 abortion to every 76 live births at Bellevue Hospital in New York; 1 to every 167 at New York Lying-In; and 1 to every 169 deliveries at Iowa University Hospital.[5] Given the state of medical knowledge and the range of medical options, as well as prevailing ideas about the physical toll pregnancy took on women, non-Catholic physicians were often willing to sacrifice the pregnancy in favor of the well-being of the woman. Medical decisions concerning these matters were often predicated upon an assumption that pregnancy itself was a physical event or a medical condition which happened to girls and women, sometimes under conditions that were not physically or medically favorable. In these cases it could be assumed that pregnancy could interact with and worsen a pre-existing condition. A prominent obstetrician reminded his colleagues in 1958 that "for years medicine has taught that pregnancy, though a normal physiologic process, is such a tremendous burden that it adds an unbearable load to any ill, diseased, or handicapped person and, therefore, the two were not compatible."[6] This perspective assumed that the woman's body was an integrated system which the pregnancy *could* undermine or disintegrate. The pregnancy itself might well take precedence over disease as the more destructive agent. Where contraindications existed, pregnancy—or the "unborn child"—was not granted precedence, or healing power, or constructed as a special condition virtually separate from the biological body or psychological mind of the impregnated female. The pregnancy was an additive, not an autonomous factor. By the twentieth century, girls and women in the United States had lost the traditional right to abortion by choice in the first trimester. But the newer association of law and medicine still sanctioned abortion under medically indicated, life-threatening conditions. In short, abortion served a function when pregnancy invaded and threatened a woman's body.

A Professional Crisis

By the early postwar years the medical consensus about the indications for abortion had fractured, and therapeutic abortion rates were plummeting in hospitals across the country. One authoritative study reported that the therapeutic abortion rate per 1,000 live births in the United States declined from 5.1 in 1943 to 2.9 in 1953, a 43 percent decline.[7] A study of legal hospital abortions in New York after the war demonstrated that the "overall frequency of therapeutic abortions declined by almost 50 percent."[8] Bellevue's ratio was moving toward the 1:362 mark it would hit in 1965. Other studies showed that hospitals attached to the universities of Virginia and California, which had ratios of 1:120 and 1:88, respectively, in the 1940s, reduced their rates by one-third to one-half over the next fifteen years.[9]

The sharp decline in legal therapeutic abortions performed in hospitals reflected the fact that by midcentury, mainstream medical opinion held that medical-technological and obstetrical advances obviated the need to interrupt pregnancy for most of the medical conditions previously considered incompatible with pregnancy.[10] The same obstetrician who recalled the era when pregnancy was considered an "added burden, extra strain, and increased load" for ill or handicapped women in the recent past pointed out that now obstetricians and their medical colleagues had access "to better understanding of such complications and a greater realization that with correct therapy the disease and pregnancy are compatible."[11]

Shared access to new technologies and treatments, however, did not mean that physicians shared a professional opinion about when and how these innovations should be applied.[12] In fact, the new medical developments gave rise to a very complicated situation for physicians; the situation could be called a *crisis* which extended over a twenty-year period, at least.

The crisis derived in part from a profoundly paradoxical relationship between medical progress, the law, and politics. On the one hand, physicians were scientific and humanitarian heroes for subduing the role of pregnancy as an "added burden" and for devising methodologies to conquer diseases threatening to pregnancy and the pregnant female. On the other hand, state laws still required that the life of the pregnant women must be medically endangered to permit abor-

tion. The legal system persisted in requiring a condition that the medical system said rarely existed. Consequently, legal demands were at odds with medical advances which claimed to have virtually removed the basis for medical judgments concerning indications for abortions.

Given their continuing legal relationship to abortion, however, and their interest in sustaining medical authority over pregnant women, physicians struggled to establish new bases for medical decision making.[13] By the early 1950s a number of physicians were airing these struggles before the medical community in the pages of the most prestigious medical journals in the United States. They described a bitterly contentious intraprofessional situation. The reports indicated that any sense of common purpose among physicians considering abortion had been severely undermined in the aftermath of medical advances. A 1952 article in the *American Journal of Obstetrics and Gynecology* referred to "considerable argument" and "this disunanimity of opinion" among physicians concerning the subject of indication for abortion.[14] The next year, the *Journal of the American Medical Association* carried an article condemning the "confusion and uncertainty" surrounding this issue within the profession.[15] Other articles chronicled specific disagreements among physicians[16] and presciently despaired that a state of harmony could ever again be attained. Two Chicago physicians asserted that no agreement among medical doctors can "be achieved regarding either individual indications [for abortion] or general principles."[17] Another physician called his attempt to study the therapeutic abortion situation "a complete disaster" because "the categories of opinion were almost equal in number to the men concerned."[18]

Many physicians did not consider these open debates a sign of health within the profession. On the contrary, there is evidence that many felt that the new disunity over the abortion issue hurt the standing of physicians as expert objec-

using the reason of psychiatric problem to get legal abortion was denied

tive practitioners of medical science.[19] Dissension also raised questions about the source and scope of medical authority. One physician observed, unhappily, that "if interruption of a previable pregnancy is requested, the law at present dictates what medical opinion should be."[20] Another put the abortion issue and his professional discontent in a larger context of "restrictive efforts in every field of medicine. . . . Qualifications and regulations and boards are limiting the scope of the practice of individual physicians. Law now directs specific methods of treatment and prophylaxis for certain diseases."[21] Others expressed deep uneasiness that they were facing pressures to look beyond their traditional subject—the physical condition of the individual pregnant woman. They were being urged, inappropriately, to include social factors in their medical diagnoses.[22]

The rise of psychiatric indications as grounds for abortion[23] solved the issue of medical authority for many practitioners but deepened the uneasiness of many others not convinced in the 1950s that psychiatry belonged within the ranks of medical science.[24] A Cleveland obstetrician identified his hospital's biggest abortion problem as "those cases done for psychiatric indications, many times questionable psychiatric reasons."[25] Another obstetrician wrote that "medical men . . . have been able to markedly reduce the therapeutic abortion rate throughout the country only to find that this least justifiable of all indications, psychiatric reasons, has been allowed to run rampant."[26] A sociologist assessed the situation this way:

> In recommending legal abortion, the psychiatrist faces the additional problem of hostility—or at least skepticism—on the part of the medical men who may be involved in ruling on his cases. Thus obstetricians may feel that psychiatric diagnoses are being used as subterfuges in instances where abortion is not really jus-

tified, and they are particularly unconvinced by assertions which the psychiatrist may feel [compelled to make] that the patient is likely to commit suicide if pregnancy is not terminated.[27]

Many essayists in the medical journals were most concerned that the use of psychiatric findings in favor of abortion further undermined the traditions and the reputation of medicine as a scientific endeavor. Psychiatry was merely a "long practiced art," at best "an infant science."[28] One physician referred to psychiatry and the application of its principles to the abortion decision as "a most nebulous, nonobjective, nonscientific approach to medicine" and pictured psychiatrists as engaged in "bedeviling" their colleagues to perform therapeutic abortions.[29] Two California obstetricians argued that physicians who relied on psychiatrists to restore medical grounds for abortion and reestablish a justification for medical participation in the decision felt their case weakened: "The extraordinary range of opinion represented among the psychiatrists is a far cry from the scientific objectivity that one hopes would apply to determinations affecting the life and health of patients."[30]

Disqualified as insufficiently scientific by many of these physicians, the psychiatrist could be identified as the "unwitting accomplice" in relation to abortion, a label with more legal than medical significance.[31] Psychiatrists were portrayed in this way as pawns of importuning women, unlike real medical doctors who initiated any abortion decision in the interest of their passive pregnant patients. One physician went a remarkable step further in associating psychiatrists with women who inappropriately initiated their own treatment.

> In some parts of the country [criminal abortions] can be obtained so easily that when patients apply for a psychiatric consultation, for the purpose, so they state, of obtaining a psychiatric recommendation

to the effect that their pregnancy be interrupted, the very fact that they make such an appointment seems to be almost presumptive proof that they do not wish the abortion, but rather psychiatric help in order to carry their child to term.[32]

In the early postwar years, then, physicians struggled with the issue of abortion sans portmanteau. Many felt compelled to argue defensively for the obsolescence of medical indications while at the same time taking the offensive, arguing to sustain their professional prerogatives as abortion decision makers. Psychiatrists alone provided capacious medical grounds for abortion, but many practitioners rejected the terrain as polluted. Women, for three generations forced by law to submit to physicians as abortion decision makers, had now begun to initiate and pressure medical doctors to provide them with abortions, sometimes on whatever shifting grounds were approved at a given time and place.[33] All physicians had to absorb the fallout that followed intramural dissension and undermined the united front of expertise. Similarly, all physicians involved in "therapeutic" abortion decisions had to adjust their personal, political, and professional judgments to the fact that the law and the law enforcement system were at least theoretically conditioning and monitoring their medical practice.

The Rise of Hospital Abortion Committees

By the mid-1950s most non-Catholic hospitals had begun to address their vulnerability in relation to abortion by finding ways to reassert medical authority over the issue and to sustain physicians' control over pregnant girls and women. Two strategies governed this process in a great many hospitals across the country. First, physicians recognized that they had to reassemble themselves as a collectivity from which professional expert diagnoses and decisions regarding individual women could be issued in one voice. In this setting psychiatrists could be team players. They could bring their special perspective on the individual into the arena of experts and thus come to the aid of the profession while validating their own standing. Second, physicians redefined pregnancy in relation to women's bodies in such a way as to efface the woman herself while giving precedence to the law and the fetus. Again, psychiatrists played a pivotal role in accomplishing the redefinition.

By the mid-1950s in many hospitals physicians assembled themselves collectively into abortion boards or committees. As a group, obstetricians, cardiologists, psychiatrists, and others considered abortion recommendations and requests and issued definitive decisions on each case. The chief of a department of obstetrics and gynecology in a large northeastern hospital described the way decision-making processes changed in many hospitals in the early 1950s.

> At Mount Sinai Hospital [in New York], before [Alan Guttmacher's innovations], a request for therapeutic abortion merely had to be signed by two senior staff members. Guttmacher established the abortion committee of five members: the chief in medicine, representatives of pediatrics and of surgery, the chief of psychiatry, and the chief of obstetrics and gynecology who acts as chairman. Requests to the committee must be supported by two consultants recommending the procedure and outlining the indications for it. One of the consultants must appear before the committee to answer additional questions. The committee must be unanimous in its approval of any request.[34]

These committees protected physicians, individually and as a profession, in a number of ways. Of paramount importance to many was the legal protection the boards provided. Four medical doctors, characterizing the therapeutic abortionist as a "fetal executioner," stressed that group review of all cases was crucial because the "legal burden" otherwise rested on the individual obstetrician.[35] The *Journal of the Indiana Medical Association* recommended group work early in the crisis: "To make as certain as human precaution can make it, that a physician might not be subjected to difficulties later on, he should have consultation with other physicians."[36] Another group of medical doctors studied the legal situation in Michigan in 1950 and determined that hospitals were compelled to establish abortion committees in order "to protect the physician" because "while abortions could be performed legally [in that state, if the mother's life was in danger], in the event of suit, the physician had no legal protection," in the absence of a committee. Rudolph W. Holmes insisted that because the law drew such a "tenuous" line of demarcation between legal and illegal abortions, "it behooves medical staffs of all reputable hospitals to institute [abortion boards]. It would be a great protection to the operator as well as a deterrent to dangerous aspersions by outsiders."[37]

For many concerned physicians insiders could be as dangerous as outsiders. These medical doctors felt that committees functioned best to mute, neutralize, or "curb liberal obstetricians" favoring too many abortions or abortions on questionable grounds. The California survey conducted in the mid-1950s and reported in the *Stanford Law Review* revealed that most physicians in the twenty-six reporting hospitals felt that the committee's central function was "to police activities of a doctor whose procedures might otherwise bring himself and his colleagues into disrepute."[38] A number of medical doctors simply described the structure as "an effective method of control."[39]

These interests in reputation and control were undoubtedly central concerns of many physicians in part because so many of them spoke and behaved one way publicly and another way privately. For example, a number of professional illegal abortionists who conducted thriving businesses in this era have reported that hundreds of medical doctors—surely among them, those who publicly claimed medical, hospital control over abortion decisions—routinely referred clients for illegal abortions.[40] By insisting on the righteousness of the mechanisms of hospital abortion committees, physicians could disassociate themselves from professional and public concerns about widespread illegal abortions, thus diminishing personal vulnerability and perhaps individual crises of conscience.

Moreover, as Carole Joffe has demonstrated so vividly, a number of respectable medical doctors—she calls them "physicians of conscience"—performed thousands of abortions in secret in these years, because of their deeply held conviction that women should be able to choose whether and when to have babies.[41] These physicians of conscience were serving a function for many of their law-abiding colleagues who regularly referred unhappily pregnant girls and women to them. They were also symbols of the broken ranks of the profession and represented a threat to that profession's probity and its safety. Because, apparently, so many hundreds of obstetricians, gynecologists, psychiatrists, and others were at least second-party participants in illegal abortions, the hospital abortion committee became important as it promoted the fictions of medical solidarity and the profession's legal compliance.

Many contemporary commentators referred to the actual legal vulnerability of physicians who performed abortions as a "phantom," and many pointed out that "no reputable physician has ever been convicted for performing an abortion in a reputable hospital."[42] This was the case both before and after abortion committees began to operate. It seems probable, then, that the most valuable service the boards actually performed was to bolster the image of physicians as members of a highly functioning professional body guided by scientific expertise and collective wisdom. The committee could transform public dissension within the medical community into public harmony and at the same time reduce the incidence of abortion. Careful study of the early functioning of one abortion committee showed that requests had fallen dramatically and "that the indications proposed during [the second year] conformed more to medical practice."[43] Two medical doctors, reflecting on their three-year experience on an abortion board in Newark, New Jersey, lauded the structure because of its "impartial, anonymous, efficient performance." Incidentally, these physicians provided an example of how such scientific expert qualities shaped the board's decision making. Two women who had contracted rubella, one at six and one at nine weeks gestation, applied for permission to abort but were rejected by the committee because in both cases the illness was "not objectively observed by a physician."[44]

Moreover, physicians could more confidently assert their right and duty to retain medical control over the abortion decision once they established the committee as a respectable forum dedicated to processing individual women in an orderly fashion. In short order the committee became a vehicle for bringing professional wisdom to bear on the issue, in part as a way to forestall the situation "where the decision for abortion may be made by legal, social or welfare groups outside of the profession."[45]

In an era when the law was increasingly positioned between the medical practitioner and the patient, many physicians recognized a need not only to reassert a proprietary role in the abortion decision but also a need to assert, through

the abortion board, medical doctors' intentions to carry out their medical responsibility judiciously, even judicially. Facing a discrepancy between hospital by-laws and state laws governing permissible abortions,[46] physicians constituted the abortion committee as a quasi-legal forum and associated themselves with the wisdom and objectivity of the law. In this way they also courted and apparently won the trust and respect of the legal profession, as well as a measure of protection against liability. Obstetricians, reporting on the success of the committee in one hospital, pointed to two outstanding achievements. First, abortion requests had almost halved since the board was established. And then, "another way of assaying the value of our committee is the informal opinion of our legal friends as well as a Judge of Probate Court that . . . all physicians [should] have the benefit of such committee approval."[47]

As physicians assumed a judicial role regarding individual requests for abortion—whether the requests originated with the obstetrician, another medical specialist, or the pregnant woman herself—inevitably, committee physicians, donning their robes in earnest, perceived the individual woman as "on trial."[48] Unfortunately, however, in many cases the cardinal principle of the U.S. legal system seems to have been inoperative. Physicians warned each other not to assume the woman's innocence. A New York medical doctor put it this way: "The physician must have a high index of suspicion for the patient who tries to pull a fast one." The source of danger was the "individual [woman] seeking to satisfy selfish needs"; the consequences of ignoring the danger were "somewhat analogous to medical opinion in any industrial compensatory action in which motivation may play a large role, and medical practice can be degraded."[49] One physician spoke for many of his colleagues when he warned of the "clever, scheming women, simply trying to hoodwink the psychiatrist and obstetrician," when they asked permission to abort.[50] Another identified "woman's main role here on earth as conceiving, delivering and raising children." Thus, he concluded, any woman who claims not to want a certain pregnancy must not be believed.[51] In this environment it is not surprising that, as one physician put it, "we have had a great many less requests for abortion [in his California hospital] since the patient and the doctor know that the patient must . . . have her case become an open trial so to speak to be decided on its merits."[52]

In order to function successfully abortion committees accepted additional assumptions about the relationship between abortion, medicine, and the role of the medical doctor. First, many physicians stressed the traditional exclusive relationship between medical science and the individual patient, a relationship that could best be honored and protected by the committee of medical scientists. In this case the individual was a simply biological or organic entity; social, economic, or other environmental factors were irrelevant to an individual pregnant woman's situation and to an abortion determination. Two physicians who felt that some of their colleagues were being inappropriately swayed by what one called the "intense [nonmedical] motivations" of importuning women,[53] cautioned:

> It would seem that a few abortions were brought about through the combined influence of economic pressure, social factors, and convenience. To deny that these forces had not influenced us would be incorrect; to accept them would be unwise; and the best course would be to view future indications in the light of strict medical principles directed toward preserving the life and health of the mother.[54]

Six years later another medical doctor referred to the "real need" to disregard any but the strictly medical indications present in the individual pregnant woman. To stray from this focus was to stray from science and from the physician's role as a medical healer. It was also an invitation for critics to impute social or political or unethical agendas to medical doctors. For example, those who exceeded their medical expertise could be accused this way: "To specify certain social indications for legal abortion is equal to legal license for the abortionist," or this way: medical doctors who granted permission liberally, on the other-than-medical grounds, perhaps did so because they "enjoy this procedure" and because of a "complete lack of professional and moral principles [which leads them to] do anything out of a desire to win a friend or to make a dollar."[55] Holding a tight focus on the individual sustained the pregnant woman as a scientific specimen which could be viewed against a neutral background. The physicians' task could be sustained as scientific. Medical doctors need not—indeed—they must not—assume the role, particularly of social critics. Although the objectivity, the neutrality, anonymity, and dignity of the law had something to offer medical doctors considering the abortion issue, in the 1950s the sociologist's arena was a minefield.

There was, however, one way that social science could support and validate therapeutic abortion practices in hospitals in the postwar years. By this time statistical information was widely recognized by sociologists, psychologists, political scientists, educators, and other academics and public policy experts as a highly valuable legitimating tool. Numbers became a valid basis for explaining, analyzing, predicting, and even justifying behavior. Inferences and conclusions, policies and politics based on statistics, became "scientifically valid."[56] As part of this trend abortion committees agreed to practice scientific medicine by statistics. Drawing on this development, one participant asserted that "the need for therapeutic abortion should be no higher than one per one thousand materni- ties."[57] A critic of this general orientation observed that many hospitals were "now practicing abortion by statistics [so] the patient is no longer a medical case but a number balanced against a quota. If she arrives after the monthly quota has been filled, she may well be rejected despite the urgency of her medical needs."[58] One report of hospital practices cited a "gynecologist [who] said his place had gotten a reputation for being easy, so tightened up and now approved one in ten."[59] Similarly, Alan Guttmacher in New York said that his hospital, Mt. Sinai, was formerly saddled by the reputation of being an easy place to get an abortion. As a consequence he set up an abortion board. "The result is that applicants for interruption of pregnancy have decreased tremendously because of the vigilance of the board and the fact that the case has to go through such a procedure."[60]

A number of studies argued for the efficacy of committees on the simple grounds that abortions decreased after the boards were established.[61] In this additional respect, then, committee-based, statistically shaped abortion decision-making bolstered the reputation of the medical profession as a collectivity of scientists. At the same time practitioners protected themselves from outside negativity. Low statistics demonstrated good scientific, nonideological practice. One commentator reported that the fear of being labeled with the reputation of "abortion mill" was so pervasive among hospital staffs in the postwar era that "many hospitals now consider a minimum abortion rate a status symbol. 'The fewer abortions, the better we look,' a Philadelphia doctor put it."[62]

Finally, as suggested earlier, although the psychiatric perspective had been initially problematic for many medical doctors involved in abortion determinations, by the late 1950s the situation had changed. By this time the abortion committees had provided psychiatrists with a rich proving ground for their specialty. According to a number of essayists, psychiatrists did rise to the aid of their colleagues by providing the expert basis for medical decision making and medical control that would have otherwise been lacking. As the biology of both disease pathology and pregnancy became less mystified and less remote because of medical-technological advances, psychiatrists stepped in, forestalling the possible empowerment of the pregnant patient. Psychiatrists constructed and drew on the unconscious as an entity which was only accessible to, and could only be decoded by, the expert. One physician observed, "If we have learned anything in psychiatry, we have learned to respect the unconscious far more than the conscious and we have learned not to take [abortion requests] at face value."[63] Another demonstrated how this observation worked in practice. "An example is a woman who comes in seemingly with an unambivalent wish to be aborted which, upon interview, turns out to be an unconscious attempt on her part to punish her husband."[64] Such a discovery, as the basis of diagnosis, could be available only to the physician.

Pregnancy Redefined

This physician and many of his colleagues were in part responding to the new pressure from many women in their offices initiating requests for legal therapeutic abortions. They were also responding from a new definition of pregnancy itself which emerged following the decline of medical indications for the interruption of pregnancy and alongside the validation of the psychiatric perspective.

Pregnancy became at this time a state inhering to the woman-as-custodian, but the pregnant woman and fetus no longer presented an integrated system. In the postwar period pregnancy was no longer viewed as an "added burden" or an "increased load," or a potentially destructive agent. It remained, under the proper circumstances only, a fundamental expression of womanhood, inexorable and transcendent but something dramatic had happened to the essence of pregnancy. After medical doctors determined that there were no longer any medical contraindications to pregnancy, pregnancy ceased to be a physical issue. Physicians now argued that "for most conditions, the natural history of the disease is not influenced deleteriously by an intercurrent pregnancy. Conversely, neither is the course of pregnancy seriously affected by a complicating medical condition."[65]

Neither did physicians consider pregnancy a psychological issue. One argued: "Statistical analysis shows that childbearing has only a small influence on the mental disorders of women and that the majority of individuals predisposed to mental disorder go through childbirth unscathed."[66] Regarding psychological treatment, medical doctors were confident that "the presence of pregnancy does not interfere with the treatment of psychiatric disease or make it less effective; in fact, we do not hesitate to administer electric shock therapy with curare while the patient is pregnant."[67] Summing up the position of many, two obstetricians wrote, "As far as a complicating disease is concerned, the expectant mother presents a problem not greatly different from that of a non-pregnant sister with the same disease, and . . . furthermore, so far as her pregnancy is concerned, she is not greatly different from other pregnant women."[68]

In essence, pregnancy was most centrally a *moral* issue, but the moral ground had shifted. As the fetus was constructed as a little person, medical doctors constructed the pregnant woman's body as a safe reproductive container. The woman, along with her physician, had the moral duty to sustain the container as fit.[69] One obstetrician explained the suitability of women for this role. "Woman is a uterus surrounded by a supporting organism and a directing personality."[70] Completely effaced, the woman-as-uterus simply housed the child. The most perfect iconic expression of this refocusing burst upon the consciousness of the general public in this country in the pages of *Life* magazine in April 1965. There were displayed the amazing photographs of Linnert Nilsson, a Swedish photographer, who had spent seven years working with surgeons in five Stockholm hospitals to capture images of the fetus in utero at many stages of its development.[71] The photographs demonstrated two startling conceptual innovations. First, in Nilsson's pictures each frame is filled entirely by a fetus in the uterine environment but no woman, no mother, no hint that the fetus is in relation to any other living entity. The fetus is ultraprivileged and apparently ultraindependent. The images suggest that if there is a woman involved in this "life before birth," she occupies another space, if not another universe, entirely. Second, the pictures aim to capture, most important, the human-ness of the "baby." Nilsson selected to focus on the eyes, the faces, the hands and feet, to stimulate the viewers' sense of sympathy and identity with the fetus. The photographer intends to portray the thirty-seven-day-old, one-half-inch fetus as baby. The embryo with a human face demands a morally nourishing environment. Providing that had become the pregnant woman's job and the meaning of pregnancy.

It is important to note here a shocking irony regarding this pictorial event. Tiny inconspicuous text accompanying the photographs indicates that almost all of the photographs were images of embryos that had been "surgically removed for a variety of reasons." The text doesn't indicate from what or whom they had been "removed." So without any clues about this fact attached to any individual picture, indeed it was the case that these fetuses were independent, if dead.

669

Drawing on the innovative notions of pregnancy and pregnant women, psychiatrists were prepared to explain the behavior of the growing number of women asking medical doctors for abortions in the postwar years. Their explanations created a broad category of women who were, by definition, in the absence of traditional medical problems, *morally* and psychologically unsuited for childbearing and certainly for motherhood because they were unwilling to serve as pregnancy vessels. Where there was an unhappily pregnant woman, there was a defective vessel. Many medical doctors agreed that an abortion *could* be performed on such a woman, but the procedure would not help as the problem was not the pregnancy. The problem was called a "psychiatric disorder" involving the woman's denial of her destiny and "amendable to treatment" as such.[72] But the tone of the diagnosis, like the tone so often used to judge women on one ground or another in these years, dripped with moral rectitude and condemnation. One psychiatrist identified the request for abortion "as proof [of the petitioner's] inability and failure to live through the destiny of being a woman."[73] Another, already cited, named motherhood as woman's "main role on earth." Arguing that abortion inevitably damaged women, he claimed that "despite protests to the contrary [and] . . . despite other sublimated types of activities," pregnancy and motherhood were "still their primary role."[74] Going a step further, a psychiatrically oriented New York obstetrician insisted that most women experiencing unwanted pregnancies—whether or not they sought abortions—were "immature, psychoneurotic, or under emotional stress," not the victims of contraceptive failures.[75]

Again, consistent with the generally misogynistic and moralistic judgments experts offered at this time, there was a broad consensus among many essayists in the medical journals and else-where that unwillingness to provide a safe environment for the fetus revealed a deeply rooted history of mental illness. One medical doctor found that "the patient, who all her life has disliked being female, found herself in conflict with men, and feared motherhood may be particularly abortion-prone." He cited the work of another specialist who identified two types of women likely to refuse pregnancy for psychological reasons: (1) "The basically immature woman who cannot accept the outstanding responsibility of mature femininity, namely becoming a mother"; and (2) "The independent, frustrated woman who has been conditioned to and yearns for the rewards of the male world and feels that maternity, the greatest reward of the female world, is much less satisfying—in fact, highly unsatisfying."[76]

A physician who responded to such a woman's expressed desire to violate her destiny was, according to many, in serious error. One highly experienced author-psychiatrist placed women who chose abortion on a sullied moral plane when he asserted that he had "never seen a patient who has not had guilt feelings about a previous . . . abortion."[77] Others felt that because the pregnancy itself was not the source of difficulty, an abortion did not solve a woman's problems but could create serious problems for her. For example,

[Abortion] coupled with ideas of guilt, self-deprecation, some recurrent preoccupation centering around the abortion and the general theme of "I let them kill my baby" might well disturb a poorly integrated personality even to psychotic proportions. Feelings of love, admiration and respect for the male partner . . . may well be distorted in the aborted woman to ideas of disgust, hate, and disrespect; "He gave me a baby then took it away." The unconscious motivation and even the flow of emotions during the readjustments to a normal sexual nonpregnancy cycle may result in deeply engrained feelings of hostility toward the husband. Abortions we may say can produce psychotic cicatrix.[78]

Indeed, husbands were often defined as the worst victims of abortion. A psychiatrist described what he had observed were its most common psychiatric sequelae:

Psychiatrists see patients who accuse themselves ... of being murderesses and then who go into very pronounced depressive reactions. We see patients who deliberately afterwards punish themselves or their husbands by forcing vasectomy upon them, or in other ways—sometimes unconsciously, but very frequently on conscious levels deliberately castrating their husbands—usually emotionally, but occasionally, even in actuality.[79]

Another physician argued that abortions were beside the point because "women who are physically vulnerable" will eventually and inevitably deteriorate. The pregnancy was beside the point, as well: "It seems to matter little with regard to future mental health whether the pregnancy is terminated or not. Those who are going to react adversely will do so irrespective of the procedure."[80] Conversely, a psychiatrist in Birmingham, England, cited the work of his colleagues as a warning to abortion-prone doctors: "Pregnancy appears to have a protective effect against the manifestations of mental disorder ... many psychotics and neurotics show quiescence of their symptoms during pregnancy itself."[81] In this case, hardly an "added burden," pregnancy becomes a variant of electric shock therapy.

In sum, the rise of hospital abortion commit-

tees and the redefinition of pregnancy in the postwar years reflected and intensified a broad cultural interest in reaffirming and reasserting male authority over women. The method of achieving male control described here was a typically insidious example of this effort, because the language created by the new insights about the nature of pregnancy required that pregnant women be disempowered. That is, the new "moral" essence of pregnancy was built upon a presupposition of judgment and control. When physicians defined pregnancy as a moral issue and counseled women to cooperate in sustaining themselves as moral and fit containers for fetuses, they demanded that pregnant girls and women cooperate in accepting the terms of their own oppression. Resistance had become a moral issue and, in effect, an immoral act.

Women Threaten Suicide

Well-known to unhappily pregnant women in the postwar era, however, was one method of resistance that sometimes cut through the language of morality: the threat of suicide. This condition alone raised the specter for medical doctors of a reintegrated mind, body, and pregnancy. A pregnant woman's threat of suicide suggested that the woman might destroy the reproductive container which gave definition to her very existence.[82] Women recognized early that they could get their medical doctors' attention by making such a threat,[83] but many physicians found it easier to believe that a woman was using her pregnancy rather than throwing away her destiny. Thus physicians proceeded very cautiously in this area. One wrote that "a mere threat of suicide or even an abortive attempt at suicide is not in itself regarded as a medical indication for therapeutic abortion; it may be nothing more than an effort to blackmail the surgeon into performing the operation."[84] An obstetrician at Columbia-Presbyterian Medical Center in New York explained his position:

I have been very much disturbed by the use of the indication of reactive depression with suicidal tendency. In cardiac disease you at least have an occasional death to validate the indication. I have not in my experience ever run across a suicide in pregnancy in a patient who was suffering from anxiety depression. . . . I think that one of the honest reasons for the reduction in the number of therapeutic abortions in the last ten years is that the obstetricians are concerned with the subterfuges that are being employed, otherwise they might be willing to be much more lenient.[85]

The abortion board at one hospital had been in operation for three years by 1960 and had adjudicated a number of requests from allegedly suicidal women. Lewis E. Savel and Irving K. Perlmutter gave examples demonstrating how the committee members were able to identify which of the petitioners should be denied permission. "One [woman] was a 40-year-old gravida v, para iv, who threatened suicide. The opinion was that such feelings were often verbalized by many women having an undesired pregnancy." This petitioner was denied an abortion. Two additional petitioners were similarly inclined but in both cases, "the psychiatric situation was judged too superficial to warrant intervention."[86] Another exemplary case described "a girl of twenty [who] was referred to a gynecologist at a large teaching hospital. Staff psychiatrists saw her to pass final approval [for the abortion]. When they rejected her . . . it was their opinion that while 'allowing the pregnancy to continue will undoubtedly cause further deterioration in her schizophrenic process, we do not think she will kill herself.'"[87]

The survey reported in the *Stanford Law Review* provides an excellent example of a suicidal pregnant woman who physicians were willing to believe deserved an abortion. An unprecedented 80 percent of reporting hospitals agreed to sanction abortion in this hypothetical case.

Mrs. C. is 32 years old and is the mother of children, aged 7, 4, and 3. Following the birth of her last child, she had what was diagnosed as a postpartum depression in which she became completely withdrawn. She was hospitalized in a state hospital for 6 months during which time she had electroshock therapy with some improvement. She has remained under psychiatric care since then but she still becomes depressed very easily and talks freely about committing suicide, saying that her family will be better off without the burden of her care.

Four weeks ago it was diagnosed that she was approximately 4 weeks' pregnant. The news of this precipitated a severe emotional crisis. This has been manifested by vomiting, spells of uncontrollable crying lasting for hours at a time, at which time the patient locked herself in her room. She threatened suicide several times in the last four weeks, saying that she could never be a "good mother" and that she was a "useless member of society."

Last night Mrs. C. was found unconscious on the floor of her living room. There was an empty bottle, which should have contained approximately eighteen sleeping pills, in her bedroom. She was taken to the hospital and has apparently responded to vigorous therapy for her barbiturate overdose.

Mrs. C.'s case evoked near consensus because this woman demonstrated her commitment to destroy the reproductive container she had become. Only in the case of such a demonstration could the moral dimension be eclipsed and the condition of pregnancy assume its previous status as an "added burden" or a destructive agent.

Doctors Threaten Sterilization

The other way that physicians frequently revealed their commitment to the new construction of women's bodies as reproductive containers was in their association of therapeutic abortion with simultaneous sterilization. As one chronicler of this era put it:

Patients actually had little or no contact with the operating physicians and often learned, only well after the fact, that the abortion had included sterilization. Because abortion patients were viewed as "psychotic," "hysterical," "depressed," "neurotic," or "guilt-laden," the symptoms associated with what psychiatrists ... term the "post-abortion hangover," the patient was considered to be in an unfit mental state to evaluate her own treatment. Early supporters of the psychiatric route believed that the abortive woman not only lost her baby, but rejected her own womanhood as well. The belief in woman-as-childbearer, a paramount function, undergirded the entire therapeutic structures.[88]

The prevalence of sterilization was widely featured in the obstetrical and psychiatric literature of the day, specifically in cases involving what one prominent expert called the "tainted individual."[89] One group of obstetricians found that "some women desiring an abortion were required to have a simultaneous sterilization operation as a condition of approval of the abortion in from one-third to two-thirds of [those]

teaching hospitals [studied] in different regions of the country. In all, 53.6 percent of teaching hospitals made this a requirement for some of their patients."[90] Another physician reported his finding of a 40 percent concomitant sterilization rate in all U.S. hospitals in the 1940s and 1950s.[91] A Chicago study of 209 aborted patients showed that medical doctors at the Lying-In Hospital in that city determined, "In the majority of cases when therapeutic abortion is indicated, the patient's medical condition warrants the prevention of future gestations"; 69.4 percent of these women were sterilized.[92]

Some physicians justified simultaneous sterilization on the grounds that any woman ill enough to warrant abortion should never again be pregnant.[93] Others shared this position but shifted the emphasis onto the medical doctor's dilemma: "A serious effort is made to control [by sterilization] the need for dealing with the same problem in the same patient twice."[94] A California psychiatrist described what he felt was a strong trend among medical doctors, "penalizing" by sterilization the patient who "needs" a therapeutic abortion. He explained the practice this way: "Often, the surgeon's stipulation for sterilization may reflect his reluctance to perform the abortion, his misunderstanding of its necessity, and his resentment of the psychiatric indications."[95] Another commentator felt that some physicians in this era resented sexual women more than they resented psychiatry: "The abortion committee [at one hospital] evaluated all patients in terms of recommendations for sterilization. Medical grounds for this 'final solution' to 'promiscuous' abortions were forcefully debated by individual members and typically included the physician's evaluation of the woman's condition and moral character."[96] The widespread use of sterilization, whatever the expressed justification, seems to suggest that many physicians in the postwar era were willing to use the sterilization option to cap the defective reproductive container. In one small midwestern hospital four requests for therapeutic abortion were presented to the committee one year. None was approved. Among them was this case: "Approval of both therapeutic abortion and sterilization was requested for a 36-year-old gravida iv, para ii for arrested pulmonary tuberculosis, thyrotoxicosis, and emotional instability. Despite the consultants' recommendations the committee did not approve the abortion, but did approve postpartum sterilization."[97]

One physician, unhappy about the coupling of sterilization and therapeutic abortion in U.S. hospitals, observed that this practice actually drove women to illegal abortionists to escape the likelihood that a legal abortion would entail the permanent loss of their fertility. He added, "I would like to point that out, because the package [therapeutic abortion–sterilization] is so frequent, I therefore consider them fortunate to have been illegally rather than therapeutically aborted, and thus spared sterilization."[98] This aspect of the discussion foreshadowed, of course, the legal institutionalization in our time of the link between abortion and sterilization, via the Hyde Amendment.

The Limits of Abortion Committees

The literature reviewed in this essay makes it clear that some influential medical doctors in the postwar era derived professional strength and ideological coherence from abortion committees and from a new disembodied definition of pregnancy. But by the middle of the 1960s it was also clear that the same factors which had pushed physicians into a defensive posture in the early postwar years continued to exert considerable pressures on the profession. These and additional factors combined to facilitate the

eclipse of medical authority over the abortion decision much sooner than many practitioners had predicted.

Over time the committees themselves could not sustain the image of professional unity and scientific purpose, even if an individual hospital could issue abortion decisions with one voice. Harold Rosen, a prominent medical doctor interested in abortion reform, noted widespread inconsistencies between hospital abortion committees in the mid-1960s which hurt the credibility of the profession.

> Not infrequently, for instance, the abortion board of one hospital, but not another, may refuse to accept a recommendation for interruption; on nine separate occasions during the past seven years, patients who have been seen in consultation in one hospital have afterwards been therapeutically aborted at adjacent hospitals with, at times, almost the same visiting staff.

At the heart of this apparent capriciousness was a continuing inability among physicians to agree on indications, even medical indications.

> If physicians do not wish to force a specific woman to carry a specific pregnancy to term, and if that woman is actually suffering from some severe physical disease then, but only then, the pathological process, provided it falls within certain categories, is in certain hospitals and by certain physicians and hospital boards considered sufficient indication for interruption. In others, it is not.[99]

In addition, Rosen noted that the medical profession continued to be rent by the abortion issue as the direct result of both medical progress in managing pregnancy and "undeclared nonmedical factors," specifically the pressure of the legal threat against physicians and restrictive legislative statutes. These factors persisted in conditioning the abortion decisions of medical doctors despite attempts to neutralize them and despite the fact that they were rarely, if ever, in fact prosecuted for performing therapeutic abortions in hospitals.

Other factors which exerted increasing pressure in the abortion arena include first, of course, women's growing insistence on breaking the link between law and medicine, so that women themselves could take the power to decide who was a mother and to decide when a woman was a mother. After the rubella epidemic and the thalidomide episode of the early 1960s, women also began to insist on a legal publicly sanctioned right to decide who was a child. The sensationally and intrusively reported plight of Sherri Finkbine in 1962 raised, above all, the specter of the pregnant woman's right to reject a fetus deeply damaged by thalidomide.[100]

Additional pressures which struck at medical authority came from the flowering of the quality of life (or "lifestyle") ethic among the middle-class in the United States which undermined the acceptability of the simple life/death dichotomy that the law mandated must govern abortion decisions. Also, in the 1960s as social criticism seeped back into mainstream public discourse, some physicians began to accept and use a definition of the purpose of medicine—in this case, of indications for abortion which placed unhappily pregnant women in desperate social and economic contexts.[101] Physicians were also involved in and influenced by the reemergence in this era of a holistic approach toward diagnostics and treatment which reflected and promoted the other two emergent trends of the 1960s. One contemporary commentator applied these trends to the abortion issue in this way:

> Distinctions between physical and mental health are meaningless in terms of modern medical thinking. Health cannot be

divorced from socio-economic factors which influence people's lives since health is a product of these conditions. In applying criteria for abortion based on maternal health, the question should be the extent to which the pregnancy threatens the general well-being of the patient.[102]

Of equal or greater importance to all these pressures undermining medical authority in the abortion arena by the mid-1960s was widespread concern and fear among whites in the United States about the "population explosion," rising welfare costs, the civil rights movement, and the "sexual revolution." Critics of these social, political, and cultural phenomena tended to target women's bodies and their reproductive capacity as a source of danger to the fabric of U.S. society. Demedicalizing and decriminalizing the abortion decision became one way to diminish the damage women's bodies could do.[103]

Conclusion

This essay leaves unexplored many issues that would shed additional light on the concerns and strategies of medical doctors sitting on hospital abortion committees in the postwar era. These include physicians' attitudes toward abortion and women of various races, ethnicities, and classes. Much research is needed in this area. The essay does not explore medical doctors' attitudes toward and relationships with illegal abortionists, a subject well worth pursuing.[104] Also left unexplored are the sources and complex nature of physicians' changing attitudes toward abortion in the 1960s and 1970s. Pregnant women themselves have not been given voice in this essay.

But the subjects of this study, a highly visible segment of the medical community, have been given voice here in order to allow us to consider what was at issue for many physicians in the immediate pre–*Roe v. Wade* decades. What is most striking in the literature reviewed for this essay is that, with the exception of the few articles prepared by Catholic medical doctors, the physicians who wrote on the abortion issue were not primarily concerned with the issue of when life begins.[105] They were, however, very concerned with what they took to be their role in the postwar cultural mandate to protect and preserve the links between sexuality, femininity, marriage, and maternity. They were also deeply concerned about their professional dignity and about devising strategies to protect and preserve the power, the prerogatives, and the legal standing of the medical profession.

An important strategy of many physicians in this era was to draw on the vulnerability of pregnant women to construct a definition of pregnancy that effaced the personhood of the individual pregnant woman. This definition created a safe place for the fetus and also for the physician forced by law to adjudicate the extremely personal decisions of women, many of whom were resisting effacement. The subordination of the pregnant woman to the fetus revitalized medical participation in the abortion decision because the medical doctor was now required to make sure that the woman stayed moral, that is, served her fetus correctly. These postwar ideas demonstrate the relationship between scientific advances and ideological positions regarding women, pregnant women, pregnancy, and fetuses. Physicians have often presented these positions as scientific, providing "evidence" for antichoice proponents. It seems clear today that if abortion decisions were again assigned by law to medical doctors, unhappily pregnant women seeking abortions would again confront a defensive profession, masking as scientists for this purpose but constrained to practice ideological medicine. Perhaps the most difficult task for prochoice advocates today, and the most crucial,

is to insist with even more vitality that they occupy the moral ground. Pregnancy is not, by definition, the moral duty of girls and women; rather, granting this population reproductive freedom *is* the moral duty of society.

Notes

1. All these cases are drawn from Herbert L. Packer and Ralph J. Gampell, "Therapeutic Abortion: A Problem in Law and Medicine," *Stanford Law Review* 11 (May 1959): 417–55.

2. Ibid., 418.

3. In fact, for many postwar commentators a woman fulfilled her destiny as mother not via pregnancy but in the proper postnatal care of her child. This is especially evident in the prevailing treatment of white unwed mothers in these years, particularly in the definition of this population as *not-mothers* and in the coercively applied prescription that these girls and women should relinquish their babies for adoption on an every-case basis. See Rickie Solinger, "Race and 'Value': Black and White Illegitimate Babies in the U.S.A., 1945–1965," *Gender and History* 4 (Autumn 1992): 343–63.

4. See, for example, Quinten Scherman, "Therapeutic Abortion," *Obstetrics and Gynecology* 11 (March 1958): 323–35. This article and others cited in this essay contradict Rosalind Petchesky's observation that a 1970 American Medical Association statement calling for the liberalization of abortion laws demonstrated that "after a century of strictly moral opposition to abortion on practically any ground, the AMA was now conceding that abortion for 'medically necessary' reasons was legitimate." See Rosalind Pollack Petchesky, *Abortion and Woman's Choice: The State, Sexuality, and Reproductive Freedom*, rev. ed. (Boston: Northeastern University Press, 1990), 124.

5. J. G. Moore and J. H. Randall, "Trends in Therapeutic Abortion: A Review of 137 Cases," *American Journal of Obstetrics and Gynecology* 63 (January 1952): 28–40.

6. Scherman, "Therapeutic Abortion," 323.

7. Mary S. Calderone, ed., *Abortion in the*

United States: A Conference Sponsored by the Planned Parenthood Federation of America, Inc. at Arden and the New York Academy of Medicine (New York: Harper & Bros., 1958), 84. Kristin Luker argues that the ratio of hospital therapeutic abortions to deliveries between 1926 and 1960 did not seem to have changed: "Abortions became neither easier nor harder to obtain over time." See Kristin Luker, *Abortion and the Politics of Motherhood* (Berkeley: University of California Press, 1984), 46. Most of the articles cited in the present essay suggest that between about 1945 and 1965, rates of hospital abortions declined in relation to births.

8. Edwin M. Gold et al., "Therapeutic Abortions in New York City: A Twenty-Year Review," *American Journal of Public Health* 55 (July 1965): 964–72.

9. Lawrence Lader, *Abortion* (Indianapolis: Bobbs-Merrill, 1966), 26–27.

10. See Calderone, ed., *Abortion in the United States*, 86–88.

11. Scherman, "Therapeutic Abortion," 325.

12. Luker, *Abortion*, 66.

13. See Calderone, ed., *Abortion in the United States*, ch. 9.

14. Moore and Randall, "Trends," 28.

15. Keith P. Russell, "Changing Indications for Therapeutic Abortion: Twenty Years' Experience at Los Angeles Community Hospital," *Journal of the American Medical Association* 151 (Jan. 10, 1953): 108.

16. Myrna Loth and H. Close Hesseltine, "Therapeutic Abortion at the Chicago Lying-In Hospital," *American Journal of Obstetrics and Gynecology* 72 (August 1956): 304–11.

17. Harry A. Pearse and Harold A. Ott, "Hospital Control of Sterilization and Therapeutic Abortion," *American Journal of Obstetrics and Gynecology* 60 (August 1950): 285.

18. Scherman, "Therapeutic Abortion," 323.

19. Roy J. Heffernan and William Lynch, "What Is the Status of Therapeutic Abortion in Modern Obstetrics?" *American Journal of Obstetrics and Gynecology* 66 (August 1953): 335.

20. Harold Rosen, "The Psychiatric Implications of Abortion: A Case Study in Hypocrisy," in *Abortion and the Law*, ed. David T. Smith (Cleveland: Press of Case Western Reserve University, 1967), 105.

21. "Comment," by W. O. Johnson, in Pearse and Ott, "Hospital Control," 299.

22. Heffernan and Lynch "What Is the Status?"; Moore and Randall, "Trends," 39; H. A. Stephenson, "Therapeutic Abortion," *Obstetrics and Gynecology* 4 (1958): 578. Also see the *New York Times*, June 22, 1965, which compares the rates of psychiatric indications for abortion in a Buffalo hospital in 1943 (13 percent) and 1963 (87.5 percent).

23. See, for example, Moore and Randall, "Trends," 34.

24. See, for example, D. I. Arbuse and J. Schechtman, *American Practitioner* 1 (October 1950): 1069.

25. "Comment," by G. K. Folger, in Pearse and Ott, "Hospital Control," 299–300.

26. Scherman, "Therapeutic Abortion," 330–31.

27. Edwin M. Schur, *Crimes Without Victims: Deviant Behavior and Public Policy* (Englewood Cliffs, N.J.: Prentice-Hall, 1965), 16.

28. Sidney Bolter, Response to Robert L. Marcus's Editorial, *American Journal of Psychiatry* 119 (February 1963): 798.

29. Alex Barno, "Criminal Abortion Deaths, Illegitimate Pregnancy Deaths, and Suicide in Pregnancy: Minnesota 1950–1965," *American Journal of Obstetrics and Gynecology* 98 (June 1967): 361.

30. Allan J. Rosenberg and Emmanuel Silver, "Suicide, Psychiatrists, and Therapeutic Abortions," *California Medicine* 102 (June 1965): 410; also see R. B. McGraw, "Legal Aspects of Termination of Pregnancy on Psychiatric Grounds," *New York State Journal of Medicine* 56 (May 15, 1956): 1605. This article carries a long case history of a woman referred for abortion on psychiatric grounds. The final vote of the abortion committee was five to five.

31. Sidney Bolter, "The Psychiatrist's Role in Therapeutic Abortion: The Unwitting Accomplice," *American Journal of Psychiatry* 119 (October 1962): 312.

32. Rosen, "Psychiatric Indications of Abortion," 90.

33. Ibid., 77, 80.

34. Lewis E. Savel, "Adjudication of Therapeutic Abortion and Sterilization," in *Therapeutic Abortion*

and Sterilization, ed. Edmund W. Overstreet (New York: Harper & Row, 1964). Alan Guttmacher gives his own description of the Mt. Sinai committee in several places, including Calderone, ed., *Abortion in the United States,* 92–93, 139; and Alan F. Guttmacher, "Therapeutic Abortion: The Doctor's Dilemma," *Journal of Mt. Sinai Hospital* 21 (1954): 111.

35. James M. Ingram et al., "Interruption of Pregnancy for Psychiatric Indications—A Suggested Method of Control," *Obstetrics and Gynecology* 29 (February 1967): 255.

36. "Laws Regulating Abortion," *Journal of the Indiana Medical Association* 40 (July 1947): 16.

37. Pearse and Ott, "Hospital Control," 290, 299.

38. Packer and Gampell, "Therapeutic Abortion," 429.

39. See, for example, Ingram et al., "Interruption of Pregnancy," 255.

40. See, especially, *The Abortionist,* by Dr. X as told to Lucy Freeman (Garden City, N.Y.: Doubleday, 1962). Elsewhere, "Dr. X," a convicted abortionist, testified, "In spite of the fact that there were 353 doctors whom I had served for many years, when the time came for those men to come forward and share the responsibility with me, there was not one in the whole group that offered to do so. . . . So actually it was the profession that convicted me, in spite of the fact that they were the very ones who had used my services." See Calderone, ed., *Abortion in the United States,* 62. Also see Zad Leavy and Jerome M. Kummer, "Criminal Abortion: Human Hardship and Unyielding Laws," *Southern California Law Review* 35 (Winter 1965): 125. Other firsthand statements about the relationship between illegal practitioners and referring physicians can be found in Dr. Ruth Barnett's *They Weep on My Doorstep* (Beaverton, Ore.: HALO Publishers, 1969).

41. Carole Joffe, "'Portraits of Three Physicians of Conscience': Abortion Before Legalization in the United States," *Journal of the History of Sexuality* 2 (July 1991): 46–67.

42. See, for example, Lader, *Abortion,* 26.

43. Pearse and Ott, "Hospital Control," 299.

44. Lewis E. Savel and Irving K. Perlmutter, "Therapeutic Abortion and Sterilization Committees: A Three-Year Experience," *American Journal of Ob-*

stetrics and Gynecology 80 (December 1960): 1198, 1194.

45. W. Joseph May, "Therapeutic Abortion in North Carolina," *North Carolina Medical Journal* 23 (December 1962): 548.

46. See, for example, Ingram et al., "Interruption of Pregnancy," 252; also Packer and Gampell, "Therapeutic Abortion," who report that 75 percent of participating hospitals felt that their abortion decisions regularly violated the law, 430. Also of interest is Jack Star, "One Million Abortions a Year: The Growing Tragedy of Abortion," *Look,* Oct. 19, 1965. Star quotes Robert E. Hall of the Department of Obstetrics and Gynecology at Columbia University's College of Physicians and Surgeons to the effect that hospital abortion boards could not do their job and stay within the law. Hall went on to give examples demonstrating that this was so.

47. Pearse and Ott, "Hospital Control," 297.

48. In England physicians called their abortion committees "tribunals." See, for example, J. V. O'Sullivan and L. Fairfield, "The Case Against Termination on Psychiatric Grounds," *Mental Health* 20 (August 1961): 97.

49. Arnold S. Levine, "The Problem of Psychiatric Disturbances in Relation to Therapeutic Abortion," *Journal of the Albert Einstein Medical Center* 6 (1958): 76.

50. Nicholas J. Eastman, "Obstetric Forward," in *Therapeutic Abortion,* ed. Harold Rosen (New York: Julian Press, 1954), xx.

51. Bolter, "Psychiatrist's Role," 315.

52. Packer and Gampell, "Therapeutic Abortion," 430.

53. Levine, "Problem," 76.

54. Moore and Randall, "Trends," 36.

55. Scherman, "Therapeutic Abortion," 323.

56. See, for example, Richard Kluger, *Simple Justice* (New York: Random House, 1975); Alfred C. Kinsey, Wardell B. Pomeroy, and Clyde W. Martin, *Sexual Behavior in the Human Male* (Philadelphia: W. B. Saunders, 1948); Alfred C. Kinsey, Wardell B. Pomeroy, and Clyde E. Martin, *Sexual Behavior in the Human Female* (Philadelphia: W. B. Saunders, 1953).

57. John Johnson, "Termination of Pregnancy

on Psychiatric Grounds," *Medical Gynecology and Sociology* 2 (1966): 2.

58. Lader, *Abortion,* 27.

59. Charles C. Dahlberg, "Abortion," in *Sexual Behavior and the Law,* ed. Ralph Slovenko (Springfield, Ill.: Charles Thomas Publishers, 1965), 384.

60. Calderone, ed., *Abortion in the United States,* 93.

61. See, for example, Ingram et al., "Interruption of Pregnancy," and Pearse and Ott, "Hospital Control."

62. Lader, *Abortion,* 28.

63. Bolter, "Psychiatrist's Role," 315.

64. "Discussion," in *Abortion Obtained and Denied: Research Perspectives,* ed. Sidney H. Newman, Mildred B. Beck, and Sarah Lewit (New York: Population Council, 1971), 77.

65. Gold et al., "Therapeutic Abortions," 969.

66. J. A. Harrington, "Psychiatric Indications for the Termination of Pregnancy," *Practitioner* 185 (November 1960): 654–58.

67. Moore and Randall, "Trends," 34.

68. Heffernan and Lynch, "What Is the Status?" 335.

69. It is difficult to tell, in this regard, whether the physician was the mother's assistant or the other way around.

70. Calderone, ed., *Abortion in the United States,* 118.

71. "The Drama of Life Before Birth," *Life,* April 30, 1965, 54.

72. Harrington, "Psychiatric Indications," 658.

73. Theodore Linz, "Reflections of the Psychiatrist," in *Therapeutic Abortion.*

74. Bolter, "Psychiatrist's Role," 314–15.

75. Hans Lehfeldt, "Willfull Exposure to Unwanted Pregnancy," *American Journal of Obstetrics and Gynecology* 78 (September 1959): 665; also see the statement of Iago Goldston defining the desire for abortion in a "so-called adult woman" as an indication of "a sick person and a sick situation . . . which could be relieved, or ameliorated [by the abortion] like cutting off a gangrenous foot" (Calderone, ed., *Abortion in the United States,* 118–19).

76. H. Flanders Dunbar, *Psychiatry in the Medical Specialties* (New York: McGraw-Hill, 1959), 279, 281.

77. Bolter, "Psychiatrist's Role," 314.

78. F. G. Ebaugh and K. D. Heuser, "Psychiatric Aspects of Therapeutic Abortion," *Postgraduate Medicine* 2 (1947): 325.

79. Calderone, ed., *Abortion in the United States,* 129.

80. Johnson, "Termination of Pregnancy," 3.

81. Harrington, "Psychiatric Indications," 655.

82. See, for example, Jacob H. Friedman, "The Vagarity of Psychiatric Indications for Therapeutic Abortion," *American Journal of Psychotherapy* 16 (April 1962): 251.

83. See, for example, Nanette Davis, *From Crime to Choice: The Transformation of Abortion in America* (Westport, Conn.: Greenwood Press, 1985), 72.

84. S. Leon Israel, "Editorials: Therapeutic Abortion," *Postgraduate Medicine* 33 (June 1963): 619–20.

85. Calderone, ed., *Abortion in the United States,* 108.

86. Savel and Perlmutter, "Therapeutic Abortion," 1194.

87. Dahlberg, "Abortion," 384.

88. Davis, *From Crime to Choice,* 73.

89. Frederick S. Taussig, *Abortion, Spontaneous and Induced: Medical and Social Aspects* (St. Louis, Mo.: C. V. Mosby Co., 1936), 79.

90. Johan W. Eliot et al., "The Obstetrician's View," in *Abortion in a Changing World,* ed. Robert E. Hall (New York: Columbia University Press, 1970), 1:93.

91. Kenneth R. Niswander, "Medical Abortion Practices in the United States," in *Abortion and the Law,* 57.

92. Loth and Hesseltine, "Therapeutic Abortion," 306; see also Pearse and Ott, "Hospital Control."

93. See, for example, Moore and Randall, "Trends," 37.

94. Savel, "Adjudication," 18.

95. Alexander Simon, "Psychiatric Indications for Therapeutic Abortion and Sterilization," in *Therapeutic Abortion and Sterilization,* 78. Also see Dahlberg, "Abortion," 383.

96. Davis, *From Crime to Choice,* 77–78.

97. Pearse and Ott, "Hospital Control," 296.

98. Calderone, ed., *Abortion in the United States,* 131.

99. Rosen, "Psychiatric Implications of Abortion," 77, 80.

100. See the *New York Times,* July 25, 1962, to Aug. 27, 1962.

101. See, for example, "Discussion by Charles S. Stevenson," in Barno, "Criminal Abortion Deaths," 364.

102. Kenneth J. Ryan, "Humane Abortion Laws and the Health Needs of Society," in *Abortion and the Law,* 68.

103. See Rickie Solinger, *Wake Up Little Susie: Single Pregnancy and Race Before Roe v. Wade* (New York: Routledge, 1992).

104. See Davis, *From Crime to Choice,* 89–90. The subject of trials in the postwar era involving women who were incarcerated for performing illegal abortions is explored in Rickie Solinger, *The Abortionist: A Woman Against the Law* (New York: Free Press, 1994).

105. See *The Abortion Problem* (Baltimore, Md.: Williams & Wilkins, 1944) for a comprehensive collection of medical doctors' concerns in this era, few of which focus on the issue of the inception of life.

A GUIDE FOR

FURTHER READING

A Guide for Further Reading

Compiled by Tomomi Kinukawa

General

Apple, Rima D., ed. *Women, Health, and Medicine in America: A Historical Handbook.* New York: Garland, 1990.

Bair, Barbara, and Susan E. Cayleff, eds. *Wings of Gauze: Women of Color and the Experience of Health and Illness.* Detroit: Wayne State University Press, 1993.

Cott, Nancy F. *The Bonds of Womanhood: "Woman's Sphere" in New England, 1780–1835.* 2d ed. New Haven, Conn.: Yale University Press, 1997.

Degler, Carl N. *At Odds: Women and the Family in America from the Revolution to the Present.* New York: Oxford University Press, 1980.

Ehrenreich, Barbara, and Deirdre English. *Complaints and Disorders: The Sexual Politics of Sickness.* Old Westbury, N. Y.: Feminist Press, 1973.

Ehrenreich, Barbara, and Deirdre English. *For Her Own Good: 150 Years of the Experts' Advice to Women.* New York: Anchor Press/Doubleday, 1978.

Epstein, Julia. *Altered Conditions: Disease, Medicine, and Storytelling.* New York: Routledge, 1995.

Fee, Elizabeth, ed. *Women and Health: The Politics of Sex in Medicine.* Farmingdale, N. Y.: Baywood, 1982.

Fee, Elizabeth, and Nancy Krieger, eds. *Women's Health, Politics, and Power: Essays on Sex/Gender, Medicine, and Public Health.* Amityville, N. Y.: Baywood, 1994.

Fellman, Anita C., and Michael Fellman. *Making Sense of Self: Medical Advice Literature in Late Nineteenth-Century America.* Philadelphia: University of Pennsylvania Press, 1981.

Kern, Stephen. *Anatomy and Destiny: A Cultural History of the Human Body.* Indianapolis: Bobbs-Merrill, 1975.

Leavitt, Judith Walzer, and Ronald L. Numbers, eds. *Sickness and Health in America: Readings in the History of Medicine and Public Health.* 3d ed. Madison: University of Wisconsin Press, 1997.

Lewin, Ellen, and Virginia Oleson, eds. *Women, Health, and Healing: Toward a New Perspective.* New York: Tavistock, 1985.

Marieskind, Helen I. *Women in the Health System: Patients, Providers, and Programs.* St. Louis, Mo.: Mosby, 1980.

Newman, Louise Michele. *Men's Ideas/Women's Realities: Popular Science, 1870–1915.* New York: Pergamon, 1985.

Price Herndl, Diane. *Invalid Women: Figuring Feminine Illness in American Fiction and Culture, 1840–1940.* Chapel Hill: University of North Carolina Press, 1993.

Risse, Guenter, Ronald L. Numbers, and Judith W. Leavitt, eds. *Medicine Without Doctors: Home Health Care in American History.* New York: Science History Publications, 1977.

Rosser, Sue V. *Women's Health—Missing from U. S. Medicine.* Bloomington: Indiana University Press, 1994.

Rothman, Sheila M. *Woman's Proper Place: A History of Changing Ideals and Practices, 1870 to the Present.* New York: Basic Books, 1978.

Shorter, Edward. *Women's Bodies: A Social History of Women's Encounter with Health, Ill Health, and Medicine.* 1982. Reprint. New Brunswick, N.J.: Transaction, 1991.

Smith-Rosenberg, Carroll. *Disorderly Conduct: Visions of Gender in Victorian America.* New York: Knopf, 1985.

Verbrugge, Martha H. *Able-Bodied Womanhood: Personal Health and Social Change in Nineteenth-Century Boston.* New York: Oxford University Press, 1988.

White, Evelyn C., ed. *The Black Women's Health Book: Speaking for Ourselves.* Seattle, Wash.: Seal Press, 1990.

Worcester, Nancy, and Mariamne H. Whatley, eds. *Women's Health: Readings on Social, Economic, and Political Issues.* 2d ed. Dubuque, Iowa: Kendall/Hunt, 1994.

Body Image and Physical Fitness

Banner, Lois W. *American Beauty.* New York: Knopf, 1983.

Bordo, Susan. *Unbearable Weight: Feminism, Western Culture, and the Body.* Berkeley: University of California Press, 1993.

Brumberg, Joan Jacobs. *Fasting Girls: The Emergence of Anorexia Nervosa as a Modern Disease.* Cambridge, Mass.: Harvard University Press, 1988.

Brumberg, Joan Jacobs. *The Body Project: An Intimate History of American Girls.* New York: Random House, 1997.

Green, Harvey. *Fit for America: Health, Fitness, Sport, and American Society.* New York: Pantheon, 1986.

Lupton, Mary J., and Emily Toth. *The Curse: A Cultural History of Menstruation.* Rev. ed. Urbana: University of Illinois Press, 1988.

Mangan, J. A., and Roberta J. Park. *From "Fair Sex" to Feminism: Sport and the Socialization of Women in the Industrial and Postindustrial Eras.* Totowa, N.J.: Cass, 1987.

Martin, Emily. *The Woman in the Body: A Cultural Analysis of Reproduction.* Baltimore, Md.: Johns Hopkins University Press, 1987.

Schwartz, Hillel. *Never Satisfied: A Cultural History of Diets, Fantasies, and Fat.* New York: Free Press, 1986.

Seid, Roberta Pollack. *Never Too Thin: Why Women Are at War with Their Bodies.* New York: Prentice-Hall, 1989.

Stanley, Gregory K. *The Rise and Fall of the Sportswoman: Women's Health, Fitness, and Athletics, 1860–1940.* New York: P. Lang, 1996.

Vertinsky, Patricia A. *The Eternally Wounded Woman: Women, Doctors, and Exercise in the Late Nineteenth Century.* New York: St. Martin's, 1990.

Sexuality

Barker-Benfield, G. J. *The Horrors of the Half-Known Life: Male Attitudes Toward Women and Sexuality in Nineteenth-Century America.* New York: Harper, 1976.

Bullough, Vern L. *Sexual Variance in Society and History.* Chicago: University of Chicago Press, 1981.

Bullough, Vern L. *Science in the Bedroom: A History of Sex Research.* New York: Basic Books, 1994.

Bullough, Vern L., and Bonnie Bullough. *Sin, Sickness, and Sanity: A History of Sexual Attitudes.* New York: New American Library, 1977.

Bullough, Vern L., and Bonnie Bullough, eds. *Human Sexuality: An Encyclopedia.* New York: Garland, 1994.

Connelly, Mark Thomas. *The Response to Prostitution in the Progressive Era.* Chapel Hill: University of North Carolina Press, 1980.

Cott, Nancy F., ed. *Prostitution.* Munich: Saur, 1993.

Cott, Nancy F., ed. *Sexuality and Sexual Behavior.* Munich: Saur, 1993.

D'Emilio, John, and Estelle B. Freedman. *Intimate Matters: A History of Sexuality in America.* New York: Harper and Row, 1988.

Ditzion, Sidney. *Marriage, Morals, and Sex in America: A History of Ideas.* New York: Octagon, 1969.

Duberman, Martin Bauml, Martha Vicinus, and George Chauncey, Jr., eds. *Hidden from History: Reclaiming the Gay and Lesbian Past.* New York: New America Library, 1989.

Faderman, Lillian. *Surpassing the Love of Men: Romantic Friendship and Love Between Women from the Renaissance to the Present.* New York: William Morrow, 1981.

Faderman, Lillian. *Odd Girls and Twilight Lovers: A History of Lesbian Life in Twentieth-Century*

America. New York: Columbia University Press, 1991.

Foster, Lawrence. *Religion and Sexuality: Three American Communal Experiments of the Nineteenth Century.* New York: Oxford University Press, 1981.

Fout, John C., and Maura Shaw Tantillo, eds. *American Sexual Politics: Sex, Gender, and Race Since the Civil War.* Chicago: University of Chicago Press, 1993.

Haller, John, and Robin Haller. *The Physician and Sexuality in Victorian America.* Urbana: University of Illinois Press, 1974.

Kern, Louis J. *An Ordered Love: Sex Roles and Sexuality in Victorian Utopias—The Shakers, the Mormons, and the Oneida Community.* Chapel Hill: University of North Carolina Press, 1981.

Leach, William. *True Love and Perfect Union: The Feminist Reform of Sex and Society.* New York: Basic Books, 1980.

Muncy, Raymond Lee. *Sex and Marriage in Utopian Communities: Nineteenth-Century America.* Bloomington: Indiana University Press, 1973.

Nathanson, Constance. *Dangerous Passage: The Social Control of Sexuality in Women's Adolescence.* Philadelphia: Temple University Press, 1991.

Pivar, David J. *Purity Crusade: Sexual Morality and Social Control, 1868–1900.* Westport, Conn.: Greenwood, 1973.

Rosen, Ruth. *The Lost Sisterhood: Prostitution in America, 1900–1918.* Baltimore, Md.: Johns Hopkins University Press, 1982.

Sears, Hal D. *The Sex Radicals: Free Love in High Victorian America.* Lawrence: Regents Press of Kansas, 1977.

Walters, Ronald G., ed. *Primers for Prudery: Sexual Advice to Victorian America.* Englewood Cliffs, N.J.: Prentice-Hall, 1974.

Fertility, Abortion, and Birth Control

Asbell, Bernard. *The Pill: A Biography of the Drug That Changed the World.* New York: Random House, 1995.

Beisel, Nicola. *Imperial Innocents: Anthony Comstock and Family Reproduction in Victorian America.* Princeton, N.J.: Princeton University Press, 1997.

Brodie, Janet Farrell. *Contraception and Abortion in Nineteenth-Century America.* Ithaca, N.Y.: Cornell University Press, 1994.

Chen, Constance. *"The Sex Side of Life": Mary Ware Dennett's Pioneering Battle for Birth Control and Sex Education.* New York: New Press, 1996.

Chesler, Ellen. *Woman of Valor: Margaret Sanger and the Birth Control Movement in America.* New York: Simon and Schuster, 1992.

Davis, Nanette J. *From Crime to Choice: The Transformation of Abortion in America.* Westport, Conn.: Greenwood, 1985.

Duden, Barbara. *Disembodying Women: Perspectives on Pregnancy and the Unborn.* Cambridge, Mass.: Harvard University Press, 1993.

Garrow, David J. *Liberty and Sexuality: The Right to Privacy and the Making of Roe v. Wade.* New York: Macmillan, 1994.

Ginsberg, Caren A. *Sex-Specific Mortality and the Economic Value of Children in Nineteenth-Century Massachusetts.* New York: Garland, 1989.

Gordon, Linda. *Woman's Body, Woman's Right: Birth Control in America.* Rev. ed. New York: Penguin, 1990.

Grant, Nicole J. *The Selling of Contraception: The Dalkon Shield Case, Sexuality, and Women's Autonomy.* Columbus: Ohio State University Press, 1992.

Hicks, Karen M. *Surviving the Dalkon Shield IUD: Women v. the Pharmaceutical Industry.* New York: Teachers College Press, 1994.

Himes, Norman E. *Medical History of Contraception.* 1936. Reprint. New York: Schocken, 1970.

Hoffer, Peter C., and N. E. H. Hull. *Murdering Mothers: Infanticide in England and New England, 1558–1803.* New York: New York University Press, 1981.

Joffe, Carole. *Doctors of Conscience: The Struggle to Provide Abortion Before and After Roe v. Wade.* Boston: Beacon, 1995.

Kaplan, Laura. *The Story of Jane: The Legendary Underground Feminist Abortion Service.* New York: Pantheon, 1995.

Keller, Allan. *Scandalous Lady: The Life and Times of Madame Restell, New York's Most Notorious Abortionist.* New York: Atheneum, 1981.

Kennedy, David. *Birth Control in America: The Career of Margaret Sanger.* New Haven, Conn.: Yale University Press, 1970.

Luker, Kristin. *Abortion and the Politics of Motherhood.* Berkeley: University of California Press, 1984.

Marsh, Margaret, and Wanda Ronner. *The Empty Cradle: Infertility in America from Colonial Times to the Present.* Baltimore, Md.: Johns Hopkins University Press, 1996.

May, Elaine Tyler. *Barren in the Promised Land: Childless Americans and the Pursuit of Happiness.* New York: Basic Books, 1995.

McCann, Carole R. *Birth Control Politics in the United States, 1916–1945.* Ithaca, N. Y.: Cornell University Press, 1994.

Mohr, James C. *Abortion in America: The Origins and Evolution of National Policy, 1800–1900.* New York: Oxford University Press, 1978.

Newman, Karen. *Fetal Positions: Individualism, Science, Vitality.* Palo Alto, Calif.: Stanford University Press, 1996.

Petchesky, Rosalind Pollack. *Abortion and Woman's Choice: The State, Sexuality, and Reproductive Freedom.* Rev. ed. Boston: Northeastern University Press, 1990.

Reagan, Leslie J. *When Abortion Was a Crime: Women, Medicine, and Law in the United States, 1867–1973.* Berkeley: University of California Press, 1997.

Reed, James. *The Birth Control Movement and American Society: From Private Vice to Public Virtue: With a New Preface on the Relationship Between Historical Scholarship and Feminist Issues.* Princeton, N.J.: Princeton University Press, 1983.

Reilly, Philip. *The Surgical Solution: A History of Involuntary Sterilization in the United States.* Baltimore, Md.: Johns Hopkins University Press, 1991.

Rubin, Eva R. *Abortion, Politics, and the Courts: Roe v. Wade and Its Aftermath.* Rev. ed. New York: Greenwood, 1987.

Solinger, Rickie. *The Abortionist: A Woman Against the Law.* New York: Free Press, 1994.

Tone, Andrea, ed. *Controlling Reproduction: An American History.* Wilmington, Del.: SR Books, 1997.

Van Horn, Susan Householder. *Women, Work, and Fertility, 1900–1986.* New York: New York University Press, 1988.

Ward, Martha C. *Poor Women, Powerful Men: America's Great Experiment in Family Planning.* Boulder, Colo.: Westview, 1986.

Wells, Robert V. *Revolution in Americans' Lives: A Demographic Perspective on the History of Americans, Their Families, and Their Society.* Westport, Conn: Greenwood, 1982.

Williams, Doone, and Greer Williams. *Every Child a Wanted Child: Clarence James Gamble, M.D. and His Work in the Birth Control Movement.* Boston: Francis A. Countway Library of Medicine, 1978.

Childbirth and Motherhood

Apple, Rima D. *Mothers and Medicine: A Social History of Infant Feeding, 1890–1950.* Madison: University of Wisconsin Press, 1987.

Apple, Rima D., and Janet Golden, eds. *Mothers and Motherhood: Readings in American History.* Columbia: Ohio State University Press, 1997.

Arnup, Katherine, Andrée Lévesque, Ruth Roach Pierson, with the assistance of Margaret Brennan, eds. *Delivering Motherhood: Maternal Ideologies and Practices in the Nineteenth and Twentieth Centuries.* New York: Routledge, 1990.

Colleau, Sophie, ed. *Childbirth: The Beginning of Motherhood, Proceedings of the Second Motherhood Symposium.* Madison, Wisc.: Women's Studies Research Center, 1982.

Corea, Gena. *The Mother Machine: Reproductive Technologies from Artificial Insemination to Artificial Wombs.* New York: Harper and Row, 1985.

Cott, Nancy F., ed. *Women's Bodies: Health and Childbirth.* Munich: Saur, 1993.

Davis-Floyd, Robbie E., and Carolyn F. Sargent. *Childbirth and Authoritative Knowledge: Cross-Cultural Perspectives.* Berkeley: University of California Press, 1997.

Donegan, Jane B. *Women and Men Midwives: Medicine, Morality, and Misogyny in Early America.* Westport, Conn.: Greenwood, 1978.

Eakins, Pamela S., ed. *The American Way of Birth.* Philadelphia: Temple University Press, 1986.

Edwards, Margot, and Mary Waldorf. *Reclaiming Birth: History and Heroines of American Childbirth Reform.* Trumansburg, N.Y.: Crossing Press, 1984.

Golden, Janet. *A Social History of Wet-Nursing in America: From Breast to Bottle.* New York: Cambridge University Press, 1996.

Gordon, Linda. *Pitied but Not Entitled: Single Mothers and the History of Welfare, 1890–1935.* New York: Free Press, 1994.

Hoffert, Sylvia D. *Private Matters: American Attitudes Toward Childbearing and Infant Nurture in the Urban North, 1800–1860.* Urbana: University of Illinois Press, 1989.

Horn, Margo. *Before It's Too Late: The Child Guidance Movement in the United States, 1922–1945.* Philadelphia: Temple University Press, 1989.

King, Charles R. *Children's Health in America: A History.* New York: Twayne, 1993.

Kunzel, Regina G. *Fallen Women, Problem Girls: Unmarried Mothers and the Professionalization of Social Work, 1890–1945.* New Haven, Conn.: Yale University Press, 1993.

Ladd-Taylor, Molly. *Raising a Baby the Government Way: Mothers' Letters to the Children's Bureau, 1915–1932.* New Brunswick, N.J.: Rutgers University Press, 1986.

Ladd-Taylor, Molly. *Mother-Work: Women, Child Welfare, and the State, 1890–1930.* Urbana: University of Illinois Press, 1994.

Leavitt, Judith Walzer. *Brought to Bed: Childbearing in America, 1750–1950.* New York: Oxford University Press, 1986.

Lindenmeyer, Kriste. *"A Right to Childhood?" The U.S. Children's Bureau and Child Welfare, 1912–1946.* Urbana: University of Illinois Press, 1997.

Loudon, Irvine. *Death in Childbirth: An International Study of Maternal Care and Maternal Mortality, 1800–1950.* New York: Oxford University Press, 1992.

Loudon, Irvine. *Childbed Fever: A Documentary History.* New York: Garland, 1995.

McMillen, Sally G. *Motherhood in the Old South: Pregnancy, Childbirth, and Infant Rearing.* Baton Rouge: Louisiana State University Press, 1990.

Morton, Marian J. *And Sin No More: Social Policy and Unwed Mothers in Cleveland, 1855–1990.* Columbus: Ohio State University Press, 1993.

Preston, Samuel H., and Michael R. Haines. *Fatal Years: Child Mortality in Late Nineteenth-Century America.* Princeton, N.J.: Princeton University Press, 1991.

Quiroga, Virginia A. Metaxas. *Poor Mothers and Babies: A Social History of Childbirth and Child Care Hospitals in Nineteenth-Century New York City.* New York: Garland, 1989.

Rich, Adrienne. *Of Woman Born: Motherhood as Experience and Institution.* New York: Norton, 1995.

Romalis, Shelly, ed. *Childbirth: Alternatives to Medical Control.* Austin: University of Texas, Press, 1981.

Rothman, Barbara Katz. *Giving Birth: Alternatives in Childbirth.* New York: Penguin, 1984.

Sandelowski, Margarete. *Pain, Pleasure, and American Childbirth: From the Twilight Sleep to the Read Method, 1914–1960.* Westport, Conn.: Greenwood, 1984.

Scholten, Catherine M. *Childbearing in American Society, 1650–1850.* New York: New York University Press, 1985.

Solinger, Rickie. *Wake Up Little Susie: Single Pregnancy and Race Before Roe v. Wade.* New York: Routledge, 1992.

Wertz, Richard W., and Dorothy C. Wertz. *Lying-in: A History of Childbirth in America: Its Technologies and Social Relations.* New Haven, Conn.: Yale University Press, 1989.

Women and Mental Illness

Chesler, Phyllis. *Women and Madness.* Garden City, N.Y.: Doubleday, 1972.

Ender, Evelyne. *Sexing the Mind: Nineteenth-Century Fictions of Hysteria.* Ithaca, N.Y.: Cornell University Press, 1995.

Geller, Jeffrey L., and Maxine Harris. *Women of the Asylum: Voices from Behind the Walls, 1840–1910.* New York: Doubleday, 1994.

Gollaher, David. *Voice for the Mad: The Life of Dorothea Dix.* New York: Free Press, 1995.

Hughes, John S., ed. *The Letters of a Victorian Madwoman.* Columbia, S.C.: University of South Carolina, 1993.

Lewin, Miriam, ed. *In the Shadow of the Past: Psychology Portrays the Sexes, A Social and Intellectual History.* New York: Columbia University Press, 1984.

Lunbeck, Elizabeth. *The Psychiatric Persuasion: Knowledge, Gender, and Power in Modern America.* Princeton, N.J.: Princeton University Press, 1994.

Micale, Mark S. *Approaching Hysteria: Disease and Its Interpretations.* Princeton, N.J.: Princeton University Press, 1995.

Small, Helen. *Love's Madness: Medicine, the Novel, and Female Insanity, 1800–1865.* New York: Oxford University Press, 1996.

Warren, Carol A. B. *Madwives: Schizophrenic Women in the 1950s.* New Brunswick, N.J.: Rutgers University Press, 1987.

A Sourcebook on Its Modern Origins. New York: Greenwood, 1986.

Logan, Onnie Lee, as told to Katherine Clark. *Motherwit: An Alabama Midwife's Story.* New York: Dutton, 1989.

Smart, Donna T., ed. *Mormon Midwife: The 1846–1888 Diaries of Patty Bartlett Sessions.* Logan: Utah State University Press, 1997.

Smith, Margaret Charles, and Linda Janet Holmes. *Listen to Me Good: The Life Story of an Alabama Midwife.* Columbus: Ohio State University Press, 1996.

Sullivan, Deborah A., and Rose Weitz. *Labor Pains: Modern Midwives and Home Birth.* New Haven, Conn.: Yale University Press, 1988.

Susie, Debra Ann. *In the Way of Our Grandmothers: A Cultural View of Twentieth-Century Midwifery in Florida.* Athens: University of Georgia Press, 1988.

Ulrich, Laurel Thatcher. *A Midwife's Tale: The Life of Martha Ballard, Based on Her Diary, 1785–1812.* New York: Knopf, 1990.

Health Care Providers: Midwives

Borst, Charlotte G. *Catching Babies: The Professionalization of Childbirth, 1870–1920.* Cambridge, Mass.: Harvard University Press, 1995.

Buss, Fran Leeper. *La Partera: Story of a Midwife.* Ann Arbor: University of Michigan Press, 1980.

DeVries, Raymond G. *Making Midwives Legal: Childbirth, Medicine, and the Law.* 2d ed. Columbus: Ohio State University Press, 1996.

Donnison, Jean. *Midwives and Medical Men: A History of the Struggle for the Control of Childbirth.* London: Historical Publications, 1988.

Ehrenreich, Barbara, and Deirdre English. *Witches, Midwives, and Nurses: A History of Women Healers.* Old Westbury, N.Y.: Feminist Press, 1974.

Leap, Nicky, and Billie Hunter. *The Midwife's Tale: An Oral History from Handywoman to Professional Midwife.* London: Scarlet Press, 1993.

Litoff, Judy Barrett. *American Midwives: 1860 to the Present.* Westport, Conn.: Greenwood, 1978.

Litoff, Judy Barrett, ed. *The American Midwife Debate:*

Health Care Providers: Nurses

Ashley, JoAnn. *Hospitals, Paternalism, and the Role of the Nurse.* New York: Teachers College Press, 1976.

Buhler-Wilkerson, Karen. *False Dawn: The Rise and Decline of Public Health Nursing, 1900–1930.* New York: Garland, 1989.

Buhler-Wilkerson, Karen, ed. *Nursing and the Public's Health: An Anthology of Sources.* New York: Garland, 1989.

Bullough, Vern L., and Bonnie Bullough. *The Care of the Sick: The Emergence of Modern Nursing.* London: Croom Helm, 1979.

Bullough, Vern L., and Bonnie Bullough. *History, Trends, and Politics of Nursing.* Norwalk, Conn.: Appleton-Century-Crofts, 1984.

Carnegie, Mary Elizabeth. *The Path We Tread: Blacks in Nursing Worldwide, 1854–1994.* 3d ed. New York: National League for Nursing Press, 1995.

Dammann, Nancy. *A Social History of the Frontier Nursing Service.* Sun City, Ariz.: Social Change Press, 1982.

Davies, Celia, ed. *Rewriting Nursing History.* Totowa, N.J.: Barnes & Noble, 1980.

Fessler, Diane Burke. *No Time for Fear: Voices of American Military Nurses in World War II.* East Lansing: Michigan State University Press, 1996.

Fitzpatrick, M. Louise, ed. *Historical Studies in Nursing.* New York: Teachers College Press, 1978.

Goldenberg, Gary. *Nurses of a Different Stripe: A History of the Columbia University School of Nursing, 1892–1992.* New York: Columbia University School of Nursing, 1992.

Hine, Darlene Clark, ed. *Black Women in the Nursing Profession: A Documentary History.* New York: Garland, 1985.

Hine, Darlene Clark. *Black Women in White: Racial Conflict and Cooperation in the Nursing Profession, 1890–1950.* Bloomington: Indiana University Press, 1989.

Jones, Anne Hudson, ed. *Images of Nurses: Perspectives from History, Art, and Literature.* Philadelphia: University of Pennsylvania Press, 1988.

Kalisch, Philip A., and Beatrice J. Kalisch. *The Advance of American Nursing.* 2d ed. Boston: Little, Brown, 1986.

Kalisch, Philip A., and Beatrice J. Kalisch. *The Changing Image of the Nurse.* Menlo Park, Calif.: Addison-Wesley Health Sciences Division, 1987.

Kalisch, Philip A., Beatrice J. Kalisch, and Margaret Scobey. *Images of Nurses on Television.* New York: Springer, 1983.

Kaufman, Martin, Joellen Watson Hawkins, Loretta P. Higgins, and Alice Howell Friedman, eds. *Dictionary of American Nursing Biography.* New York: Greenwood, 1988.

Lagemann, Ellen Condliffe, ed. *Nursing History: New Perspectives, New Possibilities.* New York: Teachers College Press, 1983.

Lewenson, Sandra Beth. *Taking Charge: Nursing, Suffrage, and Feminism in America, 1873–1920.* New York: Garland, 1993.

Mahler, Sister Mary Denis. *To Bind Up the Wounds: Catholic Sister Nurses in the U.S. Civil War.* Westport, Conn.: Greenwood, 1989.

Marshall, Helen E. *Mary Adelaide Nutting: Pioneer of Modern Nursing.* Baltimore, Md.: Johns Hopkins University Press, 1972.

Melosh, Barbara. *"The Physician's Hand": Work Culture and Conflict in American Nursing.* Philadelphia: Temple University Press, 1982.

Mottus, Jane E. *New York Nightingales: The Emergence of the Nursing Profession at Bellevue and New York Hospital, 1850–1920.* Ann Arbor, Mich.: UMI Research Press, 1980.

Norman, Elizabeth M. *Women at War: The Story of Fifty Military Nurses Who Served in Vietnam.* Philadelphia, Pa.: University of Pennsylvania Press, 1990.

Reverby, Susan. *One Strong Voice—Or An Out of Tune Chorus?: A Brief History of Women Health Workers and Their Changing Role.* New York: Bread and Roses Project, District 1199, National Union of Hospital and Health Care Employees/RWDSU/AFL-CIO, 1979.

Reverby, Susan. *Ordered to Care: The Dilemma of American Nursing, 1850–1945.* New York: Cambridge University Press, 1987.

Roberts, Joan I., and Thetis M. Group. *Feminism and Nursing: A Historical Perspective on Power, Status, and Political Activism in the Nursing Profession.* Westport, Conn.: Praeger, 1995.

Safier, Gwendolyn. *Contemporary American Leaders in Nursing: An Oral History.* New York: McGraw-Hill, 1977.

Shaw, Stephanie J. *What a Woman Ought To Be and To Do: Black Professional Women Workers During the Jim Crow Era.* Chicago: University of Chicago Press, 1996.

Staupers, Mabel K. *No Time for Prejudice: A Story of the Integration of Negroes in Nursing in the United States.* New York: Macmillan, 1961.

Thoms, Adah B. *Pathfinders: A History of the Progress of Colored Graduate Nurses.* New York: Kay Printing, 1929.

Tomblin, Barbara Brooks. *G. I. Nightingales: The Army Nurse Corps in World War II.* Lexington: University Press of Kentucky, 1996.

Health Care Providers: Physicians

Abram, Ruth J., ed. *Send Us a Lady Physician: Women Doctors in America, 1835–1920.* New York: Norton, 1985.

Alsop, Gulielma Fell. *History of the Woman's Medical College, Philadelphia, Pennsylvania, 1850–1950.* Philadelphia: Lippincott, 1950.

Dickstein, Leah J., and Carol C. Nadelson, eds. *Women Physicians in Leadership Roles.* Washington, D.C.: American Psychiatric Press, 1986.

Drachman, Virginia G. *Hospital with a Heart: Women Doctors and the Paradox of Separatism at the New England Hospital, 1862–1969.* Ithaca, N.Y.: Cornell University Press, 1984.

Glazer, Penina M., and Miriam Slater. *Unequal Colleagues: The Entrance of Women into the Professions, 1890–1940.* New Brunswick, N.J.: Rutgers University Press, 1987.

Harris, Barbara. *Beyond Her Sphere: Women and the Professions in American History.* Westport, Conn.: Greenwood, 1978.

Hill, Ruth Edmons, ed. *The Black Women Oral History Project: From the Arthur and Elizabeth Schlesinger Library in the History of Women in America, Radcliffe College.* Westport, Conn.: Meckler, 1991.

Hurd-Mead, Kate C. *A History of Women in Medicine: From the Earliest Times to the Beginning of the Nineteenth Century.* 1938. Reprint. Boston: Milford House, 1973.

Lightfoot, Sara Lawrence. *Balm in Gilead: Journey of a Healer.* Reading, Mass.: Addison-Wesley, 1988.

Lopate, Carol. *Women in Medicine.* Baltimore, Md.: Johns Hopkins University Press, 1968.

Lorber, Judith. *Women Physicians, Careers, Status, and Power.* New York: Tavistock, 1984.

Moldow, Gloria. *Women Doctors in Gilded-Age Washington: Race, Gender, and Professionalization.* Urbana: University of Illinois Press, 1987.

Morantz, Regina Markell, Cynthia Stodola Pomerleau, and Carol Hansen Fenichel, eds. *In Her Own Words: Oral Histories of Women Physicians.* Westport, Conn.: Greenwood, 1982.

Morantz-Sanchez, Regina Markell. *Sympathy and Science: Women Physicians in American Medicine.* New York: Oxford University Press, 1985.

More, Ellen S., and Maureen A. Milligran, eds. *The Empathic Practitioner: Empathy, Gender, and Medicine.* New Brunswick, N.J.: Rutgers University Press, 1994.

Rowland, Mara Canaga. *As Long as Life: The Memoirs of a Frontier Woman Doctor, Mary Canaga Rowland, 1873–1966.* Seattle, Wash.: Storm Peak Press, 1994.

Sicherman, Barbara. *Alice Hamilton: A Life in Letters.* Cambridge, Mass.: Harvard University Press, 1984.

Walsh, Mary Roth. *"Doctors Wanted: No Women Need Apply": Sexual Barriers in the Medical Profession, 1835–1975.* New Haven, Conn.: Yale University Press, 1977.

Health Care Providers: Others

Dakin, Theodora P. *A History of Women's Contribution to World Health.* Lewiston, N.Y.: Edwin Mellen Press, 1991.

Fink, Leon, and Brian Greenberg. *Upheaval in the Quiet Zone: A History of Hospital Workers' Union, Local 1199.* Urbana: University of Illinois Press, 1989.

Levin, Beatrice. *Women and Medicine: Pioneers Meeting the Challenge!* 2d ed. Lincoln, Neb.: Media Publishing, 1988.

Perrone, Bobette, H. Henrietta Stockel, and Victoria Krueger. *Medicine Women, Curanderas, and Women Doctors.* Norman: University of Oklahoma Press, 1989.

Sacks, Karen Brodkin. *Caring by the Hour: Women, Work, and Organizing at Duke Medical Center.* Urbana: University of Illinois Press, 1988.

Stepsis, Ursula, and Dolores Liptak, eds. *Pioneer Healers: The History of Women Religious in American Health Care.* New York: Crossroad, 1989.

Women, Health Reform, and Public Health

Blustein, Bonnie Ellen. *Educating for Health and Prevention: A History of the Department of Community and Preventive Medicine of the (Woman's) Medical College of Pennsylvania.* Canton, Mass.: Science History, 1993.

Carter, Richard. *The Gentle Legions.* New York: Doubleday, 1961.

Cayleff, Susan E. *Wash and Be Healed: The Water-Cure Movement and Women's Health.* Philadelphia: Temple University Press, 1987.

Donegan, Jane B. *"Hydropathic Highway to Health": Women and Water-Cure in Antebellum America.* Westport, Conn: Greenwood, 1986.

Larson, Edward J. *Sex, Race, and Science: Eugenics in the Deep South.* Baltimore, Md.: Johns Hopkins University Press, 1995.

Leavitt, Judith Walzer. *Typhoid Mary: Captive to the Public's Health.* Boston: Beacon, 1996.

Markowitz, Gerald E., and David Rosner. *Children, Race, and Power: Kenneth and Mamie Clark's Northside Center.* Charlottesville: University of Virginia Press, 1996.

Marshall, Helen E. *Dorothea Dix: Forgotten Samaritan.* Chapel Hill: University of North Carolina Press, 1973.

McCarthy, Kathleen D. *Noblesse Oblige: Charity and Cultural Philanthropy in Chicago, 1849–1929.* Chicago: University of Chicago Press, 1982.

Meckel, Richard A. *"Save the Babies": American Public Health Reform and the Prevention of Infant Mortality, 1850–1929.* Baltimore, Md.: Johns Hopkins University Press, 1990.

Numbers, Ronald L. *Prophetess of Health: Ellen G. White and the Origins of Seventh-day Adventist Health Reform.* Knoxville: University of Tennessee Press, 1992.

Pernick, Martin S. *The Black Stork: Eugenics and the Death of "Defective" Babies in American Medicine and Motion Pictures Since 1915.* New York: Oxford University Press, 1996.

Pivar, David. *Purity Crusade: Sexual Morality and Social Control, 1868–1900.* Westport, Conn.: Greenwood, 1973.

Poirier, Suzanne. *Chicago's War on Syphilis, 1937–1940: The Times, the Trib, and the Clap Doctor.* Urbana: University of Illinois Press, 1995.

Rosner, David, and Gerald Markowitz, eds. *Dying for Work: Workers' Safety and Health in Twentieth-Century America.* Bloomington: Indiana University Press, 1987.

Shapiro, Thomas M. *Population Control Politics: Women, Sterilization, and Reproductive Choice.* Philadelphia: Temple University Press, 1985.

Sklar, Kathryn Kish. *Catharine Beecher: A Study in American Domesticity.* New Haven, Conn.: Yale University Press, 1973.

Skocpol, Theda. *Protecting Soldiers and Mothers: The Political Origins of Social Policy in the United States.* Cambridge, Mass.: Belknap Press of Harvard University Press, 1992.

Smith, Susan L. *Sick and Tired of Being Sick and Tired: Black Women and the National Negro Health Movement, 1915–1950.* Philadelphia: University of Pennsylvania Press, 1995.

Tomes, Nancy. *The Gospel of Germs: Men, Women, and the Microbe in American Life.* Cambridge, Mass.: Harvard University Press, 1997.

Whorton, James C. *Crusaders for Fitness: The History of American Health Reformers.* Princeton, N.J.: Princeton University Press, 1982.

Diseases and Treatments

Apple, Rima D. *Vitamania: Vitamins in American Culture.* New Brunswick, N.J.: Rutgers University Press, 1996.

Kendall, Stephen R. *Substance and Shadow: Women and Addiction in the United States.* Cambridge, Mass.: Harvard University Press, 1996.

McGregor, Deborah Kuhn. *Sexual Surgery and the Origins of Gynecology: J. Marion Sims, His Hospital, and His Patients.* New York: Garland, 1989.

Pernick, Martin S. *A Calculus of Suffering: Pain, Professionalism, and Anesthesia in Nineteenth-Century America.* New York: Columbia University Press, 1985.

Rothman, Sheila M. *Living in the Shadow of Death: Tuberculosis and the Social Experience of Illness in America.* New York: Basic Books, 1994.

Speert, Harold. *Obstetrics and Gynecology: A History and Iconography.* 2d ed. San Francisco: Norman, 1994.

Stage, Sarah. *Female Complaints: Lydia Pinkham and the Business of Women's Medicine.* New York: Norton, 1979.

Aging

Abel, Emily K. *Who Cares for the Elderly?: Public Policy and the Experiences of Adult Daughters.* Philadelphia: Temple University Press, 1991.

Banner, Lois W. *In Full Flower: Aging Women, Power, and Sexuality: A History.* New York: Knopf, 1992.

Formanek, Ruth, ed. *The Meanings of Menopause: Historical, Medical, and Clinical Perspectives.* Hillsdale, N.J.: Analytic Press, 1989.

Lock, Margaret. *Encounters with Aging: Mythologies of Menopause in Japan and North America.* Berkeley: University of California Press, 1993.